22nd Association for the Advancement of Animal Breeding and Genetics Conference 2017

Townsville, Australia
2 – 5 July 2017

ISBN: 978-1-5108-6759-8

Printed from e-media with permission by:

Curran Associates, Inc.
57 Morehouse Lane
Red Hook, NY 12571

proceedings.com

Some format issues inherent in the e-media version may also appear in this print version.

Printed by Curran Associates, Inc. (2018)

For permission requests, please contact AAABG – Association for the Advancement of Animal Breeding and Genetics at the address below.

AAABG – Association for the Advancement of Animal Breeding and Genetics
University of New England
Armidale NSW 2351
Australia

Phone: 61 2 677 320 55
Fax: 61 2 677 332 66

AGBU@une.edu.au

Additional copies of this publication are available from:

Curran Associates, Inc.
57 Morehouse Lane
Red Hook, NY 12571 USA
Phone: 845-758-0400
Fax: 845-758-2633
Email: curran@proceedings.com
Web: www.proceedings.com

Association for the Advancement of Animal Breeding and Genetics

AAABG

Association for the Advancement of Animal Breeding and Genetics

Proceedings of the Twenty-second (22nd) Conference

Townsville, Queensland, AUSTRALIA
2nd July – 5th July 2017

Produced by:

Association for the Advancement of Animal Breeding and Genetics
C/-AGBU
University of New England
Armidale NSW 2350

Internet web site: http://www.aaabg.org

PRESIDENT'S MESSAGE

Dear AAABG Members, Delegates and Partners,
On behalf of the Organising Committee it is my pleasure to extend a warm and cheerful welcome to you all to the 22nd AAABG conference in Townsville, Queensland. We respectfully acknowledge the Traditional Owners of this land on which we are meeting, the Bindal and Wulgurukaba people, we pay respect to their Elders – past, present and emerging – and acknowledge the important role Aboriginal and Torres Strait Islander people continue to play within our community.
This is the fourth time The Sunshine State has hosted the AAABG conference with Brisbane (1982), Rockhampton (1992) and Noosa Lakes (2005) preceding Townsville. Hopefully the weather has aligned with our wishes and the tropical winter allows you to experience the lifestyle, culture and heritage that make this region so unique.
Acclaimed as the natural second capital city of Queensland, Townsville and its region have a vibrant primary sector including cattle and fisheries. It is also home to leading institutions like James Cook University (JCU), the Australian Institute of Marine Science (AIMS) and the Australian Tropical Sciences and Innovation Precinct (ATSIP), a joint venture between JCU and CSIRO.
For the 22nd AAABG conference, the Organising Committee has worked under the banner of "Science Enabling Industry Outcomes" to reflect the intent to amalgamate breeders, R&D providers and scientists in their efforts to catalyse the translation of scientific findings to industrially useful applications. In addition to the customary species- and methodology-oriented sessions, the conference program boasts two industry sessions lead by agribusiness specialists from Australia and New Zealand, as well as a Q&A session aimed at fostering a fruitful dialogue among all parties.
For their generous support, I am grateful to the sponsors. Their funding has allowed us to attract invited speakers of international calibre, hire the venue and enjoy catering that the registration fee alone could have not afforded. The conference organisers, ASN Events in general and Jennifa Vo in particular, have worked tirelessly and to the highest professional standards to ensure a successful conference.
It has been gratifying to see the enthusiasm and commitment of the session chairs supervising the reviewing process of the papers allocated to their session. Ed Charmley and Dean Jerry are accredited for organising the beef and aquaculture tours, respectively.
I hope you enjoy the program, use the opportunity to cement old friendships, establish new ones, and leave with fond, lasting memories of the 22nd AAABG conference in Townsville.

Toni Reverter
President, 2017 AAABG

ASSOCIATION FOR THE ADVANCEMENT OF ANIMAL BREEDING AND GENETICS

2017

TWENTY SECOND CONFERENCE

COMMITTEE

CITATION OF PAPERS

Papers in this publication should be cited as appearing in the Proceedings of the Association for the Advancement of Animal Breeding and Genetics (Abbreviation: Proc. Assoc. Advmt. Anim. Breed. Genet.)

For example:
Bowley F.E., Amer P.R. and Meier S. (2013) New approaches to genetic analysis of fertility traits in New Zealand dairy cattle. *Proc. Assoc. Advmt. Anim. Breed. Genet.* **20**: 37-40.

REVIEWERS and SECTION EDITORS

All papers, invited and contributed, were subjected to peer review by two referees. We acknowledge and thank those people listed below for their work in reviewing the papers (and apologise if we have inadvertently omitted any reviewer from the list).

Reviewers:

Mary Abdelsayed
Jena Alexopoulos
Othman Alqaisi
Peter Amer
Jason Archer
Robert Banks
Stephen Barwick
Irene Van Den Berg
Vinzent Boerner
Forbes Brien
Daniel Brown
Kim Bunter
Roberto Carvalheiro
Yizhou Chen
Sam Clark
Schalk Cloete
Lucy Coleman
Natalie Connors
David Cottle
Christine Couldrey
Neil Cullen
Hans Daetwyler
Akhmad Dakhlan
Dion Detterer
Ken Dodds
Sonja Dominik
Kath Donoghue
Naomi Duijvesteijn
Alison Van Eenennaam
Mohammad Ferdosi
Andrew Foote
Marina Fortes
Dorian Garrick
Masoud Soltani-Ghombavani
Evgeny Glazov
Mike Goddard
Cedric Gondro
Klint Gore
Tom Granleese
Johan Greeff
Phillip Gurman
Sarita Guy
Timothy Hancock
Rhiannon Handcock
Sue Hatcher
Ben Hayes
Fiona Hely
John Henshall
Susanne Hermesch
Andrew Hess
John Holmes
Roger Huerlimann
Dean Jerry
Gilbert Jeyaruban
Patricia Johnson
David Johnston
David Jones
Haja Kadarmideen
Mohammad Kalaldeh
Parthan Kasarapu
Mike Keehan
Octavia Kelly
Majid Khansefid
Mehar Khatkar
Brian Kinghorn
Lambros Koufariotis
Theodor Kristjansson
Sigrid Lehnert
Yutao Li
Li Li
Russell Lyons
Paul Maclean
Iona MacLeod
Hawlader Al-Mamun
Mekonnen Haile-Mariam
John McEwan
Rudi McEwin
Aaron McMillan
Lorna Mcnaughton
Thaise Pinto de Melo
Karin Meyer
Nasir Moghaddar
Amin Mohamed
Carel Muller
Laercio Porto-Neto
Baltasar Fernandes Garcia
Joanna Newton
Thuy Nguyen
Loan Nguyen
Tansyn Noble
Gertje Petersen
Wayne Pitchford
Matthew Price
Jennie Pryce
Cheryl Quinton
Fernanda Raidan
Jerome Raoul
Antonio Reverter
Fernanda Rezende
Kristian Ridley
Penny Riggs
Sara De las Heras Saldana
Marina Naval Sanchez
Marina Naval Sanchez
Lise Sandenbergh
Bruno Santos
Beth Scott
Saleh Shahinfar
Ric Sherlock
Eva Strucken
Jan Strugnell
Andrew Swan
Siriporn Tongsiri
Jose Torres Vazquez
Sam Walkom
Bradley Walmsley
Tingting Wang
Matt Wolcott
Rob Woolaston
Shernae Woolley
Ruidong Xiang
Juncong Yan
Yuandan Zhang

SPONSORS of the 22nd AAABG Conference 2017

The financial assistance of the following organisations is gratefully acknowledged.

Supported by

AAABG was formerly known as the Australian Association for Animal Breeding and Genetics. Following the 1995 OGM the name was changed when it became an organisation with a joint Australian and New Zealand membership. The Association for the Advancement of Animal Breeding and Genetics is incorporated in South Australia.

THE ASSOCIATION FOR THE ADVANCEMENT OF ANIMAL BREEDING AND GENETICS INCORPORATED

OBJECTIVES

(i) to promote scientific research on the genetics of animals;
(ii) to foster the application of genetics in animal production;
(iii) to promote communication among all those interested in the application of genetics to animal production, particularly breeders and their organisations, consultants, extension workers, educators and geneticists.

To meet these objectives, the Association will:

(i) hold regular conferences to provide a forum for:
  (a) presentation of papers and in-depth discussions of general and industry-specific topics concerning the application of genetics in commercial animal production;
  (b) scientific discussions and presentation of papers on completed research and on proposed research projects;
(ii) publish the proceedings of each Regular Conference and circulate them to all financial members;
(iii) use any such other means as may from time to time be deemed appropriate.

MEMBERSHIP

Any person interested in the application of genetics to animal production may apply for membership of the Association and, at the discretion of the Committee, be admitted to membership as an Ordinary Member.

Any organisations interested in the application of genetics to animal production may apply for membership and, at the discretion of the Committee, be admitted to membership as a Corporate member. Each such Corporate Member shall have the privilege of being represented at any meeting of the Association by one delegate appointed by the Corporate Member.

Benefits to Individual Members

- While it is not possible to produce specific recommendations or "recipes" for breeding plans that are applicable for all herd/flock sizes and management systems, principles for the development of breeding plans can be specified. Discussion of these principles, consideration of particular case studies, and demonstration of breeding programs that are in use will all be of benefit to breeders.
- Geneticists will benefit from the continuing contact with other research workers in refreshing and updating their knowledge.
- The opportunity for contact and discussions between breeders and geneticists in individual members' programs, and for geneticists in allowing for detailed discussion and appreciation of the practical management factors that often restrict application of optimum breeding programs.

Benefits to Member Organisations

- Many of the benefits to individual breeders will also apply to breeding organisations. In addition, there are benefits to be gained through coordination and integration of their efforts. Recognition of this should follow from understanding of common problems, and would lead to increased effectiveness of action and initiatives.
- Corporate members can use the Association as a forum to float ideas aimed at improving and/or increasing service to their members.

General Benefits

- Membership of the Association may be expected to provide a variety of benefits and, through the members, indirect benefits to all the animal industries.
- All members should benefit through increased recognition of problems, both at the level of research and of application, and increased understanding of current approaches to their solution.
- Well-documented communication of gains to be realised through effective breeding programs will stimulate breeders and breeding organisations, allowing increased effectiveness of application and, consequently, increased efficiency of operation.
- Increased recognition of practical problems and specific areas of major concern to individual industries should lead to increased relevance of applied research.
- All breeders will benefit indirectly because of improved services offered by the organisations which service them.
- The existence of the Association will increase appreciably the amount and use of factual information in public relations in the animal industries.
- Association members will comprise a pool of expertise – at both the applied and research levels – and, as such, individual members and the Association itself must have an impact on administrators at all levels of the animal industries and on Government organisations, leading to wiser decisions on all aspects of livestock improvement, and increased efficiency of animal production.

CONFERENCES

One of the main activities of the Association is the Conference. These Conferences will be structured to provide a forum for discussion of research problems and for breeders to discuss their problems with each other, with extension specialists and with geneticists.

ASSOCIATION FOR THE ADVANCEMENT OF ANIMAL BREEDING AND GENETICS
FELLOWS OF THE ASSOCIATION

"Persons who have rendered eminent service to animal breeding in Australia and/or New Zealand or elsewhere in the world, may be elected to Fellowship of the Association..."

Elected February 1990 R.B.M. Dun F.H.W. Morley (deceased) A.L. Rae (deceased) H.N. Turner (deceased) *Elected September 1992* K. Hammond *Elected July 1995* C.H.S. Dolling J.R. Hawker J. Litchfield *Elected February 1997* J.S.F. Barker R.E. Freer *Elected June 1999* J. Gough J.W. James *Elected July 2001* J.N. Clarke A.R. Gilmour L.R. Piper *Elected September 2005* B.M. Bindon M.E. Goddard H.-U. Graser F.W. Nicholas	*Elected September 2007* K.D. Atkins R.G. Banks G.H. Davis *Elected September 2009* N. Fogarty A. Fyfe J. McEwan R. Mortimer R. Ponzoni *Elected September 2011* B.P. Kinghorn A. McDonald *Elected October 2013* H. Burrow P. Fennessy G. Nicoll P. Parnell *Elected October 2015* P. Arthur D. Johnson K. Meyer B. Tier R. Woolaston

HONORARY MEMBERS OF THE ASSOCIATION
"Members who have rendered eminent service to the Association may be elected to Honorary Membership..."

Elected September 2009
W.A. Pattie
J. Walkley

HELEN NEWTON TURNER MEDAL TRUST

The Helen Newton Turner Medal Trust was established in 1993 following an anonymous donation to the Animal Genetics and Breeding Unit. The Helen Newton Turner Medal is awarded to provide encouragement and inspiration to those engaged in animal genetics. The Medal is named after Dr Helen Newton Turner whose career with CSIRO was dedicated to research into the genetic improvement of sheep for wool production. The Medallist is chosen by Trustees from the ranks of those persons who have made an outstanding contribution to genetic improvement of Australian livestock.

The Helen Newton Turner Medal was first awarded in 1994 to Associate Professor John James and a list of all recipients to date is given below. The recipient of the Medal is invited to deliver an Oration on a topical subject of their choice. The Oration of the 2015 Medal recipient, Dr. Arthur Gilmour, is reproduced in these proceedings.

MEDALLISTS

1994	J.W. James	*2001*	G.A. Carnaby	*2011*	R. Banks
1995	L.R. Piper	*2003*	F.W. Nicholas	*2013*	M. Goddard
1997	J. Litchfield	*2005*	K. Hammond	*2015*	A. Gilmour
1998	J.S.F. Barker	*2007*	L. Corrigan	*2017*	A. Collins
1999	C.W. Sandilands	*2009*	R. Hawker		

HELEN NEWTON TURNER AO

HELEN NEWTON TURNER MEDALIST ORATION 2015

ACCEPTANCE SPEECH
Arthur Gilmour

Introduction

I never met Helen Newton Turner but I understand she had a big influence on the Australian Sheep Industry by introducing new management and breeding strategies. This is part of a worldwide development, especially post World War II, which continues to the present: the increase in agricultural production by the simultaneous improvement of management and genetics. It had not crossed my mind that I might be considered worthy of an honour given in her memory. I would like to take this opportunity to document some of my background, my collaborators and my motivation.

Some History

I was born in Lockhart in 1949, the eldest of six children, and grew up on the family sheep/wheat farm at Boree Creek. Life was hard on those heavy soils for my parents but it was there I learned to take the initiative and to work hard at what had to be done. I attended Yanco Agricultural High School, built around Sir Samuel McCaughey's (1835-1919) mansion, and was dux in 1965.
Not having much idea what to do, after a few months at Port Kembla flat products, I secured a Commonwealth scholarship to do Agriculture at Sydney University. I transitioned to a Department of Agriculture traineeship in 1967 and so began almost 42 years service. In the fourth year, I specialised in biometry and began service as a biometrician in the State Office Block in 1970.

This was a period of expansion in agricultural research with a dozen or so biometricians supporting researchers in the design and analysis of their experiments, a role I pursued for 39 years. Computers were becoming available and I engaged in Fortran programming, to enhance the computational capability of the group. Consequently, my time was divided between research consulting and software development to meet growing demands.
When biometricians were urged to move to country research stations, I transferred to Trangie in January 1976 where I developed links with the sheep breeding team. The Australian Wool Board then gave me a scholarship to study for a PhD at Massey University under Prof Robert Anderson and Prof Al Rae between 1980-1982. After almost a year again in Sydney, I transferred to the Orange Agricultural Institute in 1984. The Department was my very supportive employer through to 2009.

Some colleagues

My PhD was on the estimation of genetic parameters for categorical traits. I met Robin Thompson in New Zealand in 1981 and visited him in Edinburgh in 1982 when I also attended the second world congress (WCGALP) in Madrid. This led to regular contact with Robin, which continues today.

Brian Cullis, a fellow biometrician, wanted to fit a Genotype x Year x Location mixed model to 10 years of wheat data but it was too large for Genstat. So, I wrote a derivative free program to perform the estimation in 1992. This was also the year I wrote BLUP software for the wool industry and the lamb industry. Robin dropped by and suggested the Average Information algorithm for Brian's analysis; it was a simple extension to the program I had. He then invited by to spend the last half of 1993 with him at Roslin Institute outside Edinburgh, extending a multiple trait BLUP program for the British Dairy Board. We three worked on the Average Information paper which was published in 1995. I then wrote a more general program which was released as ASReml in 1996. ASReml has dominated my life ever since.

Some philosophy

I learned about Jesus from my mother while at primary school, and committed myself to Him in 1961 just before going to Yanco. Therefore, I have studied and sought to share the Scriptures throughout my career. Whenever Jesus taught the crowds, he told stories based on nature, especially agriculture, and my understanding of agricultural science is underpinned by what I read in Scripture.

In the beginning God created the heavens and the earth' including all its life forms and He 'saw that it was very good'. Although it is now getting old, we still marvel at the systems God has put in place. Not least is the DNA code, the digital program specific to each kind of living thing. I conclude that the SNPs we observe today are primarily part of the initial variability built into living things so that they not only performed well at the beginning, but have been able to adapt to changing environments through history.

Critical to reconciling what we now observe with creation is the event known as Noah's flood, and the subsequent ice age. As well as explaining the origin of sedimentary layers containing fossils around the world, it also caused a genetic bottle neck for animals in particular, so that founder effects and natural selection in initially small populations have resulted in the many species we see today. This was reinforced to me when Pattie Cunningham explained the distinctions between European, African and Asian cattle at a AAABG meeting (1992?), a classic founder effect related to dispersion after the flood. And again, by a talk I heard at WCGALP in Vancouver (1994) reporting that an analysis of SNP variation indicated two genetic bottlenecks, which I associate with Creation and the Flood.

As animal and plant breeders, we are restoring a partly decayed system and I am confident there is a lot of potential for further progress. The problem for the future though is primarily a political one, not just a scientific one. The world could feed itself except for political agendas. The Scriptures anticipate the political issues will be resolved when Jesus, the Jewish Messiah, is acclaimed first by them, and then by the world, as world leader.

Conclusion

I am privileged to have been given a small part in genetic improvement and thank those who have made this possible. I encourage us all to take the initiative to do what we can each day to meet the needs around us. We each have a role, whether prominent or unnoticed, and it has been a privilege to collaborate with so many who are doing so much. I acknowledge the grace of God through Jesus Christ in giving me this role.

Arthur Gilmour
30 Sept 2015

TABLE OF CONTENTS

Proc. Assoc. Advmt. Anim. Breed. Genet. 22:1-8

FUNCTIONAL ANNOTATION OF ANIMAL GENOMES

E.L. Clark, S. Bush, R. Young, J.K. Baillie, L. Lefevre, P. Dutta, C. Muriuki, M. McCulloch, T.C. Freeman, D.W. Burt, L. Freem, C.B.A.Whitelaw, K.M. Summers, A.L. Archibald & D.A. Hume

The Roslin Institute and Royal (Dick) School of Veterinary Studies, University of Edinburgh, Easter Bush, Midlothian EH25 9RG, UK

SUMMARY

With the advent of long-read sequencing technologies, and the rapid drop in the cost of short-read sequencing, livestock geneticists have access to almost completely contiguous reference genome sequences of similar quality to human and model organisms, and massive sequence level data on variation amongst breeds and adapted populations. In livestock genomes, many protein-coding genes are marked with placeholder names, their functional orthology to human or mouse genes is ambiguous and the annotation of transcript diversity is sparse. Non-coding regulatory elements (promoters, enhancers etc) and non-coding RNAs are even less well characterised, yet available evidence from human genetics indicates that variants in these elements are enriched for trait associations. The international FAANG (Functional Annotation of Animal Genomes, www.faang.org) consortium aims to coordinate efforts to address the information gap (L. Andersson *et al.* 2015). Gene-editing technologies, combined with sequence information, offers the promise of accelerated genetic gain (Hickey *et al.* 2016). In this review, we consider some of our approaches to livestock genome annotation.

INTRODUCTION

At the previous meeting of AAABG, Perez-Enciso et al. (2015) (Perez-Encisco *et al.* 2015) reviewed the potential applications of sequence data to animal breeding; and talked of "biology-informed sequence exploitation". Since 2015, the cost of generating whole genome shotgun sequence data has continued to fall. Thus, with the most recent genotyping platforms, the $1000 genome at 30X genome coverage is not far from reality, and we and others are sequencing hundreds, and even thousands, of animals from different breeds and different adapted populations in every livestock species. The increased sequence depth increases the reliability of variant calling, including variants that impact on the function of protein-coding genes such as indels, stop gains and severely disruptive mutations (Boschiero *et al.* 2015, Telenti *et al.* 2016). These mutations are more prevalent in populations than might be expected. In a remarkable study of human populations with high levels of consanguinity, Saleheen et al. (Saleheen *et al.* 2017) reported exome sequencing of >10,000 individuals, and identified 49,000 rare predicted loss-of-function mutations of which 1317 were homozygous in at least one individual. A subset was confirmed to cause functional changes in the encoded protein, albeit clearly not lethal. An exome sequencing platform has been developed for pigs, and its application similarly predicts significant prevalence of loss-of-function alleles (Robert *et al.* 2014). This is a potential resource for functional genomics, as well as animal breeding, since the impact of such alleles can be confirmed by brother-sister mating or from prohibited homozygosity in populations (if the impact is severe). We have initiated such as backcross project in chickens, where we identified candidate loss-of-function alleles in a set of 10 founder pairs, and then mated their F1 offspring to expose homozygotes. However, even high impact functional variants are not necessarily coding. Hoff et al. (Hoff *et al.* 2017) identified seven haplotypes that were relatively prevalent in registered US Angus cattle, but were not observed as homozygotes, and used deep sequencing of >100 individuals to identify common variants within these haplotypes. None of the candidate causal variants identified was present within exons.

Another of the major impacts of deep sequencing is the improved detection of copy-number variants and sequences that are not present in the reference genome. This is somewhat constrained by the quality of the genome assembly (Couldrey *et al.* 2017) but the rapid improvement of livestock genomes, driven by the FAANG consortium, will address this issue. Indeed, the contiguity of the new goat genome, released earlier this year (Bickhart *et al.* 2017, Worley 2017), is approaching that of the completed human and mouse genomes. The recent sequencing of 10,000 human genomes at 30-40x coverage identified on average 0.7 Mb of sequence that was not present in the human reference genome (Telenti et al. 2016). Copy number and structural (e.g. inversions/translocations) variants are commonly associated with trait variation in all species. A recent study, which also reviewed some of the earlier literature, identified multiple copy number variants associated with domestication and high altitude adaptation in the Chinese Yak (Zhang *et al.* 2016). Long read sequencing provides an additional potential step-change in detection of structural variants, with an incomplete overlap between the outcomes from short-read technologies (Couldrey et al. 2017). With all of this sequence/genomic information, we have the potential to reverse the traditional information flow, and link sequence to consequence. However, there are several major challenges to overcome.

Firstly, we need much more information about the function of individual genes and regulatory sequences in a wider range of species. It is certainly the case that some functions are conserved across species. The phenotypes associated with knockouts of protein-coding genes in mice can give insights into likely functions and phenotypic consequences of loss-of-function in other species. Similarly, detailed analysis of promoter and enhancer landscapes in the liver across 20 mammalian species revealed substantial conservation of both regulatory elements and transcriptional outputs (Villar *et al.* 2015). Arguably, the liver has a rather generic "housekeeping" function in mammals that is not subject to rigorous selection. By contrast, there are radical differences between mice, pigs and humans in the response of innate immune cells to bacterial lipopolysaccharide (LPS) (Kapetanovic *et al.* 2012, Schroder *et al.* 2012) or to glucocorticoids (Jubb *et al.* 2015), associated with gain and loss of promoter and enhancer elements. It is these differences between species, and between individuals, that are of particular interest to geneticists and developmental biologists.

Secondly, we need to find a way to take account of epistasis, which manifests as variable penetrance. There are few knockout mutations in mice, or human genetic diseases, that do not exhibit some measure of phenotypic variation that is apparently a consequence of gene-gene interactions, or genetic background (Phillips 2008, Mackay 2014). Sometimes the mechanism can be disentangled based upon biological knowledge. For example, the knockout of the macrophage-specific transcription factor, PU.1, is mid-gestation lethal in homozygous PU.1 knockout inbred C57Bl/6 mice, but when the knockout allele is present in the homozygous state on a different genetic background, produces viable offspring with a neutrophil deficiency. The PU.1 protein interacts with another transcription factor, MITF, and a compound heterozygote (PU.1 +/-, MITF mi/+) phenocopies the PU.1 knockout (Luchin *et al.* 2001). Efforts to model the impact of epistasis in GWAS analysis and genomic selection have had limited success, in part due to the computational challenges (Stanislas *et al.* 2017). A subset of variable penetrance results from genomic imprinting in mammals, where the apparent heritability of a trait depends upon the parent of origin and reciprocal crosses do not produce the same outcome. The analysis of the contribution of imprinting to estimated breeding values is also computationally challenging (Nishio and Satoh 2015), but would be significantly less so if the set of imprinted loci and their functions was known in each species.

Identification of causal variants has been described as the "holy grail" for quantitative genetics (Perez-Encisco et al. 2015). Increased density of markers derived from sequence information, without functional annotation, simply approaches the tyranny of statistics. The challenge is to develop strong biological "priors" to prioritise variants that are more likely to be functionally associated with a trait. Inclusion of such biological priors clearly has the potential to enhance the power of genomic prediction in complex traits (MacLeod *et al.* 2016). So, how far have we come since 2015 in generating useful prior knowledge?

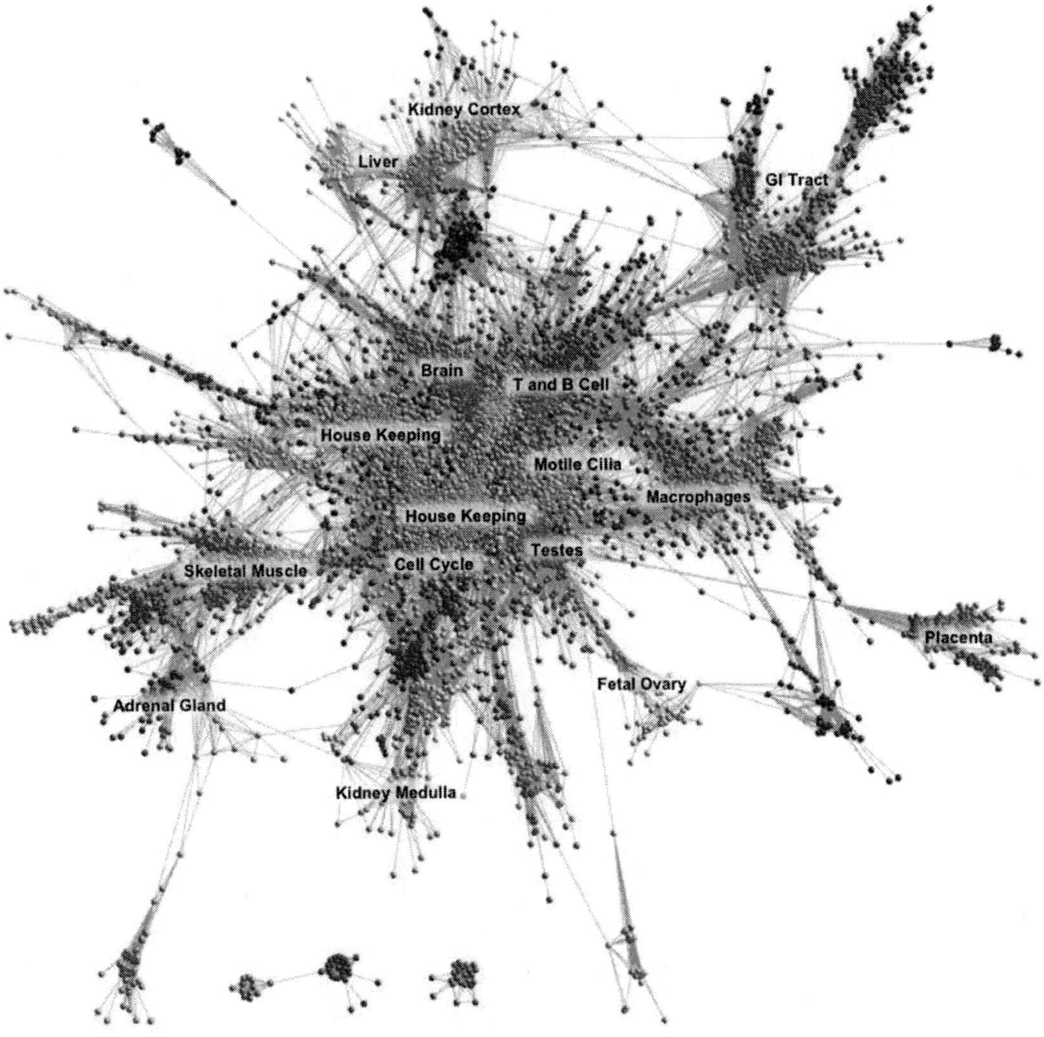

Figure 1. The transcriptional network of the sheep gene expression atlas dataset. Each node represents a single transcript, the lines between them represent correlations (edges) and the colours are shared by nodes that have correlated expression across the network (The graph is comprised of 15,192 nodes (genes) and 811,213 edges, $r = 0.75$, MCLi = 2.2).

TRANSCRIPTIONAL ATLAS PROJECTS

All of the processes that underpin development, growth, physiology and productivity depend upon the functions of numerous gene products that act together to generate pathways, macromolecular complexes, organelles, cells, organs and systems The set of genes required to deliver a cell-type, an organelle or a functional complex must share transcriptional regulation, so that their products are available in the correct place at the right time. If one samples the transcriptome of many different organ and cellular systems that differ from each other, the levels of transcripts encoding products that function together must be correlated with each other. The more physiological states that one samples, the more stringently one can determine that a pair of genes shares strict coexpression. Since the pioneering efforts that produced the Symatlas (now BioGPS, http://biogps.org) from sets of microarray data from mouse and human cells and tissues, there has been an explosion of gene expression "atlases" across multiple tissues in a number of species and within tissues across cell types and developmental time in humans and mice. The principal of guilt by association, namely that one can infer a great deal about the likely function of a gene product from its transcriptional neighbours, was clearly fulfilled in analysis of the mouse BioGPS dataset (Hume *et al.* 2010). For example, the entire set of genes encoding the lysosome was co-expressed, and specifically elevated in phagocytes. Similarly, genes involved in the cell cycle, in protein synthesis, or in extracellular matrix, clearly formed co-expression clusters because they are regulated activities and different cells and tissues engage these pathways to different extents. The exception is the set of genes that is relatively ubiquitously-expressed: the house-keeping genes. The housekeeping gene set also contains the highest proportion of genes that lack informative annotation, a reflection of the focus of biologists on differential expression. To identify and visualise transcriptional clusters in very large datasets, we utilized the network-clustering tool Biolayout *Express*^{3D}, now developed as Miru (http://www.kajeka.com). One advantage of the consistency of commercial microarray platforms was that it was possible to consolidate and integrate data from multiple laboratories, for example to generate an atlas of gene expression in human cells (Mabbott *et al.* 2013), also available as a default set on BioGPS.

The generation of transcriptional atlases for livestock species is more recent. We utilized extensive EST data to design a comprehensive microarray for the pig, and created a transcriptional atlas (Freeman *et al.* 2012). One example of the principal of guilt-by-association was the identification of a comprehensive set of transcripts associated with mitochondrial oxidative phosphorylation, and separation of the nuclear and mitochondrial-encoded transcripts (indicating that their transcription is not perfectly correlated). A bovine expression atlas was generated based upon tag sequencing of tissue from adult, juvenile and fetal tissues (Harhay *et al.* 2010) and subsequently extended in a set of 18 tissues from a single animal by RNAseq (Chamberlain *et al.* 2015). More recently, we have produced an extensive transcriptional atlas based upon direct sequencing of mRNA from six adult sheep as well as embryos and juveniles at various developmental ages (bioRxiv132696). The animals were deliberately chosen as cross breeds between the reference Texel (Jiang *et al.* 2014) and the Scottish Blackface. Figure 1 shows the overview of the transcriptional network, which clearly segregates the transcripts into tissue, cell-type and process-specific clusters. The latter clusters include a comprehensive set of genes involved in the cell cycle, protein synthesis, oxidative phosphorylation and motile cilia. Note also the close proximity of liver and kidney cortex in the network, indicating their similar expression profiles. We identified many transcripts encoding enzymes associated with gluconeogenesis and amino acid metabolism that are shared between the two organs. These data have also been made available on Biogps (biogps.org/sheepatlas). We are also currently analyzing similar projects, albeit on a smaller scale (guided by transcript diversity observed in the sheep) in commercial cross-bred goats, Indian and Mediterranean (the reference breed for the current assembly of a water buffalo genome) water buffalo and broiler and layer chickens.

These data together will produce a quantum leap in the analysis of transcript variants in each of the species and have contributed to the various genome projects to support improved annotation.

The next phase of genome/transcriptome annotation is the identification of regulatory elements. Several of the authors have had a long-term association with the FANTOM Consortium. The consortium utilized Cap Analysis of Gene Expression (CAGE) to generate a promoter-based atlas of gene expression in humans and mice (Consortium *et al.* 2014). CAGE, which is essentially genome-scale 5'RACE, also detects the short transcripts that are produced by active enhancers (R. Andersson *et al.* 2014) and the integration of information derived from detected promoter and enhancer activity can be used to infer the relationship between the two. Enhancers and promoters generated by CAGE sequencing were strongly correlated with similar elements detected by ChIP-seq analysis of the location of acetylated and methylated histones including data from the ENCODE consortium. In the analysis of a diversity of time courses of cell activation or differentiation, the transcriptional activity of enhancers in the vicinity of inducible genes was increased transiently in advance of detectable promoter activation (Arner *et al.* 2015, Baillie *et al.* 2017). The most recent FANTOM publication integrated CAGE and RNAseq data to identify 27,000 long non-coding RNAs encoded by the human genome, and to demonstrate that these transcripts derive primarily from enhancers. They further demonstrated that the lncRNAs that overlap trait-associated SNPs are expressed in cell types that are relevant to the trait in humans. The RNAseq data we have obtained from livestock species also greatly expands the diversity of lncRNAs identified and by inference, will contribute to the location of likely trait-associated regulatory elements. The FANTOM5 data from humans and mice can be usefully mapped across to other large animals such as pigs to identify conserved promoters and enhancers (Robert *et al.* 2015), in the process supporting other evidence that the transcriptome of pigs is substantially more human-like than that of mice.

APPLICATIONS OF TRANSCRIPTOMIC DATA IN GENETICS

SNPs associated with enhancers and promoters detected by the FANTOM5 consortium were more likely even than exonic SNPs to be associated with human disease susceptibilities (R. Andersson et al. 2014), mirroring evidence based upon identification of open chromatin detected as DNaseI hypersensitive sites (Maurano *et al.* 2012). More recently, genome-wide analysis of long range interactions between distal enhancers and promoters in multiple human cell types provided further links between regulatory variants and disease susceptibility traits (Javierre *et al.* 2016). The principle can be extended further. Regulatory variation in sets of genes that each contribute independently to a common pathway are likely to each contribute to a trait that depends upon that pathway. Consistent with the proposal, it is possible to identify and quantify co-expression of RNAs from trait-associated regions (bioRxiv, 095349) and from that information, to draw inferences about the likely underlying biology and to identify additional candidate susceptibility loci. Based upon that principle, we formed the hypothesis that genes involved in susceptibility to inflammatory bowel disease (IBD) were co-expressed specifically in monocytes and regulated during their differentiation. We identified a set of promoters that fulfilled that criterion and which were strongly enriched for associations with IBD, including >100 novel loci (Baillie et al. 2017).

The link between SNPs in regulatory regions and complex traits, of course has an intermediate phenotype in the form of heritable variation in the level of the regulated transcript, so-called expression quantitative trait loci (eQTL). Variation within such loci may act in cis or trans to produce differences in transcript abundance. Most evidence of eQTL to date has relied on microarray profiling of the same tissue or cell type from large numbers of individuals and conventional GWAS, or in defined crosses, an approach that has been called "genetical genomics" (de Koning *et al.* 2007, Martinez-Montes *et al.* 2017). Studies of human leukocytes have revealed that the large majority of transcripts detected on a microarray display detectable and heritable variation in expression (Fairfax *et al.* 2014, Westra *et al.* 2015).

Sequence-based analysis of the expression of each allele in individual animals has the potential to massively increase the power of detection of eQTL (Almlof *et al.* 2012), and this approach has become substantially more straightforward with the feasibility of obtaining high depth coverage of DNA and RNA sequences from the same animal(s). Chamberlain et al (Chamberlain et al. 2015) utilized RNAseq data to demonstrate the pervasive allele-specific expression of genes in 18 tissues of a single cow, including a surprising level of mono-allelic or parent of origin-specific expression and tissue-specificity. The sheep genome consortium also noted pervasive mono-allelic expression in transcriptome analysis of the pure-bred Texels (Jiang et al. 2014). In our own RNAseq data from multiple species, we have deliberately chosen to analyse cross-bred animals, and sequenced a wider diversity of tissues at greater depth than previous studies. One of the advantages of deep sequencing is that unprocessed nuclear RNAs, and lncRNA are covered at sufficient depth to detect variation in expression, and these non-coding regions have much higher density of SNVs (Barreiro *et al.* 2008). The MBASED algorithm (Mayba *et al.* 2014) can be used to integrate expression estimates from multiple SNV level RNAseq counts, to integrate allele specific expression (ASE) detection across a locus. With sufficient sequencing depth, the analysis can extend into neighbouring regulatory regions without the requirement for phasing information. We are currently utilizing this approach to identify ASE in sheep, water buffalo, pig and chicken RNAseq datasets.

One of the applications of particular interest is to begin to understand the benefits of cross-breeding or heterosis. The molecular basis for the benefits of cross-breeding is relatively poorly understood, and much of the analysis comes from plants, rather than animals (Chen 2013). In the sheep transcriptional atlas, we were able to integrate data from a smaller RNA-seq atlas derived from pure-bred Texels, produced in association with the release of the sheep genome (Jiang et al. 2014). A subset of transcripts was much more highly-expressed in the muscle and brain in the cross-bred animals than in the pure Texel animals. If most trait variation is associated with transcriptional regulation, heterosis presumably derives from some form of optimal contribution of the variant expression alleles of each parent within the cell and tissues that control the trait. Combining data from transcriptional networks and allele-specific transcription in cross-bred animals may eventually underpin the prediction of cross-bred animal performance.

GENOME EDITING

Alongside the revolution in genome sequencing, genome editing technologies provide a second revolution; the capacity to confirm predictive functions by altering the genome in model organism or in the species of interest. However, genome editing is more likely to be deployed in farmed animal species to modify or delete protein coding genes in order to generate animals with desirable genotypes that cannot readily be established by conventional selective breeding. A couple of recent examples of such desirable traits are resistance to Porcine Reproductive and Respiratory Syndrome Virus (PRRSV) and germline ablated male pigs that can serve as vehicles to increase the delivery of gametes from elite males (Burkard *et al.* 2017, Park *et al.* 2017). The use of primordial germ cells has expedited the application of germ-editing in poultry (Taylor *et al.* 2017). Perhaps more challenging is the prospect of accelerating genetic gain in breeding programmes by multiplex editing of functional variants in a single generation (Hickey et al. 2016), or even the application of so-called "gene drives" (Gonen *et al.* 2017). That prospect is certainly on the horizon, but the consequences of editing enhancer elements in mice have not been entirely predictable. Most genomic loci contain numerous apparently conserved and functional enhancers, and many others that are gained and lost between species (Villar et al. 2015). There is still some way to go before we can predict consequence from sequence in regulatory elements.

CONCLUSIONS

The availability of high throughput sequencing and its decreasing cost combined with development of new methods for modifying animal genomes has opened a wide range of approaches that will enhance genome annotation in livestock animals and lead to greater understanding of important production traits and processes such as heterosis.

REFERENCES

Almlof, J. C., et al. (2012). PLoS One **7**(12): e52260.
Andersson, L., et al. (2015). Genome Biol **16**: 57.
Andersson, R., et al. (2014). Nature **507**(7493): 455-461.
Arner, E., et al. (2015). Science **347**(6225): 1010-1014.
Baillie, J. K., et al. (2017). PLoS Genet **13**(3): e1006641.
Barreiro, L. B., et al. (2008). Nat Genet **40**(3): 340-345.
Bickhart, D. M., et al. (2017). Nat Genet **49**(4): 643-650.
Boschiero, C., et al. (2015). BMC Genomics **16**: 562.
Burkard, C., et al. (2017). PLoS Pathog **13**(2): e1006206.
Chamberlain, A. J., et al. (2015). BMC Genomics **16**: 993.
Chen, Z. J. (2013). Nat Rev Genet **14**(7): 471-482.
Couldrey, C., et al. (2017). J Dairy Sci.
de Koning, D. J., et al. (2007). Poult Sci **86**(7): 1501-1509.
Fairfax, B. P., et al. (2014). Science **343**(6175): 1246949.
FANTOM Consortium (2014). Nature **507**(7493): 462-470.
Freeman, T. C., et al. (2012). BMC Biol **10**: 90.
Gonen, S., et al. (2017). Genet Sel Evol **49**(1): 3.
Harhay, G. P., et al. (2010). Genome Biol **11**(10): R102.
Hickey, J. M., et al. (2016). J Anim Breed Genet **133**(2): 83-84.
Hoff, J., et al. (2017). BioRxiv.
Hume, D. A., et al. (2010). Genomics **95**(6): 328-338.
Javierre, B. M., et al. (2016). Cell **167**(5): 1369-1384 e1319.
Jiang, Y., et al. (2014). Science **344**(6188): 1168-1173.
Jubb, A., et al. (2015). Lancet **385 Suppl 1**: S54.
Kapetanovic, R., et al. (2012). J Immunol **188**(7): 3382-3394.
Luchin, A., et al. (2001). J Biol Chem **276**(39): 36703-36710.
Mabbott, N. A., et al. (2013). BMC Genomics **14**: 632.
Mackay, T. F. (2014). Nat Rev Genet **15**(1): 22-33.
MacLeod, I. M., et al. (2016). BMC Genomics **17**: 144.
Martinez-Montes, A. M., et al. (2017). Mamm Genome **28**(3-4): 130-142.
Maurano, M. T., et al. (2012). Science **337**(6099): 1190-1195.
Mayba, O., et al. (2014). Genome Biol **15**(8): 405.
Nishio, M. and M. Satoh (2015). Genet Sel Evol **47**: 32.
Park, K. E., et al. (2017). Sci Rep **7**: 40176.
Perez-Encisco, M., et al. (2015). Association for the Advancement of Animal Breeding Proceedings 21: 21126.
Phillips, P. C. (2008). Nat Rev Genet **9**(11): 855-867.
Robert, C., et al. (2014). BMC Genomics **15**: 550.
Robert, C., et al. (2015). BMC Genomics **16**: 970.
Saleheen, D., et al. (2017). Nature **544**(7649): 235-239.
Schroder, K., et al. (2012). Proc Natl Acad Sci U S A **109**(16): E944-953.

Stanislas, V., et al. (2017). BMC Bioinformatics **18**(1): 54.
Taylor, L., et al. (2017). Development **144**(5): 928-934.
Telenti, A., et al. (2016). Proc Natl Acad Sci U S A **113**(42): 11901-11906.
Villar, D., et al. (2015). Cell **160**(3): 554-566.
Westra, H. J., et al. (2015). PLoS Genet **11**(5): e1005223.
Worley, K. C. (2017). Nat Genet **49**(4): 485-486.
Zhang, X., et al. (2016). BMC Genomics **17**: 379.

Proc. Assoc. Advmt. Anim. Breed. Genet. 22:9-14

FROM BREEDING TO MARKET: OPPORTUNITIES WITHIN A DISRUPTED FOOD CHAIN

A.W. Campbell and P.R Amer

AbacusBio Ltd., Dunedin, New Zealand

SUMMARY

Food production and traditional agricultural systems are in a state of change due to increasing consumer demands and technological advances. In this paper, we outline international food trends as a background for creating discussion for supporting the positioning of future animal breeding programmes. We then present at a high-level, two case studies of programmes where breeding indexes could, or have been altered, to achieve changes in traits previously unselected for. We complete the paper with some discussion on the challenges of being closely aligned with industry and what that means for developing capabilities of young scientists.

THE FAST CHANGING WORLD WE LIVE IN

"Beware of the incumbent's chortle" was a line given in response to a discussion about disruption. It is of course, quite probable that the Blockbuster's former CEO chortled at the concept of live-streaming, when they turned down the opportunity to buy Netflix.

Disruption is a term in regular use and the food industry is not immune. Insect-based proteins, synthetic and plant-based meat and milk products, greenhouse gas (GHG) minimisation and consumer beliefs associated with animal welfare are challenging traditional agriculture food production systems. So too are these challenges creating opportunities and we, as scientists and technologists, have an important role to play in working closer with industry to take full advantage of them.

INTERNATIONAL FOOD TRENDS

Consumer power. Food brands have long-been established through clever marketing campaigns and product positioning. However, consumers' rising distrust of the food industry and their ability to promote or undermine companies via social media has led to a change in the balance of power. Because of this, consumers need to be at the forefront of research and development strategies of companies and industries.

For companies and industries striving to differentiate themselves from commodity producers, a sticker or label on packaging is not enough to denote where a product is from and how it has been produced. Layers of evidence as to how food has been grown and produced, fulfilling ethical, welfare and environmental considerations, and a connected value chain are critical for commanding premium food prices.

Food and health. A significant international food trend is the relationship between food and health. This is led by Chinese consumers who have been described as "the world's most health conscious," based on a long tradition of food-based medicines. In China, 73% of consumers are willing to pay a premium for healthier products (12 points higher than the global average), preferring products which treat common ailments, boost energy and strengthen immune systems (Boston Consulting Group 2014).

Aligned with the food and health trend is an increasing interest in the concept of personalised-food, and not just for humans. "Just Right by Purina" allows dog owners to order personalised-food formulations for their dogs. The formulations are derived according to the dog's breed, activity levels, coat and skin condition and the state of their stools. Similarly, for humans, Soylent is an example of formulated foods designed for specific human nutritional needs. Large food companies, such as Nestle with their "Choose Wellness," programme, are investing significantly into this area.

The changing face of food retail. The way we buy food is changing which is important for the delivery of foods in a personalised form. Factories and big companies are out of vogue. Consumers want to feel a connection to growers and the rise in popularity of Farmers' Markets is testament to this. There are increased quantities of high-end food products being sold on-line directly, or via meal-kit companies such as Blue Apron (United States) and My Food Bag (New Zealand). This ability to directly market and sell to consumers means that smaller companies can offer niche products profitably, opening-up opportunities for artisan growers.

Food as an experience. Younger consumers are increasingly seeking authentic and novel food experiences, in preference to more traditionally sought fine-dining experiences. This creates opportunities for producing novel food products derived from less traditional livestock cuts, such as offals and from other species such as crickets. It is worth noting that *Acheta domesticus* -the humble domestic cricket, is far better at converting ingested food into protein than cattle and crickets also have far greater fecundity (1,200–1,500 offspring per female).

The concept of food as an experience, also generates opportunities associated with food-tourism, of relevance to both Australia and New Zealand's significant tourism industries which are connected to our landscapes.

"You know things are changing in the food sector when you get gourmet nosh from a food truck, when your beer comes bolstered with protein, and McDonald's introduces a kale-enhanced breakfast" (Keown and Brendish, 2015).

LIVESTOCK AS A SOURCE OF PROTEIN

Protein consumption is rising internationally, especially in emerging economies. Annual meat production is projected to increase from 218 million tonnes in 1997-1999 to 376 million tonnes by 2030 (World Health Organisation). In response, emerging economies are fast-developing their own sources of protein with livestock production programmes growing in efficiency and volume throughout Asia and Africa.

In parallel, the role of ruminants in the food chain is increasingly being questioned as awareness around climate change grows. In the future, we may see political trade ramifications for high-carbon products (Ciochetto, 2016) and food producers will become more vulnerable to negative campaigns, be they political or social.

Thirty per cent of Earth's land surface is already devoted to livestock production, a practice that accounts for nearly 15% of global greenhouse-gas emissions (reviewed in Heffernan, 2017). Cows are the seen as the worst environmental culprits, not only because they emit a lot of methane, but because the production of beef uses vast quantities of water: 15,415 litres for a kilogram of beef (reviewed in Heffernan, 2017).

Alternate protein may lead to a reduction in protein sourced from livestock but it is unlikely to become an either/or situation. Livestock producers that position their products at the high-value end of the spectrum will not be as challenged by alternate proteins as those who operate in the commodity space. Fully-housed livestock, produced in commodity-style with high health and feed inputs, will increasingly be shunned by consumers.

Adding-value to livestock products, should at a very minimum, encapsulate where the product is from, how it is produced, and have sales channels which are different to traditional commodity channels. By shifting more of Australasia's production systems to this minimum value-add form we would expect increased prices and reduced volatility of those prices. This is because consumers exhibit a lower price sensitivity to products which are more expensive and further processed (Baiardi et al., 2014). A major challenge in making this shift is to ensure that the increased costs of producing a high-value product are a sound investment in the market, because almost by definition, further processing narrows the potential end-use for a product.

WEALTH OF DATA

Throughout the value chain, increasing amounts of data are being generated. Consumers are wearing devices measuring heart rates and sleep patterns. At the other end of the value chain, devices are under development for livestock to wear or be tracked by, and for land-based activities, such as irrigation and nutrient monitoring (reviewed in King 2017).

We will be moving into an era where what we eat will be defined for us by what we have done during the day, informed by internally and externally worn sensors. Similarly, farmers and animal breeders will have access to unprecedented amounts of animal performance data to strive for greater productivity with less impact on environments and creating connections with consumers.

Some of these data will have relevance for how we undertake breeding and genetic evaluation. Scientists will have access to data from greater numbers of animals and for differing traits. Traditional nucleus breeding programmes may be replaced.

In thinking about the types of capability required for positioning our industries for future success, geneticists, as both biologists and mathematicians, are in a prime position to be data integrators: adding value to inherently messy data by asking relevant questions and finding smart solutions to form the base for new technologies and applications.

CASE STUDY ONE: THE POTENTIAL TO INCLUDE GREENHOUSE GAS MITIGATION IN LIVESTOCK BREEDING INDEXES

Many livestock industries around the world are seeking good-news stories relating to environmental impact. An obvious option to reduce absolute GHG levels is to reduce livestock numbers, but this has major implications for production and economic well-being and as such, is unlikely to be taken up, unless there is considerable compliance pressure and/or economic alternatives.

An alternative is to reduce GHG intensity - GHG per unit of product. Under current selection approaches, the drive to improve production efficiencies indirectly lowers GHG intensity year-on-year. So far, modelling data has demonstrated that this is likely to be a positive news story, in that current and historic selection efforts improve livestock production efficiency substantially, and this reduces emissions intensity (Ludemann et al., 2011; Amer et al., 2017a; Amer et al 2017b; Quinton et al., 2017a, Quinton et al. 2017b).

There is a more aggressive option available for reducing GHG emissions intensity of livestock. This involves placing greater than current relative selection pressure on the traits that improve GHG emissions intensity (GHG EI) the most, and correspondingly, less relative selection pressure on traits that do not tend to improve GHG EI (Quinton et al (paper submitted to Animal) and Ludemann et al., 2011). It turns out that these indexes which extract more GHG EI gains than purely farm profit based indexes are typically efficient, in that significant improvements in GHG EI gains can be extracted with only modest reductions in the farm profitability gains expected from selecting on the modified indexes.

There is also a challenge in this approach in that placing more emphasis on the traits that reduce GHG EI the most (for example litter size in sheep (Ludemann et al., 2011) and milk production in dairy (through a dilution of emissions effect) are in reverse (Wall et al., 2010) to a directional shift in breeding goals over past decades towards traits that make animals more functional and easy to farm.

Efforts to develop novel selection criteria to improve GHG EI, including feed intake measurements, and either methane yield per unit of feed, and/or total methane yield per animal may be hampered by genetic antagonisms with functional aspects of animals. Another challenge is the cost of either trait measurement of selection candidates, or the cost of implementing genomic selection schemes, whereby the genotyping costs are more than offset by reductions in phenotyping investment.

Incentives/compliance drivers for uptake. A driver for an increased focus on breeding to reduce GHG emissions might come through audited supply chain systems. Auditing is required, because of the antagonisms and costs discussed above, so that free-riders benefiting from the supply chain story might otherwise skimp on compliance.

Interestingly, choice of breeding males by commercial farmers is potentially easier to audit than many other GHG mitigating technologies and certainly more than any sort of actual farm GHG emissions audit. This is because detailed databases already exist containing a substantial proportion of breeding sector animals. If breeders were to record the commercial farm buyers of breeding males, and/or semen, and submit these to an auditing body, then the ongoing genetic trend of the commercial farm for genetic merit for GHG emissions intensity could be predicted accurately. Occasional, or random checking, via DNA verification or genomic relationship predictions, could be deployed at low cost to ensure accuracy of the system, and miss-reporting of sire purchases. The steps required to develop such a system are quite feasible when compared with what will be required to deploy and incentivise many other GHG mitigating farm technologies.

There are opportunities to link such GHG reduction genetic programmes with national positioning programmes connected to product markets, such as Origin Green. Origin Green is an Irish national sustainability programme implemented by Board Bia, the Irish Food Board and supported by Government and private companies.

Origin Green is ostensibly a marketing effort, but the differentiation comes from the supply chain (Shelman, 2016). The programme so far has seen 90,000 farms audited and carbon footprinted. At manufacturing level, over 470 food and drink manufacturers, which represent almost 95% of their total food and drink exports, have registered to take part in Origin Green. The opportunity therefore, is for Irish producers to be incentivised to use GHG reduction indexes as part of a commitment to Origin Green, striving for a subsequent value-increase for product off those properties.

CASE STUDY TWO: THE OMEGA LAMB PROJECT

Worldwide, there is a large (>$20 billion) and rapidly growing market for omega-3 and omega-3-enhanced products, and a static or declining source of omega-3 from marine fish oils, prompting concerns of shortages. Alternative sources of omega-3s are necessary to meet demand, particularly in continental countries, like China, where there are large populations that do not eat fish regularly. The European Food Safety Authority recommends a dietary intake of 250mg of EPA plus DHA (eicosapentaenoic acid; C20:5n-3, docosahexaenoic acid; C22:6n-3), a day. The estimated average daily intake of EPA and DHA in China for example, is just 49mg. This deficiency has prompted the Chinese Nutrition Society to review its dietary guidelines in order to increase the nation's intake of DHA and EPA fatty acids. As a result, there is now a substantial volume of research investigating the enhancement of omega-3 levels in beef, lamb, pork and chicken using alternative feeds and feed-lot systems. These feed systems use fish, algal or ALA (alpha-linolenic acid; C18:3n-3) rich seed supplements to enrich the omega-3 composition and, recently, small volumes of omega-3 enhanced beef, pork and chicken products have become available in markets.

The Omega Lamb project, led by red meat processing company, Alliance Group and sheep breeding company, Headwaters, was initiated in 2011, with a view to developing a naturally differentiated lamb product. The aim of the project has been to develop value-added lamb products, high in omega-3 and also incorporating other meat quality attributes and environmental-management philosophies. In the early years of the programme, this involved analysing over 300 sire lines and 20 forage lines for their impact on fatty acid composition.

Enhanced omega-3 levels in lamb have been achieved through a combination of selective breeding - using assessment of correlated traits related to intramuscular fat levels - and diet - using a chicory-red clover finishing system. This is the basis for the development of what is now a fully commercial pipeline of products. High-health lamb products are being marketed for their health attributes and have been endorsed as high quality by chefs and independent consumer taste panel analyses. These products are currently being sold for a premium in high-end New Zealand restaurants and in Hong Kong. This season, in the first year to market, product from 30,000 lambs has been processed, with the target of processing 60,000 lambs in year two. From here, key challenges for the programme are associated with quality control and scaling to a larger volume of product.

A key driver for the success of the Omega Lamb Project has been the early involvement of people representing all parts of the value chain. This included scientists, livestock breeders, commercial farmers meat processors and marketers. Early consumer studies in three markets, the United Kingdom, Germany and China, were also important in informing where the value opportunities lay. This big picture and value-chain commitment has been challenging to manage, but has been critical to the programme's success, throughout the research and development phases and now the commercialisation phase.

CHALLENGES OF MARKET-DRIVEN BREEDING PROGRAMMES

There are many examples of market-driven breeding programmes, some of which have had limited success. Green-wash, is a term used to describe products taken to market and sold under an undeserved environmental banner. As we stated earlier, consumers have become cynical and will question the positioning of products by companies. Products and companies that are seen as inauthentic will be quickly brought-down via social media. One of the challenges with connecting breeding programmes to market is to ensure that there is legitimacy behind market claims, for example claims of superior quality. In breeding terms, such legitimacy will come from a concerted and multi-year strategic investment into understanding traits and the time taken in selection to make a measurable difference. When such programmes have failed, a factor has been that the expectations of progress have not been managed from breeder to marketer and marketers have gone out too early with product claims. Additional challenges include those of scaling breeding programmes to produce enough product for profitability, managing quality throughout the value chain and when demand is created in-market, managing year-round supply and or consumer/retailer expectations around product availability.

There are successful programmes that are managing, or in the process of managing these challenges, some additional examples include the Ora King programme (premium eating quality salmon) and Lanaco (wool-based air filtration face-masks). In these programmes, as in the Omega Project, success is underpinned by a willingness to collaborate by geneticists and industry players throughout the value chain.

CAPABILITY

Scientists need to have a genuine interest in solving industry challenges in order to engage successfully with industry. This requires a deep understanding of company and industry drivers. It is hard to develop this in a purely academic environment and scientists should be encouraged to spend considerable time outside of that environment, to the point of spending periods of time embedded in companies or industry organisations. Such time is invaluable for developing relationships and understanding why things are never as simple as they seem in terms of implementation of longer-term research and development strategies.

Performance drivers for scientists within academic environments are often in conflict with such an approach and science organisations need to strive to develop new, or align existing performance measures with such an approach.

Finally, in the experience of these authors, there is tremendous satisfaction at playing a role bringing science and industry together. It is our view that this can be done in a way which maintains scientific valour and integrity and most importantly, makes a considerable impact in taking industries forward in a changing food environment.

REFERENCES

Amer, P.R., Hely, F.S., Quinton, C.D. and Cromie, A.R. (2017a). *Submitted to Animal.*

Amer P., Quinton C., Byrne T., Cromie A., Richardson C. and Archer, J. (2017b). *AAABG*, this edition.

Baiardi, D., Bianchi, C. and Lorenzini E. (2014). *Journal of Agricultural Economics*. **66**:358-391.

Boston Consulting Group (2014). https://www.bcgperspectives.com/content/articles/center_consumer_customer_insight_globalization_insight_action_capturing_share_chinas_consumer_health_market/?chapter=2 accessed 20 April 2017.

Ciochetto, L. (2016). In: The New Zealand land and food annual. Ed Massey, C. Page 215-224.

Herreman, O. (2017). *Nature* **544**, S18–S20.

Keown, J. and Brendish, L. (2015). http://idealog.co.nz/venture/2015/12/top-global-food-trend-5-sustainability. Accessed 30th April 2017.

King, A. (2017). *Nature,* **544**, *S21–S23*

Quinton, C.D., Hely, F.S., Amer, P.R., Byrne, T., Cromie, A.R. (2017a). *Submitted to Animal.*

Quinton, C., Hely, F., Byrne T., Amer P. and Cromie, A. (2017b). *AAABG*, this edition.

Shelman, M., McLoughlin, D. and Pagell. M. (2016). In: Organizing Supply Chain Processes for Sustainable Innovation in the Agri-Food Industry, 205-233.

Wall, E., Simm, G. and Moran, D. (2010). *Animal* **4**, 366-376.

World Health Organisation, http://www.who.int/nutrition/topics/3_foodconsumption/en/index4.html accessed 26 April 2017.

Proc. Assoc. Advmt. Anim. Breed. Genet. 22:15-18

INVESTIGATING EMERGING INHERITED DISEASES IN AUSTRALIAN LIVESTOCK: A COLLABORATIVE APPROACH

S.A. Woolley[1], E.R. Tsimnadis[1], N. Nowak[1], R.L. Tulloch[1], M.R. Shariflou[1], T. Leeb[2], C.E. Willet[3], M.S. Khatkar[1], B.A. O'Rourke[4], I. Tammen[1]

[1] Sydney School of Veterinary Science, Faculty of Science, The University of Sydney, Camden, NSW, Australia
[2] Institute of Genetics, Vetsuisse Faculty, University of Bern, Bern, Switzerland
[3] Sydney Informatics Hub, Core Research Facilities, The University of Sydney, Sydney, NSW, Australia
[4] Elizabeth Macarthur Agricultural Institute, NSW Department of Primary Industries, Menangle, NSW, Australia

SUMMARY

Emerging inherited diseases can cause numerous issues for producers, including productivity loss, profit loss and animal welfare problems. Under-reporting of emerging inherited diseases can result in difficulties associated with identifying and managing these diseases. The development of a research centre between the University of Sydney and Elizabeth Macarthur Agricultural Institute, NSW Department of Primary Industries is a current collaborative effort to encourage the submission of suspected inherited disease cases. Previous collaboration has resulted in the ongoing investigation of 10 inherited diseases using SNP-based homozygosity mapping and next generation sequencing to identify positional candidate genes and causal mutations. The long-term aim is to formally develop a research centre that allows independent investigation of emerging inherited diseases in livestock that builds upon current joint research.

INTRODUCTION

Emerging inherited diseases within Australian livestock can often go unreported, either because they are misdiagnosed as non-inherited diseases or are not reported due to concerns of profit loss and reputation damage. Not reporting suspected inherited disease cases can lead to a loss of valuable sample resources and a lost opportunity to characterise the phenotype(s), thus causing a delay in investigating or monitoring these diseases. Without the assurance of a robust genotyping test to identify heterozygous individuals, the management of autosomal recessive inherited diseases can become problematic, especially if detailed pedigrees are unknown for at-risk populations (Man *et al.* 2007).

The under-reporting of suspected recessive inherited diseases can contribute to the inadvertent dissemination of deleterious alleles throughout populations. If a deleterious allele can be traced to a common ancestor within a prominent sire line, all offspring are at risk of being heterozygous for the deleterious allele and only a DNA test will be able to accurately identify true heterozygous animals. Emerging inherited disease monitoring and the implementation of management programs to avoid carrier by carrier matings are important for reducing the number of affected progeny born, as well as mitigating any production and economic losses. The importance of these management programs has been shown in the case of brachygnathia, cardiomegaly and renal hypoplasia syndrome in Merino sheep (Shariflou et al. 2013), where breeding programs have reduced the number of affected progeny born (Shariflou, personal communication).

Researchers at the University of Sydney and the Elizabeth Macarthur Agricultural Institute, NSW Department of Primary Industries (EMAI) each have a longstanding history in investigating inherited diseases in Australian livestock and have recently started to collaborate on numerous research projects. So far, 10 inherited diseases are being investigated and are likely to be inherited via a recessive mode of inheritance: congenital mandibular prognathia (CMP) in Droughtmaster cattle, pulmonary hypoplasia with anasarca (PHA) in Persian sheep, Niemann-Pick type C disease (NPC) in Angus cattle, congenital blindness (CB) in white Shorthorn cattle, cervicothoracic vertebral subluxation (CVS) in Merino sheep, a new variant of cardiomyopathy woolly haircoat syndrome (CWH) in Hereford cattle, new variants of ichthyosis fetalis (IF) in Hereford and Shorthorn cattle, suspected cases of congenital contractural arachnodactyly (CCA) in Murray Grey cattle, ovine dermatosparaxis (OD) in Merino sheep as well as the previously reported brachygnathia, cardiomegaly and renal hypoplasia syndrome (BCRHS) in Merino sheep (Shariflou et al. 2013).

A SNP-chip based homozygosity mapping approach and next generation sequencing is described with an aim to identify positional candidate genes, identify causal mutations and develop diagnostic DNA tests. The long term aim resulting from these collaborations is to develop an independent centre where producers and veterinarians can report and submit samples of suspected inherited disease cases. The centre will follow a similar approach to previous studies conducted and will benefit the Australian livestock industries through increased awareness and acceptance of reporting.

MATERIALS AND METHODS

In current collaborative research projects, SNP genotyping was performed by the Animal Genetics Laboratory (University of Queensland, Gatton, Australia) and Australian Genome Research Facility (Westmead, Australia) (Table 1). Sliding windows of 25, 50 and 100 SNPs were used to identify runs of homozygosity (ROH) for all affected animals using the bovine UMD3.1 genome assembly and the ovine Oarv1.0 genome assembly. ROH were analysed using PLINK (Purcell *et al.* 2007) and were considered to be regions of interest if these regions were shared by all of the affected animals and not with any of the carrier and control animals. These regions were scanned for positional candidate genes based on gene function.

Table 1. Number of affected and carrier DNA samples sent for SNP chip genotyping and regions of homozygosity, including species specific OMIA ID

Disease	OMIA ID[1]	Breed	Affected /Carrier	SNP chip	Region of interest
Cervicothoracic vertebral sublaxation	000077-9940	Merino	14/2	SNP50[2]	OAR10
Pulmonary hypoplasia with anasarca	000493-9940	Persian	5/5	SNP50[2]	OAR1,3,4,6,7,9,17, 25,26
Cardiomyopathy and woolly haircoat syndrome	000161-9913	Poll Hereford	2/0	SNP80[3]	BTA1,4,6,12,15,24, 25
Congenital blindness	-	Shorthorn	2/3	SNP80[3]	BTA5,14,16,22,24
Congenital contractural arachnodactyly	001511-9913	Murray Grey	5/5	SNP80[3]	BTA21
Congenital mandibular prognathia	-	Droughtmaster	9/4	SNP80[3]	BTA26
Ichthyosis fetalis	000547-9913	Hereford	1/3	SNP80[3]	multiple
Niemann-Pick disease	-	Angus	2/2	SNP80[3]	BTA3,4,16,24,29

[1]OMIA http://omia.angis.org.au, - indicates no species specific OMIA ID. [2]SNP50 = Illumina® OvineSNP50 Genotyping BeadChip (CA, USA). [3]SNP80 = GeneSeek® Genomic Profiler Bovine HD Chip 80K chip (Neogen, NE, USA).

Sanger sequencing of select candidate genes was commenced but was cost and labour intensive. Next generation sequencing (NGS) of affected animals for CMP, CVS, PHA and BCRHS using the Illumina HiSeq™ X Ten sequencing platform was performed by the Kinghorn Centre for Clinical Genomics (Garvan Institute of Medical Research, Darlinghurst, Australia) through the Ramaciotti Centre for Genomics (University of New South Wales, Sydney, Australia) with 150bp paired-end reads at 30X coverage (Table 2). This NGS data has been aligned to either the bosTau8 or oviAri3 reference genome assemblies and will be analysed for genetic variants. Samples of affected animals for IF, CWH and OD are undergoing sequencing using an in-house Illumina HiSeq® 3000 sequencing platform in Switzerland (Table 2).

Table 2. Number of affected DNA samples for next generation sequencing

Disease	Breed	Affected	Expected coverage	% of sequences with mean Q>30
Brachygnathia, cardiomegaly and renal hypoplasia syndrome	Merino	1	30X	85.84
Cardiomyopathy and woolly haircoat syndrome	Poll Hereford	2	20X	In progress
Cervicothoracic vertebral sublaxation	Merino	2	30X	92.16
Congenital mandibular prognathia	Droughtmaster	2	30X	86.58
Ichthyosis fetalis	Hereford	1	20X	In progress
Ichthyosis fetalis	Shorthorn	1	20X	In progress
Ovine dermatosparaxis	Merino	2	20X	In progress
Pulmonary hypoplasia with anasarca	Persian	2	30X	90.17

RESULTS AND DISCUSSION

Homozygosity mapping has successfully revealed and/or excluded positional candidate genes for all of the inherited diseases currently being investigated (Table 1; Shariflou *et al.* 2013; Tammen *et al.* 2016). The known mutation for CCA in Angus cattle was confirmed to be present in the Murray Grey cattle with suspected CCA. Validation of a genetic variant in a positional candidate gene for NPC is ongoing. Partial Sanger sequencing of positional candidate genes for CVS, PHA, CMP and CWH did not reveal any disease-causing mutations and affected animals were therefore re-sequenced using NGS. Previous mapping of BCRHS did not identify a clear positional candidate gene and an affected animal sample was submitted for NGS. Known candidate genes for CWH and CB were excluded and alternate candidate genes need to be investigated within the regions of interest identified (Table 1). Strong candidate genes exist for IF and OD, as these diseases have been previously characterised in different breeds (Charlier *et al.* 2008; Zhou *et al.* 2012). The affected animals tested negative for the known disease causing mutations and were re-sequenced due to suspected genetic heterogeneity.

Preliminary quality control analysis of the NGS data is positive with per base sequence quality determined by a Phred score of Q>30 ranging from 85.84% to 92.16% (Table 2) with no over-represented sequences identified. After aligning data to the bosTau8 or oviAri3 genome assemblies, allelic variations including SNPs, indels and structural variants will be identified in the regions of interest previously identified, with a focus on positional candidate genes identified by homozygosity mapping.

The results from these studies indicate that SNP genotyping and homozygosity mapping methods are highly effective in identifying positional candidate genes for a range of disorders even if sample sizes are small and phenotypes are poorly defined. Genome wide SNP genotyping and homozygosity mapping approaches have successfully identified candidate genes and causal mutations in a range of recessive inherited diseases in cattle, including ichthyosis fetalis in Chianina cattle (Charlier *et al.* 2008). The inclusion of NGS data to identify allelic variations will allow for several runs of homozygosity identified through homozygosity mapping to be further investigated.

CENTRE CONCEPT

The methodology framework and results described in the current research projects between the University of Sydney and EMAI demonstrates the success of the working relationship between both groups. The concept of an independent research centre geared towards the molecular characterisation of emerging inherited diseases in livestock could provide a central point of contact for veterinarians, breeders, producers and breed societies. It has the potential to increase confidential reporting of suspected cases and provide research services with the aim to rapidly develop low-cost diagnostic tests based on frameworks that are already implemented at both institutions. The availability of diagnostic DNA tests will allow for informed breeding decisions to be made to avoid potentially devastating profit loss and animal welfare issues.

The centre will aim to publish validated results which will increase awareness for the role of emerging inherited diseases within Australian livestock populations. The future development of the centre will be focussed on developing a stream-lined research and diagnostic service that may involve additional research and industry groups. The key driving factor behind successfully developing an independent centre will be the collaborative relationships and shared resources between numerous research groups to encourage greater surveillance of emerging inherited disease in livestock across NSW and nation-wide.

ACKNOWLEDGMENTS

The authors acknowledge and thank the producers and veterinarians for the submission of samples to EMAI and the University of Sydney. The authors thank the genetics laboratory staff at EMAI for their assistance. The University of Sydney is acknowledged for the use of services and HPC facilities at the Sydney Informatics Hub and the Faculty of Veterinary Science provided research support for 3 honours projects (NN 2013, ET 2014 & SW 2015), funding support from the Animal Welfare, FS Quiney and Dorothy Minchin Bequests. The Australian Wool Education Trust is acknowledged for the support of a research scholarship for RT (2016). The NGS is funded by a Faculty of Veterinary Science - DPI collaborative compact fund and research grant CRSII3_160738 from the Swiss National Science Foundation to TL.

REFERENCES

Charlier C., Coppieters W., Rollin F., Desmech D., Agerholm J.S., *et al.* (2008) *Nature Genet.* **40**: 449.

Man W.Y.N., Nicholas F.W. & James J.W. (2007) *Prev Vet Med.* **78**: 262.

Purcell S., Neale B., Todd-Brown K., Thomas L., Ferreira Manuel A R., *et al.* (2007) *Am J Hum Genet.* **81**: 559.

Shariflou M.R., Wade C.M., Kijas J., McCulloch R., Windsor P.A., *et al.* (2013) *Anim Genet.* **44**: 231.

Tammen I., Woolley S., Tsimnadis E., Nowak N., Tulloch R., *et al.* (2016). *J Anim Sci.* **94** (7 Supplement 4): 173.

Zhou H., Hickford J.G.H. & Fang Q. (2012) *Anim Genet.* **43**: 471.

Proc. Assoc. Advmt. Anim. Breed. Genet. 22:19-22

HIGH GENETIC MERIT DAIRY COWS CONTRIBUTE MORE TO FARM PROFIT: CASE STUDIES OF 3 AUSTRALIAN DAIRY HERDS

J.E. Newton[1], M.E. Goddard[1,2], H.N. Phuong[1], M.A. Axford[3], C.K.M. Ho[1], N.C. Nelson[1], C.F. Waterman[1], B.J. Hayes[1,4] and J.E. Pryce[1,5]

[1] Department of Economic Development, Jobs Transport and Resources, Victoria, Australia
[2] University of Melbourne, Parkville, Victoria, 3010, Australia
[3] DataGene Ltd, Bundoora,Victoria, 3083, Australia
[4] Centre for Animal Science, University of Queensland, Queensland, Australia.
[5] La Trobe University, Bundoora, Victoria, 3083, Australia

SUMMARY

One of the barriers to the adoption of Australian Breeding Values (ABVs) is not having evidence that high genetic merit dairy cows actually contribute more to farm profit in practice. Using historical financial data collected as part of the Dairy Farm Monitor (DFM) Project, and historical cow production, health and mating records, a method was developed to compare the estimated contribution to farm profit of cows of differing genetic merit. High genetic merit cows contributed between $150 and $235 per cow more to farm profit each year without compromising their productive life, or incurring higher breeding or mastitis treatment costs.

INTRODUCTION

Although the Australian dairy industry is making genetic progress, the rate of actual genetic gain, $8/year (≈0.1 genetic standard deviations) increase in the Balanced Performance Index (BPI), is less than half of what is theoretically feasible. Under optimal conditions, genetic gain is projected to increase between 0.21 and 0.5 genetic standard deviations per year for progeny-testing and genomic selection respectively (Schaeffer 2006). The ImProving Herds project was established with the goal of improving farm profit through demonstrating the value of genetics and herd improvement in the dairy industry, a key goal recognised in the national Herd Improvement 2020 Strategy. Dairy Australia recommended that increased focus be placed on case studies and regionally specific extension activities to increase knowledge, trust and use of genetic tools in the dairy industry. To incorporate this suggestion, the ImProving Herds project is centred around 34 focus farms.

An across herd study of Irish dairy herds (n= 1131) found a 1 unit increase in the Economic Breeding Index was associated with a €1.94 (≈ AU$2.76) increase in net margin per cow, after adjustment for year, stocking rate, herd size and purchased feed (Ramsbottom *et al.* 2012). This value was very close to the €2 increase in net margin per cow predicted. The Australian dairy industry is not suited to an across herd economic analysis due to climatic variability, diverse feeding and management practices and variability in milk payment systems which exacerbate between herd variation in economic performance. To control for this variability, we elected to perform a within herd analysis, with focus farms from the ImProving Herds project as case studies.

The aims of this study were to 1) develop a method to calculate the contribution an individual cow makes to farm profit over her lifetime, and 2) investigate the relationship between cow genetic merit, profit and performance at the individual farm level.

MATERIALS AND METHODS

Two historical and independent databases were used for this study of 3 Victorian dairy farms: 1) the DFM database; the DFM project is a joint initiative between Agriculture Victoria and Dairy Australia which annually collects and analyses detailed financial and farm production data from dairy farms, and 2) DataGene; the national database of cow production, pedigree and ABV records. Within-herd long term averages over the 2008 to 2016 financial years, inclusive, were calculated for farm financial data, adjusted to present day values, and herd production data. All herds had cow lactation, health and mating records and at least 2/3 of cows had ABVs. To be included in this analysis, a cow's entire productive life had to fall within the 2008 to 2016 financial years, inclusive.

The individual contribution that each cow made to farm profit over her lifetime (Cow$) was calculated using the equation:

$$Cow\$ = \$_{milk} + \$_{calf} + \$_{cull} - (\$_{rear} + \$_{feed} + \$_{mastitis} + \$_{repro} + \$_{herd})$$

Lifetime milk income ($\$_{milk}$) was calculated by multiplying total milk solids (MS) by average milk price ($/kg MS). Income from calf sales ($\$_{calf}$), and costs of mastitis treatment ($\$_{mastitis}$) and animal mating ($\$_{repro}$) were calculated by summing the number of incidences of each event and multiplying by the dollar value, in $ per cow, of one occurrence of that event. A cow's salvage value ($\$_{cull}$) was assumed to be the average within-herd cull cow price unless she was recorded as dead, then $\$_{cull}$ was $0. If more than 12 months had passed since the cow was last seen in the herd she was assumed to have been sold. The initial investment in rearing the cow to the point of entering the milking herd ($\$_{rear}$) was assumed to be $1606 (Byrne *et al.* 2016). Feed costs were calculated by multiplying the within-herd average cost of feed consumed ($/Megajoule of metabolisable energy, $/MJ ME) by each cow's energy requirements. Cow energy requirements were calculated using the equations in CSIRO (2007). They accounted for cow age and breed, lactation and pregnancy records and herd level information about distance walked each day, farm topography, liveweight and condition score loss during lactation. Dairy and general herd health costs ($\$_{herd}$) were assumed to be proportional to the cow's productive life. Day 1 was taken as the date of first calving. To account for discounting over time, all elements of the profit equation were calculated in 365 day periods, a 5% discount rate applied and then summed together.

Cow ABVs are breed specific. The 3 herds had Holstein (Herd C), Jersey (Herd A) and mixed Jersey and Holstein (Herd B) cows. DataGene presents breed specific genetic evaluations (with different bases for each breed), so the original solutions were obtained (from multi-breed models) and rescaled using the Holstein ABV parameters, enabling a within-herd, but across breed analysis to be used. The BPI is the Australia dairy industry's main index. It was developed using a bio-economic model to balance improvements in longevity, health, type, fertility and production to maximise farm profit (Byrne *et al.* 2016). For this study, within each herd each cow was classified into two sub-herds, either low or high BPI based on whether she was below or above the median BPI for her contemporary group; herd and year of first calving. A linear model weighted by cow productive life (in days) was used to test for differences in annualized physical and financial measures of cow performance in the low and high BPI sub-herds. This analysis was performed separately for each herd. The results below are presented as the estimate of the difference between the two sub-herds within each of the 3 herds from the weighted linear model.

RESULTS AND DISCUSSION

In all 3 herds, splitting the herd based on median BPI resulted in significant ($p<0.05$) differences in ABV between the high and low BPI sub-herds (Table 1). The difference in BPI between the two sub-herds ranged from $78 to $116. All high BPI sub-herds had significantly ($p<0.001$) higher BPI, milk production and survival ABVs than the below BPI sub-herds (Table 1). Two out of three high BPI sub-herds also had significantly higher cell count ABVs and lower fertility ABVs.

Cows in the high BPI sub-herds produced significantly ($p<0.05$) more litres of milk, and kilograms of fat and protein each year than their low BPI counterparts (Table 2). All high BPI sub-herds tended to have cows with a longer productive life, but this difference was only significant ($p<0.05$) for 1 herd.

Table 1 Estimated difference (s.e) in ABVs between high and low BPI sub-herds from weighted linear model. Significance of p-value (NS >0.05,* = <0.05, ** = <0.01, * = <0.001)**

Herd	BPI	Protein	Milk	Fat	Cell count	Fertility	Survival
A	78 (5)***	10 (1)***	269 (71)***	17 (2)***	3 (2)^{NS}	0 (1)^{NS}	2 (0)***
B	94 (6)***	13 (1)***	376 (66)***	18 (2)***	6 (2)**	-1 (1)*	2 (0)***
C	116 (4)***	14 (1)***	340 (45)***	21 (2)***	3 (1)*	-1 (0)**	3 (0)***

Table 2 Estimated difference (s.e) in average physical parameters between cows in high and low BPI sub-herds from weighted linear model. Significance of p-value (NS>0.05,* = <0.05, ** = <0.01, * = <0.001)**

Herd	Milk (L/yr)	Fat (kg/yr)	Prot (kg/yr)	Productive life (months)	Calving interval (days)	Lactation length (days)	No. calves (calves/yr)
A	434 (154) **	26 (6) ***	19 (5) ***	4 (3) NS	-11 (10) NS	1 (10) NS	0.0 (0.0) NS
B	411 (131) **	20 (5) ***	19 (4) ***	5 (3) NS	22 (16) NS	19 (14) NS	0.0 (0.0) NS
C	265 (125) *	27 (4) ***	19 (4) ***	4 (2) *	34 (10) ***	25 (8) **	-0.1 (0.0) ***

All high BPI sub-herds were significantly ($p<0.01$) more profitable, with the average difference ranging from $150 to $235 per cow/year (Table 3). The main source of this difference was greater yearly milk income, with cows in high BPI sub-herds generating on average between $185 and $258 more income from milk sales each year. Although feed costs were higher in the high BPI sub-herds, the extra cost of feed ranged from $30 to $42, which was more than compensated for by additional milk income. Increases to milk income were achieved without decreasing ,and in one case significantly ($p<0.05$) increasing, the average productive life of the high BPI sub-herds (Table 2) and without significantly ($p>0.05$) increasing mastitis costs (Table 3). This finding goes some way to dispel the widely-held belief that high producing animals break down earlier and are more prone to mastitis. Although cows in high BPI sub-herd C had significantly ($p<0.001$) longer calving intervals and fewer calves per year (Table 2), they also had significantly longer lactations ($p<0.01$) and a tendency ($p=0.10$) for lower AI costs each year.

Table 3 Estimated difference (s.e) in the contribution each cow makes to profit (Cow$) and Cow$ components between high and low BPI sub-herds from weighted linear model. Significance of p-value (NS >0.05,* = <0.05, ** = <0.01, * = <0.001)**

		Income ($/yr)		Costs ($/yr)				
Herd	Cow$ ($/yr)	Milk	Calf	Feed	AI	Preg test	Mastitis	Rearing
A	178 (50) ***	208 (51) ***	-2 (4) NS	-42 (16) **	6 (4) NS	0 (0) NS	-3 (3) NS	52 (44) NS
B	150 (49) **	185 (43) ***	-7 (4) NS	-34 (12) **	-4 (4) NS	0 (0) NS	-3 (3) NS	55 (39) NS
C	235 (40) ***	258 (49) ***	-10 (2) ***	-30 (12) *	6 (3) NS	0 (0) NS	1 (3) NS	31 (29) NS

At the national level the regression of profit and BPI is expected to be a $1 increase in profit for every unit increase in BPI (Byrne *et al.* 2016). In the three case study herds, the ratio between Cow$ and BPI was higher than this at $2.28, $1.60, $2.03 for herds A, B, C respectively. This differs from Ramsbottom *et al.* (2012) whose €1.94 (≈AU$2.76) increase in net margin per cow was very close to the expected increase of €2.00. A possible reason is that the Victorian herds in our study are not representative of the national average, whereas Ramsbottom *et al.*'s (2012) larger study of 1131 herds better captures the national variation in Irish dairy herds. An indication this may be the case is that average feed cost for the herds in our study ranged from $0.016 to $0.022/ MJ ME whilst the national average purchased feed cost is $0.025 /MJ ME (Byrne *et al.* 2016).

The phenotypic records that were used to calculate Cow$ have also been used in cow ABV estimation. An alternate approach that uses ABVs derived from parent average or genomic prediction could also be used. A parent average analysis was conducted, with similar results obtained. Differences in Cow$ between the sub-herds selected based on parent average BPI were significant ($p<0.05$) in two herds and approached significance ($p<0.1$) in the third herd. In choosing which set of results to present, the end goal of the ImProving Herds project needs to be considered. The goal of the ImProving Herds project is to increase knowledge, trust and usage of genetic tools, such as ABVs and the BPI index, in the Australian dairy industry. For the purposes of demonstrating that ABVs "work" to farmers it is therefore most relevant to use the ABVs in the format they appear in existing industry tools.

This analysis required in depth historical financial, pedigree, performance and management information from the case study herds which is not available on all focus farms to such a high level of detail. A simplified approach using regional historical financial information will enable a similar analysis of the project's 34 focus farms, and potentially other dairy farms, who have cow ABVs and accurate lactation records. The transferability of the approach used here to other livestock species will be determined by the availability of detailed phenotypes for key contributors to farm profit and validated financial records.

CONCLUSION

Using an independent financial data source, the DFM project, it was successfully shown that the assumption made at the national level about the positive relationship between cow genetic merit and cow contribution to farm profit holds true at the individual farm level. Although high genetic merit animals have higher feed costs, these are more than compensated for by greater milk income. Furthermore, our analysis indicates that high BPI cows do not have a shorter productive life, nor higher mastitis incidence or mating costs. These case studies provide the opportunity to contribute to localised extension activities and help build the dairy industry's trust, knowledge and use of ABVs.

ACKNOWLEDGEMENTS

The authors gratefully acknowledge the farmers who provided access to their data and Bill Malcolm, University of Melbourne, for providing feedback during method development. ImProving Herds is funded by the Gardiner Foundation and Dairy Australia through collaborative support from the Department of Economic Development, Jobs, Transport and Resources, Victoria, DataGene Ltd., Holstein Australia and the National Herd Improvement Association of Australia.

REFERENCES

Byrne T.J., et al. (2016) *J. Dairy Sci.* **99**:8146

CSIRO (2007) 'Nutrient requirements of domesticated ruminants.' CSIRO Publishing, Melbourne

Ramsbottom G., Cromie, A.R., Horan, B. and Berry, D.P. (2012) *Anim.* **6**:1031

Schaeffer L.R. (2006), *J. Anim Breed Genet* **123**:218

Proc. Assoc. Advmt. Anim. Breed. Genet. 22:23-26

THERE IS NOTHING ROUTINE ABOUT ROUTINE TESTING. A PERSPECTIVE FROM THE UNIVERSITY OF QUEENSLAND'S ANIMAL GENETICS LABORATORY

R.E. Lyons[1] and S. Buttsworth[1]

[1]The University of Queensland, School of Veterinary Sciences, Gatton, QLD 4343, Australia.

SUMMARY

The following article is a reflection on current trends and challenges in genetic testing across the livestock sector, particularly the cattle industry, from the perspective of a significant genetic testing laboratory based at The University of Queensland.

INTRODUCTION

Much has changed in genotyping technologies since The University of Queensland's Animal Genetics Laboratory (AGL) was first established in 1985. While cattle makes up the single largest species tested at AGL, we also cater for sheep, alpaca, goat and pigs, as well as services and research for the aquaculture industry, fisheries and wildlife ecology research groups. Below are insights into the operations of a successful genetics laboratory.

AGL DOES MUCH MORE THAN SIMPLY GENOTYPE CATTLE.

AGL serves a very wide client base, ranging from research organisations to breed societies, pastoral companies and small to medium-sized livestock producers. Additionally we provide support to the Gatton-based research communities, state police services and others. Hence, it is a requirement for AGL to be both nimble and adaptable. Australian farmers are a unique clientele operating a range of diverse production systems in different terrains and producing cattle for various markets, all whom have specialised requirements and expectations.

Therefore the range of services provided needs to be multi-faceted. While for some clients the experience may be purely transactional (samples in, results reported), many others are looking for a more personalised & ongoing service. AGL's clients are country people that appreciate the ability to discuss testing options and interpretations. In many cases AGL staff have built both rapport and understanding of the herds of many clients, Genotyping results are often merely the beginning, or continuation of, a long and prosperous relationship. In many cases, AGL retains critical herd-specific knowledge that spans many years, and many property managers' tenures.

GROWTH/MARKET TRENDS

The number of samples AGL receives has grown considerably (Figure 1). Looking at the last 5 years (2011-2016) alone, the growth in cattle samples, as measured by case numbers assigned per annum, has averaged 13.4% per annum. This is actually an underestimate of testing volumes given that in the last year or 2 there has been significant client-driven demand for retesting of animals already in the system, and these are not captured in Figure 1.

It is also instructive to look at testing trends over this period. From 2012 - 2016 the number of samples processed on microsatellites (MiP) has remained relatively stable at AGL, excluding a larger than normal demand in 2012 (Figure 2). During this time there has been a rapid increase in the use of genomics and SNP-base parentage (SEQ) requests. In the case of the GeneSeek Genomic Profiler low-density BeadChip (GGPLD), usage was initially for research projects, but the steadily increasing demand for the assay in 2015 and 2016 is primarily due to increased demand from livestock producers.

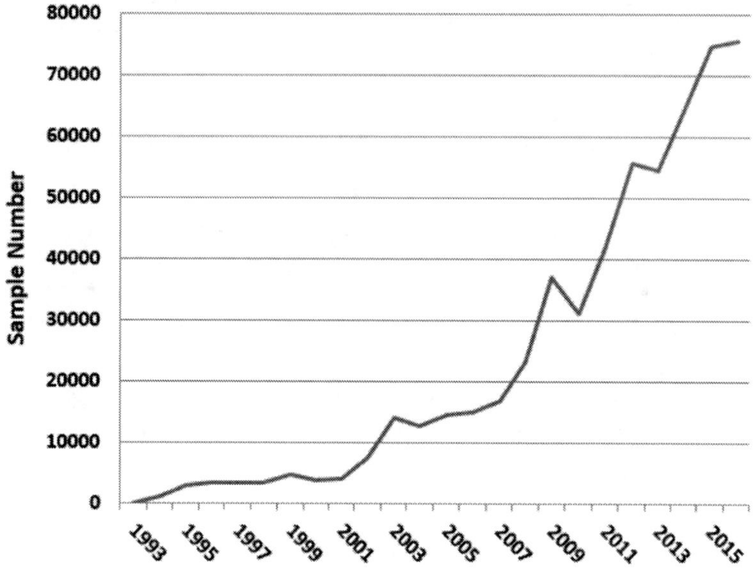

Figure 1. Cattle samples received per full year 1993 – 2016

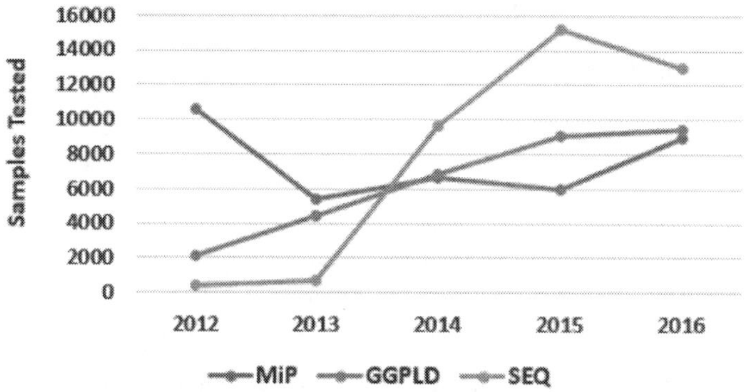

Figure 2. Count of parentage and genomic testing at AGL 2012-2016.

THE CSI EFFECT

The Crime Scene Investigation (CSI) effect is any way in which the exaggerated portrayal of forensic science on crime television shows influences public perception (Cole and Dioso-Villa 2007). It is very relevant to those working in customer-facing roles within the scientific profession.

The CSI effect manifests itself in a multitude of ways at AGL but most commonly in regards to unrealistic expectations of turnaround time or the amount and quality of sample that is required. When parentage does not immediately resolve, it is often assumed that AGL can simply run it against everything in the database to identify the correct parent. This not only assumes that the sire or dam is 'in the system", but also that AGL has the resources to develop the equivalent of a National DNA Data Bank for Australian Cattle and that sufficient markers are available to discriminate every individual. It is important to get the message out to all users and potential users of genetic and genomic testing services that 'real science' does not happen this way.

PARENTAGE CHALLENGES

From the parentage viewpoint, northern herds tend to be more complex than southern herds. This is due to a number of factors including sire-only parentage, larger overall herd and parent lists, difficulty in providing complete sire lists and a greater chance of uncaptured parents. There are also significant logistical challenges in providing resubmissions for samples that fail genotyping or produce anomalous results.

Success rates of northern parentage verification (PV) analyses can still be maximised, despite these aforementioned constraints, with open and frequent communication between AGL staff and the client. The PV success rate of a large northern herd that used this tactic was considerably improved over a 3 year period (Table 1).

Table 1. Parentage verification success rates for a large Northern herd

	Analysis 1	Analysis 2	Analysis 3
Year 1	46%	71%	89%
Year 2	61%	89%	97%
Year 3	95%	97%	

FROM MICROSATELLITES TO SNP

Much has been written about the promises of SNP-based parentage verification (SNP_PV) in livestock and animal traceability across the supply chain (Heaton et al. 2002, Van Eenennaam et al. 2007, Baruch and Weller 2008). However, costs associated with moving a breed from PV using MiP to SNP_PV are substantial, as are the logistical challenges. Retaining unused samples (with greater than 500,000 hair samples archived) at AGL has helped significantly reduce time spent sourcing new samples for animals, especially when animals are deceased. Once the decision is made to transition across to SNP_PV, experience shows us that very clear communications is essential to avoid issue of incompatible profiles between sires, dams and progeny. For smaller breeds, where there remains a lack of incentive to use genomics, then the change to SNP_PV is uneconomical and PV using MiP will probably remain part of the AGL offerings for many years to come. However as price per SNP test falls, the move to SNP will likely become attractive to even the smaller breeds.

CHALLENGES OF SNP REVOLUTION

The challenge in context of the Australian market has been trying to find the sweet spot of sufficient markers for accurate parentage at a price deemed acceptable. In an industry as diverse as the Australian cattle industry this has proven to be no simple task. AGL currently offers 2 SNP-based parentage assays: SEQ1 iPLEX panels contained a total of 138 SNP including 95 ISAG core plus 4 ISAG additional SNP, or SEQ2 consisted of 59 additional SNP for a combined total of 197 markers genotyped and total of 97 ISAG core SNP. These extra markers were developed to be informative in Brahman and Tropical Composite breeds. As reported previously (Lyons et al, 2013), we demonstrated that the ISAG-recommended core bovine SNP parentage panel is not sufficient to provide accurate parentage verification in many common Australia production systems. Further, we acknowledged that these panels were less than ideal. A number of publications over recent years has highlighted the advantage of larger numbers of SNP for parentage (Strucken et al. 2014; McClure et al., 2015), but these rarely take into account the economic reality of the market and current technologies.

PRICING CHALLENGES

Price expectations of the livestock industry do not necessarily align with commercial realities of test prices. Unlike supermarkets or other commodity-based services, and perhaps unlike standard R&D within research organisations, there is much more to be considered than the consumables' cost. Significant challenges and considerations in development and implementation of testing need to be both understood and appropriately costed. For any test performed at AGL, the samples will pass through up to 6 hands from arrival to reporting and beyond. In simple terms, there is reception, cataloguing, sample preparation, DNA extraction and QC, pre-PCR, post-PCR, data analysis and reporting data in a multitude of different formats prone to change regularly. Standardisation of reporting remains a challenge across the industry.. As already discussed, AGL prides itself on doing more than simply churning out data. AGL liaises with clients regularly and has intimate knowledge of herds and breeding regimes based on prior testing. The labour costs at AGL associated with pre- and post-testing consultations and follow-up discussions with are significant.

Other factors often overlooked, but of critical importance to the feasibility of genetic diagnostic labs include: patent and licensing considerations or costs, maintenance and depreciation costs for equipment, newer technology upgrades necessary to remain competitive, the additional costs of validation of novel platforms or assays, data and sample storage, informatics for interpretation of genomic variation, volume discounting options and commercial risk mitigation.

THE FUTURE

Much has been written about the decreasing cost per marker for genotyping and/or sequencing. The large number of high-throughput SNP genotyping technologies available are growing, but this in itself offers many challenges. Capital investments previously made will largely dictate services offered, and at AGL the reliability and reproducibility of the fixed Illumina Infinium platform has been very successful. Minimizing turnaround times and throughput variability remain important factors that have influenced AGL's model of developing in-house facilities rather than outsourcing. Genotype-By-Sequence (GBS) is often suggested as the way of the future, and certainly has a role in R&D or where flexibility is required. However, one major challenge with GBS approaches, especially for high-throughput genotyping facilities, is the considerable investment needed for bioinformatics support to properly analyse, curate and store the massive amounts of sequence data obtained from running GBS.

At the end of the day producer uptake of these technologies is not driven by cost-per-marker statistics. Producers are seeking a reproducible, highly accurate and informative result that can be translated into achieving their breeding objectives and/or a more saleable item. Reduced costs will be welcomed, but only if there is no compromise to results, and to date that has been the challenge. Attaining the 'holy grail' of 1 test per sample for everything you could need including Parentage, Recessives, Trait markers, EBVs, and ultimately the ability to make early selection decisions, is becoming a more realistic goal.

REFERENCES

Baruch, E. and J. I. Weller (2008) *Animal Genetics* **39**: 474.
Cole, S., & A. Dioso-Villa. (2007). *New England Law Review*. **41**: 435.
Heaton, M. P., G. P. Harhay, G. L. Bennett, et al. (2002) Mammalian Genome 13: 272.
Lyons, R.E., Buttsworth S., Waine D. and M. Kelly. (2013) *Proc. Aust. Assoc. Anim. Breed. Genet.*
McClure, M.C., McCarthy, J., Flynn, P., *et al.* (2015) *Proc. Int. Comm. Anim. Rec.* 175.
Strucken, E. M., Gudex, B., Ferdosi, M. H.,. *et al.* (2014) *Animal Genetics* **45**: 572.
Van Eenennaam, A. L., R. L. Weaber, D. J. Drake, et al. (2007) *J. Anim. Sci.* **85**: 3159.

Proc. Assoc. Advmt. Anim. Breed. Genet. 22:27-30

GENETIC CORRELATION BETWEEN PUREBRED AND CROSSBRED PERFORMANCE OF MERINO SHEEP FOR THREE WEIGHT TRAITS USING A GENOMIC RELATIONSHIP MATRIX.

N. Duijvesteijn[1,2] and J.H.J. van der Werf[1,2]

[1]School of Environmental and Rural Science, Univ. of New England, Armidale, NSW, Australia
[2]Cooperative Research Centre for Sheep Industry Innovation, Armidale, Australia 2351

SUMMARY

In sheep, genetic correlations between purebred and crossbred performance have not been studied extensively. The availability of genomic data on both purebreds and crossbreds makes it possible to estimate these genetic correlations for Merinos. Data of ~5000 purebred Merinos and ~ 5000 crossbred Merinos (sired by White Suffolk, Poll Dorset or Border Leicester) was used and the animals were genotyped with the Ovine 50K and phenotyped for three weight traits; weaning weight (WWT), post-weaning weight (PWWT) and carcass weight (CWT). Results showed a significant deviation from 1 for PWWT namely 0.61. While the correlation for WWT and CWT were not significant at 0.96 and 0.69 respectively. For a Merino breeding programs where emphasize is on increasing crossbred performance for PWWT (and CWT), purebred and crossbred performance should be combined in the genetic evaluation to achieve a good response to selection.

INTRODUCTION

Routine genetic evaluation of sheep in Australia is generally based on purebred performance. However, most lambs are produced as crossbreds For optimal ranking of breeding animals it is important to know whether breeding values predicted based on purebred performance, are also good predictors for crossbred performance. For example, a genetic correlation between purebred and crossbred performance (r_{pc}) of 0.8 (accurately estimated) will result in a loss in response of 20% in crossbreds when selection is based on purebred performance (Bijma *et al.* 2014). Some studies have identified moderate to high estimates for r_{pc} in Australian sheep (Ingham *et al.* 2005, Banks *et al.* 2009, Brown *et al.* 2015). These studies were all based on terminal sires having both purebred and crossbred offspring. The estimate of r_{pc} could in these cases be confounded with a potential genotype-by-environment interaction effect. It has been hard to estimate r_{pc} for Merinos as Merinos rams are rarely mated to other breeds. However, since the availability of genomic data, new opportunities arise as genetic parameters can be estimated even without structured family designs. For example, r_{pc} can be estimated through genomic relationships between purebred Merinos and crossbreds where the dam is a Merino. Such data exists abundantly in the Sheep CRC information Nucleus.

The aim of our study is to estimate r_{pc} for three weight traits, using genomic and phenotypic data on purebred Merinos and crosses between sires from terminal and maternal breeds and Merino dams.

MATERIALS AND METHODS

Animals, phenotypic and genotypic data. Data was extracted from two research datasets known as the Information Nucleus Flock (INF, (Van der Werf *et al.* 2010)) and the Sheep Genomics Flock (SGF, (White *et al.* 2012)). The data consisted of purebred Merinos (~40%) and crosses of terminal and maternal sires with Merino dams. Assigned genetic groups of base animals alongside pedigree information was used to determine the breed proportion. The sum of all Merino strains (Ultra/Superfine, Fine/Fine-medium, Medium/Strong, or undefined) was used to determine the percentage of purebred Merino. For this study, the crossbred animals should be at least 45% Merino and 45% from either Border Leicester (BL), Poll Dorset (PD) or White Suffolk (WS). The purebred

Merino were >95% Merino. Animals were genotyped using the 50k Illumina-Ovine SNP chip and 48,371 SNPs were used. Further quality control included Mendelian inconsistencies, plotting of the principal components to visually check breed assignment and removal of duplicate samples (off-diagonal relationship >0.9). To avoid that the covariance between purebred and crossbred offspring is confounded with some maternal effects, we randomly removed one of the offspring. In total the dataset consisted of 9,126 animals with 5,066 purebred Merino, 1,489 BL x Merino, 1,407 PD x Merino and 1,164 White Suffolk x Merino.

Recorded phenotypes for this study were weaning weight (WWT), post-weaning weight (PWWT) and carcass weight (CWT). Table 1 shows the number of phenotypic records for the Merino and their crosses with phenotypic information on the weight traits.

Table 1. Number of observation and phenotypic mean for purebred Merinos and their crosses.

Breed[1]	**N WWT[2]**	**Mean WWT**	**N PWWT[3]**	**Mean PWWT**	**N CWT[4]**	**Mean CWT**
Merino	5066	24.43	4623	38.67	1925	21.39
BL x Merino	1489	27.39	1095	44.24	729	22.07
PD x Merino	1407	28.47	739	45.85	1361	23.00
WS x Merino	1164	28.53	613	46.11	1128	22.89

[1]BL=Border Leicester, PD=Poll Dorset, WS=White Suffolk. [2]WWT=weaning weight. [3]PWWT=Post-weaning weight. [4]CWT=carcass weight.

Statistical analysis. Fixed effects fitted were derived from previous studies using similar data (Moghaddar *et al.* 2014) and were; birth type, rearing type, gender, age at measurement, breed and contemporary group defines as flock, birth year and management group.

The relationship matrix was constructed using genotypes to derive the genomic relationship matrix (Yang *et al.* 2010).

Linear mixed models were used to estimate the variance components and the data was fitted in the program MTG2 (Lee *et al.* 2016). Depending on the trait different random effects were fitted. The simplest model was chosen where the Likelihood Ratio Test showed no significant difference between including an extra random effect or not (results not shown).

Model 1 for WWT: $Y = Xb + Z_1a + Z_2m + e$

Model 2 for PWWT: $Y = Xb + Z_1a + Z_2m + Z_3sf + e$

Model 3 for CWT: $Y = Xb + Z_1a + Z_3sf + e$

Where **Y** is the vector with phenotypes, **b** is a vector of fixed effects, **a** is a vector of random additive genetic effects, **m** is the effect the dam, **sf** is a sire by flock interaction effects and **e** is a vector of random residual effects.

Bivariate analyses was used for all three traits, where the traits were defined by being measured either in purebred or crossbred animals, with the resulting correlation between additive genetic effects representing a correlation between purebred and crossbred performance (r_{pc}). Covariance between maternal effects in the purebred and crossbred dataset was set to zero, as dams were not allowed to have both crossbred and purebred offspring. Similarly, the covariance of the sire by flock interaction for purebred and crossbred performance was set to zero.

RESULTS

In Figure 1 the first two principle components (PC) are shown to indicate breed content of the dataset. The first PC explained 25.6 % of the genetic variance and the 2nd PC explained 19.7%. The first PC separates the Merinos from BL and the second PC separates WS and PD.

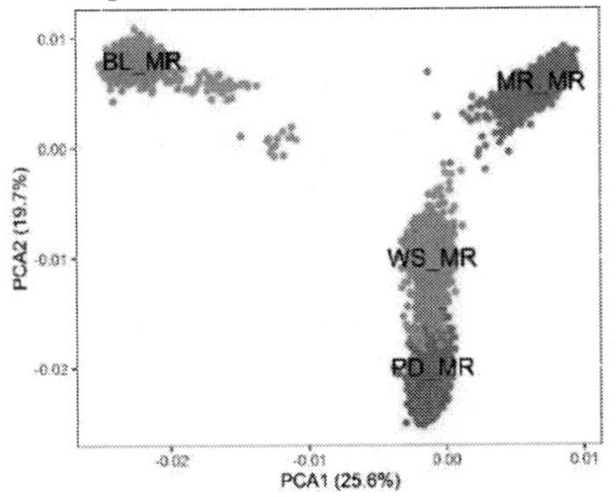

Figure 1. Plot of principal components (PC) 1 and 2 with the percentage of variance explained in brackets, where MR=Merino, BL=Border Leicester, PD=Poll Dorset, and WS=White Suffolk.

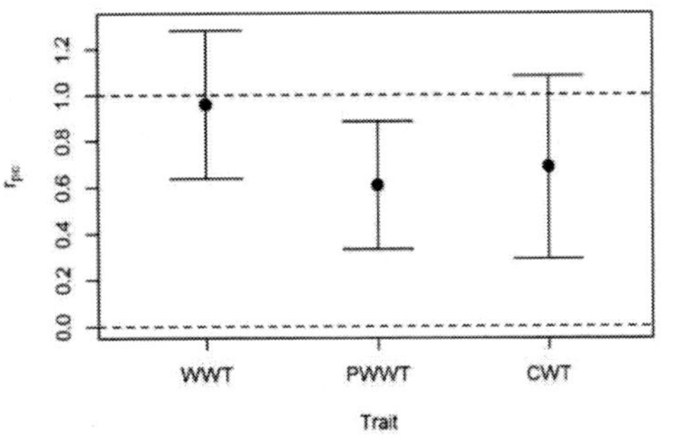

Figure 2. The genetic correlation between purebred (Merino) and crossbred performance for three weight traits. WWT= Weaning weight, PWWT=Post-weaning weight, and CWT=Carcass weight.

Table 2. Additive genetic variance (σ_a^2), maternal (σ_m^2), sire by flock interaction (σ_{sf}^2) and h^2 for each trait for purebred (PB) and crossbred (XB) performance.

		σ_a^2		σ_m^2		σ_{sf}^2		h^2	
Trait[1]	**PB/XB**	**comp[2]**	**se[3]**	**comp**	**Se**	**comp**	**se**	**comp**	**se**
WWT	PB	2.22	0.27	2.02	0.25			0.22	0.03
	XB	2.42	0.44	3.53	0.37			0.19	0.03
PWWT	PB	9.65	0.86	1.57	0.58	1.65	0.34	0.38	0.03
	XB	7.79	1.69	2.44	1.10	3.30	0.73	0.28	0.06
CWT	PB	2.65	0.41			0.60	0.20	0.38	0.05
	XB	1.16	0.24			0.37	0.11	0.20	0.04

[1]WWT=weaning weight; PWWT=Post-weaning weight; CWT=carcass weight. [2]Estimate of the variance component or ratio.[3]Approximate standard error on the estimate.

The results of the bivariate analyses are shown in Table 2 and the genetic correlation between purebred and crossbred performance (r_{pc}) with a 95% confidence interval is shown in Figure 2. The trait PWWT had the lowest r_{pc} which was 0.61 and was significantly different from one, while WWT was the highest (0.96). The trait CWT had a genetic correlation similar to PWWT (0.69), but due to lower number of records, the standard error on the estimate is larger. Results by breed group (WS X MR, BL X MR and PD X MR) showed similar trends (r_{pc} high for WWT and more variable for PWWT and CWT). In general, the r_{pc} for WS X MR and MR where lower than PD X MR or BL X MR, but due to the limited size of the data sets, standard errors were large (>0.20) and clear conclusions could not be drawn (results not shown).

DISCUSSION AND CONCLUSION

Results from the bivariate analysis show similar or slightly lower heritabilities based on genomic relationships compared to previous studies (Daetwyler *et al.* 2012, Moghaddar *et al.* 2014) ranging between 0.2 and 0.3. When the genetic correlation between purebred and crossbred performance (r_{pc}) was lower (i.e. for PWWT and CWT), the genetic variance as well as the heritability was larger in purebred animals than in crossbred animals. Brown et al (2015) found genetic correlations, which were not significantly deviating from one for similar weight traits. Their results were based on purebred Poll Dorset and their crosses. The current study focussed on a maternal contribution to crossbred performance, while other studies have often focussed on the paternal contribution to crossbred performance. A study by Moghaddar et al (2014) found a lower prediction accuracy for crossbreds for the trait PWWT for a similar dataset (genotyped Merinos including their crossbreds), lower than what was expected also after accounting for the number of haplotypes, i.e. twice the number of crossbreds gave lower accuracy than purebreds. This result could be partly explained by the r_{pc} being lower than 1. The number of studies calculating r_{pc} in sheep are limited. Other studies have mainly focused on performance traits in pigs and poultry where results seem to be very diverse in estimated r_{pc} also due to a lack of power in the datasets used (personal communication Y.C.J. Wientjes).

Generally the SE on the estimated genetic correlations were large in the current study. The SE was larger than expected when using the same size of dataset, but with paternal half sib groups (Falconer *et al.* 1996). This is likely a reflection of the smaller degree of relationship between the dam contributions and sire contributions.

To conclude, both crossbred performance and purebred performance need to be included in the estimation of the breeding values to increase crossbred performance of Merino crosses, especially for PWWT and CWT. In a Merino breeding program where both wool and meat production are selected for, the crossbred performance for production traits is relevant. If selection will be only based on purebred performance, a reduced selection response of around 40% can be expected for PWWT and CWT in the crossbreds. Therefore, genetic evaluation on traits such as PWWT and CWT should be based on both purebred and crossbred performance.

ACKNOWLEGEMENTS

The authors acknowledge the contributions of people from breeders and many CRC participant that contributed to the Sheep CRC Information Nucleus flocks. Also Nasir Moghaddar is acknowledged for his help retrieving data and helpful discussions.

REFERENCES

Banks, R., Brown, D. and Field, S. (2009) *Proc. Assoc. Advmt. Anim. Breed. Genet.* **18**: 480.

Bijma, P. and Bastiaansen, J. W. (2014) *Genet. Sel. Evol.* **46**: 1.

Brown, D. and van der Werf, J.H. (2015) *Proc. Assoc. Advmt. Anim. Breed. Genet.* **21**: 177.

Daetwyler, H. D., Swan, A.A., van der Werf, J.H. *et al.* (2012) *Genet. Sel. Evol.* **44**: 1.

Falconer, D.S. and Mackay, T.F. (1996) In 'Introduction to quantitative genetics'. Longman, Essex United Kingdom.

Ingham, V., Fogarty, N., Gilmour, A., *et al.* (2005) *Proc. Assoc. Advmt. Anim. Breed. Genet.* **16**: 227.

Lee, S.H. and Van Der Werf, J.H. (2016) *Bioinf.* **32**: 1420.

Moghaddar, N., Swan, A.A. and van der Werf, J.H. (2014) *Anim. Prod. Sci.* **54**: 544.

Moghaddar, N., Swan, A.A. and van der Werf, J.H. (2014) *Genet. Sel. Evol.* **46**: 1.

Van der Werf, J.H, Kinghorn, B., and Banks, R. (2010) *Anim. Prod. Sci.* **50**: 998.

White, J.D., Allingham, P.G., Gorman, C.M. *et al.* (2012) *Anim. Prod. Sci.* **52**: 157.

Yang, J., Benyamin, B., McEvoy, B.P. *et al.* (2010) *Nat. Genet.* **42**: 565.

Proc. Assoc. Advmt. Anim. Breed. Genet. 22:31-34

GENETIC CORRELATIONS BETWEEN PRODUCTION TRAITS AND COMPONENTS OF REPRODUCTION IN MERINO SHEEP

D.J. Brown, K.L. Bunter and A.A. Swan

Animal Genetics and Breeding Unit (a joint venture of NSW Department of Primary Industries and the University of New England), University of New England, Armidale, NSW 2351, Australia

SUMMARY

Data from the Sheep Genetics database was used in investigation of the genetic relationships between components of reproduction and traits which may be useful indirect selection criteria for reproduction rate in Merino sheep. Pre-joining weight as well as fat and eye muscle depth were favourably genetically correlated with all reproduction traits except ewe rearing ability, as were more favourable scores for maternal behaviour, wrinkle and face cover. Correlations with pre-joining condition score were only significant when unadjusted for live weight. Maternal behaviour score was favourably correlated with all reproduction traits except ewe rearing ability but the standard errors were large, and more data are required to improve the precision of estimates. These results suggest that these traits could be recorded by breeders and included in the genetic evaluation system to improve the accuracy of selection for reproduction rate in Merino sheep.

INTRODUCTION

The Sheep Genetics (SG) genetic evaluation system produces Australian Sheep Breeding Values (ASBVs) for net reproduction rate with two trait definitions, number of lambs born or weaned per ewe joined (nlb and nlw respectively). While this has worked adequately in the past there are a number of key benefits from moving towards component trait analyses, including the ability to fit different models to each trait, allowing targeted selection for components and optimal use of the data available from industry.

As reproduction traits are lowly heritable, sex-linked and expressed later in life, the accuracy of ASBVs, particularly in young animals, can be increased through the use of information on correlated traits. To make use of such correlated information, accurate estimates of the genetic and phenotypic correlations between key traits are required, many of which are not currently available in the literature. Based on the estimates presented by Hatcher *et al.* (2015), Brown and Swan (2016), Brown *et al.* (2015) and many earlier publications, important relationships exist between production and reproduction traits which could contribute to the estimation of breeding values, and the development of selection indexes.

The aim of this paper is to estimate the correlations between some key production traits and reproduction traits in Merino sheep.

MATERIALS AND METHODS

Data. Pedigree and performance data were extracted from the Sheep Genetics MERINOSELECT database (Brown *et al.* 2007). This database consists of pedigree and performance records submitted by Australian and New Zealand Merino ram breeders, and is used for genetic evaluation purposes. The database also contains information from the Sheep CRC Information Nucleus Flock (INF) and the Resource Flock. From these data all animals with at least sire known, born 2000 and later, and from flocks with a history of recording reproduction traits were included. Data were extracted for all animals with early breech wrinkle (ebwr), late body wrinkle (lbdwr), late face cover (lface), post-weaning fat and eye muscle depth (pfat and pemd) and yearling fat and eye muscle depth (yfat and yemd). Annual ewe records were also extracted for pre-joining weight (pjwt), pre-joining condition score (pjcs), number of lambs born and weaned per ewe joined (nlb and nlw), maternal behaviour of the ewe at lambing (mbs, 1 to 6, with 1 being best), ewe fertility (fert, dry or pregnant), litter size defined as the number of lambs born per ewe lambing (ls) and ewe rearing ability defined as the proportion of lambs weaned to lambs born per ewe lambing (era).

The pedigree was built using all ancestral information available. This resulted in pedigree files comprising between 78,563 and 191,392 animals for the combined dataset depending on the trait combination being analysed. A summary of the number of records available for each trait in each data set is shown in Table 1. The number of animals with records for two traits ranged from 1479 for yfat and era to 20,847 for pjwt and nlb. At the sire level, this range corresponded to 226 common sires for lface and era to 891 for pjwt and nlb.

Table 1: Summary of raw data used for each trait

Trait	Records	Animals	Sires	Flocks	Mean	SD	Min	Max
pjwt	20,847	13,315	891	27	49.97	9.21	24.00	105.50
pjcs	8,298	4,433	388	17	3.03	0.53	1.00	5.00
pfat	22,088	22,088	912	46	2.25	0.51	0.60	5.20
yfat	59,488	59,488	1,919	71	2.50	0.57	0.50	7.60
yemd	61,986	61,986	2,046	75	23.60	4.32	10.00	45.00
pemd	22,293	22,293	924	47	22.82	3.82	10.00	41.00
ebwr	85,779	85,779	1,509	55	2.27	0.99	1.00	5.00
lbdwr	35,627	35,627	928	28	2.01	0.87	1.00	5.00
lface	26,572	26,572	776	27	2.52	0.87	1.00	5.00
mbs	4,769	3,218	333	10	2.19	1.01	1.00	6.00
nlb	73,227	34,840	2,180	53	1.18	0.65	0.00	4.00
nlw	60,639	29,693	1,925	49	1.02	0.68	0.00	4.00
fert	73,227	34,840	2,180	53	0.87	0.33	0.00	1.00
ls	63,918	31,565	2,113	53	1.35	0.51	1.00	4.00
era	52,872	26,942	1,851	49	0.87	0.32	0.00	1.00

Models of analysis. Parameters were estimated in bivariate sire model analyses for each trait combination using ASReml (Gilmour *et al.* 2009). For wrinkle, weight, and condition score traits the fixed effects of contemporary group, birth type, rearing type, age of dam, and animal's age at measurement were fitted. For the body composition traits the fixed effects of contemporary group and the regression on an animal's live weight at measurement (linear and quadratic) were fitted. Contemporary group was defined as flock, year of birth, sex, date of measurement and management group subclass. For the reproduction traits the only effect fitted was the reproduction contemporary group, based on combinations of flock and year of lambing, management group, conception method (AI and Natural) and ewe age class (1, 2, and 3+ years). A random sire term for the direct genetic effects was modelled for all traits, including ancestral sire pedigree relationships. A sire model was chosen as the data structure did not support the estimation of all parameters using an animal model. An additional random term for maternal permanent environment effects was included for ebwr and pjwt. For pjcs, pjwt, mbs, and the reproduction traits repeated records were accounted for by including an additional random term to model the permanent environment of the animal. Sire by flock-year interactions were also fitted as an additional random term for all traits. Genetic groups were specified by flock of origin and fitted as random effects (Swan *et al.* 2014). As genetic groups did not significantly improve the fit of the model for mbs and the reproduction traits they were only fitted for production traits. As pre-joining weight and condition score are related an additional pre-joining condition score trait was created which included adjustment for weight at joining (pjcs2) by fitting pjwt as an additional covariate in the model.

RESULTS AND DISCUSSION

Large numbers of records were available for most traits (Table 1). The mean of 0.87 for era is slightly higher than the value of 0.81 reported by Bunter *et al.* (2016) derived from three well-recorded industry Merino flocks, suggesting that the lamb survival data may be biased upwards in this study due to incomplete recording. Heritability estimates (Table 2) for most traits were consistent with earlier publications based on MERINOSELECT data estimated predominantly using animal models. The heritabilities for pjcs and fat depth were slightly lower than previously observed as were those for most reproduction traits compared to the earlier estimates reported by Bunter *et al.* (2016).

Table 2: Phenotypic variance (σ^2_p), heritability (h^2), repeatability (r), permanent environment due to dam (dam PE), and sire by flock interaction (s^2) for each trait

Trait	σ^2_p	h^2	r	dam PE	s^2
pjwt	31.31 (2.10)	0.39 (0.11)	0.80 (0.11)	0.10 (0.11)	0.11 (0.11)
pjcs	0.12 (0.01)	0.11 (0.03)	0.22 (0.03)		0.07 (0.03)
pjcs2	0.10 (0.01)	0.11 (0.04)	0.22 (0.04)		0.06 (0.04)
pfat	0.22 (0.04)	0.12 (0.04)	.		0.03 (0.04)
yfat	0.28 (0.02)	0.10 (0.01)	.	.	0.03 (0.01)
pemd	3.68 (0.01)	0.21 (0.02)	.	.	0.04 (0.02)
yemd	3.98 (0.01)	0.22 (0.01)	.	.	0.03 (0.01)
ebwr	0.64 (0.01)	0.35 (0.03)	.	0.12 (0.03)	0.05 (0.03)
lbdwr	0.40 (0.01)	0.37 (0.05)	.	.	0.03 (0.05)
lface	0.50 (0.01)	0.35 (0.01)	.	.	0.04 (0.01)
mbs	0.81 (0.01)	0.09 (0.02)	0.22 (0.02)		0.05 (0.02)
nlb	0.33 (0.01)	0.07 (0.01)	0.16 (0.01)	.	0.01 (0.01)
nlw	0.38 (0.01)	0.04 (0.01)	0.12 (0.01)	.	0.01 (0.01)
fert	0.09 (0.01)	0.06 (0.01)	0.16 (0.01)	.	0.02 (0.01)
ls	0.21 (0.01)	0.07 (0.01)	0.16 (0.01)	.	0.01 (0.01)
era	0.09 (0.01)	0.02 (0.01)	0.11 (0.01)	.	0.01 (0.01)

Pre-joining weight and condition score were moderately correlated genetically (0.50 ± 0.09) and phenotypically (0.29 ± 0.02). Pre-joining weight, early in life fat and eye muscle depth were favourably correlated with all reproduction traits except ewe rearing ability (Table 3). These results generally agree with the earlier work of Brown and Swan (2016). However, the inconsistent correlations of body composition traits with ewe rearing ability are at odds with earlier work and may be a reflection of the incomplete recording of lamb survival, as mentioned above. Further studies with high quality data to study relationships with era are certainly warranted. Better scores for wrinkle and face cover were generally favourably associated with reproduction traits. The lack of a correlation between wrinkle and ewe rearing ability is inconsistent with results of Hatcher *et al.* (2015) who estimated significant favourable relationships between these traits in both industry and INF data. Correlations of pre-joining condition score with nlb, nlw or fert were only significant when unadjusted for live weight. Walkom and Brown (2016) estimated the correlations between these traits using just the INF data and found no significant relationship between condition score and reproduction traits unless condition score was adjusted for previous reproduction status. These results are also at odds with those observed for the fat and eye muscle depth traits which were highly genetically correlated with condition score in these data (rg between 0.68 and 0.98 across the 4 ultrasound traits) and demonstrated by earlier work of Walkom and Brown (2016). As the results for condition score appear quite inconsistent across analyses and data sets more industry data are clearly required to confirm the relationships between body composition and reproductive traits.

Maternal behaviour score was favourably correlated with all reproduction traits in absolute terms, except for ewe rearing ability. However, standard errors were large and the number of ewes recorded for mbs was relatively low, suggesting that more data are required to confirm these results.

While this study has not included other weight and wool traits, it is known that significant relationships exist between live weight, scrotal circumference, fleece weight, fibre diameter, fibre curvature and staple length with the reproductive traits and these should also be considered.

Table 3: Genetic correlations between reproduction traits, and production and visual traits

	pjwt	pjcs	pjcs2	ebwr	lbdwr	lface	pfat	yfat	pemd	yemd	mbs
nlb	0.51	0.40	-0.01	-0.32	-0.46	-0.44	0.42	0.40	0.38	0.42	-0.16
	(0.09)	(0.16)	(0.17)	(0.09)	(0.10)	(0.12)	(0.14)	(0.10)	(0.12)	(0.09)	(0.23)
nlw	0.50	0.41	-0.04	-0.43	-0.50	-0.48	0.40	0.41	0.34	0.50	-0.17
	(0.11)	(0.19)	(0.19)	(0.10)	(0.12)	(0.14)	(0.16)	(0.13)	(0.15)	(0.11)	(0.26)
fert	0.20	0.42	0.28	-0.31	-0.54	-0.09	0.59	0.34	0.45	0.37	-0.22
	(0.11)	(0.18)	(0.18)	(0.10)	(0.11)	(0.14)	(0.15)	(0.12)	(0.13)	(0.10)	(0.26)
ls	0.56	0.27	-0.25	-0.22	-0.28	-0.52	0.15	0.32	0.22	0.32	-0.06
	(0.08)	(0.17)	(0.17)	(0.09)	(0.10)	(0.11)	(0.15)	(0.10)	(0.13)	(0.09)	(0.22)
Era	0.07	0.14	0.01	-0.10	-0.09	0.04	-0.27	-0.00	-0.16	0.14	0.12
	(0.15)	(0.28)	(0.27)	(0.15)	(0.18)	(0.21)	(0.24)	(0.00)	(0.21)	(0.15)	(0.33)

CONCLUSION

These results suggest that these traits could usefully be recorded more by breeders and included in the genetic evaluation system to improve the accuracy of selection for reproduction rate in Merino sheep. More high quality data are required for maternal behaviour score, condition score and ewe rearing ability to confirm associations between these traits.

ACKNOWLEDGEMENTS

This research was funded by Meat and Livestock Australia, Australian Wool Innovation and Sheep Genetics, made possible through the support of the Australian sheep industry. The authors acknowledge the contributions of the Sheep CRC Information Nucleus, the Australia Merino Sire Evaluation Association and industry-funded research flocks.

REFERENCES

Brown D.J., Huisman A.E., Swan A.A., Graser H.-U., Woolaston R.R., *et al.* (2007) *Proc. Assoc. Advmt. Anim. Breed. Genet.* **17:** 187.

Brown D.J., Fogarty N.M., Iker C.L., Ferguson D.M., Blache D., *et al.* (2015) *Anim. Prod. Sci.* **56:** 767.

Brown D.J. and Swan A.A. (2016) *Anim. Prod. Sci.* **56:** 690.

Bunter K.L., Swan A.A., Purvis, I.W. and Brown D.J. (2016) *Anim. Prod. Sci.* **56**: 679.

Gilmour A.R., Gogel B.J., Cullis B.R. and Thompson R. (2009) '*ASREML user Guide Release 3.0*' VSN International Ltd, Hemel Hempstead, UK.

Hatcher S., Brown D.J., Brien F.D. and Hebart M.L. (2015) *Proc. Assoc. Advmt. Anim. Breed. Genet.* **21:** 354.

Swan A.A., Brown D.J. and van der Werf J.H.J (2016) *Anim. Prod. Sci.* **56:** 87.

Walkom S.F. and Brown D.J. (2016) *Anim. Prod. Sci.* **57**: 20.

Proc. Assoc. Advmt. Anim. Breed. Genet. 22:35-38

GENETIC PARAMETERS AND BREED DIFFERENCES FOR OVINE TICK COUNTS ON INDIGENOUS AND COMMERCIAL SHEEP IN AUTUMN AND SPRING

J.J.E. Cloete[1,2], A.J. Scholtz[3], S. Matthee[4] and S.W.P. Cloete[2,3]

[1]Elsenburg Agricultural Training Institute, Private Bag X1, Elsenburg 7607, South Africa
[2]Department of Animal Sciences, University of Stellenbosch, , Matieland 7602, South Africa
[3]Directorate Animal Sciences: Elsenburg, Private Bag X1, Elsenburg 7607, South Africa
[4]Department of Conservation Ecology and Entomology, University of Stellenbosch, Matieland, South Africa

SUMMARY

This paper reports body-site specific and overall tick counts as assessed during either spring or autumn for Dorper, SA Mutton Merino (SAMM) and Namaqua Afrikaner (NA) ewes maintained on natural pasture in an arid area. There seem to be a shift in the tick population challenging the hosts from autumn to spring, posing the question whether tick count in spring is genetically the same trait as tick count in autumn. The unimproved, fat tailed, indigenous NA breed had lower tick counts on all body sites compared to the two commercial breeds, the exception being tick counts on the tail of NA ewes. The other breeds have docked tails and could thus not be assessed for this site. All body-site specific tick counts were heritable, both in autumn (range 0.26-0.42) and spring (range 0.15-0.41). Ticks counts in autumn and spring were genetically very similar traits (r_g>0.88). Overall and body-site specific tick counts were heritable and should respond to selection.

INTRODUCTION

Sheep farming is very important in the South African agrarian landscape since it allows the sustainable utilization of arid rural environments (Cloete *et al.* 2014). Sheep are parasitized by ticks throughout the world, with many tick species being of veterinary and economic importance. Some ticks introduce toxins that cause paralysis (Fourie *et al.* 1989); other species can be the cause of severe tissue damage, which either results from their longer mouthparts or a tendency to form clusters (Cloete *et al.* 2016). Ticks are also responsible for anemia and production losses (Norval *et al.* 1988). Ticks are also responsible for direct damage, such as skin or hide damage, damage to udders, teats and the scrotum of livestock (Norval 1983). A variety of factors such as host type, host age or tick inter- and intraspecific interactions can affect the preferential feeding sites of ticks.

Host resistance to pathogens can be used as a component in integrated pest control programs (Walker 2011). However, research on the genetics of tick resistance is very limited in sheep. Van Marle-Köster *et al.* (2015) suggested that adapted, indigenous genetic resources have advantages over imported breeds in their response to stressful conditions, including tick infestations.

The objectives of this paper were: 1) to determine whether the tick challenge of sheep differed between seasons (autumn and spring); 2) to derive heritability estimates for body-site specific and overall tick counts within seasons; 3) to estimate genetic and phenotypic correlations between body-sites and overall tick counts; 4) to derive genetic correlations of tick counts in autumn with those in spring to determine whether tick infestation in autumn and spring are genetically similar traits.

MATERIALS AND METHODS

The experiment was carried out at the Nortier Research Farm (32°02'S and 18°20'E) in the West Coast Strandveld area of the Western Cape Province of South Africa, using a genetic resource population described by Cloete *et al.* (2013; 2016). Ewes from the indigenous fat-tailed Namaqua Afrikaner (NA) sheep breed and two commercial breeds, the Dorper and South African Mutton Merino (SAMM), were compared under marginal, extensive conditions. The Dorper is the leading

South African meat breed while the SAMM is the leading South African dual-purpose (meat and wool) breed and both breeds contribute substantial numbers of weaning weight records to the small stock improvement programme (Cloete *et al.* 2014). The NA, in contrast, is characterised by low numbers and is maintained in a few conservation flocks (Qwabe *et al.* 2012). Previous studies suggested that NA ewes were more resistant to ticks than the other breeds (Cloete *et al.* 2013; 2016).

The climate of the experimental site is Mediterranean, with 78 % of the total long-term annual precipitation of 221 mm being recorded during winter (April–September). Dry, warm summers and cool winters with an unpredictable and variable rainfall characterises the study area. The vegetation is classified as Strandveld of the West Coast (Acocks 1988). The Dorper and the SAMM were tail docked as lambs, while the fat tails of the indigenous NA were left intact. Docking was done with rubber rings applied at the third palpable joint when the lambs were approximately three weeks old.

Ticks were counted in a detailed study involving species during autumn (May) and in spring (September) of 2012 (Trial 1). Ewes (n=73) were cast and a total of 2425 ticks were removed from these animals. The detached ticks were preserved in 70% ethanol and identified according to species. Apart from this detailed study on tick species, ticks were also counted in Trial 2 on all available ewes in the autumn of 2012, 2015 and 2016 as well as in the spring of all years from 2012-2016. The total number of repeated records amounted to 914 records of 358 ewes in spring and 535 records of 341 ewes in autumn. These counts were done without considering the tick species present on the animals. Ticks were counted at three locations: the head and front legs (HFL), udder and hind legs (UHL) and perineum, including the tail of NA ewes (PT) as was described by Cloete *et al.* (2013; 2016). These counts were also summed to obtain a total tick count for each animal (TOT). All ewes were maintained in a single flock except for a six week mating period during which the breeds were kept separate. Ewes were also randomly divided into smaller groups during lambing.

The frequencies at which the respective tick species occurred in Trial 1 was compared by Chi^2-procedures. Raw tick counts in Trial 2 were extremely variable (Table 1) and needed to be suitably transformed. Individual counts were therefore transformed to square roots after 0.5 were added to individual records to reduce the difference between counts to between 0 and 1 (Dickson and Sanford 2005). ASReml (Gilmour *et al.* 2015) was used to first identify significant fixed effects (ewe breed and ewe age) then to derive genetic parameters by fitting four-trait models to all available data in the autumn and spring. The same counts in autumn and spring were then analysed together in two-trait analyses to derive genetic correlations between seasonal counts. Animal permanent environmental effects were initially modeled together with animal additive effects. Based on Log likelihood ratios, only direct animal effects were retained in the final analyses. The pedigree file contained 2713 animals, the progeny of 40 sires and 596 dams. Ethical clearance was provided by the Departmental Ethical Committee for Research on Animals (approval number R13/88).

RESULTS AND DISCUSSION

Trial 1: Ticks from the three major species differed in proportions in autumn and summer. When expressed relative to the total number of ticks recovered, the contribution of *Rhipicephalus evertsi evertsi* amounted to 0.38 in autumn and 0.44 in spring (Chi^2=19.7; degrees of freedom=1; P<0.01). *R. gertrudae* were recovered at a substantially higher proportion in autumn (0.52) than during spring (0.19; Chi^2=274.1; degrees of freedom=1; P<0.01). Corresponding proportions for *Hyalomma truncatum* amounted to 0.11 and 0.37 respectively (Chi^2=249.8; degrees of freedom=1; P<0.01). These results suggested that the tick challenge during spring and autumn was different and potentially needed different coping strategies by the host animals.

Trial 2: Raw tick counts on individual ewes were extremely variable with standard deviations often exceeding the corresponding means (Table 1). The square root transformation normalised the distributions in terms of skewness and kurtosis and reduced the observed coefficients of variation to more manageable levels, ranging from 39.5% for TOT in autumn to 66% for HFL in spring.

Table 1. Descriptive statistics for the raw and transformed tick counts analysed on ewes in autumn (n=535) and spring (n=914), namely head-front leg tick count (HFL), udder-hind leg tick count (UHL), perineum-tail tick count (PT) and total tick count (TOT)

Season	Autumn			Spring		
Trait	Raw mean ± s.d.	Range	Transformed mean ± s.d.	Raw mean ± s.d.	Range	Transformed mean ± s.d.
HFL	10.4 ± 11.1	0 – 88	2.96 ± 1.46	5.1 ± 7.4	0 – 54	1.98 ± 1.46
UHL	11.7 ± 16.0	0 – 112	2.97 ± 1.83	8.6 ± 11.4	0 – 89	2.53 ± 1.64
PT	6.7 ± 7.6	0 – 50	2.37 ± 1.26	6.8 ± 7.4	0 – 61	2.39 ± 1.27
TOT	28.8 ± 25.6	0 – 216	5.03 ± 1.98	21.0 ± 17.5	0 – 126	4.23 ± 1.76

Backtransformed means for tick counts at the HFL and UHL sites of the commercial breeds exceeded those recorded in their NA contemporaries by at least a factor of 2 (P<0.01), both during autumn and spring (Table 2). NA ewes had higher (P<0.01) PT tick counts than the Dorper in both seasons, as well as SAMM ewes during spring. Breed differences were previously reported for tick count as well as for attachment site in sheep (Fourie and Kok 1995; Cloete *et al.* 2013; 2016). The latter authors attributed the higher tick counts at the PT site in the NA to the fact that their tails were left intact. Backtransformed means for TOT in the commercial breeds exceeded those of NA ewes by between 43 and 148% (All P<0.01), suggesting a greater resistance in the indigenous breed.

Table 2. Least-squares means (±s.e.) depicting breed[1] differences between the breeds assessed for head-front leg tick count (HFL), udder-hind leg tick count (UHL), perineum-tail tick count (PT) and total tick count (TOT) recorded either in the autumn or spring with backtransformed means in brackets

Season and breed	N	Trait: HFL	UHL	PT	TOT
Autumn		**	**	**	**
NA	204	2.15 ± 0.09 (4.1)	2.15 ± 0.10 (4.1)	2.79 ± 0.08 (7.3)	4.21 ± 0.10 (17.3)
Dorper	238	2.96 ± 0.08 (8.3)	3.33 ± 0.09 (10.6)	1.95 ± 0.07 (3.3)	5.03 ± 0.09 (24.8)
SAMM	76	4.39 ± 0.14 (18.8)	3.88 ± 0.16 (14.5)	2.77 ± 0.12 (7.2)	6.59 ± 0.16 (43.0)
Spring		**	**	**	**
NA	330	1.45 ± 0.07 (1.6)	1.73 ± 0.09 (2.5)	2.88 ± 0.07 (7.8)	3.72 ± 0.09 (13.3)
Dorper	451	2.46 ± 0.06 (5.6)	3.15 ± 0.08 (9.4)	2.00 ± 0.06 (3.5)	4.69 ± 0.08 (21.5)
SAMM	133	1.98 ± 0.11 (3.4)	3.32 ± 0.15 (10.5)	2.63 ± 0.11 (6.4)	4.82 ± 0.15 (22.7)

[1] Namaqua Afrikaner (NA), Dorper and South African Mutton Merino (SAMM)
** P<0.01

Significant genetic variation was detected for all body-site specific tick counts in four-trait analyses conducted in autumn and spring (Table 3). Genetic parameters were quite similar across seasons, except for PT tick counts, where the heritability was lower in spring. These results compared well with previous heritability estimates of 0.26 for HFL, 0.53 for UHL, 0.19 for PT and 0.43 for TOT (Cloete *et al.* 2016). Grøva *et al.* (2014) accordingly reported heritability estimates of 0.37-0.52 for TOT in Norwegian lambs under conditions where another tick species, namely *Ixodes ricinus*, prevails. HFL and UHL tick counts were highly correlated to TOT on the genetic level, as would be expected for traits in a part-whole relationship. These results were also consistent with those previously reported by Cloete *et al.* (2016). Genetic correlations between tick counts recorded in autumn and spring approached, and in some cases exceeded, unity for body-site specific values (Table 3). These preliminary results suggest that resistance to ticks in autumn and spring are genetically very similar traits. Phenotypic correlations among traits were similar in sign as genetic correlations, but generally smaller in magnitude.

Table 3. (Co)variance ratios (± s.e.) for head-front leg tick count (HFL), udder-hind leg tick count (UHL), perineum-tail tick count (PT) and total tick count (TOT) recorded either in the autumn or spring based on four-trait or two-trait analyses

Component and trait	Trait			
	HFL	**UHL**	**PT**	**TOT**
(Co)variance ratios in autumn*				
HFL	**0.26 ± 0.07**	0.61 ± 0.15	0.17 ± 0.18	0.88 ± 0.08
UHL	0.20 ± 0.05	**0.39 ± 0.06**	-0.40 ± 0.14	0.81 ± 0.06
PT	0.04 ± 0.05	-0.19 ± 0.05	**0.30 ± 0.06**	0.18 ± 0.14
TOT	0.68 ± 0.05	0.68 ± 0.03	0.32 ± 0.04	**0.42 ± 0.06**
(Co)variance ratios in spring*				
HFL	**0.26 ± 0.04**	0.28 ± 0.11	0.10 ± 0.16	0.64 ± 0.08
UHL	0.20 ± 0.04	**0.41 ± 0.04**	-0.23 ± 0.14	0.85 ± 0.04
PT	0.07 ± 0.04	-0.11 ± 0.04	**0.15 ± 0.04**	0.17 ± 0.15
TOT	0.56 ± 0.03	0.74 ± 0.02	0.42 ± 0.03	**0.34 ± 0.04**
Correlations between tick counts in autumn and spring				
Genetic	0.89 ± 0.09	1.01 ± 0.02	1.00 ± 0.08	1.01 ± 0.04
Phenotypic	0.27 ± 0.05	0.48 ± 0.04	0.24 ± 0.04	0.45 ± 0.04

* Heritability in bold on the diagonal, genetic correlations above the diagonal and phenotypic correlations below the diagonal

CONCLUSIONS

The species composition of the tick challenge at the experimental site differed appreciably in species composition between autumn and spring. Notwithstanding this result, appreciable genetic variation in body site specific and total tick counts was present in both seasons. Moreover, genetic correlations between autumn and spring tick counts suggested that these traits were likely to be controlled by largely the same genes, a finding that needs to be verified in further studies.

REFERENCES

Acocks J.P.H. (1988) *3rd Ed. Memoirs Bot. Surv. S. Africa* **57**: 146.

Cloete J.J.E., Cloete S.W.P., Scholtz A.J. and Matthee S. (2013) *Proc. Assoc. Advmt. Anim. Breed. Genet.* **20**: 187.

Cloete S.W.P., Olivier J.J., Sandenbergh L. and Snyman M.A. (2014) *S. Afr. J. Anim. Sci.* **44**: 308.

Cloete S.W.P., Cloete J.J.E. and Scholtz A.J. (2016) *Vet. Parasitol.* **230**: 33.

Dickson K.A. and Sanford L.M. (2005) *Small Rumin. Res.* **56**: 189.

Fourie L.J., Petney T.N., Horak I.G. and De Jager C. (1989) *Vet. Parasitol.* **33**: 319.

Fourie L.J. and Kok D.J. (1995) *Onderstepoort J. Vet. Res.* **62**: 211.

Gilmour A.R., Gogel B.J., Cullis B.R., Welham, S.J. and Thompson R. (2015) ASREML - User Guide Release 4.1 VSN International Ltd, Hemel Hempstead, HP11ES, UK

Grøva L., Sae-Lim P. and Olesen I. (2014) *Proc. 10th World Cong. Gen. Appl. Livest. Prod.*, 16-20

Norval R.A.I. (1983) *Zimbabwe Vet. J.* **14**: 19.

Norval R.A.I., Sutherst R.W., Kurkil J., Gibson J.D. and Kerr J.D. (1988) *Vet. Parasitol.* **30**: 149.

Qwabe S.O., Van Marle-Köster E. and Visser C. (2012) *Trop. Anim. Health Prod.* **45**: 511.

Van Marle-Köster E., Visser C., Makgahlela M. and Cloete S.W.P. (2015) *Food Res. Int.* **76**: 971.

Walker A.R. (2011) *Parasitology* 15pp. DOI: 10.1017/S0031182011000709.

Proc. Assoc. Advmt. Anim. Breed. Genet. 22:39-43

GENOMIC PREDICTION USING QTL REGIONS IDENTIFIED FROM REGIONAL HERITABILITY MAPPING FOR PARASITE RESISTANCE IN AUSTRALIAN SHEEP

M. Al Kalaldeh, J.H.J. van der Werf and C. Gondro

University of New England, School of Environmental and Rural Science, Armidale 2351, Australia

SUMMARY

Genomic selection uses genomic information to predict the breeding value of animals and can achieve higher prediction accuracy than pedigree based selection. This study aimed to compare the accuracy of genomic prediction using a medium-density (50k) SNP panel, as well as an imputed high-density (600k) SNP panel, with and without including pre-selected SNPs from QTL regions identified by regional heritability mapping (RHM). The proportion of variance explained by the pre-selected SNPs combined in a genomic relationship matrix (GRM) was considerably smaller than that explained by all SNPs from the 600k panel (25% of the genomic heritability). To obtain a better estimate of the variance explained by the pre-selected SNPs, both GRMs from the pre-selected SNPs (GRM_s) and their complementary SNPs from the 600k panel (GRM_c) were fitted in a single model. The total heritability explained by both GRM_s and GRM_c when fitted together was similar to the heritability explained by fitting all SNPs in a single GRM. The GRM_s explained a smaller proportion (18%) of the total heritability, whereas the GRM_c explained 82%. Fitting either the 50k or the 600k SNP panels resulted in similar prediction accuracy for parasite resistance (~0.37). However, when both GRM_s and GRM_c were fitted together in the prediction model, genomic accuracy was increased by 10%. These results indicate that accuracy of genomic prediction can be improved by including QTL information explicitly in the prediction models.

INTRODUCTION

Traditional genetic improvement relies on the use of pedigree information and phenotypic records of farm animals to estimate their breeding values. This has led to substantial genetic gain in most livestock species, especially for the traits that are easy to measure. However, the process is often inefficient for low-heritable, expensive or difficult to measure traits. An example is parasite resistance, measured by indicator traits such as worm egg counts (WEC), which is an important health issue that affects the sheep industry worldwide. Genomic selection offers an alternative to conventional breeding programs and can increase the rate of genetic gain by using genomic information to predict the breeding values of selection animals (Hayes *et al.*, 2009).

In genomic selection, the genomic breeding values (GBV) for selection candidates are predicted based on the estimates of marker effects across the whole genome. The accuracy of predicting genomic breeding values depends on the heritability of the trait, the size of the reference population and the level of relatedness between the reference population and selection candidates (Habier *et al.*, 2010). Moreover, the accuracy is highly influenced by the level of linkage disequilibrium between the SNP markers and the QTL (quantitative trait loci) affecting the trait (Goddard 2009). Depending on the genetic architecture of the trait, the chosen statistical method used to build the prediction model will have a significant impact on prediction accuracy. Models that incorporate pre-selected SNPs from QTL regions have been shown to improve the accuracy of genomic prediction (Brondum *et al.* 2015).

The objective of this study was to compare the accuracy of genomic prediction based on a medium-density (50k) SNP panel, high-density (600k) SNP panel, and including pre-selected SNPs from QTL regions identified by regional heritability mapping for parasite resistance in Australian sheep.

MATERIALS AND METHODS

Animals. Parasite resistance, as measured by WEC, was investigated in a multi-breed sheep population from the Sheep Cooperative Research Centre information nucleus flock (INF). A total of 7,539 animals with both genotype data and WEC phenotypes were included in this analysis. Various breeds were represented in the population (Table 1) but with a significant proportion of Merino sheep, and only this breed had a substantial proportion of purebred animals. The remaining breeds were mainly represented by their crosses with Merino (van der Werf *et al.* 2010).

Table 1. Proportions of different breeds in the population

Breed	BL	COR	COOP	EF	WD	PD	TEX	AF	PS	MER
Proportion (%)	11.1	0.8	10	0.7	0.4	1.8	2.3	2	1.1	69.8

Border Leicester: **BL**, Corriedale: **COR**, Coopworth: **COOP**, East Friesian: **EF**, White Dorper:**WD**, Poll Dorset: **PD**, Texel: **TEX**, Australian Finnsheep: **AF**, Prime Samm: **PS**, Merino:**MER**

Genotypes. Animals were genotyped using the 50k Ovine marker panel (Illumina Inc., SanDiego, CA, USA). SNPs were removed if they had a minor allele frequency (MAF) < 1%, an Illumina Gentrain score (GC) less than 0.6, a call rate less than 95%, or not in Hardy-Weinberg equilibrium. Furthermore, positions of SNPs were obtained from the latest sheep genome *Ovis_aries_v3.1*, and any SNP with unknown position was removed. After applying these quality measures, 7,539 animals and 48,198 SNPs were retained. The imputation from the medium-density panel to the high-density (HD) SNP panel was performed using the Fimpute algorithm (Sargolzaei *et al.* 2014).

Cross-validation experimental design. Animals were randomly split into ten non-overlapping subsets (i.e. each subset with ~ 753 animals). For each experiment, one of the ten subsets served as a validation population and the remaining of the data served as the training population. The whole process was repeated ten times so that each subset served once as the validation population.

Regional heritability mapping (RHM). RHM was performed ten times, once for each validation set. The input to RHM consists of phenotype and genotype data (600k SNPs) on animals in the combined nine training sets. Data on animals in the validation set was not included in the RHM input. In RHM, each chromosome was divided into regions of pre-defined number of SNPs, and the variance attributable to each region was estimated. Window size of 200 SNPs was used to build genomic relationship matrix (GRM) and the window was shifted every 100 SNPs so that each two adjacent windows overlap midway. The significance was evaluated by the likelihood ratio test (LRT), comparing the RHM model which includes the regional effect with the base model composed of mean, fixed effects and random animal and error terms, but without the regional effect. The base model **(1)** and the RHM model **(2)** fitted to the data were as follows:

$$\boldsymbol{y} = \boldsymbol{Xb} + \boldsymbol{Za} + \boldsymbol{e} \quad (1)$$
$$\boldsymbol{y} = \boldsymbol{Xb} + \boldsymbol{Za} + \boldsymbol{Z_2g} + \boldsymbol{e} \quad (2)$$

where y is a vector of cube root transformed WEC records; b is a vector of fixed effects; $\boldsymbol{a}$ is a vector of random additive genetic effects, $\boldsymbol{g}$ is a vector of random regional genetic effect estimated from SNPs within each region (window), $\boldsymbol{e}$ is a vector of residuals which was assumed to be distributed as $\sim N(0, I\sigma_e^2)$, where $\boldsymbol{\sigma_e^2}$ is the residual variance. X, Z and Z_2 are incidence matrices relating fixed, additive genetic and regional genetic effects to phenotypes. $\boldsymbol{a}$ was assumed to be distributed as ~ $N(0, A\sigma_a^2)$, where $\boldsymbol{A}$ is the numerator relationship matrix (NRM) calculated from deep pedigree records and $\boldsymbol{\sigma_a^2}$ is the additive genetic variance explained by pedigree; and $\boldsymbol{g}$ was assumed to be distributed as $N(0, G\sigma_g^2)$, where G is the regional genomic relationship matrix constructed from SNPs within each region, and σ_g^2 is the regional genomic variance. The fixed effects included in the models were breed proportions, age of animals, age of dam, gender, rearing type × birth type and contemporary groups (combination of flock site, birth year and management group effects).

Selection of SNP markers. Genomic regions obtained from each of the ten-fold cross-validation RHM analyses were ranked based on their LRT and significant regions were selected. For each fold, the top five ranked regions across the ten-fold experiments were the same. SNPs located within the top five ranked regions were used to build a GRM (GRM_s) and the proportion of the variance explained by these pre-selected SNPs was estimated by replacing the NRM in model (1) by the GRM obtained from the pre-selected SNPs. Variance was not only estimated using the GRM for the selected SNPs, but also by using a complementary GRM (GRM_c) based on the remaining SNPs from the 600k SNP panel. To obtain a better estimate of the variance explained by the selected SNPs, both the GRM_s and GRM_C were fitted together in the same model.

Accuracy of genomic prediction. To evaluate the impact of the selected SNPs on prediction accuracy, genomic predictions for the validation animals was calculated and correlated with the phenotypes of the same animals. The $GRM_s$was fitted and the genomic best linear unbiased prediction (GBLUP) analysis was performed. The prediction model that includes both GRM_s and GRM_c was also evaluated. Genomic breeding values (GBV) were calculated following the ten-fold cross-validation procedure as described above. Prediction accuracy was calculated as the correlation between the predicted GBVs of the validation set and the adjusted phenotypes, which were corrected for fixed effects, divided by the square root of the trait heritability. Furthermore, the regression coefficient (slope) of the adjusted phenotypes on the GBVs was calculated to assess the bias of genomic predictions.

RESULTS AND DISCUSION

The RHM results for ten-fold experiments are shown in the Manhattan plots in Figure1. The top five ranked regions remained consistent across the ten-fold cross-validation experiments. These five regions include three windows (107 -108 Mb, 110 -112 Mb, 117 -118 Mb) on OAR2, three overlapping windows between 28 to 36 Mb on OAR6, a window between 17 to 18 Mb on OAR18, a window between 7.2 to 6.8 Mb on OAR20 and a window between 40 to 41 Mb on OAR24. 1600 SNPs located within these regions were selected to build a GRM and, the heritability explained by the pre-selected SNPs was 0.05 compared to 0.19 explained by all the SNPs from the 600k panel.

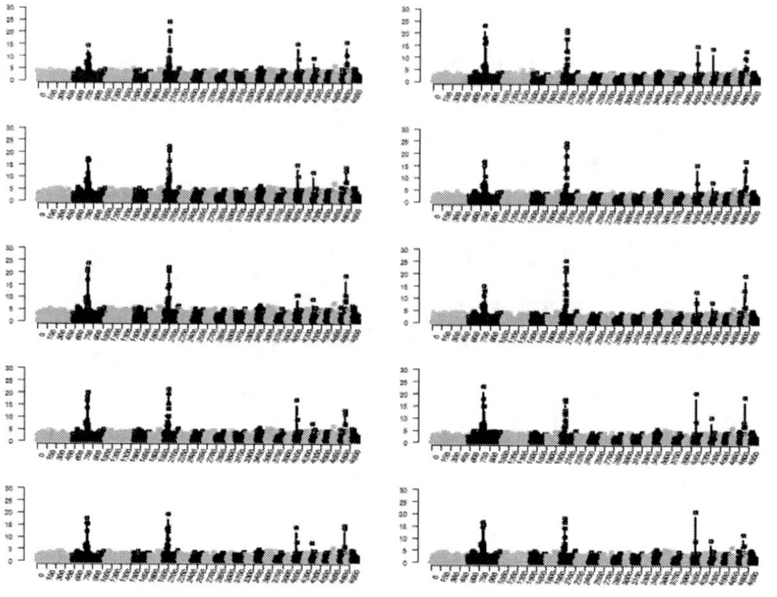

Figure 1. Manhattan plots of regional heritability mapping (RHM) results across the ten-fold cross-validation experiments. The x-axis represents the number of windows and the y-axis represents the corresponding likelihood ratio test (LRT) for each window.

Another way of testing the importance of the pre-selected SNPs was to investigate how much heritability was lost when the pre-selected SNPs were excluded from the GRM. Fitting only GRM_C, containing all SNPs in the 600k panel minus the pre-selected SNPs from the target regions, resulted in a similar heritability estimate as fitting all the SNPs. To assess the relative importance of the GRM from the selected SNPs and the GRM from the remaining SNPs, both GRM_s and GRM_C were fitted simultaneously in the same model. The proportion of variance explained when both GRM_s and GRM_C were fitted simultaneously was similar to the proportion of the genetic variance explained by fitting all the SNPs from the 600k. The GRM from the selected SNPs explained 18% of the total heritability, whereas 82% of the total heritability was explained by all the remaining SNPs (Table 2).

Table 2. The proportion of phenotypic variance (h^2) explained for parasite resistance

Selection criteria	GRM	GRM_s	GRM_C	logL
G (50k)	0.178 ± 0.020			-10673
G(600k)	0.194 ± 0.021			-10670
G(regions)		0.050 ± 0.009		-10682
GRMc			0.188 ± 0.021	-10673
G(Regions)+GRMc		0.034 ± 0.008	0.152 ± 0.021	-10638

G (50k): GRM from the 50k SNP panel, G (600k): GRM from the 600k SNP panel, G (regions): GRM_s from the pre-selected SNPs; GRMc: complementary GRM (GRMc)

Using any of the 50k and the 600k SNP panels resulted in a similar prediction accuracy for parasite resistance (~0.37, Table 3). When the GRM_s from the pre-selected SNPs was fitted alone, the prediction accuracy dropped by 18% compared to fitting all SNPs from the 600k panel. However, when both GRM_s and GRM_C were fitted together, higher prediction accuracy was observed than fitting all the SNPs in a single GRM. This is likely because a model with two components of genetic effects allows effects of the pre-selected SNPs to have larger variance than all the remaining SNPs in the panel, thus putting more weight on the pre-selected SNPs from the QTL regions. Moreover, the slopes of all models were not significantly different from 1, which indicates no significant bias in the predictions. It should however be noted that the RHM regions are not independent since they were the same across all 10-fold repeats and this can of course favourably influence the prediction accuracy. While suboptimal for a fair comparison of accuracy of prediction this lack of independence is not unexpected nor undesirable in practice since QTLs should have a real biological effect on a trait and are expected to be consistently identifiable in different datasets with similar power. If the RHM regions changed with each subset of the data, there would be greater cause for concern.

Table 3. Cross-validation prediction accuracy for parasite resistance averaged over the ten validation sets, and slope for the regression of adjusted phenotypes on the predicted breeding values

Selection criteria	Accuracy	SE(accuracy)	Slope	SE(slope)
G (50k)	0.368	0.036	0.915	0.197
G(600k)	0.374	0.036	0.916	0.193
G(regions)	0.307	0.035	0.841	0.219
G(Regions)+GRMc	0.411	0.036	0.848	0.164

CONCLUSION

The results in this study show that there is little advantage of using the imputed high density SNP panel over the medium-density panel for genomic prediction with this trait. However, by incorporating information from QTL regions explicitly into the genomic prediction model, prediction accuracy of parasite resistance increased by 10% based on the current SNP panel density. These results suggest that QTL information should be beneficial in genomic prediction, not just for parasite resistance but also for other economically important traits in sheep.

ACKNOWLEDGEMENT

This project was supported by Sheep CRC, Next-Generation BioGreen 21 Program PJ01134906 and PJ012611, Rural Development Administration, Republic of Korea and Australian Research Council (DP130100542).

REFERENCES

Hayes B.J., Bowman P.J., Chamberlain A.C., Verbyla K. and Goddard M.E. (2009) *Genet. Sel. Evol.* **41**: 51.

Habier D., Tetens J., Seefried F., Lichtner P. and Thaller G. (2010. *Genet. Sel. Evol.* **42**: 5.

Goddard, M.E., (2009) *Genetica.* **136**: 245.

Brondum R. F., Su G., Janss L., Sahana G., Guldbrandtsen B., Boichard D., *et al.* (2015) *J Dairy Sci.* **98**: 4107.

Van der Werf J.H.J., Kinghorn B.P. and Banks R.G. (2010) *Anim. Prod. Sci.* **50**: 998.

Sargolzaei M., Chesnais J. and Schenkel F. (2014) *BMC Genom.* **15**: 478.

Proc. Assoc. Advmt. Anim. Breed. Genet. 22:45-48

GENE BY BIRTH TYPE INTERACTION IN MERINO LAMB

A. Dakhlan[1], N. Moghaddar[1,2], C. Gondro[1] and J.H.J. van der Werf[1,2]

[1]School of Environmental and Rural Science, University of New England, Armidale, NSW, Australia
[2]Cooperative Research Centre for Sheep Industry Innovation, Armidale, 2351, Australia

SUMMARY

The purpose of this study was to investigate genome wide association of genetic markers with birth weight (BWT) and the interaction of significant marker effects with birth type (BT) in Merino lambs. Data used in this study consisted of 6,463 birth weight records of lambs generated from 349 sires and 4,896 dams of Merino sheep, which were genotyped using the Illumina Ovine SNP50 BeadChip (Illumina Inc., San Diego, CA, USA). After quality control 48,599 SNPs were included in the association study. We detected 11 and 15 genome-wide significant SNPs for birth weight in singles and twins, respectively, and 17 genome-wide significant SNPs were found when using all data. OAR6_41936490.1 and OAR6_41877997.1 were the most significant SNPs for single and twin birth weight, respectively. Among 17 significant SNPs detected by GWAS there were 9 that had a significant SNP by BT interaction, indicating that gene by BT interaction contributed to BWT variation.

INTRODUCTION

Birth type (BT) constitutes an environment that influences gene expression related to particular traits in sheep. In a previous study, Dakhlan *et al.* (2017) found significant genotype by environment interaction effects for birth weight (BWT) and weaning weight in Merino lambs. Twin BT is a poorer environment for BWT compared to single BT. With information on molecular genotypes it is now also possible to assess the interaction between environment and genotype at the individual gene level.

Genome Wide Association Studies (GWAS) have been widely used to identify genes that are associated with body weight in some animal species, including sheep. Jonas *et al.* (2010) detected a quantitative trait locus (QTL) on ovine (*Ovies aries*) chromosome 21 (OAR21) in Awassi-Merino backcross and Al-Mamun *et al.* (2015) identified 39 SNPs associated with body weight in Merinos, including a major QTL region on OAR6.

Birth type is one of many factors that influence growth performance of sheep and given there is a BT by growth interaction, it may be possible to differentiate between gene effects associated with BWT in single and twin BT of lambs. The purpose of this study was to investigate genotype by BT interaction at the gene level by investigating QTL associated with BWT of lambs and identify whether effects of significant markers differ between single and twin BT of lambs.

MATERIAL AND METHODS

Phenotypes for association study. Birth weight data for this study were obtained from the Information Nucleus (IN) program of the CRC for Sheep Industry Innovation in Australia. Details on this program and its design are described by Van der Werf *et al.* (2010). Birth weight records were available from 6,463 Merino lambs generated from 349 sires and 4,896 dams. These lambs were distributed over 2 BT classes: 3087 lambs were born as single and 3376 lambs were born as twins. The lambs were raised in 8 different flocks (521-2,483 lambs per flock) in up to 4 management groups per flock per year, and they were born between 2007 and 2012 (969-1,678 lambs per year).

Mixed model analysis with ASReml software (Gilmour et al., 2009) was used to generate predicted birth weight and the residual effects were used in a genome wide association study. The fixed effects in the models were birth year (2 classes), sex (2 classes), age of dam as covariate and contemporary group. As random effects in an animal model, genetic group, animal, dam, and interaction between sire and flock were fitted. There were 135 genetic groups defined. A pedigree file consisting of 17,664 animals from 11 generations was used to determine additive genetic relationships among animals and account for them in the analysis. It was assumed that dams were unrelated as limited pedigree information was available on the dams.

Genotyping and association study. Animals used in this study were genotyped using the Illumina Ovine SNP50 BeadChip (Illumina Inc., San Diego, CA, USA), and after quality control we included 48,599 SNPs. Gene annotation was done using the latest sheep genome Ovis_aries_v4.0 sequence to identify and explore candidate genes. For the association study birth weight residuals were regressed on each of the SNP genotypes individually, one at a time, using a linear model. Three analyses were undertaken with the first using all data of birth weight residuals (6,463 records), and then two analysis, one for using only records for single birth type (3,087 records) and one for twin birth type (3,376 records).

The significance threshold value ($P<2.06x10^{-7} = 0.01/48,599$) was set for genome-wide significance by applying the Bonferroni correction. To investigate gene by BT interaction for significant SNPs effect, a SNP by BT interaction term was fitted in the model used for all data.

RESULTS AND DISCUSSION

Genome-wide association study. Genome-wide significant SNPs were detected for birth weight in the combined data as well as in the data for single and twin birth types separately (Table 1). There were 11 significant SNPs (Bonferroni-corrected genome-wide association, $P<1.03x10^{-6}$) for birth weight in the single BT data set, and they were all within one region on OAR6 between 41.00 and 42.09 Mb. The most significant SNP was OAR6_41936490.1 ($P = 8.45 \times 10^{-15}$).

There were 15 significant SNPs for birth weight in the twin BT data, all but one in the same region on OAR6 as in the single BT dataset. The most significant SNP was OAR6_41877997.1 ($P = 3.02 \times 10^{-13}$). Riggio *et al.* (2013) reported that OAR6_41558126.1, OAR6_41768532.1 and OAR6_40855809.1 are associated with body weight in Scottish Blackface lambs. There were 10 significant SNPs found in this study that are the same as those SNPs found by Al-Mamun *et al.* (2015), who used post weaning weight data with a smaller (1,781 lambs) subset of the data used in this study.

According to *Ovis aries* reference genome assembly (Oar_v4.0) there were 12 genes within 17 significant SNPs that span the region between 40.45 and 42.53 Mb on OAR6, those genes are LOC105608045, LOC106991210, TRNAS-GGA (transfer RNA serine (anticodon GGA)), LOC105611897, LOC105615458, LOC106991209, TRNAW-CCA (transfer RNA tryptophan (anticodon CCA)) and LOC101104829 (60S ribosomal protein L10a pseudogene) which are both associated with body weight in Merino sheep (Al-Mamun *et al.*, 2014), KCNIP4 (Kv channel interacting protein 4) which is associated with weaning weight in cattle (Buzanskas *et al.*, 2014) and body weight aged 12 weeks in chicken (Gu *et al.*, 2011), LOC105611900, ADGRA3 (adhesion G protein-coupled receptor A3) which is associated with birth weight in pig (Wang *et al.*, 2016), and LOC101103396 (cytosolic beta-glucosidase). No information regarding the function of genes of LOC105608045, LOC106991210, LOC105611897, LOC105615458, LOC106991209, LOC105611900, and LOC101103396 have been reported in the literature. There were 10 genes (not including LOC106991210 and LOC105608045) for single BT and 11 genes (not including LOC105608045) for twin BT that span the same region.

Gene by birth type interaction. Among 17 significant SNPs detected by GWAS there were 9 SNPs that showed a significant interaction with BT (Table 1). Lambs born as a single have heavier BWT than those born as twins, indicating that a single BT provides a better environment compared to a twin BT environment. The most significant interaction was found for OAR6_41003295.1, where the allele substitution effect was -0.10 kg in singles whereas it was -0.05 kg in twins. Similar effects differences were found for other SNPs (Table 2). These result is supported by our previous study where it was found that the genetic correlation between breeding values for BWT expressed in singles and twins is less than one (Dakhlan *et al.* 2017), indicating that BWT expressed in two different BT environments is genetically not the same trait.

Table 1. SNPs that have significant association on OAR6 for single and twin birth weight and with birth weight using total data

SNP name	Position (bp)	P-value (all data)	P-value (single BT data)	P-value (twin BT data)
OAR6_40449774.1 ns	40449774	$3.01x10^{-09}$		
OAR6_40724811_X.1 ns	40724812	$1.82x10^{-11}$		$1.09x10^{-07}$
OAR6_40855809.1 ns	40855809	$2.56x10^{-11}$		$2.39x10^{-07}$
OAR6_41003295.1*	41003295	$4.96x10^{-19}$	$2.47x10^{-13}$	$1.71x10^{-09}$
s17946.1*	41384761	$4.61x10^{-18}$	$2.04x10^{-11}$	$1.37x10^{-08}$
OAR6_41476497.1*	41476497	$8.21x10^{-17}$	$1.05x10^{-10}$	$1.42x10^{-07}$
OAR6_41494878.1*	41494878	$6.26x10^{-17}$	$1.46x10^{-10}$	$9.76x10^{-08}$
OAR6_41558126.1*	41558126	$1.63x10^{-15}$	$8.88x10^{-09}$	$5.74x10^{-08}$
OAR6_41583796.1*	41583796	$8.47x10^{-15}$	$3.45x10^{-09}$	$5.67x10^{-07}$
OAR6_41709987.1*	41709987	$1.03x10^{-14}$	$1.12x10^{-08}$	$1.20x10^{-07}$
OAR6_41768532.1 ns	41768532	$8.83x10^{-17}$	$4.57x10^{-08}$	$2.82x10^{-10}$
OAR6_41850329.1 ns	41850329	$3.96x10^{-10}$		$1.12x10^{-07}$
OAR6_41877997.1 ns	41877997	$2.75x10^{-19}$	$2.13x10^{-08}$	$3.02x10^{-13}$
OAR6_41936490.1*	41936490	$4.90x10^{-25}$	$8.45x10^{-15}$	$7.26x10^{-13}$
OAR6_42094768.1*	42094768	$2.97x10^{-17}$	$2.29x10^{-11}$	$1.62x10^{-08}$
OAR6_42247197.1 ns	42247197	$2.24x10^{-07}$		$6.42x10^{-07}$
OAR6_42528741.1 ns	42528741	$5.02x10^{-08}$		

Note: *Interaction significance is based on $\alpha = 5\%$, ns = not significant interaction

CONCLUSION

In this study 11 and 15 genome-wide significant SNPs were detected for single and twin birth weight, and 17 genome-wide significant SNPs were associated with birth weight when using all data of birth weight. Twelve genes spanning the region between 40.45 and 42.53 Mb on OAR6 cause birth weight variation but 9 SNPs showed a significant interaction with birth type, indicating that the genes associated with these SNPS may have a different gene action in the two birth type environments.

Table 2. SNP effects of single and twin birth type and interaction P-value on birth weight

SNP name	SNP effect		Interaction P-value
	Single BT	Twin BT	
OAR6_40449774.1	-0.09	-0.04	5.01E-02
OAR6_40724811_X.1	-0.09	-0.06	3.23E-01
OAR6_40855809.1	0.08	0.06	4.14E-01
OAR6_41003295.1	-0.10	-0.05	6.84E-05*
s17946.1	-0.10	-0.05	3.07E-02*
OAR6_41476497.1	-0.10	-0.05	3.61E-02*
OAR6_41494878.1	0.10	0.05	3.55E-02*
OAR6_41558126.1	0.08	0.05	4.40E-02*
OAR6_41583796.1	-0.11	-0.06	3.81E-02*
OAR6_41709987.1	-0.11	-0.06	3.58E-02*
OAR6_41768532.1	0.09	0.06	1.91E-01
OAR6_41850329.1	0.09	0.07	4.54E-01
OAR6_41877997.1	-0.08	-0.06	2.99E-01
OAR6_41936490.1	0.12	0.06	2.21E-02*
OAR6_42094768.1	-0.11	-0.06	4.16E-02*
OAR6_42247197.1	-0.05	-0.05	3.32E-01
OAR6_42528741.1	-0.08	-0.03	3.91E-01

Note: *Interaction significance is based on $\alpha = 5\%$

REFERENCES

Al-Mamun H.A., Clark S.A., Kwan P. and Gondro C. (2014) In: Proceeding of 10th World Congress of Genetics Applied to Livestock Production, Vancouver Canada

Al-Mamun H. A., Kwan P, Clark S.A., Ferdosi M.H., Tellam R. *et al.* (2015) *GSE* **47**: 66.

Buzanskas M., Grossi D., Ventura R., Schenkel F., Sargolzaei M. *et l.* (2014) *Plos One* **9**: e94802.

Dakhlan A., Moghaddar N. and van der Werf J.H.J. (2017) (under review in Anim Production Sci.)

Gilmour A.R., Gogel B.J., Cullis B.R. and Thompson R. (2009) ASReml User Guide Release 3.0. VSN International Ltd, 5 The Waterhouse, Waterhouse Street, Hemel Hempstead, HP1 1ES, UK.

Gu X., Feng C., Ma L, Song C., Wang Y. *et al.* (2011) *Plos One* **6**(7): e21872.

Jonas E., Thomson P, and Raadsma H. 2010. Proceeding of the 9th World Congress on Genetics Applied to Livestock Production: 1-6.

Riggio V., Matika O, Pong-Wong R., Stear M. *et al.* (2013) *Heredity* **110**: 420.

Van der Werf J.H.J., Kinghorn B.P. and Banks R.G. (2010) *Anim. Prod. Sci.* **50**: 998.

Wang X., Liu X., Deng D., Yu M. and Li X. (2016) *BMC Genet.* **17**(Suppl 1):15

Proc. Assoc. Advmt. Anim. Breed. Genet. 22:49-52

OPTIMAL AND PRACTICAL STRATEGIES TO MANAGE AN OVULATION RATE MUTATION LOCATED ON THE X CHROMOSOME IN A FRENCH SHEEP BREED

J. Raoul[1,2,3], L. Bodin[2], J.M. Elsen[2] and A.A. Swan[3]

[1] Institut de l'Elevage, Castanet-Tolosan, France
[2] GenPhySE, INRA, Castanet-Tolosan, France
[3] Animal Genetics and Breeding Unit, University of New England, Armidale, NSW, 2351 Australia

SUMMARY

In French sheep breeding programs, several mutations affecting ovulation rate have been discovered. For mutations located on the X chromosome, the optimal management of such genes is still a challenge because nucleus flocks are small compared to Australian or New Zealand ram breeding flocks. A deterministic model was developed, and using sequential quadratic programming methodology, the combination of mating types that maximized the profit across a range of genotype costs was determined. Results show that even if losses of genetic gain were quite high compared to the gain without the major gene, the optimal use of an ovulation rate mutation located on the X chromosome was beneficial. At the current costs, the optimal strategy that gave the maximal profit was based on four different mating types. A strategy based on only the use of carrier females mated to non-carrier males gave similar results to the optimal strategy in terms of profit and genetic gain. This strategy could be adopted by French breeding programs where this kind of mutation segregates.

INTRODUCTION

The number of lambs produced per female has a large impact on profitability in meat oriented sheep production. Several mutations affecting ovulation rate, and thus number of lambs, have been identified. For example, Booroola (Piper and Bindon 1982; Davis *et al.* 1982), BMP15-Inverdale (Davis *et al.* 1982) or BMP15-Grivette (Demars *et al.* 2013), and GDF9-Cambridge (Hanrahan *et al.* 2004). Most often, these polymorphisms have a positive effect on heterozygous carrier productivity. However, in homozygous ewes, these polymorphisms lead to sterility or excessive prolificacy and high rates of neonatal lamb mortality. Therefore homozygous females are undesirable for commercial production.

Several strategies can be implemented to manage these mutations, as outlined by Amer *et al.* (1998) for mutations carried by the X chromosome (i.e. Inverdale gene) and Raoul *et al.* (2017) for mutations carried by an autosomal chromosome: the proportion of each parental genotype is defined according to the sex and matings organised. These balance high frequency of heterozygous females with genetic gain. Increasing the frequency of heterozygotes leads to a change in the proportion of available candidates which affects the overall selection differential of parents and consequently genetic gain. Amer *et al.* (1998) assessed two strategies to manage the Inverdale gene and found that depending on the strategy implemented, the loss of genetic gain was either 24%, or less than 5% compared to the gain without major gene. In the case of an autosomal polymorphism, strategies that enhance either genetic gain or heterozygous female frequency gave equal profit (Raoul *et al.* 2017) and were affected by the genotyping cost per animal.

In the French meat sheep production context, the average number of ewes per nucleus flock is about 300. With such limited flock sizes implementing a strategy which comprises a small proportion of a given mating type (less than 10%) is difficult. It is not practical at a single flock level, but could be organized via specialization of several nucleus flocks in which different flocks focus on a specific mating. This is difficult to co-ordinate, so for practical reason, French breeders would much prefer strategies based on at most two mating types. Strategies outlined for autosomal mutation management have already been discussed for French breeding programs (Raoul *et al.* 2017). The aim of this study is, for the case of a mutation carried by the X chromosome, to determine the combination of mating types that provide the maximal profit (optimal strategy) according to various genotyping costs. This optimal strategy will be compared with more practical strategies in terms of profit and genetic gain.

MATERIALS AND METHODS

A nucleus population representative of a typical French breeding program based on natural mating was modelled. A maternal production trait expressed once per year during female's reproductive life was considered as the only selected trait (*e.g.* milk production estimated through lamb weight at 30 days). Each year, 8000 ewes were mated to 200 rams. Because homozygous carrier females were not used for reproduction, 2 genotypes, non-carriers and carriers were respectively considered for males ([+] and [m]) and females ([++] and [m+]) leading to 4 mating types: 1) ♀ [++] x ♂ [+], 2) ♀[++] x ♂ [m], 3) ♀ [m+] x ♂ [+] and 4) ♀ [m+] x ♂ [m]. As the flock management was assumed to be in a steady-state, the proportion of each mating type across time was constant. The newborn candidates were divided into categories according to their parental genotypes (*i.e.* 4 matings), their sex and their own genotype (2 genotypes for males and 3 genotypes for females). Generations were overlapping and the maximum reproductive life was 6 years for males and females, with a maximum parity of 5 (i.e., from 2 to 6 years of age), leading to a replacement proportion close to 24%.

At each generation, new parents were selected within sex*genotype categories by truncation selection on EBVs: 4 truncation thresholds (2 per parental genotype) were determined across the candidate EBV distributions. For example, [++] female replacement were selected from progeny of mating types 1 and 3. Considering dam parity, these female were selected across 10 EBVs distributions. Whatever their parental genotype or dam's age, we selected females whose EBV was above the unique truncation threshold. Given those thresholds, selection differential and genetic contribution to the next generation (*i.e.* probability of gene origin) were calculated for each candidate category. Evolution of genetic values of parents and their progeny across time for the maternal trait was derived using the gene flow methodology proposed by Hill (1974): a transition matrix representing the gene flow from categories at year t to categories at year t+1 was built from genetic contributions to newborns and accounting for ageing of parents.

Discounted revenues and costs were computed for each cycle (year). The revenues were proportional to the number of lambs sold per year which was equal to the number of live lambs produced minus the number selected for replacement, and the number of live lambs produced by ewes transferred to a commercial flocks. The costs included genotyping costs made at the nucleus level and proportional breeding costs per ewe (nucleus and transferred ewes). It was assumed that 50% of newborn females would still be available after parent selection, and these surplus females would be transferred to a commercial flock where they could be retained for up to 5 parities. These female were not genotyped and only females from mating types 1, 2 and 3 were transferred. It was assumed that independently of their genotype, the selected maternal trait was related to the cost per ewe, because the trait was determined based on milk production, with higher production levels reducing feed costs per lamb. The overall profit was computed as the sum of discounted revenues minus costs over a long-term time horizon (year 5 to year 30). This overall profit was assessed for the following sets of parameters: number of lambs produced = 1.5 for non-carrier females, and +0.5 additional lambs for heterozygous females. Given the fertility, the lamb viability (higher for lambs born from non-carrier), the number of lambs weaned per ewe joined for non-carrier and carrier ewes were 1.22 and 1.44 respectively. The income per lamb sold was assumed to be constant and the production cost per lamb depended on the dam's genetic value for the selected trait and genotype. Three genotyping costs were tested: no cost, 10 and 20 € per genotyped animal.

For a given genotyping cost, the relative proportion of mating types that gave the maximum profit (the optimal strategy) was determined using an algorithm based on sequential quadratic programming methodology. The gain in the absence of the major gene and two simplified strategies was also assessed based on 1 mating only, ♀ [m] x ♂ [++] (S1, corresponding to the "self-sustaining scheme" outlined by Amer (1998)) or 2 mating types, ♀ [++] x ♂ [m] and ♀ [m+] x ♂ [+], (named S2). The proportion of each mating types of these strategies is shown in Table 1.

Table 1: Proportion (%) of each mating type of alternative strategies assessed for the management of an ovulation rate mutation[1] located on the X chromosome.

Mating type	♀[++]x♂[+]	♀[++]x♂[m]	♀[m+]x♂[+]	♀[m+]x ♂[m]
Gain without major gene	100	0	0	0
S1	0	0	100	0
S2	0	60	40	0

[1] Biallelic locus (X chromosome) influencing the number of lambs per female (1.5 for [++] and 2.0 for [m+]).

RESULTS AND DISCUSSION

Table 2 gives the proportion of each mating type in the nucleus that maximizes profit according to the genotyping cost. Results show that when genotyping costs were not included (cost=0), the best strategy was to bred only carriers females and mate them to non-carrier males.

Table 2: Percentage of each mating type in the optimal strategy to manage an ovulation rate mutation[1] located on the X chromosome, according to three genotyping costs (€).

		Mating type			
	genotyping costs	♀[++]x♂[+]	♀[++]x♂[m]	♀[m+]x♂[+]	♀[m+]x ♂[m]
optimal strategy	0	0	0	100	0
	10	21	49	12	18
	20	39	57	0	4

[1] Biallelic locus (X chromosome) influencing the number of lambs per female (1.5 for [++] and 2.0 for [m+]).

For a genotyping cost equal to 10 €, the optimal strategies combined the 4 mating types. The main mating type was non-carrier females mated to carrier males (49% of all matings). In this strategy 30% of the nucleus females were carriers. For a genotyping cost equal to 20, the proportion of non-carrier females mated to carrier males reached 57%. The proportion of carrier females in the nucleus reduced to 4% which corresponded to the minimum requirement to replace carrier males and produced heterozygous females transferred to commercial flocks.

Table 3 shows the genetic gain achieved by the nucleus for all strategies assessed, the genotyping requirements, the frequencies of heterozygous females (nucleus and transferred) and the profit. Apart from the heterozygous frequencies, all results are expressed relative to values obtained for the optimal strategy when there was no genotype cost (=100 in the first row of Table 3).

Results show that when genotyping costs were not included, the optimal strategy maximized the heterozygous female frequency in the nucleus. In this case, a proportion of m+ females were selected for the nucleus, whereas all ++ females were available for transfer. This lead to a reduction in the heterozygous frequency of transferred females to 24%. When genotyping costs were included, the strategy maximized the heterozygous frequency of transferred females. In this case, mating type 2 (♀[++]x ♂[m]) which produces m+ females without genotyping was used, allowing production of heterozygous females to be transferred to a commercial flock. For a moderate genotyping cost (10€), the number of genotyping remained at a significant level and allowed implementation of a strategy providing a substantial genetic gain. For a high genotyping cost, the number of genotypes was very low and limited to genotyping male progeny of the mating ♀[m+]x ♂[m] only, implemented to replace male carriers. Even if this mating produced homozygous carrier females which were culled, it allowed a higher genetic value of carrier males and a higher genetic gain compared to the use of the mating ♀[++]x ♂[m] This strategy maintained the high proportion of heterozygous females transferred to a commercial flock and limited losses in genetic gain.

Table 3: Genetic gain, genotyping needs, heterozygous female frequencies and profit of various strategies according to the genotyping costs (€).

	Geno. Costs	Genetic gain[1]	Genotyping requirements[2]	Het. freq (nucleus)	Het. freq (transferred)	Profit[3]
	0	100.0	100	1.00	0.24	100.0
Optimal strategy	10	100.4	27	0.29	1.00	79.1
	20	85.1	4	0.04	1.00	74.6
Gain without major gene	-	125.4	0	0.00	0.00	72.5
S1[4]	0	100.0	100	1.00	0.20	100.0
	10	100.0	100	1.00	0.20	77.1
	20	100.0	100	1.00	0.20	54.2
S2[5]	0	103.5	59	0.40	1.00	85.8
	10	103.5	59	0.40	1.00	72.4
	20	103.5	59	0.40	1.00	58.9

[1] 100=genetic gain obtained for the optimal strategy at null genotyping costs
[2] 100=number of genotype for the optimal strategy at null genotyping costs
[3] 100= profit obtained for the optimal strategy at null genotyping costs
[4] Simplified strategy based on one mating type ♀ [m] x ♂ [++]
[5] Simplified strategy based on two mating types ♀ [++] x ♂ [m] and ♀ [m+] x ♂ [+]

The genetic gains for the S1 and S2 strategies were similar to those obtained for optimal strategies, and losses of genetic gain ranged from 22 to 25%, compared to gain without the major gene, similar to the results obtained by Amer *et al.* (1998). Profit obtained for S1 was higher than S2 except at the high genotyping cost. In this case, simple management of the mutation gave lower profit than its eradication. Given the current genotyping cost, approximately 10 €, S1 is a strategy which could be considered for French breeding programs. This strategy has quite high genotyping requirements (two genotyped animals per selected replacement) but results in profitability similar to the optimal strategy and a high productivity in the nucleus flocks. The use of a tool combining parentage assignment and mutation genotyping, which is available in France, would decrease the genotyping cost and make application the S1 strategy more attractive.

REFERENCES

Amer P.R., McEwan J.C., Dodds K.G. and Davis G.H. (1998) *Proc. NZ Soc. Anim. Prod.*, **58**: 157

Piper L.R. and Bindon B.M. (1982) *In the Booroola Merino*, **9**:20. CSIRO, Melbourne

Davis G.H., Montgomery G.W., Allison A.J., Kelly R.W. and Bray A.R. (1982) *NZ J. Agric. Res.* **25**: 525.

Demars J., Fabre S., Sarry J., Rossetti R., Gilbert H., Persani L., Tosser-Klopp G., Mulsant P., Nowak Z., Drobik W., Martyniuk E. and Bodin L. (2013) *PLoS Genetics* **9**:4

Hanrahan J.P., Gregan S.M., Mulsant P., Mullen M., Davis G.H., Powell R., Galloway S.M (2004) *Biol. Reprod.* **70**:900

Hill W.G. (1974). *Ani. Prod.*, **18**:117

Raoul J., Palhière I., Astruc J.M., Swan A.A. and Elsen J.M. (2017) submitted to *Animal*

Proc. Assoc. Advmt. Anim. Breed. Genet. 22:53-56

INCLUDING FEED INTAKE AND GREENHOUSE GAS EMISSIONS INTENSITY INTO EXISTING SELECTION INDEXES – THEORY AND PRACTICE

P. Amer[1], C. Quinton[1], T. Byrne[1], A. Cromie[2], C. Richardson[3] and J. Archer[1]

[1] AbacusBio Ltd, PO Box 5585, Dunedin
[2] ICBF, Bandon, Ireland
[3] CGIL, University of Guelph, Canada

SUMMARY

Theoretical approaches used to calculate economic weights for feed intake and greenhouse gas emissions intensity that can be used to augment existing selection indexes are outlined. The approaches are discussed. Evaluation of these traits and their index weightings in a way that makes them independent of other traits already in the breeding objective is highly desirable to industry practitioners wishing to minimise disruption to current systems. However, this requires both biological and statistical concerns associated with definition of residual traits to be overcome.

INTRODUCTION

While collecting records, and undertaking genetic/genomic evaluations of merit for feed intake and greenhouse gas (GHG) emissions traits is costly, estimated breeding values for these traits are of growing interest for inclusion in national selection indexes, because of their current and potential high relative economic importance (Archer et al. 1997; Wall et al. 2010).

Substantial investment has been dedicated towards recording of phenotypes, but the specific options for explicitly defining the estimated breeding value traits and applying weighting to them remains contentious. The complexity of the issues is exacerbated by the fact that existing selection criteria and estimated breeding values are already linked to both the amount of feed consumed and the amount of GHG emissions by animals in a commercial farm system. This means that double counting must be avoided, either through the choice of novel estimated breeding value trait definitions, or through adjustments to the weightings applied to existing traits in the index.

This paper introduces the options for implementation of selection criteria for novel traits addressing GHG emissions and feed efficiency into selection indexes and discusses their strengths and weaknesses.

EXISTING SELECTION CRITERIA

Existing selection indexes for farmed ruminant livestock commonly affect GHG emissions intensity, defined as the amount of GHG emitted from the farm system per unit of product generated. Methods have recently been developed to show how output-increasing traits dilute the "fixed" GHG emissions that do not increase in proportion to the extra output. For example higher output per animal without any increase in animal rearing and maintenance feed requirements improves the biological efficiency of the farm system (Wall et al. 2010). This concept has been formalised recently into a methodological framework that can be applied across multiple livestock species (Amer et al. 2017). It can be expected that the relative importance of existing traits in selection indexes based on farm profitability will be different to their relative importance based on GHG emissions intensity. Thus, there is an opportunity to divert the selection direction slightly away from selection solely for farm profitability, so as to achieve greater than current gains in GHG emissions intensity (Quinton et al, 2017). For example, increasing breeding female survival improves GHG emissions intensity through a reduction in feed and associated emissions for rearing of replacements. In contrast, the trait maternal milk production in meat production systems, while desired by farmers, typically does not increase market output, and is less feed efficient than converting feed into meat directly by the growing meat animal. Thus, weightings on emissions intensity traits do not fully align with farm profitability weightings. Because traits that improve GHG emissions intensity tend in most cases to also increase farm profitability, trade-offs from profitability weighting to emissions intensity weighting will in most cases be very efficient, in that significant improvements in emissions intensity gains may be achieved with only small reductions in profitability gains. These gains can be achieved with confidence because of the strong association between production efficiency (through both genetic and non-genetic means) and GHG emission intensity.

Existing selection indexes also typically account for correlated increases in feed intake associated with genetic changes in traits which are not direct measures of feed intake. Examples include the reductions in the economic values for milk component yields and young animal carcase output to account for well-known associations of these traits with feed intake. Mature breeding female weight is also often used as a proxy to predict the higher maintenance feed intake, and higher rearing feed costs associated with larger mature size individuals. While these proxy associations only account for a proportion of genetic variation in feed intake and production system efficiency, the proportion that they do account for is achieved with considerable prediction accuracy, because live weight and dairy system milk traits are typically evaluated with considerable accuracy, in a very large proportion of selection candidates and at modest recording expense.

DEFINITION OF NEW SELECTION CRITERIA

New measurement technologies may soon make it possible to record the feed intake and greenhouse gas emissions of animals on a sufficient scale for either conventional or genomic prediction of estimated breeding values, such that these traits can be included in selection indexes. It is important to consider though that many selection candidates are likely to be evaluated with only very modest accuracy. Even with genomic selection approaches, the accuracy of prediction of these novel selection criteria will at best be modest. This is due to the small size of training populations, and the risk of selection candidates having insufficient numbers of phenotyped close relatives in the training population.

Classical selection criteria typically predict fractional components of feed efficiency and GHG emissions intensity with high accuracy. Examples include milk yield and growth rate. It is well known that animals that produce more milk, or which grow faster, will on average require more feed than their contemporaries. The accuracy of novel selection criteria to predict the whole genetic variation is typically much lower (see Figure 1).

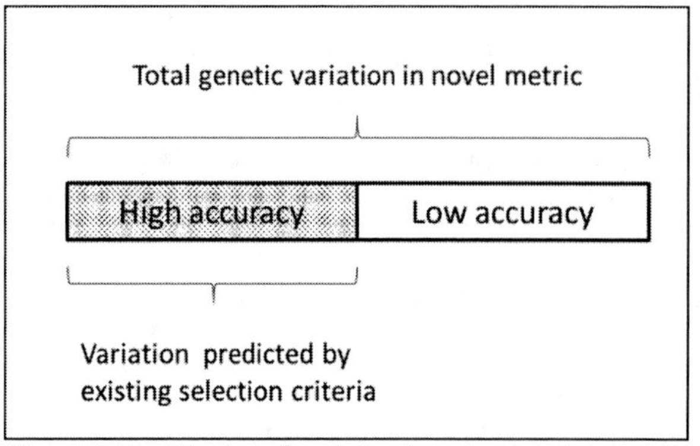

Figure 1. Partitioning of variation in a novel metric into the component predicted at high accuracy by existing selection criteria, and the component predicted at lower accuracy by phenotypes for the

novel metric.

Therefore, care must be taken to maximise the efficiency of prediction of the full breeding objective, while maintaining clarity and transparency of the index to users in industry. There are three options. Firstly, it is possible to adjust the phenotype for the novel metric so that the estimated breeding value predicts only variation that is not already predicted by existing selection criteria. The second and third options involve a more substantial rebuild of the existing selection index. These options are discussed further below.

OPTION 1. RESIDUAL TRAIT DEFINITION

Residual feed intake is a classic example of implementation of option one. From a breeding objective perspective, this approach is very elegant, as no adjustment is required to existing selection criteria and the weightings that are applied to them in the existing breeding objective. The economic value for feed intake also does not require any major rework. The cost of an extra unit of feed is the same, irrespective of whether it is an extra unit of total feed intake, or an extra unit of feed required after accounting for the feed intake expected to be associated with maintenance, growth and/or milk production (i.e. other traits in the index). Furthermore, it is likely that feed intake is a different genetic trait (i.e. correlations less than one) across different life stages of the animal, but the selection criterion breeding value is likely to be based on phenotypes measured at only one life stage (e.g. feed intake measured in young males for a small test window, used to predict feed intake in both growing and mature females, with the mature females split into both dry and lactating states). It is reasonable to hypothesise that the residual feed intake trait will be more highly genetically correlated across different life stages than total feed intake, but this will be hard to prove definitively in practice. Never-the-less, some estimation of these correlations is required for appropriate selection index construction, irrespective of the option taken.

It is important to note that many quantitative geneticists and biologists have reservations about the definition and use of residual traits, particularly for feed intake. The arguments put forward are beyond the scope of this paper, but relate to both statistical properties of the resulting traits, and potential detrimental biological consequences for fitness traits resulting from selection for a narrow definition of the trait.

OPTION 2. GROSS TRAIT DEFINITION – USE OF CORRELATED PREDICTORS

An alternative to option one is to accommodate the novel metric in the breeding objective without any adjustments. For example, a gross methane estimated breeding value might be included with a negative economic value associated with methane output. Similarly, a breeding value for total feed intake could be included in the breeding objective, either with or without the gross methane trait. Because output traits such as growth rate, body size and milk production are highly genetically correlated with both feed intake and GHG emissions associated with methane output, then these output traits should in principle be included as correlated predictor traits in the genetic evaluation of the methane or total feed intake breeding values.

The only advantage of this approach is that the need to define an estimated breeding value for a residual or adjusted trait is circumvented. There are many disadvantages. Any accounting for feed costs or associated emissions in the existing trait economic values must be removed. This will result cosmetically in much larger economic weightings for output-increasing genetic traits. The extent to which existing selection criteria are useful as correlated predictors of feed intake and gross methane emissions will differ substantially for animals in different life stages. For example, milk yield potential will be an important predictor of total feed intake in lactating cows, but much less so in young growing animals. The reverse would be expected for body weight traits recorded in young animals. This makes the process of estimating breeding values quite arduous, whereby complex multi-trait predictions are required with many parameters. Many of these parameters will not be well estimated. While the same parameters are likely required in option one, in that case they are more likely to be applied in the definition of the breeding objective, than in the computation of the estimated breeding values. Finally, for GHG traits, finding the balance of weighting on gross methane, gross feed intake and conventional traits that achieves a quantified and understood improvement in emissions intensity will be highly complex. For example, milk yield would be penalised heavily in an index including gross methane because of the strong association between milk production and GHG emissions in a lactating dairy cow, and it is well known that gain in milk production per cow is a substantial contributor to gains in GHG emissions intensity on dairy farms.

OPTION 3. GROSS TRAIT DEFINITION – IGNORING CORRELATED PREDICTORS

Option two is potentially complex to deploy, and so it might be tempting to completely ignore the fact that existing selection criteria are useful predictors of a component of the genetic variation in a target trait of interest. This would circumvent the problem stated above in terms of the implied complexity for the genetic evaluation system. For the situation of total feed intake as a trait with an economic value in the selection index, some weighting penalty should be left on existing selection criteria to account for their association with feed intake. Otherwise, it is likely that animals with low accuracy estimated breeding values for feed intake would have excessive weighting placed on output- and feed intake-increasing traits. However, the optimal amount of weighting would depend on the information sources available for both production traits and feed intake for each animal. Thus, such an implementation would likely be inefficient for many animals.

DISCUSSION AND CONCLUSION

Many quantitative geneticists are currently considering how novel selection criteria for feed intake and GHG emissions could be included in selection indexes. Two groupings exist. Firstly, there are researchers focused on development of the new traits, who view the options in terms of their biological interpretation. Secondly, there are practitioners who see the novel traits as a complex extension of systems and processes already in place and accepted by industry users. This paper has presented some options for deployment, and discussed their advantages and disadvantages. In general, option one, whereby the novel phenotypes are evaluated in a way that makes them independent of other traits already in the breeding objective is highly desirable to industry practitioners wishing to minimise disruption to current systems. However, this requires both biological and statistical concerns associated with definition of residual traits to be overcome. Failure to do so is likely to delay industry deployment, and so the challenge to overcome the objections will need to be addressed by those advocating for inclusion of the new information in industry selection indexes.

REFERENCES

Amer, P, Quinton, C, Hely, F. and Cromie, A. (2017) *Animal* In submission

Archer, J.A., Arthur, P.F., Herd, R.M., Parnell, P.F. and Pitchford, W.S. (1997) *J Anim Sci*, **75**:2024-2032

Quinton, C.D., Hely, F.S., Amer, P.R., Byrne, T. and Cromie, A.R. (2017) *Animal* In submission

Wall, E., Simm, G. and Moran, D. (2010) *Animal*, **4**(3):366–376.

Proc. Assoc. Advmt. Anim. Breed. Genet. 22:57-60

GENOMIC ANALYSIS OF HEALTH TRAITS USING AN AUSTRALIAN GENOTYPED COW REFERENCE POPULATION

M. Abdelsayed[1], M. Haile-Mariam[1] and J.E. Pryce[1,2]

[1]Agriculture Victoria, Department of Economic Development, Jobs, Transport and Resources
[2]La Trobe University, Agribio, 5 Ring Road, Bundoora, VIC 3083, Australia

SUMMARY

In recent years there has been increasing interest internationally in estimating breeding values for traits that can reduce farm costs, such as health traits in livestock. One of the limitations in developing breeding values for health traits in Australia has been lack of data. In this study, we have estimated reliabilities of genomic breeding values for health traits when only clinical records on health disorders are used that are collected from a genomic reference population (Ginfo). Reliabilities for bulls with daughters in the reference population are 27%, and 25% for mastitis and an "all-disease" trait, respectively. For bulls with no daughters in the reference population, reliabilities are 4% and 12% for mastitis and the "all-disease" trait, respectively. In contrast, reliabilities for reproductive disorders and metabolic diseases were much lower (<15%). Mastitis and "all-diseases" have higher incidences and also higher heritability estimates than the other diseases, which is likely to be the reason for higher reliability estimates. Although estimates are still regarded as low, they are in line with expectations for a newly-recorded trait. Investigation into the improvement of reliabilities through the use of predictor traits through multi-trait analysis is the next step for this research.

INTRODUCTION

In the past, genetic selection for milk production was the main focus for the driver of dairy farm profitability. While making great genetic gains in milk production, an unfavourable relationship between production and disease resistance has become apparent (Pryce *et al.* 1997; Rauw *et al.* 1998; Koeck*et al.* 2012). Dairy cow health will continue to deteriorate if disease traits, or their predictors, are not included in breeding objectives. Healthy cows are more productive, easier to manage, require less intervention, have improved animal welfare and contribute to profitability by reducing production costs.

Health and fertility traits generally have low heritability estimates (<5%) compared to production traits (>30%) (Egger-Danner *et al.*2015). However, there is sufficient genetic variation to still make selection feasible for low heritable traits, and this has been evident in the dairy industry with the improvements made with selecting directly on fertility (Pryce *et al.* 2014).

Traits like health and fertility have large impacts on the dairy industry but sometimes data availability is low. One option is to obtain records from a dedicated reference population of genotyped cows with phenotypes of interest. This has already started in Australia with the establishment of the first 100 Genomic Information Nucleus herds (Ginfo). Ginfo was a large-scale genotyping project (103 herds and 32,386 cows) to increase the size of the Australian dairy reference population to improve the reliability of Australian genomic breeding values.

The objectives of this study were to estimate 'clinical' genomic health breeding values for the major disease traits such as mastitis, reproductive disorders, lameness, metabolic disorders and an overall "all-disease" trait using the health data collected from the Ginfo herds and secondly to determine the reliability of those estimated breeding values.

MATERIALS AND METHODS

Health data and genotypes. A total of 487,503 electronic health records were accessed from 90 (of 103) Ginfo herds. Genotypes were available on 15,632 cows that also had health records. Genotypes of 2,984 bulls with daughter health records were also obtained from DataGene.

Disease categories. The major disease traits (mastitis, reproductive disorders, lameness, and metabolic disorders) were converted into binary traits. Each disease was coded with a 0 or 1 for every cow-lactation record, where 1 corresponds to a cow having a particular disease at any time in a lactation period and 0 if it does not have that disease. For the "all-diseases" category, if a cow has any record of any disease event, it was coded 1, or otherwise 0 as healthy.

Genomic analysis. The reference dataset contained 11,458 genotyped Holstein cows (out of the total 15,632). The validation dataset contained 494 genotyped bulls, with 6,989 daughters that had health records (n = 22,276) but were not genotyped themselves, so not included in the reference set. Bulls with less than 5 daughters were excluded from the analysis.

For the estimation of genomic breeding values the following linear mixed animal model was used:

$$y = \mu + HYS + Parity + MOC + \beta_1 Agecalving + \beta_2 Agecalving^2 + CowID + GRM + e,$$

where y= observable health traits (binary trait 0 or 1), μ = trait mean, HYS = Herd-Year-Season contemporary group, Parity = 4 levels of parity (1, 2, 3, > 4), MOC = month of calving 1 to 12, Agecalving = age at calving from 18 months to 220 months (calving date – birth date) fitted as a covariate and 2nd order polynomial, CowID = random permanent environmental cow effect to account for repeated measures, GRM = random term for the genetic markers (SNPs), and e = random error term. The model was fitted using ASReml Version 4 (Gilmour *et al.*, 2015).

Reliability of genomic prediction. Two methods were used to estimate the reliability of genomic prediction:

1. Theoretical (expected) reliability (R) = $1 - \frac{PEV}{\sigma_g^2}$,

where, the prediction error variance (PEV) = squared standard error of the direct genetic value (DGV) for each animal in the dataset, and σ_g^2 is the additive genomic variance, obtained from the REML estimate.

2. Empirical (observed) reliability using cross-validation

$$= r(DGV, DTD)^2$$

Cross-validation was performed by predicting DGVs for the 494 genotyped bulls that had daughters with health records but were not genotyped. Reliability was then estimated as a simple Pearson's squared correlation between the direct genomic breeding value (DGV) and the corrected phenotypes (residuals) which were used to calculate the daughter trait deviations (DTD) for each bull. The reliability was adjusted by dividing it by the average reliability of DTDs (h^2*average effective number of daughters for the genotyped bulls) (Haile-Mariam *et al.*, 2012).

RESULTS AND DISCUSSION

A summary of the number of records used in the genomic analysis for each health trait is reported in Table 1 for Holsteins.

Table 1. Summary of the number of cow-lactations, cases of disease (n) recorded for each health trait (MAST = mastitis, REPRO = reproductive disorders, LAME = lameness, METAB = metabolic diseases, ALL DIS = "all-diseases") and heritability estimates ($\hat{h}^2$ ±standard errors) for Holsteins using all parity records

Traits	n	$\hat{h}^2$ ±S.E
Cow-Lac	33,000	
MAST	3,735	0.03 ± 0.004
REPRO	2,498	0.01 ± 0.002
LAME	248	0.00 ± 0.00
METAB	241	0.002 ± 0.002
ALL DIS	6,085	0.02 ± 0.004

Mastitis and the all disease category had the largest number of records followed by reproduction, lameness and metabolic disorder categories. The same patterns were also evident with the reliabilities of genomic predictions with the highest being mastitis and the all disease category, followed by reproductive and metabolic disorders (Table 2).

Table 2. Average expected reliabilities (R) of genomic breeding values for cows and bulls with daughters in the reference dataset and bulls in the validation dataset (V) and Cross-validation accuracy and reliability (r^2) for each health trait (MAST = mastitis, REPRO = reproductive disorders, LAME = lameness, METAB = metabolic diseases, ALL DIS = all diseases)

	Expected Reliability			Cross-Validation	
Traits	Bulls*	Cows	Bulls_V^	Accuracy	r^2
MAST	0.33	0.23	0.18	0.12	0.04
REPRO	0.15	0.09	0.05	0.02	0.004
METAB	0.04	0.01	-0.01	-0.01	0.003
ALLDIS	0.31	0.20	0.16	0.18	0.12

*Bulls with daughters in the reference set (n= 948); ^Bulls with no daughters in the reference set (n= 494)

The prediction error variance and cross-validation methods produce similar reliability estimates. The reliabilities are low but are comparatively higher for mastitis and the all disease category (Table 2). Bulls generally had higher reliabilities than cows, due to bulls having greater than 5 daughters in the data.

The lower reliability for metabolic disease is associated with fewer records in comparison to mastitis and the all disease trait. Further, mastitis and the "all-diseases" trait had higher heritabilities and incidences than the other disease traits (Table 2), possibly an indication of why their reliabilities are higher. There is still potential for improving these traits' reliabilities by two means: 1) by including DTDs in the reference set for the 948 genotyped bulls, and 2) by incorporating predictor traits, for example inclusion of both mastitis and SCC data is expected to improve the reliabilities of GEBVs for mastitis.

The reliability estimate for lameness was unsatisfactory to report (R=0) due to the low number of records associated with this trait, and zero heritability. However, there may be merit in recording different types of lameness (e.g. laminitis, etc.) and developing new ways of recording, such as using a phone app. We expect that collection of more data and distinguishing between types of lameness may help to develop genomic breeding values for this trait.

CONCLUSIONS

Overall the results from this study are in line with expected reliabilities for new traits with comparatively small amounts of data and provide a good foundation for further improvement of reliabilities for health traits. It is encouraging that reasonable reliabilities were achieved for diseases such as mastitis and the all disease trait. Having more health event data being identified and made available to the dairy industry, and further investigation in combining predictor traits, will assist in providing genomic breeding values with greater reliability.

REFERENCES

Egger-Danner C., Cole J. B.,Pryce J. E., GenglerN., HeringstadB.,Bradley A. and StockK. F.(2015) *Anim.* **9**: 207.

Gilmour A. R., Gogel B. J.,Cullis B.R., Welham S. J. and Thompson R. (2015). ASReml User Guide Release 4.1 Structural Specification, VSN International Ltd, Hemel Hempstead, HP1 1ES, UK

Haile-Mariam M.,Nieuwhof G. J., Beard K.T., Konstatinov K.V.and Hayes B. J. (2012) *J. Anim. Breed. Genet.* **130**:31.

Koeck A., Miglior F., Kelton D. F. and Schenkel F. S. (2012) *J. Dairy Sci.* **95**: 4099.

Pryce J.E., Veerkamp R.F., Thompson R., Hill W.G. and Simm G. (1997) *Anim. Sci.* **65**: 353.

Pryce J.E., Woolaston R., Berry D.P., Wall E., Winters M., Butler R. and Shaffer M. (2014) World Trends in Dairy Cow Fertility. *Proc. 10th World Congress Genet. App.Livest. Prod.* **154**

RauwW. M., Kanis E.,Noordhuizen-Stassen N. and GrommersF. J. (1998) *Livest. Prod. Sci.* **56**:15.

Proc. Assoc. Advmt. Anim. Breed. Genet. 22:61-64

GENETIC TRENDS IN THE ESTIMATED FEED INTAKE OF ANGUS CATTLE

B.J. Walmsley, A.L. Henzell and S.A. Barwick

Animal Genetics & Breeding Unit (AGBU*), Univ. of New England, Armidale, NSW, 2351

SUMMARY

Genetic trends are presented for the estimated feed intake of young Angus animals at pasture and in the feedlot, and of Angus cows at pasture for a self-replacing, 100d-finished production system. Increases in feed intake over time, both at pasture and in the feedlot, are estimated to have accompanied genetic gains in productivity traits in Angus cattle. The estimated increases are both in feed requirement and residual feed intake, with the latter being smaller in magnitude. The need for industry to record feed intake to facilitate selection for feed efficiency and, in the absence of this, for stocking rate to be managed in commercial herds to offset increases in feed intake, are factors briefly discussed in connection with industry realising benefits from genetic improvement.

INTRODUCTION

Feed intake has a major influence on beef production profitability, but it is difficult to measure in the grazing animal and consequently it is not easily included in genetic evaluation. In Australia, there is a protocol (eg. Exton 2001) for industry recording of residual or 'net' feed intake (ie. feed intake at the same liveweight and gain). The high cost of measuring feed intake has so far limited its recording. This paper examines genetic trends since 1985 in the estimated feed requirement and residual feed intake of young Angus cattle at pasture and in the feedlot, and in the feed requirement of Angus cows at pasture. Some implications for whether benefits from genetic gain are being realised in industry are briefly discussed.

METHODS

Breeding objectives. Breeding objectives for net return per cow were derived with BreedObject (Barwick *et al.* 2005) for pasture finished, 100d feedlot finished (self-replacing cow herd at pasture, steers finished at 640kg at 22m), and 220d feedlot finished animals. Results are presented only for the 100d-fed system, as patterns in results for other systems were similar. Traits in the breeding objective were sale weight, dressing %, saleable meat %, rump fat depth, marbling score, feedlot entry weight, weaning weight (direct & maternal), mature cow weight, cow weaning rate, residual feed intake-pasture, residual feed intake-feedlot, and cow condition score. The general form of the economic value for traits is Δ returns – Δ feed requirement cost – Δ non-feed management cost. The feed requirement associated with a unit change in each objective trait was estimated using the equation systems described by Freer *et al.* (2007).

Genetic trends in productivity traits. EBVs for the breeding objective traits were predicted from the January 2017 BREEDPLAN EBVs of 1,895,481 Angus animals born from 1985 through to 2015, and summarised by year of birth. Predictions used the relation $\hat{g} = \hat{\mu}\,\mathbf{G}_{11}^{-1}\mathbf{G}_{12}$, where $\hat{g}$ and $\hat{\mu}$ are EBVs for breeding objective traits and from BREEDPLAN, and $\mathbf{G}_{11}$ and $\mathbf{G}_{12}$ are genetic covariances among BREEDPLAN EBVs and between these and the objective traits, respectively. Genetic parameters employed were derived from industry and literature estimates and are those used for developing Angus indexes in Australia. The trends in Figure 1 are for selected objective traits of those listed above for the young animal or cow.

---

* AGBU is a joint venture of NSW Department of Primary Industries and University of New England

Genetic trends in feed intake. Genetic trends in feed intake were obtained as index trends by restricting the prices received and costs incurred in the breeding objective to zero except those for feed. Feed requirement and residual feed intake trends were obtained by omitting or retaining the residual trait in the objective. In principle, total feed intake is the sum of feed requirement and the residual trait. Because these components can be correlated, feed intake trends were derived with both components in the objective. The trends in Figure 2 are in terms of the estimated total feed intake (excluding any period of surplus feed) per animal (young animal, cow or cow/calf unit) for that segment of the production system (cow herd, backgrounding at pasture or feedlot finishing).

RESULTS AND DISCUSSION

Figure 1 demonstrates estimated genetic trends occurring in selected objective traits of Angus.

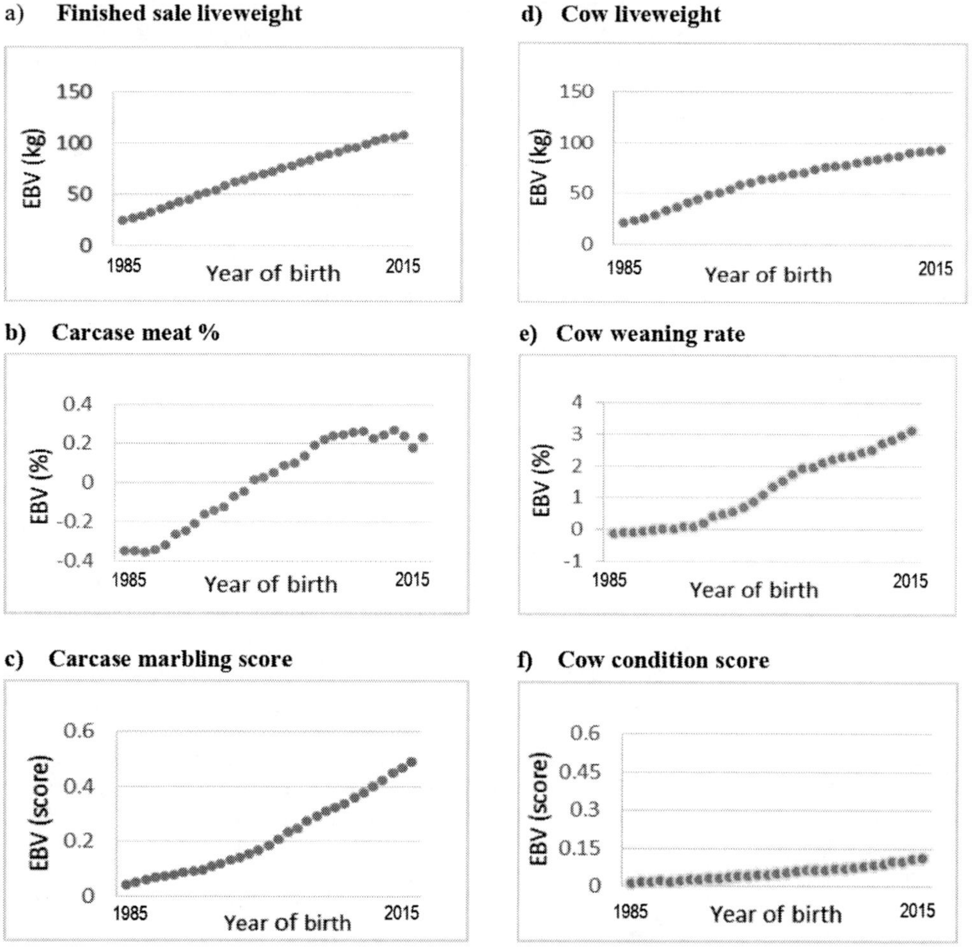

Figure 1. Genetic trends in breeding objective traits for the young animal or cow in Angus cattle for a self-replacing cow herd with steers 100-d feedlot finished after backgrounding.

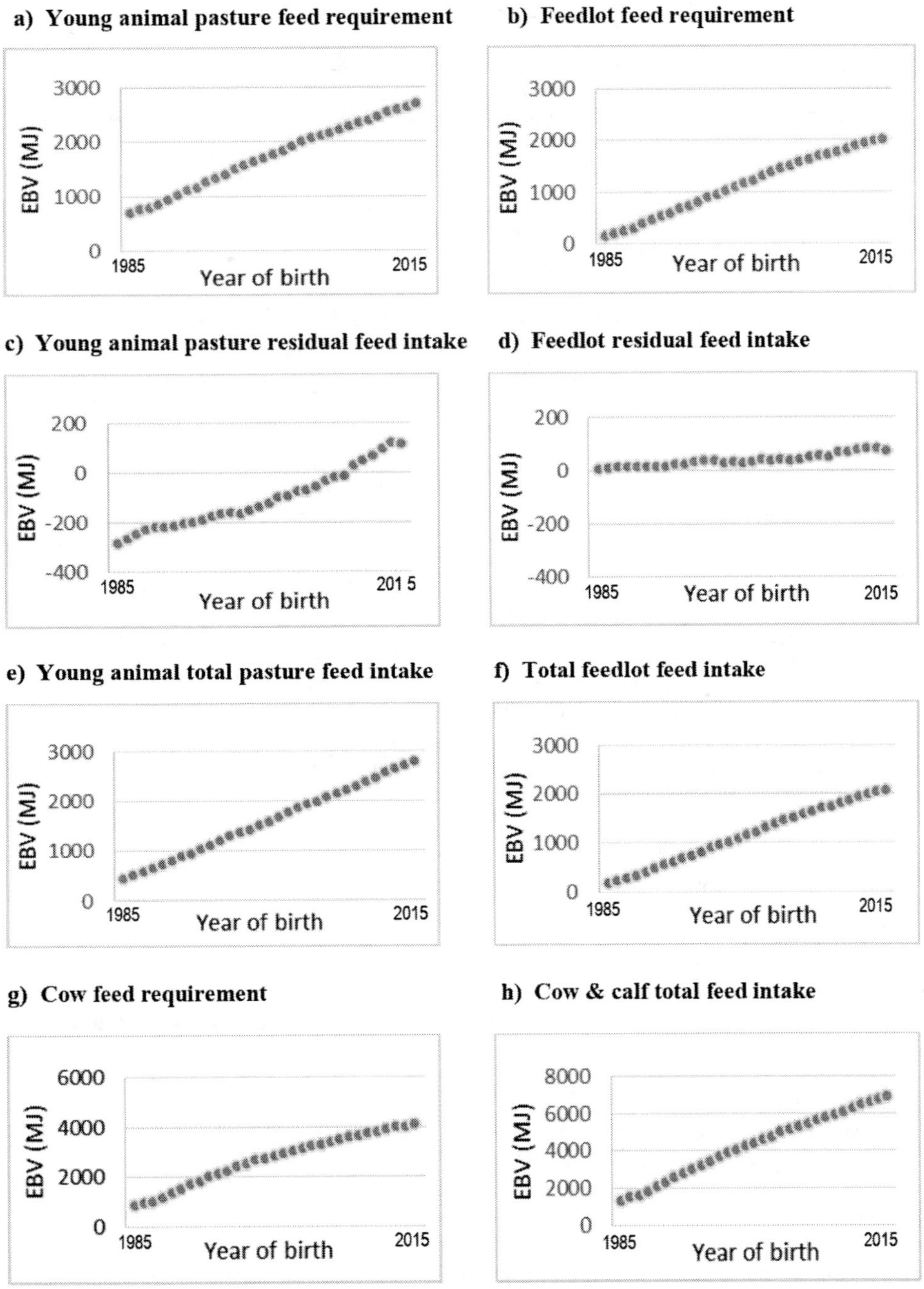

Figure 2. Genetic trends in the estimated feed intake of Angus cattle for a self-replacing cow herd with steers 100-d feedlot finished after backgrounding. The trends are in terms of total feed (excluding any period of surplus feed) for that production system component (cow plus calf to weaning, backgrounding or feedlot finishing).

Figure 2 shows the gains in productivity traits in Figure 1 have been accompanied by increases in estimated feed intake, involving both the animals' requirement for production and its residual. In the 100d-fed system, feed intake is estimated to have increased both at pasture and in the feedlot. In the 30 years between 1985 and 2015 the increase in the intake of cows at pasture (about 3000 MJ, Figure 2g) and the cow and calf unit at pasture (about 5000 MJ; calves at pasture from weaning at 7m until feedlot entry at 18.5m), means the expected DSE rating of Angus cows has also increased.

The estimated increases in feed intake (Figure 2), in particular residual feed intake, illustrate the need for industry recording of feed intake so feed efficiency can be improved along with productivity. Selection indexes derived for industry in the past with BreedObject (Barwick and Henzell 2005), that have increased over time (not presented), take account of the cost of the increased feed requirement but residual feed intake has only recently been included (released 2016). Figures 2c and 2d show residual feed intakes of Angus are increasing rather than decreasing (decreases are needed to increase feed efficiency), reflecting the existence of underlying low positive genetic correlations between feed requirement and residual feed traits. Given this correlation not recording feed intake to estimate residual feed intake EBVs and continued selection for increased growth and mature size will allow beef feed efficiency to continue to decrease.

The results also suggest that animal genetic improvement and pasture stocking rate management need to be considered jointly. In an earlier illustration (Barwick *et al.* 2011) it was shown that genetic improvement was likely to have the extra benefit of improving pasture utilisation when stocking rates are low. At high stocking rates, it was shown that benefits from genetic improvement may not be realised unless stocking rate is reduced or other feed is provided. Without this management change, there is environmental decline from the point of view of the animal, as individual feed demands have increased. This situation could also be occurring in other production systems and other grazing species. Graham *et al.* (2015) drew attention to the possibility of other forms of environmental decline limiting benefits from genetic improvement being realised.

Though data are scarce, it is commonly held that industry pasture utilisation rates are low. Anecdotal evidence from industry suggests this may be changing, though it is not clear if this is only at particular times of the year and in lower-rainfall seasons. The beef industry needs more recording of feed intake so feed efficiency can be improved. In the absence of efficiency improvement, when pasture utilisation is high, it is critical for benefits to be realised from genetic improvement that commercial producers are aware of the trends in feed intake that accompany genetically higher-performing animals. It may also help for industry selection indexes to be derived at two or more levels of feed availability/cost (eg. supplementary feed; $100/tonne vs $300/tonne).

ACKNOWLEDGEMENTS

We thank the Angus Society of Australia for data access, and NSW Dept. of Primary Industries and Meat & Livestock Australia for financial support.

REFERENCES

Barwick S.A. and Henzell A.L. (2005) *Aust. J. Exp. Agric.* **45**: 923.

Barwick S.A., Swan A.A., Hermesch S. and Graser H-U. (2011) *Proc. Assoc. Advmt. Anim. Breed. Genet.* **19**: 23.

Exton S. (2001) 'Testing beef cattle for net feed efficiency'. NSW Agriculture, Trangie NSW.

Freer M., Dove H. and Nolan J.V. (Eds) (2007) 'Nutrient requirements of domesticated ruminants'. CSIRO Publishing, Collingwood Vic.

Graham, P., Swan A.A. and Banks, R.G. (2015) *Proc. Assoc. Advmt. Anim. Breed. Genet.* **21**: 245.

Proc. Assoc. Advmt. Anim. Breed. Genet. 22:65-68

RELATIONSHIP BETWEEN FEED INTAKE, ENERGY EXPENDITURE AND METHANE EMISSIONS: IMPLICATIONS FOR GENETIC EVALUATION

S. Dominik[1], D.L. Robinson [2], A.J. Donaldson[2], M. Cameron[2], K.L. Austin[2], and V.H. Oddy [2]

[1] CSIRO Agriculture and Food, Armidale, Australia
[2] Beef Centre, NSW DPI, Armidale, Australia

SUMMARY

Portable accumulation chambers (PAC) enable enteric gas emissions of sheep to be measured under field conditions. Feed intake is highly correlated with methane emission and should be accounted for in models for parameter estimation of methane emissions, but it cannot be measured in the field. In this study, different linear mixed models were fitted to methane and carbon dioxide emissions and oxygen consumption to investigate the consequences of not adjusting for feed intake, as well as adjusting for effects that indirectly account for feed intake, such as live weight, carbon dioxide or oxygen. The significance of permanent environmental effects was also tested. The results demonstrate that feed intake accounts for a considerable amount of the variance in methane emissions. In this animal house experiment, where sheep were fed at 1.5 x maintenance, much of the variation in feed intake appeared to be related to non-genetic effects of the animal. Consequently, fitting a permanent environmental effect yielded similar heritability estimates to those of models that adjusted for feed intake. Repeated measures of greenhouse gas emission in PAC require more complex models including permanent environmental effects to produce acceptable estimates.

INTRODUCTION

Enteric methane emissions are strongly correlated with feed intake. Criticism has been raised, that, without appropriate measures of production, selection to genetically reduce methane emissions could lead to decreased production because of decreased feed intake (Arthur *et al.* 2009). One approach is therefore to adjust methane emissions for feed intake. Technologies to measure methane and other enteric gas emissions of sheep include respiration chambers (RC) and portable accumulation chambers (PAC). The advantage of PAC is that they can be used in the field; the disadvantage is that under field conditions, it is not possible to measure feed intake.

The aim of this study was quantify the differences in variance components and heritability estimates for enteric gas emissions and oxygen consumption from models with and without adjustment for feed intake, or proxies for feed intake that can easily be measured. In addition, the outcomes of fitting permanent environmental effects were explored.

MATERIALS AND METHODS

Data. Enteric gas emission traits were measured on 512 Information Nucleus Flock (INF) follower ewes at Armidale, New South Wales. The ewes were born between August 2007 and October 2013. Data were collected in an indoor facility using PAC with two measurement protocols that differed in time off feed prior to measurement. Protocol PAC0 measured animals immediately off feed and PAC1 kept animals 1 hr off feed prior to measurement. Methane, CO_2 and O_2 (ml/min), live weights (kg) and feed intake (g) were recorded. Measurements from the two PAC protocols were highly correlated, with genetic correlations ranging from 0.75 to 1.00. Therefore, records for PAC0 and PAC1 were regarded as repeat measures, resulting in two PAC measurements per animal. Ewes were tested from mid-April 2015 to mid-March 2016.

Feed was offered in the mornings at 1.5 x maintenance requirements and feed intake recorded from 8 am on the day prior to PAC measurements to 8 am on the day of measurement (FIDP) and from 8 am on the measurement day until the time the animal entered the PAC (FIOD).

Statistical analysis. Variance components and heritabilities for gas emission traits were estimated using ASReml (Gilmour *et al.* 2009). An extensive back-pedigree with 13 genetic groups was used. Univariate mixed animal repeatability models were run to estimate parameters. Fixed effects included test batch, birth year, measurement date, measurement protocol, testing run (RUN, 7 levels, with 4 for PAC0 and 3 for PAC1), and PAC (from 1 to 12). Ten models were tested for CO_2 and O_2, and twelve models for CH_4. For each gas trait, the first model fitted all significant fixed effects, but not direct or indirect adjustment for feed intake (Model no adj). Other models fitted either feed intake (FIOD and FIDP and their interaction with RUN) as Model FI, live weight (Model LWT), feed intake and live weight (Model FI+LWT), CO_2 (Model CO_2) or O_2 (Model O_2). Only significant fixed effects and interactions were retained in the final models. All models were fitted with and without permanent environmental effect (PE). Random effects included animal ID to estimate the genetic variance and a permanent environmental effect, fitted as an identity matrix of the animal ID.

RESULTS AND DISCUSSION

Basic features of the dataset and the distribution of their raw phenotypes are shown in Table 1. Table 2 shows the variance components and resulting heritability estimates for CH_4, CO_2 and O_2 from the different models, with and without adjustment for feed intake or a substitute (LWT, CO_2 or O_2) without and with permanent environmental effect (+PE). For all traits, the phenotypic variances decreased after fitting FI, LWT, FI and LWT or CO_2 or O_2, as might be expected. For CH_4, feed intake accounted for the most variation, whereas O_2 accounted for most of the variation in CO_2 and vice versa. As a consequence of the reduction in phenotypic variances, genetic and environmental variances were also reduced, with environmental variance being less affected than genetic variance.

Table 1. Mean ($\pm$sd: standard deviation), minimum (Min) and maximum (Max) of methane (CH_4), carbon dioxide (CO_2) and oxygen (O_2) (in ml/min)

	Mean (+ sd))	Min	Max
CH_4	36.27 $\pm$ 9.35	4.97	75.31
CO_2	422.30 $\pm$ 82.56	207.40	734.90
O_2	-451.60 $\pm$77.43	-732.50	-257.80

The change in heritability estimates also reflects the substantial amount of variance related to the covariates fitted. Previously reported heritabilities for CH_4 from field measurements of sheep in PAC ranged from 0.05 – 0.19 (Robinson *et al.* 2014a; Goopy *et al.* 2016). As might be expected, the results from this controlled animal house study were higher than published estimates from field measurements. Results from the different models in this study support the conclusion of Robinson *et al.* (2014b), that a substantial proportion of the variation in CH_4 emissions is related to variation in feed intake. In fact, economic modelling of breeding objectives suggests that methane measurements can be used as a proxy for feed intake, and that the resulting improvements in feed efficiency will often be more valuable than the reductions in greenhouse gas emissions (Robinson and Oddy 2016).

Robinson *et al.* (2014b) highlighted the importance of PE effects in regards to CH_4 emission traits. They noted significant effects of twins being reared as singles and hypothesised about other causes, such as diet, rumen volume and their impacts on short or long-term variation in rumen microbial composition. In our study, the effect of fitting a permanent environmental effect was tested for all models (+PE). As assessed by likelihood ratio tests, the significance of PE was not associated with a particular trait, but appeared to depend on the covariates that were fitted. The more variance could be captured by the covariate, i.e. FI and also CO_2 for O_2 emissions, the less variance was due to the PE effect. Interestingly, fitting a permanent environmental effect in model no adj yielded similar heritability estimates for CH_4, CO_2 and O_2 to those from model FI. The repeated measures in this dataset allowed both permanent environmental effects and measurement errors to be estimated. Another approach would be to explore the measurements from PAC0 and PAC1 in a bivariate analysis as correlated traits.

Table 2. Genetic (V_G), residual (V_E), phenotypic (V_P) and permanent environmental (V_{Pe}) variance component (including significance), log likelihood (Logl) for each model with and without permanent environmental effect and heritability estimates (h^2) for CH_4, CO_2 emission and O_2 consumption

			CH_4			
	V_G	V_E	V_P	V_{Pe}	Logl	h^2
no adj	45.25	27.10	72.35	--	-2697.32	0.63 (0.03)
no adj + PE	25.77	26.88	70.62	17.98	-2696.89	0.36 (0.14)
LWT	36.64	27.32	63.97	--	-2666.19	0.57 (0.03)
LWT + PE	9.54	26.88	61.70	25.28**	-2660.79	0.15 (0.13)
CO_2	20.62	17.64	38.26	--	-2416.55	0.54 (0.03)
CO_2 + PE	15.06	17.52	37.85	5.27	-2415.14	0.40 (0.13)
O_2	17.54	18.34	35.88	--	-2404.53	0.49 (0.03)
O_2 + PE	12.70	18.20	35.56	4.66	-2403.16	0.36 (0.13)
FI	5.09	14.32	19.41	--	-2974.91	0.26 (0.04)
FI + PE	3.78	14.22	19.35	1.35***	-2173.48	0.20 (0.11)
LWT+FI	4.35	14.17	18.53	--	-2158.48	0.23 (0.04)
LWT+FI+PE	2.98	14.06	18.48	1.44	-2157.50	0.16 (0.10)
			CO_2			
	V_G	V_E	V_P	V_{Pe}	Logl	h^2
no adj	2805.21	1750.00	4555.20	--	-4866.12	0.62 (0.03)
no adj + PE	1554.93	1734.66	4455.60	1156.03*	-4863.86	0.35 (0.14)
LWT	1775.60	1772.57	3548.20	--	-4793.18	0.50 (0.03)
LWT + PE	63.73	1734.82	3424.70	1626.17***	-4783.88	0.02 (0.12)
O_2	48.67	584.76	633.44	--	-3990.31	0.08 (0.04)
O_2 + PE	15.66	577.65	633.13	39.83	-3988.94	0.02 (0.08)
FI	765.72	1309.08	2074.80	--	-4578.01	0.37 (0.04)
FI + PE	232.22	1288.01	2044.80	524.61*	-4574.84	0.11 (0.12)
LWT+FI	670.32	1262.78	1933.10	--	-4547.54	0.35 (0.04)
LWT+FI+PE	224.89	1243.27	1911.00	442.86*	-4544.99	0.12 (0.12)
			O_2			
	V_G	V_E	V_P	V_{Pe}	Logl	h^2
no adj	1985.11	1099.85	3085.00	--	-4666.59	0.64 (0.03)
no adj + PE	1366.19	1093.60	3028.40	568.65	-4665.34	0.44 (0.15)
LWT	1200.57	1113.36	2313.90	--	-4583.37	0.52 (0.03)
LWT + PE	50.39	1092.20	2227.00	1084.39***	-4574.16	0.02 (0.12)
CO_2	182.16	543.77	725.93	--	-4058.93	0.19 (0.04)
CO_2 + PE	93.33	535.94	722.72	93.33	-4057.45	0.13 (0.10)
FI	621.80	887.77	1509.60	--	-4420.35	0.41 (0.04)
FI + PE	212.64	876.54	1482.50	393.31**	-4417.26	0.14 (0.13)
LWT+FI	543.05	851.49	1394.50	--	-4686.23	0.39 (0.04)
LWT+FI+PE	116.01	837.05	1369.90	416.81***	-4382.35	0.08 (0.12)

Significance of log likelihood ratio test: $P < 0.05$ *;$P < 0.01$ **, $P < 0.001$***

Despite relatively small numbers of animals (total of 512), the PE was more often significant than not. Live weight, is of course highly heritable, and when it was accounted for in Model LWT, the estimate of genetic variation was small for CO_2, but there was still variation due to PE effects. One possible explanation is that in some animals both CO_2 production and O_2 consumption have a different relationship with live weight, and, to a lesser extent, feed intake. What this might be is yet to be determined, but could include learned behaviour such as stress responses that might contribute to additional PE variation in O_2 consumption and CO_2 emissions.

Robinson *et al.* (2016) noted that the repeatability of methane measurements diminishes over time, falling from an average of 0.48 for measurements in the same week to 0.20 for the average of 6 repeated measurements on the same animals from 2009-2014. This suggests that some of the variation attributed to PE effects could in fact be temporary and (perhaps to a greater extent than genetic effects) relate to factors affecting the animal during the particular month each batch of sheep spent in the animal house.

CONCLUSION

Ideally feed intake is accounted for in models for genetic parameter estimation of CH_4 emission, however feed intake measures are difficult to obtain in the field. Repeated measures of enteric gas emission in sheep provide an opportunity to estimate both measurement errors and non-genetic animal environmental effects. The latter were usually significant and accounted for some variation in feed intake and other factors that, in models ignoring the PE effect, would be included in estimates of the genetic variance and result in inflated estimates of heritability.

ACKNOWLEDGEMENT

This research was supported by funding from the Australian Government, Department of Agriculture as part of Filling the Research Gaps program of the Carbon Farming Initiative and from Meat and Livestock Australia.

REFERENCES

Arthur P.F., Donoghue K.A., Herd R.M. and Hegarty R.S. (2009) Proc. Assoc. Advmt. Anim. Breed. Genet. **18**:472-475.

Gilmour A.R., Gogel B.J., Cullis B.R. and Thompson R. (2009) ASReml User Guide Release 3.0 VSN International Ltd, Hemel Hempstead, HP1 1ES, UK.

Goopy J.P., Robinson D.L, Woodgate R.T, Donaldson A.J., Oddy V.H, Vercoe P.E. and Hegarty R.S. (2016) *Anim. Prod. Sci.* **56**: 116.

Robinson D.L.,Goopy J.P., Hegarty R.S., Oddy V.H., Thompson, A.N., Toovey, A.F., Macleay, C.A., Briegal, J.R., and Woodgate R.T, Donaldson A.J. and Vercoe P.E. (2014a) *J. Anim.Sci.* **92**: 4349.

Robinson D.L.,Goopy J.P., Donaldson A.J., Woodgate R.T, Oddy V.H. and Hegarty R.S. (2014b) *Animal* **8** (12): 1935.

Robinson D.L. and Oddy, V.H. (2016). Journal of Animal Science doi: 10.2527/jas.2016-0469.

Proc. Assoc. Advmt. Anim. Breed. Genet. 22:69-72

A WEB-BASED APPLICATION TO ASSIST SELECTION FOR HEAT TOLERANCE IN COMBINATION WITH THE BALANCED PERFORMANCE INDEX IN AUSTRALIAN DAIRY CATTLE

T.T.T. Nguyen[1], B.J. Hayes[1,2], J.E.Pryce[1,3]

[1] Agriculture Research Division, Department of Economics, Jobs, Transport and Resources, 5 Ring Road, Bundoora, Victoria 3083, Australia

[2] Queensland Alliance for Agriculture & Food Innovation, Queensland Alliance for Agriculture & Food Innovation, Brisbane, St Lucia, Queensland 4072, Australia

[3] School of Applied Systems Biology, La Trobe University, Bundoora, Victoria, 3083, Australia

SUMMARY

The aim of this study was to develop a future-scenarios selection tool to assist farmers in making selection decisions, that combines the current national dairy selection index, known as the Balanced Performance Index (BPI) with a proposed heat tolerance (HT) genomic estimated breeding value (GEBV). Heat tolerance GEBV was estimated for 12,062 genotyped cows and 10,981 bulls, using an established genomic-prediction equation. Publicly available future daily average temperature and humidity data were used to calculate mean daily temperature–humidity index for each dairy herd. One way to ascertain heat tolerance is the rate of decline in milk production traits to rising heat loads, this definition was the basis of the heat tolerance breeding values (BV_HT). An economic estimate of an individual cow's BV_HT was calculated by multiplying HT GEBVs for milk, fat and protein yields by their respective economic values that are used in the BPI. This was scaled for each region by multiplying BV_HT by the heat load, which is the temperature–humidity index (THI) units exceeding the threshold per year at a particular location. BV_HT were incorporated into the BPI as: BPI_HT = BPI + BV_HT; where BPI_HT is the 'augmented BPI' breeding value including HT. A web-based application was developed enabling farmers to predict the future heat load of a herd and take steps to aim at genetic improvement in future generations by selecting bulls and cows that rank high for the 'augmented BPI'.

INTRODUCTION

It is widely recognised that heat stress has significant impacts on the performance of dairy cows. When heat stressed, animals consume less feed, followed by a decline in milk yield (St-Pierre et al., 2003). In Australia, it is projected that major dairying regions will experience an increase in daily average temperatures as well as more frequent heat waves (CSIRO and BoM, 2015). Therefore, there is a need for the industry to develop strategies to mitigate the impacts of a warming climate on animal performance.

Apart from providing cooling devices and managing diets for cows on hot days, selection for more heat tolerant animals is an approach worthy of investigation. In this regard, Nguyen *et al.* (2016) developed genomic estimated breeding values (GEBVs) for heat tolerance (HT) for Australian Holsteins and Jerseys, which is the rate of decline in milk production traits to rising heat loads. The study found that using high-density single-nucleotide polymorphism (SNP) genotypes, HT GEBV can be predicted with an accuracy ranging between 0.42–0.61. The HT GEBV has unfavourable correlations with production traits, but a favourable correlation with fertility. In addition, the HT GEBVs were validated through an experiment where genomically predicted heat-susceptible and predicted heat-tolerant animals show a significant difference in milk yield losses, and rectal and intra-vaginal temperatures when experiencing a mild simulated heat wave (Garner *et al.*, 2016). A breeding value for HT is planned to be released to the dairy industry in the near future.

Given the complexity in the relationships between HT and other traits in the current selection indices, one relevant question is how farmers can balance the selection for HT with their existing priorities. Farmers in regions where heat stress is more of an issue may prioritise selection for HT to a greater extent than those in cooler climates. In the present study, we developed a future-scenarios selection tool that enables farmers to make informed decisions so as to balance the selection of current economic drivers traits in the BPI with HT simultaneously by varying the weight applied to HT_BV for individual farms by heat load.

MATERIALS AND METHODS

Projected future climate data. The Commonwealth Scientific and Industrial Research Organisation (CSIRO) and the Bureau of Meteorology (BoM) have provided details of projected future climate-change scenarios in Australia over the 21st century (CSIRO and BoM, 2015). Appropriate climate projection models used in the present study were selected following the advice of CSIRO climate scientists. We chose medium and high emission scenarios (RCP4.5 and RCP8.5 (carbon dioxide level of 540 x10^{-6} μmol/mol by 2100) and RCP8.5 (carbon dioxide concentration of 940 x 10^{-6} μmol/mol by 2100) as examples.

On the basis of the selected models, projected average daily temperature and humidity for weather stations (namely weather station data) were downloaded from the 'Climate Change in Australia' website (http://www.climatechangeinaustralia.gov.au/, 01 March 2016). In addition, gridded average daily temperature and humidity data (namely gridded data) were also obtained directly from the Climate Research and Services, CSIRO Oceans and Atmosphere (Aspendale, Victoria). We used data from the nearest grid (≤1 km distance to weather station) to patch missing weather-station data. Weather data were matched to the nearest postcode provided the distance between the weather station and centroid of the postcode was no more than 60 km.

Daily average THI was calculated for each day from 2020 to 2035, as per Nguyen *et al.* (2016). According to Hayes *et al.* (2003), averaged THI of the test day and 1, 2, 3 and 4 days before the test day of exceeding 60 could result in a decline in milk yield. Therefore, we defined heat load of a given year as the total of five-consecutive-day-average THI units exceeding 60 in that year, which is referred to as THI hereafter.

HT and BPI breeding values. In order to calculate the future profitability of a herd with and without selection for HT, and under different climate-change scenarios, the current genetic merit of a herd is required, as well as the genetic merit of the bulls on offer. So as to have a reasonably large group of cows and bulls that would span many herds and many bull-selection possibilities, HT GEBVs of genotyped cows and bulls were predicted using the equation developed by Nguyen *et al.* (2016) for all genotyped cows and bulls. BPIs for both cows and bulls of the February 2016 release were obtained from DataGene (formally Australian Dairy Herd Improvement Scheme).

The heat-tolerance breeding value (BV_HT) in dollars (so it can be readily combined with the BPI) was expressed as:

$$\mathrm{BV\_HT} = (\mathrm{EW_m GEBVht_m} + \mathrm{EW_f GEBVht_f} + \mathrm{EW_p GEBVht_p})\,\mathrm{HL},$$

where BV_HT is the breeding value of heat tolerance in monetary term; $\mathrm{EW_m} = -0.10$, $\mathrm{EW_f} = 1.79$, $\mathrm{EW_p} = 6.92$ are economic weight of milk, fat and protein respectively, which are currently used in the BPI (Byrne *et al.*, 2016); $\mathrm{GEBVht_m}$, $\mathrm{GEBVht_f}$ and $\mathrm{GEBVht_p}$ are genomic breeding values of heat tolerance in relation to milk, fat and protein respectively; HL is the total number of THI units exceeding 60 in a year.

We combined BPI and BV_HT for each animal as follows:

BPI_HT = BPI + BV_HT,

Where BPI_HT is the 'augmented BPI' breeding value with heat tolerance included; and BV_HT is breeding value of heat tolerance.

Data visualisation. The application HOTdAIRy v.01 developed in R (R Core Team 2015), using the 'shiny' package (Chang *et al.*, 2016) in RStudio (RStudio Team, 2015). We obtained the postal area shape file from the Australian Bureau of Statistics for postcode boundaries (http://www.abs.gov.au/, 01 March 2016).

RESULTS AND DISCUSSION

We successfully obtained daily average temperature and humidity for 58 weather stations in Australia. Of these, we were able to match 57 stations with 1861 postcodes covering 3836 herds (85.4% of the total number of Australian dairy herds). The average number of days per year with THI exceeding the threshold of 60 were 313, 235, 242, 176, 164, 120, 121 in Queensland, New South Wales, Western Australia, South Australia, Northern Victoria, Gippsland and Western Victoria, respectively. The average number of THI units exceeding the threshold (THI > = 60) per year ranged from 2,587 (Gippsland) to 2,676 (Western Victoria), 3,240 (Northern Victoria), 3,445 (South Australia), 4,338 (Western Australia) and 6,019 (Queensland), indicating that all major dairying regions will be affected by excessive heat load, but at different levels.

For demonstration purposes, we have included only information from genotyped cows that currently belong to the Genomic Information Nucleus (Ginfo) herds, and genomic bulls, in our tool. We successfully estimated HT GEBV for 12,062 genotyped cows (10,680 Holsteins and 1,382 Jerseys from 80 Ginfo herds), and 10,981 genomic bulls (9,306 Holsteins and 1,675 Jerseys). The BV_HT significantly varied according to the level of heat load. For example, if the heat load of year 2025 was applied, BV_HT among the 10,981 bulls analysed ranged between AU$–29 to A$21 per cow (mean set at zero) under the conditions in Johanna, Victoria, but the range of BV_HT changes to AU$–174 to AU$126 per cow at the conditions in Rockhampton, Queensland. The correlation between BPI and BPI_HT for bulls was, therefore, higher (0.99) if the heat load in Johanna was applied, than it was under Rockhampton (0.95) conditions. Figure 1 shows an example scatter plot of BPI vs BPI_HT for the bulls under the conditions in Rockhamton.

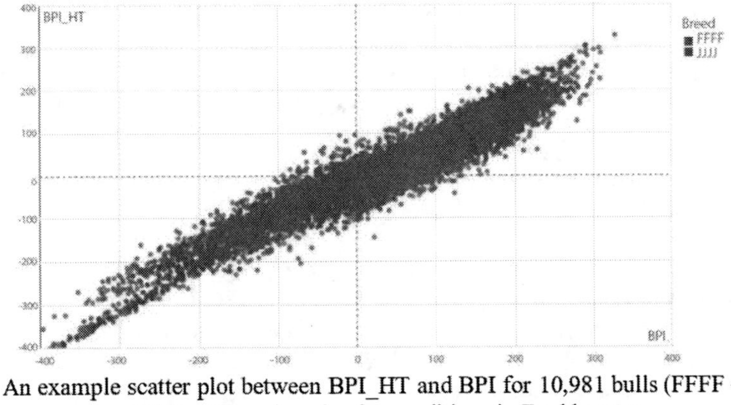

Figure 1. An example scatter plot between BPI_HT and BPI for 10,981 bulls (FFFF = Holsteins and JJJJ = Jerseys) under the conditions in Rockhamton.

A typical workflow in the web-based application HOTdAIRy v.01 (https://tnshinyr.shinyapps.io/app12) begins with providing inputs, including a herd postcode, a herd ID, a future year (2020, 2025, 2030 or 2035) and a greenhouse gas-emission scenario. The outputs include the amount of heat load calculated based on the inputs, scatter plots between BPI_HT and BPI for cows and bulls, and relevant tables which can be sorted and downloaded. The ranked cow and heifer list can then be used to make selection decisions on which animal to keep in the herd, and which to cull, on the basis of predicted performance with the projected future heat load. The highest-ranked bulls maximising the profit under the given projected future climate conditions on the farm.

One special characteristic of HT is that its breeding value depends on the amount of heat load animals are expected to experience. Heat load varies between regions and our approach was to use heat load as a weight for the trait, i.e. in regions with high heat load, emphasis on selection for HT is higher and *vice versa*. That means BV_HT for an animal depends on the herd locality. In the study, we were able to use projected climate data from CSIRO and BoM (2015) to determine levels of heat load for most dairying regions in Australia, which also serves as a weight for HT in the 'augmented BPI' index. This method of inclusion of a new trait opens opportunities for the inclusion of other traits of this nature in the index.

Our demo version of future-scenarios selection tool is currently a standalone web-based application. However, it is also flexible in terms of incorporation into other existing tools that farmers are currently using. Once possible option is to integrate it into the Good Bulls app (http://www.datagene.com.au/); thereby, BPI_HT breeding values and ranking can be viewed along with BPI, HWI and TWI.

In summary, we have created a practical future-scenarios selection tool that can be used by dairy farmers and breeders to make informed decisions in selecting for HT and BPI, that is customised to their dairy region and includes options for various future climate-change scenarios. The tool will become particularly relevant given the continuing increase in average temperature and frequency of heat-wave events. Our study is the first attempt to incorporate HT into selection indices for dairy cattle. It is important because profitability and animal welfare can be improved simultaneously through identifying animals that are able to cope with current and future climate change in a way that is consistent with the impact of HT on local farm profitability.

ACKNOWLEDGEMENTS

We thank the Department of Agriculture and Water Resources (Canberra, Australia) for funding this work, Dairy Futures Cooperative Research Centre for overall support, and DataGene (Melbourne, Australia) for providing breeding values for BPI. Special thanks go to Dr John Clark of the Climate Research and Services, CSIRO Oceans and Atmosphere (Aspendale, Victoria) for advice on selection of climate models and for sharing gridded projected future-climate data.

REFERENCES

Byrne T. J., Santos B. F. S., Amer P. R., Martin-Collado D., Pryce J. E., and Axford M. (2016). *J Dairy Sci* **99**:8146-8167.

Chang W., Cheng J., Allaire J. J., Xie Y., and McPherson J. (2016). http://CRAN.R-project.org/package=shiny.

CSIRO and BoM (2015). CSIRO and Bureau of Meteorology, Australia.

Garner J. B., Douglas M., Williams R. S. O., Wales W. J., Nguyen T. T. T., and Hayes B. J. (2016). *Nature Scientific Reports* **6**:1-8.

Hayes B. J., Carrick M., Bowman P., and Goddard M. E. (2003) *J Dairy Sci* **86**:3736-3744.

Nguyen T. T. T., Bowman P., Haile-Mariam M., Pryce J. E., and Hayes B. J. (2016). *J Dairy Sci* **99**:2849-2862.

RStudio Team. 2015. http://www.rstudio.com/.

St-Pierre N. R., Cobanov B., and Schnitkey G. (2003). *J Dairy Sci* **86**:52-77.

Proc. Assoc. Advmt. Anim. Breed. Genet. 22:73-80

CAN WE FRAME AND UNDERSTAND CROSS-VALIDATION RESULTS IN ANIMAL BREEDING?

A. Legarra[1], A. Reverter[2]

[1] UMR 1388 GenPhySE, INRA, Castanet Tolosan, France
[2] CSIRO Agriculture and Food, 306 Carmody Rd., St. Lucia, QLD 4067, Australia

SUMMARY

Performance of genomic selection is typically evaluated by cross-validation. In this work we review and point out some problems and features of the cross-validation metrics. Then we propose a semiparametric alternative using statistics derived from the "Method R".

INTRODUCTION

Genomic prediction of breeding values via genomic BLUP (GBLUP) is expensive and requires initial and continuous investments in genotyping. State of the art theory so far does not yield convincing *a priori* estimates of the increased accuracy of genomic prediction vs. pedigree-based predictions. Thus, cross-validation has been extensively used (e.g. Legarra *et al.* 2008; VanRaden *et al.* 2009; Mantysaari *et al.* 2010; Christensen *et al.* 2012). The theory of cross-validation is poorly understood in the context of heavily related and selected data (but see (Gianola and Schön, 2016)). For instance, how to evaluate accuracy for maternal traits is very unclear. Here we provide a brief review of this topic and suggest some options.

CROSS-VALIDATION BIAS AND ACCURACY

What cross-validation? Forecasters such as pedigree-BLUP and GBLUP may behave differently according to what the "forecasted" target is. Breeders have a difficult task, namely, to forecast the best reproducers in order to select them. In this, they are different from *machine learners*, whose objective is (from our perspective) to forecast present phenomena. Thus, it is rather obvious that for breeders the best method is such that allows taking the best selection decisions, that it is, the method that best predicts future performance of an individual knowing its genetic background.

We will call this *forward cross-validation*. Its features are three-fold: (1) It needs the definition of a cut-off date; (2) It needs the construction of "Full" and "Reduced" data sets (Mantysaari *et al.* 2010; Olson *et al.* 2011); and (3) In its crudest form, it does not provide any form of randomisation and therefore a point estimate of goodness of prediction is obtained, without any associated measure of uncertainty.

In contrast, the classical *random folding* k-fold cross-validation in its most classic form splits randomly the data into k distinct sets and predicts one set from the remaining $k-1$ sets. Its key features include: (1) Extremely simple to implement; (2) Provides estimates of standard error of metrics of cross-validation; (3) Not realistic in an animal breeding setting and the ranking of methods is not suitable for practical purposes; and (4) Tends to overfit (case of leave-one-out)

Some more esoteric forms of cross-validation exist. Legarra *et al.* (2008) *split folds "across" or "within" families*, obtaining very different results. But this is undoable (and little useful) for regular animal breeding data. The *k-means for cross-validation* (Saatchi *et al.* 2011) separates individuals into "most distinct" folds, and the i-th fold is predicted from the remaining k-1 folds. This does not answer the breeder's question, which most often wants to predict from *close*, not from *far* animals.

Which metrics? To assess the *predictive ability* of the different forecasters, animal breeders are highly formatted by Henderson's BLUP, which in turn was highly dependent upon dairy cattle genetic improvement. Metrics commonly used come from linear regression, named in this paper *predictive abilities*, are:

$$\text{Bias: } b_0 = E(u - \hat{u}); \qquad \text{Slope: } b_1 = \frac{Cov(u,\hat{u})}{Var(\hat{u})}; \quad \text{Accuracy: } r = \frac{Cov(u,\hat{u})}{\sqrt{Var(u)Var(\hat{u})}}$$

Sometimes mean squared error is used ($MSE = b_0^2 + \sigma_u^2(1 + r^2/b_1^2 - 2r^2/b_1)$). Properties of BLUP in absence of selection are no bias, slope of 1, and maximum accuracy. Henderson defined this at the individual level on a frequentist basis (over conceptual repetitions). Bias=0 and slope=1 ensure fair comparisons across old and young animals. This is important if the scheme mixes proven and young animals, like dairy cattle. It seems less relevant in schemes were reproducers are culled quickly (pigs, chicken) with beef species falling someone in the middle, we believe. Deviations may exist if there is selection, because bias and slope are related to genetic gain and dispersion (see Figure 1).

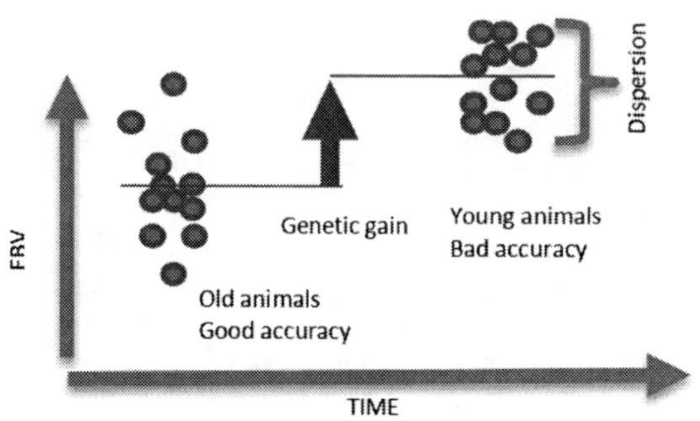

Figure 1 Typical scenario for retrospective analysis.

What is it meant by classical bias? Animal breeders probably agree to Henderson's (1973) sentence "*most users would, I think, be reluctant deliberately to bias comparisons between different groups, for example to underevaluate young sires as compared to older ones*". Here we have an operational definition of bias. In formal terms this implies that at a given point in time:

$$b_0^{[Henderson]} = \left(\mathbf{1}'\hat{\mathbf{u}}_{group1} - \mathbf{1}'\hat{\mathbf{u}}_{group2}\right) - \left(\mathbf{1}'\mathbf{u}_{group1} - \mathbf{1}'\mathbf{u}_{group2}\right)$$
$$= \left(\mathbf{1}'\hat{\mathbf{u}}_{group1} - \mathbf{1}'\mathbf{u}_{group1}\right) - \left(\mathbf{1}'\hat{\mathbf{u}}_{group2} - \mathbf{1}'\mathbf{u}_{group2}\right)$$

This definition has practical implications: if the candidates are chosen *across* groups, selection decisions are optimal if there is no bias. Thus, it is expected that $b_0^{[Henderson]} = 0$. There may be several definitions of groups: (1) Different conditions (grazing vs. indoor fed cattle). This case should be addressed by the model used for evaluation; (2) Within country, different amounts of information that cumulate in time (progeny-tested vs. genomic bulls). This case is strongly affected by within-country genetic trend (see below); (3) Same amount of information, but different origins (US vs. FR). This case is most affected by wrong estimates of the difference in genetic level across countries (Bonaiti *et al.* 1993; Powell and Wiggans 1994).

The Interbull definition. Interbull uses retrospective tests (Boichard *et al.* 1995; Mantysaari *et al.* 2010) that compare EBV's *before* and *after* progeny testing.

$$b_0^{[Interbull]} = \mathbf{1}'\hat{\mathbf{u}}_t - \mathbf{1}'\hat{\mathbf{u}}_{t-1}$$

If progeny testing gives exact EBVs, then $\hat{\mathbf{u}}_t = \mathbf{u}_t$ and $b_0^{[Interbull]} = \mathbf{1}'\mathbf{u} - \mathbf{1}'\hat{\mathbf{u}}_{t-1}$.Note that $b_0^{[Henderson]} \neq b_0^{[Interbull]}$, but if group1 is "very old" proven bulls and $\hat{\mathbf{u}}_t = \mathbf{u}_t$ and group2 is genomic bulls (then becoming proven bulls) then $b_0^{[Henderson]} = b_0^{[Interbull]}$. This may be rather obvious, but it only holds for progeny testing data.

What happens under selection? Assume that we want to compare selection candidates with "proven" animals. If there is *no* selection, then $\mathbf{1}'\boldsymbol{u}_{group1} = \mathbf{1}'\boldsymbol{u}_{group2}$ and there is actually no need to make the test. Alas, if *there is* selection, then

$$b_0^{[Henderson]} = (\mathbf{1}'\widehat{\boldsymbol{u}}_{group1} - \mathbf{1}'\widehat{\boldsymbol{u}}_{group2}) - (\mathbf{1}'\boldsymbol{u}_{group1} - \mathbf{1}'\boldsymbol{u}_{group2}) = n(\hat{\Delta} - \Delta)$$

in other words, unbiasedness requires a correct (unbiased!) estimate of the realized genetic trend.

What is overdispersion, a.k.a {Interbull, genomic} bias? Is it affected by selection?

Dairy cattle breeders are much concerned by overdispersion of genomic proofs. If there is too much dispersion of $\widehat{\boldsymbol{u}}_{genomic}$, the retained candidates will have unfairly high $\widehat{\boldsymbol{u}}_{genomic}$. This could be staten more formally as "the mean of the EBVs of the selected candidates should be equal to the mean of the TBVs". If selection is by truncation and under multivariate normality, the true mean *after* selection is $\mu_T = (\mathbf{1}'\mathbf{u})/n + ir\sigma_u$, but this mean is (implicitly) predicted before selection as $\mu_E = (\mathbf{1}'\hat{\mathbf{u}})/n + i\sigma_{\hat{u}}$.

For $\mu_T = \mu_E$ to hold, we need the first unbiasedness condition (b_0 above), plus a second condition, $\sigma_{\hat{u}} = r\sigma_u$. But this condition *only* holds if $Cov(u, \hat{u}) = Var(\hat{u})$, which amounts to the regression coefficient to be 1:

$$b_1 = \frac{Cov(u, \hat{u})}{Var(\hat{u})}$$

This is the Interbull official, and most put forward, test of unbiasedness and nowadays more often called as "bias". It is easy to see why $b_1 = 1$ may not hold, namely, because selection modifies variances in rather unpredictable manners. The expected $Cov(u, \hat{u}) = Var(\hat{u})$ holds under quite restrictive conditions (Henderson 1982).

Evaluations can easily be biased. Unbiasedness of current genetic evaluations is more wishful thinking than an established fact. Unbiasedness exist only if several conditions hold:

- The model is correct (linear model, effects, heritabilities…)
- The selection process is described by the data
- Multivariate normality

Thus, there are many reasons why there is wrong estimate of the genetic trend and thus there will be bias:

- Collinearity of contemporary groups and genetic trend (this is the usual case)
- Genetic groups in the model
- Heritability is wrong (or changes with time)
- Analysis are single trait whereas selection is multiple trait
- Selection decisions not based on data.

In addition, genetic gain can be estimated one generation forward (but no more) unless an explicit selection model is included. In other words, retrospective analysis cannot be done deleting two generations of records. This would need explicit introduction of the selection process.

Why some species/traits seem biased where others do not? Basically, if there is *no* selection then *automatically* $b_0 = 0$ holds (i.e., all possible sets of candidates have 0 average value), and most likely $b_1 = 1$ holds, because selection does not change variances, and if a decent estimator of genetic variance is used, then genetic parameters are such that $b_1 = \frac{Cov(u,\hat{u})}{Var(\hat{u})} = 1$ by construction, in particular in a BLUP context. So, bias is expected to increase more with higher genetic gains.

An example is *pigs*. Christensen *et al.* (Christensen *et al.* 2012) found slopes below 1 (~0.9) for a heritable, selected trait (daily gain), whereas Xiang *et al.* (Xiang *et al.* 2016) found regressions nearly one for hard-to-select trait litter size.

In Lacaune dairy *sheep* (Baloche *et al.* 2014), we can put together the following. Figure 1 shows the regression slopes vs. the expected genetic gain or the expected loss of genetic variance based on Robertson (1977) . In theory, the reduction in variance is accounted for by genetic evaluation (Bijma 2012). In practice, this does not seem to be the case. A possible solution may be to reestimate this variance in each cycle of selection.

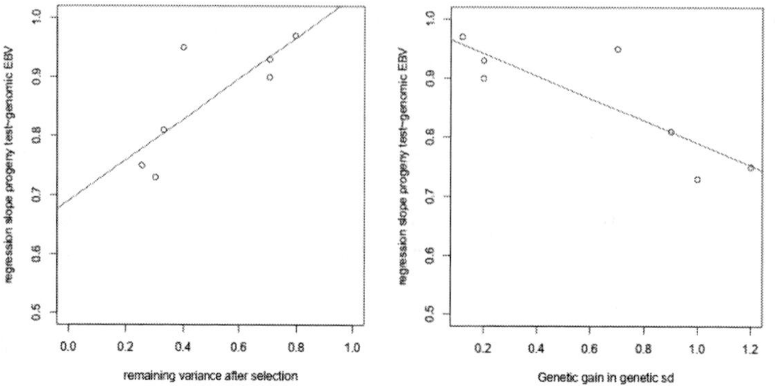

Figure 2 Slope b_1 vs. expected reduction in genetic variance (left) or genetic gain (right) by trait in Lacaune dairy sheep.

Vitezica *et al.* (2011) compared by *simulation* several predictors in selected populations in a SSGBLUP context. Statistic b_1 generally indicated bias, that was higher with less heritability. High heritability increases the selection differential and reduces variances, but it also gives more information. Interestingly, the only method which provided unbiased $b_1 = 0.99$ resulted in strong bias $b_0 = 1.38\sigma_u$. Thus, *both* bias should be checked.

What do we mean by accuracy? In animal breeding textbooks, accuracy (r, with reliability r^2) is presented twice: first, as a component of $\Delta_G = ir\sigma_u$ (so, a populational parameter) and, second, as a measure of uncertainty of $\hat{u}$ (an individual parameter). However, when selecting from real populations, EBVs are correlated across individuals, so the individual accuracies may be meaningless. In other words: it is pointless to obtain $r_i = 0.70$ and $r_j = 0.70$ if $r(\hat{u}_i, \hat{u}_j) = 0.69$.

Cross-validation accuracies are computed as correlations $r^2 = \frac{Cov(u,\hat{u})}{Var(u)Var(\hat{u})}$. They indicate our ability to rank individuals *within* a cohort. The fact that these accuracies are computed regardless of the correlated structure of both u and $\hat{u}$ has unclear implications. In fact, it can be shown that, if Hendersonian conditions hold, $E(r)^2 = 1 - \frac{\left(diag(C^{22}) - \overline{C^{22}}\right)}{\left(diag(G) - \overline{G}\right)}$ is the expectation of the observed reliability. This reliability takes into account the "classical" reliability contained in the diagonal terms but also the relationships a priori (in $\boldsymbol{G}$) and a posteriori (in $\boldsymbol{C}^{22}$) across individuals. If the evaluation method cannot rank correctly *within* the validation sample, then diagonal and off-diagonal values of $\boldsymbol{C}^{22}$ are similar and reliability drops down. This is a desirable behaviour.

Selection also affects observed cross-validation accuracy (Edel et al., 2012; Bijma 2012). If the cross-validation test uses elite animals, accuracies are underestimated. In other words, it is easy to rank all animals, but more difficult to rank elite animals. The reduction is such that

$$r^2_{selected} = 1 - \left(1 - r^2_{unselected}\right)\frac{\sigma^2_{\hat{u}_{unselected}}}{\sigma^2_{\hat{u}_{selected}}} \quad .$$

ISSUES OF CROSS-VALIDATION METRICS

The accuracy of cross-validation metrics. After an experiment has been carried out, the breeder wants to know if the genomic accuracy is really different from the parents average accuracy. A simple method is to use the theoretical standard error of the estimates; for b_0 and b_1 these are from classical regression theory. For the correlation, this is a bit more convoluted, but an option is to use Fisher's z-transform: $z = \frac{1}{2} ln \frac{1+r}{1-r}$ has approximate s.e. $1/\sqrt{n-3}$ where n is the number of data points used. From this a confidence interval can be worked out. For instance, in the Basco-Bearnaise breed genomic predictions of 87 rams were 0.06 more accurate than parent averages (Legarra *et al.* 2014); this implies a rather symmetric 95% confidence interval of $[-0.15, 0.27]$.

There is a source of bias and two sources of randomness in cross-validation metrics. The source of bias is that individuals are related both at the stage of prediction (parent average and genomic) and later, at the stage of validation (moment at which they have data; except for the case of progeny-tested animals for which proofs can be assumed uncorrelated). This has been discussed above. The two sources of randomness are: (1) Sampling of the reference population, (2) Sampling of the validation population. Fisher's z-transform and Hotelling-Williams test include both. However, they do not consider that individuals are related, and therefore the accuracy is likely to be overestimated. Again, a theoretical equation can be worked out to estimate $Var(r)$.

(Re)Sampling of the validation population. A more practical approach involves using (re)sampling techniques. In k-fold cross-validation this is immediate but, as discussed before, the setting is not realistic. In (Mäntysaari and Koivula 2012; Legarra *et al.* 2014; Cuyabano *et al.* 2015), sampling of the validation population was addressed by *bootstrapping*, i.e. sampling n individuals with replacement from the original n individuals in the validation data set. This method main virtue is that it avoids strong influence of outliers in the validation data set. It also allows formal comparisons of accuracies. Its main drawback is that it does not addresses the sampling of the reference population.

(Re)sampling of the reference population. Recently, (Mikshowsky *et al.* 2016) bootstrapped, not the validation, but the *reference* population. This also provides distribution of metrics. However, it may be argued that, in a dairy cattle reference population, including a sire twice (what the bootstrapping actually does) is like including it once, because the accuracy of the sire pseudo-phenotype is close to 1 in dairy cattle. Thus, including it twice will not change much the solution for the sire – or the contribution of the sire to SNPs solutions. Therefore, randomness comes from *removing* sires more than by *overrepresenting* sires. In that sense, Mikshowsky *et al.* (2016) bootstrap corresponds to Tukey's jackknife with more than one data point removed.

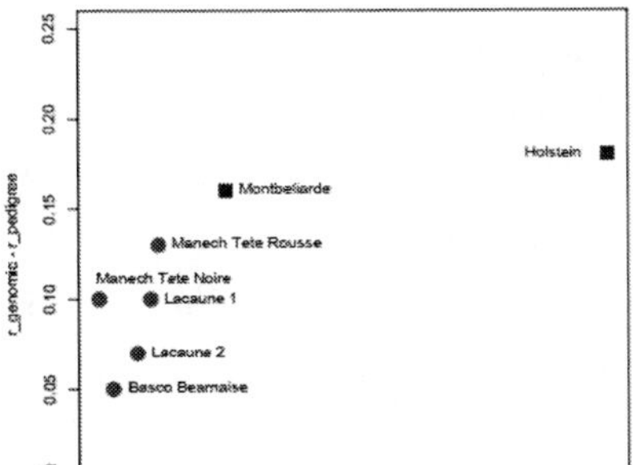

Figure 3 Genomic accuracy and family size.

Superiority of genomic on pedigree predictions is a function of family structure of the validation data set. Consider a set of two generations, a generation of parents and one of descendants: n full-sib families with k offspring each. Parents have information (say, own weight) but there is not information for the offspring. We can ask: is it worth doing genomic prediction?

Families can be easily ranked based on parent average, but there is not possibility to rank within families with pedigree information. However, genomic information *can* rank *within* family as well as *across* families. Thus, the observed benefit of GBLUP by retrospective analysis will be larger in a set composed of *few* families with a large number of candidates *within* families. In the limit, if there is one big family, pedigree prediction has 0 accuracy, whereas if there are n families with 1 offspring each, pedigree and genomic predictions should behave similarly.

This is supported by Figure 3 in which we plot the genomic vs pedigree accuracy for milk yield for five dairy sheep and two dairy cattle breeds in France, as a function of family size. Clearly, the larger the family size, the larger the benefit because genomic selection allows distinguishing sibs. This raises several questions: (1) Do comparisons reflect "genetic architecture" or merely data structure in the validation? (2) Do selection schemes that select across families get less benefit from genomic selection? (3) Is Holstein gaining a lot from genomic selection because it has higher LD than other breeds or just as an artefact of its family structure?

Which variables to use on the metrics? In the dairy industry, sires do not have phenotypes, so that comparisons are between (G)EBV's and the "true" progeny proofs or deregressed proofs. In other species, it is more common to compare (G)EBV's to "true" phenotypes, say $\boldsymbol{y}$, using an approximation $r = Corr(GEBV, y)/h$ where h^2 is the heritability (Legarra *et al.* 2008). This is unsatisfactory, for conceptual and practical reasons:

- The equation above for r assumes uncorrelated individuals and GEBV's
- Records $\boldsymbol{y}$ are typically pre-corrected to $\boldsymbol{y}^* = \boldsymbol{y} - \boldsymbol{X}\hat{\boldsymbol{b}}$, and the results are sensitive to precorrection. It is unclear what happens if there are contemporary groups in $\boldsymbol{b}$ that are not present in the training data.
- If the whole data set is used for precorrection, then a relationship structure is fit (e.g. pedigree relationships) as $\boldsymbol{y}^* = (\boldsymbol{I} - \boldsymbol{X}(\boldsymbol{X}'(\boldsymbol{ZAZ}\sigma_u^2 + \boldsymbol{I}\sigma_e^2)^{-1}\boldsymbol{X})^{-})\boldsymbol{y}$ where $\boldsymbol{A}\sigma_u^2$ is assumed to be "correct". If the assumed relationship is biased or incorrect, so will be $\hat{\boldsymbol{b}}$ and $\boldsymbol{y}^*$, and the bias will be toward the assumed relationship. This may explain some puzzling results, e.g. poor performance of genomic prediction in low heritable traits such as fertility (Hayes *et al.* 2009).
- Even after precorrection, there will be a remaining covariance structure across pre-corrected $\boldsymbol{y}^*$. This structure is notoriously hard to model (and rarely modelled). This may explain phenomena such as $\frac{Corr(GEBV, y^*)}{h} > 1$.
- Some precorrected $\boldsymbol{y}^*$ are too clumsy (Ricard *et al.* 2013) to be believed or computed in practice, for instance maternal effects.

CROSS-VALIDATION ACCURACIES FROM METHOD R

Description of the method. We propose to use the properties of method R to construct metrics of cross-validation. Reverter *et al.* (1994) observed that the regression of EBVs obtained with "whole" (w) data on EBVs estimated with "partial" (p) data, $b_{w,p} = \frac{Cov(\hat{u}_w, \hat{u}_p)}{Var(\hat{u}_p)}$ is 1, and this checks bias (in the sense b_1 before). The correlation of partial on whole (eq. 7-9 in their paper) $\rho_{p,w} = \frac{Cov(\hat{u}_p, \hat{u}_w)}{\sqrt{Var(\hat{u}_w)Var(\hat{u}_p)}}$ is a function of respective accuracies. Invoking exchangeability, both equations can be extended to multivariate forms, and expectations can be taken in both the numerator and the denominator, resulting in:

$$b_{w,p} = \hat{\mathbf{u}}_w' \boldsymbol{K}^{-1} \hat{\mathbf{u}}_p / \hat{\mathbf{u}}_p' \boldsymbol{K}^{-1} \hat{\mathbf{u}}_p$$

where $\boldsymbol{K}$ is a matrix of relationships, $b_{p,w}$ with an expected value of 1, and

$$\rho_{w,p} = \hat{\mathbf{u}}_p' \boldsymbol{K}^{-1} \hat{\mathbf{u}}_w / \sqrt{\hat{\mathbf{u}}_p' \boldsymbol{K}^{-1} \hat{\mathbf{u}}_p \hat{\mathbf{u}}_w' \boldsymbol{K}^{-1} \hat{\mathbf{u}}_w}$$

with an expected value $E(\rho_{w,p}) = \sqrt{\frac{\mu_{acc_p^2}}{\mu_{acc_w^2}}}$ that is, proportional to the relative increase in average reliabilities. As more data cumulates, $\hat{\mathbf{u}}$ tends towards the true breeding values, thus $\hat{\mathbf{u}}_w$ is more accurate than $\hat{\mathbf{u}}_p$. The empirical covariance $\hat{\mathbf{u}}'_w\mathbf{K}^{-1}\hat{\mathbf{u}}_p$ measures the strength of the association between the two, whereas $\hat{\mathbf{u}}'_p\mathbf{K}^{-1}\hat{\mathbf{u}}_p$ measures the extent of shrinkage due to lack of information. In other words, the theoretical prediction error covariances are replaced by empirical ones (Thompson 2001). By combining cross-validation and theory from mixed models, we hope to retain the best of both worlds: a measure of accuracy that corresponds to reality and that is little affected by the existence of related, unbalanced data. Therefore, an algorithm to estimate accuracy of (say) PBLUP and GBLUP is:

1. Compute EBV's with all data ("whole") using, say, GBLUP (which method should not be critical if all animals have data or progeny)
2. Choose cutoff date
3. Create "partial" data: Set values after cutoff date to missing
4. Compute EBVs based on "partial" and GBLUP
5. Compute statistic $b_{w,p}^{GBLUP} = \frac{\hat{\mathbf{u}}'_p\mathbf{K}^{-1}\hat{\mathbf{u}}_w}{\hat{\mathbf{u}}'_p\mathbf{K}^{-1}\hat{\mathbf{u}}_p}$
6. Compute statistic $\rho_{p,w}^{GBLUP} = \frac{\hat{\mathbf{u}}'_p\mathbf{K}^{-1}\hat{\mathbf{u}}_w}{\sqrt{\hat{\mathbf{u}}'_w\mathbf{K}^{-1}\hat{\mathbf{u}}_w\hat{\mathbf{u}}'_p\mathbf{K}^{-1}\hat{\mathbf{u}}_p}}$
7. Compute EBVs based on "partial" and PBLUP
8. Compute statistic $b_{w,p}^{PBLUP} = \frac{\hat{\mathbf{u}}'_p\mathbf{K}^{-1}\hat{\mathbf{u}}_w}{\hat{\mathbf{u}}'_p\mathbf{K}^{-1}\hat{\mathbf{u}}_p}$
9. Compute statistic $\rho_{p,w}^{PBLUP} = \frac{\hat{\mathbf{u}}'_p\mathbf{K}^{-1}\hat{\mathbf{u}}_w}{\sqrt{\hat{\mathbf{u}}'_w\mathbf{K}^{-1}\hat{\mathbf{u}}_w\hat{\mathbf{u}}'_p\mathbf{K}^{-1}\hat{\mathbf{u}}_p}}$

For forward cross-validation, the statistics should be computed for the focal individuals (i.e., candidates to selection). On exit, $b_{w,p}^{GBLUP}$ should be 1 (unbiased method) and is equivalent to b_1 and $\rho_{p,w}^{GBLUP}$ and $\rho_{p,w}^{PBLUP}$ describes the respective accuracies of GBLUP and PBLUP. An extra statistic is bias $\mu_{wp} = b_0 = (\mathbf{1}'\mathbf{K}^{-1}\hat{\mathbf{u}}_w - \mathbf{1}'\mathbf{K}^{-1}\hat{\mathbf{u}}_p)/n$. Matrix **K** should be the "true" relationship matrix across individuals but there should be no great difference in using either genomic or pedigree relationships as far as they are correct. The procedure has several advantages: is completely general (it can be used e.g. for maternal traits or random regression), it is semi-automatic, and can, at least potentially, provide estimates of the accuracy of the cross-validation metric. There are though many points that need to be addressed: robustness to misspecification, the role of selection (and how to avoid biases in the estimates of the different $b's$), how to sample efficiently, etc.

TEST WITH REAL LIFE DATA SETS

In beef cattle, we used genetic and phenotypic resources from Brahman cows (N = 995) and bulls (N = 1,116) outlined in (Porto-Neto *et al.* 2015). The phenotype was yearling body weight. A procedure "method R" as above was introduced to assess accuracy of GBLUP, and random (1000 replicates) splits of the data set in training and validation was used, as animals are quite unrelated and belong to a single generation. We only present very briefly the results. The statistic $b_{w,p} = 0.96 \pm 0.08$ (in the whole population) showed that evaluation was nearly unbiased, whereas $\rho_{p,w} = 0.67 \pm 0.02$ has a correlation of 0.81 with conventional cross-validation accuracy estimated as $\frac{Corr(GEBV,y^*)}{h}$.

In dairy sheep, we used a large data set (Manech Tete Rousse) of 1,700,000 milk yield performances, 500,000 animals in pedigree and 2,111 sires with 50K genotypes. Data was split at 2011 in training and validation. For *all* individuals, unbiasedness of (SSG)BLUP was checked with results $\mu_{w,p} = b_0 = 0.2\sigma_g = 5$ (liters), $b_{w,p} = b_1 = 0.996$, so genetic evaluation is virtually unbiased for b_1 (slope) but not for b_0 (genetic trend), which is unsurprising because the model includes Unknown Parent Groups. Later, candidates to selection were compared, with $\rho_{w,p}^{SSGBLUP} = 0.55$ vs. $\rho_{w,p}^{BLUP} = 0.39$, and both evaluations where notoriously biased ($b_1^{SSGBLUP} = 0.77$, $b_1^{BLUP} = 0.70$), possibly due to selection not well accounted for. All these results agree well with previous analysis (Legarra *et al.* 2014).

ACKNOWLEDGEMENTS

Toni Reverter and Andrés Legarra benefit funding from the INRA-CSIRO linkage proposals 2016/2017. AL also financed by grants GENOMIA (Poctefa, Feder) and GenoPyr (Feder).

REFERENCES

Baloche G., Legarra A., Sallé G., Larroque H., Astruc J. M., Robert-Granié C., Barillet F. (2014) *J. Dairy Sci.* **97**: 1107–1116.

Bijma P. (2012) *J. Anim. Breed. Genet.* **129**: 345–358.

Boichard D., Bonaiti B., Barbat A., Mattalia S. (1995) *J. Dairy Sci.* **78**: 431–437.

Bonaiti B., Boichard D., Barbat A., Mattalia S. (1993) *Interbull Bull.* **8**

Christensen O., Madsen P., Nielsen B., Ostersen T., Su G. (2012) *Animal* **6**: 1565–1571.

Cuyabano B. C. D., Su G., Rosa G. J. M., Lund M. S., Gianola D. (2015) *J. Dairy Sci.* **98**: 7351–7363.

Edel, C., Neuner, S., Emmerling, R. and Goetz, K.U., 2012. *Interbull Bull.* **46**

Gianola D., Schön C.-C. (2016) *G3 GenesGenomesGenetics* **6**: 3107–3128.

Hayes B. J., Bowman P. J., Chamberlain A. J., Goddard M. E. (2009) *J Dairy Sci* **92**: 433–443.

Henderson C. R. (1973) *J Anim Sci* (**Symposium**) 10-41

Legarra A., Robert-Granié C., Manfredi E., Elsen J.-M. (2008) *Genetics* **180**: 611–618.

Legarra A., Baloche G., Barillet F., Astruc J., Soulas C., Aguerre X., Arrese F., Mintegi L., Lasarte M., Maeztu F. (2014) *J. Dairy Sci.* **97**: 3200–3212.

Mantysaari E., Liu Z., VanRaden P. (2010) *Interbull Bull* **41**.

Mäntysaari E. A., Koivula M. (2012) *Interbull Bull.* **46**

Mikshowsky A. A., Gianola D., Weigel K. A. (2016) *J. Dairy Sci.* **99**: 3632–3645.

Olson K., VanRaden P., Tooker M., Cooper T. (2011) *J. Dairy Sci.* **94**: 2613–2620.

Porto-Neto L. R., Barendse W., Henshall J. M., McWilliam S. M., Lehnert S. A., Reverter A. (2015) *Genet. Sel. Evol.* **47**: 84.

Powell R. L., Wiggans G. R. (1994) *Interbull Bull.* **10**.

Reverter A., Golden B. L., Bourdon R. M., Brinks J. S. (1994) *J. Anim. Sci.* **72**: 34–37.

Ricard A., Danvy S., Legarra A. (2013) *J. Anim. Sci.* **91**: 1076–1085.

Robertson A. (1977) Z. Für Tierz. Zücht. **94**: 131–135.

Saatchi M., McClure M. C., McKay S. D., Rolf M. M., Kim J., et al. (2011) *Genet. Sel. Evol.* **43**: 40.

Thompson R. (2001) *Livest. Prod. Sci.* **72**: 129–134.

VanRaden P. M., Tassell C. P. V., Wiggans G. R., Sonstegard T. S., Schnabel R. D., Taylor J. F., Schenkel F. S. (2009) *J Dairy Sci* **92**: 16–24.

Vitezica Z., Aguilar I., Misztal I., Legarra A. (2011) *Genet. Res.* **93**: 357–366.

Xiang T., Nielsen B., Su G., Legarra A., Christensen O. F. (2016) *J. Anim. Sci.* **94**: 936–948.

Proc. Assoc. Advmt. Anim. Breed. Genet. 22:81-84

A COMPARISON OF RELATEDNESS ESTIMATES FROM SNP CHIP GENOTYPES AND FROM GENOTYPING-BY-SEQUENCING RESULTS

K.G. Dodds, J.C. McEwan, T.C. Van Stijn, R. Brauning and S.M. Clarke

AgResearch, Invermay Agricultural Centre, Mosgiel, New Zealand

SUMMARY

Estimates of genomic relatedness derived from either SNP chip (two different densities) or genotyping-by-sequencing (GBS) resources were compared in a group of 95 sheep. The estimates were highly correlated (r = 0.983-0.992 for relatedness between individuals) although GBS estimates were slightly higher than chip-based estimates. These results provide evidence that GBS is a useful technique for genomic studies.

INTRODUCTION

Genomic information is increasingly being used in animal breeding. Many livestock industries have SNP chips available at a range of densities and at a cost where they are being used in breeding programmes. The SNP chip results are used either directly or indirectly, often after imputation to a higher density, to estimate genomic relatedness between animals in breeding programmes. An alternative technology is to use genotyping-by-sequencing (GBS), based on sequencing a fraction of the genome, possibly at low depth (to reduce costs). GBS can be applied in species without extensive genomic resources (such as SNP chips and reference genome assemblies). Methods have been developed to estimate relatedness using GBS results (Dodds *et al.* 2015). Here we compare relatedness estimates in a sub-flock of 95 sheep genotyped using both genotyping technologies.

MATERIALS AND METHODS

Animals. A group of sheep that had previously been genotyped using SNP chips were chosen for GBS genotyping to allow a comparison of methods. This group were a set of 89 male and female progeny from a single cohort (born in 2014), 5 of their sires and a control sample; 80 of the progeny had their sire in these data. Two of the sires were Primera, two were predominantly Texel, and the other was predominantly Texel x Coopworth. The control animal was a Texel x Coopworth. The dams were unrecorded, but were a maternal type (predominantly Romney).

SNP chip genotypes. The set of animals had been previously genotyped. All animals except for 12 of the progeny had been genotyped with the Illumina ovine HD beadchip (Kijas *et al.* 2014). Although this chip assays over 600,000 SNPs, only 41,020 of those SNPs (referred to as 41k) are used here, being those that are also on the Illumina ovine SNP50 beadchip and which passed quality control (including being autosomal) on both chips using the criteria in Auvray *et al.* (2014). All progeny had been genotyped with a custom Illumina BovineLDplusovine SNP chip which assays 5283 ovine SNPs; this study used 4015 (referred to as 4k) of those SNPs, being those that were also on both the HD beadchip and the SNP50 beadchip, and which passed quality control. For some animals, genotypes for these SNPs from a higher density chip were used as the 4k genotypes.

GBS genotypes. The animals were genotyped by GBS using the methods described by Dodds *et al.* (2015) and based on the GBS protocol of Elshire *et al.* (2011). Briefly, DNA samples and a negative control were digested with *Pst*I; a different barcode adaptor was added to each sample, along with a common adapter. Samples were then combined and fragments in the range 150-500bp were selected and single-end sequenced on one lane of an Illumina HiSeq2500 resulting in approximately 2 million reads per sample. The resulting sequence reads were adapter-trimmed and then UNEAK (Lu *et al.* 2013) was used to detect variants (without the use of a reference genome) and report allele counts for each variant and sample.

Estimation of relatedness. Relatedness between each pair of individuals, and self-relatedness for each individual were estimated by the methods of Dodds *et al.* (2015) which accounts for the read depth in a genotype call. This included estimation from SNP chip genotypes, where the depth was taken to be infinite. This is then equivalent to the first method of vanRaden (2008), except that only SNPs with data for the individual or pair of individuals involved are used for that estimate (i.e., missing genotypes are not imputed). The allele frequencies used were taken as the sample allele frequencies using allele counts. For chip data the allele counts were the usual counts (e.g. AA has 2 A alleles). All SNPs reported by UNEAK were used for the GBS-based analysis. Methods are compared by correlation and by regressions of the differences on the means (Altman and Bland 1983) for each pair of methods. Standard errors for the regressions using pairs of individuals were calculated using the number of individuals rather than the number of pairs as an approximate method to account for the non-independence of the pairs.

RESULTS AND DISCUSSION

The GBS process resulted in calls for 68,293 SNPs with a mean read depth of 6.1. The 41k SNPs had 407 with a minor allele frequencies (MAF) of 0 in these data, and these were removed before further analysis. Summary statistics are shown in Table 1; for GBS, having at least one read at a SNP is taken as a call. Call rates were high for the chip data, but lower for GBS due to the randomness of the sequence reads. The MAFs were highest for the 4k chip, where SNPs were highly selected to be informative, and lowest for GBS where SNPs were not pre-selected.

Table 1. Summary statistics for the different genotyping methods

Marker set	Number of SNPs used	Mean call rate	Mean minor allele frequency	Mean inbreeding estimate	Mean relatedness
41k chip	40,613	99.96%	0.289	-0.037	-0.012
4k chip	4,014	99.37%	0.367	-0.035	-0.010
GBS	68,293	86.73%	0.225	0.058	-0.003

Table 2. Summary statistics for relatedness comparisons including correlations of the estimates and regressions of the differences (first marker type minus second marker type) on the means

Marker comparison	Relatedness	Number compared	Correlation (r)	Mean difference (SE)	Slope (SE)
41k – 4k	Self	83	0.844	-0.002 (0.002)	0.093 (0.065)
41k – GBS	Self	83	0.769	-0.095 (0.003)***	0.060 (0.080)
4k – GBS	Self	95	0.662	-0.094 (0.003)***	-0.068 (0.093)
41k – 4k	Between	3403	0.992	-0.001 (0.002)	-0.012 (0.014)
41k – GBS	Between	3403	0.989	-0.008 (0.002)***	-0.055 (0.016)**
4k – GBS	Between	4465	0.983	-0.007 (0.002)**	-0.047 (0.019)*

* $P<0.05$, ** $P<0.01$, *** $P<0.001$

Comparisons of relatedness estimates are shown in Figure 1 and Tables 1 and 2. In general, the estimates appear to be quite similar across methods. GBS produced higher ($P<0.001$) inbreeding estimates and they were less consistent with the chip estimates than the two chip results were with each other. The breeding design for the progeny set tends to involve breed crosses, so we would expect inbreeding to be low (with low variation) in the flock. The differences in inbreeding between GBS and chips appeared to be uniform over the observed range; the regression slopes for the differences were not significant. One possible reason for GBS giving higher inbreeding estimates is that the SNPs have not been pre-selected, and in particular are likely to include non-autosomal SNPs.

This could elevate the results for males, as they would appear homozygous for X-linked and Y chromosome markers. The inbreeding in the male progeny was higher than in the females, but by only a small amount (0.005, SE = 0.006, NS). These regions would be expected to have around half the average read depth (in males) and the method of estimating inbreeding adjusts for un-observed heterozygosity with low depth (assuming autosomal markers), which would dampen any increase in estimated inbreeding due to these regions.

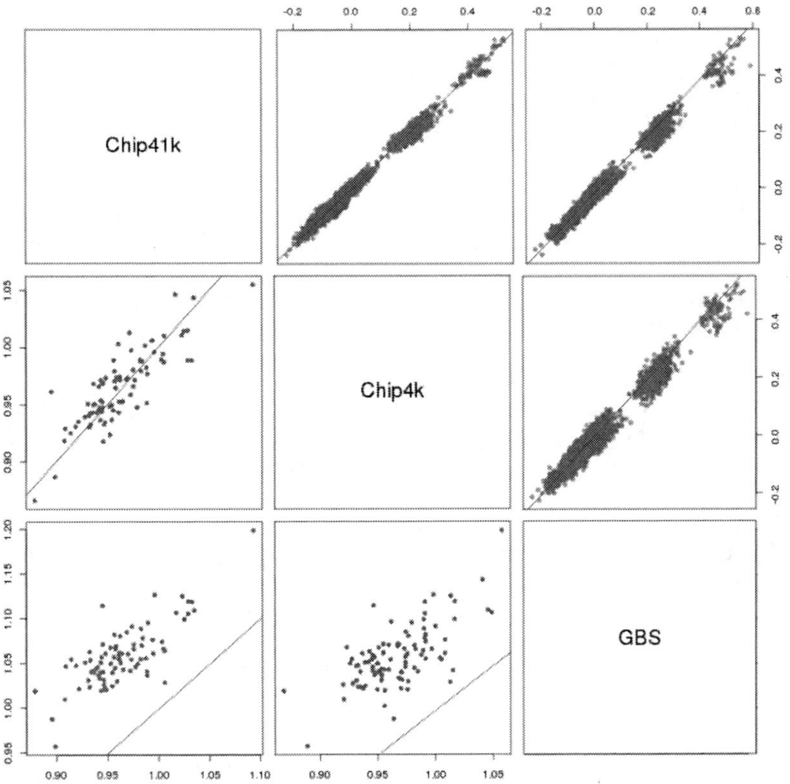

Figure 1. Comparison of relatedness estimates using different genotyping methods. Plots below the diagonal are for self-relatedness of individuals and those above the diagonal are for relatedness between all pairs of individuals. Diagonal labels show the method for the horizontal axis in that column and vertical axis in that row. Lines of equality are also drawn.

The relatedness values were all highly correlated (Figure 1, Table 2). Once again, GBS produced higher ($P<0.01$) values overall, but only by a small amount (0.007 or 0.008 on average). There was also a significant ($P<0.05$) slope for these two comparisons, meaning that there was a larger difference between GBS-based estimates and chip-based estimates for higher values of relatedness. The relatedness estimates form three main groups. The group with higher values are mainly sire-progeny pairs, but there are also pairs from within the progeny group, presumably full-sibs. The middle group contains a pair of sires, while all other pairs are within the progeny group, presumably half-sibs.

The relatedness estimates average close to zero, a by-product of estimating allele frequencies within the dataset, rather than having ancestral frequencies (Yang *et al.* 2010). As GBS SNPs were not pre-selected, and the methods gave similar estimates, it suggests that there is not a large ascertainment bias on the chips, in terms of estimating relatedness. It is also interesting to note that the estimates appear to be similarly correlated for low values compared to high values of relatedness. This suggests that the rankings of relatedness when the estimates are negative are still meaningful (pairs with more negative values are less related than pairs with negative values close to zero).

One of the main reasons for estimating relatedness in agricultural species is to allow genomic selection, for example these estimates can be used directly in a GBLUP model. Having the relatedness estimates for the three methods correlate well suggests that they would perform similarly for genomic prediction, but further work is needed to verify this. For example, it is generally accepted that at least 10,000 SNPs are needed for genomic prediction, suggesting that the high correlation (0.992) between the 4k and 41k sets seen here may not be enough to guarantee satisfactory predictions from the 4k set. If GBS is to be adopted in resources were many individuals have been genotyped with SNP chips, there will need to be an investigation on how to combine relatedness estimates from different methods as has been required for combining pedigree and genomic-based relatedness (Aguilar *et al.* 2010).

We have shown that there is good agreement between relatedness estimates from GBS and from SNP chips, especially in terms of their correlation. There were some small differences in the mean levels of relatedness, so that adjustments would be required if combining data using different methods. It would be useful if this comparison could be extended to genomic relatedness estimation across divergent breeds and also to examine different GBS protocols, i.e. different enzymes, to check the robustness of these results. In summary, GBS is a promising method for genomic analyses using relatedness estimates and can be rapidly deployed, even for species with poor genomic resources.

ACKNOWLEDGEMENTS

This work was funded by MBIE (NZ) through the "Genomics for Production & Security in a Biological Economy" programme. The Illumina ovine HD BeadChip was developed by the International Sheep Genomics Consortium with FarmIQ (http://www.farmiq.co.nz/), a joint NZ Primary Growth Partnership programme. The animals involved in this work were part of the FarmIQ project and thanks is extended to Focus Genetics for access to this resource.

REFERENCES

Aguilar I., Misztal I., Johnson D.L., Legarra A., Tsuruta S. and Lawlor T.J. (2010) *J. Dairy Sci.* **93**:743.

Altman D.G. and Bland J.M. (1983) *J. Royal Stat. Soc. Ser. D (The Statistician)* **32**:307.

Auvray B., McEwan J.C., Newman S.A.N., Lee M. and Dodds K.G. (2014) *J. Anim. Sci.* **92**:4375.

Dodds K.G., McEwan J.C., Brauning R., Anderson R.A., Van Stijn T.C., Kristjánsson T. and Clarke S.M. (2015) *BMC Genomics* **16**:1047.

Elshire R.J., Glaubitz J.C., Sun Q., Poland J.A., Kawamoto K., Buckler E.S. and Mitchell S.E. (2011) *PLoS ONE* **6**:e19379.

Kijas J.W., Porto-Neto L., Dominik S., Reverter A., Bunch R., McCulloch R., Hayes B.J., Brauning R., McEwan J. and the International Sheep Genomics Consortium (2014) *Anim. Genet.* **45**:754.

Lu F., Lipka A.E., Glaubitz J., Elshire R., Cherney J.H., Casler M.D., Buckler E.S. and Costich D.E. (2013) *PLoS Genet.* **9**:e1003215.

VanRaden P.M. (2008) *J. Dairy Sci.* **91**:4414.

Yang J., Benyamin B., McEvoy B.P., Gordon S., Henders A.K., Nyholt D.R., Madden P.A., Heath A.C., Martin N.G., Montgomery G.W., Goddard M.E. and Visscher P.M. (2010) *Nat. Genet.* **42**:565.

Proc. Assoc. Advmt. Anim. Breed. Genet. 22:85-88

ACCURACY OF GENOMIC PREDICTION WHEN QTL EFFECTS ARE NOT NORMALLY DISTRIBUTED

M.E. Goddard[1,2] and T.H.E. Meuwissen[3]

[1] Faculty of Veterinary and Agricultural Sciences, University of Melbourne, Victoria, Australia
[2] AgriBio, Department of Economic Development, Jobs, Transport and Resources, Victoria, Australia
[3] Norwegian Life Sciences University, Norway

SUMMARY

In this paper we examine, using simulation and an analytical method, the factors that control the accuracy of genomic prediction when the effects of chromosome segments are not normally distributed, for instance, because many chromosome segments do not contain a QTL. In this situation non-linear methods of analysis give higher accuracy than GBLUP but the advantage is small unless the distribution of chromosome segment effects departs markedly from a normal distribution and the distribution assumed by the method of analysis also departs markedly from a normal distribution. The effect of sample size on accuracy of non-linear methods is similar to that with GBLUP but the advantage of non-linear methods over GBLUP increases with sample size when accuracy is low.

INTRODUCTION

Before implementing genomic prediction of breeding values (genomic selection), it would be useful to be able to predict the accuracy that might be achieved or at least to understand the factors controlling accuracy so that the optimum combination could be used. If genomic estimated breeding values (GEBVs) are estimated using GBLUP (Meuwissen et al 2001), there is good theory to predict the accuracy (Daetwyler et al 2008, Goddard 2009). In this case, the accuracy or correlation between EBV and true breeding value (r) is approximately given by MacLeod et al (2014)

$$r^2 = \theta c/(1+\theta - h^2r^2) \quad (1)$$

where c = the proportion of genetic variance explained by markers
h^2 = heritability
$\theta = Nh^2c/M_e$
N= number of records in the training population
M_e = effective number of independent chromosome segments in the genome.

This is not an explicit formula for r^2 because r^2 appears on both sides of the equation. However, we choose to present the formula in this way because it makes clear the way in which increasing accuracy decreases the unexplained variance and so further increases accuracy. If the causal variants or QTL have similar properties to the markers, then $c = M/(M + M_e)$ where M is the number of markers. However, c is often less than this presumably because the QTL have lower linkage disequilibrium (LD) with the markers than the markers do amongst themselves.

Estimation of breeding values using GBLUP, as above, is a Bayesian prediction if it is assumed that the effects of the markers are all drawn from a normal distribution with mean zero and constant variance. That is, a model in which the genomic relationships between the animals is estimated from the markers (GBLUP) is equivalent to a model in which SNP effects are assumed to be normally distributed (SNP-BLUP). Other assumptions about the distribution of marker effects lead to other methods of estimation of which some have been called Bayes A, B, C or R. Although BLUP is a linear estimate in the phenotypic values (y), these other Bayesian methods are non-linear in y. These non-linear Bayesian methods give higher accuracy than BLUP in some cases (MacLeod e al 2014) but there is no theory that predicts how much more accurate and in what circumstances. As well as the parameters that affect GBLUP accuracy, the accuracy of non-linear methods could be affected by the true distribution of marker effects and the distribution assumed by the method of analysis. The aim of this paper is illustrate how these parameters affect the accuracy of non-linear Bayesian methods of predicting breeding value. We use simulation and a simplified analytical model.

MATERIALS AND METHODS

Analytical method. Here we assume that the markers and QTL are identical and there are M_e independent QTL so that the accuracy of estimating a single QTL effect (r) is equal to the accuracy with which the combined value of all QTL is estimated. This can then be calculated using numerical integration. That is, $r^2 = V(\hat{q})/V(q)$ and $V(\hat{q}) = \int f(q)E(\hat{q}|q)^2\, dq$, where q is the effect of a QTL assumed to have a mean of zero, f(q) is the distribution of QTL effects, $E(\hat{q}|q)$ is the expectation of the estimate of q ($\hat{q}$) given q.

Simulation. We simulated a genome of length 1M in a population of $N_e = 1000$ until it reached mutation-drift equilibrium. At this point there were approximately 33,000 SNPs segregating of which between 3 and 290 were designated as QTL and their effect sampled from a distribution that was either exponential or gamma (shape parameter = 0.09) or t-distribution (degrees of freedom = 4.1 or 4.2). The scale of the effects was adjusted so that a fixed heritability was reached after adding normally distributed environmental effects. The linkage disequilibrium among the markers means that the effective number of chromosome segments (M_e) is approximately 300. The simulated data on 200 animals were analysed with BLUP, Bayes A, Bayes B (Meuwissen et al 2001) and Bayes R (Erbe et al 2012) and the correlation between true breeding value and EBV calculated in an independent set of animals. Because the results depend to θ, the simulation approximately corresponds to a genome of 30 M but with a sample size of 30 * 200 = 6000.

RESULTS AND DISCUSSION

Simulation results. Table 1 lists the accuracy achieved when $h^2 = 0.5$ and the all 33,000 markers were used so that all genetic variance is explained by the markers (c=1 in equation 1).

Table 1 Effect of distribution of QTL and distribution assumed by the method of analysis on accuracy (%) of EBVs.
For Bayes R Sim. = simulation results, anal. = analytic approximation, all other results are from simulation.

No. QTL	Distribution	Method of analysis GBLUP	 Bayes B	Bayes R sim.	 anal.	 Bayes A
3	exponential	51	97	95	98	67
30	exponential	49	83	82	85	54
30	gamma	48	88	89	96	65
30	t (df = 4.105)	54	81	82	81	57
290	t (df = 4.225)	52	57	55	61	51

When GBLUP is used, assuming a normal distribution of marker effects, the accuracy is nearly the same (~0.5) regardless of the true distribution of QTL effects. Although there are 33,000 SNPs, there are only about 300 effective independent chromosome segments. Therefore the last distribution in table 1 with 290 QTL with effects drawn from a t distribution does not differ greatly from a distribution in which all chromosome segments have an effect drawn from a normal distribution. Consequently the Bayesian methods have little advantage over GBLUP. When there are less than or equal to 30 QTL, many chromosome segments have zero effect and the distribution differs more markedly from a normal distribution. In these cases Bayes B and Bayes R have similar accuracy and an advantage over GBLUP. Bayes B and Bayes R assume a distribution of marker effects in which some effects are zero and this agrees with the true distribution in the first 4 cases in table 1. Bayes A assumes no effects are zero but all SNP effects follow a t-distribution. The accuracy it achieves is in between that of GBLUP and Bayes B or R.

The accuracy of the non-linear methods (e.g. Bayes B and R) depends in part on the kurtosis of the distribution of effects of chromosome segments. If many segments have zero effect (i.e. no QTL in the segment) the kurtosis is increased. However, the kurtosis is not the only parameter of the distribution that affects the accuracy of EBVs. In table 1 the gamma distribution with 30 QTL and the exponential distribution with 3 QTL have similar kurtosis but the exponential distribution leads to higher accuracy. This is because the gamma distribution with shape parameter of 0.094 has some large effects but also many very small effects that are hard to estimate accurately.

The results in table 1 can be summarised by

- the true distribution must differ greatly from a normal before non-linear methods have an advantage over GBLUP,
- it is not worthwhile to use a non-linear method of analysis unless it assumes a distribution of marker effects that differ greatly from a normal distribution.

Analytical method. Here we calculated the accuracy of estimating the effect of a single QTL assuming that the method of analysis used the same distribution of QTL effects as used to generate true QTL effects. Table 1 shows that the analytical method overestimates the accuracy found by simulation. This is expected. The analytical method assumes there is only one marker per effective chromosome segment, whereas in the simulation there are approximately 100. The GBLUP analysis shrinks estimates of marker effects but the amount of shrinkage is not effected by the size of the estimated effect. Consequently, the effect of a chromosome segment can be shared among several markers with little loss of accuracy. But the non-linear methods shrinks apparently large effects less than small effects (Figure 2) and so, if the effect of a single QTL is shared among several markers, the effect is shrunk too much and this reduces the accuracy.

Apart from this over prediction of accuracy, the analytical method does predict the differences in accuracy between distributions (Table 1) and, although not shown here, it also predicts the effect of changing θ reasonably well. In figure 1, we use the analytical method to examine the effect of θ on accuracy. The y-axis of the graph is $T = r^2/(1-r^2)$. For GBLUP analysis this is almost equal to θ but differs from it due to the $-h^2r^2$ term in equation 1. This term corrects for the reduction in error variance when estimating the effect of one marker due to the simultaneous prediction of the effects of all other markers (Daetwyler et al 2008). Consequently, T is slightly greater than θ for GBLUP and this disparity increases slightly with θ. For the non-linear methods, T increases faster than linear in θ and the advantage over GBLUP increases with θ at first and then reaches a constant ratio.

In real data within one breed, the distribution of QTL effects may be most similar to the t-distribution with 290 QTL in 300 effective chromosome segments corresponding to 8100 QTL in a 30M genome. This would explain why non-linear methods enjoy only a small advantage over BLUP in many cases. The advantage of non-linear methods would be expected to increase if multiple breeds were analysed or the population had a high effective population size e.g. in humans.

REFERENCES

Daetwyler, H.D., Villanueva, B. & Woolliams, J.A. 2008 Accuracy of Predicting the Genetic Risk of Disease Using a Genome-Wide Approach. *PLoS ONE* **3**, e3395.

Erbe M., Hayes B.J., Matukumalli L.K., Goswami S., Bowman P.J., Reich C.M., Mason B.A. and Goddard M.E. (2012) *J. Dairy Sci.* **95**: 4114.

Goddard, M. 2009 Genomic selection: prediction of accuracy and maximisation of long term response. *Genetica* **136**, 245-257.

MacLeod IM, Hayes BJ, & Goddard ME (2014) The Effects of Demography and Long-Term Selection on the Accuracy of Genomic Prediction with Sequence Data. *Genetics* 198(4):1671-1684.

Meuwissen, T., Hayes, B. and Goddard, M. 2001 Prediction of total genetic value using genome-wide dense marker maps. *Genetics* **157**, 1819 - 1829.

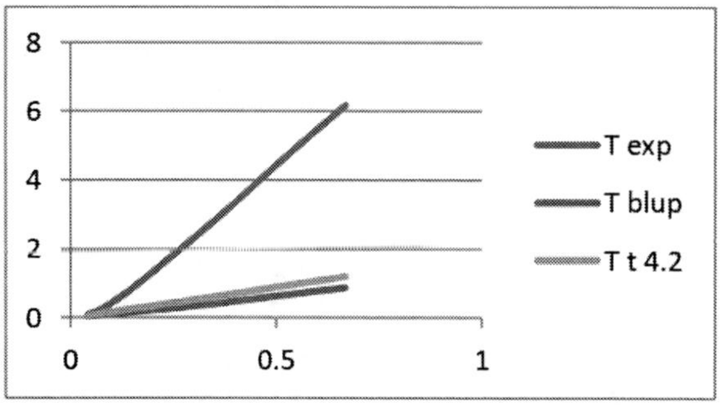

Figure 1. The effect of θ on $T = r^2 / (1-r^2)$. The graphs show the effect of θ on accuracy from the analytical method for the exponential distribution of 30 QTL effects (T exp), the normal distribution of 300 QTL effects (T blup) and the t-distribution with degrees of freedom = 4.225 of 290 QTL effects.

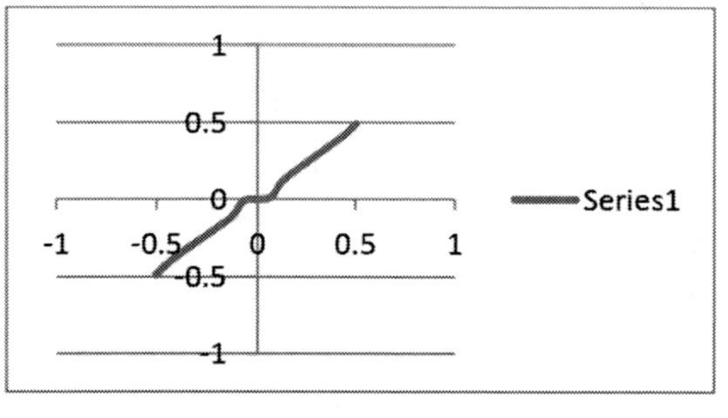

Figure 2. Estimated QTL effect size vs true QTL effect size from the analytical method under the exponential distribution of 30 QTL in 300 effective chromosomal segments (arbitary scale of effect sizes).

Proc. Assoc. Advmt. Anim. Breed. Genet. 22:89-92

APPROXIMATING THE ACCURACY OF SINGLE STEP EBVS

L. Li, A.A. Swan and B. Tier

Animal Genetics and Breeding Unit*, University of New England, Armidale, NSW 2351, Australia

SUMMARY

To accompany the implementation of multi-trait Single Step Genomic BLUP (SS-GBLUP) in the BREEDPLAN and OVIS genetic evaluation systems, an algorithm to approximate accuracy with genomic information has been developed and is presented in this paper. Data from full terminal sire OVIS and Brahman BREEDPLAN runs were processed using this new method. Results demonstrate that the approximated accuracy of SS-GBLUP estimated breed values (EBVs) is highly correlated (R^2 >0.96) with exact accuracies in several small example analyses for both beef and sheep. SS-GBLUP EBV accuracies increase more for traits with a larger reference population and for traits with higher heritabilities. Animals with low pedigree-only (ABLUP) EBV accuracies benefit more from genomic information than animals with high ABLUP EBV accuracies.

INTRODUCTION

Single Step Genomic BLUP (SS-GBLUP, e.g. Legarra *et al.* 2014) was implemented in the Australian sheep and beef cattle evaluation systems OVIS and BREEDPLAN during 2016, simultaneously combining phenotypic, pedigree, and genomic information. Conceptually, SS-GBLUP is compatible to the existing pedigree BLUP models and is relatively straightforward to implement by replacing the traditional inverse pedigree relationship matrix ($\mathbf{A}^{-1}$) in the mixed model equations (MME) with $\mathbf{H}^{-1}$ (Christensen and Lund, 2010):

$$\boldsymbol{H}^{-1} = \boldsymbol{A}^{-1} + \begin{pmatrix} 0 & 0 \\ 0 & \boldsymbol{G}^{-1} - \boldsymbol{A}_{22}^{-1} \end{pmatrix}$$

where $\mathbf{G}$ and $\mathbf{A}_{22}$ are genomic and pedigree relationship matrices for genotyped animals, respectively. This make modification of models and software to estimate breeding values (EBVs) relative straightforward, although computational requirements can increase significantly.

Accuracies of EBVs are also an important output of genetic evaluation systems, and these have traditionally been approximated using "effective progeny numbers" (EPN) as a basis which accumulate information from animals' own performance, progeny, parents, and from correlated traits (Graser and Tier 1997). In this paper, we present a modification to this algorithm to account for EPN from genomic information, allowing the calculation of accuracies for SS-GBLUP EBVs. We also investigate the impact of genomic information on the improvement of accuracy of EBV for real examples.

MATERIALS AND METHODS

Algorithms to derive "genomic EPN". In order to ensure compatibility with the current accuracy algorithm, information from the genomic relationship matrix needs to be expressed in the form of an EPN for each animal. This "genomic EPN" must be accumulated with existing sources of EPN to derive approximations of the total accuracy for multi-trait SS-GBLUP analyses. The steps required are described below.

Step 1. Calculate a prediction error variance (PEV) using a series of single trait GBLUP pseudo-analyses. For each trait, we construct the MME for genotyped animals with additive genetic effects

*AGBU is a joint venture of NSW Department of Primary Industries and the University of New England

considered in the model, ignoring all the other fixed and random effects. The diagonal of the inverse of the MME then represents the genomic PEV for the trait. Because the pedigree relationship matrix $\mathbf{A}_{22}$ for these animals has already been used to contribute accuracy from pedigree and performance information, and also because a proportion of **G** is used to build $\mathbf{H}^{-1}$ in SS-GBLUP, an adjusted PEV must be used to derive the contribution of genomic information to accuracy. This adjusted PEV for the ith animal is calculated as:

$$PEV_i^* = wt * PEV_i + (1 - wt)\sigma_a^2$$

where σ_a^2 is the additive variance, and *wt* is a tuning parameter (referred to as the "genomic PEV weight" below) determined empirically by comparing approximate accuracies calculated across a range of *wt* values from 0.1 to 1.0 with exact accuracies calculated by direct inversion of the SS-GBLUP mixed model equations for a range of examples reported below.

After PEV^* for each trait is calculated with appropriate values of *wt*, accuracy is calculated as:

$$acc = \sqrt{1 - PEV_i^* / g_{ii}\sigma_a^2}$$

where g_{ii} is the diagonal of **G** for the ith animal. This is assumed to be the gain in accuracy due to genomic information for genotyped animals.

Step 2. Propagate genomic accuracy to un-genotyped ancestors and descendants so that the impact of genomic information on close relatives is included. Propagation is performed upwards first (to ancestors) and then downwards (to descendants). If an un-genotyped animal has its parents and progeny genotyped, accuracy is calculated from the progeny, except for the case where only one progeny and both parents are genotyped, in which accuracy is calculated from the parents. The accuracy of un-genotyped parents with genotyped progeny is given by:

$$acc = \overline{acc} \times (1 - 0.5^n)$$

where $\overline{acc}$ is the average accuracy over n genotyped progeny for the sire or dam. The accuracy of the un-genotyped progeny is given by:

$$acc = \sqrt{(acc_{sire}^2 + acc_{dam}^2)/4}$$

Step 3. Accuracy for genotyped animals and progeny and parents of genotyped animals is transformed to the equivalent number of effective progeny as:

$$EPN = \delta \times acc^2/(1 - acc^2)$$

where $\delta = (4 - h^2)/h^2$ and h^2 is the heritability of the trait.

Step 4. For each animal with genomic EPN derived from the above single trait analyses, multiple trait EPN are derived by constructing multiple trait MME with additive genetic effects as follows: 1) Accumulating the residual matrices based on the common minimal EPN across traits based on the phenotypes observed into a trait by trait matrix; 2) The additive genetic co-variance matrix is added to the accumulated residual matrix; 3) Multiple trait PEV are then calculated by inverting this matrix and then converted to EPN following the procedures above.

Step 5. Because EPN due to genomic information for each animal are confounded with EPN arising from phenotypic own-performance information, the final step is to calculate the difference between the genomic EPN of an animal and the EPN arising from its own phenotype, as calculated from the current algorithm. Only when this difference is positive is the genomic EPN accumulated with EPN from all other sources to derive the final accuracy.

Note that when calculating final EBV accuracies following the formula in equation above, rather than using g_{ii} for genotyped animals, we use $\lambda g_{ii} + (1 - \lambda)a_{ii}$, following the specification of the

H matrix used for SS-GBLUP, where λ is a weighting factor for genomic and pedigree information as described by McMillan and Swan (2017).

Selection of genomic PEV weight. Data from OVIS and BREEDPLAN runs were used to investigating the genomic prediction error variance weight. Traits considered for sheep were intra-muscular fat (IMF, $h^2 = 0.56$) and shear force at day 5 (SF5, $h^2 = 0.32$), with the data including 11,416 genotyped animals from terminal sire evaluation. Traits considered for beef were beef 600 day weight (FWD, $h^2 = 0.49$) and days to calving (DTC, $h^2 = 0.08$), with the data including 5,847 genotyped animals from the Brahman BREEDPLAN evaluation. These data were analysed repeatedly with the new accuracy algorithm, fitting a range of genomic PEV weights from 0 to 1 in increments of 0.1. The approximate accuracies derived were then compared to exact accuracies calculated from PEVs derived by inversion of the mixed model coefficient matrix for each data set, and varying the value of λ used to construct the **G** matrix from 0 to 1 in increments of 0.1.

Application to industry data. The new SS-GBLUP accuracy algorithm was applied to data from full sheep terminal sire evaluation and full Brahman BREEDPLAN runs. The numbers of genotyped animals were 11,832 in sheep and 7,166 in beef data.

RESULTS AND DISCUSSION

Selection of the genomic PEV weight. Based on comparison of approximate accuracy calculated over a series of genomic PEV weights to exact accuracy with a series of λ values, results showed that the means and standard deviations of true accuracies increased with λ from 0 to 1. When $\lambda = 0.5$, the value currently chosen to run SS-GBLUP analyses in OVIS and BREEDPLAN, the closest genomic PEV weight for the new accuracy algorithm was 0.3 for all sheep and beef traits, based on comparison of means and standard deviations. Across all four traits, high R-squared values (>0.95) and regression coefficients (from 0.96 to 1.1) were observed for the regression of approximate accuracies with genomic PEV weight = 0.3 on true accuracies with $\lambda = 0.5$, indicating a genomic PEV weight of 0.30 is appropriate to tune the genomic prediction error variances for the current implementation of SS-GBLUP.

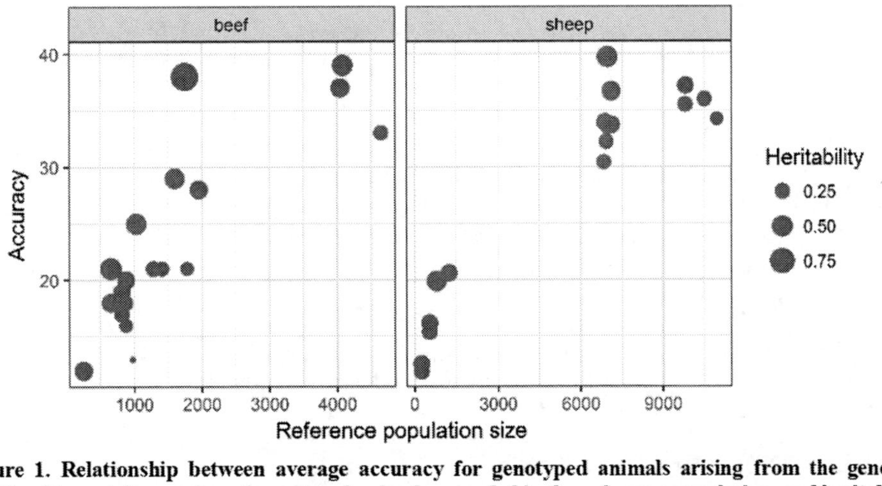

Figure 1. Relationship between average accuracy for genotyped animals arising from the genomic relationship matrix, number of genotyped animals recorded in the reference population, and heritability (size of points) for different sheep and beef traits.

The impact of genomic information on accuracy. The relationship between average accuracies for genotyped animals arising from the genomic information, number of genotyped animals recorded in the reference population, and heritability for sheep and beef traits are shown in Figure 1. The average accuracies of genotyped animals as calculated in Step 1 above varied from 0.12 to 0.40 in both sheep and beef across different traits. The accuracies were positively related to the number of animals with records and heritability for each trait.

Figure 2 shows the distribution of average accuracy improvement for SS-GBLUP relative to ABLUP for beef and sheep. For animals with low starting ABLUP accuracies (<30%), the SS-GBLUP accuracy was on average 18% points higher for sheep (ranging from 11 to 24% points), and on average 13% points higher for beef (ranging from 3 to 29% points). For medium starting accuracies (30 to 50%), the improvements were 6% (2 to 8%) for sheep and 4 (1 to 9%) for beef, while very little improvement in accuracy was observed for high starting accuracies (>50%). These trends confirm expected benefits of accuracy from genomic information.

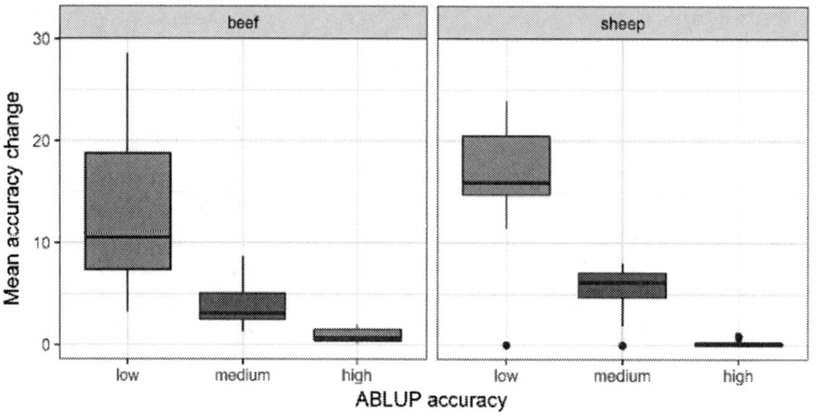

Figure 2. Distribution across beef and sheep traits of improvement of SS-GBLUP accuracies over ABLUP accuracies within bands of ABLUP accuracy from low (<30), medium (30 – 50) and high (>50).

CONCLUSIONS

An algorithm to approximate SS-GBLUP EBV accuracies was developed, and shown to be consistent with exact accuracies in several small example analyses for beef and sheep. SS-GBLUP EBV accuracies increase more for traits with a larger reference population (numbers of animals phenotyped and genotyped), and for traits with higher heritabilities. Animals with low pedigree-only (ABLUP) EBV accuracies gain more improvement in accuracy from genomic information than animals with high ABLUP EBV accuracies.

ACKNOWLEDGMENTS

This project was funded by the MLA through projects B.SGN.0127 and B.BFG.0050

REFERENCES

Christensen O.F., and Lund M.S. (2010) *Gen. Sel. Evol.* **42**: 2.

Graser H.-U and Tier B. (1997) *Proc. Assoc. Advmt. Anim. Breed. Genet.* **12**: 547.

Legarra A., Christensen O.F., Aguilar I. and Misztal I. (2014) *Livest. Sci.* **166**: 54.

McMillan, A.J. and Swan, A.A. (2017) *Proc. Assoc. Advmt. Anim. Breed. Genet.* **21:** these procedings.

Proc. Assoc. Advmt. Anim. Breed. Genet. 22:93-96

APPROXIMATE GBLUP FOR EFFICIENT ROUTINE EVALUATIONS

T. P. Hancock[1,2]

1 DEDJTR, Victoria, Australia
2 DataGene Limited

SUMMARY

We present a computationally efficient approach to GBLUP which approximates inverse reference set matrix by optimally selecting the most informative animal cohort. The optimal animal cohort, named core reference animals, are identified through a Partial Incomplete Cholesky Decomposition (PICD) and selected such that the reconstruction error is at a specified user percentage. Our application of PICD on the Australian Holstein and Jersey reference sets shows that allowing a small error halves the effective size of reference set, resulting in significant gains in performance with only minor differences between exact and approximate breeding values and reliabilities ($r > 0.99$). Overall our results show that application of methods like PICD aimed at eliminating redundancy within large reference sets, significant performance gains can be made without sacrificing accuracy.

INTRODUCTION

Genomic evaluations are routinely used to evaluate the performance of dairy cattle world-wide. These genomic evaluations impose a significant and ever increasing computational burden on the evaluation organisations. This computational burden must be offset by the requirement to maintain a meaningful animal reference set to ensure that accurate and reliable predictions are made for the young animals entering the system. Up to now the focus has been on increasing the accuracy and reliability of genomic evaluations with projects such as GINFO (Pryce et al, in press) succeeding in increasing the overall reliability of the Australian genomic evaluations between 2 and 7 percent, by doubling the number of animals in the reference set. The cost of doubling the size of the reference set results in a dramatic increase in computational burden. GBLUP (Van Raden, 2008) like algorithms can be solved for breeding values using gradient techniques highly efficiently, however the reliability computation requires the explicit inverse of the genomic reference set matrix which scales at cubic complexity. With reference sets continuing to grow, and now including more than 35000 Australian dairy animals, more efficient solutions for genomic evaluations are required.

The accuracy and reliability of a genomic breeding value for a young, non-reference animal, is not based on the size of the reference set, but how related that animal is to the reference set. Additionally, the genomic relationship structure within the reference set animals are not related to the quality of their phenotypic information. Therefore simply adding animals to the reference set based on the quality of their phenotype alone will not ensure more reliable predictions into the future and is likely to make routine evaluations computationally infeasible.

In this paper we investigate the feasibility of a Partial Incomplete Cholseky Decomposition PICD (Foster et al, 2009) to identify a smaller cohort of reference set animals, named core reference set animals, which can be used to optimally represent the structure within full reference set. PICD has been shown in kernel regression literature to provide a robust approximate solution to a related model to GBLUP (Foster et al, 2009). In this paper we extend PICD for application to the GBLUP model by accounting for the diagonal weighting of all reference set animals to ensure that phenotype accuracy information is included in the evaluation of all animals. We show that application of PICD with a small degree of error can significantly reduce computational time without dramatically moving from the estimated breeding values or reliability from the full model.

MATERIALS AND METHODS

The equations for the GBLUP breeding values $\hat{a}$ and reliabilities *rel* are as follows (Van Raden, 2008),

$$\hat{\boldsymbol{a}} = \boldsymbol{G}_{cr}(\boldsymbol{G}_{rr} + \boldsymbol{R})^{-1}\boldsymbol{y} \qquad \text{and} \qquad \boldsymbol{rel} = \frac{\mathbf{diag}[\boldsymbol{G}_{cr}(\boldsymbol{G}_{rr} + \boldsymbol{R})^{-1}\boldsymbol{G}_{cr}^T]}{\mathbf{diag}[\boldsymbol{G}]} .$$

where $\boldsymbol{G}_{rr}$ is the genomic relationship matrix of the reference set animals, $\boldsymbol{R}$ is a diagonal matrix of observation weights and $\boldsymbol{G}_{cr}$ is the genomic covariance matrix of all animals with the reference set animals. The cost of a GBLUP model is in the evaluation of $(\boldsymbol{G}_{rr} + \boldsymbol{R})^{-1}$ where the number of required operations scales cubically, $O(r^3)$, as the number of reference set animals, *r*, increases.

Partial Incomplete Cholesky Decomposition (PICD) (Foster et al, 2009) is a variant of the Cholesky decomposition which employs both row pivoting and a diagonal error tolerance to create a rank-reduced decomposition. The purpose of PICD is to select from $\boldsymbol{G}_{rr}$ a reduced cohort of animals, called core reference animals, which are representative of the entire population. This cohort can then be used to reconstruct $\boldsymbol{G}_{rr}$ by,

$$\boldsymbol{G}_{rr} = \boldsymbol{L}^T\boldsymbol{L} \approx \boldsymbol{L}_k^T\boldsymbol{L}_k \, ,$$

where *k* is the set of core reference animals, $k < r$, and $\boldsymbol{L}_k$ is the Cholesky complement only including the currently selected *k* animals.

The PICD algorithm identifies the core reference animal by performing single Cholesky updates to $\boldsymbol{L}_k$, animal-by-animal in a stage-wise and greedy fashion where the next animal to be added $\boldsymbol{L}_k$ is selected such that it maximally reduces the reconstruction error. The reconstruction error is a measurement of how well $\boldsymbol{L}_k^T\boldsymbol{L}_k$ predicts $\boldsymbol{G}_{rr}$. The addition of all *r* animals completely reconstructs the full Cholesky complement with no error. Therefore the reconstruction error can be measured as a percentage of complete reconstruction.

The algorithm requires as input the acceptable amount of error as a percentage, and from this will create a Cholskey complement, $\boldsymbol{L}_k$, of size (*N, k*) where *k* number of animals required to approximate the original matrix at that error percentage. The advantage of using this approach to others such as Singular Value Decomposition (SVD) is its ability to pick the specific animals required for the reconstruction, whereas SVD projects each animal onto every eigenvector. Therefore PICD is a means of selecting the most informative animals from the reference set.

PCID when used in the kernel regression setting reduces the cost complexity from order $O(r^3)$ to $O(kr^2)$ (Rasmussen and Williams, 2006). However, within the reference set of GBLUP there are also observation weightings defined. To allow for all reference set animals to have their observation weight applied we must derive a subset-of-regressors approximation of $(\boldsymbol{G}_{rr} + \boldsymbol{R})^{-1}$ using the Nystrom approximation of $\boldsymbol{G}_{rr}$ (Rasmussen and Williams, 2006). The Nystrom approximation of $\boldsymbol{G}_{rr}$ is the approximation of the $\boldsymbol{G}_{rr}$ using a subset of rows and can be expressed as,

$$\widehat{\boldsymbol{G}}_{rr} = \boldsymbol{G}_{rk}\boldsymbol{G}_{kk}^{-1}\boldsymbol{G}_{kr}$$

where the *k* animals are selected from the reference set using PICD. From this representation of $\widehat{\boldsymbol{G}}_{rr}$ we can apply the Woodbury matrix identity to gain an approximation of the whole system inclusive of the observation weights,

$$(\boldsymbol{G}_{rr} + \boldsymbol{R})^{-1} \approx \left(\widehat{\boldsymbol{G}}_{rr} + \boldsymbol{R}\right)^{-1} = \boldsymbol{R}^{-1} - \boldsymbol{R}^{-1}\boldsymbol{G}_{rk}(\boldsymbol{G}_{kk} + \boldsymbol{G}_{kr}\boldsymbol{R}^{-1}\boldsymbol{G}_{rk})^{-1}\boldsymbol{G}_{kr}\boldsymbol{R}^{-1}$$

where $\boldsymbol{G}_{rk}$ is the covariance between the all reference animals and the core reference animals. This approximation to GBLUP allows for a selection of core animals from the reference set, without losing any phenotypic information from the model. Once the solution to the approximate GBLUP is attained the pre and post multiplication by $\boldsymbol{G}_{cr}$ is still required to compute the breeding values and reliabilities respectively. If no error tolerance is specified the approximation will yield exactly the same results as solving the system directly. It is suggested that this be treated like a heritability analysis and run once annually, out of scope of an evaluation.

PICD is also similar in idea to the sparse inverse of $\boldsymbol{G}$ with the APY algorithm of (Misztal, 2014) however PICD is a reduced rank approximation where as APY is a sparse approximation. The main advantage of PICD over APY is reducing the size of the entire system required to be solved through the efficient use of the Woodbury matrix identity above. APY on the other hand approximates only $\boldsymbol{G}$ or $\boldsymbol{G}^{-1}$ which still requires the addition of observation weights, $\boldsymbol{R}$, and solution of the entire system to be computed.

MATERIALS AND METHODS

To evaluate our proposed PICD approximated GBLUP we perform a simple parameter sweep on the percent error for the PICD algorithm and evaluate three different metrics.

1. The computational elapsed time.
2. The number of animals in the core reference set.
3. The correlation between breeding values and reliabilities as compared to the exact solutions.

The PICD program was developed in-house and implemented in R using Rcpp and compiled using the Intel MKL library. The datasets under consideration are the 58961 non-duplicated Holstein bulls and cows as well as the 11768 non-duplicated Jersey bulls and cows from the December 2016 ABV ADHIS release. Of these animals 32481 Holstein and 8846 Jersey bulls and cows were found in the full Protein GEBV reference set. The parameter sweep is run between 0 and 50% allowable error in increments of 5%.

RESULTS AND DISCUSSION

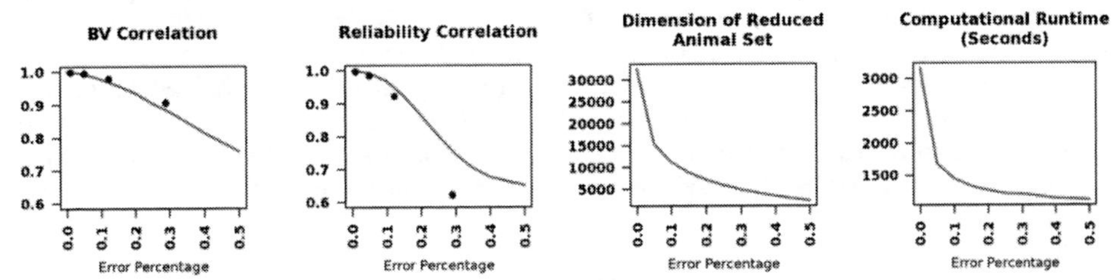

Figure 1: Holstein parameter screening results. Green line is the correlation between the exact solution and the PICD algorithm and the blue dots are the average correlation of 10 repeats of randomly selecting rows at four specified error tolerance.

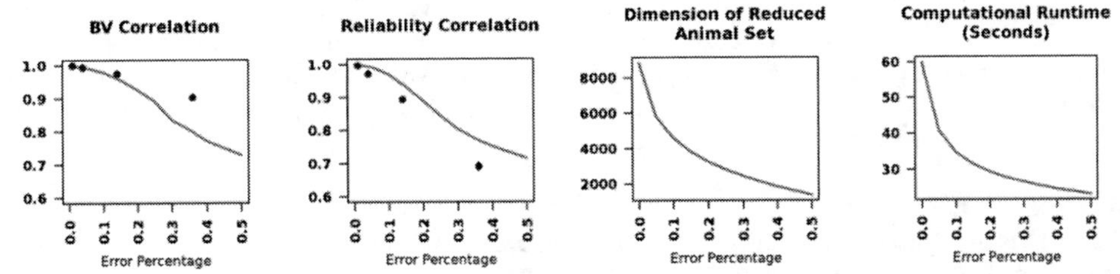

Figure 2: Jersey parameter screening results. Green line the correlation between the exact solution and PICD algorithm and the blue dots are the average correlation of 10 repeats of randomly selecting rows at four specified error tolerance.

Figure 1 and 2 present the parameter sweep results for the Holstein and Jersey analyses respectively. The results include the computation of breeding values and reliability for all animals in the analysis, including non-reference animals with no phenotype. From left to right, the first two plots are the correlation between approximate breeding values and reliabilities compared to exact GBLUP calculation, the dimension of the core reference animal set, *k*, and the run time.

Both Holstein and Jersey sets share the same profile, where at small amount of acceptable errors the approximate methods correlate very well (r > 0.99, % error = 0.05) with the exact solutions. The animals removed are predominantly bulls rather than cows. In of the 7754 cows and 1092 bulls in the Jersey reference set 2464 (32 %) cows and 591 (54 %) bulls were removed by PICD at 0.05 error tolerance. Of the 28228 cows and 4253 bulls in the Holstein reference set 13761 (49 %) cows and 3295 (78 %) bulls were removed PICD at 0.05 error tolerance. The removal of bulls from the reference is likely due to the selection of bulls results in stronger relationships between them, and therefore they produce more redundant set in terms of genotypic variation. The surprising result from these parameter sweeps by imposing only a small error the amount of animals in the core reference set is approximately.

The observed massive reduction in the reference set size is a result of the genomic redundancy within the reference set created by one-sided selection of animals. Reference set inclusion is based bulls having more than 10 daughters or cows in specific projects with phenotypic records, not on how related the animal is to the existing reference set. This approach is likely to select a reference set with a large number of highly related animals who collectively contribute very little to the performance of the overall evaluation. Algorithms like PICD are able to parse this redundant set and capture the key animals required to maintaining accuracy and reliability. The availability of such algorithms therefore encourages the continued collection of phenotypes and from the ever increasing pool of reference set animals timely evaluations are still possible.

At larger amounts of acceptable errors we observe that the PICD approximated reliabilities are significantly closer to the exact reliabilities than those computed from a random sample. However, the breeding values estimated by PICD are more poorly estimated, in particular within the Jersey analysis. This drop in performance is because PICD seeks to remove all redundancy within the genomic relationship matrix, without any knowledge of the phenotype. This style of selection may inadvertently remove animals with phenotypes that are highly informative for the trait under analysis because their relatives are already included in core reference set. This reduces the accuracy of breeding value estimation, but not reliability estimation, as the reliability is a function only of the relationship matrix (the target of PICD) not the phenotypic importance. This problem is well known and could potentially be overcome by selecting an animal subset using more complex objective functions which seek to balance the contributions from both left and right hand side GBLUP equations (Rasmussen and Williams, 2006).

In conclusion we have shown that it is possible to dramatically decrease the running time of genomic evaluations, without a significant impact on accuracy or reliability, by defining a smaller set of core reference animals. The implementation PICD with only small amount of error will reduce the computational burden on evaluation organisations allowing them to screen more animals, faster and more often.

REFERENCES

Foster L. et al, (2009) *Journal of Mach. Learn. Research*, **10**:857.

Misztal et al, (2014) *J Dairy Sci*, **97**:3943–3952.

Pryce J.E. (2016) *Interbull Bulletin*, **50**: in press.

Rasmussen C.E. and Willams C.K.I , (2006) 'Gaussian Processes for Machine Learning'. MIT Press.

VanRaden P.M. (2008) *J Dairy Sci*, **91**:4414.

Proc. Assoc. Advmt. Anim. Breed. Genet. 22:97-100

ON BREED COMPOSITION ESTIMATION OF CROSS-BRED ANIMALS USING NON-LINEAR OPTIMISATION

Vinzent Boerner

Animal Genetics and Breeding Unit, University of New England, Armidale, NSW 2351, Australia

SUMMARY

Genetically admixed animals are common in most quantitative genetic analysis, and usually are a result of intended crosses between two or more pure breed populations to enhance productivity. Disregarding the genetic heterogeneous architecture of admixed individuals may lead to poor or even wrong inference about the quality, quantity and genome location of genetic factors affecting phenotypes, and it could reduce the accuracy of estimates of genetic merit. In this article a non-linear optimisation approach (constrained genomic regression, CGR) is presented to describe the marker genotype of a focus animal as a linear function of marker allele frequencies of possible populations of origin. The algorithm was tested on a beef cattle data set consisting of 11639 animals from 11 different breeds with marker genotypes of 4022 single nucleotide polymorphisms, which were used to generate 5000 artificially cross-bred animals. For comparison the data set was also analysed with the ADMIXTURE software (ADM). CGR outperformed ADM with a maximum difference between the true and estimated breed proportion of 0.25 and 0.28 for the 5 and 25 cross-over data set respectively. For ADM this parameter was 0.83 and 0.66. The mean squared estimation error was 15 and 5 times larger for ADM compared to CGR for the 5 and 25 cross-over data set respectively. In addition, CGR always outperformed ADM in terms of speed by factor 20.

INTRODUCTION

The quantification of pure breed proportions of cross-bred animals' genomes is of relevance for genome wide association studies, estimation of population parameters, breeding value estimation and cross-breeding program optimisation. The most widely used methodology for marker based breed proportion estimation is likelihood formulation of the animals' genotype probability conditional on the pure breed population allele frequencies, where the latter are estimated in turn from the animals' genotypes and the assigned breed proportion (Pritchard et al. 2000). The whole system is evaluated using Gibbs Sampling (Pritchard et al. 2000; Raj et al. 2014), expectation maximisation (Tang et al. 2005), or, as a sped-up version, a block relaxation algorithm (Alexander et al. 2009). Since often the allele frequencies of pure breed populations can be estimated from animals of known pure breed origin, Alexander et al. (2009) shortcut their method to facilitate quicker breed proportion estimation for cross bred animals. However, the likelihood based method has two major shortcomings: a) the likelihood formulation assumes the absence of linkage disequilibrium between markers and orthogonality of pure breed population allele frequency vectors, and b) processing time becomes an obstacle if there are many marker genotypes (e.g. 700k or full genome sequences). This article describes a non-linear optimisation method (constrained genomic regression, CGR) for the estimation of pure breed proportions of cross-bred animals' maker genotypes, which overcomes both the limitations of the likelihood based method and allows a meaningful interpretation of the results even if the number of possible pure breeds is huge (see Chiang et al. 2010; Kuehn et al. 2011, for an unconstrained version of this approach). The algorithm was applied to 4k single nucleotide polymorphisms (SNP) genotypes of 5000 cross-bred animals artificially generated from real genotypes of 11639 animals from 11 different breeds. Result were compared to results from ADMIXTURE (ADM) (Alexander et al. 2009).

METHODS

Model. The problem to solve can be written as $argmin_b f(b) = y'y - 2y'Xb + b'X'Xb(1)$ subject to $b_i \geqslant 0\{i = 1,..,N\}(2)$ and $\sum b_i = 1(3)$where y is the marker genotype vector of the cross-bred animal, X is a column matrix of pure breed population allele frequency vectors, and N is the number of pure breeds. Note that equations (2) and (3) comprise constraints to the solutions of equation (1). Values in vector b are regression coefficients regressing y on the columns of X. Minimising equation 1 with respect to equation 2 and 3 will yield a vector b of which values will not only explain the genotype in y as a linear function of population allele frequencies in X, coefficients also have the straight forward interpretation of what proportion of Xb is explained by each column in X.

Data. The cattle data set consisted of 11639 animals from 11 different cattle breeds (Brahman (1492), Angus (1473), Murray Grey (316), Limousin (1395), Charolais (899), Hereford (1500), Simmental (337), Shorthorn (1126), Wagyu (1497), Santa Gertrudis (1474) and Drought Master (130)). Since genotypes of these animals were from various SNP panels, the 4022 SNP were selected which all panels had in common. The SNP genotypes were randomly phased to obtain haplotypes. Cross-bred animals were generated over five rounds. In round one the sex was randomly assigned to the 11639 pure-bred animals and 1000 males and 1000 females were randomly chosen (with replacement) to serve as parents. From each pair of parents one offspring was generated by joining their gametes generated from their haplotypes assuming 25 or 5 randomly located cross-overs. In the subsequent four rounds the 2000 parents were selected among previous 1000 offspring implying more than one offspring per parent. Thus, the total number of artificial admixed offspring was 5000. Table 1 summarises the number of cross-bred animals with 1 to 11 pure breed proportions in their genome.

Table 1: Summary of number of cross-bred animals with genome proportions of 1 to 11 pure breeds.

Number of cross-overs	Number of pure breeds contributing to a cross-bred genome										
	1	2	3	4	5	6	7	8	9	10	11
5	121	970	465	594	424	529	584	618	477	205	13
25	120	968	453	610	394	478	465	559	576	312	65

Note that table rows sum up to 5000, which is the number of cross-bred animals.

Result evaluation. Let bT be the row matrix of true breed proportions, and bE its estimated equivalent, with row dimension equal to the number of cross-bred animals and column dimension equal to the number of possible pure breeds. Results were evaluated by a parameter M calculated as the maximum of $|bT - bE|$, and by a parameter S calculated as the mean of $(bT - bE)^2$.

Software. CGR was implemented in a FORTRAN program which called the NLopt library (Johnson 2011). The optimisation solver used the augmented Lagrangian algorithm as global solver and the method of moving asymptotes as a local solver. All computations were carried out on a desktop computer with an Intel(R) Core(TM) i7-3770 processor and 32GB of memory.

RESULTS

Table 2 summarises the results for the cross-bred animals when the number of cross-overs during gametogenesis was 5 and 25 respectively. Invariably of the number of cross-overs CGR always performed better than ADM. The greatest absolute difference between the true and estimated breed proportion estimated by CGR was 0.24 and 0.28 for the 5 and 25 cross-over data set respectively, whereas for ADM that parameter was 0.85 and 0.67. The parameter S for the ADM results was 15 times larger than that for CGR results when the 5 cross-over data set was used. This difference to shrunk to 5 times larger when the 25 cross-over data set was used.

Table 2: Statistics of the breed proportion estimation error subject to the number of cross-overs when generating cross-bred animals and the used algorithm, where M is the maximum absolute error across all cross-bred animals and all possible breeds, and S is the mean of the squared estimation error calculated across all animals and possible breeds.

Number of cross-overs	CGR		ADM	
	M	S	M	S
5	0.24691	0.00103	0.85393	0.01578
25	0.28217	0.00107	0.67077	0.00566

CGR needed about 16 real time seconds for estimating the pure breed proportions of all 5000 cross-bred animals, whereas ADM needed 292 and 336 real time seconds for the 5 and 25 cross-over data set, respectively, which is an increase in processing time by a factor of 20. Note that the processing time was obtained without exploiting the parallel processing capabilities of both algorithms.

DISCUSSION

Results show that when pure breed population allele frequencies are known, the less elaborate modelling approach of CGR performs better than the ADM approach. Both algorithms do not account for linkage disequilibrium between marker. However, in addition to not assuming any LD between markers, the likelihood formulation of the ADM algorithm assumes also orthogonality between pure breed population allele frequency vectors. While this might be the case between very distant breeds having diverged many generations ago, it is unlikely to be the case for commercial beef cattle breeds. While CGR in its current formulation is not accounting for LD explicitly, it accounts for non-orthogonality between pure breed allele frequency vectors which might be one reason for the better performance. However, CGR could also account for LD by reformulating formula 1 to a generalised least square problem with the co-variance matrix of vector y reflecting the LD between markers, although this approach is limited by the number of markers. Beside better performance CGR generated more accurate results in a processing real time of only 5 % of that of ADM. This will becoming even more relevant when the number of marker used increases to 50k or more.

CONCLUSION

The results show that the simple modelling approach implemented in CGR provides accurate estimations of breed proportions in cross-bred animals. Moreover, CGR proved to be robust against LD, accounts for non-orthogonality of allele frequency vectors of founder breeds and is fast enough to deliver results for tens of thousands of animals in a reasonable time.

REFERENCES

Alexander D. H., Novembre J. and Lange K. (2009) *Genome Res.* **19**(9):1655.
Chiang C., Gajdos Z., Korn J. M., Kuruvilla F. G., Butler J. L., Hackett R., Guiducci C., Nguyen T. T., Wilks R., Forrester T. et al. (2010) *PLoS Genet.* **6**(3):e1000866.
Johnson S. G. (2011) The NLopt nonlinear-optimisation package.
Kuehn L., Keele J., Bennett G., McDaneld T., Smith T., Snelling W., Sonstegard T. and Thallman R. (2011) *J. Anim. Sci.* **89**(6):1742.
Pritchard J. K., Stephens M. and Donnelly P. (2000) *Genetics* **155**(2):945.
Raj A., Stephens M. and Pritchard J. K. (2014) *Genetics* **197**(2):573.
Tang H., Peng J., Wang P. and Risch N. J. (2005) *Genet. Epidemiol.* **28**(4):289.

Proc. Assoc. Advmt. Anim. Breed. Genet. 22:101-104

POSSIBILITIES OF BINOMIAL PROBABILISTIC PRINCIPAL COMPONENT MODELS TO IDENTIFY GROUPS IN GENOTYPED POPULATIONS

J.B. Holmes[1], K.G. Dodds[2], and M.A. Lee[1]

[1] Department of Mathematics and Statistics, University of Otago, Dunedin, New Zealand
[2] AgResearch, Invermay Agricultural Centre, Mosgiel, New Zealand

SUMMARY

There has been extensive research, particularly in human genetics, devoted to the development of methods that use genotype data for the identification of distinct genetic sub-populations within the population of interest. Some of these methods have also been incorporated in the field of animal breeding in order to improve the accuracy of predicted breeding values through their use as genetic group effects. In this paper, we compared a method of finding sub-populations based on a decomposition of a normalised matrix derived from genotype data, to a modified probabilistic PCA model that took into account the non-normal nature of the genotype data. In an initial study, where we used a dataset from the New Zealand sheep industry with a known breed composition, we found that the modified probabilistic PCA model gave equivalent separation between breeds to EIGENSTRAT.

INTRODUCTION

Livestock programs aim to optimise long-term genetic gain. To do this the ideal is for breeding values to be as accurate as possible. One method of improving breeding value accuracy is through the fitting of genetic groups. However in practice, genetic groups often prove difficult to define (Kuehn *et al.* 2007).

With the increased availability of genotype data, there has been a move towards replacing pedigree records with genotype data for the construction of the relationship matrix to improve breeding value accuracy. In addition there have been attempts to use genotype data to define structure within the population of interest, which is then fitted in the model, usually as a fixed effect. An example of this is EIGENSTRAT (Patterson *et al.* 2006), which in practice is very similar to the eigen-decomposition of the second genomic relationship matrix proposed in VanRaden (2008). This method ignores the non-normal nature of the genotype data and has been shown to reduce across breed accuracy when used as a genetic group (Daetwyler *et al.* 2012).

To deal with the issues outlined, we propose a probabilistic PCA model that explicitly takes into account the ideal conditions of binomially distributed genotypes. We then compared the two methods, focusing on their respective ability to distinguish between genetic groups, which we took to correspond to the recorded breed.

MATERIALS AND METHODS

Data. The genotype data (5K Illumina SNP Chip) available was from 8,902 animals born from 2000 to 2014, each with up to 5,283 markers recorded. Genotypes which were missing for more than 1 % of animals or monomorphic for all animals were omitted from analysis. The removal of animals with any missing genotypes reduced the dataset to 1,672 animals with 5,170 markers recorded. Breed composition data was obtained from Sheep Improvement Limited (SIL). The distribution of breeds in the dataset is indicated on Table 1.

EIGENSTRAT. This method of identifying population structure was introduced in Patterson *et al.* (2006). It assumes a $n \times m$ matrix of genotypes $\boldsymbol{Z}$ with rows corresponding to individuals and columns to markers and coded 0, 1, 2 where the numbers correspond to the number of copies of the A allele. Each column j of $\boldsymbol{Z}$ was then normalised by subtracting by twice the allele frequency p_j and dividing the result by $\sqrt{p_j(1-p_j)}$ to form the matrix $\boldsymbol{M}$. Eigen-decomposition (Principal Component analysis) was then performed on the matrix $\frac{1}{m}\boldsymbol{MM'}$. Determination of population structure was then made using the resulting eigenvectors (Principal components).

Table 1. Breed distribution of genotyped animals as recorded in SIL

Breed distribution of animals			
Breed	Number of animals	Breed	Number of animals
Unknown	11	Perendale	133
Romney	495	Highlander	31
Coopworth	67	Composite	2
Overall distribution of breeds where known			
Breed	% in population	Breed	% in population
Romney	48.13	Poll Dorset	1.28
Coopworth	14.87	East Friesian	1.04
Perendale	13.88	Highlander	3.37
Finnish Landrace	1.12	Composite	3.57
Texel	6.70	Other Breeds	2.53
Suffolk	3.51	(less than 1 % of population)	

Binomial probabilistic principal component analysis (BPPCA). Under ideal conditions of Hardy-Weinberg equilibrium and no linkage disequilibrium, each of the markers *j* observed from individual *i* can be regarded as realisations of a binomial random variable.

$$\boldsymbol{Z}_{ij} \sim Bin(2, p_{ij}) \quad [1]$$

BPPCA assumes that the individual-marker specific allele frequency p_{ij} can be modelled using the link function $\boldsymbol{\theta}_{ij} = \log(p_{ij}/(1-p_{ij}))$ as a function of a marker specific intercept μ_j, f principal components, where f was pre-determined, and an error term. This results in the following model for the observed genotype pattern, where $\boldsymbol{\theta}$ is a $n \times m$ matrix of link functions, $\boldsymbol{L}$ a $n \times f$ matrix of components, $\boldsymbol{F}$ a $f \times m$ matrix of scores, and $\boldsymbol{e}$ is a $n \times m$ matrix of residuals.

$$\boldsymbol{Z}_{ij} \sim Bin(2, (1+\mathrm{e}^{-\theta_{ij}})^{-1})$$
$$\boldsymbol{\theta}_{ij} = \mu_j + \textstyle\sum_k \boldsymbol{L}_{ik}\boldsymbol{F}_{kj} + \boldsymbol{e}_{ij},\ \boldsymbol{F}_{kj} \sim N(0,1),\ \boldsymbol{e}_{ij} \sim N(0,\sigma^2) \quad [2]$$

To fit the model, we used Pólya-gamma data augmentation as outlined in Polson et.al (2013) and previously implemented for a similar model in Klami (2014). This allowed closed form conditional posteriors to be obtained for all model parameters. Based on the eigenvalue scree plot obtained from implementing the EIGENSTRAT method, the number of components to fit was fixed at five. Estimates were obtained from the posterior means found by using a blocked Gibbs sampler based on the conditional posteriors. The Gibbs sampler was stopped once the relative change in $\overline{\boldsymbol{\theta}}$ dropped below 1×10^{-5}. Spectral value decomposition was then applied to the initial estimates to ensure orthogonal components. This ensured comparability of components to those extracted using EIGENSTRAT.

RESULTS AND DISCUSSION

Ability to separate breeds based on principal components. Figure 1 plots the first two principal components obtained from EIGENSTRAT and BPPCA with pure breed animals highlighted. Both methods were able to distinguish between different pure breed populations.

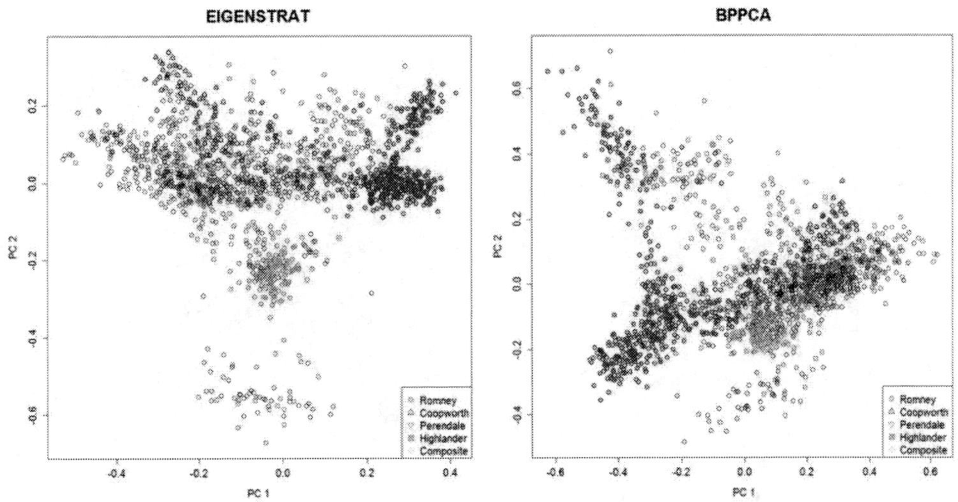

Figure 1. First two principal components obtained from EIGENSTRAT and BPPCA.

Possible uses of the principal components (PC) to represent population structure. Since it is established that principal component analysis on normalised genotype data can distinguish between sub-populations, the fitting of PC has been used extensively to account for population structure in models. The PC are usually fitted as fixed effects. Since EIGENSTRAT extracts PC from the decomposition of the genomic relationship matrix, we suggest that it is more appropriate to fit the PC as random effects. In addition, fitting a decomposition of the genomic relationship matrix in addition to the genomic relationship matrix could be regarded as over-fitting.

In BPPCA, PC are constructed at the link function level, not directly from the observed data. This means that the relationship between the PC and the genomic relationship matrix is indirect. This can be demonstrated by the law of total variance and noting that $E(\boldsymbol{p})$ and $Var(\boldsymbol{p})$, where $\boldsymbol{p}$ is the vector of latent probabilities for each animal, are both functions of the BPPCA PC. It may also mean representing population structure using PC from the BPPCA model is less prone to the reduction of across breed accuracy seen in Daetwyler *et al.* (2012).

$$Var(\boldsymbol{Z}) = E\big(Var(\boldsymbol{Z}|\boldsymbol{p})\big) + Var\big(E(\boldsymbol{Z}|\boldsymbol{p})\big) \quad [3]$$
$$= diag\{E\big(2\boldsymbol{p}(1-\boldsymbol{p})\big)\} + Var(2\boldsymbol{p}) = 2diag\{E(\boldsymbol{p}) - E(\boldsymbol{p})^2 - Var(\boldsymbol{p})\} + 4Var(\boldsymbol{p})$$

If the genotype data can be represented by a low rank matrix factorisation at the link function level, the correlations between animals implied by the PC would be higher (if correlation is positive) or lower (if correlation is negative) than the corresponding correlations in the genomic relationship matrix. However EIGENSTRAT extracts a reduced number of PC, which contain more information about covariance than variance elements. Therefore the implied correlation between random structure effects of different animals is similar between the two methods. This is shown in Figure 2, which shows heat maps of the implied between animal correlation.

Figure 2 shows if PC are used as a classification tool to distinguish between breeds, similar results were obtained from EIGENSTRAT and BPPCA. In our dataset, both clearly identify each pure breed population, sub-groups within the Romneys and classify the animals of unknown breed as Perendale. Corresponding PC extracted by the two methods were highly correlated, except for component 2 and 3, as seen in Table 2. The high negative correlation seen in component 1 and 5 is due to the sign invariance property of estimated loadings in latent factor models.

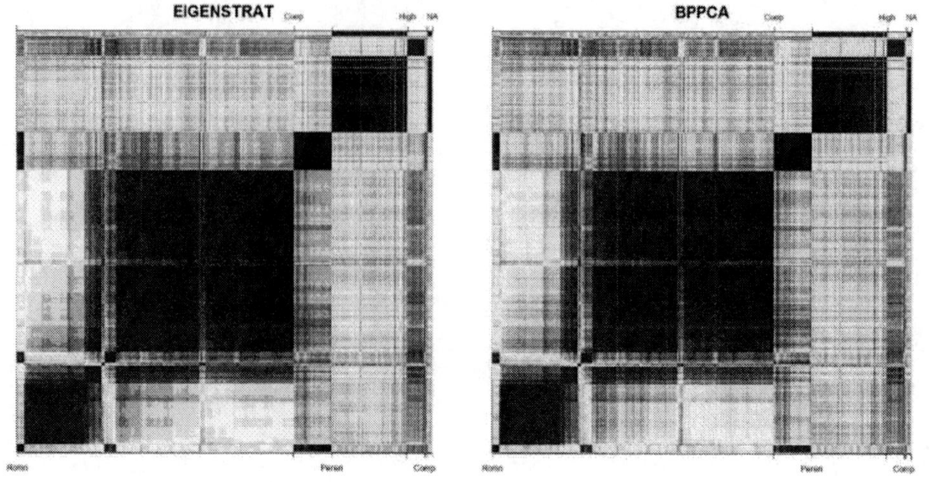

Figure 2. Heat maps of implied correlations between animals that were either of pure or unknown breed. (Dark Red: High positive correlation, Dark Blue: High negative correlation)

Table 2. Correlation between EIGENSTRAT and BPPCA principal components

EIGENSTRAT component	BPPCA component 1	2	3	4	5
1	-0.9922	-0.0342	-0.0062	0.0194	0.0537
2	-0.0299	0.7297	0.6536	0.1353	-0.0483
3	0.0159	-0.6636	0.7352	0.0157	-0.0345
4	0.0130	-0.0895	-0.1086	0.9677	-0.1509
5	-0.0509	-0.0023	-0.0406	-0.1534	-0.9683

Conclusions. BPPCA can be shown to successfully distinguish between different breeds and identify the breed of unknown animals but we did not find substantial differences to EIGENSTRAT for either property. Currently BPPCA is much slower to implement and the challenge will be to determine if the method has advantages in populations with different sub-structure than the example given. In the future, the fitting of principal components from EIGENSTRAT and BPPCA as random effects in a BLUP model can be compared for their efficacy in the prediction of breeding values with respect to accuracy and bias.

ACKNOWLEDGEMENTS

The authors would like to acknowledge the generous support of Beef + Lamb New Zealand Genetics, the Ministry of Business, Innovation and Employment and NZ Sheep breeders.

REFERENCES

Daetwyler H.D., Kemper K.E., van der Werf J.H.J., and Hayes, B.J. (2012) *J. Anim. Sci.* **90**:3375.
Klami A. (2014) *J. Mach. Learn. Res.* **39**: 112.
Kuehn L.A., Lewis, R.M. and Notter, D.R. (2007) *Genet. Sel. Evol.* **39**:225
Patterson N., Price A.L. and Reich D. (2006) *Plos. Genet.* **2(12)**: 2074.
Polson N.G., Scott J.G., and Windle J. (2013) *J. Am. Stat. Assoc.* **108**: 1339.
VanRaden P.M. (2008) *J. Dairy Sci.* **91**: 4414.

Proc. Assoc. Advmt. Anim. Breed. Genet. 22:105-112

THE NEXT WAVE IN SELECTIVE BREEDING: IMPLEMENTING GENOMIC SELECTION IN AQUACULTURE

K.R. Zenger[1,2], M.S. Khatkar[2,3], D.R. Jerry[1,2] and H.W. Raadsma[2,3]

[1]Centre for Sustainable Tropical Fisheries and Aquaculture, College of Science and Engineering, James Cook University, Queensland, Australia.
[2]ARC Research Hub for Advanced Prawn Breeding, James Cook University, Townsville QLD, Australia.
[3]Sydney School of Veterinary Science, Faculty of Science, The University of Sydney, Camden, NSW, Australia.

SUMMARY

Advanced animal breeding in aquaculture has reached a tipping point where the commercial implementation of genomic selection to improve productivity and disease resistance is becoming reality. However, the success of practical implementation of genomic selection depends on the specific aquaculture species, production system and available phenotyping and genetic resources. Using the experience learned from commercial programs for pearl oysters and marine shrimp, we highlight current benefits and options in cost-effective high-throughput genotyping and phenotyping technologies for genomic selection applications relevant to aquaculture species, followed by discussion of some of the lessons learnt when dealing with its practical implementation, including what is needed to build adequate genotype resources for non-model species; confounded breeding objective verse trait measurements; complex traits and unknown interactions; multi-family breeding schemes; multi-stage selection schemes, and transition to a genomic selection breeding program incorporating minimisation of inbreeding.

INTRODUCTION

Classical breeding programs for farmed plant and animal species are based on phenotypic selection of individuals in conjunction with knowledge on genetic relationships and quantitative genetic principles. Breeders have enhanced production traits of farmed species by selecting superior individuals as parents for succeeding generations. However, the efficiency of this method is limited when traits are difficult-to-measure, can only be measured late in life, are sex limited, or have low heritability. Over the past two decades, rapid developments in genomics have resulted in breeders incorporating genetic marker technology in the form of Marker Assisted Selection (MAS) to aid in the animal selection process. Although this technique can be useful for some simple traits, application of MAS to improve complex traits controlled by many genes of small effect is limited. Genetic improvement in these traits can only be achieved through more advanced genomic methods (Eggen 2014).

With recent advances in molecular biotechnology and quantitative analysis methods, it is now possible to accurately predict and use genome-wide molecular breeding values for improved animal selection. This approach is termed Genomic Selection (GS) and was first proposed by Meuwissen *et al.* (2001), and has gained significant application within the animal genetics community. In this approach, animal selection decisions are based on genomic breeding values (GBVs) predicted from genome-wide loci. GS is based on the theory that with sufficiently high numbers of loci across the genome, most quantitative trait loci will be in strong linkage disequilibrium with at least one marker. GS simultaneously estimates the combined genetic effects of all relevant genes and provides accurate predictions of genetic merit for a trait. Furthermore, genome wide markers can be directly use to compute the genomic relationship matrix (GRM), which can then be used to compute genomic breeding values using standard mixed model equations. GRM, even based on a smaller subset of markers, can provide an accurate estimate of the proportion of the genome shared by related individuals and hence provides higher accuracy of estimation of breeding values as compared to estimates based on pedigree information alone (Forni *et al.* 2011).

Integration of GS methods into aquaculture breeding programs promise to rapidly increase genetic gains through improved accuracy of breeding value estimation. GS has the highest potential for traits that cannot be directly measured on the selection candidates and can be used to capture both within- and between-family genetic variances (Nielsen *et al.* 2009). This makes genomic selection a powerful approach in aquaculture, since many traits (eg., disease resistance, carcass quality and pearl quality traits) must be measured on the siblings of the actual selection candidates, rather than the selected candidates themselves. Furthermore, GS can minimise inbreeding while maximising genetic gain beyond that of current practices (Daetwyler *et al.* 2007). This is of particular benefit to aquaculture where species are often highly fecund and the number of contributing families reared in closed farms is low, resulting in rapid inbreeding if pedigree is not tracked (Gjedrem 2005). Despite all of these advantages, a limited number of aquaculture breeding companies and associated research programs are attempting to implement GS into commercial operations for long-term genetic gain (eg., Tsai *et al.* 2017; Khatkar *et al.* 2017a; Jones *et al.* 2017).

The success of the practical implementation of GS in aquaculture production systems depends on the breeding objectives, selection criteria, infrastructure, genomic resources and phenotypic recording / analysis systems. Each of these aspects can have different challenges depending on the specific aquaculture species and production system. Here we aim to provide an overview of the opportunities for the adoption of genomic selection within aquaculture, with particular focus on the challenges of implementation and long-term use in aquaculture commercial systems.

VALUE OF GENOMIC SELECTION IN AQUACULTURE

The breeding design of aquaculture species is primarily governed by the biology of the animal and available farm resources. Commercial selective breeding programs have recently expanded to a diverse range of species (eg., crustaceans such as shrimp, oysters and finfish). Primarily, most aquaculture selection programs have focused on growth, which can be selected easily based on either simple individual, or pedigree family-based selection approaches (eg., between, within and combined). For disease traits or other traits that require destructive sampling, family-based sib-selection is more commonly practised. Sib-selection, whilst allowing family average breeding values to be calculated, only exploits half of the available additive genetic variance (ie., exploits the between family variance), which limits genetic gains, and can also lead to increased inbreeding as not all families are selected to contribute to the next generation stocks.

In aquaculture, GS has been theoretically shown to simultaneously increase genetic gains, while decreasing inbreeding by up to 81% when compared with traditional selection programs (Sonesson and Meuwissen 2009). Although, the monetary value of individual animals of most aquaculture species is generally low (eg., compared to livestock), they are highly fecund and have a relatively short generation interval. This not only provides the ability for varied selective breeding strategies, but also for generating the thousands of phenotypic records required for accurate GS predictions. Furthermore, with a limited number of discrete broodstock capable of producing offspring for the entire production system, the farm effective population size is relatively small. This characteristic enables GS to be implemented on a family-based, or farm-wide basis, utilising a lower density of genome-wide loci compared to outbred populations (see genomic information section below). In aquaculture, GS improvement programs can have a rapid impact on genetic improvement particularly through the use of a structured nucleus breeding scheme. As with traditional selective breeding programs, the potential of GS will vary across different species depending on differences in life cycle, fecundity, effective population size and breeding objectives.

To date the successful application of GS in aquaculture has been limited to a handful of research projects. For example, sea lice resistance in Atlantic salmon (Tsai *et al.* 2016), bacterial cold-water disease resistance in rainbow trout (Vallejo *et al.* 2017) pasteurellosis resistance in gilthead sea bream (Palaiokostas *et al.* 2016) and shell size in scallops (Dou *et al.* 2016). For commercial aquaculture applications of GS, there is limited public information available, and progress is reported here on optimisation and implementation within the authors own programs. Here, GS is being directly integrated into shrimp breeding programs for multiple production traits (eg., size, disease resistance, colour, survival, Khatkar *et al.* 2017a these proceedings), as well as pearl oyster breeding programs for both host oyster and donor oyster traits (eg., shell size and pearl quality traits, Jones *et al.* 2017 these proceedings). Within these programs, the feasibility of successfully applying GS has relied on the availability of high-quality genomic resources, comprehensive information on genetic parameters for all traits and extensive trait phenotype records in the reference population.

COST-EFFECTIVE GENOMIC INFORMATION

In aquaculture breeding, the number of individuals to genotype can be large (particularly for traits with low heritability). Apart from optimising the number of training or selection candidates for routine genotyping (ie., based on GS modelling and farm breeding scheme, eg., Sonesson and Meuwissen 2009), reducing the cost or number of genome-wide markers is a viable solution. Our own data show that derivation of genomic relationships can be achieved with relatively low-density SNP panels (Figure 1; 1,000-3,000 SNPs;) compared to those derived from medium-to high density SNP panels (eg., 50,000+ SNPs; see also Ødegård *et al.* 2014). However, such accuracies deteriorate rapidly if very low-density SNP panels are used (<1,000 SNPs).

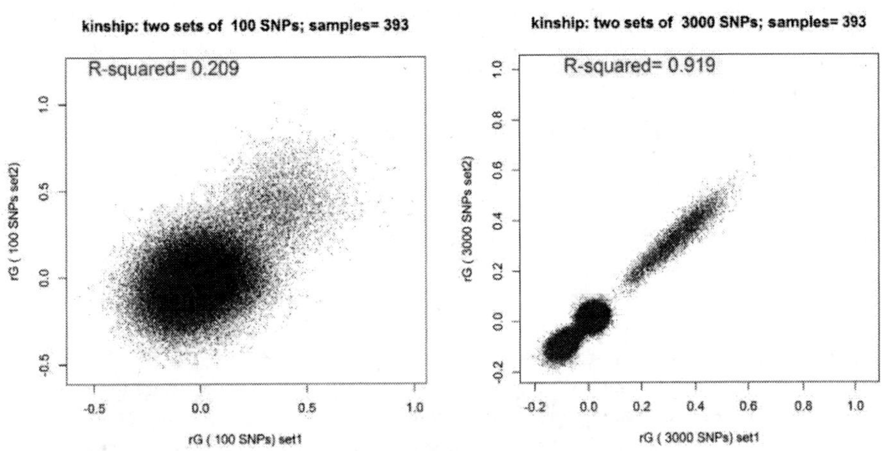

Figure 1. Comparison of SNP based kinship estimates (rG) computed using two independent sets of (a) 100 SNPs and (b) 3,000 SNPs, calculated on 393 shrimp samples.

To our knowledge, there are only a handful of aquaculture species that have commercially available SNP genotyping arrays available (ie., Affymetrix Axiom Salmon genotyping array, Affymetrix Axiom Trout genotyping array and Illumina Infinium ShrimpLD-24 genotyping array). The lack of commercially available genotyping SNP arrays for aquaculture adds significant additional cost to GS genotyping, as these resources need to be first development and tested. However, the recent development of high-throughput and cost-effective genotyping by sequencing (GBS) technologies has significantly reduced both the cost of developing and genotyping SNPs for non-model species (eg., Lind *et al.* 2017). As such, GBS is rapidly becoming the methodology of choice for aquaculture species (Robledo *et al.* 2017). Compared to SNP array based genotyping platforms, GBS requires significantly more quality control (QC) measures to ensure robust genotype data is produced. This is primarily a result of the molecular technique itself, which can introduce spurious and missing data when proper control and data filtering methods are not put in place. Aquaculture species can be particularly sensitive to these anomalies given their sometimes highly polymorphic and repetitive genome structures, a problem particularly observed for crustaceans and oysters (eg., Yu *et al.* 2015; Lal *et al.* 2016).

Another method to reduce the cost of genotyping is through imputation of genotypes, where most of the animals can be genotyped with a low-cost, low-density SNP panel. The genotypes of these animals can be imputed up to high-density by using information on a smaller number of reference individuals (typically broodstock) genotyped with a larger high-density SNP panel that also captures the same SNP as represented on smaller arrays. Such imputed *in-silico* genotypes can then be used for GS and other genomic analyses. Such strategies have been shown to improve the accuracy of GS in livestock (Khatkar *et al.* 2012) and aquaculture species (Tsai *et al.* 2017).

The number of individuals in the reference panel and number of markers in the low-density panel depends on the effective population size of the breeding stock and relationship between reference and test populations. A small effective population size, as present in many aquaculture stocks, will require smaller number of animals in the reference panel and can be imputed with high accuracy with smaller number of SNPs in the low density panel. Moreover, if all the contributing broodstock are genotyped with the high-density panel, the accuracy of imputation in the progeny, genotyped with even smaller SNP panel, could be quite high using a pedigree based imputation approach (Hickey *et al.* 2012). However, accurate imputation requires knowledge about the precise location of SNPs across the genome. For most aquaculture species genetic linkage maps and / or genome assemblies are in the early stages of development (Abdelrahman *et al.* 2017).

NEXT-GENERATION PHENOTYPING

Accurate phenotypes on commercially important traits are critical for any breeding program. This becomes especially challenging in aquaculture where large numbers of animals need to be recorded. Any error in the trait recording will reduce effective estimated heritability and hence realised genetic gain. High-throughput and precise phenotyping strategies are required to supply the large amount of trait data required for commercial scale GS applications. Within this framework, the objective is to increase the accuracy, precision and throughput of phenotypic assessment while reducing costs and minimising labour in an intensive production system. Today, phenotyping is quickly emerging as the major operational bottleneck limiting the power and speed of commercial GS programs (eg., Cobb *et al.* 2013). This problem is compounded in aquaculture where fecundity, progeny numbers from breeding pairs and variable survival rates create circumstances where individual phenotypes and traceability are nearly impossible to obtain without new methodologies. Furthermore, aquaculture does not have the benefit of standardised global phenotyping programs such as in livestock (eg., dairy cattle). Designing effective on-farm phenotyping strategies requires integrated solutions incorporating biologists, computer scientists, statisticians and engineers.

More recently, automation, imaging and software developments have paved the way for many quantitative phenotyping studies. Within these developments, digital imaging has emerged as a cornerstone to capturing quantitative phenotypic information. Visual imaging has already allowed many production traits to be measured efficiently and accurately across different production industries including aquaculture (Cobb *et al.* 2013; Saberioon e*t al.* 2016). For example, fish length has been estimated in Rainbow trout (Miranda and Romero 2017) and fish mass in Jade Perch (Viazzi *et al.* 2015) with very low residual errors using automated computer vision techniques. Furthermore, fish skin colour and pearl quality traits (eg., colour, lustre, completion), which traditionally are recorded as categorical traits, can now be recorded as highly-reliable continuous quantitative traits based on UV-Vis spectrophotometry (eg., Kustrin and Morton 2015), which ultimately will improve GS predictions. Other emerging aquaculture phenotyping techniques are Near Infra Red (NIR) spectroscopy and Hyperspectral imaging (HIS) which combines spectroscopy with imaging technology. These techniques are able to quantify and evaluate the chemical (eg., fat, protein, moisture) and physical (eg., freshness, texture, colour) attributes of aquatic animals with relatively high accuracies of prediction ($r > 0.8$, see Liu *et al.* 2013; Saberioon *et al.* 2016). All of these machine vision systems (MVS) are able to extract and analyse quantitative information from digital images and have the ability to improve the accuracy of the phenotype by electronically analysing the data at a pixel level across spectral regions not always visible to the human eye.

MVS usually consists of two components, the image acquisition system hardware (ie., UV-Vis, NIR and HIS) and data extraction system software. The latter typically incorporates computer based processing and optimised statistical methods and algorithms specific for the trait of interest, which is often the limiting factor in applying MVS. The development of advanced image analysis software including artificial neural network (ANN) algorithms based on machine learning approaches has been an important step forward in the development of analysis systems for automated MVS phenotyping (eg., Grys *et al.* 2016).

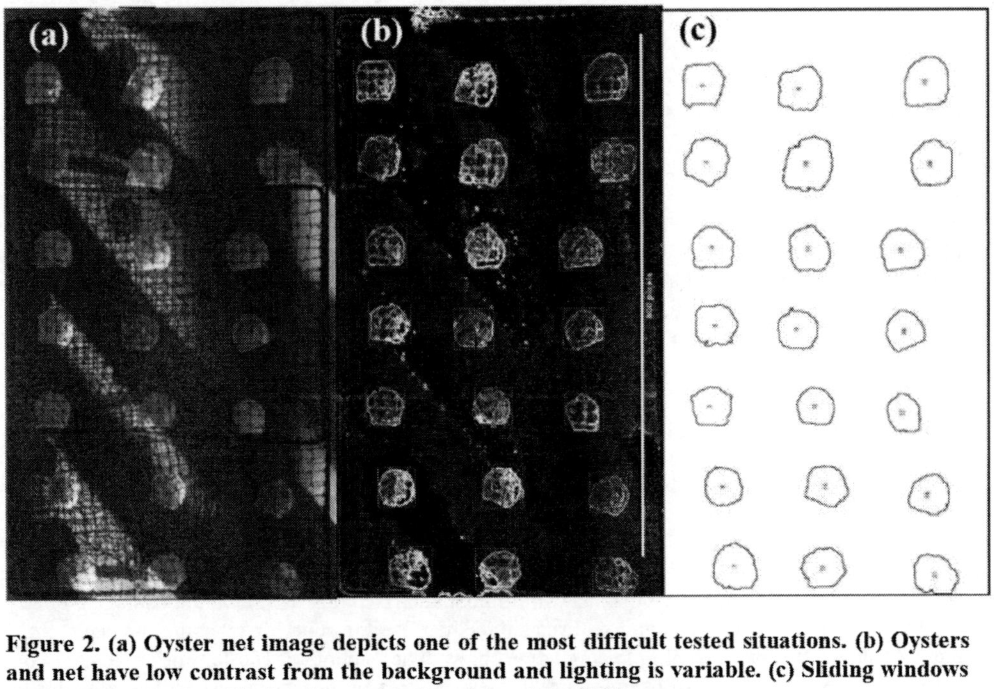

Figure 2. (a) Oyster net image depicts one of the most difficult tested situations. (b) Oysters and net have low contrast from the background and lighting is variable. (c) Sliding windows CNNs correctly identified and measured oysters with >93% accuracy.

Within our own research programs (ie., for marine shrimp and pearl oyster), machine learning algorithms have allowed precise inexpensive phenotyping across diverse production traits. For example, MVS systems have been used for pearl oyster growth data as well as pearl quality traits (eg., colour, size, lustre, completion). Although still in development, sliding window algorithms and Convolutional Neural Network (CNN) with rule-association based clustering yielded high accuracy (exceeding 93%) in Object Character Recognition (OCR) for the oysters in nets within the full spectrum of commercial situations (Figure 2; P. Toole unpub. data). By definition, CNN learning algorithms get more precise when presented with more data. This supervised learning approach has been undertaken with developing methodologies on how to automate the entry of commercial data into a noSQL or graph-based database.

IMPLEMENTING GENOMIC SELECTION ON FARM – LESSONS LEARNT

Greatest immediate value from genomic selection is realised where genomic breeding values can be targeted against traits that drive economic returns to commercial farmers. Typically such traits are based on yields of harvested product. Although this sounds straightforward enough, practical limitations become immediately apparent in situations where traits under commercial grow-out conditions vary substantially from performance recording environments in often pathogen-free central nucleus breeding facilities (as used in specific pathogen free shrimp breeding programs for instance). For most aquaculture systems the Genotype by Environment (GxE) interactions are largely unknown and limit the value of GS training data if the genetic correlation between the central nucleus breeding values and on-farm breeding values is significantly less than unity (ie., < 0.6). Fortunately, genomic selection platforms allow for field data to be linked to nucleus broodstock through DNA derived genomic relationships and on-farm phenotyping. Secondly, genomic selection programs become increasingly more complex when harvest yields are determined by diverse genomes, as is the case of pearl oyster, with a host recipient seeded with the saibo of a donor. The need to have accurate breeding values for both host and donor oyster may eventually result in the need of separate breeding lines for both. Unknown interactions between host and donor further complicate the application of genomic selection if such epistatic effects are significantly greater than zero. In the case of pearl oyster the multi-factorial nature of pearl value adds to the complexity of setting up multi-trait genomic selection. Thirdly, and potentially of greatest commercial appeal for genomic selection is to build disease resistance into the genetic improvement program as has been highlighted above. Most central nucleus breeding programs are pathogen free and breeding decisions are based on family sib-selection, but commercial grow out environments are under constant disease challenge. It is unlikely that simply screening commercial stocks will yield data of sufficient quality to obtain genomic breeding values for disease resistance, since most disease field challenges are uncontrolled, and often resistance to multiple pathogens is of interest. One potential solution is to expose large mixed-family progeny cohorts to standardised disease challenge and ascertain survival statistics from pooled genotype data pre- and post-challenge. Finally, it is almost certain that for most genomic selection programs, there will be a need for ongoing phenotyping to update the training sets, and cross validate data collected under diverse commercial environments and to monitor unfavourable genetic correlated responses.

Perhaps one of the greatest advantages offered by application of genomic selection over conventional breeding programs, is that large-scale multi-family data can be resolved retrospectively through genomic relationships. This has two immediate and highly significant advantages. Firstly, the predicted genetic response and realised inbreeding are far superior over the management of multiple single-family lines. Simple simulation shows that a cohort of 100 families in a single line outperforms the average of 100 single-family lines and creates long-term sustainable value for the industry (Khatkar *et al.* 2017b, these proceeding). Secondly, the enormous costs in establishing and maintaining single-family mating, spawning and rearing facilities are not required under a genomic selection program using a large scale multi-family breeding program. In many cases the commercial infrastructure for propagation is sufficient, and the cost saving outweighs the cost incurred for genotyping.

In our experience, the transition from existing/traditional selection programs into a genomic selection program is challenging since most mating and infrastructure designs in central nucleus breeding facilities do not capture the advantages offered by genomic selection programs. In the case where simple mass produced commercial stocks are produced, or where no genetic improvement programs are in place, imposing a genomic selection program is potentially straightforward. The main requirement is that the species is domesticated, since lifecycles need to be closed for ongoing selection and capture of genetic gain. Where source broodstock has been harvested from wild stock, the base generation needs to be adequately represented in the foundation stocks, and inclusion of "new" ongoing sampling of wild stocks limited. Once an adequate training data set against commercially well-defined breeding objectives has been completed, a robust test-set and validation phase is required to determine the accuracy of the genomic predictions. For easy to measure traits of moderate to high heritability, this is relatively easy to achieve; however, for most, if not all diseases, and complex multi-factorial traits, the development of adequate training data sets will remain a logistical challenge. Of practical concern is also how best to use available information. For most applications, genotyping potential candidates under selection remains a significant cost. The use of multi-stage selection, based on simple phenotypic selection as a primary selection, followed by genomic sampling (DNA sampling genotyping and tracking tagged individuals) and selection is likely the most cost-effective application of this technology (Khatkar *et al.* 2017b these proceeding). Other applications of genomic selection include the genomic management to minimize inbreeding by candidate selection and mate allocation to maximize genomic diversity. Genomic selection also offers an additional commercial benefit, to pre-screen females and males in the current generation for production of commercial animals, given that relatively few females are needed to generate the many millions of larvae for commercial production. The exact benefits of GS breeding programs will be dependent on the species and nature of the aquaculture enterprise.

REFERENCES

Abdelrahman H., ElHady M., Alcivar-Warren A., Allen S., et al. (2017). *BMC Genomics*, **18**:191, doi:10.1186/s12864-017-3557-1.

Cobb J.N., Declerck G., Greenberg A., Clark R., McCouch S. (2013). *Theoretical Applied Genetics*, **126**:867–887.

Daetwyler H.D., Villanueva B., Woolliams J.A. (2007). *Journal of Animal Breeding Genetics*, **124**:369-376.

Dou J., Li X., Fu Q., Jiao W., Li Y., et al. (2016). *Scientific Reports*, **6**:19244. doi:10.1038/srep19244.

Eggen A. (2012). Animal Frontiers 2:10–15.

Forni, S., Aguilar I., Misztal I. (2011). *Genet Sel Evol*, **43**:1, doi:10.1186/1297-9686-43-1.

Gjedrem, T. (2005) In: 'Selection and Breeding Programs in Aquaculture', editor Gjedrem T., Springer, Berlin, Heidelberg.

Grys B.T., Lo D.S., Sahin N., Kraus O.Z., Morris Q., Boone C., Andrews B. (2016). *Journal of Cell Biology*, doi:10.1083/jcb.201610026.

Hayes B.J., Pryce J., Chamberlain A.J., Bowman P.J., Goddard M.E. (2010). *PLoS Genet* **6**(9): e1001139.

Hickey, J.M., Kinghorn, B.P., Tier, B., van der Werf, J.H.J., Cleveland, M.A. (2012). *Genet Sel Evol*, **44**:9, doi:10.1186/1297-9686-44-9.

Jones D.B., Toole P.B., Khatkar M.S., Raadsma H.W., Jerry D.R., Zenger K.R. (2017). Proceedings of the AAABG 22nd Conference 2017, 2nd-5th July 2017, Townsville, QLD, Australia.

Khatkar, M. S., G. Moser, B. J. Hayes and H. W. Raadsma (2012). *BMC Genomics* **13**: 538, doi:10.1186/1471-2164-13-538.

Aquaculture

Khatkar M.S., Zenger K.R., Jones D.B., van der Steen H.A.M., Jerry D.R., Raadsma H.W. (2017a). Proceedings of the AAABG 22nd Conference 2017, 2nd-5th July 2017, Townsville, QLD, Australia.

Khatkar M.S., Coman,G.J., Thomson, P.C., Raadsma H.W. (2017b). Proceedings of the AAABG 22nd Conference 2017, 2nd-5th July 2017, Townsville, QLD, Australia.

Kustrin S.A., Morton D.W. (2015) *Modern Chemistry Applications*, **3**:152, doi:10.4172/2329-6798.1000152.

Lal M.M., Southgate P.C., Jerry D.R., Zenger K.R. (2016). *Marine Genomics* **25**:57–68.

Lind, C. E., Kilian A., Benzie J.A.H. (2017). *Animal Genetics* **48**:362-364.

Liu D., Zeng X.A., Sun D.W. (2013). *Applied Spectroscopy Reviews*, **48**:609-628.

Ødegård, J., Moen T., Santi N., Korsvoll S.A., Kjøglum S., Meuwissen T.H.E. (2014) *Frontiers in Genetics*, **5**: 402

Meuwissen T.H.E., Hayes B.J., Goddard M.E. (2001). *Genetics*, **157**:1819-1829.

Miranda J.M. and Romero M. (2017). *Aquaculture Engineering*, **76**:41-49.

Nielsen H.M., Sonesson A.K., Yazdi H., Meuwissen T.H.E. (2009). *Aquaculture*, **289**:259-264.

Palaiokostas C., Ferraresso S., Franch R., Houston R.D., Bargelloni L. (2016) *G3 (Bethesda)* **6**: 3693–3700.

Robledo D., Palaiokostas C., Bargelloni L., Martínez P., Houston R. D. (2017). *Reviews in Aquaculture*. **0**: 1-13. doi:10.1111/raq.12193.

Sonesson A.K., Meuwissen T.H.E. (2009). *Genet Sel Evol* **41**:37, doi:10.1186/1297-9686-41-37

Tsai H.Y., Hamilton A., Tinch A.E., Guy D.R., Bron J.E., Taggart J.B. et al. (2016). *Genet Sel Evol* **48**:47, doi:10.1186/s12711-016-0226-9.

Tsai, H.Y., Matika O., Edwards S. M., Antolin-Sanchez R., Hamilton A., Guy D. R., Tinch A. E. K. Gharbi, M. J. Stear, J. B. Taggart, J. E. Bron, J. M. Hickey and R. D. Houston (2017). *G3 (Bethesda)* **7**(4): 1377-1383.

Vallejo R.L., Leeds T.D., Gao G., Parsons J.E., Martin K.E., Evenhuis J.P., Fragomeni B.O., Wiens G.D., Palti Y. (2017). *Genet Sel Evol.* **49**:17, doi:10.1186/s12711-017-0293-6.

Saberioon, M., Gholizadeh, A., Cisar, P., Pautsina, A. and Urban, J. (2016). *Reviews in Aquaculture*, doi:10.1111/raq.12143.

Viazzi S., Van Hoestenberghe S., Goddeeris B.M., Berckmans D. (2015). *Aquacultural Engineering*, **64**: 42–48.

Yu Y., Zhang X., Yuan J., Li F., Chen X., et.al. (2015). *Scientific Reports*, **5**:15612, doi:10.1038/srep15612.

Proc. Assoc. Advmt. Anim. Breed. Genet. 22:113-116

UNDERSTANDING THE TRANSCRIPTIONAL CHANGES ASSOCIATED WITH ONSET OF MATURATION IN ATLANTIC SALMON

A. Mohamed[1], A. Reverter[1], P. Kube[2], H. King[2], B. Evans[3] and J. Kijas[1]

[1]CSIRO Agriculture and Food, Queensland Bioscience Precinct, St Lucia, QLD 4067, Australia
[2]CSIRO Agriculture, CSIRO, Hobart, Tasmania, Australia
[3]Tassal Ltd., Hobart, Tasmania, Australia

SUMMARY

Sexual maturation of Atlantic salmon, *Salmo salar*, is a complicated process that involves many variables that can act to activate and/or inhibit sexual development. Unwanted early onset of sexual maturation of Atlantic salmon is a challenge for the salmon aquaculture industry, as it has negative impacts on growth rate and product quality. Consequently, there has been a significant amount of research aiming to understand the biological mechanisms driving early salmon maturation. We present the description of a proposed animal trial, coupled with RNA-Seq based transcriptomics, designed to elucidate the earliest triggers which commit animals to sexual maturation. Our approach has two major components. First, animals will be photoperiod manipulated to artificially narrow the time window during which maturation is initiated. Tissue samples will be collected before, during and after the initiation event. The second component involves next generation sequencing to obtain detailed gene expression profiles. We will target the brain, pituitary and gonad tissues as the brain-pituitary-gonad (BPG) axis is central to regulating sexual maturation. We anticipate our approach has the potential to both identify the genes involved, and open new approaches to control the timing of maturation in this important production species.

INTRODUCTION

Sexual maturation is the process by which organisms become mature and are capable of reproducing. In Atlantic salmon, the development of sexual maturation is complex, with extreme variability in age and size at maturation (Good and Davidson 2015). Moreover, the variability in timing of maturation is considered a significant problem to Atlantic salmon aquaculture, specifically Atlantic salmon that mature at an early age are more susceptible to opportunistic microbes (St-Hilaire *et al.* 1998), exhibit decreased feed conversion efficiency and lower than normal growth rate (McClure *et al.* 2007), and have reduced product quality (Aksnes *et al.* 1986). In salmon industry, photoperiod management is the general practice to control animal maturation. The brain-pituitary-gonad (BPG) axis is a key regulator of sexual development in vertebrates. Activation of neurons in the hypothalamus leads to production of gonadotropin releasing hormones (GnRH), which stimulate the release of gonadotropins such as follicle stimulating hormone (FSH) and luteinising hormone (LH) from the pituitary gland. In the gonads, gonadotropins induce the production of gonadal steroids (e.g., testosterone, estrogen and progesterone), which in turn affect various aspects of sex-related physiology, secondary sexual characteristics and behaviour. Consequently, analysing transcriptomic changes in the BPG axis during the early stages of sexual maturation in Atlantic salmon could identify differentially expressed genes and gene co-expression networks operating to control the process.

It is possible maturation is inhibited during the juvenile life stages until specific physiological/biochemical thresholds are attained. The thresholds include, for example, levels of adipose tissue (Rowe *et al.* 1991) and energy reserves (Kadri *et al.* 1996), which provide information about the optimal fitness and triggers a developmental switch towards maturation. These thresholds are influenced by environmental factors (Taranger *et al.* 2010) of which photoperiod is considered an essential determinant for initiating sexual maturation in teleosts including Atlantic salmon

(Bromage *et al.* 2001). Photoperiod effects facilitate optimal timing of conditions that favour growth and survival of young animals. Moreover, those physiological thresholds are genetically determined to some extent. For example, Barson *et al.* (2015) identified a single locus in the Atlantic salmon genome that is associated with age at maturity through a genome wide association study. The causal gene is likely to be the vestigial-like family member 3 gene (*VGLL3*), which has a role in adiposity, however its precise role is yet to be determined.

The mechanisms underlying the onset of maturation are not understood in Atlantic salmon. This is primarily because it is difficult to sample animals as they commit to the maturation pathway. This project describes an animal experiment designed to identify the genes, gene expression differences and gene networks driving initiation of sexual maturation in Atlantic salmon.

MATERIALS AND METHODS

Experimental design. In order to maximise the probability of sampling animals during the earliest stages of the maturation process, well before the appearance of the phenotypic changes associated with maturing fish, the decision window for animals to initiate maturation should be as short as possible. Consequently animals will be managed via photoperiod manipulation to synchronise the timing of commitment into maturation. We will study a population of female broodstock that will be approximately 36 months post fertilization in April 2017 (~3.1 to 3.4 yrs at sampling). The proposed management of the animals and associated timeline is given in Fig.1.

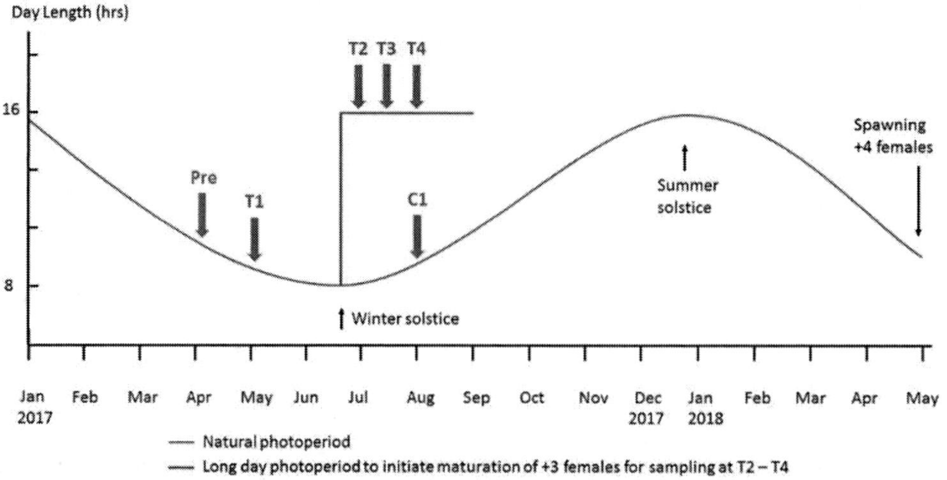

Figure 1. Induction of maturation through photoperiod manipulation and suggested time points for tissue sampling and RNA isolation.

Sampling. Activation of the brain-pituitary-gonad axis is central to reproductive development and prioritises the three target tissues for examination in the experiment. An expanded set of tissues (liver, spleen and muscle) may be harvested at sampling, however these three are the focus for RNA-Seq data generation. In order to measure and control for variation between individuals, we propose to sample 4 fish at each of the T1 – T4 and the C1 time point (total of 20 fish). This will enable variation within tissues and time points to be evaluated. The maturation status of animals (leading up to the long day photoperiod initiation) is currently being monitored by ultrasound. Ultrasound data and update on the T1 samples will be presented at the conference.

Transcriptomic data generation and high-throughput sequencing. A total of 60 RNA libraries will be generated arising from 5 (time points) x 3 (tissues; brain, pituitary, ovary) x 4 (biological replicates). RNA-Seq libraries will be prepared using the Illumina TruSeq RNA Sample Preparation Kit. RNA-Seq libraries will be sequenced on four lanes on an Illumina HiSeq platform. Sequencing should produce (at least) 25 million individual 100-bp paired-end reads per library. RNA-Seq data will be processed and analysed for differential expression in response to the onset of maturation Fig.2.

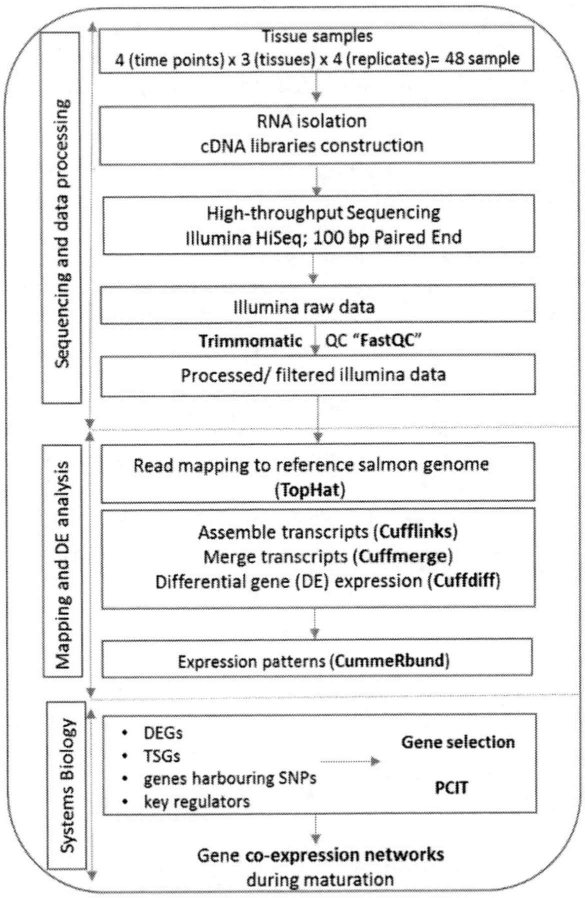

Figure 2. Flow chart of the pipeline for RNA-Seq (transcriptomic) data generation, identification of differentially expressed genes and subsequent gene co-expression networks analyses.

RNA-Seq data processing and differential expression analysis. Illumina raw reads will be checked using FastQC, a quality control tool for NGS data. Illumina universal and indexed adapters will be removed and data will be filtered based on quality using Trimmomatic software (Bolger *et al.* 2014). Illumina reads will be analysed according to the Tuxedo protocol (Trapnell *et al.* 2012). Briefly, the processed Illumina reads will be mapped separately against the salmon reference genome (Lien *et al.* 2016) using TopHat, a gapped/ spliced mapper, in order to generate alignment (accepted_hits.bam) files. Then the Cufflinks suite will be used for differential expression. First transcripts will be assembled and quantified using cufflinks, then transcripts will be merged into a single transcriptome using cuffmerge and differential expression will be calculated using cuffdiff. The R package CummeRbund will be then used to explore the gene expression data and create volcano plots and heatmaps to visualise the differential expression. The list of differentially expressed genes (DEGs) will be analysed using GO and KEGG databases for pathway enrichment among the gene lists.

Gene co-expression network analysis. Gene co-expression networks will be analysed as described in (Canovas *et al.* 2014). Briefly, in addition to the list of DEGs, tissue-specific genes (TSGs), genes harbouring SNPs reported to be associated with maturation traits and key regulators such as transcription factors (TF) will be used to generate and analyse gene co-expression networks. The DEGs, TSGs, key TF and SNP harbouring genes will be used as nodes and significant connections will be identified using the partial correlation and information theory (PCIT) algorithm (Watson-Haigh *et al.* 2010) in the R environment. The PCIT ascertains the correlation between genes and network nodes after taking into account all other genes present in the dataset. The PCIT output will be viewed with Cytoscape, a software for analysis and visualisation of gene co-expression networks (Shannon *et al.* 2003). The highly interconnected gene clusters and significantly overrepresented Gene Ontology terms will be identified. Those clusters may be of biological significance to maturation in Atlantic salmon. The analysis flowchart is summarized in Fig.2.

CONCLUSIONS

Execution of the proposed experiment will generate a tissue collection and a large transcriptomic dataset that has not yet been obtained by the research community. The project is focused to investigate the biological mechanisms driving the onset of sexual development, with a view to developing novel approaches to assist management of unwanted early maturation within the Atlantic salmon industry.

REFERENCES

Aksnes A., Gjerde B. and Roald S. (1986). *Aquaculture* **53**:7–20.

Barson J., Aykanat T., Hindar K., *et al.* (2015). *Nature.* **528**:405–408.

Bolger M., Lohse M., Usade B. (2014). *Bioinformatics.* **30**:15.

Bromage N., Porter M., and Randall C. (2001). *Aquaculture.* **197**:63–98.

Cánovas A., Reverter A., DeAtley L., *et al.* (2014). *PLoS ONE.* **9**:7.

Good C. and Davidson J. (2015). *JWAS.* **47**:5.

Kadri S., Mitchell F., Metcalfe N., Huntingford F. and Thorpe J. (1996). *Aquaculture,* **142**:245–257.

Lien S., Koop B., Sandve S., *et al.* (2016). *Nature.* **533**, 200–205.

McClure A., Hammell L., Moore M., Dohoo R. and Burnley H. (2007). *Aquaculture.* **272**:370–379.

Rowe, K., Thorpe J., and Shanks A. (1991). *Can J Fish Aquat Sci.* **48**:405–413.

Shannon P., Markiel A., Ozier O., et al. (2003). *Genome Res.***13**: 2498-2504

St-Hilaire S., Ribble C., Whitaker D. and Kent M. (1998). *Aquaculture.* **162**: 69–77.

Taranger L., Carillo M., Schulz R., *et al.* (2010). *Gen Comp Endocrinol.* **165**: 483–515.

Trapnell C, Williams A, Pertea G, *et al.* (2010). *Nature Biotechnology.* **28**, 511–515.

Watson-Haigh N., Kadarmideen H. and Reverter A. (2010). *Bioinformatics.* **26**:3.

Proc. Assoc. Advmt. Anim. Breed. Genet. 22:117-120

DEVELOPMENT OF A LOW-DENSITY COMMERCIAL GENOTYPING ARRAY FOR THE WHITE LEGGED SHRIMP, *LITOPENAEUS VANNAMEI*

D.B. Jones[1], K.R. Zenger[1,2], M.S Khatkar[2,3], H.W. Raadsma [2,3], H.A.M. van der Steen[4], J. Prochaska[4,5] , S. Forêt[6] and D.R. Jerry[1,2]

[1] Centre for Sustainable Tropical Fisheries and Aquaculture, College of Science and Engineering, James Cook University, Queensland, Australia
[2] ARC Hub for Advanced Prawn Breeding, James Cook University, Townsville QLD, Australia
[3] Sydney School of Veterinary Science, Faculty of Science, The University of Sydney, Camden, NSW, Australia
[4] Global Gen, Jalan Raya Narogong Km 14, Desa Cikiwul Bantar Gebang Bekasi, 17310 Indonesia
[5] Amity Aquaculture, LLC. 1603 Capitol Ave., Suite #310 A317, Cheyenne, WY 82001 USA
[6] ARC Centre of Excellence for Coral Reef Studies, James Cook University, Townsville, QLD, Australia

SUMMARY

The Pacific whiteleg shrimp, *Litopenaeus vannamei*, is the most farmed shrimp species globally. The development of high quality genomic resources including a dense array of genetic markers and genetic maps are pivotal to integrating genomic selection in this species. We describe the development and utility of an Illumina low-density single nucleotide polymorphism (SNP) array (Infinium ShrimpLD-24 v1.0) which is now commercially available. These resources set the foundation for investigating the architecture of complex traits and genomic selection.

INTRODUCTION

The whiteleg shrimp, *Litopenaeus vannamei*, is an intensively farmed species with global production exceeding 3 million tonnes annually (GLOBEFISH 2016). Current breeding programs for *L. vannamei* use traditional phenotypic selection to produce shrimp with enhanced growth and that exhibit-lowered susceptibility to various viral pathogens like Taura syndrome virus (TSV) and White spot syndrome virus (WSSV). While this traditional approach has been moderately successful in producing more productive shrimp strains, genetic progress using multi-trait phenotypic selection in *L. vannamei* is in some cases significantly impeded by an unfavourable genetic correlation between growth and disease, as well as a poor correlative response in susceptibility to multiple diseases (Gitterle *et al.* 2007, Huang *et al.* 2012, Gjedrem 2015). *L. vannamei* is an aquaculture species that would benefit substantially from the integration of genomic information into traditional breeding programs, particularly for disease and growth traits. Recent increased research effort has yielded a number of genome-wide SNP and genome map resources for *L. vannamei* (Ciobanu *et al.* 2010, Du *et al.* 2010, Yu *et al.* 2015). However, none have yet to be made commercially available. Herein, we present a large transcriptome sequence reference assembly with utility for mining over 26,662 high quality SNP markers and a commercially available Illumina Infinium ShrimpLD-24 v1.0 genotyping array with 8,967 SNPs for *L. vannamei*.

MATERIALS AND METHODS

Sequencing, assembly and annotation

To enable the identification and development of genome-wide Type I SNPs, high-quality total RNA was extracted from the pleopod tissue of 30 *L. vannamei* individuals (provided by Global Gen, Indonesia) using TRIZOL® Reagent (Life Technologies). Equimolar pooled RNA was converted to cDNA using the Mint cDNA synthesis kit (Evrogen) and sequenced using an Illumina GA-IIX at 76 bp paired-end resulting in approximately 25 gigabases of paired-end EST sequence data (~10x genome coverage). Sequences were screened using the software Seqclean (https://sourceforge.net/projects/seqclean/) and MOTHUR (Schloss *et al.* 2009). The cleaned sequence data was assembled using Velvet V1.0 (Zerbino *et al.* 2008) and OASES (Schulz *et al.* 2012). Transcript assemblies were conducted at kmer lengths of k39, k41, k43, k45, k47, k49, k51 and k53 before being clustered together at a 90% sequence identify threshold using the software CD-HIT (Li *et al.* 2006). Assembly of the cleaned-up sequence data produced 76,963 contigs (N50 = 2,375 bp and average contig length = 1,429 bp).

SNP Discovery and Filtering

Genome-wide SNPs were identified within SAMTOOLs (Li *et al.* 2009). The varFilter option in SAMTOOLs was employed to filter SNPs, keeping only the most informative (i.e. minor allele frequency (MAF) >0.25, read depth >10 reads, minor allele reads >2, SNP mapping quality >25, flanking sequence quality >25). Any SNP identified within 50 bp of a candidate SNP was excluded to ensure a conservative flanking region for probe design. SNPs with the highest MAF and read depth were submitted for assay development analysis using Illumina's Assay Design Tool (ADT) and included if their ADT score was greater than 0.7. To ensure no unintentional duplicate SNPs were included on the array, probes for each SNP were mapped to the initial assembly using NOVOCRAFT (Novocraft Technologies) and only the probes that mapped uniquely were included.

Infinium Array Genotyping

To validate the performance of the Illumina ShrimpLD-24 v1.0 genotyping array, 1,134 female and 193 male parents of families (produced by Global Gen, Indonesia) were genotyped. To ensure all genotype calls were genuine and to identify aberrant SNP and DNA samples, strict data integrity was undertaken in GenomeStudio V2011.1 following methods outlined in Jones *et al.* (2013). Genotype reproducibility between batches was tested using 52 replicate samples and 26 replicate SNPs. SNPs with a MAF greater than 0.01 were considered polymorphic. SNPs were investigated for conformation to Hardy-Weinberg Equilibrium (HWE) and Mendelian Inheritance (MI) patterns.

To demonstrate the utility of the SNPs included on the Infinium ShrimpLD-24 v1.0 array, we generated a preliminary linkage map using 30 grand-maternal and 19 grand-paternal families containing 15 progeny on average. The linkage map was constructed in Carthagene V1.3 (de Givry *et al.* 2005) using an iterative *buildfw*, *annealing*, *flips 6* and *polish* method until the best map were produced. Finally, genomic relationship matrixes (GRMs) were calculated with subsets of SNPs and the full array to determine the minimum number of SNPs required for genomic selection (GS).

RESULTS AND DISCUSSION

Sequencing and assembly of transcripts

In total, over 25 Gb of sequence data (329 million raw EST sequences, 76 bp paired-end, ~15x genome coverage) was produced from an Illumina GA-IIx run. After sequence trimming, 19.7 Gb of high-quality data was retained. Assembly of remaining sequence data produced 76,963 contigs (N50 = 2,375 bp and the average contig length = 1,429 bp). The average read depth over all contigs was 210 reads with a median of 29. The assembled contig sequences and mapped raw reads have been submitted to GenBank (Accession number: SRP094129). This significant genomic resource enables the mining of over 17,000 additional SNPs not included within any commercial SNP array.

SNP discovery and filtering

From the assembled sequence dataset, 234,452 putative SNPs were identified *in-silico* before strict filtering parameters were applied. By filtering out all SNPs with a read depth less than 10 reads and a MAF of less than 0.25, a total of 26,662 high-quality SNPs were identified. A total of 1,142 SNPs did not return ADT values > 0.7 and 1,006 SNPs did not map to unique contigs and were removed. A further 7,003 SNPs were excluded due to being located within the flanking region of another SNP resulting in a final list of 9,447 high-value SNPs. Of these, the highest scoring 8,967 SNPs [8,616 novel; and 351 developed in Ciobanu *et al.* (2010) and mapped in Du *et al.* (2010)] were incorporated into the Illumina ShrimpLD-24 v1.0 array enabling high throughput, cost effective and accurate genotyping. The average MAF and ADT score of these high-value SNPs was 0.37 and 0.95 respectively. SNPs included on the custom array have been submitted to dbSNP on NCBI [ss2137297825-ss2137306471 (the current study); rs159816077-rs159831399 (Du *et al.* 2010); and rs142459135-rs142459627 (Ciobanu *et al.* 2010)]. The ShrimpLD-24 v1.0 array is available at https://www.illumina.com/products/by-type/microarray-kits/infinium-shrimp-ld.html.

Infinium array genotyping and validation

In total, 1,327 individuals were genotyped on the ShrimpLD-24 v1.0 array. From these samples, 70 (5.3%) individuals produced call rates of less than 90% and were removed from further analysis leaving 1,257 unique individuals to investigate SNP array performance. Analysis of the resulting genotypic data revealed that 6.0% of the SNPs did not amplify successfully (probe did not bind to the DNA) and 13.0% of the SNPs returned ambiguous clusters. From the resulting 7,259 SNPs, the SNP conversion and validation rates were 80.9% and 95.6% respectively (Table 1). Further filtering (i.e. excluding SNPs with a MAF < 0.01, SNP duplication, low call rates, or deviations from HWE or MI expectations) resulted in a final dataset of 6,379 high quality SNPs with an extremely high call rate (98.9%). The average minor allele frequency of these high-value SNPs was 0.37.

Table 1: SNP array performance indicating the number of SNPs retained throughout filtering.

SNP Exclusion Category	# SNPs excluded	#SNPs remaining
Total Number of SNPs:		8,967
Probe Didn't Bind	539	
Ambiguous Clusters	1169	
Number of SNPs producing genotypes (conversion rate):		7,259 (80.95%)
Monomorphic	318	
Number Validated SNPs (validation rate):		6,941 (95.62%)
HWE deviations (Heterozygous Excess / Deficit)	163	
Mendelian Inheritance Errors	399	
Number of SNPs with minimal errors:		6,379 (87.88%)
Mendelian Inheritance Errors (< 0.01), or MAF < 0.01	90	
Duplicated SNPs	43	
Call rate < 90%, or Only 2 Clusters	190	
Number of SNPs with no errors:		6,056 (83.43%)

A total of 52 replicate samples were included to evaluate array performance with concordance between replicate samples exceeding 99.9%. This provided strong support for highly reliable genotypic data across all validated SNPs. Furthermore, we reliably constructed a moderate density linkage map of 44 linkage groups containing 4,370 SNPs. These SNPs span 98.12% of the estimated genome size of 4619.3 cM at an average interval of 0.97 (map data to be revised and presented in subsequent publication). The number of markers placed within each linkage group ranged from 22 – 169 and linkage group distances ranged from 24.9 – 159.5 cM. By assigning positional information to these SNPs, not only we demonstrate their utility, but improve their value within ongoing studies.

In the current breeding program, 3,000 highly informative SNPs provided adequate power for accurate GRM calculations when compared to the 6,379 high quality filtered SNPs [Figure 1;

correlation value of r^2 = 0.99; see Khatkar et al. 2017 (these proceedings) for GS analysis]. The minimum number of SNPs for GRM analysis is also supported in similar studies of closed farm populations including Atlantic Salmon (Tsai *et al.* 2015).

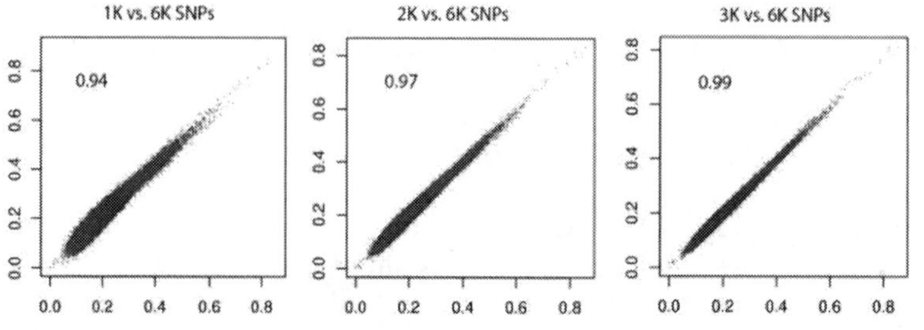

Figure 1: GRM comparisons of different subsets of SNPs.

The development and validation of a large EST-derived SNP resource is pivotal for ongoing research including identifying the major genes underlying important commercial traits, predicting production performance and developing genetic selective breeding programs for *L. vannamei*. If further SNPs are required these can be sourced from the SNP *in-silico* database. High SNP conversion rates are anticipated since the observed conversion rate within this array was > 80%.

REFERENCES

Ciobanu, D. C., J. W. M. Bastiaansen, J. Magrin, J. L. Rocha, D. H. Jiang, N. Yu, B. Geiger, N. Deeb, D. Rocha, H. Gong, B. P. Kinghorn, G. S. Plastow, H. A. M. Van Der Steen and A. J. Mileham (2010) Anim Genet **411**: 39-47.

de Givry, S., M. Bouchez, P. Chabrier, D. Milan and T. Schiex (2005) Bioinformatics **218**: 1703-1704.

Du, Z. Q., D. C. Ciobanu, S. K. Onteru, D. Gorbach, A. J. Mileham, G. Jaramillo and M. F. Rothschild (2010) Anim Genet **413**: 286-294.

Gitterle, T., H. Johansen, C. Erazo, C. Lozano, J. Cock, M. Salazar and M. Rye (2007) Aquaculture **272**, Supplement 1: S262.

Gjedrem, T. (2015) J Mar Sci Eng. **31**: 146.

GLOBEFISH, F. (2016). FAO GLOBEFISH Highlights, 1/2016: 48

Huang, Y. C., Z. X. Yin, S. P. Weng, J. G. He and S. D. Li (2012) Aquaculture 364: 111-117.

Jones, D. B., D. R. Jerry, S. Forêt, D. A. Konovalov and K. R. Zenger (2013) Mar Biotechnol **156**: 647-658.

Li, H., B. Handsaker, A. Wysoker, T. Fennell, J. Ruan, N. Homer, G. Marth, G. Abecasis and R. Durbin (2009) Bioinformatics **2516**: 2078-2079.

Li, W. and A. Godzik (2006) Bioinformatics **2213**: 1658-1659.

Schloss, P. D., S. L. Westcott, T. Ryabin, J. R. Hall, M. Hartmann, E. B. Hollister, R. A. Lesniewski, B. B. Oakley, D. H. Parks and C. J. Robinson (2009) Appl Environ Microbiol **7523**: 7537.

Schulz, M. H., D. R. Zerbino, M. Vingron and E. Birney (2012) Bioinformatics **288**: 1086-1092.

Tsai, H.-Y., A. Hamilton, A. E. Tinch, D. R. Guy, K. Gharbi, M. J. Stear, O. Matika, S. C. Bishop and R. D. Houston (2015) BMC Genomics **161**: 969.

Yu, Y., X. Zhang, J. Yuan, F. Li, X. Chen, Y. Zhao, L. Huang, H. Zheng and J. Xiang (2015) Sci Rep **5**: 15612.

Zerbino, D. R. and E. Birney (2008) Genome Res. **185**: 821-829.

Proc. Assoc. Advmt. Anim. Breed. Genet. 22:121-124

QUANTITATIVE GENOMIC ANALYSES IN THE PACIFIC WHITELEG SHRIMP *LITOPENAEUS VANNAMEI*

M.S Khatkar[1,2], K.R. Zenger[2,3], D.B. Jones[2,3], Jeff Prochaska[4], H.A.M van der Steen[4], D.R. Jerry[2,3], and H.W. Raadsma[1,2]

[1] Sydney School of Veterinary Science, Faculty of Science, The University of Sydney, Camden, NSW, Australia

[2]ARC Hub for Advanced Prawn Breeding, James Cook University, Queensland, Australia

[3] Centre for Sustainable Tropical Fisheries and Aquaculture, College of Science and Engineering, James Cook University, Queensland, Australia

[4] Global Gen, Jalan Raya Narogong Km 14, Desa Cikiwul Bantar Gebang Bekasi, 17310 Indonesia

SUMMARY

Traditional genetic improvement programs for Pacific white-leg shrimp (*Litopenaeus vannamei)* rely on family selection to improve growth and disease resistance traits. DNA technologies can help in simplifying breeding schemes and increasing genetic gains particularly for complex or difficult to measure traits. Here we present the results of genome-wide association and whole genomic prediction analyses using average family allele frequencies and the family mean of a growth trait in a genetic resource population consisting 1,934 animals and 690 families of *L. vannamei* genotyped with 8,967 genome-wide SNPs. After correcting for FDR, no significant SNPs were detected for growth. The accuracy of DGV in mirror prediction is much higher (0.65-0.69) as compared to forward prediction. A SNP that may be closely linked to the sex locus was identified with the female being the heterogametic sex.

INTRODUCTION

The Pacific white-leg shrimp (*Litopenaeus vannamei)* is an important aquaculture species and the most widely farmed shrimp globally. Traditional genetic improvement programs for *L. vannamei* rely on family selection to improve growth and disease resistance traits. Recent advances in high-throughput genotyping and analytical methods can help to simplify breeding schemes and increase genetic gain, particularly for complex or difficult to measure traits. In particular the mapping of quantitative trait loci (QTL), or genes with large effect may have an immediate application in marker assisted selection (MAS). We conducted a genome-wide association analysis for growth and a sex associated trait in *L. vannamei* by genotyping a resource population with a purpose built genome wide SNP panel and explored the possibility of genomic selection in *L. vannamei*.

MATERIALS AND METHODS

We built a resource database for *L. vannamei* by genotyping a total of 1,934 samples with 8,967 genome-wide SNPs on the Illumina Infinium ShrimpLD-24 v1.0 genotyping array (Jones et al. 2017 - these proceedings). These included 1,134 female and 123 male parents along with 677 nauplii (larval shrimp) pools. Following SNP quality control (QC), 5,893 SNPs were used for all analyses. An integrated linkage and LODE map was constructed using 631 progeny from 30 grand maternal and 19 grand paternal traced families (Jones et al 2017 - these proceedings). In total, 4,817 SNPs were mapped to 44 linkage groups that span a total of 4552.5 cM and cover an estimated 98.12% of the *L. vannamei* genome. The average interval, excluding intervals of 0 cM, was 2.67 cM. This map was utilised for all subsequent GWAS analyses and presenting results as Manhattan plots.

For the GWAS, average family allele frequencies were used for 690 families. For an additional 94 families, the genotype of the parents were available and for these the realised family-mean allele frequencies were computed as the mean of parental alleles. Out of these, based on the availability of genotypic and phenotypic data, family-wise mean frequencies on 416 families were finally used for conducting SNP association with the traits.

Phenotypic data on the family mean of 416 families on one growth trait G.d2All (Growth rate in grams per day for all tanks) were used for the current analyses. The family mean value of G.d2All ranged from 0.16 to 0.45 with a mean and standard deviation of 0.31 and 0.05, respectively. The overall distribution of G.d2All (Figure 1a) indicated that this trait is normally distributed. Figure 1b shows batch-wise distribution. In addition, individual genotypes of 1,963 animals and their sex-status were analysed to detect any sex associated SNP association.

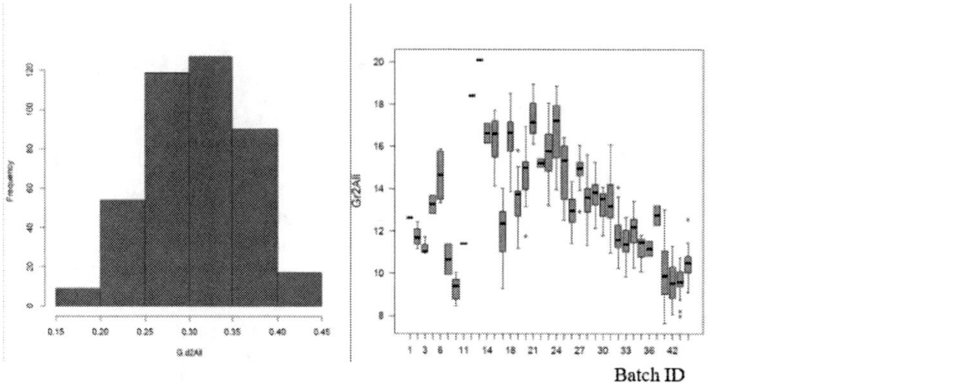

Figure 1. a) Overall distribution of the growth trait, G.d2All, presented as histogram. b) Batch-wise mean and distribution of the growth trait. The x axis represents batch id in a chronological order.

Genome-wide association (GWA) analyses. The association analysis was conducted using the allele frequencies and mean phenotypic value of the traits for the families. A realized additive relationship matrix (**K**) (Endelman, 2011) was computed to calculate molecular kinship among all families using scaled mean allelic frequencies. The regression of the mean family phenotype on SNP genotypes were conducted by fitting the mean allele frequency as a covariate and adjusting for across family relationships using the following linear mixed model:

$$\mathbf{y} = \mathbf{X}\boldsymbol{\beta} + \mathbf{Zu} + \boldsymbol{\varepsilon}$$

where **y** is a vector of the phenotypic value (trait), **X** is the incidence matrix incorporating mean and SNP allele frequency; **β** is a vector representing coefficients of the fixed effects, **Z** is an incidence matrix mapping phenotype records to families, **u** is a vector of polygenic genetic effects such that $\mathrm{var}(\mathbf{u}) = \sigma_g^2\mathbf{K}$, where **K** is the kinship matrix as described above, and **ε** is vector of residual random errors with $\mathrm{var}(\boldsymbol{\varepsilon}) = \sigma_e^2\mathbf{I}$. The model was fitted using ASReml (Gilmour, Gogele, Cullis, & Thompson, 2009). Genome-wide false discovery rate was computed using the q-value package in R (www.r-project.org).

Accuracy of genomic prediction. Genomic selection uses information from all SNP to derive Direct Genomic Values (DGV). Accuracy of direct genomic values using SNP genotypes was investigated by dividing the data on 416 families into a training set and a validation/test set. Three different sets of training and test sets were investigated by using a different proportion of the families in validation and test sets viz. 1) 75 % in training and 25 % in test; 2) 67% in training and 33 % in test; 3) 50 % in training and 50 % in test. In forward prediction, the training set consisted of older families and the test set consisted of recent families. In the mirror prediction the families were allotted randomly to the training and the test sets across all batches.

DGV were estimated using a best linear unbiased prediction (BLUP) method which used a Gaussian kernel prediction based on the Euclidean distance matrix for K where K is the kinship matrix as described above. This model is implemented in R package rrBLUP (Endelman, 2011). The accuracy of DGV prediction was computed as the Pearson's correlation coefficient between DGV and the mean phenotypic value of the families in the test set. The bias was computed as the regression coefficient of DGV on the phenotypic value.

RESULTS AND DISCUSSION

Genome-wide association (GWA) analyses for growth. Genome-wide associations expressed as log-P value for each marker are presented as a Manhattan plot in Figure 2. The unmapped SNPs are shown without any chromosome label on right hand side of the plot. There were 83 SNPs significant at P <0.05. However, after correcting for FDR no significant SNPs were detected for growth. A few clusters of SNPs with P-value <0.001 were identified, however, due to the high FDR, these could only be considered as suggestive at best. Overall these GWAS results suggest that no gene of large effect regulates this growth trait. In order to detect significant SNPs of moderate or small effect, a substantially larger sample size and a higher SNP density would be required.

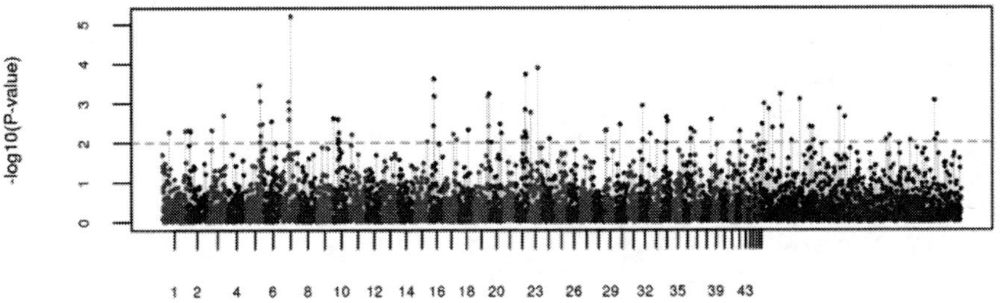

Figure 2: Genome wide SNPs associations with growth trait, G.d2All, presented as Manhattan plot.

Sex-associated SNPs. The genome-wide associations of SNPs with sex status of the animals presented in Figure 3 as Manhattan plot shows one very significant cluster of SNPs on linkage group 44. The most significant SNP was associated with sex status of the animals with -log10 (p)=294 at the start of LG44. Minor allelic frequency for this SNP was 0.3 indicating that this is a common SNP. The strong association and frequency of males and females genotypes suggest that this SNP may be closely linked to the sex locus. Most females (95%) were heterozygous whereas most males (95%) were homozygous for the major allele of the most significant SNP. These results are in agreement with earlier studies which suggest that the sex in penaeid species is mainly genetic and determined by a WZ–ZZ chromosomal system where the female is the heterogametic sex (Staelens et al., 2008). This also raised possibility of monosex culture. Potentially homogametic females and males can be used as parents to yield sexually uniform heterogametic female offspring. Monosex sex culture in prawn has been reported more profitable as compared to rearing of mixed sex animals (Mohanakumaran Nair, Salin, Raju, & Sebastian, 2006). In addition, monsex culture system can provide some protection to genetically superior stock.

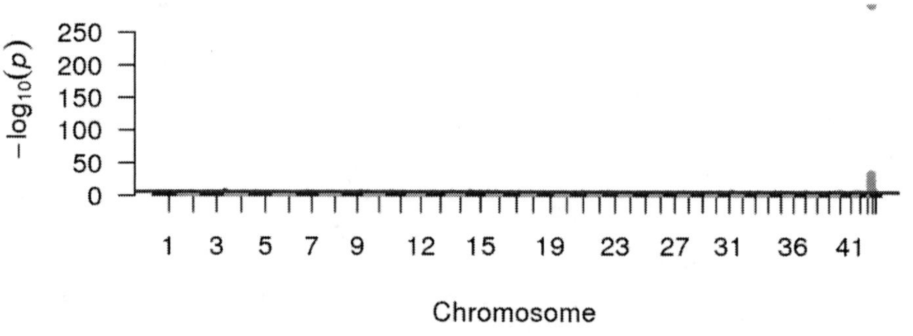

Figure 3. Genome-wide associations for sex status of animals.

Accuracy of genomic prediction of growth trait (G.d2All). The data on a growth trait G.d2All on 416 families were analysed for this analysis. The mean of each family was obtained by pooling data across tanks as described in the methods section. The genotypic data on family-wise mean allelic frequencies for 4,686 QC SNPs were included in this analysis.

The accuracy of DGV in mirror prediction (randomly dividing families in training and test set, Table 1) is much higher (0.65-0.69) as compared to forward prediction (0.17-0.32) (Table 1). The prediction accuracies in the mirror prediction indicate the potential level of accuracies of genomic selection in shrimp. It seems that declining trend with a very large batch effect of G.d2All (Figure 1b) hampered the accuracy of genomic prediction in forward prediction.

Partitioning 50 % families in training and 50 % in test gave higher accuracy as compared to other partitions in forward prediction (Table 1). Inconsistent accuracies in different partition/sets indicate that the current sample size for genomic prediction is too small which is further complicated by the large batch effect confounded with families.

Table 1. Accuracy of genomic prediction for a growth trait (G.d2All).

% in training	Number of families		Mirror prediction		Forward prediction	
	Training	Test	Accuracy	Bias	Accuracy	Bias
75	312	104	0.693	1.256	0.168	0.413
67	277	139	0.632	1.039	0.279	0.757
50	208	208	0.647	1.478	0.315	0.671

CONCLUSION

This study identified a major region associated with sex, and demonstrated that genomic selection has potential application with moderate number of SNPs, family average phenotypic records, and based on family DNA pool frequency data for commercially important traits in *L. vannamei*.

REFERENCES

Endelman, J. B. (2011). The Plant Genome, **4**(3), 250-255.

Gilmour, A. R., Gogele, B. J., Cullis, B. R., & Thompson, R. (2009). ASReml User Guide Release 3.0. VSN International Ltd, Hemel Hempstead, HP1 1ES, UK.

Mohanakumaran Nair, C., Salin, K. R., Raju, M. S., & Sebastian, M. (2006). Aquaculture Research, **37**(9), 949-954.

Staelens, J., Rombaut, D., Vercauteren, et al. (2008). Genetics, **179**(2), 917-925.

Proc. Assoc. Advmt. Anim. Breed. Genet. 22:125-128

SELECTION FOR *Piscirickettsia salmonis* (SRS) RESISTANCE IN ATLANTIC SALMON (*Salmo salar*) USING GENOTYPING BY SEQUENCING (GBS)

Theódór Kristjánsson[1], Ken G Dodds[2], John C McEwan[2], Rudiger Brauning[2], Rayna M Anderson[2], Tracey C van Stijn[2], and Shannon M Clarke[2]

[1]Stofnfiskur, Staðarberg 2-4, Hafnarfjörður IS-221, Iceland; theodor@stofnfiskur.is
[2]AgResearch, Invermay Agricultural Centre, Private Bag 50034, Mosgiel, New Zealand

SUMMARY

The genetic variance for SRS resistance in Atlantic salmon was estimated based on the challenge test performed in VESO (Norway). Data was obtained from 1881 juveniles tested at average weight of 38.7 grams. The juveniles belong to full and half sib mating structure comprising 100 full sib and 100 and paternal half-sibs families. The challenge test, which lasted for 47 days, had a cumulative mortality of 82%. Two statistical models were used to estimate genetic parameters: test survival model (TS) and day of death (DD). Estimated heritabilities for the models differ from 0.23 (TS) to 0.41 (DD). A tissue sample was taken from each juvenile for DNA extraction prior to Genotyping-by-Sequencing (GBS) using *Pst*I for the restriction digest. Subsequent filtering of GBS SNPs resulted in 22.917 SNPs (~23k) derived from the diploid region of the genome for further analysis. Using the Kinship using GBS with Depth adjustment (KGD) method to estimate a genomic relationship matrix (GRM) allowed a Genomic Best Linear Unbiased Prediction (GBLUP) evaluation of breeding value for SRS resistance. The results suggest that by using GBS with GBLUP in genotyped but non-challenged half and full sib candidates, both the accuracy and genetic gain, would increase 21-22% compared with conventional pedigree based BLUP methodology.

INTRODUCTION

Piscirickettsia salmonis (SRS) is caused by the intracellular Gram-negative bacterium, *Piscirickettsia salmonis*, first identified in Chile and later in Canada and several European countries (Corbeil and Crane 2005). SRS has been reported to infect a wide range of Salmonidae pink salmon (*Oncorhynchus gorbuscha*), chinook salmon (*Oncorhynchus tshawytscha*) and rainbow trout (*Oncorhynchus mykiss*) (Corbeil and Crane 2005). Although SRS has wide geographic range, it has caused larger outbreaks in South America than in Europe.

SRS is epizootic in Chile and losses due to SRS are significant and have severely hit the Chilean Atlantic salmon and Coho salmon industry (Cvitanich *et al.* 1991). Mortality rates have been reported to be 30-90% among Coho salmon (Corbeil and Crane 2005). Treatments with antibiotic and vaccination have provided some advantage, but do not give control of the disease.

In recent years there has been an increased focus on genetic improvement programs to select more robust and resistant individuals towards diseases. To date a number of studies have been conducted to determine additive genetic variation for disease resistance for both bacterial and virus diseases in Atlantic Salmon (Ødegård *et al.* 2011; Gjedrem *et al.* 2012). For the last ten years these studies have been supported by extensive genomic research including the use of genomic selection (GS). Studies in Atlantic salmon breeding have shown that genetic gain and accuracy can be improved substantially with GS, even with sparse SNPs (4K) (Sonesson and Meuwissen 2009; Villanueva *et al.* 2011; Ødegård *et al.* 2014). Most of the genotyping in salmon breeding has used SNP-chips, however, more recently high throughput, low cost GBS genotyping and analysis methods have been developed (Elshire *et al.* 2011; Dodds *et al.* 2015). These methods offer several advantages albeit at the expense of more complicated bioinformatics analysis.

The aim of this study was to estimate genetic variance of salmon towards SRS resistance. GBS together with KGD analysis were utilized for SNP filtering and later GBLUP was used to estimate breeding values for both challenged and non-challenged test groups, which were compared with conventional pedigree based BLUP methodology.

MATERIALS AND METHODS

Animals for both challenged and non-challenged test groups were chosen from Stofnfiskur breeding population and hatched in November 2012, year class 2012-2 (YC12-2). A nested mating design was used in which, one male was used to fertilize eggs from two females, creating groups of full-sibs and paternal half-sibs. In total, 100 females were mated with 50 males. Each family was reared in a one cubic meter tank until the fish were tagged with a PITtag. The average tagging weight was 15.8 (SD = 6.6) g. After tagging the fish were pooled and reared in a single communal tank for 6 months. The challenge group was then moved to VESO in Norway. In total, 2.400 fish (20 fish per family) were transported in bags from Stofnfiskur to VESO Vikan by plane. The fish were kept in two separate tanks until challenge testing. After 4 weeks of acclimatization at 12°C in freshwater, the fish were acclimatized to 15°C freshwater for one week before the challenge. The challenge weight was 38.7 (SD = 9.7) g. Out of 2.400 fish, 400 were used as challenge carriers (shedders) and marked by adipose fin. The cohabitation challenge was performed in one tank by injecting the shedder fish with *Piscirickettsia salmonis* and adding these fish directly to the same tank as the tested fish. Mortality was observed throughout a 47-day period after challenge.

In January 2016, 2.846 fish from were selected from YC12-2 as a non-challenged test group and future breeding candidates in the Stofnfiskurs breeding nucleus. Fin clips were taken from both challenged and non-challenged test groups and stored in 96 % ethanol for DNA analyzing. The tissue samples were sent to AgResearch, New Zealand, for DNA extraction and GBS using *Pst*I and the protocol and subsequent processing was as described in Dodds *et al.* (2015) except that 190 bar-coded samples were sequenced per lane.

GBLUP and BLUP were fitted in mixed linear models using DMU 6, software package for animal breeding (Madsen and Jensen 2013). Two models were used for the analysis. Model one was Test survival (TS) where the individuals are scored 0 if it dies within challenge test time and 1 otherwise. The second model was Day of death (DD) where individuals were scored at the day of death in the challenge ranging from day 1- 47 and individuals which survive the challenge test were considered censored. Non-challenged individuals were given missing values. The model is as follows: $\boldsymbol{y} = \boldsymbol{X}\boldsymbol{b} + \boldsymbol{Z}\boldsymbol{a} + \boldsymbol{e}$, where $\boldsymbol{y}$ is the vector of the survivals score either as 0/1 or day of death, and $\boldsymbol{b}$ is a vector of fixed effects, which included sex and rearing tanks. The vector $\boldsymbol{a}$ is a vector of random additive genetic effects of individual animals. KGD method was used to estimate a GRM in GBLUP and pedigree information was used in BLUP.

RESULTS AND DISCUSSION

The mean survival at day 47 of the SRS disease challenge test was 18% and ranged from 0 to 60% (Figure 1). A total of 29.671putative SNPs were identified using GBS methodology. After filtering by using the KGD method, the 22.917 remaining SNPs were used to create the GRM for GBLUP evaluation.

Running the TS and DD models in DMU 6 gave in both cases higher estimate of heritability and accuracy of the estimated value compared to conventional pedigree based BLUP methodology (see Table 1).

Table 1. The estimates of genetic parameters. BLUP refers to pedigree based BLUP methodology and GBLUP refers to genomic based BLUP methodology.

BLUP methods	Models	h^2	Accuracy in challenged group	Accuracy in non-challenged group	Increased accuracy
BLUP	DD	0.35	0.71	0.56	
GBLUP	DD	0.41	0.82	0.68	21%
BLUP	TS	0.23	0.69	0.54	
GBLUP	TS	0.26	0.76	0.66	22%

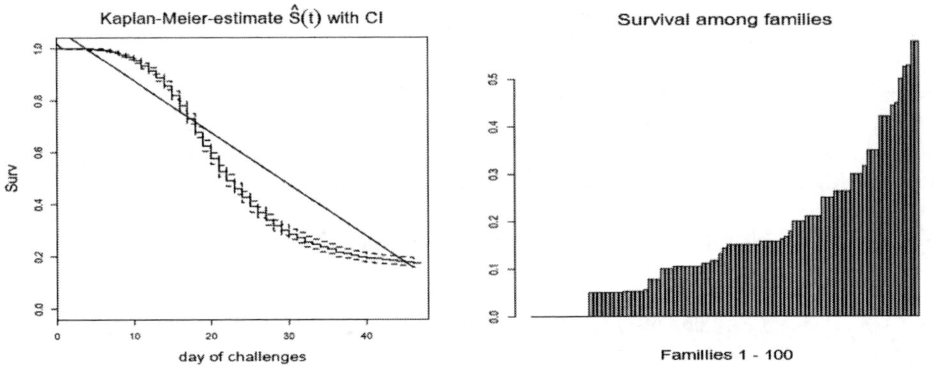

Figure 1. On the left is the Kaplan–Meier mortality curves for 47 days of challenge, on the right is the variation among 100 families tested.

This study shows a substantial increase in accuracy by applying GBS with GBLUP where the KGD method is applied to create the GRM. This is in line with other studies in salmon breeding. Ødegård *et al.* (2014) showed that a considerable improvement can be gained even from sparse SNPs (4k) but increased accuracy starts to converge rapidly from 22k to 220k, confirming the 23k SNPs from this present study would be sufficient to utilize the full potential of GBLUP.

Estimated heritabilities indicate that there is moderate additive genetic variance of SRS resistance. Moreover, heritability of DD model was higher than estimated in the TS model for both BLUP and GBLUP (Table 1). However, the estimates from the two models give different results in heritability. It should be noted that traits are defined very differently in these two models. These heritability estimates are similar to Yáñez *et al.* (2013). In both models GBLUP gives an increased accuracy and heritability compared to pedigree based BLUP methodology. Where GRM is created with SNPs, such as in GBLUP, random deviations from relationships caused by Mendelian sampling terms can be quantified more accurately.

In salmon, breeding for SRS disease resistance is difficult because breeding companies don't use infected challenged fish for breeding. Instead non-challenged sibs are used as breeding candidates (sib testing). Such evaluation has many drawbacks in relation to the amount of genetic progress that can be realized within a breeding program when depending only on pedigree information to predict breeding values by using conventional BLUP. When the predicted breeding values are not based on an individual's own performance, selected accuracy would be lower. Moreover, variation of Mendelian sampling effects within a family cannot be used to select superior animals within the best family.

Using genomic information, such as GBLUP, increases the importance of the Mendelian sampling term, or the within family variance, and reduces the importance of family compared to traditional BLUP valuation. Thus, breeding programs for traits with low heritability and relatively few records per trait measured, such as carcass and disease resistance, are those which can benefit from GS.

REFERENCES

Corbeil S., Hyatt, A.D. and Crane M.S.J. (2005) *Dis. Aquat. Organ.* **64**: 37.
Cvitanich J., Garate N. and Smith C. (1991) *J. Fish Dis.* **14**: 121.
Dodds K.G., McEwan J.C., Brauning R., Anderson R.M., van Stijn T.C., Kristjánsson, T. and Clarke S.M. (2015) *BMC Genomics* **16**:1047.
Elshire R.J., Glaubitz J.C., Sun Q., Poland J.A., Kawamoto K., Buckler E.S. and Mitchell S.E. (2011) *PLoS One* **6**: 1.
Gjedrem T., Robinson N. and Rye, M. (2012) *Aquaculture* **350**: 117.
Madsen P. and Jensen J. (2013) *http://dmu.agrsci.dk/DMU.*
Ødegård J., Baranski M., Gjerde B. and Gjedrem T. (2011) *Aquac. Res.* **42**: 103.
Ødegård J., Moen T., Santi N., Korsvoll S.A., Kjøglum S. and Meuwisse T.H.E. (2014) *Front. Genet.* **5**: 1.
Sonesson A.K. and Meuwissen T.H.E. (2009) *Genet. Sel. Evol.* **41**: 37.
Villanueva B., Fernández J., García-Cortés L.A., Varona L., Daetwyler H.D. and Toro M.A. (2011) *J. Anim. Sci.* **89**: 3433.
Yáñez J.M., Bangera R., Lhorente J.P., Oyarzún M. and Neira R. (2013) *Aquaculture* **414**: 155.

Proc. Assoc. Advmt. Anim. Breed. Genet. 22:129-132

BREEDING FOR DISEASE RESISTANCE IN AUSTRALIAN SHRIMP: HOW DO WE GET THERE?

T.H. Noble[1,2,3], N. Wade[1,3], G.J. Coman[1,3], D.R. Jerry[1,2]

[1]ARC ITRH for Advanced Prawn Breeding, James Cook University, Queensland, Australia
[2] College of Science and Engineering, James Cook University, Queensland, Australia
[3] Integrated Sustainable Aquaculture Production, CSIRO Agriculture & Food, Queensland, Australia

SUMMARY

Shrimp farming is a highly valuable aquaculture industry globally. Domesticated and selectively bred stocks of *Litopenaeus vannamei* are farmed throughout Asia and South America, however, selective breeding in Australian farmed shrimp (*Penaeus monodon*) is currently severely underutilised. Disease is the biggest threat to shrimp production globally and selective breeding is thought to be a more effective long term disease management strategy. Breeding resistant shrimp has been accomplished for very few diseases using laboratory disease challenge tests, sib-selection and conventional breeding methodologies. Genomic selection offers the potential to significantly advance shrimp selective breeding particularly for complex traits like disease resistance through increased accuracy and selection intensity. In Australia, a breeding program is currently underway developing and applying new and improved methods for selection for disease resistance in shrimp.

INTRODUCTION

Selective breeding plays an important role in increasing farming productivity and helping to meet the increasing global demand for animal protein. Aquaculture is the fastest growing primary production industry, yet less than 10% of world aquaculture production is based on selectively bred and genetically improved stocks (Gjedrem *et al.* 2012). Within the global aquaculture industry, farming of penaeid shrimp is a highly valuable sector, with most production taking place in Asia and South America using the species *Litopenaeus vannamei* (Pacific White Shrimp). Domesticated specific pathogen free (SPF) and recently selectively bred populations have been developed for this species, largely in response to the widespread disease problems the industry has faced and the catastrophic losses that result when a disease manifests in a new region (Lightner 2005). However, in the Australian shrimp farming context, the major species farmed is *Penaeus monodon* (Black Tiger Shrimp) and production is based nearly exclusively on unimproved seed derived from wild caught broodstock (although there are smaller scale domestication and breeding programs currently being developed).

Disease is perhaps the most significant issue for shrimp production globally (Stentiford *et al.* 2012) and until recently Australia has been fortunate to remain free of the major pathogens that have resulted in catastrophic production losses in Asia and Latin America. Over the last decade losses due to disease are thought to have cost the industry at least \$20bn (Shinn 2016). For example, White Spot Syndrome Virus (WSSV) is estimated to have cost at least \$8bn, however, some estimates make it closer to \$15bn since its emergence in South East Asia in the early 1990's (Lightner *et al.* 2012). Acute Hepatopancreatic Necrosis Disease (AHPND), a more recent disease impacting shrimp farming, is estimated to cause losses in production in the Thai shrimp industry alone between \$1.7 and \$2bn annually (Shinn 2016).

In December 2016 the first outbreak of WSSV was detected in Australia in South East Queensland and has had a significant immediate impact on production, brought about uncertain consequences for future production in the area, as well as having ramifications to seafood products in Australia more broadly. Additionally, an AHPND-like disease was detected in 2 Australian shrimp farms in early 2016, which was found to be caused from a similar acting, but different pathogenic strain of bacteria than that found in Asia (Nick Moody, CSIRO pers. comm.). These examples highlight how exotic diseases pose a great threat to Australian shrimp farming; however, Australian farms are also often exposed to endemic pathogens, such as gill-associated virus (GAV), that have a less devastating, but nonetheless persistent impact on production (Munro *et al.* 2011). This is because these viruses are highly prevalent in wild and farmed stocks; prevalence of GAV for example approaches 100% in some cases of *P. monodon* populations (Walker & Winton 2010).

As shrimp lack an adaptive immune system, common disease management strategies such as vaccination are not an option for shrimp. The most common management strategy used in regions where highly pathogenic diseases are present is the use of specific pathogen free (SPF) stocks that are tested and certified free of major disease causing pathogens. Whilst not selected for resistance or tolerance to the pathogen, SPF shrimp have allowed the industry to operate in regions where pathogens are present through the stocking of "clean" shrimp into ponds. However, SPF shrimp are still naïve to infection with massive losses due to disease continuing to occur and there is evidence they perform poorly in the presence of disease compared to wild stocks (Moss *et al.* 2001). Improving disease resistance through selective breeding is seen to be a more sustainable, long term strategy for the industry and as a result instigation of selective breeding programs for shrimp that capitalise on additive genetic variability in disease tolerance within farmed populations are underway.

MEASURING DISEASE RESISTANCE

The ability to accurately and reliably measure a trait under selection is core to any breeding program. For shrimp disease, measuring resistance is largely based on survival, either on-farm during grow out, or in laboratory challenge tests. Laboratory challenges tests are most commonly used because inoculation of the pathogen and environmental conditions can be more easily controlled. Challenge methods in shrimp include; injection of the pathogen into abdominal muscle, ingestion of infected material and waterborne exposure. Breeding programs that utilize disease challenge tests to measure disease resistance are based on family selection. Here a subset of progeny from a family are removed from the core breeding nucleus facilities and disease challenged. Family survival estimates are then calculated after a specified amount of time post inoculation and families are ranked on their survival performance. Families to perpetuate into the breeding program are then selected based on the family's performance. This approach means the breeding candidates themselves are never exposed to the disease, but rather chosen based on the estimated breeding values (EBV) of their disease challenged sibs (i.e. sib selection). This allows breeding companies to not only improve disease tolerance through accumulation of additive genetic variability, but practice SPF management strategies. One disadvantage of the approach, however, is that family selection only utilises the between-family genetic variance within a population and ignores 50% of the available genetic variance that is represented within-family. This, coupled with the phenotypic performance of the selected candidate having never been evaluated can lead to inaccuracies in EBV, reduced selection intensity, and therefore can lower the genetic gains realised.

Another characteristic of shrimp disease challenge tests is that resistance is often only measured as a single trait, survival. However, survivorship is complex, can be influenced by many non-disease related factors and may not manifest predominantly, or entirely through survivorship, instead causing issues with growth or deformities (e.g. runt deformity syndrome caused by Infectious Hypodermal Hematopoietic Necrosis Virus (Lightner 1999)). Therefore, alternative methods such as measuring viral load, or presence of disease associated genetic markers, may be useful in evaluating disease resistance.

A large assumption made when using controlled challenge tests in breeding programs is that resistance measured during challenge testing accurately reflects resistance under grow out farm conditions. This is largely untested for shrimp breeding programs. If there are differences then significant genotype-by-environment (GxE) interactions may be occurring which will reduce the efficiency of selection and genetic gains realised. The only known correlation published on this issue in shrimp was a phenotypic correlation between TSV challenge survival and commercial pond survival in *L. vannamei* (Moss *et al.* 2005). Here moderate positive correlations were reported (0.55 and 0.68), however, phenotypic correlations are insubstantial as there is no inclusion of the genetic effects; this information is still lacking in shrimp.

LESSONS FROM OVERSEAS GENETIC IMPROVEMENT PROGRAMS

There are few published studies that have investigated the quantitative genetics of disease resistance in shrimp. However, information on the implementation and success of disease resistance traits being incorporated into breeding programs is variable and very limited. Nearly all work has been carried out on *L. vannamei* and the most well-known success story in shrimp has been selecting *L. vannamei* for resistance against Taura Syndrome Virus (TSV). This trait has been incorporated in several breeding programs (Cock *et al.* 2009), as it has high phenotypic variation (14.6 - 93.8%) and genetic variance is moderate to high; heritability estimates across the different breeding programs range between 0.2 – 0.4 (Argue *et al.* 2002, Odegard *et al.* 2011). Response to selection has also been very good, with survival rates shown to increase by at least 18.4% per generation (Argue *et al.* 2002, White *et al.* 2002). Unfortunately, TSV disease resistance was found to be negatively correlated with growth (Argue *et al.* 2002), therefore both growth and resistance to TSV were incorporated in the breeding programs as separate traits selected for in individual breeding lines (Argue *et al.* 2002, Odegard *et al.* 2011). Despite this impediment, selecting for TSV resistance has been so successful that TSV resistant shrimp are widely used throughout the shrimp farming industry and TSV is no longer considered a major threat to production.

Conversely, breeding for resistance to WSSV has had limited success. This can in part be due to the highly virulent nature of this virus and very small genetic variation often observed both under field and controlled challenge conditions (>90 % mortality is commonly found). Estimates of heritability for resistance to WSSV under controlled challenge conditions were found to be <0.1 (Gitterle *et al.* 2005). Similar to TSV, resistance to WSSV was also negatively correlated (- 0.55 & - 0.64) with harvest weight (Gitterle *et al.* 2005). More recently, however, there have been reports of significant improvement of resistance to WSSV: For example 3 families of *L. vannamei* from a Panamanian breeding program had significantly higher survival compared to the unselected "Kona" shrimp breeding line (Cuellar-Anjel *et al.* 2011). It is difficult to get a full appreciation of how successful breeding for resistance to WSSV has been, most likely due to the commercial sensitivities of genetically improved stocks; however, this virus continues to be a major problem for shrimp farming worldwide which would suggest breeding for improved resistance has had little success so far.

OPPORTUNITIES FOR AUSTRALIAN SHRIMP FARMING

Australia has been somewhat fortunate that until recently it has been free of many of the highly virulent and devastating diseases that have occurred in overseas shrimp farms. The only known example in Australia whereby a breeding program has directly incorporated disease testing was via viral screening of wild and domesticated *P. monodon* broodstock to identify individuals with natural high GAV loads that were then removed from the spawning group (Coman *et al.* 2013). It is unknown how effective this strategy was in reducing the impact of GAV on production and there is no evidence that the approach leads to significant accumulation of advantageous additive genetic variance for GAV tolerance. Moving forward, GAV will likely continue to be an important virus affecting Australian shrimp farms, as this virus is highly prevalent in the wild and in shrimp farms.

Conventional methods of quantitative genetics used so far for shrimp breeding programs, while successful at improving growth rate, have been less effective for improving disease resistance as evidenced by an absence of resistant strains to most virulent diseases. Possibly this lack of progress is a consequence of the selection models used (i.e. sib selection) and/or laboratory challenge tests which don't accurately estimate disease additive genetic variation as it manifests itself on-farm under complex environmental interactions. Genomic selection, however, offers the potential to increase the accuracy and selection intensity of complex traits like disease resistance (Castillo-Juarez *et al.* 2015), along with more readily accessible integration of on-farm performance. This is because genomic selection allows individual phenotype data from both laboratory and on-farm performance trials to be linked with predictive genome-wide markers which can then be applied to select unchallenged individuals through genotyping only (i.e. thereby maintaining SPF status in the breeding nucleus). Genomic selection under this model would increase genetic gain as it utilizes both between and within-family variance and is able to estimate individual EBVs to use for selection of breeding candidates. Furthermore, the identification of SNPs associated with disease resistance may also be applied through quantitative trait loci (QTL) and marker assisted selection. All of this combined should allow for greater accuracy of genetic merit estimates, increased selection intensity and hence genetic gains for disease resistance traits (Castillo-Juarez *et al.* 2015). Developing and applying these new technologies are currently underway for *P. monodon* in a developing breeding program in Australia.

REFRENCES

Argue B.J., Arce S.M., Lotz J.M., Moss S.M. (2002) Aquaculture 204: 447

Castillo-Juárez H., Campos-Montes G.R., Caballero-Zamora A., Montaldo H.H. (2015) Front. Genet. 6: 93

Cock J., Gitterle T., Salazar M., Rye M. (2009) Aquaculture 286: 1

Coman G., Sellars M., Henshall J., & Preston N. (2013) Proc. World Aquac. Soc.

Cuéllar-Anjel J., White-Noble B., Schofield P., Chamorro R., Lightner D. V. (2012) Aquaculture 36: 368

Gitterle T., Salte R., Gjerde B., Cock J., Johansen H., Salazar M., Lozano C., Rye M. (2005). Aquaculture 246: 139

Gjedrem T., Robinson N., Rye M. (2012) Aquaculture 350–353: 117

Lightner D.V. (1999) J. Appl. Aquaculture 9: 27

Lightner D.V. (2005) J. World Aquac. Soc. 36: 229

Lightner D.V., Redman R.M., Pantoja C.R., Tang K.F.J., Noble B.L., Schofield P., Mohney L.L., Nunan L.M., Navarro S.A. (2012) J. Invertebr. Pathol. 110: 174

Moss S.M., Arce S.M, Argue B.J., Otoshi C.A, Fernanda C., Tacon A.G.J. (2001) The New Wave Proc. Spec. Session Sustainable Shrimp Culture

Moss S.M., Doyle R.W., Lightner D. V. (2005) In 'Diseases in Asian Aquaculture V', pp. 379-393, editors Walker P., Lester R. and Bondad-Reantaso M.G., Asian Fisheries Society, Manila.

Munro J., Callinan R., Owens L. (2011) J. Fish Dis. 34: 13

Odegard J., Gitterle T., Madsen P., Meuwissen T.H., Yazdi M.H. Gjerde B., Pulgarin C., Rye M.(2011) Genet. Sel. Evol. 43: 14

Shinn (2016). The Aquaculture Round Table Series

Stentiford G.D., Neil D.M., Peeler E.J., Shields J.D., Small H.J., Flegel T.W., Vlak J.M., Jones B., Moradog F., Mossh S., Lotz J., Bartholomay L., Behringer D.C., Hautonl C., Lightnerm D.V. (2012) J. Invertebr. Pathol. 110: 141

Walker P.J. and Winton J.R. (2010) Vet. Res. 41: 51

White B.L., Schofield P.J., Poulos B.T., Lightner D. V. (2002) J. World Aquac. Soc. 33: 341

Proc. Assoc. Advmt. Anim. Breed. Genet. 22:133-136

RELIABILITIES OF AUSTRALIAN DAIRY GENOMIC BREEDING VALUES INCREASE THROUGH THE ADDITION OF GENOTYPED FEMALES WITH EXCELLENT PHENOTYPES

J.E. Pryce[1,2], P. Douglas[3], C.M. Reich[1], A.J. Chamberlain[1], P.J. Bowman[1,2], T.T.T. Nguyen[1], B.A. Mason[1], C.P. Prowse-Wilkins[1], G.J. Nieuwhof[1,4], T. Hancock[1,4], M. Shaffer[4] and B.J. Hayes[1,5]

[1] Department of Economic Development, Bundoora, VIC 3083, Australia
[2] La Trobe University, Bundoora, VIC 3083, Australia
[3]Holstein Australia, Hawthorn, VIC 3122, Australia
[4]DataGene, Bundoora, VIC 3083, Australia
[5]University of Queensland, Brisbane, QLD, Australia

SUMMARY

Ginfo is a large-scale genotyping project to increase the size of the Australian dairy reference population. In total, there were 32,386 cows from 103 herds with excellent records located across Australia's main dairy regions. The increase in the reliabilities of breeding values for young genomic bulls (without daughters) was between 5% and 7% in Holsteins and between 2% and 3% in Jerseys. For example, in Holsteins, the reliability of daughter fertility breeding values increased from 41% to 46%. Incorporating genotypes from herds selected on the quality of their phenotypes has increased the reliability of genomic breeding values.

INTRODUCTION

Phenotypic data underpins the calculation of both traditional and genomic breeding values. A reference population of genotyped individuals with phenotypes is required to calculate associations between genetic markers and phenotypes and form a genomic prediction equation. Without sufficient data, the relationship between the reference population and the general population weakens and so does the relevance of the genomic prediction equations.

In Australia, the male reference populations comprise around 4000 Holsteins and 1000 Jerseys. Previous research investments have already resulted in female populations of approximately 10,000 Holstein 4000 Jersey females being added to the national reference populations of the respective breeds. These data were from projects that focused on cows with large quantities of phenotypes. Instead, the aim of Ginfo was to select herds that had high quality phenotypes.

The aim of this study was to quantify the change in reliability of genomic breeding values for Australian breeding values through adding the Ginfo population to the reference population.

MATERIALS AND METHODS

Herd Selection. To qualify for the reference population, known as Ginfo (Genomic Information Nucleus), Australian dairy herds were scored according to the quality of the records contributing to the national database using an index that rewards cows with fertility, conformation, survival, workability, somatic cell count and milk yield data; in the scoring system, the maximum score was 25 and having complete fertility phenotypes can make up 10 of these points. The highest scoring herds (n=103) were invited to participate in the project.

The 103 Ginfo herds have been contributing records on 32,386 daughters of 2,917 bulls to the Ginfo project. Tail hair samples were collected from all the cows in the recruited herds for genotyping and data on milk production traits, somatic cell count, mating, pregnancy and calving data for multiple parities were provided to DataGene. First parity cows from Ginfo herds were also type classified by Holstein Australia. The herds were from across Australia's main dairy regions with an intention to proportionally represent dairy cow populations. For example, two thirds of the herds were in Victoria aligning with the distribution of dairy cows across Australia.

Genotyping by sequencing (GBS). Genotyping by sequencing (GBS) was used to procure genotypes. The GBS methodology used has previously been described by Chamberlain *et al.* (2015). Briefly, probes were designed to the flanking sequencing of 9,102 target SNP, of which 5,119 were part of the Illumina Infinium Bovine SNP50 beadchip. The HiSeq2000 and HiSeq3000 genome analyser platforms using single read chemistry were used for sequencing.

Quality control steps of sequence reads were as follows: 1) poor quality bases (qscore <20) were removed using scripts developed in house; 2) alignment was with BWA v0.7.7; 3) Samtools v0.1.19 (Li *et al.*, 2009) mpileup tool was used to create vcf files and allele counts at the 9,102 target SNPs and 4) Allele counts were used to call genotypes, where the total count was >=6 and a heterozygote had a minor allele frequency > 0.167. The genotypes in UMD 3.1 forward format were converted to Illumina's top-top format. The next step was imputation of GBS genotypes to those used by DataGene in routine genomic evaluations (Nieuwhof *et al.*, 2010).

All animals were imputed to a 50K evaluation panel using Fimpute (Sargolzaei *et al.*, 2014). The Ginfo project also enhanced the DataGene evaluation SNP panel to include new variants which were identified by whole genome sequence (WGS) analysis which were found to be located near new QTLs for the traits within the evaluation. These WGS variants were added to the DataGene evaluation panel through the Ginfo GBS genotypes and all other animals were imputed for these WGS SNPs.

Impact of Ginfo population on reliability of genomic selection. The Ginfo cows and their associated phenotypes were added to the genomic reference population. In April 2016, when our comparisons were done, the existing reference populations comprised 4,172 bulls and 10,254 cows for Holsteins and 1,097 bulls and 4,232 cows for Jerseys. The cows that were already included in the reference population were selected using similar selection criteria for phenotype quality, as described already for Ginfo.

Reliabilities were estimated for all traits evaluated by DataGene using software developed in house for genomic selection (Nieuwhof *et al.*, 2010) implementing the mixed model equations for genomic selection as described by Garrick (2007). The reliabilities of genomic bulls with no daughters were compared when estimated with and without Ginfo cows in the reference population.

RESULTS AND DISCUSSION

The number of Ginfo cows added to the reference population was 17,108 and 3,347 for Holsteins and Jerseys respectively. At the time the Ginfo population was added to the Australian national reference population, they represented 54% and 39% of the Holstein and Jersey populations respectively.

On average the increase in reliability from adding Ginfo to the reference population was 5.8% and 2.5% for young genotyped Holstein and Jersey bulls respectively (Figure 1). The impact varied by trait, with gains of between 5% and 7% for Holsteins and between 2% and 3% for Jerseys. For example, in Holsteins, the reliability of daughter fertility increased from 41% to 46%, while overall type increased from 42% to 49%. This is similar to approximations derived by applying the equation of Daetwyler *et al.* (2008) to predict the reliability of genomic prediction for varying reference population sizes. The scores for herds that are in Ginfo are on average higher for Holsteins than Jerseys, which is a consequence of the relative population sizes. This could have partly contributed to the smaller increase in reliability for Jerseys compared to Holsteins.

One of the main questions in the design of future reference populations is whether to focus on increasing reliabilities through genotyped bulls with large progeny groups, or on genotyped cows with their own phenotypes (Gonzalez-Recio *et al.*, 2014; Chesnais *et al.*, 2016). Another alternative, discussed by Plieschke *et al.* (2016) is genotyping and phenotyping a fixed number of first crop daughters, as this increases the reliability of the sire. The general conclusions of Chesnais *et al.* (2016) are that when phenotypes are inexpensive and easy to measure on a large scale for key traits of interest, bull reference populations are better, while for expensive or difficult to measure traits, it is preferable to have a reference population of genotyped females. However, there is also a case for female reference populations, where the usual source of new phenotypes (i.e. the number of progeny-tested bulls) is in decline. In Australia, the number of bulls with sufficient daughters with publishable proofs for production traits by year of birth has gradually been declining, by around 60 per year. Consequently, a genomic reference population that does not solely rely on progeny-tested sires is important. When large male reference populations are already available, the impact of adding females on reliabilities is comparatively small; so the value for these sorts of reference populations is more around the new traits that can be measured in dedicated reference populations.

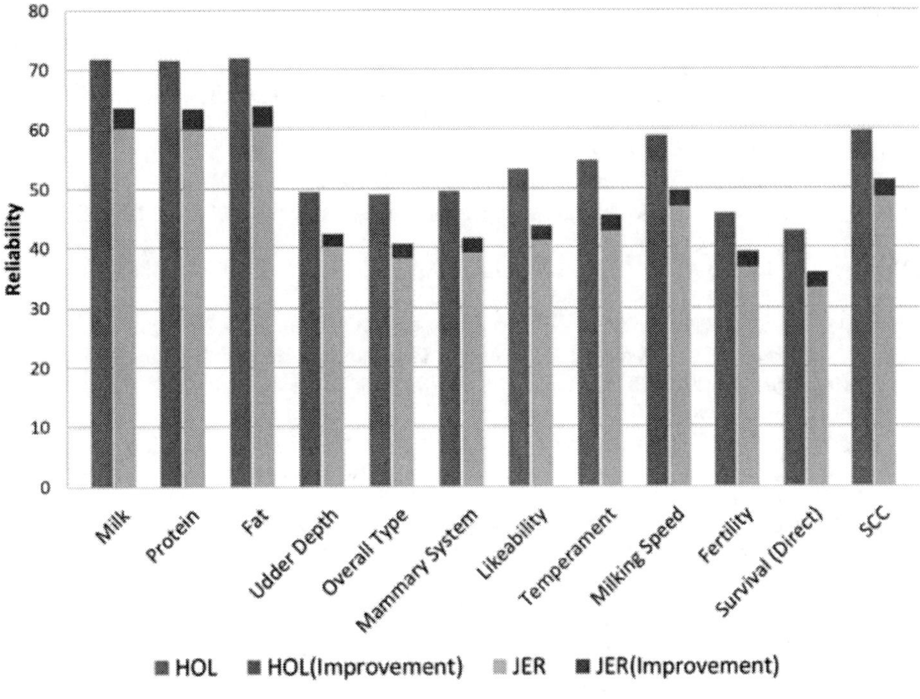

Figure 1. Reliabilities of traits with and without the Ginfo population

The Ginfo reference population is projected to encompass approximately 60,000 milking animals in 200 herds to reflect the genetics, location and farm systems in the broader Australian dairy population. Ginfo is anticipated to become a primary source for the Australian industry's ongoing evaluation of the current suite of genomic Breeding Values. In addition we are also investigating the collection of emerging and new phenotypes of interest to farmers particularly for animal health traits and traits associated with resource availability and efficiency (Abdelsayed *et al.*, 2017).

Dairy

One of the philosophies in establishing Ginfo was to develop relationships with Australian dairy farmers who have a shared interest in the value of high quality phenotypes and genotype results. Although the model we have used to date included all genotyping costs being covered through research funding, we envisage that this will change as we move to a model where farmers pay for a much larger proportion of the genotyping cost themselves. While the genotyping results (breeding values) of lactating cows may have limited use for decision making, there is considerable value in genotyping results for heifers, most notably in selecting the best replacements (Pryce and Hayes, 2012 Calus *et al*, 2013). Therefore, the investment strategy needs to balance the benefits to the farmer versus the benefits to the broader dairy industry.

ACKNOWLEDGEMENTS
The Ginfo research project was funded by Dairy Futures CRC (Melbourne, Australia). Ginfo is now an industry initiative and is part of DataGene (Melbourne, Australia).

REFERENCES

Abdelsayed, M., Haile-Mariam M and Pryce, J.E. (2017) *Proc. Assoc. Advmt. Anim. Breed. Genet.* **22** (submitted)

Calus, M.P.L., Bijma, P., van Arendonk, J. and Veerkamp, R.F. (2013) *Ann. Mtg Europ Assoc. Anim. Prod.* **26**.

Chamberlain A.J., Beatson P., Mason B.A., Prowse-Wilkins C, Reich CM., Schrooten C, Hayes B.J., Tibbits J., and M. Hayden (2015) *Proc. Assoc. Advmt. Anim. Genet.* **21**: 69-72.

Chesnais, J.P., Cooper, T.A., Wiggans, G.R., Sargolzaei, M., Pryce, J.E. and Miglior, F. (2016) *J. Dairy Sci.* **99**: 2413-2427.

Daetwyler, H.D., Villanueva B. and Woolliams J.A., (2008). *PLoS One*, **3**: e3395.

Garrick, D. J. (2007) *J. Anim. Sci.* **85**: 376.

Gonzalez-Recio, O., Coffey, M.P. and Pryce, J.E. (2014). *J. Dairy Sci.* **97**: 7905-7915.

Li, H., Handsaker, B., Wysoker, A., Fennell, T., Ruan, J., Homer, N. and Durbin, R. (2009) *Bioinformat.* **25**: 2078-2079.

Nieuwhof, G.J., Beard, K.T., Konstantinov, K.V., Bowman, P.J., and Hayes, B.J. (2010) *Proceedings of 2010 Interbull meeting, Riga, Latvia*

Plieschke, L., Edel, C., Pimentel, E.C., Emmerling, R., Bennewitz, J. and Götz, K.U. (2016) *Genet. Sel. Evol* **48**: 73.

Pryce, J.E. and Hayes, B.J. (2012). *Anim. Prod.Sci.* **52**: 180-184.

Sargolzaei, M., Chesnais, J. P., and Schenkel, F. S. (2014). *BMC genomics*, **15**: 478.

Proc. Assoc. Advmt. Anim. Breed. Genet. 22:137-140

DOES EXPERIENCING BIRTH DIFFICULTY INFLUENCE PERFORMANCE AS AN ADULT (LATER LIFE) IN HOLSTEIN CATTLE?

M. Haile-Mariam[1] and J.E. Pryce[1,2]

[1] Department of Economic Development, Jobs, Transport and Resources, Bundoora, VIC 3083, Australia
[2] School of Applied Systems Biology, La Trobe University, Bundoora, VIC 3083, Australia

SUMMARY

The effects of experiencing difficulty at birth on the performance of animals as adult were estimated using data of Holstein cows that calved over two decades. Calves that experienced difficulty at birth showed a reduced fertility and fitness relative to those born without difficulty as expressed by late calving for the first time, longer calving interval and lowered survival from first to second lactation. However, experiencing difficulty at birth did not reduce the milk yield of the animal as an adult. Although the effect of birth difficulty on performance of the animal as an adult is small, relative to subsequent performance of cows that experienced difficulty themselves it should serve as an additional incentive to improve calving performance and management of calves born with difficulty.

INTRODUCTION

Calving is a key event in any cattle production operation and is essential for the sustainability of the herd. In dairy industries where statistics are available, phenotypic dystocia rates appear to have increased (Mee, 2008) which means that the economic and welfare implication of calving difficulty (CD) is also increasing. A number of studies have quantified the effect of CD on the productivity of cows that experienced difficulty. For example, Dematawewa and Berger (1997) estimated that the financial cost of dystocia to be 41% due to production losses, 31% due to poor fertility and 25% due to cow and calf morbidity and mortality. Several others have reported that the effect of CD on subsequent milk yield of cows is insignificant (Rajala and Gröhn, 1998; McClintock, 2004). On the other hand, the effect of difficult birth on the performance of the calf over its lifetime is not well documented, although a few studies exist (e.g. Eaglen et al. 2011). Evidence from other mammalian species including cattle (Lombard et al., 2007, Dwyer, 2008), shows that experiencing difficulty at birth could affect the health and development of offspring. The study by Eaglen et al. (2011) based on data from the UK, showed that the production and fertility of calves born following a difficult birth is reduced. Eaglen et al. (2011) observed that the milk yield of cows that experienced extreme difficulty at birth with veterinary assistance amounted to only 91% of those born without any difficulty. They also showed that calves that experienced difficulty at birth were less fertile as adults, but their estimates were associated with large standard errors (Eaglen et al. 2011).

Quantifying the effect of birth difficulty is important because it can serve as an additional incentive to adopt both genetic and non-genetic approaches to improve calving performance. Therefore the aim of this study is to estimate the effect of experiencing difficulty at birth on performance traits such as age at first calving, milk yield, fertility and survival in Holstein cows.

MATERIALS AND METHODS

Data on calving difficulty (CD) and other performance traits including fitness and milk yield traits of cows that calved between 1995 and 2016 were extracted from the national dairy genetics database operated by DataGene Ltd. First, Holstein cows with valid CD (i.e. single female) and service sire and date of calving were selected from data extracted for genetic evaluations of CD.

Then, from the pedigree database, where all births are recorded, animals that were born on the same date, from the same cow (dam) and mating bull (sires) were selected. For female calves these data were then matched based on their national identification number to their performance as adults. The age at first calving for the animals selected for this study varied from 18 to 40 months. The number of cows with data for age at first calving (AFC), survival (Surv), calving interval (CI) and 305-day milk yield (305 MY) by level of birth difficulty is shown in Table 1.

Four levels of difficulty, as suggested by McClintock (2004), were defined. The effect of level of birth difficulty (i.e. 4 levels) on AFC, CI, Surv and 305 MY was assessed by fitting a model including herd-year-season of calving as an adult, month and year of birth as a calf and age at first calving for all traits except AFC. For AFC, herd-year-season of birth instead of calving was fitted. The effect of level of CD on milk yield traits was also estimated using test-day milk yields in the first 150-day of lactation. For this analysis the fixed effects fitted were herd-test date and year-season of calving, instead of herd-year-season. In the test-day model the interaction of days in milk (DIM) as a covariate with the 4 levels of CD were fitted in addition to cow and sire as random effects. The random effect of cow was fitted to account for repeated test-day record of cows and the random effect of sire was fitted to estimate the effect of CD on milk after accounting for genetic differences among sires. To further explore possible reasons for the effect of CD on fitness and production traits covariance analyses were performed using multi-trait models. These analyses provided estimates of correlations between CD levels and MY, CI, Surv and AFC using a sire model with additive genetic relationships. The pedigree used included sires of animals with information on CD and performance and their parents going back to 1950s. All data analyses were performed using ASReml (Gilmour et al. 2009).

Table 1. Number of calves with their level of birth difficulty and their performance information until the beginning of the second lactation in Holstein

Birth difficulty & observations		Traits			
Level	Observations (%)	Age at first calving	305-day milk yield	Survival	Calving interval
Normal	311951(92.98)	311775	291872	281793	216257
Slight	15849 (4.72)	15843	14858	14468	10891
Moderate	7256 (2.16)	7250	6828	6606	4962
Extreme	442 (0.13)	442	384	342	245

RESULTS AND DISCUSSION

Table 2 shows the deviation in AFC, CI, Surv and 305 MY for CD levels from those born with no difficulty. The effect of CD on AFC, Surv and CI are significant but small in magnitude. On the other hand, the effect of CD on 305 MY is insignificant. Table 2 also shows that cows born with the extreme level of difficulty of 4 produced more milk in absolute terms than those born with no difficulty but the difference was not statistically significant because the number of cows was small. The lack of a clear effect of CD on MY was confirmed by analysing the total test-day milk yield data over the first 150-days. The 150-day milk yield analyses showed that cows that experienced slight and moderate difficulty produced less than those born normally (Table 2) suggesting that early milk yield is better suited to estimate the effect of CD. In these data we also observed that CD did not have significant effect on fertility traits such as pregnancy rate, first service non return rate and calving to first service interval mainly because the number of cows with data on these traits was lower than those for AFC and CI, for example. However, there was a trend that in all cases cows experiencing some birth difficulty showed a reduced fertility compared to calves born normally. The effect of experiencing birth difficulty in later parities were not larger than those observed in first parity cows, so these results are not tabulated.

Table 2. Effect of birth difficulty on age at first calving, survival, calving interval and 305 and 150 day milk yield as deviations from normal births

Trait	Level of birth difficulty			
	Normal	Slight	Moderate	Extreme
Age at first calving, months	0.0^a	2.44±0.79^b	1.75±1.17^{ab}	10.17±4.36^b
Calving interval, days	0.0^a	2.36±0.68^b	4.14±1.01^b	4.82±4.17^{ab}
Survival (%)	0.0^a	-0.92±0.36^b	-1.81±0.54^b	-1.59±2.17^{ab}
305-day milk, Litre	0.0^a	55.0±18.1^b	63.45±26.8^b	121.2±98.9^{ab}
150 test-day milk, Litre	0.0^a	-53.2±17.0^b	-75.3±25.2^b	155.4±95.5^{ab}

^{a,b} Solutions designated with different letters are significantly different (P<0.05) from each other.

Table 3. Correlations between calving difficulty at birth and subsequent performance as adults

Traits	Genetic correlation	Residual correlation
Age at first calving	0.22±0.10	0.01±0.0
Calving interval	0.30±0.08	0.01±0.0
Survival	-0.25±0.08	-0.01±0.0
305-day milk	-0.05±0.07	0.01±0.0
150 test-day milk	-0.05±0.07	0.01±0.0

The results in Table 2 show that the effect of experiencing CD as a calf on all traits are small and may have little economic significance. In particular the effect of experiencing difficulty at birth is small compared to the effect on subsequent fertility and survival of cows that experienced CD themselves. In the current data, CI of cows following CD category of 2, 3 and 4 increased by 6.8, 12.3 and 24.4 days, respectively, relative to cows that did have a normal calving. Similarly survival from 1st to 2nd calving was reduced by 2.7, 7.2 and 13.9% when CD increased from category 2 to 4, respectively, compared to normal calving. On the other hand, the subsequent milk yield of cows was not affected by CD level of cows. Our results on the effect of CD on the subsequent performance of cows agree with those reported by McClintock (2004) who, using part of these data, observed that survival and fertility of cows was reduced following CD but that MY was not affected. Further analyses using test-day data also showed that the effect of CD on subsequent milk yield of cows that experienced difficulty is small even when observed in the first 150-days of lactation. Cows that had extreme CD produced 77 litres less milk over the first 150-days than cows that calved without difficulty. This limited or no losses of MY following difficulty agrees with some studies (Rajala and Gröhn, 1998) but disagrees with others (Dematawewa and Berger, 1997; Eaglen et al. 2011).

The effect of CD on subsequent performance of cows that experience CD is well documented but the effect on calves born with difficulty is less well known (Eaglen et al. 2011). A few studies have looked at the effect of experiencing CD on the health and development of calves (Lombard et al. 2007; Lundborg et al. 2003). The effect of experiencing difficulty at birth on performance (e.g. growth) up to first calving age could not be established in the current study because we do not have information on decisions after birth until age at 1st calving. Obviously a large part of the economic loss due to birth difficulty is the death of calves around calving. Calf deaths at about birth time in these data were 3%, 11%, 34% and 49%, respectively, in female calves that were born without, with slight, moderate and extreme difficulty, suggesting that a large number of animals that experienced moderate to severe CD ended up being excluded from this study. The selected nature the data that is available for this sort of analysis means that economic losses of difficulty at birth are hard to measure and cannot be compared to performance in cows that experienced CD.

Eaglen et al. (2011) observed MY and fertility of animals that experienced difficulty at birth was reduced. Their results with regard to fertility traits were confirmed by our analyses and we also found that both fitness and AFC was affected by CD, suggesting the possible long-term effect of CD at birth on performance to at least second calving. A bigger effect of CD on AFC (Heinrichs et al. 2005) and MY (Heinrichs and Heinrichs, 2011) was observed in US Holsteins where imputation techniques were used to avoid bias due to missing data. The reasons for such long-term effects of CD at birth on the performance of as adult was related to epigenetic processes or other so-called developmental programming (Eaglen et al. 2011). Heinrichs et al. (2005) suggests that calves that experienced CD are likely to grow slower and calve at an older age than those born with no difficulty. The implication of this is that, if calves born with CD are to be used as replacements, they should perhaps be provided with better management.

The results in Table 3 on correlations agree with those in Table 2 and they show that there is a significant genetic component to the observed reduction in fertility and survival with the increase in level of CD.

CONCLUSIONS

Although the effect of birth difficulty on the performance of the animal as an adult is small, it should serve as an additional incentive to improve calving performance and management of calves born with difficulty. However, both quantifying the effect of events such as CD and developing an overall herd improvement strategy requires data from birth to 1st calving age, including information on recruitment of replacements from dairy herds, which is currently unavailable.

ACKNOWLEDGEMENTS

We thank DataGene Ltd. for provision of data and Dr Phil Bowman for extraction of the data.

REFERENCES

Dematawewa C.M.B. and Berger P.J. (1997) J. Dairy Sci. **80**:754–761.

Dwyer C. (2008) Small Rumin. Res. **76**:31–41.

Eaglen S.A.E., Coffey M.P., Woolliams J.A., Mrode R. and Wall E. (2011) J. Dairy Sci. **94**:5413–5423.

Gilmour A.R., Gogel B.J., Cullis B.R. and Thompson R (2009) ASReml User Guide Release 3.0.

Heinrichs A.J. and Heinrichs B.S. (2011) J. Dairy Sci. **94**:336-341.

Heinrichs A.J., Heinrichs B.S., Harel O., Rogers G.W. and Place N.T. (2005) J. Dairy Sci. **88**:2828-2835.

Lombard J.E., Garry F.B., Tomlinson S.M. and Garber L.P. (2007) J. Dairy Sci. 90:1751–1760.

Lundborg G.K., Oltenacu P.A., Maizon D.O., Svensson E.C. and Liberg P.G. A. (2003) Prev. Vet. Med. **60**:175–190.

McClintock S. (2004) PhD Dissertation, the University of Melbourne.

Mee J. E. (2008) *Vet. J.* **176**:93-101.

Rajala P.J. and Gröhn Y.T. (1998) *J. Dairy Sci.* **81**:3172–3181.

Proc. Assoc. Advmt. Anim. Breed. Genet. 22:141-144

PITFALLS OF PRE-SELECTING SUBSETS OF SEQUENCE VARIANTS FOR GENOMIC PREDICTION

I.M. MacLeod[1], S. Bolormaa[1], C. Schrooten[4], M.E. Goddard[1,2] and H. Daetwyler[2,3]

[1] AgriBio, Department of Economic Development, Jobs, Transport and Resources, Victoria.
[2] Faculty of Veterinary and Agricultural Sciences, University of Melbourne, Victoria, Australia
[3] School of Applied Systems Biology, La Trobe University, Victoria, Australia
[4] CRV, Netherlands

SUMMARY

Genomic prediction (GP) in farm livestock generally exploits SNP array genotypes. Now it is possible to impute from SNP chip genotypes to whole genome sequence. However, in an industry setting it is impractical to implement GP using millions of sequence variants. Livestock industries are therefore keen to leverage sequence data by selecting subsets of variants to develop custom SNP arrays. In this study we demonstrate that there are potential pitfalls in this approach that can lead to considerable bias in GP and can underestimate the potential advantages of sequence.

INTRODUCTION

Genomic prediction is becoming a popular tool for livestock breeding, and in commercial settings generally exploits SNP array genotypes. Recently, large numbers of animals have been sequenced, enabling imputation to whole-genome sequence for any animal with SNP array genotypes. In theory all imputed sequence variants (> 20 million) could be used for genomic prediction and this should include the causal mutations. However, in practice this is computationally impractical for livestock industries. Furthermore, prediction models that include many millions of imputed sequence variants have not yet increased genomic prediction accuracy relative to SNP array genotypes (van Binsbergen *et al.* 2015; Calus *et al.* 2016). This may be a result of: 1) exacerbated "large p small n" problem leading to an over-saturated model, 2) difficulty in precisely estimating SNP effects due to long distance linkage disequilibrium (LD) and 3) imputation errors. A practical solution is to discover important sequence variants associated with key traits and then design custom SNP arrays that combine the selected variants with SNP from existing commercial arrays (e.g. Wiggans *et al.* 2016). This reduces industry problems associated with large genotype data sets, reduces the "large p small n" analytical issue and increases genotyping accuracy of important sequence variants.

In dairy cattle, several studies have attempted to gain advantage from imputed whole-genome sequence by running a single SNP regression analysis (GWAS) to identify a subset of the most significant sequence variants, and then combining these with lower density SNP array genotypes for genomic prediction (Brøndum *et al.* 2015; van den Berg *et al.* 2016; Veerkamp *et al.* 2016). Similarly, Wiggans *et al.* (2016) demonstrated a small advantage in genomic prediction accuracy by pre-selecting the most informative SNP from high density SNP array genotypes and then using this SNP subset to train the prediction equations. In all these studies, the analysis to select the top variants and their subsequent analysis to train the genomic prediction equations was carried out with the same reference population.

Here, we demonstrate that when pre-selected variants are discovered in the same reference population that is used to train subsequent genomic predictions, this approach can result in significant bias in the predictions. Furthermore, our results suggest that this approach may underestimate potential gains from using subsets of sequence variants in both accuracy and persistency of genomic prediction. We demonstrate that these pitfalls can be avoided by pre-selecting sequence SNP from a population that is independent from the reference population used to train the genomic prediction equations.

MATERIALS AND METHODS

We chose a data set of 21,879 dairy cattle with real genotypes and simulated phenotypes from the same data described in MacLeod *et al.*(2016). Briefly, the genotypes included 2.785 million imputed sequence variants and Illumina 800K Bovine HD beadChip genotypes. Sequence variants included only those in gene coding regions or in putative regulatory regions 5 Kb up- and downstream of genes. After pruning out one of all SNP pairs in perfect LD and SNP with minor allele frequency < 0.002, a total of 994,019 variants remained ("SEQ"). Three trait phenotypes were simulated for all animals by selecting 4000 of these variants to be causal mutations (QTN) with three different genetic architectures and a heritability of 0.6 (details in MacLeod *et al.* 2016). For each trait 3485, 500 and 15 additive QTN effects were sampled from three different normal distributions with a mean of zero and variances of $0.0001\sigma_g^2$, $0.001\ \sigma_g^2$ and $0.01\ \sigma_g^2$ respectively, where σ_g^2 is the additive genetic variance. Breeding values (BV) for all animals were calculated as: $BV_j = \sum_{i=1}^{4000} x_{ij}\alpha_i$, where a_i is the ith QTL effect and x_{ij} represents the ith genotype (coded 0, 1 or 2 for genotypes aa, Aa and AA) of animal *j*.

The animals included 16,133 Holstein, 4861 Jersey and 885 Australian Red breed. The 885 Australian Red and the youngest 584 Holstein were used as two separate validation populations (one distantly related and one closely related). The remaining animals were divided into two separate mixed breed reference sets: Refl with 7991 Holstein and 2323 Jersey, and Ref2 with 7558 Holstein and 2538 Jersey. Pedigree records were available for both Refl and Ref2. We applied two methods of genomic prediction: GBLUP and BayesR, with the standard model described in MacLeod *et al.* (2016). In the BayesR analyses, variant effects were sampled from four normal distributions with mean of zero and variances as described above for simulated QTN effects. BayesR is a useful method for QTN discovery (e.g. MacLeod *et al.* 2016) so we used BayesR rather than GWAS to identify a subset of putative QTN.

First we undertook QTN discovery separately in Refl and Ref2 using the SEQ genotypes (included the surrogate QTN) and then chose the top 500 putative QTN from each analysis. Then we created two custom SNP chips: the first combined the top putative QTN from Refl with the 50K Illumina BovineSNP50 chip genotypes (Chip_Refl) and the second combined the top 500 putative QTN from Ref2 with the 50K set (Chip_Ref2). These custom chips were then used for genomic prediction in reference population Refl. Thus genomic prediction with Chip_Refl mimics the approach taken by several recent studies mentioned above: i.e. the QTN discovery population (Refl) was not independent of the reference population used to train the genomic predictions. In contrast, for Chip_Ref2 the selected putative QTN were discovered in a population (Ref2) that was independent of the one used to train the genomic prediction equations (Refl). Finally, the two validation populations were used to test accuracy and bias of prediction equations derived from Refl with the custom SNP chips as well as the full SEQ, 800K and 50K genotypes. BayesR results are presented as the average of five MCMC chains and results for both GBLUP and BayesR were averaged across the three trait phenotypes (trends being similar). The accuracy of genomic prediction was calculated as the correlation between predicted and true breeding values, and bias was assessed by the regression of the true breeding value on the predicted value.

RESULTS AND DISCUSSION

The accuracy of genomic prediction was highest for SEQ genotypes (Fig 1) as expected because SEQ included all surrogate QTN variants. The relative advantage of SEQ was greater for the Australian Red validation compared to the Holsteins. This reflects the extra precision of the prediction which is only apparent when validation animals are not strongly related to the reference set (Aust. Red breed animals were not in Ref1 or Ref2). The BayesR accuracy was always higher than GBLUP. This was not surprising because we simulated a mixture model with many small effects and a few large effects and Bayesian models are generally superior to GBLUP for this scenario. For BayesR and GBLUP there was an increase in accuracy using either Chip_Ref1 or Chip_Ref2 compared to 50K only. However, for BayesR this advantage was greater for Chip_Ref2 where the putative QTN were discovered in a population that was independent of the reference population that was subsequently used to train the genomic prediction equation. For Chip_Ref2 the accuracy of prediction exceeded the accuracy of the HD 800K and the relative increase was higher for Australian Reds than Holsteins. For GBLUP there was little difference in the accuracy from the two custom chips. However, when we created custom chips by combining the top 5000 SNP from the SEQ analyses with the 50K set, the accuracy of GBLUP markedly improved with Chip_Ref2 compared to Chip_Ref1 (results not shown).

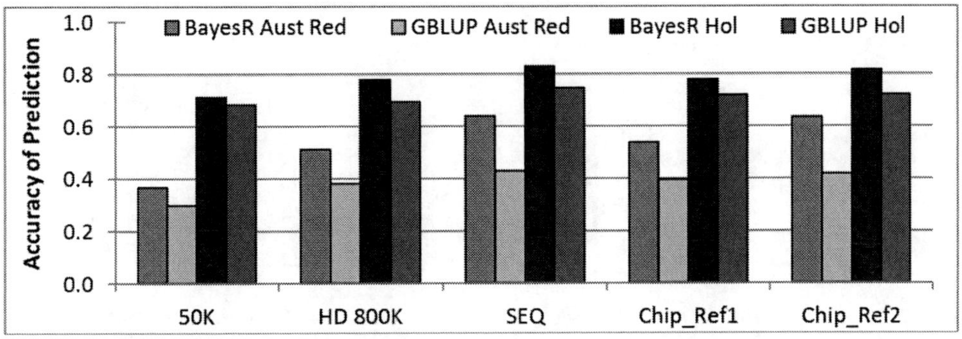

Figure 1. Accuracy of genomic prediction equations trained in Ref1 using a range of SNP genotypes and validated in Holsteins and Australian Reds. SEQ represents ~1million sequence variants, including the SNP chosen as surrogate QTN. Chip_Ref1 is a custom chip of 50K + 500 top putative QTN discovered in the same Ref1, while Chip_Ref2 is a custom chip with 50K + 500 top putative QTN discovered independently in Ref2.

Overall, the bias of genomic prediction (Fig 2) was largest for Chip_Ref1 where the top SNP were discovered in the same set as subsequently used to train the genomic predictions (Ref1). The regression was < 1 which indicates that genomic breeding values were over-predicted. This over-prediction can cause problems for the industry because genomic breeding values would be biased upwards compared to traditional breeding values. We were able to correct the bias (BayesR and GBLUP) in both validation sets, by using custom Chip_Ref2. We also investigated the proportion of variance explained by SNP in each analysis, and found that this variance was considerably over-estimated in the case of Chip_Ref1, compared to Chip_Ref2 where the variance was more accurately estimated.

This indicates that the bias is mainly due to a form of the "winner's curse" or "Beavis effect". That is, a proportion of the selected putative QTN from Ref1 were estimated to have a larger effect than the real effect, and when Chip_Ref1 was used for genomic prediction in the same Ref1 set, these effects are again overestimated. In BayesR the bias was more serious than GBLUP possibly because the BayesR mixture model allows for some large QTN effects, while GBLUP assumes all SNP effects are sampled from a single distribution so that larger effects are regressed more towards the mean. This phenomenon of bias was also reported by Veerkamp *et al.* (2016) using dairy cattle data. However, in the studies by Brøndum *et al.* (2015) and van den Berg *et al.* (2016) the bias was less apparent, most likely because their putative QTN discovery population did not exactly overlap with the genomic prediction reference populations. Wiggans *et al.* (2016) did not test for bias in their study. It might be expected that bias and reduced accuracy may be exacerbated if a GWAS is used to select the top putative variants because the Beavis effect is likely to be more pronounced with SNP effects fitted as fixed effects.

In conclusion, it is important to recognise the pitfalls of pre-selecting subsets of SNP for genomic prediction and to take steps to mitigate them, such as using independent reference populations for QTN discovery and genomic prediction. A potential alternative which does not require two independent populations is a new analytical approach (van den Berg *et al* 2017 - these proceedings) derived from a hybrid method of Expectation-Maximisation with BayesR (HyB_BR) developed by Wang *et al.* (2016).

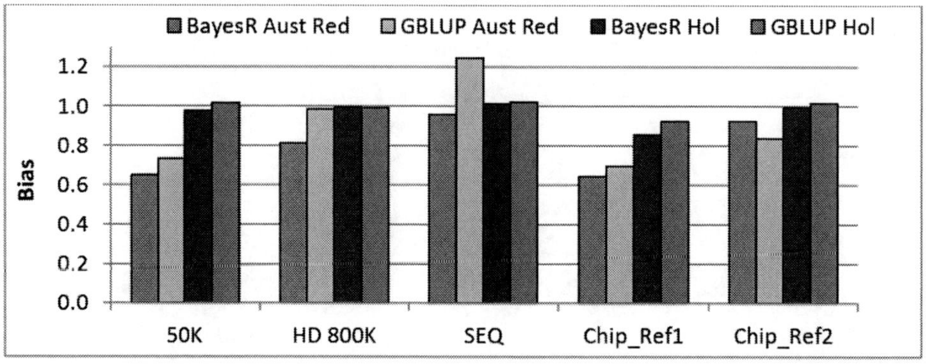

Figure 2. Bias of genomic prediction equations trained in Ref1 and validated in Holsteins or Australian Reds using a range of SNP genotypes. SEQ represents ~1million sequence variants and includes the SNP chosen as surrogate QTN. Chip_Ref1 is a custom chip of 50K + 500 top putative QTN discovered in the same Ref1, while Chip_Ref2 is a custom chip with 50K + 500 top putative QTN discovered independently in Ref2.

ACKNOWLEDGEMENTS

We acknowledge DataGene (and CRV Netherlands for providing access to some of the genotypes used in this study. We acknowledge our partners in the 1000 Bull Genomes Project for access to the reference genomes.

REFERENCES

Brøndum R.F., Su G., Janss L., et all. (2015) *J. Dairy Sci.* **98**: 4107.

Calus M.P.L., Bouwman A.C., Schrooten C. and Veerkamp R.F. (2016) *Gen. Sel. Evol.* **48**: 49.

MacLeod I.M., Bowman P.J., Vander Jagt C.J., et all. (2016) *BMC Genomics* **17**: 144.

van Binsbergen R., Calus M.P.L., Bink M.C.A.M., van Eeuwijk F.A., Schrooten C. and Veerkamp R.F. (2015) *Gen. Sel. Evol.* **47**: 71.

van den Berg I., Boichard D. and Lund M.S. (2016) *Gen. Sel. Evol.* **48**: 83.

Veerkamp R.F., Bouwman A.C., Schrooten C. and Calus M.P.L. (2016) *Gen. Sel. Evol.* **48**: 95.

Wang T., Chen Y.-P.P., Bowman P.J., Goddard M.E. and Hayes B.J. (2016) *BMC Genomics* **17**: 1.

Wiggans G.R., Cooper T.A., VanRaden P.M., Van Tassell C.P., Bickhart D.M. and Sonstegard T.S. (2016) *J. Dairy Sci.* **99**: 4504.

Proc. Assoc. Advmt. Anim. Breed. Genet. 22:145-148

GENOME-WIDE UNCORRELATED TRAIT ANALYSIS IDENTIFIES PLEITROPIC MARKERS FOR DAIRY CATTLE IN AUSTRALIA

R. Xiang[1,2], I.M. MacLeod[2], S. Bolormaa[2,3] and M.E. Goddard[1,2]

[1] Faculty of Veterinary & Agricultural Science, University of Melbourne, Parkville, VIC3010, Australia

[2] AgriBio, Dept. Economic Development, Jobs, Transport & Resources, Bundoora, VIC3083, Australia

[3] Cooperative Research Centre for Sheep Industry Innovation, Armidale, NSW 2351, Australia

SUMMARY

Selection should favour alleles which increase profitability considering their effects across all important traits. Therefore, understanding pleiotropy is an important aim. Obviously if traits are genetically correlated they must share some causal variants but it is possible that even uncorrelated traits share some causal variants. Here we analyse 25 traits on Australian dairy cattle. The 25 raw traits (RTs), covering milk production, fertility, behaviour, somatic cell count and conformation, of 2841 bulls were used to calculate uncorrelated principal components (PCs) and Cholesky transformation traits (CT). Multi-trait meta-analyses of single-trait genome-wide association studies (GWAS) for RT, PC and CT in these bulls were validated in 6821 cows. We observed a positive relationship between heritability estimates and the number significant SNPs detected in RTs and CTs. However, there was no relationship between the phenotypic importance of PCs and the number of significant SNPs detected. The major dairy cattle locus DGAT1 not only affected dairy production traits, also had validated small effects on fertility, milk speed and temperament. Our results highlight the importance of using genetic information of all traits to maximise pleiotropy detection and prioritise multi-trait genetic markers for the dairy industry.

INTRODUCTION

The profitability of dairy farming depends on many traits including milk production, fertility, diseases, workability and conformation or type traits (Byrne et al., 2015). Therefore, genomic selection should target genetic variants that increase an economic combination of traits such as the balanced performance index (BPI). When identifying genetic markers, such as single nucleotide polymorphisms (SNPs), associated with economic traits, we need to know the effect of the marker on all economic traits not just those where the marker has the biggest effect. That is, we would like to understand the pleiotropic effects of genes across all important traits.

Widespread pleiotropic effects of SNPs have been observed in beef cattle (Bolormaa et al., 2014) and sheep (Bolormaa et al., 2016). If traits are genetically correlated there must be some genes that affect both traits. However, it is also possible that uncorrelated traits share some causal variants. Principal component (PC) analysis, producing a small number of uncorrelated traits, has been proposed for conducting multi-trait genetic analysis (Klei, Luca, Devlin, & Roeder, 2008). If genes act through a limited number of physiological pathways, principle component analysis might capture the most important pathways in the first few PCs leading to a simple picture of pleiotropy.

To further understand pleiotropy in the dairy cattle population, a dataset from the Australian Dairy Herd Improvement Scheme (ADHIS) with 25 traits recorded on 9662 animals was retrieved. These 25 raw traits (RTs), including milk production, survival, fertility, temperament and linear type traits, were used to construct uncorrelated PCs and Cholesky transformed traits (CTs) (Golub & Van Loan, 2012). RTs and generated PCs and CTs were analysed with multi-trait genome-wide association studies (GWAS).

MATERIALS AND METHODS

Analyses included genotype of 2841 bulls as the discovery population and 6821 cows as the validation population from the breeds Holstein, Jersey and Australian Red. The distribution of genomic relatedness of bulls and cows in three breeds were shown in Figure 1. SNPs were genotyped by Illumina BovineLD BeadChip (7K), Illumina Bovine SNP array (54K) and Illumina Bovine HD genotypes (777 K). All animals were imputed to HD genotypes using Fimpute (Sargolzaei, Chesnais, & Schenkel, 2014) and in total, 632,002 SNPs were used. SNPs with minor allele frequency <0.01 or significant departure from Hardy-Weinberg equilibrium ($p<0.001$) were filtered out. The 25 phenotypic traits of these animals (trait deviations for cows and daughter trait deviations for bulls) were from the April 2016 genetic evaluations from the DataGene. Daughter trait deviations were the average trait deviations of a bull's daughters and all phenotypes were pre-corrected for known fixed effects.

The generation of PCs for the nth animal (u_n) was based on eigen-decomposition of $k=25$ RTs (g_n): $u_n = T' g_n$; Where u_n was a $k \times 1$ vector of PC scores for the animal n; T was an $k \times k$ matrix of eigenvectors such that the variance matrix of the PC Var(T'g) = D, a diagonal matrix of eigenvalues; g_n was an $k \times 1$ vector of RT for animal n. The CT scores for the nth animal (c_n) were calculated based on the Cholesky decomposition: $c_n = L^{-1} g_n$; where; L was the $k \times k$ matrix of the Cholesky factors which satisfied LL'= V(g), the $k \times k$ covariance matrix (Golub & Van Loan, 2012); g_n was a k × 1 vector of RT for the animal n. Single-trait GWAS was performed in GEMMA (Zhou & Stephens, 2014) using data from the discovery population: $y = mean + fixed\ effects + SNP_i + GRM + e$; where y= vector of k RTs, PCs or CTs for bulls; fixed effects= breeds; SNP_i = that each SNP genotype was fitted as a covariate one at a time; a polygenic random effect described by the GRM= genomic relatedness matrix calculated from GEMMA based on all SNPs; e = error. A multi-trait meta-analysis based on either the 25 RTs, 25 PCs or 25 CTs followed previous procedures (Bolormaa et al., 2016; Bolormaa et al., 2014). SNPs that were significant in the discovery sample were tested in the validation sample using an index of traits that maximises the effect of the SNP (Bolormaa et al., 2016; Bolormaa et al., 2014). Single-trait GWAS in the validation population was also used to confirm SNP effects on individual RTs.

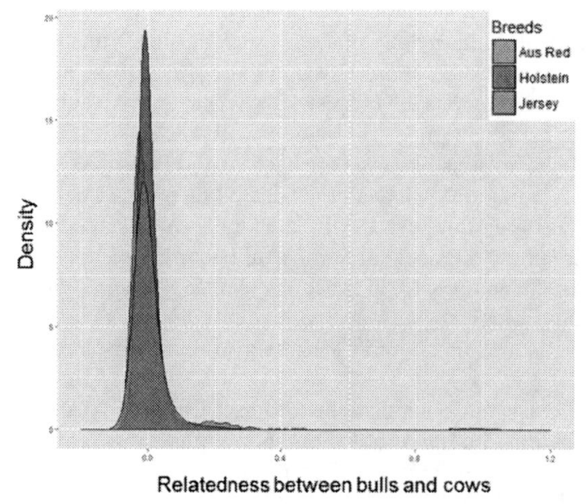

Figure 1. Density plot of the genomic relationship matrix between bulls and cows.

RESULTS AND DISCUSSION

For both RTs and CTs, the number of significant ($P<1\times10^{-5}$) SNPs detected by single-trait GWAS generally increased with the estimated heritability of the phenotype because the power to detect effects increases with h^2 (Figure 2A,B). (The heritability of bull phenotypes is the proportion of variation in daughter trait deviation explained by all SNPs jointly). Consistent with previous reports (Kemper et al., 2015; MacLeod et al., 2016), the RT of milk, protein and fat yield had the highest heritability estimates (all $h^2>0.8$ and se=0.02) and the largest numbers of significant SNPs (more than 100) detected. Survival and fertility as reproductive complex traits had mid-range heritability estimates (both $h^2>0.5$ and se=0.03) with 27 and 31 significant SNPs detected, respectively. Mid-range heritability was also estimated for temperament and milk speed (both $h^2>0.5$ and se=0.03). However, single-trait GWAS only detected 6 and 13 significant SNPs for temperament and milk speed, respectively. The h^2 of likeability is 0.48 (se=0.03) with only four significant SNPs detected. The heritability estimates of dairy type traits ranged from 0.35 (rear legs set, se=0.04) to 0.69 (front teat placement, se=0.03). However, all type traits had a small number of significant SNP detected. Rear legs set had 15 significant SNPs and front teat placement had only 1 significant SNP. This is likely to be due to the complexity of the type traits, i.e. a large number of causal variants each with a very small effect. Our discovery sample size may not be large enough to capture highly significant SNPs.

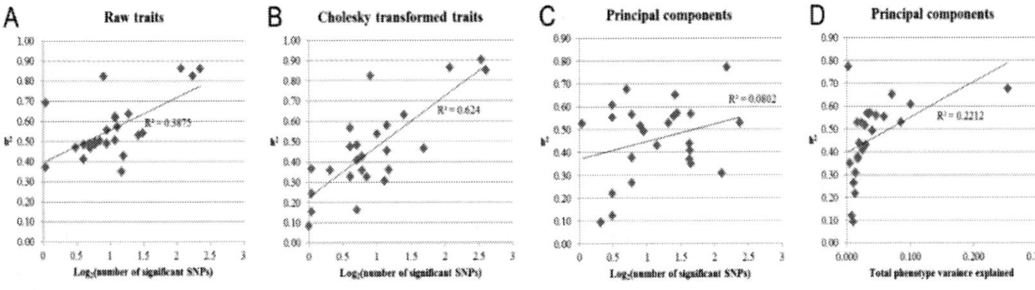

Figure 2.The relationship between heritability estimates and the number of significant SNPs detected by single-trait GWAS for RTs (A), CTs (B) and PC (C). D: The relationship between the heritability estimates of each PC and the total phenotypic variance explained by each PC.

The 25 PCs, accumulatively explained 93% of the total phenotypic variances, showed a complicated pattern (Figure 2C,D). The first PC, which explained 25% of total phenotypic variances for all RTs, had a high estimation of heritability (0.67, se=0.03) but only 5 significant SNPs. This PC had loadings from many traits and perhaps this generates a very complex trait affected by many genes. On the other hand, PC18 with top factor loading related to milk fat yield, only contributed 1.7% of the variances to all traits, had a modest heritability (0.53, se=0.03), but had the largest number of significant SNPs (241) amongst the PCs. The last PC (PC25) with high positive factor loading for protein yield and high negative factor lading for milk yield, explained 0.03% of the variances in all traits, had a modest heritability (0.50±0.03) but 153 significant SNPs. Our results are consistent with a previous simulation study in humans where the genetic information of all PCs are important (Aschard et al., 2014). Thus, only considering a small number of PCs might cause loss of power for genetic analysis.

Thus the results do not support the hypothesis that genes act through a small number of common, physiological mechanisms. This is exemplified by SNPs within and near DGAT which have significant effects on several PCs. This occurs because the effects of DGAT do not follow the pattern described by the overall genetic correlations. For instance, milk and fat yield are positively correlated but the allele of DGAT which increases milk decreases fat yield.

Three multi-trait meta-analyses were performed based on either all RTs or PCs or CTs. The 3 meta-analyses largely detected the same significant SNPs as they are all approximations to a full multi-trait analysis. They also detected many more significant SNPs than single-trait GWAS using the same threshold ($P<1\times10^{-5}$ and FDR<0.01) (Figure 3A,B).

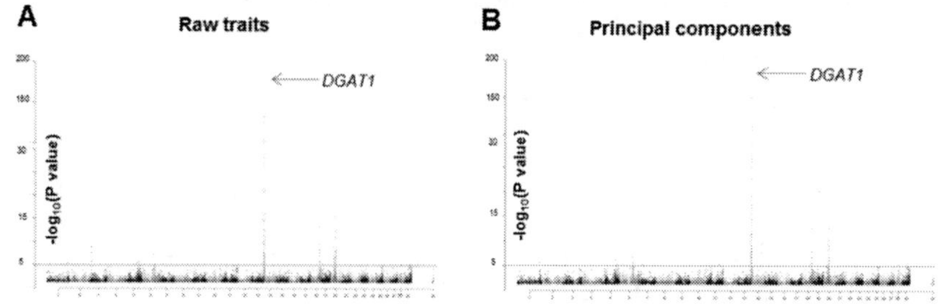

C

DGAT1.Chr14:1.8M

RT.Prot	RT.Fat	RT.Milk	RT.SCC	RT.SurvD	RT.Fert	RT.Temp	RT.MSpeed	RT.Stat	RT.Like	RT.Angul	RT.Bone	RT.CentL
[illegible]	[illegible]	[illegible]			2.9	1.8	-1.5					
PC1	PC2	PC3	PC4	PC5	PC6	PC7	PC8	PC9	PC10	PC11	PC12	PC13
	2.0			1.1	1.5	2.2	-2.0	-3.1	2.8	4.9	5.0	7.3
CT.Prot	CT.Fat	CT.Milk	CT.SCC	CT.SurvD	CT.Fert	CT.Temp	CT.MSpeed	CT.Stat	CT.Like	CT.Angul	CT.Bone	CT.CentL
[illegible]	[illegible]	-1.9	-1.3		4.5	1.6	-1.2	3.7			1.4	1.4

RT.ChestW	RT.ForeA	RT.MuzW	RT.PinSet	RT.PinW	RT.RSet	RT.RearAH	RT.RearAW	RT.TeatPF	RT.UdTex	RT.OType	RT.Mamm
	3.0			-1.1		-2.1	-2.2				
PC14	PC15	PC16	PC17	PC18	PC19	PC20	PC21	PC22	PC23	PC24	PC25
-2.6	-7.1	5.1	-7.1	[illegible]	11.8		-4.1		2.5	7.7	3.3
CT.ChestW	CT.ForeA	CT.MuzW	CT.PinSet	CT.PinW	CT.RSet	CT.RearAH	CT.RearAW	CT.TeatPF	CT.UdTex	CT.OType	CT.Mamm
-2.2	1.7	-1.1				-1.5		2.0		-4.1	2.0

Figure 3. A-B: Manhattan plot of multi-trait meta-analysis. C: t values with absolute values >1 of DGAT1 across traits.

DGAT1 was the most significant locus in the multi-trait analysis (Figure 3A,B) with effects on many RT, PCs and CTs. Along with the strong effects on production traits, DGAT1 also had small but validated effects on fertility, milk speed, temperament and type RTs, which are important information for the breeders (Figure 3C). This highlights the advantage of conducting multi-trait analysis in extending knowledge for unknown effects of known loci. Most SNPs did not have a significant effect on many traits as DGAT did but this may indicate a lack of power rather than a lack of pleiotropic effects. If the example of DGAT is repeated for other loci it is important because it indicates that SNPs with a small effect on one trait may be detected by their large effect on another trait.

REFERENCES

Aschard, H., Vilhjálmsson, B. J., et al. (2014). *Am. J. Hum. Genet.*, ***94***(5), 662-676.
Bolormaa, S., Hayes, B. J., et al. (2016). *BMC Genomics*, ***17***(1), 1.
Bolormaa, S., Pryce, J. E., et al. (2014). *PLoS Genet*, ***10***(3), e1004198.
Byrne, T., Martin-Collado, D., et al. (2015). *Proc. Assoc. Advmt. Breed. Genet*, ***21***, 21-24.
Golub, G. H., & Van Loan, C. F. (2012). *Matrix computations* (Vol. **3**): JHU Press.
Kemper, K. E., Reich, C. M., et al. (2015). *Genet. Sel. Evol.*, ***47***(1), 1.
Klei, L., Luca, D., et al. (2008). *Genet. Epidemiol.*, ***32***(1), 9-19.
MacLeod, I., Bowman, P., et al. (2016). *BMC Genomics*, ***17***(1), 1.
Sargolzaei, M., Chesnais, J. P., et al. (2014). *BMC Genomics*, ***15***(1), 1.
Yang, J., Lee, S. H., et al. (2011). *Am. J. Hum. Genet.*, ***88***(1), 76-82.
Zhou, X., & Stephens, M. (2014). *Nat. Methods*, ***11***(4), 407.

Proc. Assoc. Advmt. Anim. Breed. Genet. 22:149-152

EXPRESSION OF HETEROSIS FOR LIVE WEIGHT IN GROWTH CURVES OF NEW ZEALAND DAIRY HEIFERS

R.C. Handcock[1], N. Lopez-Villalobos[1], L.R. McNaughton[2], G.R. Edwards[3] and R.E Hickson[1]

[1]Institute of Veterinary, Animal and Biomedical Sciences, Massey University, Private Bag 11-222, Palmerston North, New Zealand
[2]Livestock Improvement Corporation, Private Bag 3016, Hamilton 3240, New Zealand
[3]Faculty of Agriculture and Life Sciences, PO Box 85084, Lincoln University, Lincoln 7647, Christchurch

SUMMARY

The objective of this study was to model the growth curves of New Zealand dairy heifers and to estimate breed and heterosis effects between Holstein Friesian (F), Jersey (J) and Holstein Friesian-Jersey crossbreed (FxJ) heifers before first calving at 2 years of age. Data consisted of 1,653,214 liveweight (LWT) records obtained from 189,936 spring-born dairy heifers located in 1,547 herds. A fourth order Legendre polynomial was fitted to the LWT data to model growth curves. At all ages F heifers were heavier than FxJ which were heavier than J heifers. The difference among the breeds varied over the growth period studied; F heifers were not a constant percentage heavier than FxJ or J heifers. This demonstrates that the breeds in this study exhibited different growth patterns. Breed effects, defined as F-J, were positive and ranged from 12.8 kg to 55.3 kg. Heterosis effects, expressed in kilograms, were positive and increased up to 18 months of age. Expressed as a percentage of the average of the parental breeds, heterosis was greatest at nine months of age (3.6%) and least at 22 months of age (2.0%). In conclusion, in New Zealand dairy heifers, heterosis effects were different throughout the growth period and F, J and FxJ heifers exhibited different growth patterns.

INTRODUCTION

The predominant dairy breeds in New Zealand are Holstein-Friesian (F), Jersey (J) and Holstein-Friesian x Jersey crossbred (FxJ) (Livestock Improvement Corporation and DairyNZ 2016). Holstein-Friesian is a later maturing and heavier breed compared with the lighter and early maturing J (Leche 1971; Hickson *et al.* 2012). Jersey heifers attained puberty at a younger age compared with F heifers (Hickson *et al.* 2011), further emphasising their earlier maturity. Heterosis is present in FxJ animals for mature liveweight (LWT) and ranges from 7.2 to 10 kg (Harris *et al.* 1996; Harris 2005), but has not been documented in growing heifers in New Zealand. Current target LWTs for dairy heifers are 30%, 60% and 90% of mature LWT at six, 15 and 22 months of age (Burke *et al.* 2007). There are differences in the proportion of target achieved between breeds (McNaughton and Lopdell 2013; Handcock *et al.* 2016), suggesting a potential difference in growth pattern and therefore, indicating that appropriate target percentage may be different among breeds. Due to the pasture-based farming systems in New Zealand, dairy heifers tend to follow a seasonal pattern of growth that matches pasture quality and quantity (Litherland *et al.* 2002; Handcock *et al.* 2016). The objective of this study was to use the model of the growth curves for New Zealand dairy heifers to estimate breed and heterosis effects.

MATERIALS AND METHODS

Liveweight records of New Zealand dairy heifers were extracted from the Livestock Improvement Corporation database. Heifers that were spring-born between the 2006-07 and 2013-14 dairy seasons, had at least two LWT records between birth and 12 months of age and two LWT records between 13 months of age and first calving at approximately two years of age, or 24 months of age if the heifer did not have any recorded calving dates were considered. Initial data cleaning was completed by calculating the mean and standard deviation of LWT for each age (in months), and for each breed. Liveweight records that were more than four standard deviations from their corresponding breed-age mean were removed (Pietersma *et al.* 2006). This method was iterated until no more records were deleted (Pietersma *et al.* 2006). This left a dataset comprised of 1,656,433 LWT records obtained from 189,936 dairy heifers located in 1,547 herds. Of these heifers, 48,026 were F; 12,407 were J and 129,503 were FxJ. Only heifers with known dam and sire and less than 2/16 (12.5%) of breeds other than F or J were included in the dataset.

Based on recorded pedigree and sire and dam breed proportions; individual animal's breed proportions were known, and were used to calculate coefficient of specific heterosis.

Statistical analysis. A Legendre polynomial of order four was fitted to LWT data using random regression to obtain regression coefficients for each heifer using ASReml (Gilmour *et al.* 2015). To remove outlier observations the relative measurement error (RME) was calculated as:

$$\text{RME} = \left(\frac{Predicted\ LWT - Actual\ LWT}{Predicted\ LWT}\right) \times 100$$

Any actual LWT between three and 23 months of age that had an absolute RME greater than 18% (mean + four standard deviations) was considered an outlier and removed from the dataset. The RME calculates the percentage deviation of the actual LWT from the predicted LWT by assuming that the predicted LWT is the "true" value. At birth, one, two and 24 months of age the accuracy of the fourth-order polynomial was low (data not shown). Records were not removed at these ages as the predicted LWT was not accurate enough to be defined as the "true" value. The cleaned dataset included 1,653,214 observations (0.2% of data removed) on the same 189,936 animals. An order-four Legendre polynomial was fitted to the cleaned dataset and was used for subsequent analysis. The individual regression coefficients were used to estimate LWT at 3, 6, 9, 12, 15, 18 and 22 months of age for each heifer.

Breed and heterosis effects for LWT at the different ages were estimated using a linear mixed model in SAS version 9.4 (SAS Institute Inc). The mixed model included the fixed effect of birth year, dam age (2 years old, or 3 years old and older), and island (North vs South) as class effects, the deviation from median birthdate (within-herd), proportion of F, proportion of Other breeds (O), heterosis FxJ, heterosis FxO and heterosis JxO fitted as covariates, and the random effect of herd of birth. The estimates of the regression coefficients were used to predict the LWT of F, J and F_1 F×J cows at different ages.

RESULTS

Predicted means and standard errors of LWT at different ages for the three breed groups are presented in Table 1. At all ages, F heifers were heavier than FxJ which in turn were heavier than J heifers. Figure 1 displays the deviation of F and FxJ from J, as well as the expected average of the parental breeds ((F + J)/2). Holstein-Friesian heifers were 15.8% heavier than J heifers at three months of age; decreasing to 14.9% at nine months of age and increasing to a maximum at 17 months of age. First cross FxJ heifers were between 10.5 and 12.2% heavier than J heifers from three to 19 months of age; decreasing to 7.6% by 22 months. The difference between FxJ and the expected average of the parental breeds ranged from 2.1 to 3.9% with the greatest difference occurring between 8 and 12 months of age.

The estimates for breed and heterosis effects for LWT are shown in Table 1. Breed differences between F and J was estimated to be greatest at 18 months of age. Heterosis, in absolute values was positive at all ages and greatest at 18 months of age. Heterosis (as a proportion of parent average) was greatest at 9 months of age.

Table 1. Live weight of Holstein-Friesian (F), Jersey (J) and first cross (F_1) Holstein-Friesian-Jersey crossbred (FxJ) heifers, and estimates of breed and heterosis effects at different ages

Age	Live weight (kg)			Breed effect	Heterosis (FxJ)	
(months)	F	F_1 FxJ	J	F-J (kg)	kg	% †
3	93.5 ± 0.3	89.2 ± 0.3	80.8 ± 0.3	12.8* ± 0.2	2.1* ± 0.1	2.4%
6	156.5 ± 0.5	150.9 ± 0.5	135.8 ± 0.5	20.7* ± 0.2	4.8* ± 0.2	3.3%
9	193.2 ± 0.6	187.1 ± 0.6	168.1 ± 0.6	25.1* ± 0.3	6.5* ± 0.2	3.6%
12	238.8 ± 0.7	230.0 ± 0.7	205.7 ± 0.7	33.1* ± 0.3	7.8* ± 0.3	3.5%
15	304.6 ± 0.7	291.1 ± 0.7	259.5 ± 0.8	45.2* ± 0.4	9.0* ± 0.3	3.2%
18	380.2 ± 0.8	362.2 ± 0.8	324.8 ± 0.8	55.3* ± 0.4	9.7* ± 0.3	2.8%
22	430.4 ± 0.7	417.4 ± 0.7	388.0 ± 0.8	42.4* ± 0.4	8.2* ± 0.3	2.0%

*Mean is significantly different from zero (P<0.0001).

† Expressed as a percentage of heterosis effects relative to the phenotypic average of the parental breeds ((F+J)/2).

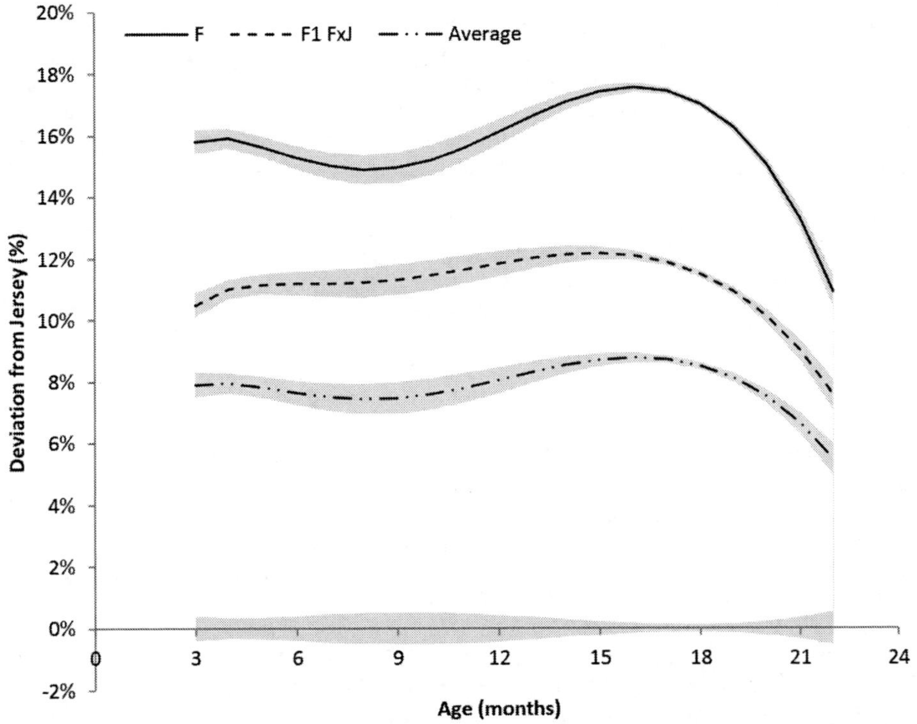

Figure 1: Deviation of estimated live weight of Holstein Friesian (F), Holstein Friesian-Jersey crossbred (F_1 FxJ) and the expected average of the parental breeds ((F + J)/2; Average) from Jersey (J) heifers ("0" line) derived from the fourth-order Legendre polynomial. Grey shading represents 95% confidence intervals for each breed.

DISCUSSION

The LWT of F heifers was consistently heavier than FxJ and J heifers, as expected based on the difference in mature size (Hickson *et al.* 2012; Livestock Improvement Corporation and DairyNZ 2016). However, the growth pattern was different; F heifers were not a constant percentage heavier than FxJ or J heifers (Figure 1). If the growth pattern was similar among the breeds the slope of the lines in Figure 1 would be zero for both F and FxJ. From four to nine months of age the difference between F and J decreased; over this same period, the difference between FxJ and J was constant. Furthermore, heterosis exhibited by FxJ heifers was the greatest (3.6%) at nine months of age. The difference indicates that at different ages, one breed has a greater potential for growth compared with the other and heterosis significantly contributes to the difference in growth pattern.

Heterosis estimates for mature FxJ cows range from 7.2 kg to 10 kg (Harris *et al.* 1996; Harris 2005); similar to the values in the current study from nine months of age onwards. Heterosis varied throughout the growth period in first generation FxJ heifers in the USA (Hilder and Fohrman 1949); and were similar at 3 (2.2%), 6 (3.2%) and 18 months of age (2.5%) to what was reported in the current study. At 9, 12 and 15 months of age, heterosis estimates for the current study were greater than those reported by Hilder and Fohrman (1949) (1.3, 0.4 and 1.4% respectively). The estimates reported by Hilder and Fohrman (1949) are from a total of 18 calves that were reared in a predominantly indoor system. In contrast, the results reported in the current study are from 189,936 heifers reared in a pasture based system, which may explain the differences in heterosis estimates between the two studies.

In conclusion, New Zealand dairy heifers exhibited heterosis effects throughout the growth period and F, J and FxJ heifers displayed different growth patterns. The weight for age targets may therefore be different for these breeds, due to the differing growth patterns.

ACKNOWLEDGEMENTS

The authors would like to acknowledge Katie Eketone for extracting the data provided by Livestock Improvement Corporation (Hamilton, New Zealand).

REFERENCES

Burke C., Blackwell M. and Little S. 2007. The InCalf Book for New Zealand dairy farmers. DairyNZ, Hamilton.

Gilmour A.R., Gogel B.J., Cullis B.R., Welham S.J. and Thompson R. 2015. ASReml User Guide Release 4.1 Structural Specification. VSN International Ltd, Hemel Hempstead, HP1 1ES, UK.

Handcock R.C., Lopdell T. and McNaughton L.R. (2016). *Proc. N.Z. Soc. Ani. Prod.* **76**: 3.

Harris B.L. (2005). *N Z Vet J.* **53**: 384.

Harris B.L., Clark J.M. and Jackson R.G. (1996). *Proc. N.Z. Soc. Ani. Prod.* **56**: 12.

Hickson R.E., Balcomb C.C., Fraser K.R., Lopez-Villalobos N., Kenyon P.R. and Morris S.T. (2011). *Proc. Assoc. Advmt. Anim. Breed. Genet.* **19**: 51.

Hickson R.E., Kenyon P.R., Lopez-Villalobos N., Laven R.L., Fraser K.R., Balcomb C.C. and Morris S.T. (2012). *Proc. Soc. Sheep & Beef Cattle Vet. NZVA.* **2012**: 4.22.1.

Hilder R.A. and Fohrman M.H. (1949). *J. Agri. Res.* **78**: 457.

Leche T.F. (1971). *Aus. J. Agri. Res.* **22**: 829.

Litherland A.J., Woodward S.J.R., Stevens D.R., McDougal D.B., Boom C.J., Knight T.L. and Lambert M.G. (2002). *Proc. N.Z. Soc. Ani. Prod.* **62**: 138.

Livestock Improvement Corporation and DairyNZ. 2016. New Zealand Dairy Statistics 2015-16. Livestock Improvement Corporation, DairyNZ, Hamilton, New Zealand.

McNaughton L.R. and Lopdell T. (2013). *Proc. N.Z. Soc. Ani. Prod.* **73**: 103.

Pietersma D., Lacroix R., Lefebvre D., Cue R.I. and Wade K. (2006). *Can. J. Anim. Sci.* **86**: 325.

Proc. Assoc. Advmt. Anim. Breed. Genet. 22:153-156

INCREASING THE ACCURACY OF GENOMIC PREDICTION FOR RESIDUAL FEED INTAKE IN DAIRY CATTLE BY USING SNPs ASSOCIATED WITH RFI IN BEEF CATTLE

M. Khansefid[1,2], J.E. Pryce[2,3] and M.E. Goddard[1,2]

[1] Department of Agriculture and Food Systems, The University of Melbourne, Melbourne, Victoria 3010, Australia
[2] Agriculture Victoria, AgriBio, Centre for AgriBioscience, Bundoora, Victoria 3083, Australia
[3] School of Applied Systems Biology, La Trobe University, Bundoora, Victoria 3083, Australia

SUMMARY

Residual feed intake (RFI) is a measure of the efficiency of animals in converting feed to products. Improving RFI in dairy cattle can reduce the costs of raising heifers and producing milk. However, calculating RFI requires expensive equipment to measure the feed intake for each individual. Since April 2015, a "Feed Saved" breeding value has been available in Australia that combines RFI with maintenance requirements. However, the size of the reference population used for genomic prediction of RFI is comparatively and consequently the accuracy of predictions are modest. To improve the prediction accuracy for RFI, the current reference population consisting of 843 heifers and 236 Australian cows and 954 European cows (357 British and 597 Dutch) was extended by including RFI measurements of 206 Australian cows. Furthermore, information from markers which were associated with RFI in 4,772 beef cattle ($p < 0.001$) was used to construct a genomic relationship matrix (GRM) for dairy cattle. We also compared the use of imputed whole genome sequence (WGS) data with 800K SNP-chip genotypes and 2 methods of calculating a GRM described by VanRaden (2008) and Yang *et al.* (2010). The use of the SNPs from the 800K SNP-chip which were associated with RFI in beef cattle improved the accuracy of genomic estimated breeding values (GEBVs) in dairy cattle. However, the use of imputed WGS data did not improve prediction accuracy, especially when the Yang *et al.* method of calculating the GRM was used. The Yang *et al.* method gives extra weight to rare alleles and these SNPs have low imputation accuracy. So it is likely that errors in imputation affect the results when using WGS and this effect is magnified when Yang *et al.* is used to construct the GRM. The best model tested was the GRM built using the Yang et *al.* method with SNP-chip genotypes and when extra weight was given to the SNPs associated with RFI in beef cattle. The accuracy of GEBVs for RFI in the best model for heifers and cows were 0.67 and 0.46, respectively.

INTRODUCTION

The efficiency of dairy cattle in utilizing feed to grow and produce milk is one of the main factors influencing the profitability of production (Berry and Crowley 2013). Residual feed intake (RFI) is one of the criteria for measuring feed efficiency. RFI is the difference between actual and predicted feed intake for each individual (Koch *et al.* 1963) which has high to moderate heritability in growing heifers and low to moderate heritability in milking cows (Berry and Pryce 2014). Hence, improving the efficiency of animals in converting feed to products is feasible by selecting and breeding cattle which need less feed than average to gain the same weight, or produce the same amount of milk. However, RFI is expensive to measure because it requires precise measurements of individual feed consumption, weight gain, and also milk yield and its components in milking cows and this has limited direct selection for feed efficiency in dairy cattle (Beever and Doyle 2007). Moreover, due to the polygenic architecture of feed efficiency, it is hard to find major genes influencing RFI. Genomic selection, using single nucleotide polymorphisms

(SNP) genotypes to estimate breeding values without measuring feed intake on selection candidates, could overcome this limitation (Meuwissen *et al.* 2001). However, genomic selection still requires a genotyped reference population with phenotypes for RFI. The limited size of reference populations, especially for RFI in milking cows, results in modest prediction accuracies. Since April 2015 the "Feed Saved" breeding value has been available in Australia and is also part of the national selection index (Pryce *et al.* 2015). Feed Saved includes genomic breeding values for RFI in heifers and cows and maintenance requirements in lactating cows and uses genomic predictions of RFI using a reference population of Australian cows and heifers and European cows. The aim of this research is to increase the accuracy of genomic estimated breeding values (GEBVs) for RFI in Australian heifers and milking cows through 1) including more animals in the reference population, 2) RFI information from non-dairy breeds and 3) using whole genome sequence (WGS) instead of SNP-chip genotypes.

MATERIALS AND METHODS

Animals and RFI measurements. The RFI measurements used for this study were from 843 Australian heifers and 440 Australian cows (139 animals had RFI measurements as heifers and cows), 954 European cows (357 British and 597 Dutch) and 4,772 beef cattle (Khansefid *et al.* 2014; Pryce *et al.* 2015). In this study, the RFI measurements in Australian cows were recalculated after including 206 new animals to the model described by Pryce *et al.* (2015).

Genotypes. The Australian heifers had 800K (Illumina HD Bovine SNP chip) genotypes (Pryce *et al.* 2012) and the rest of the dairy cattle had 50K (Illumina BovineSNP50K) genotypes which were imputed to HD (Pryce *et al.* 2014). For beef cattle, the SNP genotypes were either from HD or imputed from lower density (7K, 10K or 50K) to 800K (Khansefid *et al.* 2014). Moreover, for all datasets, the SNP-chip genotypes were imputed to WGS genotypes using FImpute (Sargolzaei *et al.* 2014) and RUN4 of 1000Bulls project as the reference.

Genome-wide association study (GWAS). The GWAS was conducted using beef cattle data to find associations between each SNP and RFI measurements using the model described by Khansefid *et al.* (2014) but using WGS genotypes in addition to SNP-chip genotypes.

Genomic relationship matrix (GRM). The GRMs were constructed using 2 methods (Yang *et al.* (2010) and VanRaden (2008)) for SNP-chip ($GRM_{SNP\text{-}chip}$) and WGS genotypes (GRM_{WGS}). Separate GRMs were also calculated from the SNP-chip ($GRM_{SNP\text{-}chip}$*) and WGS genotypes (GRM_{WGS}*) using the SNPs that were associated with RFI in beef cattle (p <0.001).

Statistical model. RFI measurements for heifers, Australian cows and European cows were considered to be 3 separate traits and were fitted in a multi-trait model (Equation 1) to calculate GEBVs using ASReml (Gilmour *et al.* 2009), where **y** is a T×1 vector consisting of RFI measurements on 1 or more of the 3 traits for each animals, **Z** is an incidence matrix associating observations to animals and traits, **g** contains the breeding values for each of 3 traits for all animals distributed as N(0, **G**⊗**K**), **G** is the genomic relationship matrix and **K** is a matrix of additive genetic variances and covariances between RFI in 3 datasets and **e** is a vector of residual terms.

$$\mathbf{y}=\mathbf{Zg}+\mathbf{e} \quad [1]$$

To give extra weights to the SNPs associated with RFI in beef cattle, the average of $GRM_{SNP\text{-}chip}$ and $GRM_{SNP\text{-}chip}$* (i.e. $GRM_{SNP\text{-}chip\ \&\ SNP\text{-}chip*}$) and also the average of GRM_{WGS} and GRM_{WGS}* (i.e. $GRM_{WGS\ \&\ WGS*}$) were calculated and fitted in equation 1.

Accuracy of genomic prediction. The accuracy of genomic predictions was calculated with a 5 fold cross-validation strategy. The dataset was divided into 5 subsets, 4 of the subsets were used as a reference population and the 5th subset was used as a validation sample. The animals in the 5 subsets were selected randomly except paternal half sibs were always placed in the same subset. Then, the GEBVs of validation animals, whose phenotypes were not included in the analysis, were estimated by genomic BLUP. The accuracy of each validation set was calculated as the correlation between GEBVs and RFI phenotypes divided by the square root of estimated heritability (h^2 for RFI in heifers and Australian cows were estimated 0.33 and 0.26, respectively) and the average across 5 validation sets was reported as the accuracy of prediction.

RESULTS AND DISCUSSION

Genotypes. In SNP-chip genotypes, 569,179 SNPs were in common between the datasets and had minor allele frequency (MAF) greater than 0.001. In WGS genotypes, 24,352,503 SNPs had MAF > 0.001.

GWAS. Among the common SNPs that had MAF > 0.001, 1,739 SNPs in SNP-chip genotypes and 60,646 SNPs in WGS genotypes were significantly associated with RFI in beef cattle ($p < 0.001$). So, about 0.3% of the SNPs were associated with RFI in beef cattle ($p < 0.001$) for both the SNP-chip or WGS genotypes.

Accuracy of genomic prediction. The accuracies of genomic predictions using different GRMs in Equation 1, are shown in Table 1. Substituting $GRM_{SNP\text{-}chip}$ with GRM_{WGS} did not improve the prediction accuracies. The accuracies were actually reduced when the Yang *et al.* (2010) method was used in GRM_{WGS} construction. So, the assumption of Yang *et al.* (2010) that rare alleles are more informative for WGS data seems to be incorrect. In WGS data there are many more SNPs with low MAF distributed across genome than in SNP-chip markers. Moreover, the accuracy of imputation is lower for rare alleles and therefore giving extra weight to these SNPs could reduce the accuracy of genomic predictions.

When the SNP-chip genotypes were used to construct the GRMs, there was no noticeable difference between the GRMs according to Yang *et al.* (2010) or VanRaden (2008). However, using WGS genotypes to make the GRMs, there was a slight superiority in constructing the GRMs according to VanRaden (2008).

Giving extra weight to the SNPs associated with RFI in beef cattle by using $GRM_{SNP\text{-}chip\ \&\ SNP\text{-}chip*}$ improved the accuracy of predictions in heifers and Australian cows. However, using $GRM_{WGS\ \&\ WGS*}$ in the model did not improve the prediction accuracy. When using WGS, the SNPs associated with RFI ($p < 0.001$) in beef cattle were distributed across the genome and included many low MAF SNPs. The rate of false positive associations seems to be higher for SNPs with low MAF because the imputation has more errors for these SNPs.. This problem could potentially be solved if a more stringent p-value was used to choose the SNPs for GRM_{WGS}*, however due to polygenic architecture of RFI, some SNPs with small effects would also be excluded.

Table 1. The accuracy of RFI predictions for Australian heifers and cows using different GRMs constructed according to Yang *et al.* (2010) and VanRaden (2008) (in parenthesis)

GRM	Accuracy of RFI prediction for heifers	Accuracy of RFI prediction for cows
$GRM_{SNP\text{-}chip}$	0.57 (0.56)	0.34 (0.35)
$GRM_{SNP\text{-}chip\ \&\ SNP\text{-}chip*}$	0.67 (0.65)	0.46 (0.41)
GRM_{WGS}	0.52 (0.55)	0.31 (0.39)
$GRM_{WGS\ \&\ WGS*}$	0.50 (0.53)	0.32 (0.34)

$GRM_{SNP\text{-}chip}$ is constructed from 569,179 SNP-chip genotypes.
$GRM_{SNP\text{-}chip\ \&\ SNP\text{-}chip*}$ is the average of $GRM_{SNP\text{-}chip}$ and the GRM built from 1,739 SNPs in SNP-chip genotypes which were significantly associated with RFI in beef cattle ($p < 0.001$).
GRM_{WGS} is constructed from 24,352,503 WGS genotypes.
$GRM_{WGS\ \&\ WGS*}$ is the average of GRM_{WGS} and the GRM built from 60,646 SNPs in WGS genotypes which were significantly associated with RFI in beef cattle ($p < 0.001$).

The best model tested was the GRM built using the Yang et *al.* method with SNP-chip genotypes and when extra weight was given to the SNPs associated with RFI in beef cattle. The accuracy of GEBVs for RFI in the best model for heifers and cows were 0.67 and 0.46, respectively.

CONCLUSION

This study shows that giving extra weight to SNPs that were associated with RFI in beef cattle increased the accuracy of GEBVs in dairy cattle. However, using imputed WGS data instead of 800K SNP-chip genotypes did not improve the prediction accuracy of genomic BLUP especially when the Yang *et al.* (2010) method was used to build GRM. The poor performance of WGS could be due to imputation errors and the use of BLUP rather than a non-linear method of calculating GEBVs. So, in order to benefit from using WGS genotypes, we need to: 1) use more accurate imputation, or direct genotyping of sequence variants, 2) find suitable statistical models such as Bayesian models, which allow a large proportion of SNPs to have zero effects and 3) use knowledge about the functionality of sequence variants. However, the current solution is to use SNP chip genotypes and to give extra weight in the GRM to the SNPs associated with RFI in beef cattle.

ACKNOWLEDGEMENTS

The authors would like to acknowledge Wageningen University and Scotland's Rural University College for data contribution, and also Dairy Futures CRC and DairyBio for funding this research.

REFERENCES

Beever, D.E. and Doyle, P.T. (2007) *Anim. Prod. Sci.* **47**: 645.

Berry, D.P. and Crowley, J.J. (2013) *J. Anim. Sci.* **91**: 1594.

Berry, D.P. and Pryce, J.E. (2014) *10th World Congr. Genet. Appl. Livest. Prod., Vancouver, Canada. Am. Soc. Anim. Sci., Champaign, IL.*

Gilmour, A.R., Gogel, B.J., Cullis, B.R., Thompson, R. and Butler, D. (2009) *VSN International Ltd, Hemel Hempstead, UK.*

Khansefid, M., Pryce, J.E., Bolormaa, S., Miller, S.P., Wang, Z., Li, C. and Goddard, M.E. (2014) *J. Anim. Sci.* **92**: 3270.

Koch, R.M., Swiger, L.A., Chambers, D. and Gregory, K.E. (1963) *J. Anim. Sci.* **22**: 486.

Meuwissen, T.H., Hayes B.J. and Goddard M. E. (2001) *Genetics* **157**: 1819.

Pryce, J.E., Gonzalez-Recio, O., Nieuwhof, G., Wales, W.J., Coffey, M.P., Hayes, B.J. and Goddard, M.E. (2015) *J. Dairy Sci.* **98**: 7340.

Pryce, J.E., Johnston, J., Hayes, B.J., Sahana, G., Weigel, K.A., McParland, S., Spurlock, D., Krattenmacher, N., Spelman, R.J., Wall, E. and Calus, M.P.L. (2014). *J. Dairy Sci.* **97**: 1799.

Pryce, J.E., Arias, J., Bowman, P.J., Davis, S.R., Macdonald, K.A., Waghorn, G.C., Wales, W.J., Williams, Y.J., Spelman, R.J. and Hayes, B.J. (2012) *J. Dairy Sci.* **95**: 2108.

Sargolzaei, M., Chesnais, J.P. and Schenkel, F.S. (2014) *BMC genomics* **15**: 1.

VanRaden, P.M. (2008) *J. Dairy Sci.* **91**: 4414.

Yang, J., Benyamin, B., McEvoy, B.P., Gordon, S., Henders, A.K., Nyholt, D.R., Madden, P.A., Heath, A.C., Martin, N.G., Montgomery, G.W., Goddard, M.E. and Visscher P.M. (2010) *Nat. Genet.* **42**: 565.

Proc. Assoc. Advmt. Anim. Breed. Genet. 22:157-160

CORRELATIONS AND GENETIC TRENDS FOR SELECTION INDICES ASSESSED USING AUSTRALIAN AND NORTH AMERICAN BULL PROOFS

B.A Scott[1], M. Abdelsayed[1] and J.E. Pryce[1,2]

[1] Agriculture Victoria, AgriBio, Centre for AgriBioscience, Bundoora, Victoria 3083, Australia
[2] School of Applied Systems Biology, La Trobe University, Bundoora, Victoria 3083, Australia

SUMMARY

The aim of this research was to evaluate correlations between national selection indices and estimate the rates of genetic gain within and between countries, using bull breeding values from Australia, USA and Canada. High ranking sires in the USA and Canada do not necessarily rank highly in Australia. The correlations between bull proofs in Australia and either Canada or USA ranged between 0.74 and 0.86 for the indices assessed, implying that national breeding objectives and genotype by environment interactions are important. Since 2010, which is similar to the start of widespread use of genomic bulls, there has been considerable increase in the rate of genetic gain in all three countries.

INTRODUCTION

The concept that animals do not always rank the same in different environments, or that there is an advantage to a genotype in one environment that is not seen in another environment is known as a genotype by environment (GxE) interaction (Falconer and Mackay, 1996). Typically, animal breeders are more concerned about the re-ranking of animals than the differences in scale between environments. If re-ranking is substantial, then specific genotypes are required for specific environments, a correlation of >0.8 is often considered to be a threshold of importance, although it is somewhat arbitrary (Falconer and Mackay, 1996).

International exchange of genotypes is very common in dairy cattle breeding and therefore a bull can sire cows in more than one country at the same time. For these bulls and their relatives, it is possible to calculate correlations of their proofs between countries, which is indicative of GxE. Interbull, the international bull evaluation service already provides some of this information for traits such as milk production and somatic cell count, however for national selection indices there are no comparisons. In Australia, in addition to the Balanced Performance Index (BPI), there are two other national selection indices available from DataGene (the Health Weighted Index and Type Weighted Index) that align with farmer philosophies (Martin-Collado *et al.*, 2015). Similarly, in the USA there are five indices for farmers to choose between. The combination of traits within an index and their respective weights varies by country, which will reduce correlations between indices.

Within country, the success of a breeding programme is often assessed as the rate of genetic gain achieved, especially in the primary selection tool, such as a selection index. Genomic selection was predicted to double the rate of genetic gain mainly through the shortening of the generation interval (Shaeffer, 2006). Since 2010, genomic selection programs have been widely adopted in genetic evaluations around the world (Pryce and Daetwyler, 2011). To date, there have been relatively few studies that have compared the realised rate of genetic gain before and after the implementation of genomic selection.

The aim of this study was to evaluate correlations between Australian, USA and Canadian indices and rates of genetic gain in these indices. For comparison, a selected number of traits (stature, milk yield and overall type) genetic correlations between countries were also estimated.

MATERIALS AND METHODS

Selection indices and breeding values of predicted transmitting abilities (PTAs) from Holsteins in Australia, USA and Canada were used in the analyses. The bull breeding value file from DataGene was used for the Australian analysis (accessed April 2015; n=9,470). The data, included both Australian Breeding Values (ABV) for bulls (n=7,423) and bulls that had an international proof determined by Interbull (ABV(i)) (n=2,047). The American data was provided by the Council on Dairy Cattle Breeding (CDCB) (n=287,207). Total Performance Index (TPI) is another USA index, that was accessed directly from Holstein USA (Tom Lawlor personal communication, 2015) (n=4,080). The Canadian bull proof file (April 2015) was downloaded from the Canadian Dairy Network (CDN) in June 2015 (n=12,269).

The bull files were merged based on their international IDs, where only bulls born after 1990 were considered. The number of bulls that had dual proofs with Australia was 8,226 with USA indices (NM, CM, FM and GM), 2,981 with TPI data and 1,874 with the Canadian index.

Table 1: List of indices used in the evaluation and their country of origin

Abbreviated index	Index name	Country	Source	No. of Bulls
BPI	Balanced performance index	Australia	DataGene	9,470
HWI	Health weighted index	Australia	DataGene	9,470
TWI	Type weighted index	Australia	DataGene	9,470
TPI	Total performance index	USA	Holstein USA	4,072
NM	Net merit	USA	CDCB	151,246
CM	Cheese merit	USA	CDCB	151,246
FM	Fluid merit	USA	CDCB	151,246
GM	Grazing merit	USA	CDCB	151,246
LPI	Lifetime profit index	Canada	CDN	9,217

Pearson correlations between indices were calculated using merged data using the statistical package R (R Core Team, 2013).

The genetic trends were calculated as regressions of breeding values (or PTAs) on year of birth for bulls born between 1990-2000; 2000-2010 and from 2010. To make comparisons between countries, the genetic trends were transformed into genetic standard deviations using the genetic standard deviation associated with each time period. The standard deviation for each interval (e.g. between 1990-2000) was calculated by taking the mean standard deviations per year over the period, then calculating the mean of the SD values within each time interval. The regression was divided by this number to give the rate of genetic gain in standard deviation units.

RESULTS AND DISCUSSION

Correlations within countries were high and reasonably strong correlations exist between Australian indices and all the American indices (Table 2). Removing bulls that only have an Interbull proof had minimal effects on the correlations (below the diagonal). Correlations of the Canadian LPI with the Australian indices ranged between 0.83 and 0.86 (Table 3). BPI seems to be more closely related to LPI than NM or TPI (0.86, 0.81 and 0.77 respectively).

The correlations presented in Tables 2 and 3 indicate the relative response to selection that could be expected when selecting based on a foreign index. Ranking bulls using any of the Australian indices will result in similar sires being selected, as the correlations between BPI, HWI and TWI are very high (0.98, 0.95 and 0.94). When selecting bulls in Australia using their North American index, sire re-ranking is expected, as the correlations between the Australian index and North American indices range from 0.77 for BPI and TPI, 0.81 for BPI and NM to 0.86 for BPI and LPI. However, the correlations between indices depend on three factors; firstly, the traits in the indices and their respective weights. It is very likely that genuine economic drivers differ between countries. Secondly, whether true genotype by environment interactions are occurring and thirdly, the differences in trait definition between countries. Sires that rank highly for their respective indices in the USA or Canada do not necessarily rank highly in Australia.

Table 2: Correlations of Australian indices (Balanced Performance Index (BPI), Health Weighted Index (HWI) and Type Weighted Index (TWI)) with USA indices (Net Merit (NM), Cheese Merit (CM), Fluid Merit (FM), Grazing Merit (GM) and Total Performance Index (TPI)), above the diagonal includes both domestic and interbull proofs, below the diagonal is domestic proof only.

	BPI	HWI	TWI	NM	CM	FM	GM	TPI
BPI		0.98	0.95	0.81	0.83	0.75	0.81	0.77
HWI	0.98		0.94	0.82	0.83	0.77	0.82	0.78
TWI	0.96	0.95		0.81	0.82	0.77	0.79	0.81
NM	0.80	0.80	0.78		1.00	0.99	0.99	0.97
CM	0.81	0.81	0.79	1.00		0.97	0.99	0.97
FM	0.74	0.76	0.74	0.99	0.98		0.97	0.97
GM	0.79	0.80	0.76	0.99	0.99	0.98		0.96
TPI	0.76	0.76	0.78	0.98	0.97	0.97	0.97	

SEs<0.02

Table 3: Correlations of Australian indices (Balanced Performance Index (BPI), Health Weighted Index (HWI) and Type Weighted Index (TWI)) with the Canadian index Lifetime Profit Index (LPI; CAN) above the diagonal includes both domestic and interbull proofs, below the diagonal is domestic proof only.

	BPI	HWI	TWI	LPI
BPI		0.98±0.01	0.95±0.01	0.86±0.02
HWI	0.97±0.01		0.94±0.01	0.83±0.02
TWI	0.94±0.01	0.93±0.01		0.86±0.02
LPI	0.83±0.02	0.80±0.03	0.83±0.02	

It should be noted, that when correlations between traits instead of indices were estimated, those that were objectively scored had strong correlations (stature 0.94 AUS-CAN), suggesting little to no GxE. Similarly, there was a moderate correlation with milk yield across all three countries (0.83 AUS-USA and 0.88 AUS-CAN). Composite traits, that are more subjectively measured, typically had lower correlations with AUS, such as overall conformation, for example for overall type the correlations were 0.56 AUS-CAN and 0.59 AUS-USA although there are differences in trait definition between countries and increased error variance (subjectivity) in some traits may also be driving weak correlations, there is likely to be GxE as well.

For all indices, the rate of genetic gain has increased dramatically since 2010 (Table 4). Rates of genetic gain were higher when all bulls were included and analysed based on their country of origin compared to bulls with dual proofs (Table 4 vs. Tables 5 and Table 6) and reflects the overall increase in genetic gain in these countries. The rate of genetic gain for bulls with proofs in Australia and a North American country was faster for TPI and LPI (Tables 5 and 6), implying that sires in Australia are being selected based on their international proof. There has been an increase in the number of international bulls used in recent years, with around 45% of daughters of registered bulls being sired by North American bulls since 2010, which compares to 28% from 2000-2010. The reduced rate of genetic gain seen in the Australian indices with dual proofs compared to North American indices could be explained by the GxE interaction that exists and bulls that rank highly on the USA or Canadian indices are not necessarily well suited to the Australian environment reflecting the lower rate of genetic gain. These rate of genetic gain since

2010 should be treated cautiously, as the number of years of data in the analysis was comparatively small. Rates of genetic gain should be re-estimated as more data becomes available.

Table 4: Holstein genetic trends in genetic standard deviations calculated as the regression of indices on year of birth for the following time intervals; 1990-2000, 2000-2010, 2010-now; for all available bulls with progeny in the country of origin (Australia – BPI, HWI, TWI (n=7,412); USA – TPI (n=4,072), NM (n=151,246); Canada - LPI (n=5,663))

	BPI	HWI	TWI	TPI	NM	LPI
SD of index*	66.4	56.2	66.3	220	98.2	261
1990-2000	0.20±0.002	0.17±0.002	0.20±0.002	0.29±0.004	0.25±0.000	0.22±0.003
2000-2010	0.22±0.005	0.22±0.005	0.24±0.005	0.30±0.007	0.32±0.001	0.32±0.007
2010-now	0.42±0.037	0.44±0.038	0.48±0.036	0.68±0.039	0.40±0.003	N/A

*The overall standard deviation

Table 5: Holstein genetic trends calculated in genetic standard deviations as the regressions of indices on the following time intervals; 1990-2000, 2000-2010; for bulls with dual proofs in Australia and USA for all indices except TPI, the number of bulls used was 8,548. For TPI 2,981 bulls were used that had dual proofs

	BPI	HWI	TWI	TPI	NM	CM	FM	GM
1990-2000	0.15	0.14	0.16	0.25	0.16	0.15	0.15	0.15
2000-2010	0.15	0.16	0.17	0.30	0.25	0.07	0.23	0.24

SEs<0.01

Table 6: Holstein genetic trends in genetic standard deviations calculated as regressions of the following time intervals; 1990-2000, 2000-2010; for bulls with dual proof in Australia and Canada (n=1,874)

	BPI	HWI	TWI	LPI
1990-2000	0.14	0.13	0.14	0.19
2000-2010	0.14	0.13	0.12	0.21

SEs<0.01

CONCLUSIONS

Sires that rank highly for their respective indices in the USA or Canada do not necessarily rank highly in Australia, with correlations between BPI and NM (USA), TPI (USA) and LPI (Canada) being 0.81, 0.77 and 0.86 respectively. Weak correlations are driven by GxE, different trait weightings and definitions and the degree of subjectivity of measuring traits in the indices. Since 2010, there has been a considerable increase in the rate of genetic gain in all countries. This could be a result of the introduction of genomics, the increase in the number of bulls being genomically tested and shorter generation intervals.

ACKNOWLEDGEMENTS

This research was funded by Dairy Futures CRC.

REFERENCES

Falconer, D.S. and Mackay T.F.C. (1996) 4th ed. Longman Group, Essex, UK.
Martin-Collado, D., Byrne, T. J., Amer, P. R., *et al.* (2015). *J. Dairy Sci. 98*(6), 4148-4161.
Mathur, P.K. (2002) *Proc. of the 7th World Congress Genet. Appl. Livest. Prod.* **18**: 1-8
Pryce, J.E., Haile-Marian, M., Goddard, M., & Hayes, B. (2014). *J. Genet. Sel. Evol. 46(1),* 71.
Schaeffer, L. R. (2006) *J. Anim. Breed. Genet. 123(4),* 218-223.
R Core Team (2013). Vienna, Austria.

Proc. Assoc. Advmt. Anim. Breed. Genet. 22:161-166

THE NEW BOVINE REFERENCE ASSEMBLY AND ITS VALUE FOR GENOMIC RESEARCH

Juan F. Medrano

Department of Animal Science, University of California Davis, Davis, CA, USA

The development of the assembly has been a major joint effort of several groups making efficient use of very limited resources. **Participants in this effort have been:** T.P.L. Smith (USDA, ARS, USMARC, Clay Center, NE), B.D. Rosen (Animal Genomics and Improvement Laboratory, ARS USDA, Beltsville, MD), S. Koren (National Human Genome Research Institute, Bethesda, MD), A. Zimin (University of Maryland, College Park, MD), R.D. Schnabel (Division of Animal Sciences, University of Missouri, Columbia), D. Bickhart (Cell Wall Biology and Utilization Laboratory, ARS USDA, Madison, WI), R. Hall (Pacific Biosciences, Menlo Park, CA), S.J. Schultheiss and C. Dreischer (Computomics GmbH, Tuebingen, Germany). Funding for the project was provided by USDA/NRSP-8 Animal Genome, USDA-ARS Meat Animal Research Center, Neogen and Zoetis.

SUMMARY

There are two public cattle genome reference assemblies (UMD3.1.1 and Btau5.0.1) that were based primarily on the same set of data. Both assemblies used sequences of a minimum tiling path of BAC clones from the CHORI-240 library (prepared using DNA from L1 Domino 99375), augmented by low coverage whole genome shotgun sequencing (WGS) from his daughter, L1 Dominette 01449. Updates and new assembly releases through the years have led to significant improvements, but as confirmed by the recently developed cattle genome optical map (BtOM1.0), there are numerous differences between these assemblies that have produced ambiguities that continue to impact and hamper genomic analysis in cattle. Recent advances in long-read sequence technology, combined with new scaffolding technologies, have made it possible to create a completely new *de-novo* Dominette assembly. An approximately 80X PacBio FALCON based de-novo assembly, followed by scaffolding with Dovetail Genomics Chicago library/HiRise technology, the BtOM1.0 Optical Map of Dominette and a recombination map of 59K autosomal SNPs. The scaffolded assembly was then refined with independent *de-novo* assemblies from CANU and MaSuRCA, yielding chromosome length scaffolds. Preliminary assembly statistics include an N50 contig size of 22 Mb and an N50 scaffold size of 104 Mb representing several fold improvements over UMD3.1 (contig N50=0.97Mb, scaffold N50=6.4Mb). Additionally, full-length transcripts from 30 Dominette tissues have been sequenced with PacBio using the Iso-Seq method to support improved annotation. A public version of the new ARS-UCD assembly is expected to be released in mid 2017. An update on the status of the long-read based assembly of Dominette will be presented here, providing some perspective on the value of having an improved bovine reference sequence.

INTRODUCTION

The availability of accurate well-annotated genome assemblies in agricultural species have become essential tools to enable the understanding of phenotypic variation and practical applications of DNA technologies. In cattle, numerous opportunities exist in the application of genomic selection and of new technologies, like gene editing to improve production efficiency. For the human and mouse genomes, enormous efforts and resources have been spent to develop what has been referred to as Gold (targeted finishing with haplotype resolution of critical regions) and Platinum (contiguous haplotype-resolved representation of the entire genome) level genome sequence assemblies. In order to approach some of these advanced states of genome resolution in cattle a significant effort has been placed towards developing a new improved Dominette assembly.

There are currently two public cattle genome reference assemblies (UMD3.1.1 and Btau5.0.1), that were based primarily on the same set of data. Both assemblies used sequences of a minimum tiling path of BAC clones from the CHORI-240 library (prepared using DNA from L1 Domino 99375), augmented by low coverage whole genome shotgun sequencing (WGS) from his daughter, L1 Dominette 01449. Subsequently, the Btau5.0.1 assembly was improved by gap filling with low coverage long-read WGS, and more recently with P5 PacBio reads. Scaffolding used combinations of radiation hybrid map and genetic linkage map data, as well as a large number of BAC end sequences. Recently, a cattle genome optical map (BtOM1.0) was developed (Zhou et al. 2015), which confirmed that there are numerous differences between these assemblies that have produced ambiguities that continue to impact and hamper genomic analysis in cattle.

Two animals (Domino and Dominette) were used to produce the assemblies that resulted in an increased amount of diversity between haplotypes. Both assemblies, although having used practically the same sequenced data, are significantly different, appearing as assembly errors, genome segmental inversions, chromosomal placements, sequence gap numbers and discrepancies of sequence coverage of the bovine genome. Although there have been periodic updates of both sequence assemblies, many issues still remain in both. It is very difficult when one encounters discrepancies between assemblies to know what is correct, and this impacts genomic studies. Table 1 shows a comparison of the genomic statistics of both assemblies, UMD3.1.1 and BTAU 5.0.1.

Table 1: Comparison of current cattle genome assemblies (NCBI report)

UMD3.1.1 (Reported April 2009 (Genome Biol)	**BTAU 5.0.1** (Released (11/19/2015)
Based on 9x Sanger coverage WGS Dominette	Based on 9x Sanger coverage of Dominette
BAC path Domino	BAC ends, + 19x coverage P5 PacBio
RH map and human-cow synteny map	BAC end, RH map, PBJelly2
Genomic statistics:	Genomic statistics:
75,618 contigs (97 kb contig N50) 42,267 contigs (276 kb contig N50)	42,267 contigs (276 kb contig N50)
6337 scaffolds (6.4 Mb scaffold N50)	5,998 scaffolds (6.8 Mb scaffold N50)
3,193 gaps between scaffolds	2,856 gaps between scaffolds

GENOME ASSEMBLY PROCESS

Creating a genome assembly is the process of reconstructing a genome to develop a database of DNA sequences that represent an example individual of the species, from a collection of short or long sequence reads. A de-novo assembly is performed without the aid of a reference genome and the genome is reconstructed by directly reconstructing the puzzle of sequence reads. One complicating factor in the reconstruction is the presence of repetitive sequences, particularly when using short read lengths that do not span the length of the repeats. Recent advances in Pacific Biosciences Smrt single-molecule sequencing technologies, with the generation of 20-50 kb have allowed resolving repetitive sequences and the creation of accurate genome assemblies (Berlin et al. 2015).

The repetitive content of genomes on both large and small scales, including structures near centromeres and telomeres, large paralog gene families, like zinc fingers, and the distribution of interspersed nuclear elements such as LINEs and SINEs are the cause of many of the incorrect assembly problems we have had in the past. Such difficult-to-assemble content composes large portions of eukaryotic genomes, about 60-70% of the human genome (de Koning et al 2011).

Although PacBio reads are error prone, errors are at random and can be overcome by sufficient coverage producing highly accurate assemblies, and it has been demonstrated from assemblies in humans and other organisms that single-molecule sequencing can produce de-novo near complete eukaryotic assemblies that are 99.99% accurate compared to the available references. In addition to the technical quality of the assemblies, the time to produce an assembly has been reduced by 5x and cost by more than 200 orders of magnitude.

The creation of the final assembly is an iterative process that evolves as scaffolds and super-scaffolds are built. The initial assembly process used in the creation of the Dominette *de-novo* assembly is shown in Figure 1.

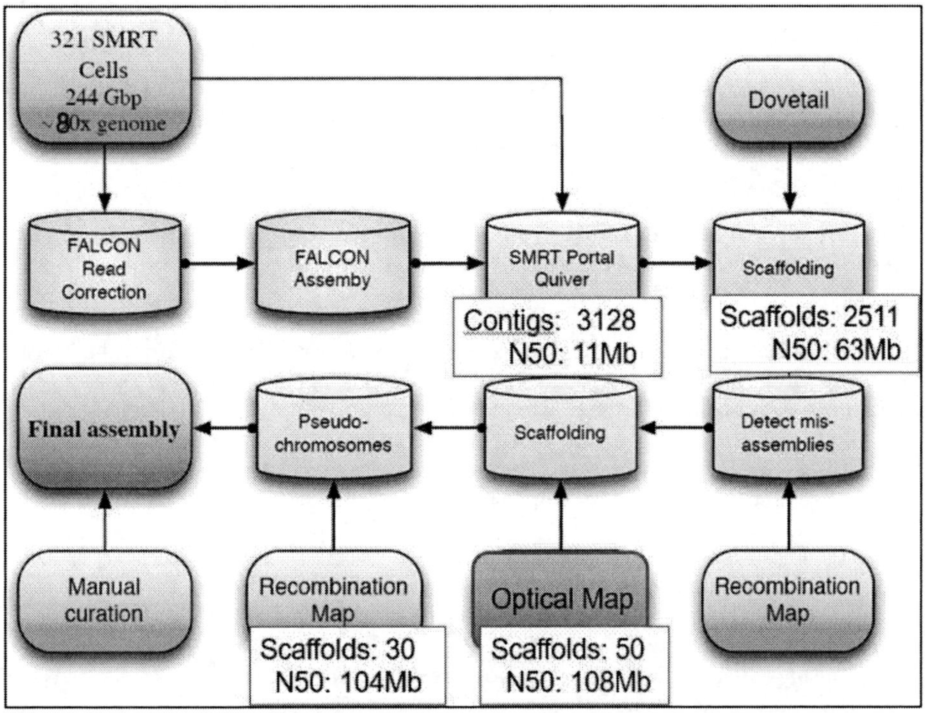

Figure 1: Initial Dominette *de-novo* assembly (January 2017). Approximately 321 PacBio SMRT cells with an average size of 20 kb were produced for a ~80x genome coverage followed by a hierarchical genome assembly process of PacBio long reads using FALCON, and Quiver for polishing. Quiver generated consensus contigs using local realignment of reads to the assembly to correct short insertions, deletions and substitutions errors. Following the construction of contigs from pre-assembled reads, the true assembly process is dependent on the correct orientation of contigs into scaffolds and super-scaffolds. This required other data types, like Dove Tail Chicago libraries, recombination map, optical map, alignment of short read sequences and manual curation.

Among the initial scaffolding resources used in the Dominette assembly were a Dovetail Chicago Hi-Rise library, the optical map and a genetic map. The Dovetail method is based on producing DNA linkages of up to several hundred kb to make sequencing libraries that can link distant fragments. This long-range mate pair data can be used to orient contigs and largely improve scaffolding in *de-novo* assemblies (Putnam et al. 2016). The Optical Map is a high throughput system that produces ordered restriction maps from individual molecules of genomic DNA (Zhou et al. 2015). The approach is ideal for identifying structural variants and studying genome structure. The recombination map used is a sex-specific recombination map of 59K autosomal SNP (Ma et al. 2015). The map was used primarily for breaking, ordering and orienting contigs.

Depending on how all the data is used and how the scaffolding resources are applied one ends up with several assemblies in which some super-scaffolds are better assembled or one assembly captures longer scaffolds and contigs. All this needs to be carefully examined in order to develop a stable assembly for further improvement. Early statistics after the initial scaffolding of the new Dominette assembly are shown in Table 2.

Table 2: Early progress statistics of the new Dominette assembly ARS-UCD v1.0 (Jan, 2017)

Based on ~80x PacBio data P6 chemistry, Falcon assembly - Quiver (PacBio)
Scaffolding: Dovetail HiRise, Optical Map, Rec Map
Genomic statistics:
2816 contigs (22.6 Mb contig N50)
30 scaffolds (104 Mb scaffold N50, L50 12)
Largest scaffold length 211 Mb (Chr 1)
460 gaps

For sequence annotation, full-length transcripts spanning entire isoforms using the PacBio Iso-Seq method from approximately 30 Dominette tissues are being developed. We expect to release an assembly with haplotype-resolved chromosomes.

CONCLUSIONS

- Long single-molecule reads, despite higher per-read error rate, create higher quality reference genomes at a fraction of the cost of earlier technologies.
- The improvement in quality of the cow assembly will have substantial impact on many genetic and molecular genetic studies.
- Many studies would benefit from re-mapping reads, and/or analysis of GWAS with improved marker order.
- The improvements in the cow assembly are substantial enough that it is worth considering waiting for them for ongoing GWAS and WGR studies.
- We expect to have a version of the new ARS-UCD assembly available this summer-2017, through NCBI.

REFERENCES

Berlin K, Koren S, Chin CS, Drake JP, Landolin JM, Phillippy AM (2015). *Nat Biotechnol.* **33**: 623.

de Koning AP, Gu W, Castoe TA, Batzer MA, Pollock DD (2011). *PLoS Genet.* 7(12):e1002384

Ma, L., O'Connell, J.R., VanRaden, P.M., Shen, B., Padhi, A., Sun, C., Bickhart, D.M., Cole, J.B., Null, D.J., Liu, G.E., et al. (2015). *PLOS Genet* **11**: e1005387.

Putnam, N.H., O'Connell, B.L., Stites, J.C., Rice, B.J., Blanchette, M., Calef, R., Troll, C.J., Fields, A., Hartley, P.D., Sugnet, C.W., et al. (2016). *Genome Res.* **26**: 342–350.
Zhou, S., Goldstein, S., Place, M., Bechner, M., Patino, D., Potamousis, K., Ravindran, P., Pape, L., Rincon, G., Hernandez-Ortiz, J., et al. (2015). *BMC Genomics* **16**: 644.

Proc. Assoc. Advmt. Anim. Breed. Genet. 22:167-170

EXTENSIVE SEQUENCING OF A TROPICALLY ADAPTED BREED – THE BRAHMAN SEQUENCING PROJECT

L. Koufariotis[1], B.J. Hayes[1], M. Kelly[2], B. Burns[3], R. Lyons[4] and S. Moore[1]

[1] Centre of Animal Science, Queensland Alliance for Agriculture and Food Innovation, The
[2]University of Queensland, Brisbane, 4072, QLD, Australia
[2]Australian Agricultural Company (AACo), Brisbane, QLD, Australia
[3]Queensland Department of Agriculture and Fisheries, Rockhampton, 4702, Queensland, Australia
[4]School of Veterinary Science, University of Queensland, Gatton, Queensland, 4343, Australia

SUMMARY

Brahman cattle are well adapted to tropical environments and are extensively used for beef production in Northern Australia. Identifying mutations in Brahman genomes associated with adaptation, fertility, meat quality and growth rates would facilitate genome selection and therefore accelerate genetic gain for these traits, in both Brahman cattle and composite cattle with Brahman ancestry. In this paper, 36 million high quality variants (SNP and Indels) were discovered from 46 whole genome sequenced Brahman bulls that represent key ancestors of the breed in Australia. As some infusion of *Bos taurus* into Brahman cattle has occurred during breed formation, we investigate regions of the Brahman genome that have high *Bos taurus* introgression. We identified multiple genome regions in Brahmans that were *Bos taurus* in origin, and investigated the roles, pathways and trait associations the of genes found in these regions.

INTRODUCTION

Brahman cattle are a breed of *Bos indicus* cattle, developed in the southern United States that are well adapted to tropical environments. They are a cross breed between four different types of Zebu cattle; Gir, Gujarat, Ongole and Krishna Valley. In northern Australia, Brahman cattle have a major impact on the Australian beef industry and are widely used in beef production due to their suitability to these harsh environments. To increase genetic diversity, Brahman cattle have been "graded up" from existing *Bos taurus* breeds in both Australia and United States.

Identifying mutations in Brahman genomes associated with adaptation, fertility, meat quality and growth rates would facilitate genome selection and therefore accelerate genetic gain for these traits, in both Brahman cattle and composite cattle with Brahman ancestry. With this ultimate aim, 46 Brahman cattle that were key ancestors of the breed were whole genome sequenced in this study. We first identified all the variants in these genomes such as single nucleotides (SNPs) and insertion/deletions (Indels), then annotated the variants into functional classes based on their locations on the genome. Finally, we used the sequence information to identify, for each bull, whether chromosome segments were indicine or taurine in origin.

MATERIALS AND METHODS

Bulls for sequencing were selected using an algorithm that identified 46 bulls that captured the highest proportion of genetic variation in the breed, based on an analysis of an extensive Brahman pedigree and a stepwise regression procedure to avoid double counting of ancestral genomes, and took into account whether DNA, extracted from semen straws or Ampules, was available for a bull or not (Daetwyler 2014; Druet 2014). The selected bulls were sequenced on an Illumina HiSeq sequencer, at an average of 12.5 times genome coverage, and a range of 10 times genome coverage to 30 times genome coverage.

Reads were mapped to the bovine genome (UMD3.1) with BWA, and variants were detected in the sequence with a GATK pipeline (McKenna *et al.* 2010). The variants included single nucleotide polymorphisms (SNP) and small insertion deletions (indels).

SNP variants were filtered based on modified filtering thresholds recommended by the GATK Best-Practices guideline (DePristo *et al.* 2011) to remove SNP that had poor quality scores.

Bos taurus and Gir variants were obtained from the 1000 bulls genomes project (Daetwyler *et al.* 2014) and from the study by (Liao *et al.* 2013), respectively. All SNP were obtained from each study and the allele frequencies for each SNP was calculated using in-house scripts and common SNP found in Brahman cattle were selected for. The population structure between Brahman and *Bos taurus* and between Brahman and Gir cattle was determining by calculating the Fixation value (FST) for each SNP in both taurine and Gir with the SNP in each individual Brahman animal, based on the methodology by the study (Bolormaa *et al.* 2011), using the following formula:

$$FST = \frac{Ht - Hs}{Ht}$$

Where:

$$Hs = PBT(1 - PBT) + Pbrai(1 - Pbrai)$$

And

$$Ht = \frac{2 * (PBT + Pbrai)}{2} \times 1 - (\frac{PBT + Pbrai}{2})$$

PBT is the SNP allele frequency of the alternative allele in either *Bos taurus* or Gir and *Pbrai* is the zygosity information of the Brahman SNP for that animal. *Hs* is to calculate heterozygosities in the subpopulation and *Ht* is to calculate the overall heterozygosity. This resulted in two datasets, one is the FST between each individual Brahman animal and the Gir SNPs and the other is between individual Brahman animals with *Bos taurus* SNPs. Following this, we grouped SNPs into fixed windows of 250 kilobases (kb) and calculated the average FST for all SNP within each fixed window across the genome. This was done by simply by adding the FST of all SNP found in the 250 kb fixed windows and dividing by the total number of SNP, as shown in the following formula:

$$FSTavg = \frac{\Sigma(FSTfw)}{n}$$

Where *FSTfw* is the FST for the SNP found within a fixed window, and *n* is the total number of SNP found in that fixed window. Animals were then sorted based on date of birth (from oldest to youngest), and the average FST in each fixed window across all animals was calculated.

SNP Annotation: Brahman SNPs were annotated into intergenic, intragenic, introns, exons, CDS, UTR (both 5'&3'), 5 kb upstream of TSS, 5 kb downstream of genes, missense, synonymous and splice site classes by querying the Ensembl variant database version 87 (Yates *et al.* 2016). CpG Isles annotations were from the study by (Su *et al.* 2014).

RESULTS AND DISCUSSION

Initial analysis of the sequence data revealed the Brahman genomes had a much higher rate of polymorphism than that observed in *Bos taurus* breeds. This is likely a reflection of a larger ancestral population size for *Bos indicus* cattle than *Bos taurus* cattle (pre-domestication), and the fact there was some infusion of *Bos taurus* breeds into Brahmans during breed formation. Additionally, as Brahman is the synthesis of 4 different indicus breed types, the impact on the breed formation is reduced.

Following SNP annotation, we find that, as expected, the majority of the variants are found within the intergenic regions of the genome, as shown in Table 1, with close to 74 % of all SNP found in this class.

Table 1. Total number of Brahman SNP found in each annotation class

Annotation	No. of SNP	Percent of Genome
Intergenic	26,505,585	73.39 %
Intragenic	9,608,525	26.61 %
Intron	9,088,608	25.17 %
Exon	433,758	1.20 %
CDS	349,093	0.97 %
3' UTR	69,733	0.19 %
5' UTR	16,426	0.05 %
5kb Downstream of TTS	692,447	1.92 %
5kb Upstream of TSS	699,424	1.94 %
Synonymous	128,397	0.36 %
Missense	83,069	0.23 %
CpG Meth	995,319	2.76 %
Splice Sites	21,789	0.06 %
Total	**36,114,110**	

Just over 25% of Brahman SNPs are found within the intragenic class (Table 1), the majority of which (95%) are actually intronic variants. The percent of variants that are found within coding genes (Exon and CDS class) is 1.2%. Similar results have been observed in *Bos Taurus* (Koufariotis *et al.* 2014).

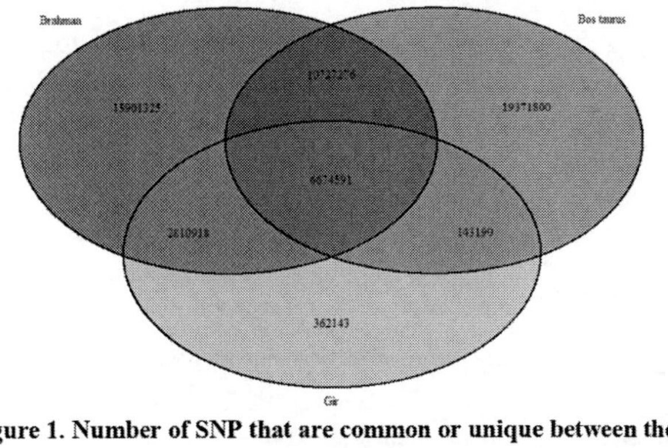

Figure 1. Number of SNP that are common or unique between the breeds

Figure 1 shows a Venn diagram of the overlap of SNP between the breeds. We see that there are more SNP in common between Brahman and Gir (95%) as opposed to Brahman and *Bos taurus* (48%). This is expected as Gir, which are a indicus breed, is one of the 4 ancestors to Brahman. Further 6,674,591 SNP in Brahman are found in both Gir and *Bos taurus*, which is a total of 18.5% of all Brahman SNP.

We examined the population structure between the Brahman cattle and taurine/Gir cattle to determine how much of the genome in Brahman is influenced, or infused with taurine and Gir. Overall, we find that Brahman and Gir cattle are the most similar (average FST over the whole genome is 0.19) compared with *Bos taurus* (average FST is 0.26) across the whole genome, as the average FST remains relatively low (is closer to 0).

However, we do observe some interesting findings on some chromosomes in where *Bos taurus* is more similar to Brahman than Gir, based on the FST window results. One example of this is on Chromosome 23 (Figure 2), in which we see a large region that is very similar to Brahman. This region is most likely the bovine lymphocyte antigen (BoLA) region. The BoLA regions is part of the MHC complex that display foreign peptides within a cell to cytotoxic T cells, triggering an immune response.

This raises the question, why is Brahman acquiring *Bos taurus* alleles in this region? One suggestion is that this could be due to MHC diversity in the alleles and their heterozygosities, which has been observed in many other vertebrate species (Salmier *et al.* 2016). It must be noted though, that this could also be due to a miss-assembly of that region in the UMD3.1, leading to these results.

Finally, we identified a region of *Bos taurus* introgression around the PLAG1 gene that was previously described by (Fortes et al., 2013). We find that the *Bos taurus* introgression in this region increases in frequency in younger animals, which could reflect a selection pressure on age at puberty.

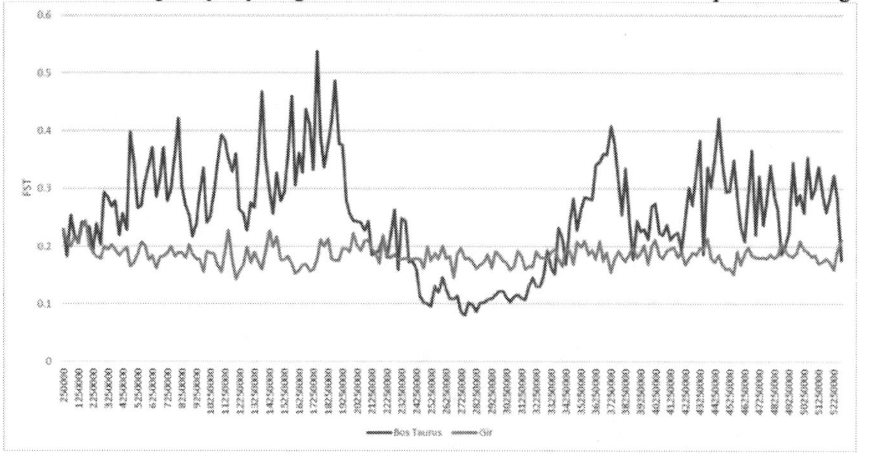

Figure 2: The average FST in each fixed window on chromosome 23 between Gir cattle (Orange line), and *Bos taurus* (blue line).

The next step in this project is to link genome variation amongst the bulls to variation in key traits such as fertility and meat quality and to examine if variants in certain genomic regions (such as coding regions, non-coding regions, regulatory regions) are more likely to influence complex beef trait variation.

REFERENCES

Bolormaa, S. et al. 2011. *Journal of animal science* **89**: 2050-2060.

Daetwyler, H. D. et al. 2014. *Nature Genetics* **46**: 858-865.

DePristo, M. A. et al. 2011. *Nature Genetics* **43**: 491-498.

Druet, T., I. M. Macleod, and B. J. Hayes. 2014. *Heredity* **112**: 39-47.

Fortes, M. R. et al. 2013. *Animal Genetics* **44**: 636-647.

Koufariotis, L., Y. P. Chen, S. Bolormaa, and B. J. Hayes. 2014. *BMC Genomics* **15**: 436.

Liao, X. et al. 2013. *Genome* **56**: 592-598.

McKenna, A. et al. 2010. *Genome research* **20**: 1297-1303.

Salmier, A., B. de Thoisy, B. Crouau-Roy, V. Lacoste, and A. Lavergne. 2016. *BMC Evolutionary Biology* **16**: 229.

Su, J., Y. Wang, X. Xing, J. Liu, and Y. Zhang. 2014. *BMC genomics* **15**: 12.

Yates, A. et al. 2016. *Nucleic Acids Research* **44**: D710-716.

Proc. Assoc. Advmt. Anim. Breed. Genet. 22:171-174

GENOMICS CAN CONTRIBUTE TO SELECTION TO IMPROVE BOTTLE TEATS IN TROPICAL BEEF GENOTYPES

M.L. Wolcott, D.J. Johnston and Y.D. Zhang

Animal Genetics and Breeding Unit*, University of New England, Armidale, NSW 2351, Australia

SUMMARY

Beef CRC research showed that a subjective score of teat size (small (1) to large (5)) was heritable in tropically adapted Brahman (BRAH) and Tropical Composite (TCOMP) cows, and that higher teat scores (bottle teats) were genetically associated with higher calf losses from birth to weaning. Teat traits are only expressed in females, and the research showed that they tended to display more variation in later life; making them ideal candidates for genomic selection. Front and rear teat scores (TSF and TSB respectively) were recorded in cows at calving through up to 6 matings. From these, a trait was created which described a cows maximum lifetime teat score (TSM), as well as a binary trait which distinguished cows that received a teat score of 4 or 5 at any time through their lives (1) from those which did not (0) (MSB). Results confirmed the heritability of TSF and TSB (h^2 = 0.30 to 0.40), and variation in both TSM and MSB was also shown to have a genetic basis (h^2 = 0.49 and 0.46 respectively for BRAH, and 0.29 and 0.22 for TCOMP). Genome wide association analyses identified large numbers of significant SNPs but did not suggest a likelihood of identifying a small number of SNPs of large effect. It is unlikely therefore, that a simple diagnostic test (based a small number of SNPs) could be developed for the traits. Conventional genomic selection, however, is likely to present opportunities to improving teat traits by selection in tropically adapted beef genotypes, with accuracies of genomic prediction of 0.23 to 0.35 for TSM and MSB across both genotypes.

INTRODUCTION

Results from the Co-operative Research Centre for Beef Genetic Technologies' northern breeding project (Beef CRC) showed that a subjective score of teat size (1 (smaller) to 5 (larger)) was heritable (h^2 = 0.30 to 0.38) in Brahman and Tropical Composite cows (Bunter *et al.* 2014). That study also showed that higher teat scores (bottle teats) were genetically associated with increased calf losses from birth to weaning (r_g = 0.54 ± 0.18), and that teat score tended to increase with cow age in both genotypes. This, combined with the sexually dimorphic nature of the trait makes it a prime candidate for genomic selection. This study aimed to determine whether SNPs of large effect for variation in teat score could be identified in the Brahman and Tropical Composite females, and to estimate the accuracy of genomic prediction for teat traits.

MATERIALS AND METHODS

Cow management and trait definitions. A comprehensive account of cow herd management is provided by Johnston *et al.* (2009) and Johnston *et al.* (2014), and Bunter *et al.* (2014) presented a description of teat score measurement in the Beef CRC Brahman (BRAH: n = 969) and Tropical Composite (TCOMP: n = 1085) females. In brief, females were transported from their properties of origin (5 BRAH and 4 for TCOMP) to on one of four research properties. Defining features of each location were described in detail by Barwick *et al.* (2009), and were selected to be representative of the major production environments of northern Australia. Cows were first mated, in multiple sire groups, to calve as 3-year-olds, and were re-mated annually thereafter, unless culled for failing to wean a calf from consecutive matings.

---

* AGBU is a joint venture of NSW Department of Primary Industries and the University of New England.

At calving, calves and their dams were located within 24 hours of birth, and calves were individually identified, and their gender, date of birth and dam's teat scores recorded. Teat scores were assigned for front (**TSF**) and back (**TSB**) teats on a 1 – 5 scale, with higher scores representing greater teat size. For this study, a trait describing a cows maximum teat score (across all years for which cows had a record) was created for each animal (**TSM**), and a binary trait (**MSB**) was also generated which identified whether cows had a TSM of less than or equal to 3 (coded 0) or greater than 3 (coded 1). As a cow could have teat scores recorded at multiple calvings, there were a total of 3661 TSF and TSB, and 939 TSM and MSB for Brahman cows; with 5006 TSF and TSB, and 1053 TSM and MSB for Tropical Composite females (Table 1).

Modelling and variance component estimation for teat score traits. Fixed effect models for all teat traits initially included terms which described cohort (the location the cows were managed and their year of birth), the cows property of origin, cow month of birth, previous mating outcomes and, for TCOMP, the genetic groups of their sire and dam. For both genotypes, all first order interactions were also tested in the initial models. Terms were dropped sequentially from the models in order of non-significance ($P < 0.05$) to produce the final models for each trait-by-genotype combination. Following the methods described by Bunter *et al.* (2014), variance components for teat score traits were estimated in ASReml, with animal fitted as random and relationships between animals described using a three generation pedigree. For TSF and TSB, which included repeated records from animals, a permanent animal genetic effect was also fitted as random in the models.

Genotyping and quality control for genotype data. The genotype data used for study was a subset of Beef CRC genomic dataset. The database included high density Illumina genotypes (HD: 729,068 SNPs) for 1137 animals, with a further 14, 110 imputed to this level from the results of Illumina 50K or 80K SNP chips using the BEAGLE software package (Browning and Browning, 2011), with an accuracy of 90% (as described by Zhang *et al.* 2014). Of the 969 BRAH and 1085 TCOMP cows with records for teat score traits, 939 and 1053 respectively had SNP genotypes which could be analysed for this study. SNPs with low minor allele frequencies (< 0.05) were excluded from the analyses, as were those which deviated significantly ($P < 10^{-5}$) from Hardy-Weinberg Equilibrium, resulting in a total of 567,445 analysable SNPs.

Genome wide association study. The magnitude of individual SNP effects were estimated as a fixed effect in a mixed model that included animal fitted as random and all significant descriptors of environmental variation, as described by Hawken *et al.* (2012). The expected false discovery rate (**FDR**) was calculated as: $FDR = p\ (1 - s/t)\ /\ [\ (s/t)\ (1-p)\ ]$, where **p** represents the threshold significance level tested (e.g. 0.01), **s** is the number of significant markers, and **t** is the total number of markers evaluated. To account for the multiple testing inherent in the GWAS methodology, further rigor was applied to the testing of significance of SNPs by applying a Bonferroni correction. This much more stringent evaluation of significance in multiple testing experiments divides the significance level applied by the number of tests carried out, and sets this as the threshold at which significance was evaluated.

Genomic prediction and five-fold internal cross validation. Genomic estimated breeding values (GEBV) for teat score traits were calculated using genomic best linear unbiased prediction (GBLUP). GEBVs were estimated with the genomic relationship matrix fitted as random (Zhang *et al.* 2014), and inverted using the Wombat software package (Meyer 2007). GEBV Accuracy (**ACC**) was calculated as: $ACC = r\ /\ h$, where **r** is the correlation between GEBVs and phenotypes (adjusted for fixed effects) and **h** is the square root of the heritability for the trait when estimated in a model which contained all phenotypes and significant fixed effects, and with relationships described using a 3 generation pedigree. The accuracy reported for this study is the mean of five estimates from a five-fold internal cross validation of GEBV estimates (Zhang *et al.* 2014).

RESULTS AND DISCUSSION

Descriptive statistical and genetic parameters for teat traits. Summary statistics, additive variances and heritabilities for teat score traits are presented in Table 1. Means show that on average, BRAH cows had higher TSF, TSB and TSM than TCOMP, and that it was more common for BRAH cows to record a teat score of 4 or 5 than it was for TCOMP (MSB = 0.30 and 0.21 for BRAH and TCOMP). For the subset of BRAH and TCOMP females with genotypes, heritabilities for TSF and TSB were consistent with those reported by Bunter *et al.* (2014), which was expected, as the estimates were based on very similar datasets (h^2 TSF and TSB = 0.38 and 0.30 for BRAH and 0.37 and 0.31 for TCOMP from that study). Heritabilities for TSM and MSB were comparable with those for TSF and TSB. These results suggest that if breeders of tropically adapted beef cattle wished to apply selection to reduce teat size, or to select to reduce the incidence of high teat scores, this could be undertaken sucessfully.

Table 1. Number of records analysed (N), mean and standard deviation (SD), with additive variance (σ_a^2) and resultant heritability (and associated standard error (s.e.)) for teat score traits in Brahman and Tropical Composite cows.

Traits	N	Mean	SD	σ_a^2	h^2	s.e.
Brahman						
TSF	3661	2.81	0.89	0.27	0.40	0.08
TSB	3661	2.70	0.82	0.18	0.32	0.07
TSM	939	3.39	0.88	0.39	0.59	0.12
MSB	939	0.30	0.46	0.10	0.46	0.12
Tropical Composite						
TSF	5006	2.65	0.84	0.17	0.30	0.08
TSB	5006	2.53	0.79	0.16	0.30	0.08
TSM	1053	3.22	0.77	0.15	0.29	0.10
MSB	1053	0.21	0.41	0.04	0.22	0.09

Genome wide association study (GWAS) for teat score traits. Tables 2 presents the number of significant SNPs identified at levels of $P \leq 0.05$, 0.01, 0.001 and 0.0001 for each of the teat score traits evaluated. Results for both genotypes suggest that there were high numbers of significant SNPs identified for the teat score traits evaluated. Across the four teat traits evaluated, the expected FDR averaged 0.41, 0.41, 0.42 and 0.52, for significance levels of 0.05, 0.01, 0.001 and 0.0001 respectively. FDR tended to be higher for TCOMP with results suggesting that for TSF and TSB, none of the SNPs identified as significant were beyond the expectations of chance. For the remaining teat traits, however, these results demonstrate a capacity to successfully identify significant SNPs, and suggest that genomic selection could be undertaken sucessfully in the genetic evaluation for tropically adapted beef genotypes to improve teat score. After accounting for the multiple testing associated with GWAS analyses through the application of a Bonferroni correction ($P \leq 0.05$ = 8.8 E-8), there were far fewer significant single SNPs associated with the teat score traits evaluated. For BRAH, none retained significance, while only a small proportion of SNPs identified for TCOMP TSM and MSB traits were still significant (n = 0 to 24).

Table 2. Number of significant SNPs identified from genome wide association studies and accuracy of genomic prediction based on 5-fold cross validation for teat score traits in Brahman and Tropical Composite cows.

Breed	Trait	Number of significant SNPs at P ≤				Accuracy of genomic prediction
		0.05	0.01	0.001	0.0001	
Brahman	TSF	60352	11955	1063	13	0.12
	TSB	75518	17730	2015	13	0.08
	TSM	62820	13659	1458	12	0.30
	MSB	64107	14253	1656	8	0.27
Tropical Composite	TSF	67045	13971	1416	5	0.09
	TSB	49633	8433	579	3	0.10
	TSM	77891	20038	3181	136	0.23
	MSB	66195	15211	2068	88	0.35

Genomic prediction for teat score traits. Table 2 also presents the accuracies of GEBVs estimated from five-fold internal cross-validation for teat score traits. The results suggest that GEBVs for TSF and TSB would provide only limited opportunities to change the trait by genomic selection, due to the low accuracies of GEBVs (ACC = 0.08 to 0.12). TSM and MSB displayed higher accuracies of genomic prediction (0.23 to 0.35), suggsting that genomic information could make a useful contribution to genetic evaluation for these traits.

CONCLUSIONS

This study confirmed the heritability of repeated measures of teat scores, and found that new traits, describing maximum lifetime teat score and whether cows ever recorded a high score (4 or 5), displayed similar heritabilities ($h^2 = 0.22$ to 0.59). A genome wide association study showed that variation in teat traits was associated with a large numbers of genes, and that development of a genomic test, based on small numbers of genes of large effect, was unlikely to be successful. Results indicated however, that GEBVs for teat traits had accuracies between 8 and 30 percent. Beef CRC research showed that high teat scores (bottle teats) were significantly genetically associated with early calf survival in tropical beef genotypes. The results of this study suggest that genomics could contribute to the genetic evaluation for teat traits, and correlated female reproduction traits, if they were included in the evaluation for tropical beef breeds.

REFERENCES

Barwick S.A., Johnston D.J., Burrow H.M., Holroyd R.G., Fordyce G., *et al.* (2009) *Anim. Prod. Sci.* **49**: 367.

Bunter K. L. and Johnston D.J. (2014) *Anim. Prod. Sci.* **54**: 50.

Browning B.L. and Browning S.R. (2011) *Am. J. Hum. Genet.*, **88,** 173-182.

Hawken R.J., Zhang Y.D., Fortes M.R.S., et al. (2012) *J. Anim. Sci.*: 1398.

Johnston D.J., Barwick S.A., Corbet N.J., Fordyce G., Holroyd R.G., *et al.* (2009) *Anim. Prod. Sci.* **49**: 399.

Johnston D.J., Barwick S.A., Fordyce G., Holroyd R.G., Williams P.J., *et al.* (2014) *Anim. Prod. Sci.* **54**: 1.

Gilmour A.R., Gogel B.J., Cullis B.R. and Thompson R. (2009) 'ASReml user guide. Release 3.0.' VSN International: Hemel Hempstead, UK.

Meyer K. (2007) *J Zhejiang Uni. Sci.* **B 8**: 815.

Zhang Y.D., Johnston D.J., Bolormaa S, Hawken R.J. and Tier B. (2014) *Anim. Prod. Sci.* **54:** 16.

Proc. Assoc. Advmt. Anim. Breed. Genet. 22:175-178

POSSIBILITIES OF SHORTENING THE NUMBER DAYS ON FEED FOR CALCULATING NFI IN CATTLE

S.A. Clark and J.H.J. van der Werf

School of Environmental and Rural Science, University of New England, Armidale, 2351

SUMMARY

Feed efficiency is an important trait in many beef cattle breeding programs. The current measure of feed efficiency used in beef cattle is net feed intake (NFI). Current measurement protocols stipulate a 70 day test period to obtain a reliable NFI phenotype, incurring a significant economic cost. This study examined the phenotypic and genetic implications of shortening the number of days on feed in the calculation of NFI phenotypes. Phenotypic and genetic parameters were estimated for 5 trial lengths (14, 28, 42, 56 and 70 days) for NFI and its component traits; average daily gain (ADG), metabolic mid weight (MMWT) and daily feed intake (DFI). For NFI, 56 days on feed was highly genetically correlated with 70 day estimates. For shorter periods correlations reduced and variance components changed substantially. In contrast, daily feed intake could be measured well in short periods of time with genetic correlations of > 0.95 for lengths greater than 14 days. To substantially reduce the time on feed it is suggested that breeders consider collecting DFI information rather than NFI. If this was seen to be desirable an alternative way of balancing feed intake and weight gain would need to be explored.

INTRODUCTION

The cost associated with feeding animals is one of the major expenses to all livestock production systems. How efficiently this feed is converted into animal products is often termed feed efficiency (Archer 1999). Feed efficiency (FE) is an important breeding objective trait in many beef cattle breeding programs as breeders attempt to find an optimum balance between increased production levels and costs of production. Many authors have suggested different ways of measuring feed efficiency which range from ratio traits like feed conversion ratio (FCR) to traits corrected for production like net (or residual) feed intake (Arthur *et al.* 2001). The current measure of feed efficiency used in beef cattle is NFI, which describes the difference between actual feed intake and the expected feed intake required for maintenance and growth (Arthur *et al.* 2001). This process makes NFI phenotypically independent of growth and maintenance; however, genetic correlations between each of the traits often remain (Archer 1999).

Current industry protocols require that daily feed intake (DFI) and live weight be recorded for 70 days so that an accurate NFI phenotype can be estimated (Archer *et al.,* 1999). Given the large economic cost associated with this, most recording has been limited to small groups of animals recorded at central testing sites (i.e. Tullimba feedlot). This study aimed to examine the phenotypic and genetic implications of shortening the number of days required to obtain a reliable phenotypes associated with feed efficiency.

MATERIALS AND METHODS

The phenotypic data examined in this study included live weights, and DFI measures from 1725 Angus Steers collected from 2012 to 2016 at Tullimba Feedlot. On entry to the feedlot, the animals in this study ranged from 470-700 days of age and weighed 400-520 kg. Initially animals were conditioned for 21 days and fed for an additional 70 days over which time all data was collected (NFI test period). All animals were weighed 6 times over the 70-day test period (fortnightly). Average daily gain was calculated as the regression of weight on time (days) while MMWT was obtained as the mid-point weight to the 0.73 power (Arthur *et al.*, 2001; Berry and Crowley, 2013).

Net feed intake was estimated as the residuals from the following regression; FI = b_1(ADG) + b_2(MMWT) + NFI (Arthur and Herd, 2008; Berry and Crowley, 2013). The regression coeffecients (b) were estimated across cohorts of steers. NFI, ADG, MMWT and DFI were calculated each for 5 trial lengths (14, 28, 42, 56 and 70 days), hence we considered 20 traits in total. Each reduced trial length was compared to the current 70 day period using a series of pairwise bivariate animal models using ASReml (Gilmour *et al.*, 2009). Fixed effects of mean, cohort, Pen (within cohort), Age and Dam age where fitted for each comparison.

For each pair of traits, the following bivariate animal model was used:

$$[1]\ y = Xb + Za + e$$

where **y** is the vector of the phenotypes for two traits; **X** is a matrix relating phenotypes to fixed effects; **b** is vector of fixed effects for the traits analysed; **Z** is a matrix relating animals to the data; **a** is a vector which contains random additive genetic effects of animals; and **e** is a vector with residuals for the analysed traits. Furthermore, variance structures of random effects are described as:

$$var = \begin{bmatrix} a \\ e \end{bmatrix} = \begin{bmatrix} A \otimes G_0 & 0 \\ 0 & I \otimes R_0 \end{bmatrix}$$

Where G_0, and R_0 denote 2x2 matrices containing additive genetic and residual covariance components, respectively; **A** is the numerator relationship matrix derived from pedigree information; **I** is the identity matrix; and $\otimes$ is the direct product of the matrix.

RESULTS AND DISCUSSION

Genetic correlations for all traits were high (>0.90) when comparing 70 day test period with 56 day test period (Table 1, 2 and 3). Genetic (σ^2_A) and phenotypic (σ^2_p) variance similar when comparing traits recorded at 56 days and 70 days, respectively. The lowest genetic correlation, between 56 and 70 days on feed, was NFI whilst the highest correlations were observed for DFI. Phenotypic correlations between each time period were lower than the genetic correlations for all traits.

Table 1. Variance component and heritability (h^2) estimates for NFI (kg/day) across all trial lengths and the phenotypic (r_P) and genetic (r_A) correlations of the reduced days with the full 70 days trial

Days on feed	σ^2_P	σ^2_A	h^2	*SE*	r_A	*SE*	r_P	*SE*
70	0.25	0.05	0.20	*0.06*				
56	0.25	0.06	0.22	*0.06*	0.92	*0.07*	0.80	*0.01*
42	0.37	0.09	0.23	*0.06*	0.87	*0.09*	0.54	*0.02*
28	0.64	0.11	0.17	*0.06*	0.88	*0.13*	0.40	*0.02*
14	0.94	0.20	0.21	*0.06*	0.75	*0.16*	0.30	*0.02*

Table 2. Variance components and heritability (h^2) estimates for ADG (kg/day) across all trial lengths and the phenotypic (r_P) and genetic (r_A) correlations of the reduced days with the full 70 days trial

Days on feed	σ^2_P	σ^2_A	h^2	*SE*	r_A	*SE*	r_P	*SE*
70	0.09	0.02	0.27	*0.07*				
56	0.11	0.03	0.27	*0.07*	0.98	*0.02*	0.89	*0.01*
42	0.17	0.04	0.22	*0.06*	0.94	*0.05*	0.76	*0.01*
28	0.31	0.06	0.19	*0.06*	0.88	*0.09*	0.57	*0.02*
14	1.30	0.03	0.02	*0.04*	0.82	*0.65*	0.23	*0.02*

Table 3. Variance components and heritability (h^2) estimates for DFI (kg/day) across all trial lengths and the phenotypic (r_P) and genetic (r_A) correlations of the reduced days with the full 70 days trial

Days on feed	σ^2_P	σ^2_A	h^2	*SE*	r_A	*SE*	r_P	*SE*
70	1.48	0.71	0.48	*0.08*				
56	1.56	0.75	0.48	*0.08*	1.00	*0.00*	0.98	*0.00*
42	1.69	0.83	0.49	*0.08*	0.99	*0.01*	0.95	*0.00*
28	1.85	0.85	0.46	*0.08*	0.97	*0.01*	0.89	*0.01*
14	2.12	0.81	0.38	*0.08*	0.95	*0.03*	0.78	*0.03*

For all traits, as the number of days was reduced the amount of genetic and phenotypic variance that was observed increased substantially. This was most evident for ADG where phenotypic variance increased such that for 28 days and 14 days on feed σ^2_P was 3 times or 15 times greater than 70 day estimates, respectively. In contrast, MMWT was very consistent across all NFI test periods, with r_A >0.97 and r_p>0.96 across all test lengths (results not shown). The inability to estimate ADG accurately was a major limitation to reducing the time in which NFI can be recorded. The impact of this declining accuracy with reduced NFI test length can be observed in Figure 1a which shows that, as NFI test period is reduced, the regression coefficient relating to the adjustment of DFI for ADG (b1), in the calculation of NFI, is greatly reduced. Furthermore, Figure 1b shows that NFI gradually becomes more like DFI with phenotypic and genetic correlation between NFI and DFI increasing as the number of days is reduced.

The time taken to precisely estimate ADG and therefore NFI has previously been reduced from 140 to 112 now to 70 days (McPeake, C. A., and D. S. Buchanan. 1986, Archer *et al.* 1997). Results from this current study suggest, given the high genetic correlation between 56 and 70 days, it may be possible to reduce the testing period further (to 56 days). The results in this study were similar to those presented by Archer *et al.* 1997. Differences between the results of the current study and that of Archer *et al.* (1997) may be a result of test animals in that study being in an earlier growth stage (250 days of age) compared to a finishing stage (~500 days of age) in the current study. This may explain why the genetic correlation for 56 days is higher in this present study than those presented by Archer *et al.* 1997.

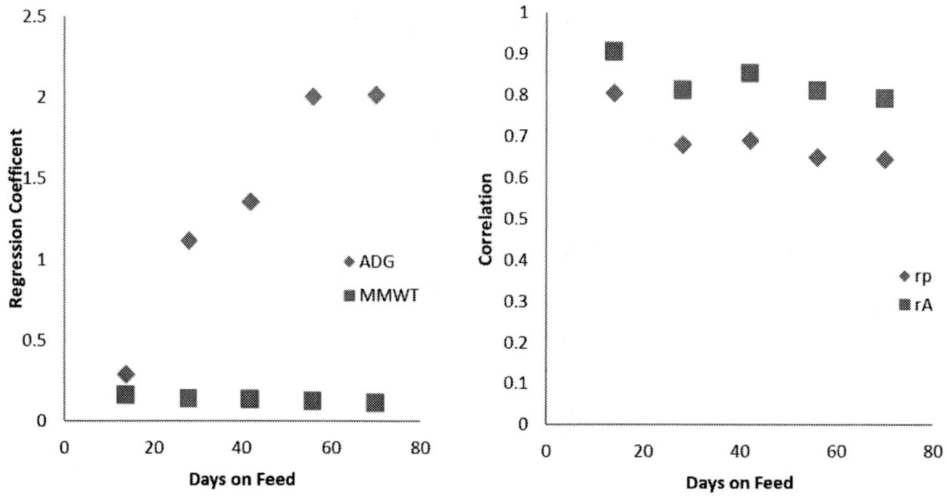

Figure 1. a) The regression coefficients (b) for ADG and MMWT used to calculate NFI, and b) the phenotypic and genetic correlations between NFI and DFI for reduced NFI test periods.

In contrast to NFI and ADG, it appears that testing length for MMWT and DFI could be reduced as genetic and phenotypic correlations indicate that MMWT and DFI were very consistent over time. The time taken to record DFI could be reduced to 28 days and even short periods could be examined as larger amounts of data become available. Possible alternatives to collecting NFI routinely may be to collect a series of DFI measurements for a single animal or test more animals for shorter test periods. The later would reduce the costs associated with collecting feed intake information. It is likely that it would also increase overall response to selection as more animals could be recorded. This would only be possible if growth traits (like ADG) were measured at different stages (additional records) in the breeding program. Some consideration is needed, prior to the reduction of the number of days on feed, to understand the relationship between growth and feed intake to ensure breeders achieve what is desired in their given breeding program. A key question is: are we happy with the current definition of NFI, or would a different definition of feed efficiency better serve breeders, feeders and processors? It could be plausible that only recording feed intake or creating an disconnect between feed intake and growth, by measuring both traits at different times, may in fact completely change the weight placed on each trait when selecting for feed efficiency.

REFERENCES

Archer, J.A., Arthur, P.F., Herd, R.M., Parnell, P.F. and Pitchford, W.S. (1997) *J. Anim. Sci.* **75**:2024

Archer J. A., Richardson E. C., Herd R. M., and Arthur. P. F. (1999) *Aust. J. Agric. Res.* **50**: 147.

Arthur P.F., Archer J. A., Johnston D. J., Herd R. M., Richardson E. C. and Parnel P. E. (2001) *J Anim Sci.* **79**(11), 2805.

Arthur, J. P. F., and Herd R. M. (2008) *Revista Brasileira de Zootecnia.* **37**: 269-279.

Berry D. P., Crowley J. J. (2013). *J Anim Sci.* **91**: 1594.

Gilmour A. R., Gogel B., Cullis B., Thompson R., and Butler D. (2009). VSN International Ltd, Hemel Hempstead, UK.

McPeake, C. A., and D. S. Buchanan. 1986. *J. Anim. Sci.* ***63****(Suppl. 1):209.*

Proc. Assoc. Advmt. Anim. Breed. Genet. 22:179-182

USE OF GENOMIC SELECTION IN A TROPICALLY ADAPTED COMPOSITE BEEF PROGRAM

W.S. Pitchford[1], G.I. Popplewell[2] and R. Terle[1]

[1] School of Animal and Veterinary Sciences, University of Adelaide, Roseworthy Campus SA 5371, Australia

[2] Popplewell Composites Pty Ltd, 33 Tom Schmidt Crt, Mt Samson, Qld 4520, Australia

SUMMARY

Popplewell composites objectively breed tropically adapted multi-breed composite bulls for beef production. They recently genotyped the whole herd and this paper reports analysis of this data. The data was analysed using G-BLUP using a genomic relationship matrix based on 23,094 polygenic markers for 1,104 animals. Preliminary estimates of heritabilities and variances were close to published estimates for similar cattle from northern Australia. Heterozygosity effects were substantial for reproduction and growth.

INTRODUCTION

The Popplewell Composite program was established in 2008 using objectively selected genetics from Angus, Belmont Red / Bonsmara, Senepol and Brahman population. The objectives of the program are to deliver continuous additive genetic improvement in meat production and quality, and female fertility improvement through replacement of traditional *Bos indicus* dominated herds with Taurus / Sanga / Indicus tropically adapted composites (Burrow *et al.* 2003) in addition to introgression of favourable qualitative alleles such as Poll and slick coat.

Genetic evaluation of livestock has traditionally been based on information on genetic relationships between animals (pedigree) and performance of animals or their relatives. Initially this was using sire models, then all known relationships could be modelled using the relationship matrix and analysing the data using best linear unbiased prediction based on the so called animal model (Quaas and Pollak 1980). There have been numerous developments to this method over the years (Graser *et al.* 2005). However, the system has limitations when animals with limited pedigree information are included, especially in tropical beef populations with large use of multiple sire mating systems before the availability of parentage testing technology. Genomic selection as proposed by Muewissen *et al.* (2001) with further developments (e.g. Hayes and Goddard 2011) enables breeding value estimation based on DNA rather than pedigree information. Furthermore, for composite herds a "genetic groups" effect (Gilmour *et al.* 2009) is often included but a genomic relationship matrix can simultaneously account for both between and within-breed genetic variation.

Female reproductive performance is an important profit driver for northern Australian beef production systems. The aim of this paper is to report preliminary genetic parameters for reproduction, growth and carcass quality traits using a genomic relationship matrix in a tropically adapted composite herd. Heterozygosity effects which reflect heterosis or dominance effects which are commonly large for female reproduction traits in taurine x indicine hybrids (Pitchford *et al.* 1993) are also reported.

MATERIALS AND METHODS

Herd management. The Popplewell Composite nucleus cow herd is run in coastal South East Queensland, rotationally grazed on Seteria, Kikuya and Rhodes grass based pastures and exposed to tropical parasites. The herd is phenotyped for fertility, birth weight, growth, flight speed, tick resistance and live-ultrasound carcass traits. Semen tested yearling bulls are sold to commercial and bull multiplier herds in Tropical and Subtropical regions of Australia. All heifers born into the program are first mated as yearlings which is not typical of tropical breed seed-stock herds.

Prior to G-BLUP, hair and or semen samples for DNA extraction had always been collected and stored on all nucleus animals and DNA technology use had been limited mainly to parentage determination and introgression of favourable Poll genes. The commitment to storing tissue and collecting economic relevant phenotypes provided a bank of DNA and data ideal for whole herd G-BLUP without the need for blending of pedigree and genomic relationships. Pedigree data allowed for comparison of pedigree BLUP and G-BLUP models.

Processing marker data. Animals were genotyped on either the Illumina GeneSeek GGP Bovine LD chip (versions 3 and 4) or Illumina BovineHD chip. A matrix of AB genotype calls for 1,119 animals and 29,464 SNPs were extracted from text output files and the minor alleles counted for each genotype (i.e. 0, 1, 2), where the minor allele was calculated across the 1,119 animals. Duplicate animals were removed, monomorphic SNPs and those with minor allele frequency less than 0.01 were also removed, leaving 23,094 SNPs on 1,104 animals. Heterozygosity for each animal was calculated by summing the number of heterozygous genotypes as a proportion of all called genotypes. Heterozygosity is a measure of dominance and reflects heterosis. The values ranged from 25-47%.

A standardised matrix of counts for each SNP was generated by subtracting its mean and dividing by its standard deviation. Missing values were replaced by the standardised mean (0). This starting matrix was multiplied by its transpose and divided by the number of SNPs to generate a relationship matrix which was then inverted ready for analysis.

Statistical analysis. Phenotypes were available for up to 3,934 animals depending on the trait but only 1,104 were genotyped. This paper reports analysis of a subset of phenotypes for animals present in the relationship matrix. The data was analysed using a linear mixed model in ASREML-R (Butler *et al.* 2009). Fixed effects were birth year (2008-2015), sex (male, female), dam age (2-10 years but coded as heifer or mature), age (by fitting birth date as a covariate within year), and heterozygosity (Het%). Contemporary group was defined as management group within birth year and sex. Management groups for later ages were comprised of current management group and previous management groups as described by Graser *et al.* (2005). Ultrasound traits included day of measurement in the contemporary group definition and included weight as a covariate within contemporary group. Scrotal size included a covariate of age within contemporary group. Lastly, the random animal effects were fitted as the inverse of the genomic relationship matrix.

The traits analysed were birth weight, weights at 200, 400 and 600 days (kg), ultrasound loin eye muscle area (cm^2), P8 fat depth, rib fat depth (mm) and intramuscular fat content (%). Maternal genetic effects were not included in initial analyses but will be for birth and 200 day weights in future.

Fertility was measured only on naturally mated females as days from joining to calving with yearling heifers (HDC) separate from those joined from 2 years old (mature, MDC). Those that failed to calve had a 32 day penalty added to the maximum DC value in their management group. Sex, dam age and heifer age effects were not included in the analysis of HDC or MDC. Mature weight was analysed using fixed effects of age in years, lactation number and heterozygosity.

RESULTS AND DISCUSSION

The population is a composite of Africander (Bonsmara and Belmont Red), Senepol, Red Angus and Brahman. A summary of the genetic variation is presented based on principal component analysis of the SNP genotypes (Figure 1). The G-BLUP performed well at describing both between and within breed variation in a single step. Fitting calculated heterozygosity avoided bias in BLUP estimates resulting from heterosis, especially for fertility.

The combination of breeds during the development was expected to lead to large variation in traits that differ between breeds. However, for most traits the variances and heritabilities (Table 1) were very similar to those reported by Wolcott *et al.* (2014) and Johnston *et al.* (2014) for tropical composite cattle measured as part of the CRC for Beef Genetic Technologies. A small exception would be that herein the cattle were younger when ultrasound scanning so the mean and variance in the fat traits was lower than the CRC cattle.

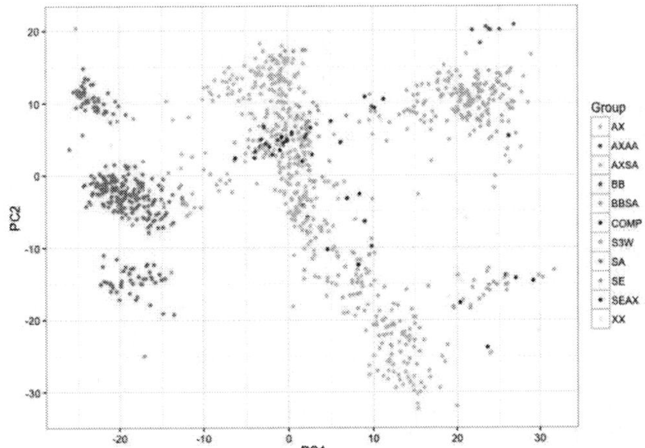

Figure 1. Genetic variation coded by breed of origin. AX is Africander (right), AXSA is Africander x Senepol/Angus (middle), BB is Brahman (top right) and SA is Senepol x Red Angus (left). Other combinations are minor.

Numbers of cattle were a limitation for accurate heritability estimation (Table 1). The numbers for growth and carcass traits was around 800 but for male (scrotal size) and female fertility traits, numbers were very small. Despite this and the fact that a genomic rather than an animal relationship matrix was used, heritability estimates were very close to published values for equivalent breeds and traits (Barwick *et al.* 2009). It is especially encouraging that the preliminary heritability estimates herein for days to calving for first parity and mature cows were almost identical to those presented by Johnston *et al.* (2014). However, a difference herein is that heifers were joined at 15 rather than 27 months. To conceive to calve at 2 years, heifers need to be cycling by around 400 days. Johnston *et al.* (2009) reported that composite heifers averaged 650 days at puberty. Thus, the program herein is putting substantial phenotypic and genetic selection pressure on heifer puberty because it is such a large profit driver and given the number that conceived, it must be working.

Those with greater heterozygosity were bigger and had better female reproduction (conceived faster, Table 1). All of these are as expected based on heterosis in taurine x indicine crosses (e.g. Pitchford *et al.* 1993). This would likely have a significant effect on profitability of commercial herds.

The practical outcome of this work is that this breeding program should achieve significant gains for commercial clients. A selection index was developed based on a combination of approximate economic values and desired gains. The 2016 mating decisions will lead to cattle with higher growth, more fat and improved fertility through both increased scrotal size and decreased days to calving. In addition, there will be small decreases in birth weight and mature cow weight as well as a small increase in eye muscle area. There was no direct selection for fat but this was a correlated response resulting from positive correlations with growth and fertility. There is expected to be ongoing improvement due to the extensive measurement program, all animals genotyped and mating allocations based on optimising breeding value and genetic diversity outcomes. In addition, the program will further accelerate in scale through strategic partnerships with bull customers using genotypes and phenotypes from their bull multiplier and commercial tier herds.

Table 1. Summary of data, phenotypic variance, heritability and heterozygosity estimates.

Trait	No.	Mean	SD	Min	Max	σ_P^2	h^2	Het%
Birth weight (kg)	892	36.8	5.3	21	55	16.9	0.41	0.32**
200 d weight (kg)	883	204	51	75	415	460	0.11	1.65**
400 d weight (kg)	801	320	64	152	528	905	0.35	3.17**
600 d weight (kg)	351	374	66	232	694	1078	0.56	2.70**
Eye muscle area (cm^2)	790	55.3	12.8	23	96	27.9	0.39	0.31**
Rump P8 fat (mm)	790	3.8	1.6	1	10	1.18	0.23	0.060
Rib fat depth (mm)	790	2.9	1.1	1	7	0.57	0.15	0.021
Intramuscular fat (%)	790	3.5	1.0	1	6	0.45	0.20	0.017**
Scrotal size (cm)	409	29.8	4.0	20	41	8.03	0.62	0.13
Heifer DC	255	348	33	271	393	1021	0.21	-2.77*
Mature DC	503	333	21	271	368	1099	0.14	-2.43**
Mature weight (kg)	433	486	68	324	666	2510	0.60	2.26*

DC is days to calving from date of joining to calving with a 32 day penalty for non-calvers.
Het% is regression of trait on percentage of polymorphic SNPs that were heterozygous.
Approximate standard errors of preliminary heritability estimates were large for all traits and >1 for some. * P<0.05, **P<0.01

In conclusion, this tropical composite breeding program has been innovative in storing DNA and then genotyping all animals. This has enabled genomic analysis of both traditional BREEDPLAN and new traits important for reproduction. Preliminary estimates of heritabilities are similar to other studies and important heterozygosity effects have also been reported.

REFERENCES

Barwick, SA, Wolcott, ML, Johnston, DJ, Burrow, HM and Sullivan, MT (2009) *Anim. Prod. Sci.* **49**:351-366.

Burrow H.M., Griffith G,R.,Barwick S.A. and Holmes W.E. (2003) Where to From Brahmans, *AAABG* 294-297

Gilmour, AR, Gogel, BJ, Cullis, BR and Thompson, R (2009) *ASReml User Guide*

Graser, H-U, Tier, B, Johnston, DJ and Barwick, SA (2005) *Aust. J. Exp. Agric.* **45**:913-921.

Hayes, BJ and Goddard, ME (2011) *J. Anim. Sci.* **86**:2089-2091.

Johnston, DJ, Barwick, SA, Corbet, NJ, Fordyce, G, Holroyd, RG, Williams, PJ and Burrow, HM (2009) *Anim. Prod. Sci.* **49**:399-412.

Johnston, DJ, Barwick, SA, Fordyce, G, Holroyd, RG, Williams, PJ, Corbet, NJ and Grant, T (2014) *Anim. Prod. Sci.* **54**:1-15.

Meuwissen, THE, Hayes, BJ and Goddard, ME (2001) *Genetics* **157**:1819-1829.

Pitchford, WS, Barlow, R and Hearnshaw, H (1993) *Livest. Prod.Sci.* **33**:141-150.

Quaas, RL and Pollack, EJ (1980) *J. Anim. Sci.* **51**:1277-1287.

Wolcott, ML, Johnston, DJ, Barwick, SA, Corbet, NJ and Williams, PJ (2014) *Anim. Prod. Sci.* **54**:37-49.

Proc. Assoc. Advmt. Anim. Breed. Genet. 22:183-186

FACTORS AFFECTING PUBERTY ATTAINMENT IN PUREBRED AND CROSSBRED HEIFERS

J. G. Alexopoulos[1,2], M.L. Hebart[1] and W.S. Pitchford[1]

[1]School of Animal and Veterinary Sciences, University of Adelaide, Roseworthy Campus SA 5371, Australia
[2]Present address: SunPork Farms South, Coleman Road, Pinkerton Plains SA

SUMMARY

This paper reports breed differences for puberty attainment in Angus and Hereford sired heifers from the first cohort of a larger trial. Pre-joining, there were no statistically significant differences between the breeds in age, weight, height, fat and net weight although crossbreed heifers were 17 kg heavier (5%), 0.2 mm fatter (5%), 15 mm taller (1%) with 8 kg more net weight than purebreds. A greater number of crossbred heifers had reached puberty prior to joining (21%; P=0.062). Age and fat depth were significant for attainment of puberty, but height and weight were not.

INTRODUCTION

The reproductive performance of female breeding stock is one of the largest economic drivers in Southern Australian beef production systems (Wathes *et al.* 2014). The aim was to utilise a crossbreeding system to optimise fertility traits, by utilising heterosis and breed effects. These mechanisms effectively increase performance traits and adaptability of genetic resources to the climate, environment and nutritional availability (Gregory and Cundiff 1980).

Genetic selection in temperate beef genotypes has primarily focused on the selection and improvement of feedlot production traits using BREEDPLAN estimated breeding values (EBVs) for, weight at endpoint, and more recently meat quality and yield (Hebart *et al.* 2016). Strong selection pressure on improving feedlot performance and carcass quality traits have been accompanied by a negative trend in fertility in some herds (Wathes *et al.* 2014).

Heifer conception rates are primarily determined by the age at which puberty is attained (Day 2015). The optimum age to reach puberty is by 13 months for joining at 15 months and calving at 24 months (Patterson *et al.* 1992). Age at puberty varies largely within and between breeds, and is dictated by the environment (Chenoweth 1994), plane of nutrition and the genetics of both sire and dam (Patterson *et al.* 1992). Body weight is the key determinate to attain puberty (Wathes *et al.* 2014), literature recommends *Bos taurus* breeds reach 60% of mature cow weight (MCW), 30-45 days prior to joining to ensure conception rates >85% (Ahmadzadeh *et al.* 2011). A recent study by Jones *et al.* (2016) reports that both fat and weight are important for a successful conception and heavier heifers can succeed with lower levels of fat.

The aim of this paper was to assess the factors that affect puberty attainment in a cohort of Angus and Hereford sired beef heifers from Angus dams. It was hypothesised that crossbred heifers will achieve a minimum proportion of mature cow weight earlier, and reach puberty at a younger age than the purebreds.

MATERIALS AND METHODS

Animals and heifer management. The 208 heifers for this study came from Angus dams joined to either 11 Hereford or four Angus sires using artificial insemination, resulting in 135 Hereford and 72 Angus sired heifers in the first cohort of the "Black Baldy" trial. The heifers were located at, Musselroe Bay in the north east of Tasmania. The Angus dams came from two management groups that differed in age of dam (maiden-2 year old or mature-3 year old), insemination dates, paddock location and sire line.

Heifer calves were born late June to early August 2015. Calves were weaned from both management groups in March 2016 and joined as one cohort and grown out to joining together under the same management.

Ovarian assessment was performed to define portion pubertal, by transrectal real-time ultrasonography at three time points when heifers averaged 300, 387 and 448 days of age, heifers that had reached puberty were excluded from subsequent ovarian measurements. Scanning involved detection of a corpus luteum (CL) or formation of a corpus albican (CA) on either the left or right ovary, confirming that the animal had ovulated (Monteiro *et al.* 2013).

At each ovarian measurement all heifers were weighed on a scale placed under a crush, height was recorded by measuring the distance from the hip to the top of the crush, this number was subtracted from the height from the top to the base of the crush, P8 fat depth was obtained using ultrasonography.

Body condition score is related to body weight and fat coverage, both which can be used to assess reproductive performance (Jones *et al.* 2016). This trial did not record body condition thus generated a new trait to define heifer condition, where weight was regressed on height at each measurement and the residuals were interpreted as net weight. Heifers with a positive net weight were considered to have a better or greater condition than expected for a given height, and therefore were expected to reach puberty at an earlier age.

Statistical analysis. All data was analysed using GenStat 15th Edition Sp2 (VSN Int 2016) statistical programme. Two primary models were fitted:

1. To determine whether there was a significant difference between the breeds in the attainment of puberty and other traits (age, weight, height, P8 fat depth and net weight), a general (generalised for puberty) linear mixed model was fitted to the data. The fixed effects included management group and breed, the interaction between these effects were also tested. A random effect of sire nested within breed was also included.
2. To determine what traits were significantly influencing the attainment of puberty, a generalised linear mixed model was fitted to the data with the fixed covariates of age, weight, height and P8 fat depth. Age by weight, height and P8 fat interaction terms were also included but were not significant ($P>0.05$) and so were not included in the final model.

RESULTS AND DISCUSSION

There were no statistically significant differences between the breeds in age, weight, height, P8 fat and net weight (model 1), although crossbred heifers were 15 mm taller (1%), 0.2 mm fatter (5%), 17 kg heavier (5%) with 8 kg more net weight. An additional 21% of crossbreed heifers had reached puberty at the final scan, prior to joining, this included heifers that were pubertal at scan point one and two (Table 1, Figure 1). Management group effects were large for all traits (not presented).

Table 1. Predicted means and standard errors (S.E.) for sire breed pubertal by scan three (pre-joining)

Heifer trait (units)	Angus	S.E.	Hereford	S.E.	P-value
Age (days)	449	3	448	2	0.540
Weight (kg)	329	7	346	4	0.075†
Height (mm)	1197	7	1212	4	0.066†
P8 fat (mm)	3.7	0.2	3.9	0.1	0.594
Net weight (kg)	-8	4	0	3	0.410
Pubertal (%)	35	7	56	4	0.062†

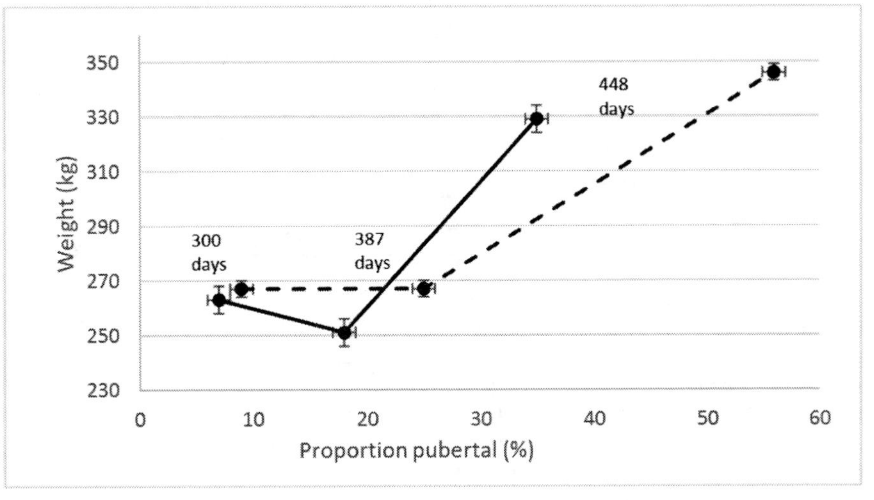

Figure 1. Breed differences in mean body weight and proportion pubertal (Model 1) at scans one, two and three (300, 387 and 448 days of age respectively). Angus heifers are represented by the solid line and Hereford cross Angus by the broken line.

Previous studies have suggested that getting heifers to a minimum proportion of MCW 30-45 days prior to joining, will increase first season conception rates (Patterson *et al.* 1992), through younger attainment of puberty. A complementarity advantage acquired through crossbreeding is heterosis, the additive genetic merit and growth inherited by the progeny from the terminal sire, which effects the age of puberty in offspring (Gregory and Cundiff, 1980). Wiltbank *et al.* (1969) demonstrated this in Angus and Hereford reciprical crosses reporting the Angus x Hereford and Hereford x Angus reached puberty earlier (29 and 55 days respectivley) than their purebred counter parts. Breed was almost significant in this trial for weight (P=0.075), this non-significant result may be due to the loss in body weight in both the Angus and Hereford x Angus between scan one and two when harsh environmental conditions were encountered. Average daily growth (ADG) was not recorded in this trial although at each measurement crossbred heifers were heavier with higher portions of cycling individuals. This concurs with recent literature reporting that ADG before 15 months can stimulate growth paths leading to early onset of puberty (Wathes *et al.*

2014). Gasser *et al.* (2006) demonstrated this, showing heifers with high ADG (1.2 kg/day) reached puberty by 271 days, compared to heifers with low ADG (0.7 kg/day) that reached puberty at 331 days weighing 282 kg and 367 kg respectively. The results from the current trial suggest that crossbreed heifers were able to cope with harsher environmental conditions, maintaining weight and condition more efficiently than their purebred counterpart, resulting in higher portions of heifers reaching puberty by the final ovarian scan prior to joining.

Age and fat depth were associated with attainment of puberty (Table 2), but weight and height were not. Two-way interactions between variables were tested and were not significant and were not retained in the final model. Acquiring a critical fat depth was more important for heifers to attain puberty than age alone. Age at puberty is strongly correlated to first season conception rates, Byerley *et al.* (1987) demonstrated the importance of age reporting that heifers that attain puberty earlier in life, and that had cycled multiple times prior to mating, had pregnancy rates of 78% in comparison to 57% in heifers that had reached puberty but where only in first oestrus at joining. A recent study by Jones *et al.* (2016) concluded that both fat and muscle were important for heifer conception. To achieve 85% conception rates under a six week joining period, Jones *et* al. (2016) predicted that Angus heifers needed to be 52% of MCW with 8 mm of rib fat or 69% of MCW with 4 mm of fat.

Table 2. Type 3 tests of significance (P-value) for traits affecting puberty at scan three

	Age	Weight	Height	P8 fat
Pubertal (%)	0.021*	0.428	0.457	0.015*

CONCLUSION

In conclusion, age and fat depth were important for attainment of puberty. Interestingly despite an almost significant difference in portion pubertal between breeds prior to joining there was no difference in net weight or fat depths. Although not significant crossbred heifers were heavier, taller and fatter with a higher portion pubertal at joining, this may increase first season conception rates. Lastly, the proportion of pubertal heifers at different weight and ages presented should guide future studies investigating attainment of puberty in southern Australian heifers.

REFERENCES

Ahmadzadeh A., Carnahan K., Autran C. (2011) *ARSBCO.* **31**: 45-60.

Byerley D.J., Staigmiller R.B., Berardinelli J.G., Short R.E. (1987) *J. Anim. Sci.* **65**: 645-650.

Chenoweth P. (1994) *Aust. Vet. J.* **71**: 422-426.

Day M.L. (2015) *2015 Florida Ruminant Nutrition Symposium*.

Gasser C., Behlke E., Grum D., Day M. (2006) *J. Anim. Sci.* **84**: 3118-3122.

Gregory K., Cundiff L. (1980) *J. Anim. Sci.* **51**: 1224-1242.

Hebart M.L., Accioly J.M., Copping K.J., Deland M.P.B., Herd R.M., *et al.* (2016) *Anim. Prod. Sci.* (In press)

Jones F.M., Accioly J.M., Copping M.P.B., Graham J.F., Hebart M.L., *et al.* (2016) *Anim. Prod. Sci.* (In press)

Monteiro F., Mercadante M., Barros C., Satrapa R., Silva J., *et al.* (2013) *Theriogenology* **80**, 10-17.

Patterson D., Perry R., Kiracofe G., Bellows R., Staigmiller R., *et al.* (1992) *J. Anim. Sci.* **70**: 4018-4035.

Wathes D., Pollott G., Johnson K., Richardson H., Cooke J. (2014) *Anim.* **8**: 91-104.

Wiltbank J., Kasson C., Ingalls J. (1969) *J. Anim. Sci.* **29:** 602-605.

Proc. Assoc. Advmt. Anim. Breed. Genet. 22:187-190

PREDICTION OF GENOMIC BREEDING VALUES OF PRIMAL CUT WEIGHTS IN KOREAN HANWOO CATTLE FROM DIFFERENT GROWTH AND CARCASS TRAITS

H.A. Al-Mamun[1], S. Kim[2], B. H. Park[2], M. N. Park[2] and C. Gondro[1]

[1] School of Environmental and Rural Science, University of New England, Australia
[2] Animal Breeding & Genetics Division, National Institute of Animal Science, Republic of Korea

SUMMARY

In Korean beef industry, selection indices are currently limited to carcass weight (CWT), marbling score (MS), eye muscle area (EMA) and back fat thickness (BFT), which are the four traits used to determine the grade of a carcass. However, other important traits have received less attention; for example, yearling weight which influences both meat quality and the yields of the primal cuts that command premium prices. In this paper, we evaluate how well genomic prediction based on routinely measured phenotypes (body weight at different ages 6, 12, 18 and 24 months, CWT and EMA) can predict other commercially important traits (MS, BFT, various primal cuts and total percentage of meat yield) which are not usually recorded. We also compare the prediction accuracy of the primal cuts and yield derived from body weight and carcass weight predictors with the prediction accuracy using the trait itself. Our results suggest that, direct genomic prediction of primal cuts and yield had a higher accuracy, and in the future some consideration should be given to better account for primal cuts and yield in the breeding program.

INTRODUCTION

Hanwoo is the most important cattle in Korea and its history traces back 5,000 years (Jo *et al.* 2012). Hanwoo beef has unique marbling characteristics which makes it highly sought after by consumers at premium prices (Han and Lee 2010; Kim *et al.* 2010; Jo *et al.* 2012). Korean cattle breeding policies are primarily focused on increasing marbling and body weight. These two traits, particularly marbling score, are the key determinants of the carcass' grade and, consequently, its price (Park *et al.* 2002; Kim *et al.* 2010; Alam *et al.* 2013). Since marbling drives most of the profit in the Korean beef industry, producers often prolong feeding periods to achieve better marbling, even if at the expense of increased backfat thickness (BFT) which incurs a grading penalty.

Considerable effort to select superior Hanwoo bulls based on the genetic parameter estimates of carcass traits has already been made (Lee *et al.* 2000; Baik *et al.* 2003; Choy *et al.* 2008). However, selection indices are currently limited to carcass weight (CWT), marbling score (MS), eye muscle area (EMA) and backfat (BFT), which are the four traits used to determine the grade of a carcass. However, other important traits have received less attention; for example, yearling weight which influences both meat quality and quantity (Lopez-Campos et al. 2012), and the yields of the primal cuts that command premium prices. Differences in price exist between different primal cuts (Morris *et al.* 2010) and large variation in yield of the primals within each grade has been reported (Moon et al. 2003). This variation affects the accuracy of the estimates of grading and consequently there is significant averaging out in the payment system. Thus, the current grading scheme based on CWT, MS, EMA and BFT may not accurately reflect the differences within the carcass primal cuts and the actual *realized* sales price in the retail market.

The broad adoption of molecular technologies for genomic selection in livestock species (Hayes *et al.* 2009; Goddard *et al.* 2010) has significantly increased the rate of genetic progress. Genomic selection can provide more accurate estimates of breeding values earlier in the life of breeding animals, higher selection accuracy and shortening of generation intervals. Additionally, hard or expensive to measure traits can be improved more effectively by predicting EBVs of un-phenotyped animals directly from their genotypes (Gondro *et al.* 2013). Thus, genomic selection allows new traits to be selected on and provides the information needed for better indexes and payment/reward systems. The ability to better align the grading system with the actual retail value of the carcass can provide significant benefits to the Korean beef industry.

In this paper, we evaluate how well genomic prediction based on routinely measured phenotypes (body weight at different ages 6, 12, 18 and 24 months, carcass weight and eye muscle area) can predict other commercially important traits (MS, BFT, various primal cuts and total percentage of meat yield) which are not usually recorded. We also compare the prediction accuracy of the primal cuts derived from body weight and carcass weight predictors with the prediction accuracy using the trait itself.

MATERIALS AND METHODS

Animals and Traits: The present study analysed the records of 1,092 Hanwoo males raised under the Korean National Hanwoo Cattle Improvement System from 1997 to 2013. The growth and carcass traits considered in the present study included body weights at different ages (6, 12, 18 and 24 months), cold carcass weight (CWT), eye muscle area (EMA), back fat thickness (BFT), and marbling score (MS). Primal-cut yield (percent of carcass weight composed of both unique and composite meat cuts from the forequarters and hindquarters) included the yields of chuck (CHK), shoulder (SLD), brisket and flank (BAF), ribs (RIB), tenderloin (TLN), striploin (STLN), sirloin (SLN), top round (TRND), round (RND), fore- and hind-shins (FHS), and total primal cut (TPC, sum of all primal cuts) and percentage of meat yield (Meat %). Summary data on different weights, carcass traits and primal-cut yields are shown in Table 1.

Table 1. Phenotypic mean, standard deviation and heritability with SE

Trait	Mean	SD	h^2 (±SE)
WT6m	169.07	31.08	0.24±0.07
WT12m	320.91	41.27	0.29±0.07
WT18m	483.93	52.08	0.39±0.08
WT24m	634.86	67.66	0.48±0.08
CWT	362.33	41.14	0.56±0.08
EMA	81.28	8.72	0.49±0.07
BFT	8.48	3.3	0.48±0.08
MS	3.38	1.56	0.56±0.08
CHK	12.94	3.71	0.34±0.07
SLD	22.84	2.84	0.62±0.07
BAF	27.92	4.95	0.38±0.08
RIB	55.68	7.59	0.41±0.08
TLN	5.8	0.79	0.49±0.08
STLN	34.8	4.55	0.51±0.08
SLN	7.46	1.08	0.50±0.08
TRND	19.52	2.31	0.70±0.07
RND	31.87	3.75	0.71±0.07
FHS	14.46	2.61	0.32±0.08
TPC	233.28	26.15	0.58±0.08
Meat %	64.46	2.72	0.43±0.07

Statistical Analysis: Heritability of each trait was estimated using a univariate model in MTG2 software (Lee and van der Werf 2016). As multi-trait (3 x 3 and more) analyses failed to converge, a series of bivariate analyses using MTG2 was used to calculate the genetic correlations between the traits. Relationship among the animals were accounted for using a genomic relationship matrix (GRM) obtained from SNP data and was fitted as a random effect in the model. Phenotypic correlations were calculated as the Pearson correlations between the residuals of the phenotypes after removing the fixed effects using a liner model in R.

Prediction of genomic breeding values were obtained from the genomic best linear unbiased prediction (GBLUP) method in MTG2. Prediction accuracy was calculated as the Pearson correlation between the adjusted phenotypes (residuals of the phenotypes after accounting for the fixed effects) divided by the square root of the heritability of the trait. The average of 10-fold cross validation with 10 replicates is reported herein.

RESULTS AND DISCUSSION

Heritabilities for the traits were all moderate to high, ranging from 0.24 for WT6m to 0.71 for RND. Comparing the rest of the traits, top round and round have very high heritability. Standard errors for the heritabilities ranged between 0.07 and 0.08.

Genetic correlations between body weight, carcass weight and EMA with different primal cuts are shown in Table 2. Genetic correlations between weights and different primal cuts increase as the cattle become older. Although all primal cuts and the total primal cuts have medium to very high genetic correlations with body weight, carcass weight and EMA, Meat % has very low or negative correlations with these traits except for EMA. EMA has moderate generic correlations with Meat % (0.47±0.12) and selection for EMA can increase percentage of meat yield. WT12m had a relatively higher correlation with TLN and CWT had a higher correlation with STLN compared to other loin cuts.

Table 2. Genetic correlations with SE between weight at different ages, carcass traits and primal-cut yields

	CHK	SLD	BAF	RIB	TLN	STLN
WT6m	0.37±0.18	0.45±0.13	0.44±0.16	0.5±0.13	0.59±0.13	0.52±0.13
WT12m	0.61±0.14	0.68±0.09	0.61±0.12	0.71±0.08	0.8±0.08	0.68±0.08
WT18m	0.61±0.12	0.73±0.06	0.73±0.09	0.85±0.05	0.76±0.07	0.76±0.06
WT24m	0.59±0.11	0.76±0.05	0.85±0.06	0.94±0.03	0.76±0.06	0.83±0.04
CWT	0.67±0.09	0.82±0.04	0.86±0.06	0.96±0.02	0.76±0.06	0.87±0.03
EMA	0.58±0.11	0.70±0.07	0.60±0.10	0.57±0.10	0.60±0.09	0.81±0.05
	SLN	TRND	RND	FHS	TPC	Meat %
WT6m	0.45±0.14	0.4±0.13	0.45±0.12	0.4±0.18	0.51±0.12	-0.21±0.16
WT12m	0.73±0.09	0.62±0.09	0.67±0.08	0.7±0.12	0.75±0.07	-0.12±0.16
WT18m	0.77±0.07	0.75±0.06	0.76±0.06	0.74±0.1	0.85±0.04	-0.15±0.14
WT24m	0.77±0.06	0.81±0.05	0.8±0.05	0.78±0.08	0.91±0.02	-0.1±0.13
CWT	0.80±0.05	0.86±0.04	0.85±0.03	0.89±0.06	0.96±0.01	-0.08±0.13
EMA	0.85±0.05	0.77±0.06	0.70±0.06	0.83±0.10	0.76±0.06	0.47±0.12

Prediction accuracy of growth traits and carcass traits using weights at different ages, CWT and EMA are given in Table 3. Weight traits, not surprisingly, are good predictors of each other but poor predictors of BFT and marginally better for MS. EMA could be predicted with reasonable and increasing accuracy from the body weights as the animal ages and from CWT.

Table 3. Prediction accuracies of growth traits and carcass traits using body weights, CWT and EMA

	WT6m	WT12m	WT18m	WT24m	CWT	EMA	BFT	MS
WT6m	0.34	0.31	0.25	0.22	0.23	0.18	0.03	0.09
WT12m	-	0.38	0.34	0.29	0.30	0.22	0.06	0.10
WT18m	-	-	0.39	0.37	0.36	0.26	0.09	0.11
WT24m	-	-	-	0.41	0.41	0.29	0.08	0.12
CWT	-	-	-	-	0.45	0.33	0.09	0.13
EMA	-	-	-	-	0.30	0.46	-0.08	0.13

Prediction accuracies of primal cuts and meat yield percentage using weights at different ages, CWT and EMA are summarized in Table 4. On average, accuracies of primal predictions increase as age increases (0.205 WT6m; 0.279 WT12m, 0.322 WT18m, 0.365 WT24m, 0.405 CWT) but always lower than the accuracies derived from the primal traits themselves (average 0.456). Interestingly all body weights at different ages and CWT failed completely to predict the percentage of meat yield. This could be explained with the fact that, we observed very low or negative genetic and phenotypic (data not shown) correlations between these traits and Meat% trait. However, in comparison to the other traits, EMA performed quite well to predict the percentage of meat yield (accuracy was 28%).

The last row of Table 4 shows the prediction accuracies for the primal cuts when the trait itself was used in the prediction model. On average, prediction of primal cuts from the primal cuts phenotypes themselves increased prediction accuracies in 122.8% in relation to WT6m, 63.27%

WT12m, 41.29% WT18m, 24.87% WT24m, 12.78% CWT and 34.42% EMA. Prediction accuracy of percentage of meat yield from itself had a 46% accuracy.

CONCLUSIONS

Genomic predictions from weights measured later in life and CWT are useful correlated traits to select on primals but come at an increase in generation interval (and CWT is essentially nonsensical in practice). The highest accuracies of selection were obtained directly from the primal cuts themselves but adoption requires investment in phenotyping and genotyping. There is good potential to make better use of genomics to improve selection for high valued cuts and redesign the selection indexes as well as the grading system to better reflect the *true* value of a carcass. EMA was somewhat useful to predict yield but weights were very poor predictors; here again, direct genomic prediction of yield had a high accuracy, in the future some consideration should be given to better account for yield in the breeding program.

Table 4. Prediction accuracy of primal cuts weights from weights at different ages, CWT and EMA

	CHK	SLD	BAF	RIB	TLN	STLN
WT6m	0.18	0.23	0.18	0.18	0.23	0.22
WT12m	0.26	0.31	0.23	0.25	0.31	0.28
WT18m	0.28	0.34	0.29	0.32	0.32	0.33
WT24m	0.29	0.39	0.35	0.39	0.35	0.38
CWT	0.34	0.43	0.38	0.42	0.36	0.42
EMA	0.32	0.37	0.27	0.26	0.30	0.38
^	*0.41*	*0.52*	*0.38*	*0.42*	*0.45*	*0.47*
	SLN	**TRND**	**RND**	**FHS**	**TPC**	**Meat %**
WT6m	0.20	0.21	0.22	0.18	0.22	0.00
WT12m	0.29	0.28	0.29	0.27	0.30	0.03
WT18m	0.32	0.34	0.34	0.30	0.36	0.01
WT24m	0.35	0.39	0.38	0.34	0.41	0.03
CWT	0.38	0.43	0.43	0.40	0.46	0.05
EMA	0.38	0.39	0.36	0.36	0.37	0.28
^	*0.46*	*0.52*	*0.51*	*0.38*	*0.49*	*0.46*

^ *Prediction of the trait from itself*

ACKNOWLEDGEMENT

This project was supported by a grant from the Next-Generation BioGreen 21 Program PJ01134906, Rural Development Administration, Republic of Korea and Australian Research Council (DP130100542).

REFERENCES

Alam M., Cho K.H., Lee S.S., Choy Y.H., Kim H.S., et al. (2013) *Asian-Australasian Journal of Animal Sciences* **26**, 1388-98.

Baik D.H., Hoque M.A., Park G.H., Park H.K., Shim K.S. et al. (2003) *Asian Australas. J. Anim. Sci* **16(1)**.

Choy Y.H., Lee C.W., Kim H.C., Choi S.B., Choi J.G. et al. (2008) *Asian-Australasian Journal of Animal Sciences* **21**, 1227-32.

Goddard M.E., Hayes B.J. and Meuwissen T.H. (2010) *Genet Res (Camb)* **92**, 413-21.

Gondro C., van der Werf J.H. and Hayes B. (2013). Springer Sciences + Business Media, LLC.

Han S.W. and Lee B.O. (2010) *J. Agric. Life Sci.* **22**, 73:89.

Hayes B.J., Bowman P.J., Chamberlain A.J. and Goddard M.E. (2009) *J Dairy Sci* **92**, 433-43.

Jo C., Cho S.H., Chang J. and Nam K.C. (2012) *Animal Frontiers* **2**, 32-8.

Kim J., Kim D., Lee J. and Lee C. (2010) *Asian-Aust. J. Anim. Sci.* **23**, 848:54.

Lee J.W., Choi S.B., Kim J.S., Keown J.F. and Van Vleck L.D. (2000) *Journal of Animal Science* **78**, 1181-90.

Lee S.H. and van der Werf J.H. (2016) *Bioinformatics* **32**, 1420-2.

Morris C.A., Cullen N.G. and McCall D.G. (2010) *New Zealand Journal of Agricultural Research* **42**, 415-21.

Park G.B., Moon S.S., Ko Y.D., Ha J.K., Lee J.G., et al. (2002) *J Anim Sci* **80**, 129-36.

Proc. Assoc. Advmt. Anim. Breed. Genet. 22:191-196

CUTTING AND PASTING: THE FUTURE OF GENETIC IMPROVEMENT FOR FOOD ANIMAL GENOMES

T.S. Sonstegard[1], D. Carlson[1], P.B. Hackett[2], A. Watson[1], and S.C. Fahrenkrug[1]

Recombinetics, Inc. St. Paul, MN U.S.A.
[2]College of Biological Sciences, University of Minnesota, Minneapolis, MN U.S.A.

SUMMARY

Gene editing technologies based on site-directed nucleases continue to improve at a rapid pace and have evolved to a point where they can be useful for direct introgression of high effect alleles into naïve populations of food animals. Herein, we review basic mechanics of site-directed nuclease action and how this activity is deployed to produce precision bred alleles into animal genomes. We also discuss the variety of traits being deployed, and differences between introducing alleles already found in nature versus gene knock out and other rationale design approaches for genetic improvement. Finally, perspectives for regulatory approval and commercialization are summarized to highlight some of the obstacles, which may hinder the widespread adoption of gene editing technology as one of the primary tools of animal breeding.

BACKGROUND

Since the first livestock domestication events approximately 10,000 years ago, the efficiency of animal production in the developed world has continued to improve through selection for desirable traits related to protein yield. The long tradition of selective breeding has relied on superior production traits emerging from the natural genetic flux. Many other advances in animal husbandry, like advanced reproductive techniques (ART), new animal medicines, and feed additives have supported production increases derived from selective breeding outcomes. Although modern genetic techniques like genome selection are increasing the accuracy with which we can find and select for these valuable alleles, genetic improvement is still limited by the availability and frequency of beneficial alleles in our current breed populations and slowed by linkage disequilibrium (LD) and long generation intervals. Furthermore, antagonistic effects are problematic in breeding practice due to tight linkages of alleles with opposite (pleiotropic) effects, e.g., the antagonistic effects of dairy fertility and/or disease resistance with milk production.

The importance of precision in animal breeding is further underscored by challenges related to global food security (FAO 2017). A burgeoning middle class of consumers estimated to be growing from 1.8 to 4.9 billion by 2030 will significantly contribute to the increasing human demand for animal protein. Furthermore, the geographical regions where this growth in consumption is highest underscore the inefficiencies of local adapted livestock varieties. These unrefined varieties are not capable of sustainably meeting demands in rapidly expanding markets, unless rapid improvements are made in average production output per animal. Crossbreeding provides an alternative to rapidly improve production in these adapted varieties, but this strategy has historically provided only short term bursts of increased performance due to heterosis and the introgression of beneficial alleles. The long term downside of such admixture is that locally adapted or purpose-bred genetics can be diluted, requiring additional, lengthy backcrossing to reach the breeding objectives of a more productive, adapted animal. Therefore, new technologies that augment current selection methods for genetic improvement must be used as part of the solution across production systems. There are thousands of yet to be discovered, desirable traits in animals that allow them to survive well in their current environments. Ultimately, solutions based on animal breeding are hampered by generation interval times, and economic feasibility and practicality for low input production systems.

The opportunity provided introgression of alleles initiated by gene editing. Advanced breeding techniques based on genome editing offer an alternative method for rapid acceleration of genetic improvement in a single generation that is directed and sustainable. The results can be disseminated in subsequent generations through traditional breeding methods. This technology allows for adaptive breeding of elite production lines, rapid production improvement of adapted lines, and conservation of diversity by introduction of heirloom alleles lost through intensive selection for a single production trait. Thus, precision genome editing as an animal breeding tools has the potential to be a "Game Changer". The possible applications and precision of site-directed nucleases (SDN)s could not, and were not predicted or made possible until very recently. By breaking a DNA molecule at a specific site, we can induce a designed genetic change by instructing the cell's repair mechanisms. The various molecular scissors provide the power previously used by plant breeders to introduce new traits by double stranded DNA breaks (DSB) made through treatment with radiation or mutagenic compounds. SDNs, like TAL effector nucleases (TALENs) and clustered regularly interspaced short palindromic repeats (CRISPR)/Cas9, have improved upon the accuracy of gene targeting by 10,000,000,000-fold, so there is no need to wade through all the random undesirable mutant outcomes of random mutational breeding. Therefore, the precision of genome editing presents a unique, asymmetric opportunity, with negligible risk and the potential for relatively very high genetic gain.

The process of editing and its deployment. Foreign DNA is often touted by the anti-biotech advocates as a primary reason for safety concerns, even though the average human consumes 10^{14} unknown genes per day. From a mechanistic perspective, the molecular components used to trigger the editing process (post-DSB) do not actually introduce or transfer any foreign material into target genome (Jasin and Haber 2016). Rather, a competition for the selected repair mechanism takes place that is dependent on the availability of cellular factors to initiate non-homologous end joining (NHEJ) or homology directed repair (HDR). HDR is facilitated but still not always favoured when a DNA template is provided to direct allelic information for gene conversion, whereby specified nucleotides can be copied in reverse-complement into the DSB site (Bozas et al. 2009; Jasin and Haber, 2016). There is firm experimental evidence suggesting template information can be provided as either ssDNA or dsDNA with potentially no size limitation just differences in efficiency for the gene conversion (Paix et al. 2016). Conversion by HDR takes place by one of two components: invasion-mediated synthesis-dependent strand annealing (SDSA) pathway or by single-strand annealing (SSA). This underscores that editing based on DSB followed by HDR is a completely natural process, because there is no transferred or introduced recombinant DNA constructs or synthetic DNA placed into the target genome.

There are currently only two proven ways to deploy gene editing technology in food animals. These are through transfection of fibroblasts destined for nuclear transfer cloning (primordial germ cells in poultry) or by microinjection of mammalian one celled embryos (Tizard et al. 2016). The transfection/injection of the "molecular scissors" into animal germplasm (fibroblasts/one-cell embryos) only increases the frequency of DNA breaks at the target locus billions fold over the natural processes of mutation. The DSBs stimulate gene conversion, which is not something novel or unnatural. In the case of sister chromatid repair, the template is copied, not physically acquired. This is analogous to gene conversion processes during meiosis, which is supported by evidence in yeast and *Drosophila* that somatic repair goes by the SDSA mechanism and involves invasion, copying, then withdrawal of the extended strand and re-pairing with the other end of the break (Bozas et al. 2009). There is also evidence in *C. elegans* that indicates that the whole of a large insert is copied *de novo* (Paix et al. 2016). This type of recombineering blurs the distinctions laid out between categorizing gene edits as something mechanistically different when considering differences in sequence length post gene conversion (e.g.,. a 10 bp versus a 1000 bp increase in allele size). The only way to separate an introduction of natural alleles from trans or cis-genesis events is to define these events as "could not be obtained by conventional breeding" or not, respectively. The opportunities now exist to introduce any sequence without the need for a recombinant DNA construct.

The deployment of editing tools for precision breeding has a spectrum of efficiencies depending on the input parameters of the editing tools, such that in most cases, many of the resultant animals from microinjection of IVF embryos may have no edits or be mosaic for an edit (Wei et al. 2015). This means some animals produced from IVF embryo injection may need to go through a Mendelian transmission test to confirm commercial viability as a germ plasm product. For regulatory considerations and commercial viability of the technology, it would be beneficial if any animals produced by injection treatment of IVF embryos, which retain no edits or cannot transmit an edit by sexual reproduction, would be treated as conventionally bred animals. In essence, the mutagenic treatment failed as if there were no treatment applied. Furthermore, any recipient animals carrying edited clones or IVF embryos should have no restrictions under conventional animal quality measures relative to entering the food chain for human consumption.

Some concern has also been raised relative to the stability of an edit and other unintended edits at off-target sites in the genome. This concern is raised based on past observations of some transgenes being lost over time from modified genomes. However, the terms used by regulators regarding "genetic stability" & "genome integrity" are meaningless phrases in the context of genome editing. For example, deep sequencing reveals that in a typical human genome there is an average of 1 mutation every 1000 bp ($6X10^6$ total), >50 Loss-Of-Function mutations in disease-related genes, hundreds sequence (gene) duplications and translocations, 10^{15} active transposons in a single human (>100/cell), and 60-100 new *de novo* mutations not from the parents ($1/10^8$ bp) with mutations varying from cell-to-cell in a single person. The rates of natural mutagenesis have been shown to be similar in cattle (Kadri et al. 2016), and negative outcomes relative to phenotype can occur from such "normal" mutagenic events even though the animals are predicted to be of superior genetic merit (Schutz et al. 2016).

Current traits and demonstration animals. Besides the need to integrate with existing systems of selection, we believe commercial deployment of genome editing should initially be focused on those traits that are beneficial along the entire value chain from animal to consumer. Such a strategy promotes animal welfare, sustainability, and consumer acceptance of the technology. For example, one of the first traits deployed in cattle was introgression of the *celtic* polled allele into horned dairy animals (Carlson et al. 2016), where changes were made using natural occurring alleles; consistent with conventional breeding principles. The resultant animals demonstrated that genome editing could benefit animal welfare by eliminating stressful management practices (dehorning), while achieving acceptance from animal advocacy groups that influence consumer food product decisions. The only remaining challenge for commercial deployment of *polled* by gene editing is regulatory approval.

In addition to consumer acceptance and regulatory approval, another limiting factor for commercial deployment is the availability of known sequence variants for target traits that add substantial value to offset the costs of trait deployment by ART. The current number of known variants that have both major effects on production, health and/or welfare and exist in low allele frequency in the most popular breeds is severely limited, especially with respect to poultry and swine. Some of this can be attributed to the fact that most livestock traits are "complex", i.e., variation in phenotype is due to effects from numerous loci with additive gene action, usually in the range of 10-100, sometimes thousands of genes. Whole genome selection in a breeding program iteratively enriches for desirable production alleles at all these loci, and the molecular markers used for guiding selection decisions rarely correspond to causal polymorphisms. Identification of causal alleles is not necessary for substantial genetic progress by genome selection, but is necessary for moving alleles to new genetic backgrounds.

The historical emergence and selection for major effect alleles indicates that the right mutation in the right gene can have a dramatic effect on a complex trait. Indeed, our appreciation of genetic potential is likely limited by epistasis, pleiotropy, and small effective population sizes. Most variants of this type as targets for editing have been reported in cattle. These variants have major effects on traits like thermotolerance (*SLICK*), muscling and tenderness (*double muscling*), coat colour, milk components (*DGAT1*), and fertility (multiple recessive lethals like HH1 and JH1). To date, these variants have only been introduced or corrected in bovine fibroblast cell lines using TALENs and HDR templates, and only a single Nelore bull was made with a *myostatin* knock out by IVF injection of TALENs into single cell embryos (Carlson et al. unpublished). Eventually, these bovine traits will be deployed for commercialization, but they may not represent the best opportunities for the use of precision breeding in food animals.

A strong argument can be made that genome editing for disease resistance traits represent the best commercial opportunity, because these traits improve animal well-being while changing industry inputs through reduction of reliance on antibiotics, vaccination, and other physical biosecurity protocols and surveillance. Again, there are practically no known variants that contribute to a substantial proportion of the phenotypic variance for resistance to a specific pathogen. Discovery of such variants may improve with revised reference genomes that contain more accurate assemblies of immune complex gene clusters. Better SNP tools and sequence alignments can then be applied to improve past and future variant discovery efforts. However, until then, we must rely on our limited knowledge of host:pathogen interactions for editing by rationale design of candidate genes. Other methods for informing rationale design and testing hypothesized causal alleles, especially for disease resistance, are needed. Recent methods by Yueng et al. (2017) demonstrate the power of using CRISPR/Cas9 to edit candidate immune genes in stem cells that can be differentiated into macrophages for pathogen challenge testing to compare how variants change response to infection and disease. Also, new gene targets for resistance can be identified using CRISPR library screening methods to interrogate gene function across an animal genome in a systematic and comprehensive manner (Zhou et al. 2014).

Although bovine traits based on "rationale design" have received publicity for conferring resistance to TB (Gao et al. 2017) and bovine respiratory disease, neither case has demonstrated resistance through a pathogen challenge of the edited animals. In pigs, Carlson et al. (unpublished), have potentially developed pigs with natural variant edits that are resistant to foot and mouth disease virus (FMDV). FMDV is a member of the picornaviruses, which replicate after infection by taking over the host cell's protein synthesis machinery. The viral proteses expressed early in the viral life cycle target and disable cap dependent translation of mRNA to shift protein synthesis to its own uncapped mRNA. The edits copied into these potentially FMD resistant pig were based on a two amino acid change to *EIF4G1* at a predicted viral cleavage site targeted by FMDV proteases (L^{pro}). L^{pro} cleavage of EIF4G1 disables the mRNA bridging function that helps allow cap dependent proteins synthesis. The amino acid variation in edited sequence was derived from *EIF4G2*; and thus, represents a natural variant found in swine. We showed these edits in *EIF4G1* gave pig embryonic fibroblasts protection from cell death caused by active L^{pro}, and the edited fibroblast cells with altered *EIF4G1* genotypes were also resistance to L^{pro} cleavage. This result provides support for further investigation to test mutant *EIF4G1* cells and animals with live FMDV challenges in a biosecure facility.

To date, probably the best case of gene editing to make an animal resistance to a pathogen was reported by Whitworth and colleagues (2015), where a knockout of swine *CD163* resulted in protection from infection by porcine reproductive and respiratory syndrome virus (PRRSV). However, subsequent efforts by Wells et al. (2016) to swap domains with a human paralog of *CD163* and by Burkard et al. (2017) to only delete exon 7 of *CD163* suggest that the other functions of this gene must be maintained to make a commercially viable, healthy animal. Thus, the continued focus to only alter a specific portion of *CD163* that facilitates infection of PRRSV into the host macrophage.

Regulation and impediments to commercialization. Gene editing has emerged as a powerful research tool that can be used to systematically test hypothesized genotype/phenotype associations, particularly for major effect alleles. Furthermore, editing can be a powerful tool to study epistatic and pleiotropic effects, enabling the comparison of phenotypes presented when a polymorphism resides in the context of original versus a comparator genome. It seems very likely that gene editing may serve not only as a research tool, but also a way to achieve non-meiotic introgression of high value polymorphisms into commercial populations, breeds, or elite individuals where they don't already exist or are present at frequencies too low for effective enrichment by selection.

So what is impeding widespread activity of such research towards commercialization of gene edited animals? The answer is probably the uncertainty of the regulatory approval systems, which inhibits investment and negatively affects innovation and commercialization. Even though molecular genetics is a highly precise science (an investment of over $300 Billion dollars in knowledge about DNA since the 1980s), our regulators in the US insist that SDN mutagenesis of DNA, where the cell's DNA repair mechanisms act naturally to produce the genetic change, is not a breeding technique but rather a drug treatment. Furthermore, the parameters laid out for approval of gene edited animals mirror those for testing efficacy of chemical compounds that may have residues in food products. Editing has no residues. Unintended mutations are possible, but requiring a measurement of a "Durable Genome" and the purity of the enzyme used to induce the DSB is scientifically a nonsensical concept. Genome editing is very predictable (more so than sex), and the technological innovation based on SDN activity is exponential. It is inevitable that over-regulation will always be based on outdated concepts, which opens up opportunities to amplify public fears by the big business of anti-biotech advocates (e.g., Greenpeace). The consequences of lengthy, expensive regulation have the potential to greatly diminish the widespread use of editing in livestock. If regulation is expensive, in a commercial endeavour with narrow profit margins for the genetics provider; then only a few "blockbuster" products will make it through regulatory approval. Expensive regulation leaves out countries that need gene edited animals the most, and supports a few multinational companies (e.g., Bayer) that can eventually overcome cost barriers to profit. However, if costs go down, then even farmers in small ecosystems can benefit from optimized agricultural animals.

We have the power to improve the world –will we use it? Today the conversation remains at a 1980s level, due to wilful ignorance by well-meaning, intelligent people. Challenges for acceptance are public "concerns" and resistance to the technology is more apparent than real. So even though the science is straight forward; regulatory approval is not. In the current schema and based on previous approval of a single GMO animal for consumption, one could predict that regulation will be 99% of financial costs of precision breeding by genome editing in the US. This will extinguish the use of gene editing for animal improvement. The fact-based argument, that ever-increasing changes in micro-climates and emerging disease threats around the world supports the application of single generation re-tooling of animals to accommodate global demand for protein in future generations, seems to have no resonance with the anti-biotechnology advocates. Their position blocks the intuitive need for genetic retooling of agricultural products, to better suit production in new environments world-wide to accommodate population increases and environmental constraints on agriculture. That can only happen if new, climate-adapted strains of animals are available at reasonable cost in poor countries. Ultimately, most rationale well intended people want the same outcomes from the use of new technology - healthy people, animals and environment – worldwide. Genome editing offers risk-free solutions – worldwide.

REFERENCES

Bozas A., Beumer K.J., Trautman J.K., Carroll D. (2009) **182**:641-51.

Burkard C., Lillico S.G., Reid E., Jackson B., Mileham A.J., Ait-Ali T., Whitelaw C.B., Archibald A.L. (2017) *PLoS Pathog.* **13:** e1006206.

Carlson D.F., Lancto C.A., Zang B., Kim E.S., Walton M., Oldeschulte D., Seabury C., Sonstegard T.S., Fahrenkrug S.C. (2016) *Nat Biotechnol.* **34:** 479-81.

FAO. (2017) The future of food and agriculture – Trends and challenges. Rome.

Gao Y., Wu H., Wang Y., Liu X., Chen L., Li Q., Cui C., Liu X., Zhang J., Zhang Y. (2017) *Genome Biol.* **18:** 13.

Jasin M., Haber J.E. (2016) *DNA Repair* **44:** 6-16.

Kadri N.K., Harland C., Faux P., Cambisano N., Karim L., Coppieters W., Fritz S., Mullaart E., Baurain D., Boichard D., Spelman R., Charlier C., Georges M., Druet T. (2016) *Genome Res.* **26:** 1323-1332.

Schütz E., Wehrhahn C., Wanjek M., Bortfeld R., Wemheuer W.E., Beck J., Brenig B. (2016) *PLoS One.* **11:**e0154602.

Tizard M., Hallerman E., Fahrenkrug S., Newell-McGloughlin M., Gibson J,. de Loos F., Wagner S., Laible G., Han J.Y., D'Occhio M., Kelly L., Lowenthal J., Gobius K., Silva P., Cooper C., Doran T. (2016) *Transgenic Res.* **25**: 575-95.

Wei J., Wagner S., Lu D., Maclean P., Carlson D.F., Fahrenkrug S.C., Laible G. (2015) *Sci Rep.* **5:**11735.

Wells K.D., Bardot R., Whitworth K.M., Trible B.R., Fang Y., Mileham A., Kerrigan M.A., Samuel M.S., Prather R.S., Rowland R.R. (2017) *J Virol.* **91:** e01521-16.

Whitworth K.M., Rowland R.R., Ewen C.L., Trible B.R., Kerrigan M.A., Cino-Ozuna A.G., Samuel M.S., Lightner J.E., McLaren D.G., Mileham A.J., Wells K.D., Prather R.S. (2015) *Nat Biotechnol.* **34:** 20-2.

Yeung A.T.Y., Hale C., Lee A.H., Gill E.E., Bushell W., Parry-Smith D., Goulding D., Pickard D., Roumeliotis T., Choudhary J., Thomson N., Skarnes W.C., Dougan G., Hancock R.E.W. (2017) *Nat Commun.* **8:** 15013.

Zhou Y., Zhu S., Cai C., Yuan P., Li C., Huang Y., Wei W. (2014) *Nature.* **509:** 487-91.

Proc. Assoc. Advmt. Anim. Breed. Genet. 22:197-200

DETECTION AND ASSESSMENT OF COPY NUMBER VARIATION USING PACBIO LONG READ SEQUENCING IN NEW ZEALAND DAIRY CATTLE

C. Couldrey[1], M. Keehan[1], T. Johnson[1], K. Tiplady[1], A. Winkelman[1], C. Thresher[1], M. D. Littlejohn[1], A. Scott[1], K. E. Kemper[2], B. Hayes[3], S.R. Davis[1] & R.J. Spelman[1]

[1]Research and Development, Livestock Improvement Corporation, Hamilton, New Zealand
[2] Institute for Molecular Bioscience, University of Queensland, St Lucia, Queensland, Australia
[3] Centre for Animal Science, University of Queensland, St Lucia, Queensland, Australia

SUMMARY

Structural variants (SVs) have eluded easy detection and characterisation, particularly in non-human species. However, there is increasing evidence that SVs not only contribute a substantial proportion of genetic variation but have significant influence on phenotypes. Here we present discovery of copy number variants (CNVs) (a subset of SVs) in a prominent New Zealand dairy bull using long read PacBio sequencing technology. Validation of CNVs was undertaken utilising whole genome Illumina sequencing of 557 cattle representing the wider New Zealand dairy cattle population. The ability to utilise CNVnator to "genotype" the 557 cattle for copy number across all regions identified as putative CNVs, allowed a genome-wide assessment of transmission level of copy number based on pedigree. The more highly transmissible a putative CNV region was observed to be, the more likely the distribution of copy number was multi-modal across the 557 sequenced animals. This transmission based approach was able to confirm a subset of CNVs that segregates in the New Zealand dairy cattle population. Genome-wide identification and validation of CNVs is an important step towards their inclusion into genomic selection strategies.

INTRODUCTION

The introduction of genomic selection to dairy cattle breeding has increased the rate of genetic gain. To date, genomic selection has largely focused on the utilisation of SNPs and very small insertions or deletions. Very little regard has been given to larger variations such as CNVs. While SVs (including CNVs) account for the greatest amount of total polymorphic content among individual genomes (Weischenfeldt et al. 2013), the focus on SNPs and small indels is presumably due to the ease with which such variation can be genotyped at a minimal cost. However, advances in genomic technologies are resulting in an increasing amount of evidence indicating that these larger sequence variations make important contributions to genetic and phenotypic variation (MacDonald et al. 2014, Zarrei et al. 2015, Sudmant et al. 2015, Weischenfeldt et al. 2013). No single technology, detection strategy, or algorithm can capture the entire spectrum of SVs in the genome. The collective effort of the human 1000 Genomes Project has utilised both a variety of SV detection platforms and algorithms to generate an integrated map of 68,818 SVs in unrelated individuals (Sudmant et al. 2015). This is now considered the gold standard SV list in humans, yet the authors still state that "SV discovery remains a challenge nonetheless, and the full complexity and spectrum of SV is not yet understood" (Sudmant et al. 2015).

The desire to have a comprehensive list of SVs in a population is not unique to human genomics, however, SV detection is critically dependent on the quality of genome assembly, which for species such as cattle, lags behind the quality of the human genome. Furthermore, while SV/CNV detection algorithms invariably report the presence of large numbers of CNVs in each individual, these detection algorithms are plagued with a high rate of false discovery. Without a gold standard with which to compare detected variants, case by case validation is a lengthy process and not suited for genome-wide analysis.

In animals such as cattle, a desire to understand the genome is driven by production traits and the desire to predict animal performance at an early age through genomic selection. As widespread genotyping and imputation of genotypes to sequence level (Druet, Macleod, and Hayes 2014) becomes more common, there is an increasing need to not only capture SNP variation, as CNVs may severely impact imputation (LIC unpublished data), and also be associated with, or contribute to important production trait phenotypes (Kadri et al. 2014, Xu et al. 2014).

The recent availability of long read single molecule sequencing (up to 80 kilobases (kb)) provides a new technology for the identifications of CNVs. This technology offers the possibility of single reads that span complex CNVs (Sedlazeck et al. 2015). We have utilised long read single molecule sequencing of a New Zealand Holstein Friesian bull with the vision of improving imputation and ultimately genomic selection and association studies.

MATERIALS AND METHODS

PacBio Sequence and SV Detection: PacBio long read sequences were generated from a Zealand Holstein-Friesian bull by Cold Spring Harbor Laboratories. The PacBio SMRT pipeline was used to generate filtered sub reads in fastq format. Alignment of reads to the UMD 3.1 bovine genome assembly was undertaken using BWA-MEM (v0.7.12; https://arxiv.org/abs/1303.3997) with options "–M –x pacbio".

SVs were called using Sniffles (v0.0.1 https://github.com/fritzsedlazeck/Sniffles). Structural variants displaying > 95% reciprocal overlap with a UMD3.1 contig were removed as these likely represent genome assembly errors. Further filtering retained only SVs present in a single contig.

Illumina Sequence and CNV Genotyping: Illumina HiSeq sequencing of 557 animals representing the population structure of New Zealand dairy cattle and phenotypes of interest has previously been described (Littlejohn et al. 2016). Read-depth-based CNV genotyping analysis was undertaken across the genome of animals sequenced on the Illumina HiSeq platform using CNVnator v0.3 (Abyzov et al. 2011) using a bin size of 150bp. Based on breakpoints identified by Sniffles, copy number was determined for each CNV greater than 100bp in length in each of 557 animals. Mendelian inheritance of copy number was assessed using a mixed linear model. The independent variables were the fixed effect of the mean and the random effect of the animal. The dependent variable was the copy number. The variance of the additive genetic effect of animal was based on a pedigree of each animal and their sire and dam, traced for seven generations. ASREML-r (version 3.0) (Gilmour et al. 2009) was used for estimation of variance components. The variance associated with the animal effect is analogous to the additive genetic variance and heritability is additive genetic variation/phenotypic variation, however, in terms of CNV inheritance, "transmission level" is used instead of the term heritability. A transmission level of 0 indicates either a denovo mutation in Esteem, or a sequencing artefact, or alternatively a transmission level of 1 indicates that the copy number is inherited in a Mendelian fashion.

Effect of CNVs on phasing allelic R^2: Using sequence data from all 557 animals, phasing allelic R^2 (AR^2) was determined for, each SNP within the 936 CNV regions found to have high transmission levels, each SNP outside the CNVs, and each SNP 50, 100, 500, 1000, 3000bp either side of the CNVs.

SNP tagging of CNVs: Correlations between copy numbers for each for the final 2661 CNVs and genotypes from 50K Illumina SNPchip or full sequence were determined.

RESULTS AND DISCUSSION

A total of 32x PacBio coverage of the bovine genome (UMD3.1) was generated. Sniffles software identified a total of 38,709 putative SVs of which 19,797 were CNVs (deletions n=18,577, duplications n=1220). Of the 3532 CNVs (deletions n=3055, duplications n=477), that remained after filtering, sizes ranged from 1 – 79,450bp with a median size of 321bp (mean size of 818bp). CNVs smaller than 100bp (n=869) were excluded from further analysis as copy number could not be accurately predicted by CNVnator (2661 CNVs remained).

Using CNVnator in genotyping mode we were able to determine copy number at all putative CNV locations identified by PacBio sequence in all 556 animals. These CNV genotypes were used as 'phenotypes' in order to allow the copy number transmission level to be estimated using ASREML in an attempt to make a distinction between real CNVs and the many false positives detected when calling CNVs from short read sequencing. Putative CNVs showed a wide range of transmission levels. Approximately 30% of CNVs called from PacBio sequence showed high transmission level (936 CNVs > 0.70). Sorting CNVs by level of transmission and plotting distribution of copy number in the population indicated a trend of increasing multimodality of copy number with increasing transmission level. Many of the CNVs with a calculated transmission level of greater than 0.6 showed a clear bi- or trimodal distribution of copy number across the 557 animals. The multimodality of copy number, together with visual observation of bam files containing sequencing read-depths, insert size, and the presence of split reads are all consistent with the detection of bona fide CNVs, provided strong evidence that these highly tranmissible CNVs were likely to be present in our population. The observation that many of these trimodal distributions represented deletions (0, 1, vs 2 copies) reflects, at least in part, the relative ease with which deletions are able to be detected relative to duplications, due to the large proportional differences in sequence content for deletions (Abyzov et al. 2011).

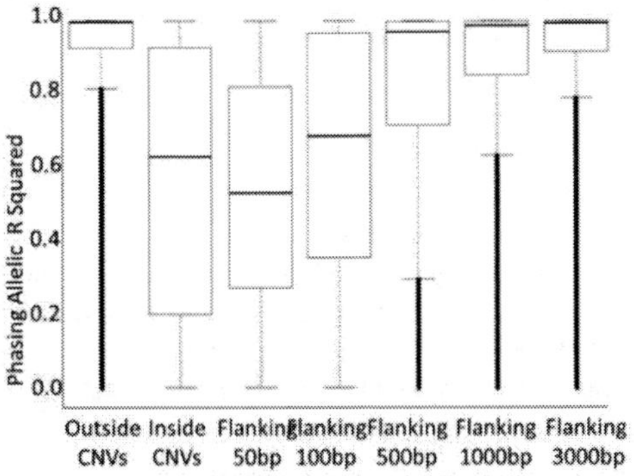

Figure 1 Phasing allelic R2 for SNPs outside, within, and flanking 936 CNVs with high transmission levels

Figure 1 illustrates the detrimental effect of CNVs on the ability to correctly phase SNP genotypes, not only within the CNV itself, but also in the surrounding sequence. Given the vast number of CNVs even in this one individual, it is expected that accuracy of imputation will be negatively affected by the inability to phase the reference sequence accurately.

While the data presented here is not a comprehensive list of CNVs in the New Zealand dairy cattle population, it does illustrate the potential of long read single molecule sequencing as an additional valuable source for identification of CNVs. Furthermore, long read sequence information, combined with independent short read sequencing and pedigree information in 557 animals representative of the population provide compelling evidence of the existence of CNVs in our dairy cattle population, and are not simply false positive results and allows us to begin a catalogue of CNVs. Characterisation of population CNVs has two major benefits to the cattle breeding industry. Firstly, once identified, CNVs may be cheaply identified alongside SNPs by simply adding appropriately designed probes to existing SNP chip genotyping platforms and including CNV genotype information as an additional source of genetic variation in genomic prediction models. Secondly undertaking imputation in a CNV aware manner to bypass poor phasing and increase imputation accuracy.

It could be argued that much of the CNV variation is already captured by SNP in linkage disequilibrium with CNVs. However, it is unlikely that multi-allelic CNVs would be accurately tagged using bi-allelic SNP, and initial reports indicate that around 20% of large CNVs identified from SNP chip platforms are not well tagged (Xu et al. 2014). Figure 2 illustrates the correlation

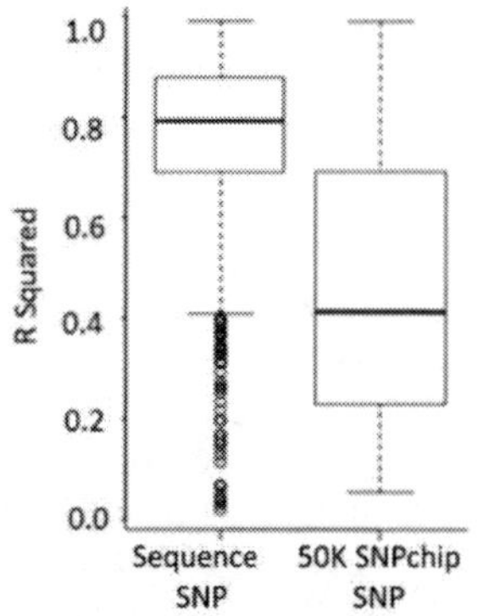

Figure 2 Correlation between copy number and genotype from sequence and 50K SNPchip for 936 CNVs with high transmission levels

between CNV and SNP genotypes on the 50K SNPchip as well as from sequence. Our results indicate that very few of the 936 highly transmissible CNVs are tagged well by SNP on the 50K SNPchip, and unsurprisingly many more CNVs are well tagged by sequence derived SNPs. Our current genomic selection protocols utilise only SNPs present on the 50K SNPchip, and therefore, to date, only a very limited amount of genetic variation from CNVs is being captured and utilised. As a move towards including sequence derived SNPs that tag CNVs could help in improving the accuracy of genomic selection

From a practical perspective, the presence of CNVs may have implications for phasing and imputation of other classes of variants. Given the increasing use of imputation of SNP chip genotypes to whole genome sequence, understanding where CNVs are located in the genome and ideally devising strategies for their correct imputation are of great importance for accurate genome-wide imputation and the generation of accurate genotype information to be utilised in genomic prediction models.

CONCLUSION

We present here the first step towards a gold standard list of CNVs in dairy cattle by utilising both long and short read sequencing technologies together with conservative filtering steps and an easy genome-wide strategy for assessing the Mendelian inheritance. Collectively this provided compelling evidence that these SVs do segregate in the population. Given the increasing use of imputation strategies being used in cattle breeding, identification and characterisation of CNVs (and all classes of SVs) will lead to improved imputation accuracy and will ultimately contribute to improved genomic prediction.

REFERENCES

Abyzov, A., A. E. Urban, M. Snyder, and M. Gerstein. (2011)Genome Res 21 (6):974-84.

Druet, T., I. M. Macleod, and B. J. Hayes. (2014). Heredity (Edinb) 112 (1):39-47.

Duan, J., J. G. Zhang, H. W. Deng, and Y. P. Wang. (2013) PLoS One 8 (3):e59128.

Gilmour, A. R., B. R. Gogel, B.R. Cullis, and R Thompson. (2009). ASREML User Guide Release 3.0.: VSN International Ltd, Hemel Hempstead, UK.

Kadri, N. K., G. Sahana, C. Charlier, T. Iso-Touru, B. Guldbrandtsen, *et al.*. (2014) PLoS Genet 10 (1):e1004049.

Littlejohn, M. D., K Tiplady, T. A. Fink, K. Lehnert, T. Lopdell, *et al.* (2016) Scientific Reports 6.

MacDonald, J. R., R. Ziman, R. K. Yuen, L. Feuk, and S. W. Scherer. (2014). Nucleic Acids Res 42 (Database issue):D986-92.

Sedlazeck, F.J., P Rescheneder, M Nattestad, and M.C. Schatz. (2015) Genome Informatics, Cold Spring Harbor, New York.

Sudmant, P. H. and 1000 Genomes Project Consortium. (2015) " Nature 526 (7571):75-81.

Weischenfeldt, J., O. Symmons, F. Spitz, and J. O. Korbel. (2013) Nat Rev Genet 14 (2):125-38.

Xu, L., J. B. Cole, D. M. Bickhart, Y. Hou, J. Song, *et al.* (2014) BMC Genomics 15:683.

Zarrei, M., J. R. MacDonald, D. Merico, and S. W. Scherer. (2015Nat Rev Genet 16 (3):172-83.

Zhang, X., R. Du, S. Li, F. Zhang, L. Jin, and H. Wang. (2014) BMC Bioinformatics 15:50.

Proc. Assoc. Advmt. Anim. Breed. Genet. 22:201-204

1000 BULL GENOMES AND SHEEPGENOMEDB PROJECTS: ENABLING COST-EFFECTIVE SEQUENCE LEVEL ANALYSES GLOBALLY

H.D. Daetwyler[1,2,3], R. Brauning[4], A.J. Chamberlain[1], S. McWilliam[5], A. McCulloch[4], C.J. Vander Jagt[1], B. Sunduimijid[1,3], B.J. Hayes[1,6] and J.W. Kijas[5]

[1] Agriculture Victoria, AgriBio, Centre for AgriBioscience, VIC, Australia
[2] La Trobe University, VIC, Australia
[3] CRC for Sheep Industry Innovation, Armidale, NSW, Australia
[4] AgResearch Ltd, Mosgiel, New Zealand
[5] CSIRO, QLD, Australia
[6] Queensland University, QLD, Australia

SUMMARY

Whole-genome sequence data has several potential uses for animal breeding, including accelerated detection of mutations with deleterious and beneficial effects, as well as increasing the accuracy of genomic selection. It is cost-effective to share data in global consortia to enable more powerful imputation of sequence into animal populations that have been genotyped at lower density. This then facilitates more powerful downstream analyses such as genome-wide association and genomic prediction. Here we describe two such projects, namely the 1000 Bull Genomes Consortium and SheepGenomesDB.

INTRODUCTION

Whole-genome sequence provides detailed information of an individual's genetic make-up, which can be used to pinpoint genetic variants and genotypes for all animals in the sample. The accuracy of the analyses is dependent on the read depth, which is the average number of short reads (usually about 100 basepairs (bp) in length) aligned per base on the reference genome (e.g. 10 fold coverage).

Sequence data has several important uses. It can be used to track down Mendelian disorders and recessive alleles in affected individuals, using sequence variants from unaffected animals as controls enabling powerful filters to reduce the number of candidate mutations. Secondly, it enables more powerful genome-wide association studies, because either the causative mutations are themselves a sequence variant or would be in high linkage disequilibrium (LD) with a genotyped variant. Thirdly, it could improve the accuracy of genomic prediction, a benefit that can likely only be harnessed in multi-breed reference populations due the small effective population size of most commercial livestock breeds. For the latter two applications, the cost of sequencing has been prohibitive to sequencing the tens of thousands of individuals needed for powerful genome-wide association and genomic prediction. An alternative option is to impute sequence genotypes into animals that are already genotyped at lower density (preferably with a high density SNP chip). While the accuracy of imputation is not perfect, especially for lower minor allele frequency variants, it still has been demonstrated to increase the power of analyses. The need therefore arises for large reference populations of whole-genome sequenced animals for imputation.

The 1000 Bull Genome Project (1000 Bulls) and SheepGenomesDB Project are meeting this need in cattle and sheep, respectively. Many cattle breeds use the same sires across the globe, which makes populations very genetically connected and, therefore, using one reference population for imputation is advantageous. In addition, because sequence information captures even the short haplotypes shared across breeds, there is a benefit to using all breeds as a combined reference for imputation. Both projects have grown quickly with the 1000 Bulls now close to

2800 animals in Run6 and SheepGenomesDB with 935 animals in Run2. Described here is the organisation principles of both projects, bioinformatic pipelines, and the animals included and number of variants discovered in the latest analyses.

MATERIALS AND METHODS

In the 1000 Bulls the lead institution is Agriculture Victoria (AgVic) and each partner contributes their sequence data to the project. In turn, each partner receives all genetic variants and sequence genotypes discovered from animals in the project. There are currently 36 partners from 22 countries. The data is only available to participating partners and partners are expected to share identifying information, pedigrees and metadata for all animals where possible. Public cattle sequences not already included are also downloaded from the NCBI sequence read archive (SRA) and incorporated. The SheepGenomesDB project is organised differently and is jointly managed by CSIRO, AgResearch (AgR) and Agriculture Victoria (AgVic). It requires all included raw sequence data to be public at NCBI SRA, but allows animals to have anonymous identifiers with meta-data giving information on breed and sex (if known). As all input data is public, all variants and genotypes found are also made public.

Processing of Sequence. 1000 Bulls – Partners are responsible for processing and aligning sequences, which are then transferred to AgVic in BAM format for inclusion in analyses. Sequence quality scores must be Phred+33 encoded. The recommended quality control (QC) and processing of whole genome sequences in fastq format is as follows: 1) remove Illumina reads that fail the chastity filter; 2) remove adaptor sequence from reads; 3) trim low quality bases (Phred <20) from 5' and 3' ends of reads; 4) discard reads with a mean Phred quality score of <20 and >=3 bases not called (i.e. N); 5) remove known artifacts (e.g. Illumina NextSeq are known to have erroneous strings of A and/or G at 3' ends); 6) after trimming, discard reads that are too short (<50% of original read length). Reads that are left unpaired after QC may be aligned. Helpful programs for processing are quadtrim (https://bitbucket.org/arobinson/quadtrim), Picard (http://picard.sourceforge.net/index.shtml), Samtools (Li *et al.* 2009), GATK (DePristo *et al.* 2011), seqtk (https://github.com/lh3/seqtk), and FastQC (http://www.bioinformatics.babraham.ac.uk/projects/fastqc/). Sequences are then aligned to the UMD3.1 *bos taurus taurus* reference genome downloaded from the 1000 Bull Genomes website (www.1000bullgenomes.org) with the Burrows-Wheeler Aligner (*bwa align or bwa mem*) using default parameters (Li & Durbin 2009). BAM files must contain sample identifiers in Interbull format if available. One BAM files per animal per partner should be locally realigned, PCR duplicates removed, sorted, and indexed. Mean read depth should be calculated using GATK *DepthOfCoverage* and provided. Upon receipt of BAM files from partners, AgVic performs format and QC checks on files and amends them if needed where possible. Partners are encouraged to submit Bovine HD genotypes and any meta-data including pedigrees along with sequences.

SheepGenomesDB – The pipeline in use for this project is the same as for the 1000 Bulls except that all sequences must be publicly available at NCBI sequence read archive. Raw fastq files are downloaded by AgR and AgVic and then processed as above. Alignments are done to OAR3.1, ftp://ftp.ensembl.org/pub/release-78/fasta/ovis_aries/dna/Ovis_aries.Oar_v3.1.dna_sm.toplevel.fa.gz. Both AgR and AgVic, have a full set of BAM files for the project. For each animal, a standard SheepGenomesID is created which provides country of origin, breed, and sex of the animals (naming convention document on www.sheepgenomesdb.org).

Variant Calling, Filtering, and Refinement. 1000 Bulls – Samtools (currently version 1.3) *mpileup* is used to call single nucleotide polymorphisms (SNP) and short insertions and deletions (indels). This results in a variant call format (VCF) file which contains the following information: variant position, reference and alternative allele, quality metrics and read depth, genotypes for all animals at that positions, and genotype probabilities for all possible genotypes per animal per position. The set of variants called at this stage will include low confidence variants for which there may not be enough evidence. Filtering of variants has been shown to improve the quality of the variant set as judged by the concordance of sequence and SNP chip genotypes at overlapping positions as well as the rate of opposing homozygotes (OppHom) found in parent-offspring pairs (there should not be any). Filtering is done with custom python scripts that use the VCF parser PyVCF (https://github.com/jamescasbon/PyVCF). Variants are removed if they have: 1) >1 alternative alleles; 2) no alternate allele observations in both forward and reverse direction reads; 3) overall quality score QUAL <20 and mapping quality score <30; 4) < minimum read depth of 10 or >3 standard deviations from the median read depth; 5) failed OppHom (>10% of parent-offspring pairs were OppHom); 6) the same bp position; 7) a proximity of <10 bp between indels or <3 bp between SNP in which case the lower QUAL variant was removed.

The resulting VCF files still contain a proportion of genotypes that are missing or called with high uncertainty. Imputation programs that are able to use genotype probabilities can be used to impute missing and refine uncertain calls using the haplotypes found in the collective set. In the 1000 Bulls we use Beagle 4.0 (Browning & Browning 2009) for this purpose. In cattle two separate analyses (Runs) are performed, one includes only *taurine* cattle and the other includes all animals.

SheepGenomesDB – The sheep pipeline is as above with the following differences. AgVic runs Samtools and AgR calls variants with GATK UnifiedGenotyper. Variants from both callers are then independently filtered as above, excluding the OppHom filter because only few Parent-Offspring pairs exist in the sheep sample to date. In addition, Samtools and GATK calls are merged to create two sets: 1) a unison set of calls with filtered overlapping variants, and 2) a complete set of calls that contain all unfiltered variants from both callers. Both sets are made public at the European Variant Archive (http://www.ebi.ac.uk/eva), which also annotates all variants and connects them to dbSNP.

Quality Control of Variants and Genotypes. Concordance of bovine HD SNP chip and sequence genotypes is performed for all animals in the 1000 Bulls if available. This concordance is expected to be >95% for good quality sequence (depending on read depth) and can typically be improved using Beagle. If concordance is <80% it may indicate that the SNP chip and sequence have not originated from the same animal and may highlight sample tracking issues. Parent-Offspring OppHom are checked for all pairs and should be less 0.1%. Furthermore, the number of singletons and heterozygosity per animal are calculated. If an animal has a very large number of singleton variants and it has breed contemporaries, it would indicate an issue with its data. Similarly, if heterozygosity is very high, it indicates that DNA has been mixed at some point during the generation of the sequence. Finally, all animals are checked whether they have genotypes in all genomic regions.

RESULTS AND DISCUSSION

The 1000 Bulls has grown fast over time starting with 238 *taurine* animals from 4 breeds in Run2 (2012) (Daetwyler *et al.* 2014) and 1756 animals from 55 breeds in Run5 (2015) across *taurine* and *indicine* sub-species. Run5 identified 67.3 million variants, of which 64.8 million were SNP and 2.5 million were indels. Run6 is currently underway and includes close to 2800 animals across more than 70 breeds (Figure 1). A new feature of Run6 is the inclusion of related species such as the Gaur, Yak, *turano mongolicus*, ancient cattle, and an auroch. It also has much expanded collection of African and *indicine* breeds.

SheepGenomesDB Run1 (2016) discovered 50 million variants and contained 453 sheep and included many New Zealand breeds, the International Sheep Genomics Consortium global diversity set, and Moroccan as well as Iranian sheep from the NextGen project (available at http://www.ebi.ac.uk/eva/?eva-study=PRJEB14685). Run2, for which the Samtools variant calling at AgVic has concluded, contains 935 sheep with the SheepCRC contributing a large number of animals from the main four Australian breeds (Merino, Polled Dorset, White Suffolk and Border Leicester, Figure 1) and the USDA contributing their Sheep Diversity Panel animals.

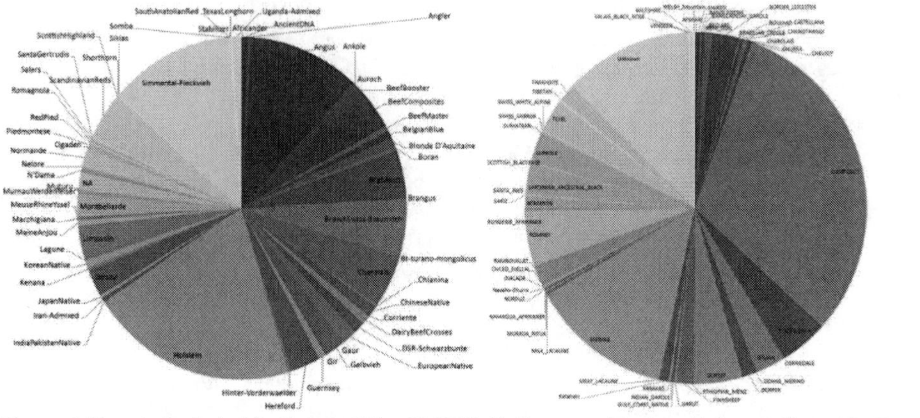

Figure 1 Breeds included in Run6 of the 1000 Bull Genomes Project (left panel) and in Run2 of the SheepGenomesDB Project (right panel). Sheep composites are primarily crosses of Australian and New Zealand breeds.

The 1000 Bulls data has been the basis for several studies that detected causative mutations, where its genomes served as controls (e.g. Daetwyler *et al.* 2014; Murgiano *et al.* 2015). It has been the basis for imputation at many consortium partner institutions, which have then used the data to perform imputation, GWAS and genomic selection (e.g. Bouwman & Veerkamp 2014; van den Berg *et al.* 2016). Similar benefits are expected to be realised in sheep. The two consortia are the most complete inventory of cattle and sheep genetic variants globally and will form the basis for sequenced-based animal breeding research in many countries.

ACKNOWLEDGEMENTS

The authors would like to thank all 1000 Bull Genomes Consortium partners and SheepGenomesDB contributors for sharing their data. Funding from the USDA, the DairyBio project (a collaboration between Dairy Australia and the Victorian Government), and the Cooperative Research Centre for Sheep Industry Innovation is acknowledged.

REFERENCES

Bouwman A.C. and Veerkamp R.F. (2014) *BMC Genetics* **15**, 105.
Browning B.L. and Browning S.R. (2009) *Am. J. Hum. Genet.* **84**, 210-23.
Daetwyler H.D., Capitan A., Pausch H., et al (2014) *Nat. Genet.* **46**, 858-65.
DePristo M.A., Banks E., Poplin R., et al. (2011) *Nat. Genet.* **43**, 491-8.
Li H. and Durbin R. (2009) *Bioinformatics* **25**, 1754-60.
Li H., Handsaker B., Wysoker A., et al (2009) *Bioinformatics* **25**, 2078-9.
Murgiano L., Shirokova V., Welle M.M., et al (2015) *PLoS Genet.* **11**, e1005427.
van den Berg I., Boichard D. and Lund M.S. (2016) *Genet. Sel. Evol.* **48**, 83.

Proc. Assoc. Advmt. Anim. Breed. Genet. 22:205-208

TRANSCRIPTOMICS PROFILING REVEALS CANDIDATE GENES TO IMPROVE FEED EFFICIENCY IN DAIRY CATTLE

M.S. Salleh[1], G. Mazzoni[1], J.K. Höglund[2], D.W. Olijhoek[2,3], P. Lund[3], P. Løvendahl[2], and H.N. Kadarmideen[4]

[1]Dept. of Veterinary and Animal Sciences, Faculty of Health and Medical Sciences, University of Copenhagen, Frederiksberg C, Denmark
[2]Dept. of Molecular Biology and Genetics, Aarhus University, Tjele, Denmark
[3]Dept. of Animal Science, Aarhus University, Tjele, Denmark
[4]Department of Bio and Health Informatics, Technical University of Denmark, DK-2800, Kgs. Lyngby, Denmark

SUMMARY

Feed is the largest variable cost in milk production industries, thus improving feed efficiency (FE) will give better use of resources. To identify and select animals with high FE, it might be helpful to understand the biological mechanisms and the role of gene expression patterns across the whole genome (transcriptomics). In the present study, RNA sequencing data was used to detect differentially expressed (DE) genes in Danish Holstein and Jersey dairy cows having either a high or low FE (assessed as residual feed intake (RFI). Functional analysis was performed on these genes to identify molecular pathways involved in FE. Ten Jersey and nine Holstein cows were used in the experiment and divided into two RFI groups depending on their calculated RFI. The two RFI groups received a Control (C) and High Concentrate (HC) diet containing 68:32 and 39:61 ratio of forage:concentrate, respectively. This enabled us to compare the interaction between RFI status and diet. The mRNA samples extracted from liver biopsies were paired end sequenced. The RNA-Seq gene expression data was then analyzed using a statistical-bioinformatics pipeline to identify DE genes and perform functional enrichment. We compared gene expressions of the RFI groups, and identified 70 and 19 DE genes in Holstein and Jersey, respectively. An interaction term (RFI x diet) detected two significantly DE genes in Jersey cows. The functional enrichment analysis of the DE genes showed involvement in pathways that might regulate RFI, such as primary immunodeficiency, retinol metabolism, starch and sucrose metabolism, arachidonic acid metabolism and cytochrome P450 drug metabolism. In conclusion, the transcriptomics approach was effective in identifying DE genes and understanding their biological functions. These findings could contribute to the development of biomarkers for RFI and to improving augmented genomic selection procedures that make use of functional information.

INTRODUCTION

Improving feed efficiency of dairy cattle can mean big savings for milk producers. One way to improve feed efficiency is by genetic selection for cows producing more milk for the same amount of feed. Residual feed intake which is the difference between actual and predicted feed intake has been used widely as a measure of feed efficiency in livestock (Berry & Crowley 2012; Connor *et al.* 2013; Lin *et al.* 2013; Tempelman *et al.* 2015).

To understand the mechanisms of action affecting feed efficiency, we suggest the use of system genetics approach including transcriptomics techniques. This might be helpful in supporting genomic selection in the future (Kadarmideen 2014; Zhang *et al.* 2014). The liver plays an important role in the metabolism of nutrients (Partridge *et al.* 2014). Hence, liver transcriptomics might give insight into feed efficiency in dairy cows.

The objective of the study was to identify potential regulatory genes and molecular pathways involved in feed efficiency of dairy cattle by characterising the liver transcriptome based on RNA-Seq technologies.

MATERIALS AND METHODS

Ten Jersey and nine Holstein cows were selected from the research herd of 200 animals in Danish Cattle Research Centre (DCRC), Aarhus University, Denmark. The data from this herd have previously been used in quantitative genetic studies regarding feed or dry matter intake (Li *et al.* 2016). Animals of both breeds were divided into two groups: high- or low-RFI. Residual feed intake was defined using a random regression model (Tempelman *et al.* 2015). Here, the random animal solutions were extracted from a random regression model in which dry matter intake was regressed on the following fixed effects: weeks of lactation, the management group in which the cows were held, and the interaction between weeks of lactation, breed and parity. Fixed linear regressions were applied to adjust for metabolic body weight, daily live weight change and daily body condition score change (fitted with a Legendre polynomial), and energy corrected milk yield. The random effects were cow within the breed and cow within the breed and parity. All cows received a low-concentrate [control (C)] and a high-concentrate (HC) diet in a crossover design with two periods. There was approximately a 30% difference in concentrate proportion on a dry matter (DM) basis between the C and HC diets which were 68:32 and 39:61 ratio of forage:concentrate, respectively.

Approximately 10–20 mg of liver tissue were collected from all the experimental cows at the end of each feeding trial. mRNA was extracted from the liver tissue samples using the Qiazol, RNeasy® Mini Kit and MaXtract High Density and sequenced with Illumina HiSeq 2500. The quality of all mRNA samples was above 8 RIN (RNA Integrity Number).

RNA-Seq data of each cow were analyzed to identify differentially expressed genes. The DE analysis was performed separately for each breed using R package DESeq2 setting all the parameters to default values (Love *et al.* 2014). Two different models were fitted:

Model 1 Y=Parity number + Diet + RFI
Model 2 Y=Parity number + Diet + RFI + Diet*RFI

where: Y is the gene expression counts, RFI is a dummy variable that represents the feed efficiency of the animals (high- and low-RFI), and Parity number was included as a dummy variable to control for potential confounding effects. In Model 1, we assumed an additive effect without interaction between two treatment diet and two RFI groups. In Model 2, we assumed an interaction between two treatment diets by two RFI groups. Differentially expressed genes were considered at a False Discovery Rate (FDR) < 5%.

Finally, functional enrichment analysis on the entire expression profile was performed using Gene Set Enrichment Analysis (GSEA). It has been demonstrated previously that GSEA provide insights into the biology behind a set of genes in terms of how the DEGs interact.

RESULTS AND DISCUSSION

On average, 91% of the read pairs (26,067,856 read pairs) were uniquely mapped to the bovine reference genome UMD 3.1 from Ensembl database release 82. On average, 62% of the read pairs mapped to exonic regions, 20% to intronic regions and almost 18% to intergenic regions.

In total, 12,025 genes in the Holstein breed and 11,905 genes in the Jersey breed were used after removing low expression genes to identify the DEGs. A total of 70 Holstein and 19 Jersey DEGs (Table 1) were identified by comparing between high- and low-RFI directly without accounting for any interaction. The interaction analysis showed low numbers of DEGs in both diet groups (Table 1). Among the top DEGs in Holsteins were *ACACA, CYP2C9, CYP7A1, CYP11A1, ELOVL6, FOSL2, HCLS1, IFI6, NR1H4, RYR1, SOCS2, TBC1D8, CR2, CTH, DGAT2, FGFR2, SLC20A1* and *TAF6*. The top DEGs in Jerseys were *CYP3A4, EXTL2, TMEM102, FDXR, GIMAP4, GIMAP8, GNG10, HLA-B* and *ZNF613*. Most of the genes identified as DEGs in both breed were also found as DEGs in other RFI divergent study by (Weber *et al.* 2016). In total, 22 Holstein genes and 14 Jersey genes were detected as significant DEGs (p values < 0.05) for the interaction analysis. No significant genes were identified in Holstein cows for the interaction (Table 1). However, two Jersey genes, SEC24 Homolog D (*SEC24D*) and FLT3-Interacting Zinc Finger 1 (*FIZ1*), were differentially expressed (p values < 0.05) in the RFI groups depending on the two diet types.

We identified seven overrepresented pathways for the set of downregulated genes and none for the upregulated genes in high-RFI group Holsteins. In Jerseys, two pathways were overrepresented for genes with negative-fold changes and three pathways for genes with positive-fold changes. The top KEGG pathways for the genes downregulated in the high-RFI group in Holsteins and in Jerseys is the primary immunodeficiency pathway, while the significant pathways identified for genes upregulated in the high-RFI group were only detected in Jerseys. We also identified, that most of the pathways within the strong indications thresholds (FDR q-value <0.05), were related to the metabolism of retinol, starch and sucrose, ether lipid and cytochrome P450 drug metabolism.

Table 1. Number of differentially expressed genes between high- and low-RFI in a separate diet group in the model with interaction term, and without interaction term according to the corrected p values < 0.05

	Control	High Concentrate	With Interaction	Without interaction
Holstein	9	13	0	70
Jersey	6	6	2	19

The functional enrichment and pathway analysis of the DEGs contribute towards understanding the function of these genes in relation to feed efficiency. The steroid hormone biosynthesis pathway was one of the top KEGG pathways identified in the analysis of negative energy balance in dairy cows (McCabe *et al.* 2012). We also discovered that this pathway was overrepresented in the set of genes upregulated in high-RFI group in Jersey cows (FDR < 0.05). Steroid hormone biosynthesis should always occur in the adrenal glands and gonads, while the liver is the site of steroid hormone inactivation. The upregulation of steroid hormone biosynthesis pathway indicated that steroid hormone was inactivated in high-RFI group. Therefore, we would conclude that this pathway plays an important role in RFI. In support, both *CYP11A1* and *CYP7A1* that were upregulated in high-RFI group in Holstein, which function in cholesterol homeostasis, were identified as DEGs in our experiments and they are part of this pathway in KEGG.

Primary immunodeficiency pathway is a heterogeneous group of disorders. This pathway was the top overrepresented pathway detected by GSEA and was significantly enriched in both cattle breeds. The downregulation of the primary immunodeficiency pathway in both breeds of high-RFI cows suggests that the immunity may affect feed efficiency. (Ozuna *et al.* 2012) observed that primary immunodeficiency disorder is consistently inherited by low-feed efficient pigs. Consistently, Kogelman *et al.* (2014) and Do *et al.* (2013) reported a correlation between genes related to immunodeficiency function disorders or immunity-related diseases and low-feed efficiency in pigs.

Notably, the genes identified for the interaction between RFI and diet, were also associated with immunodeficiency. The impact of the diet on genes belonging to the immunodeficiency pathway and it paves the way for future studies to determine how to improve diet in relation to the genetic background of the animals. Two protein-coding genes, *SEC24D* and *FIZ1*, were differentially expressed in response to diet and were associated with pathways including immune system and transport to the golgi and subsequent modification as well as in transcriptional regulation (www.genecards.org). The lack of a more extensive differential gene expression response to diets indicate that differences in the concentrate proportions between the diets, as tested in this study, may not be able to disturb gene expression levels.

In conclusion, the results reveal differences in biological mechanisms related to residual feed intake in Holsteins and Jerseys. The study provided 70 and 19 candidate genes involved in the regulation of residual feed intake pathways in Holstein and Jersey cattle, respectively. The functional enrichment analysis of the DE genes showed involvement in pathways that might regulate feed efficiency, such as primary immunodeficiency, retinol metabolism, starch and sucrose metabolism, arachidonic acid metabolism and drug metabolism cytochrome P450. The relationship between retinol metabolism and the feed conversion ratio phenotype in Nellore beef cattle has been previously described (de Almeida Santana *et al.* 2016). The candidate genes identified in this study might be useful for explaining biological effects of genomic markers in genomic selection methods utilizing functional information.

REFERENCES

Berry D. and Crowley J. (2012) *Journal of animal science* **90**: 109.

Connor E., Hutchison J., Norman H., Olson K., Van Tassell C., Leith J. and Baldwin R. (2013) *Journal of animal science.* **91**: 3978.

de Almeida Santana M.H., Junior G.A.O., Cesar A.S.M., Freua M.C., da Costa Gomes R., e Silva S.d.L., Leme P.R., Fukumasu H., Carvalho M.E. & Ventura R.V. (2016) *Journal of applied genetics* **57**, 495-504.

Do D.N., Strath, A.B., Ostersen T., Jensen J., Mark T. and Kadarmideen H.N. (2013). *PloS one.* **8**: e71509.

Kadarmideen H.N. (2014) *Livestock Science.* **166**: 232.

Kogelman L.J.A., Cirera S., Zhernakova D.V., Fredholm M., Franke L. and Kadarmideen H.N., (2014) *BMC Med Genomics.* **7**: 57.

Li B., Fikse W., Lassen J., Lidauer M., Løvendahl P., Mäntysaari P. & Berglund B. (2016) *Journal of dairy science* **99**, 7232-9.

Love M.I., Huber W. & Anders S. (2014) *Genome biology* **15**, 550.

Lin Z., Macleod I. and Pryce J. (2013) *Journal of dairy science.* **96**: 2654.

McCabe M., Waters S., Morris D., Kenny D., Lynn D. and Creevey, C. (2012) *BMC genomics.* **13**: 193.

Ozuna A.C., Rowland R.R., Nietfeld J.C., Kerrigan M.A., Dekkers J. and Wyatt C.R. (2012) *Veterinary Pathology.* **50**: 144

Partridge C.G., Fawcett G.L., Wang B., Semenkovich C.F. and Cheverud J.M. (2014) *BMC genomics.* **15**: 1.

Tempelman R., Spurlock D., Coffey M., Veerkamp R., Armentano L., Weigel K., De Haas Y., Staples C., Connor E. and Lu Y. (2015) *Journal of dairy science.* **98**: 2013.

Weber K.L., Welly B.T., Van Eenennaam A.L., Young A.E., Porto-Neto L.R., Reverter A. & Rincon G. (2016) *PloS one* **11**, e0152274.

Zhang Z.H., Jhaveri D.J., Marshall V.M., Bauer D.C., Edson J., Narayanan R.K., Robinson G.J., Lundberg A.E., Bartlett P.F. and Wray N.R. (2014). *PloS one.* **9**: e103207.

Proc. Assoc. Advmt. Anim. Breed. Genet. 22:209-212

PREDICTION OF GENOME-WIDE REGULATORY REGIONS IN SHEEP

M. Naval-Sanchez, Q. Nguyen, B.P*. Dalrymple, T. Vuocolo, R. L. Tellam, L.R. Porto-Neto, S. McWilliam, A. Reverter and J. Kijas**

CSIRO Agriculture & Food, 306 Carmody Road, St. Lucia, 4067, QLD, Australia; *Current address: Institute for Molecular Bioscience, 306 Carmody Road, St. Lucia, 4067, QLD, Australia ; **The UWA Institute of Agriculture, The University of Western Australia, 35 Stirling Highway, Crawley, WA 6009, Australia

SUMMARY

The annotation of the regulatory genome is essential to investigate the link between genotype and phenotype. In this study, we applied the computational method "Human Projection of Regulatory Sequences" (HPRS) (Nguyen *et al.* 2016) to project human regulatory information to sheep coordinates and provide a predictive sheep regulatory genome.

Firstly, we selected human large-scale publicly available human datasets as a reference for promoter and enhancer regions. Secondly, we converted the human regulatory information into sheep coordinates. We successfully mapped 70% and 65% of human promoter and enhancers regions into the sheep genome. Finally, we evaluated whether the predicted sheep regulatory genome captures sheep-regulatory information by assessing its overlap with in-house H3K27ac and H3K4me3 ChIP-seq data from sheep brown adipose tissue. We find that our predicted regulatory elements are enriched for sheep regulatory regions and present high sensitivity and specificity to discern between promoters and enhancers.

INTRODUCTION

The human regulatory genome has been extensively characterized by large-scale genomic Consortiums such as the ENCODE (ENCODE Project Consortium 2012) and Epigenomics Roadmap (Roadmap Epigenomics Consortium *et al.* 2015). Meanwhile, the functional annotation of livestock species, specifically sheep, is lagging behind. Projects such as the Functional Annotation of Animal Genomes (FAANG) (Consortium *et al.* 2015) aim to resolve this issue in the near future. Alternatively, computational approaches, in particular the "Human Projection of Regulatory Sequences" (HPRS) (Nguyen *et al.* 2016) pipeline has been successfully used to project human regulatory information into cattle coordinates providing high confidence regulatory information at the promoter and enhancer level.

Here, we apply the HPRS pipeline to predict the sheep regulatory genome and use sheep-specific regulatory data to show that the method captures sheep regulatory information with high sensitivity and specificity.

MATERIALS AND METHODS

Human genomic databases. Human regulatory information was obtained from three distinct databases:

1) FANTOM5 promoters and enhancer atlas detected by CAGE (Forrest *et al.* 2014; Andersson *et al.* 2014).
URL:http://fantom.gsc.riken.jp/5/datafiles/latest/extra/CAGE_peaks/hg19.cage_peak_phase1and2combined_coord.bed.gz; http://enhancer.binf.ku.dk/presets/permissive_enhancers.bed.

2) Epigenomics Roadmap enhancers from 88 human primary tissues (Roadmap Epigenomics Consortium *et al.* 2015). We use the chromatin states defined as enhancer, enhancer genic and enhancer bivalent.URL: http://egg2.wustl.edu/roadmap/web_portal/chr_state_learning.html.

3) ENCODE Transcription Factor Binding Sites (TFBSs): ENCODE proximal and distal TFBSs by ChIP-seq for 163 TFs. URL: https://www.encodeproject.org/data/annotations/v2/.

Human Projection of Regulatory Regions pipeline (HPRS). We followed the same procedure as (Nguyen *et al.* 2016) (https://bitbucket.csiro.au/users/ngu121/repos/hprs/browse/). In brief, the program liftOver (minMatch=0.2) (Hinrichs *et al.* 2006) was applied to convert human regions into sheep coordinates. Unmapped regions or not reciprocally mapped were allowed multiple mapping (liftOver, minMatchMulti >=0.80). The results from different datasets were then combined into a single dataset with non-overlapping regions.

Sheep Experimental ChIP-seq. Chromatin immunoprecipitation followed by next generation sequencing (ChIP-Seq) of the histone chromatin modification H3K4me3 and H3K27ac was performed on perirenal brown adipose tissue at 130 days post conception from three and two animals respectively. Sequence reads were mapped to the unmasked ovine genome sequence (*Ovis aries* Oar_v3.1.74) using the NGS core tool mapping application in CLCBIO (Peak calling comparing the H3K4me3 or H3K27ac ChiP-Seq versus the input control was performed using MACS (Zhang *et al.* 2008). Only peaks found in both replicates per chromatin mark, either H3K4me3 or H3K27ac, were further considered.

Validation of the sheep regulatory information. We produced 1,000 randomizations for each genomic feature using bedtools shuffle (-noOverlapping) (Quinlan 2014) set. Next, we calculated an empirical p-value per feature and overlap by counting how many times an equal or greater overlap observed in the original features was observed in the 1,000 randomizations.

RESULTS AND DISCUSSION

In order to annotate the sheep regulatory genome we selected human promoter and enhancer information from large-sequencing international consortiums such as FANTOM5, RoadMap Epigenomics and ENCODE (Table 1). Data from different databases differ in the biochemical process used to define enhancers and promoters, number of detected features, feature length and genome coverage (Table1). For example, RoadMap chromatin marks provide larger genome coverage (4.93% and 35.79% for promoters and enhancers, respectively) due to the capture of regulatory information from a larger number of conditions, namely 88 distinct human primary tissues.

To depict a potential sheep regulatory genome we converted human regulatory information into sheep coordinates. Table 1 shows that for each database we were able to successfully recover from 58.28% to 72.56% of their human regulatory information. It also shows that the recovery of proximal or promoter elements is higher (70%) compared to distal or enhancer elements (62%) in agreement with higher sequence conservation at the promoter than at the enhancer level. Based on these steps we captured 21.35% of the sheep genome as potentially regulatory (4.40% promoter and 16.95 % enhancer-like) (Table 2).

Next, we performed H3K27ac and H3K4me3 ChIP-seq in sheep late gestation perirenal brown adipose tissue. These chromatin marks indicate active chromatin and promoter regions respectively. A total of 35,366 regions were identified by H3K27ac, whereas 16,098 regions were identified as promoters using H3K4me3. 26,496 regions only enriched with H3K27ac and not H3K4me3 were defined as enhancers.

We assessed the recovery of sheep brown adipose H3K27ac for each converted dataset (Figure 1A) ranging from 12% recovery from FANTOM enhancers to 93% recovery from RoadMap Enhancers. Finally, to evaluate if the converted datasets were enriched for sheep regulatory information we performed 1,000 randomizations per dataset and compared their H3K27ac recovery with the original features (Figure 1B). Promoter and enhancer databases showed a clear enrichment for sheep brown adipose regulatory regions compared to random (Figure 1B). However, Enhancers Roadmap dataset presented a much lower enrichment probably caused by presenting a higher number of features from multiple tissues that appear as false positives once compared to a single tissue, namely, brown adipose regulatory information. Next, ENCODE TFBSs (proximal and distal) are depleted for general brown active chromatin. This can be explained because these datatypes present higher sensitivity and specificity for promoters and enhancers (Figure 1 C-D) rather than general open-chromatin (H3K27ac). Thus, although depleted for the overlap with the ensemble of H3K27ac signal they are enriched for H3K4me3 and enhancer signal respectively (data not shown).

Table 1. Summary statistics of regulatory sequences

	Human				Sheep				
Database	# Features	# Merged	Avg bp	% Genome	% Mapped	# Features	# Merged	Avg bp	% Genome
Promoters FANTOM	201,802	198,710	20	0.13	70.43	142,140	137,779	18	0.09
Promoters RoadMap	1,771,836	146,860	1053	4.93	68.15	1,207,522	85,378	1099	3.62
Enhancers FANTOM	43,011	43,011	288	0.39	64.13	27,583	27,532	288	1.63
Enhancers RoadMap	9,928,635	494,583	2270	35.79	58.28	5,786,318	3,80,785	2180	32.09
ENCODE Proximal TFBSs	384,343	384,343	150	1.84	72.56	278,883	271,308	153	1.61
ENCODE Distal TFBSs	1,122,364	1,122,364	150	5.37	65.05	730,112	723,645	155	4.34

Table 2. Predicted sheep regulatory sequences

	# Features	Average bp	Total bp	% Sheep genome
Promoters	258472	441	113987479	4.40
Enhancers	387613	1131	438586881	16.95

To assess the sensitivity and specificity of each converted datatype we calculated their overlap with sheep promoters and enhancers (Figure 1 C-D). In this case we showed that promoters and ENCODE proximal TFBSs databases clearly recover most sheep adipose H3K4me3 peaks (Figure 1C). Thus, concluding that promoter datatypes recover mostly promoter regions rather than enhancers.

Alternatively, the same analysis at the enhancer level clearly showed that ENCODE Distal TFBSs is specific for enhancers recovering 71% of enhancers and only 33% of sheep promoters. However, the rest of enhancer databases do not only recovery enhancer regions but promoters as well (Figure 1D). For example, RoadMap enhancers recovered 90% of sheep enhancers and 92% of sheep promoters. To solve that issue we only considered enhancers with no overlap to converted promoter datasets. This resulted in 62% of sheep enhancer recovery and only 13% promoter recovery.

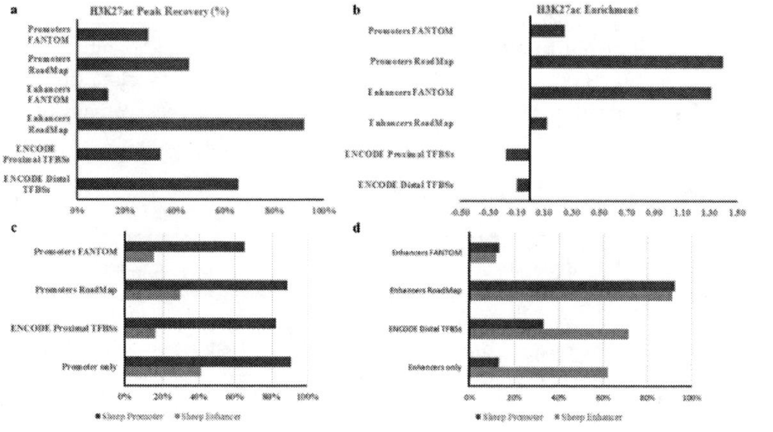

Figure 1. Recovery of experimentally defined sheep active chromatin, promoters and enhancers. (a) Percentage of recovery of sheep H3K27ac peaks from sheep brown adipose by the distinct sheep converted datatypes. (b) Fold enrichment compared to 1000 randomizations. Recovery of sheep promoters and enhancers, as a measure of specificity and sensitivity, by Promoter (c) and enhancer datasets (d).

CONCLUSIONS

Altogether, we show that the application of the HPRS pipeline successfully converts human regulatory information into sheep coordinates with potential regulatory function. This predicted regulatory map will allow the prioritization of trait-associated genetic variants, as well as further investigation and understanding between genetic variants, functional impact and phenotype. Further filtering of the dataset will be performed to increase the signal-to-noise ratio as performed in the original study (Nguyen *et al.* 2016) and then we will make this resource available to the sheep community.

REFERENCES

Andersson R., Gebhard C., Miguel-Escalada I., Hoof I., Bornholdt J., *et al.* (2014) *Nature* 507: 455.

Consortium T.F., Andersson L., Archibald A.L., Bottema C.D., Brauning R., *et al.* 2015. (2016) *Genome Biol* 16.

ENCODE Project Consortium. (2012) *Nature* 489: 57.

Forrest A.R.R, Kawaji H., Rehli M., Baillie J.K., de Hoon M.J.L., *et al.* (2014) Nature 507:462.

Hinrichs A.S., Karolchik D., Baertsch R., Barber G.P., Bejerano G., *et al.* (2006) *Nucleic Acids Res* 34: D590.

Nguyen, Q., Tellam, R.L., Kijas, J., Barendse, W. and Dalrymple B.P. (2016) *35th Proc. Int. Society for Animal Genetics Conference*, P1037

Quinlan A.R. (2014) *Curr Protoc Bioinforma Ed Board Andreas Baxevanis Al* 47: 11.12.1.

Roadmap Epigenomics Consortium, Kundaje A., Meuleman W., Ernst J., Bilenky M., *et al.* (2015). *Nature* 518: 317.

Zhang Y., Liu T., Meyer C.A., Eeckhoute J., Johnson D.S., *et al.* (2008) *Genome Biol* 9: R137.

Proc. Assoc. Advmt. Anim. Breed. Genet. 22:213-216

DEPOT-SPECIFIC GENE EXPRESSION DURING DIFFERENTIATION OF HANWOO MUSCULAR SATELLITE CELLS

S. De las Heras-Saldana[1], K. Y. Chung[2], S. H. Lee[3] and C. Gondro[1]

[1]School of Environmental and Rural Science, University of New England, Armidale, Australia
[2]Hanwoo Research Institute, National Institute of Animal Science, RDA, South Korea
[3]Division of Animal and Dairy Science, Chungnam National University, South Korea

SUMMARY

The Korean Hanwoo cattle is highly regarded for its high marbling ability (intramuscular fat) and this trait is the key drive of profitability for the industry. The study of muscle development and differentiation in Hanwoo is important to improve understanding of the genes and pathways that regulate these processes and to identify markers for genomic selection. In this study, we performed a culture cell experiment using bovine muscle satellite cells combined with a time-series RNA-seq analysis to measure the transcriptome expression levels during the development of satellite cells from *Longissimus dorsi* (LD) and *Semimembranosus* (SM) muscle. RNA-seq data was collected on days 0, 1, 2, 4, 7 and 14 after differentiation treatment with an average of 35,727,746 cleaned reads per sample. Between 77% and 85% of the reads were mapped to the reference genome (*Bos taurus* UMD3.1 from Ensembl). The genes Hoxc11, Sim2, Hoxc8, Hoxb9, Zic2, Zic4, Tbx4 and Hoxb4 were differentially expressed between LD and SM across time, suggesting that they could drive specific characterization of each muscle. The levels of expression vary vastly between time points according to the stage of muscle differentiation and development. At the beginning of the experiment, the genes involved in proliferation were enriched while their expression reduced drastically after day 2. However, at day 4 and until day 14 there was an enrichment in the genes involved in actin cytoskeleton, muscle cell differentiation and structural constituents of muscle.

INTRODUCTION

Biochemical, proteomic and gene expression characterization of the various muscle depots in cattle can assist our efforts to find improved markers for meat traits such as marbling or tenderness. The differentiation of bovine muscle satellite cells is a good model for muscle development studies since their nuclei contribute to postnatal muscle growth remodelling of pre-existing fibres and can provide insight into the genes involved in muscle growth and depot differentiation.

In muscle development, the Myogenic Regulatory Factors (MRFs) are well known to control myogenesis by the modulation of the myoblast proliferation, migration and fusion (Braun and Gautel 2011). There are four MRFs (Myf5, MyoD, Mrf4, and MyoG), however several other genes contribute to the balance of growth and differentiation (Eng *et al.* 2013). Genes for myogenesis (MYL2, MYH3) and adipogenesis (PPARY, and fabp4) of muscular satellite cells into myotubes-formed cells and adipocyte-like cells were identified in Hanwoo using microarrays (Lee *et al.* 2012).

Global RNA profiling of myogenesis in satellite cells is a good model to understand how changes in gene expression over time determine muscle proliferation and differentiation. However further work is needed to elucidate the molecular mechanisms involved in muscle differentiation and to understand differences between muscle types. In this study, two muscles (*Longissimus dorsi* -LD and *Semimembranosus* -SM) were sampled from three Hanwoo calves to extract muscle satellite cells (MSC). These cells were cultured and allowed to differentiate into myotubes, this process was studied using RNA-seq to characterize the transcriptional changes during myogenesis and how the gene expression profiles change between the differentiation of LD and SM.

MATERIALS AND METHODS

The satellite cells were isolated from the *Longissimus dorsi* (LD) and *Semimembranosus muscle* (SM) of three Korean Hanwoo new born calf as described previously by Frey *et al.* (1995) and Johnson *et al.* (1998). LD and SM satellite cells were cultured on DMEM containing DMEM and 10% FBS until they reached 60% to 70% confluence. We then induced differentiation using DMEM with 3% horse serum for 14 days. Cell samples were collected at confluence point (day 0) and at days 1, 2, 4, 7 and 14 after differentiation treatment to perform histological and RNA-seq analysis. The stain Hematoxylin and Hoechst was used to determine the muscle differentiation stages.

mRNA libraries were prepared and sequenced with the Illumina HiSeq 2000 sequencing system. The quality of resulting paired-end 100bp reads were assessed with FastQC v0.11.3, the adaptors and bad quality bases were removed with Trimmomatic v0.33. Bowtie2 v2.2.6 was used to map the reads to the reference genome *Bos taurus* (UMD3.1 from Ensembl). Bioconductor packages GenomicFeatures 1.22.13 and GenomicAlignments 1.6.3 performed the assembly and read count. We used edgeR 3.12.0 in the analysis of differentially expressed (DE) genes and considered significant DE genes those with a false discovery rate <0.05 and the logarithm fold change (logFC) $\geq$ 2. The functional enrichment of GO terms was performed by ClusterProfiler 2.5.5 and the pathway analysis was done with Pathview 1.10.1.

RESULTS AND DISCUSSION

The bovine MSC reached confluence approximately after 4 to 5 days of culture and it was at days 1, 2, after the differentiation treatment, when the myoblast initiated the terminal differentiation and multinucleated myotubes started to form, however they were not notably visible until day 4. In the last stage of differentiation myotubes went through significant morphological changes while they fused to form mature multinucleated myotubes.

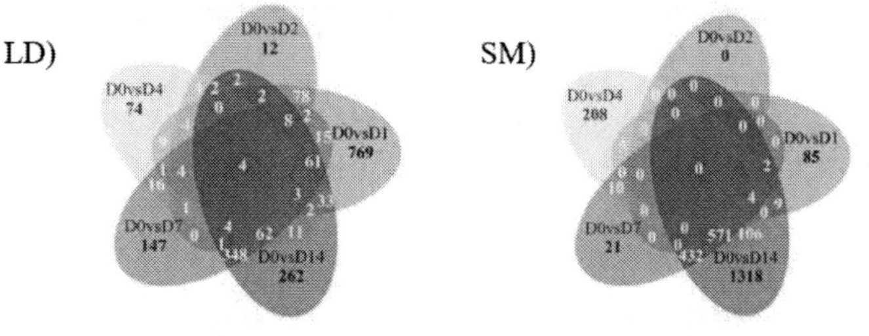

Figure 1. Number of common DE genes between time-point contrasts for LD and SM muscle.

From the sequencing data, between 77% and 84% of the reads were mapped (from average 35,727,746 total reads per sample) to the reference genome. We identified the genes that overlapped in each contrast of day 0 *versus* the sampled time and the genes that were differentially expressed exclusively at each time point (Figure 1). In LD, there were more DE at the beginning of the experiment with 769 DE genes expressed just in day 1, while the opposite pattern was observed in SM were there were 1318 DE genes at day 14. These results could reflect differences between muscle depots due to different rates of proliferation and differentiation during myogenesis.

The genes Hoxc11, Sim2, Hoxc8, Hoxb9, Zic2, Zic4, Tbx4 and Hoxb4 were differentially expressed between LD and SM at different time points, however some of these genes have not been previously directly associated with bovine muscle development.

Table I. GO functional annotation for the enriched terms in A) LD and B) SM muscle at days 1, 7, and 14. MF: molecular function; CC: cellular components; BP: biological process. Number of DE genes that are up-regulated (↑) and down-regulated (↓).

		d1 vs d0		d7 vs d0		D14 vs d0	
		Term	*DE*	*Term*	*DE*	*Term*	*DE*
LD	MF	Calcium ion binding	↓19	Cytoskeletal protein binding	↑7↓8	Receptor activity	↑15↓2
		Transmembrane receptor activity	↓15	Double-stranded DNA binding	↑4↓6	Cytoskeletal protein binding	↑9↓8
		Hormone activity	↑1↓7	Microtubule binding	↓7	Protein kinase binding	↑6↓5
						Peptidase inhibitor activity	↑4↓2
	CC	Extracellular space	↑ 4↓31	Cytoskeletal part	↑5↓20	Cytoskeletal part	↑7↓16
		Myofibril	↓25	Myofibril	↑12	Myofibril	↑13
		Sarcomere	↓22	Microtubule	↑12	Sarcomere	↑12
		Actin cytoskeleton	↓17	Sarcomere	↑11	Microtubule organizing center	↑1↓11
		I band	↓13	I band	↑7	I band	↑6
	BP	Regulation of multicellular organismal process	↑1↓37	Cell cycle	↑1↓33	Cell cycle	↑3↓29
		Immune system process	↑1↓27	Cytoskeleton organization	↑5↓15	Phosphorylation	↑13↓15
		Muscle structure development	↓17	Protein phosphorylation	↑6↓13	Cell proliferation	↑9↓16
		Actin filament-based process	↓14	Cell proliferation	↑5↓13	Muscle structure development	↑13↓3
		Muscle cell differentiation	↓11	Microtubule-based process	↓15	Negative regulation of proteolysis	↑6↓4
SM	MF			Cytoskeletal protein binding	↑12↓10	Receptor binding	↑15↓16
				Peptidase inhibitor activity	↑5↓4	Receptor activity	↑27↓3
				Protease binding	↑4↓2	Cytoskeletal protein binding	↑17↓12
	CC	Contractile fiber	↓10	Cytoskeletal part	↑23↓13	Extracellular space	↑34↓13
		Sarcomere	↓9	Extracellular space	↑24↓9	Cytoskeletal part	↑18↓24
		Myofibril	↓9	Myofibril	↑20	Myofibril	↑23↓1
		Actin cytoskeleton	↓9	Sarcomere	↑18	Sarcomere	↑21↓1
		I band	↓5	Microtubule	↑1↓12	I band	↑12
	BP	Muscle structure development	↓7	Cell cycle	↑5↓34	Immune system process	↑36↓10
		Regulation of muscle system process	↓5	Cytoskeleton organization	↑10↓16	Cell cycle	↑6↓36
		Striated muscle tissue development	↓5	Inflammatory response	↑12↓3	Cell proliferation	↑15↓23
		Regulation of muscle contraction	↓4	Muscle organ development	↑12	Cytoskeleton organization	↑16↓18
				G2/M transition of mitotic cell cycle	↑1↓4	Muscle structure development	↑21↓5

The Homeobox (HOX) family genes seem to be crucial for correct development and regulate muscle-specific genes (Houghton and Rosenthal 1999). In a study in mouse, the knockdown of Zic2 resulted in a delay in the activation of Myf5 with a subsequent delay in MyoD but the expression of Pax3 was not affected (Pan *et al.* 2011). Our results suggest that these eight genes could be involved in muscle type differentiation leading the muscle to develop characteristics specific for one depot or another.

With respect to the expression of the MRFs, we found that genes MyoG, MyoD and Myf6 showed higher expression in myoblasts in the process of differentiation and maintained high expression levels in myotubes which agrees with previous studies (Rajesh *et al.* 2011; Tripathi *et al.* 2014). At the protein level, Tripathi *et al.* (2014) reported that genes Myf5 and MyoD are highly expressed 10 days after differentiation treatment, however, in this study Myf5 did not present significant change in expression during differentiation.

In the enriched GO terms, we observed differences during the end of proliferation (day 0) compared with the myoblast differentiation (7 and 14 days after the differentiation treatments) (Table I). At the beginning of the experiment (day 0) the enriched terms were cell cycle, proliferation and G2/M transition of mitotic (Table I) in concordance with the proliferation events occurring in the myoblast. However, the expression of genes involved in these terms started to decrease in the subsequent days due to the shift from cell division to differentiation similar to previously reported in other bovine studies (Lee *et al.* 2012; He and Liu 2013). From the KEGG analysis some of the enriched pathways were dilated cardiomyopathy, cardiac muscle contraction, calcium signalling pathway, cell adhesion molecules (CAMs), cell cycle and adrenergic signalling in cardiomyocytes.

CONCLUSIONS

The differential expression of the genes Hoxc11, Sim2, Hoxc8, Hoxb9, Zic2, Zic4, Tbx4 and Hoxb4 suggest their implication in muscle depot differentiation during early development.

ACKNOWLEDGEMENT

This project was supported by a grant from the Next-Generation BioGreen 21 Program PJ01134906 and PJ012611, Rural Development Administration, Republic of Korea.

REFERENCES

Braun T. and Gautel M. (2011) *Nature reviews Molecular cell biology* **12:** 349-61.

Eng D., Ma H.Y., Gross M.K. and Kioussi C. (2013) *ISRN Developmental Biology 2013* **2013**:1-8.

Frey R.S., Johnson B.J., Hathaway M.R., White M.E. and Dayton W.R. (1995) *Basic Appl Myol.* **5**: 71-9.

He H. and Liu X. (2013) *PloS one* **8:** e64356.

Houghton L. and Rosenthal N. (1999) *Developmental Dynamics* **216:** 385-97.

Johnson B.J., Halstead N., White M.E., Hathaway M.R., DiCostanzo A. and Dayton W.R. (1998) *Journal of Animal Science* **76:** 2779-86.

Lee E.J., Bajracharya P., Lee D.-M., Kang S.W., Lee Y.S., Lee H.-J., Hong S.K., Chang J., Kim J.W. and Schnabel R.D. (2012) *Genes and Genomics* **34**: 133-48.

Pan H., Gustafsson M.K., Aruga J., Tiedken J.J., Chen J.C.J. and Emerson C.P. (2011) *Developmental biology* **351**: 120-7.

Rajesh R.V., Jang E.-J., Choi I.-H., Heo K.-N., Yoon D.-H., Kim T.-H. and Lee H.-J. (2011) *Asian-Australasian Journal of Animal Sciences* **24**: 1288-302.

Tripathi A.K., Patel A.K., Shah R.K., Patel A.B., Shah T.M., Bhatt V.D. and Joshi C.G. (2014) *Mechanisms of development* **132**: 79-92.

Proc. Assoc. Advmt. Anim. Breed. Genet. 22:217-220

GENOTYPING-BY-SEQUENCING FOR GENETIC IMPROVEMENT IN HONEYBEES

G. E. L. Petersen[1, 2], P. F. Fennessy[2], T. C. Van Stijn[3], S. M. Clarke[3] and P. K. Dearden[1]

[1] Department of Biochemistry, University of Otago, Dunedin, New Zealand
[2] AbacusBio Limited, PO Box 5585, Dunedin, New Zealand
[3] AgResearch, Invermay Agricultural Centre, Mosgiel, New Zealand

SUMMARY

The polyandrous mating habits of honeybee queens, the small size of the animals themselves and the eusocial organisation of honeybee colonies present unique challenges for the establishment of genetic improvement programs.

Here we present a method to genotype honeybee queens using pools of their male offspring as proxies in a Genotyping-by-Sequencing (GBS) protocol.

GBS makes use of restriction site-associated short reads of DNA sequencing, effectively sampling the genome. Aligning these short reads to the reference genome provides a reliable and repeatable, but cost-effective genotyping protocol for honeybee queens. We found contamination of drone pools with unrelated drones to be an issue which can be overcome by using drone larvae directly from the brood comb.

INTRODUCTION

The Western Honeybee, *Apis mellifera*, has shifted into the focus of applied animal science due to the increase in importance of pollination services and rising prices for monofloral honeys. Simultaneously, honeybee breeding efforts have become more sophisticated, leading to the development of specialised elite queen breeding operations. While elite queen breeders have been able to adopt a number of breeding techniques from other livestock species, such as performance testing and artificial insemination (AI), their work is subject to a number of biological limitations.

Honeybees are organised in colonies of up to 60,000 or so individuals, with only one reproducing female per colony, the queen, and a handful of reproducing males, the drones. The bulk of a honeybee colony is made up of functionally sterile female workers. As descendants of the same queen mother, workers are either half- or full sisters that are constantly being replenished. Honeybee queens produce up to 2,000 eggs per day, both fertilized and unfertilized. Fertilized eggs develop into females, while unfertilised eggs produce haploid males. The sperm stores are accumulated during one virginal mating flight, where the queen mates with 6 to 25 drones.

Polyandry and the resulting diversity within the work force have been shown to be crucial factors for colony fitness (Mattila *et al.* 2007). However, they severely limit the accuracy of hive/queen pedigree, making it impossible to record pedigree beyond maternal line and a list of potential drone sources (queens presiding over colonies in the same area or used to gather drones for AI). AI can be a valuable tool to control mating, but if from a single drone source, it will diminish the performance of the resulting colony. Thus, queen genotyping is the only way to accurately determine a queen's ancestry without compromising colony performance. Unfortunately, the continuous policing of the physical state of the queen by her workers complicates sampling for genotyping, since the sampling of hemolymph or body parts (e.g. tarsus, antennae) will result in death, either directly or by enticing workers to overthrow the impaired queen.

In light of these obstacles, a new approach to the genotyping of breeder queens is necessary. With the haploid male drones arising from unfertilised eggs, they potentially provide a sample of the queen's genome without interfering with her physically. This paper describes results of a pilot study to evaluate the efficacy of assessing the DNA of a queen through genotyping her haploid drone offspring.

MATERIALS AND METHODS

Honeybee samples. 7 samples of multiple drones and corresponding queens were obtained from a commercial beekeeping operation in Marlborough, New Zealand, between January 2016 (drones) and April-May 2016 (queens). In January, 10-14 adult drones were removed from each colony, immediately put on ice in the field and subsequently frozen at -20° C. Simultaneously, basic information about the colonies was collected.

At arrival in the laboratory, drones were transferred into a -80° freezer until required for DNA extraction. The same collection and storage procedures were followed for the queens. In addition, 1 old breeder queen (sample 8) was sampled with no corresponding drones, resulting in 8 samples.

DNA extraction and Drone Pooling. Genomic DNA (gDNA) was extracted using the "ZR-96 Tissue & Insect DNA Kit" (Zymo Research, Irvine, CA, USA). Due to known issues with PCR amplification of DNA extracted from honeybee heads (Boncristiani et al. 2011), thorax was chosen as the standard substrate for gDNA extraction from both drones and queens.

The standard protocol for the ZR-96 Tissue & Insect DNA Kit was followed, with a 2010 GenoGrinder® (SPEX® SamplePrep, Metuchen, NJ, USA) serving as tissue homogenizer, set to 1,200 rpm for 8 minutes.

For each of the 7 drone/queen samples, DNA from 5, 6, 7 and 8 drones was pooled, as well as from all drones in the sample. DNA concentrations for the pools via Qubit® dsDNA HS Assay Kit (ThermoFisher Scientific, Waltham, MA, USA) ranged from 11.1μg/ml to 62.9μg/ml.

Evaluation of different body parts. For sample 8, DNA was extracted from thorax, legs, wings and head, to evaluate the option of using a part of the body that would require a less time-consuming insect dissection protocol.

Genotyping-by-Sequencing. A 96-well plate, containing DNA from the following was prepared for subsequent genotyping by GBS: 7 samples with 5 drone pools (5x, 6x, 7x, 8x, all drones) and 1 (sample 1), 2 (sample 4) or 3 (samples 2-3, 5-7) repeats of corresponding queens, as well as a set of DNA from queen sample 8 (3x legs, 1x head, 1x thorax, 1x wings). The plate also contained DNA from 20 individual drones from samples 6 and 7.

Following enzyme selection and adapter optimisation (data not shown), GBS was carried out using a double digest with ApeK1 and Msp1, following an optimised version of the original GBS protocol (Elshire *et al.* 2011, Dodds *et al.* 2015). 100 ng of DNA per "sample" was utilised to prepare an 80-sample indexed GBS library that was subsequently further purified using the Pippin Prep (SAGE Science, Beverly, MA, USA) with size selection to generate a 150-500bp DNA sequencing library. Single-end sequencing (1x100) was performed on an Illumina HiSeq2500 with v4 chemistry, yielding approximately 25Gb of raw sequence data per lane. Raw fastq files were quality checked using FastQC v0.10.1 (Andrews 2010).

Data analysis, SNP calling and alignment. Approximately 260 Million raw reads were processed with UNEAK, Tassel version 3.0.170, (Lu *et al.* 2013) to detect variants and report reference and alternative allele counts at variant sites. The resulting ~27,800 called SNPs were further processed to construct a relationship matrix using R software (KGD; Dodds *et al.* 2015). KGD unbiased estimates of relatedness were calculated via method 1 of VanRaden (2008) adjusted to account for sequence read depth at each individual SNP location including SNPs with zero/missing reads.

In order to create a more consistent methodology that can be evaluated across GBS runs with potentially different restriction enzyme cut patterns, the sequencing data were additionally analysed using TASSEL 5 GBSv2 (Glaubitz *et al.* 2014) and the Burrows-Wheeler Aligner (Li *et al.* 2010). In this process, the short reads were aligned to the *A. mellifera* reference genome (Weinstock *et al.* 2006) before SNP calling. After alignment, ~46,400 SNPs could be fed into the KGD R software.

RESULTS AND DISCUSSION

GBS of honeybee drone pools and corresponding queens. Samples 1 and 2 showed consistent internal relationships (> 0.98), while for sample 3, the relationship dropped markedly with the addition of drone #6 (to ~0.86) and recovered with the addition of more drones (to ~1.1). In all three cases, the observed relationships between drone pools and queen were similar to the relationships between queen repeats.

Two samples (4 and 5) showed strong relationships both within the drone pools and within the queen repeats (~1.1), but not between the drone pools and the queen from the respective hives (~0.6).

For samples 6 and 7, all drones contributing to the pools were genotyped individually to determine if they had been sampled correctly. These individual assessments showed that for sample 6, only 6 out of the 10 drones were in fact sons of the queen from the corresponding hive; for sample 7, this was true only for 4 out of the 10 drones. As a result, relationships between drone pools and the queen diminished with the addition of more drones.

Drone pools as queen proxies. Evaluation of the relationships within and between samples consisting of pools of varying numbers of adult drones and corresponding queen mothers showed that there is considerable variation in the accuracy with which the drone pools reflect the genome of the hive queen.

Under ideal conditions, when the sampled drones are descendants of the targeted queen, drone pools appear to be a valid way to genotype their queen mother (see Table 1).

Table 1. Relationships between drone pools and their mother queen (results shown for sample 2)

	5x	6x	7x	8x	total	queen	queen	queen
5x	1.33	1.00	0.98	0.99	1.00	0.98	1.01	0.98
6x		1.31	0.99	0.99	0.99	0.98	0.99	0.99
7x			1.31	0.99	0.99	0.97	0.98	0.97
8x				1.25	0.99	0.99	0.99	0.99
total					1.30	0.98	1.02	1.00
queen						1.28	1.00	0.99
queen							1.32	1.00
queen								1.27

Most of the variation in the results could be traced back to the accidental sampling of unrelated drones as outlined below.

Sampling of adult drones in a recently re-queened colony. Samples 2 and 3 showed a pattern of strong relationships between drone pools, but weaker relationships between drone pools and queen. Queens in these colonies had been replaced prior to sample collection, and sampled drones were descendants of the old queen, not the one presiding over the colony at collection. This highlights the importance of knowing the history of the hive before sampling.

Displacement of adult drones. Samples 5 and 7 showed limited relationships between drone pools and corresponding queens due to the fact that only 6 of 10 and 4 of 10 drones respectively were sons of their putative mothers. Both of these colonies were situated in a very tightly-packed yard in which adult drones returning to the hive could potentially drift over to another hive and end up in foreign colonies.

These two problems with the accidental sampling of mismatched drones can be overcome by switching from adult drones to drone brood as the source of DNA.

GBS of different honeybee body parts. There were 3 repeats of DNA extracted from the legs of queen 8; one of these failed to give a GBS result for unknown reasons. However other than this, relationships between different genotypes generated based on DNA extracted from different body parts were consistently over 0.87 (see Table 2).

Table 2. Internal relationships for queen 8

	legs	legs	head	wings	thorax
legs	1.17	0.90	0.88	0.90	0.91
legs		1.21	0.87	0.90	0.90
head			1.30	0.90	0.89
wings				1.24	0.92
thorax					1.20

These results are consistent with repeats of the same DNA extract. Moreover, the good GBS result based on DNA extracted from the queen's head suggests that with the use of the ZR-96 Insect & Tissue DNA Kit, DNA of a sufficient purity can be generated to avoid the PCR-inhibiting effects of honeybee compound eyes.

The use of wings as a non-lethal way to genotype bees has been suggested previously (Chaline *et al.* 2004). If sampling of an unmated queen is desired (e.g. to plan an AI mating), wings could present a valuable alternative to the proxy-based protocol presented here. Wing-origin DNA from queen 8 showed similar relationships to the other body parts both within and between samples (data not shown), but only 3.4 µg DNA per ml. Due to the low DNA concentration, wing clippings (~1/3 of the wing, removed to mark a queen) are unlikely to be a reliable source of DNA for GBS, but our findings show that it is possible to use whole sets of wings in cases where wings are not necessary for a queen's success (e.g. virgin AI queens).

CONCLUSION

GBS of pooled drones is a reliable non-invasive genotyping method for honeybee queens, provided that a certain time has passed after a new queen is introduced into the hive and drones are sampled before emerging from the brood comb. Furthermore, by aligning the short reads generated via GBS to the *Apis mellifera* reference genome, this method can be used across populations. These findings will enable breeders of elite honeybees to take their breeding programs to the next level; for example, by controlling inbreeding without the use of restrictive breeding techniques such as single-drone AI.

ACKNOWLEDGEMENTS

G. Petersen is currently a Callaghan Innovation PhD fellow in the Department of Biochemistry at the University of Otago, in cooperation with AbacusBio Limited and Taylor Pass Honey Company, Blenheim, New Zealand, with additional funding being provided by Taylor Pass Honey Co. This work was carried out in collaboration with the 'Genomics for Production & Security in a Biological Economy' C10X1306 MBIE AgResearch program.

REFERENCES

Andrews S. (2010). Available online at: http://www.bioinformatics.babraham.ac.uk/projects/fastqc

Boncristiani, H. F., Li, J., Evans, J. D., Pettis, J. S. and Chen, Y. (2011) *Apidologie* **42**(4): 457–60.

Chaline, N., Ratnieks, F. L. W., Raine, N. E. and Badcock, N. S. (2004). *Apidologie* **35**: 311–18.

Dodds, K. G., McEwan, J. C., Brauning, R., Anderson, R. M., van Stijn, T. C., Kristjansson, T. and Clarke, S. (2015) *BMC Genomics* **16**(1047): 1–15.

Elshire, R. J., Glaubitz, J. C., Sun, Q., Poland, J. A., Kawamoto, K., Buckler, E. S. and Mitchell, S. E. (2011) *PloS ONE* **6**(5): e19379.

Glaubitz, J. C., Casstevens, T. M., Lu, F., Harriman, J., Elshire, R., Sun, Q. and Buckler, E. S. (2014) *PLoS ONE* **9**(2).

Li, H. and Durbin, R. (2010) *Bioinformatics*, Epub. [PMID: 20080505].

Lu, F., Lipka, A. E., Glaubitz, J. C., Elshire, R., Cherney, J. H., Casler, M. D., Buckler, E. S. and Costich, D. E. (2013) *PLoS Genetics* **9**(1).

Mattila, H. R. and Seeley, T. D. (2007) *Science* **317**: 362–64.

Weinstock, G. M. *et al.* (2006) *Nature* **443**(7114):931-949.

Proc. Assoc. Advmt. Anim. Breed. Genet. 22:221-228

GENETICS AND GENOMICS OF SWINE LEAN GROWTH AT THE INTERFACE BETWEEN HOST AND COMMENSAL GUT BACTERIA

C. Maltecca[1,2], B. Lu[1] and F. Tiezzi[1]

[1] Department of Animal Science, North Carolina State University, Raleigh, NC 27695-7627, USA.
[2]Genetics Program, North Carolina State University, Raleigh, NC 27695-7627, USA.

SUMMARY

The main objectives of this paper will be to quantify the impact of gut microbiome composition on lean growth in swine and to quantify the heritability of relative taxa abundance in swine. The gut microbiome absorbed a significant portion of the phenotypic variation, ranging from approximately 3% for microbiome composition at weaning on ADG15 to more than 65% for microbiome composition at week 15 on ADG15. Point estimates for the heritabilities of the 57 taxonomical families ranged from low to moderately high ranging from less than 5% to almost 50%, according to family and time point. Different patterns of h^2 (from low to high and vice versa) were observed across time for different families possibly reflecting the overall abundance of a particular family across the trial.

INTRODUCTION

Efficiency of producing saleable meat products is largely determined by costs associated with feed and by the amount of and quality of lean meat produced(Hoque et al., 2009)·(Hoque et al., 2008)·(McGlone and Pond, 2003). Utilizing feed resources more efficiently has become a clear challenge that faces the livestock industry. Recent efforts have been devoted to identify and exploit the genomic variability of individual pigs in increasing feed efficiency(Jiao et al., 2014a)·(Jiao et al., 2014b) (Howard et al., 2015). While partially successful this approach presents limitations. First, feed efficiency is not a directly measurable trait. Instead it must be obtained from its components and it includes all traits associated with the efficiency of feed utilization, typically feed conversion ratio (FCR) or its reciprocal (feed:gain ratio) or RFI (Koch et al., 1963). These commonly used measures have inherent flaws (Arthur and Herd, 2008). More importantly, a continued effort concentrating only on the pig variability for efficiency will inevitably result in diminished marginal gains, incurring in concomitant losses of overall fitness and diversity over time(Colleau and Tribout, 2008). The amount and type of bacteria present in the gut of individuals represent a key part of all mammalian organisms (Gill et al., 2006). The makeup of the microbiome represents a vast pool of genomic diversity that contributes to the individual physiology and health (Pflughoeft and Versalovic, 2012). Particularly, the intestinal microbiome directly affects the degradation of carbohydrates, provides short chain fatty acids, mitigates and alter the effect of potential toxic compounds and produce essential vitamins(Gill et al., 2006). Different composition of the gut population in humans has been linked to the ability of degrading enzymes, maintain a certain population balance and influence the overall health status(Cho and Blaser, 2012). Relatively few full microbiome sequencing studies have been conducted in swine to date (Isaacson and Kim, 2012), while many studies have focused on either humans or model organisms. There is nonetheless a striking physiological similarity between the human and the swine intestine such that the second is currently successfully employed as model for the first (Odle et al., 2014)· (Heinritz et al., 2013)· (Zhang et al., 2013). Several studies comparing different geographical populations of humans and studies comparing different animal species have found that host genetic differences play a significant role in the composition of the microbiome. One study of tilapia, toads, geckos, quail, and mice tested changes in the microbiota of the colon and cecum after periods of fasting (Kohl et al., 2014). The study found that in most species, there was more genetic diversity in the colon microbiome during a fast compared to a regular diet (Kohl et al., 2014). This suggests how environmental factors such as diet are not the only features keeping the microbiota balance but that other factors are at play most likely related to genetic. In the same study, results from the cecum found that in tilapia and toads, although there were initial changes in genetic diversity after the start of the fast, the microbiotic species returned to normal later in the fast (Kohl et al., 2014), again suggesting how the microbiota seem able to "self-regulate" without input from the environment. In the same study, mice showed no changes in the microbiotic composition during fasting (Kohl et al., 2014). In this case it appears that the microbiota might be completely controlled by the host genetic. Similar studies have been conducted in humans. For example Goodrich et al.(Goodrich et al., 2014a) found that twins' fecal samples have a more similar microbiota composition than unrelated individuals, with monozygotic twins having a more similar composition than dizygotic. This again suggests that genetics might play a significant role in the microbiome composition. A study of samples of Columbian gut microbiome found that samples of people with a higher BMI had less *Firmicutes* while European gut microbiome did not show decreased *Furmicutes* (Escobar et al., 2014). Differences in microbiome between individuals of different BMIs seem to indicate a direct genetic influence. A study of data from a twin study (Goodrich et al., 2014b) further linked human genotype and the composition of the gut microbiome. The study identified *Christensenellaceae* group as central to a network of co-occurring heritable microbes that has been associated with lean body mass index (BMI)(Ley, 2015). Numerous studies of rodents suggest that the gut microbiota populations are sensitive to genetic, and can produce or influence signals that directly or indirectly impact energy balance (weight gain or loss) and energy stores (Parks et al., 2013). Thus, the microbiota is certainly implicated in the development of obesity, and with tissue deposition in general. There are compelling arguments for the existence of a genetic control over the abundance of taxa in different species and the link of these with energy balance and growth. Currently some evidence has been presented in pigs.

The main objectives of this paper will be to quantify the impact of gut microbiome composition on lean growth in swine and to quantify the heritability of relative taxa abundance in swine.

MATERIALS AND METHODS

From a Duroc closed-nucleus population 28 boars were selected to be sires of the individuals used in this trial. Sires were mated to crossbred sows to generate terminal-cross piglets. These were weaned at an average of 19 days of age and grouped in single-sire-gender pens (groups). During the nursery, growth and finish period, all pigs will be fed standard diets. End of test was declared on a pen-specific basis, entire pens of pigs were taken off test and sent for harvest at a pen mean live weight of 304.6 ± 5.51 lb.

Live weight measurements were taken on individual pigs at the start (weaning) and end of the study and weeks 15, 18 and 22 post-weaning. Ultrasound back-fat depth and *Longissimus* muscle depth and area at approximately the 10th rib were measured on the right side of the pig on a transverse ultrasound scan taken at weeks 15, 18 and 22 post-weaning and at the end of the study. Fecal samples were taken for a total 1300 individual pigs at three time points. After editing, there were 3,783 fecal samples collected, including 15-24 days of age (1205 individuals), 115-124 days old (1295 individuals), and 180-217 days old (1283 individuals). Microbiome composition was obtained by amplifying the V6-V8 region of the 16S rRNA genes of the stool samples through pyrotag sequencing. After sequence processing and QC, there were 10,000 sequence reads per sample. Reads were organized into 2,026 phylotypes (operational taxonomic units, OTUs). Any taxonomic identifier with a confidence score below 80% was grouped as "unassigned". The 2,026 phylotypes were classified into 14 known phyla, 57 families, 112 genera and 213 species.

The bacterial composition of 3,783 samples was determined in each taxonomic level according to the read counts of the 2,026 taxonomically-annotated OTUs in each sample. The 3,783 samples were statistically compared according to their age range using the Kruskal-Wallis test. Further discrimination of grouping of the different taxonomical units was performed through principal component analysis (PCA).

To investigate the divergence of microbial community at taxonomic species level, samples were clustered by age group and sex. All taxonomic units with unassigned genus and/or species were removed, leaving a total of 380 OTUs for this analysis.

To highlight potential association between particular taxa combinations and growth/composition phenotypes, pseudo-enterotypes were obtained for growth and carcass composition through clustering of individuals and families.

The overall contribution of microbiome to phenotypic variability was investigated through linear mixed models. Two traits were considered, average daily gain at market weight (ADGM), as well as average daily gain at 15 weeks (ADG15). For each of the traits a model that included fixed effects of sex, dam-line, contemporary group, back fat at market weight and random effects of permanent environmental effect, animal additive genetic effect (A), and residual, were fitted. This base line model was compared to a model that a random Microbiome (M) effect. Three microbiome compositions were fitted separately to the models representing the populations present at weaning 15 weeks of age and off-test.

The host genetic control over microbiome composition was investigate at the family level. Second-degree polynomial random regression models utilizing 57 family abundance as the dependent variable were fitted. The models included time and sex and their interaction as fixed effect and random regression on animal and permanent environmental effects. All models were run with ASREML v.4.0.

RESULTS AND DISCUSSION

The bacterial composition of 3,783 samples was determined in each taxonomic level according to the read counts of the 2,026 taxonomically-annotated OTUs in each sample. The 3,783 samples were statistically compared according to age range. The Kruskal-Wallis test for differences in bacterial composition among the three age groups showed that 55 out of 57 bacterial families had significantly different ($P<0.005$) abundance counts between 15-24d and the rest of the samples. The proportion of the 10 most different families is presented in Figure 1. Similarly, the 115-124d and 180-217d groups were significantly different by 45 out of 57 families. The bacteria proportion of the 6 most different families is shown in Figure 2. At 15-24 d, the fecal bacteria were presented by three main phyla, *Firmicutes* (39.38%), *Bacteroidetes* (29.93%) and *Proteobacteria* (22.16%). Over time, the proportion of bacteria in the two phyla *Bacteroidetes* and *Proteobacteria* decreased, while the proportion of bacteria in the phylum *Firmicutes* pronouncedly increased to 72.71% and 77.26% at 115-124 d and 180-217 d, respectively. Our findings agree well with the reports by Kim et al. (2011), Ivarsson et al. (2011), Dicksved et al. (2015).

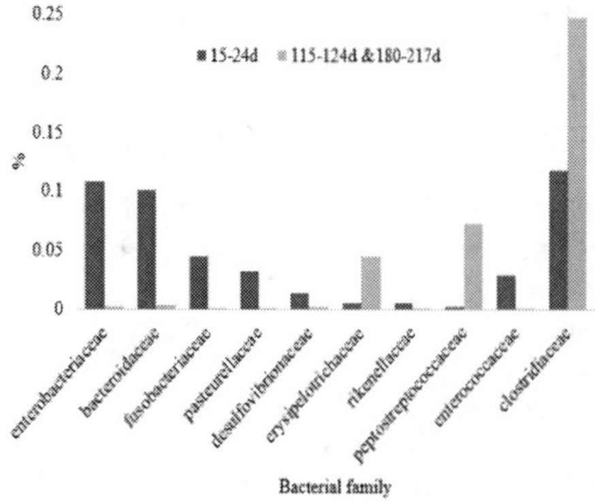

Figure 1. Bacteria proportion of the 10 most different families between 15-24d and 115-124d & 180-217d

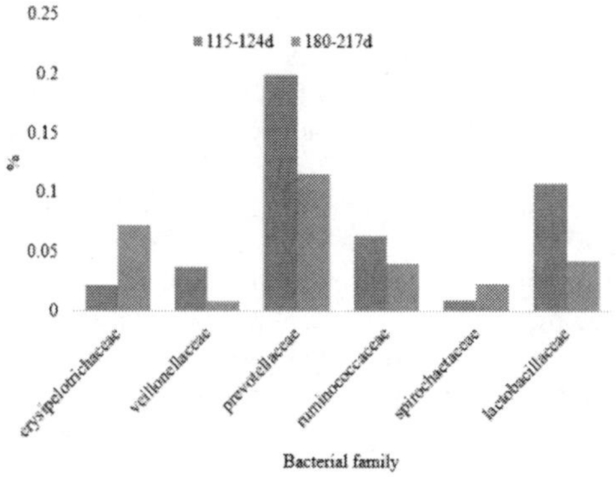

Figure 2. Bacteria proportion of the 6 most different families between 115-124d and 180-217d groups

Using principal component analysis (PCA), family-level bacterial composition data of 3,783 samples over 3 time points were decomposed into two factors that explained 44.03% of the variance (Figure 3). Principal component 1 (PC1), which explained 31.26% of the variance, was heavily negatively loaded with *Enterobacteriaceae, Bacteroidaceae, Fusobacteriaceae, Enterococcaceae, and Pasteurellaceae.* Principal component 2 (PC2) was heavily loaded with *Clostridiaceae* and *Enterobacteriaceae*, and negatively loaded with *Prevotellaceae* and *Fusobacteriaceae*.

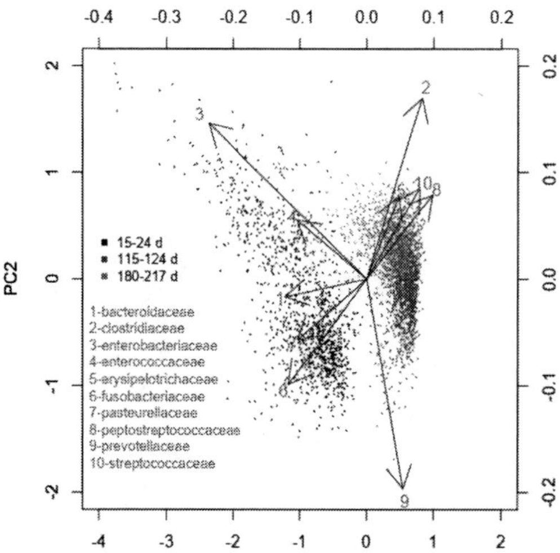

Figure 3. Principal component analysis of bacterial families and the 10 largest loadings of bacterial families for PC1 and PC2.

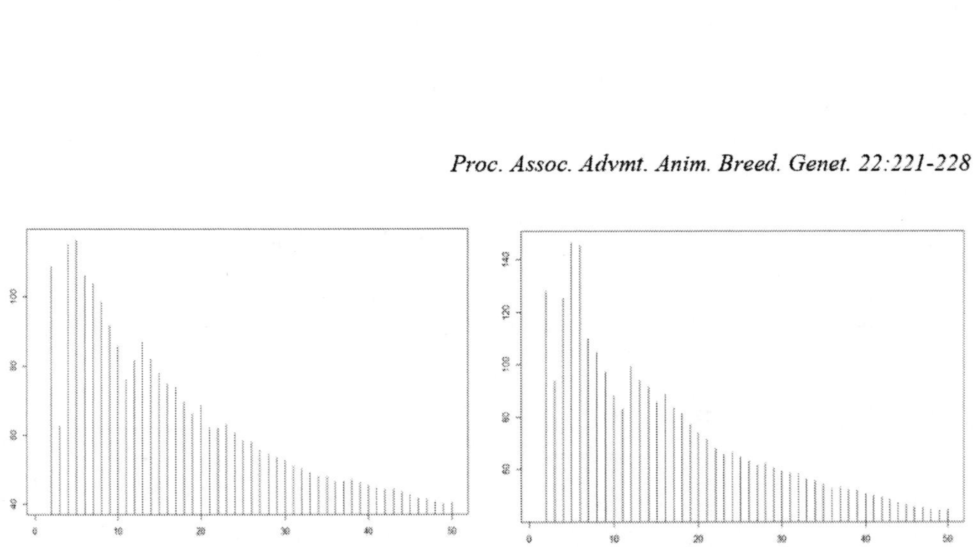

Figure 4. Calinski-Harabasz indexes for number of clusters of samples from 115-124 d males (left) and 180-217 d males (right). Number of clusters on X-axis. The index on Y-axis.

Samples of 115-124 d and 180-217 d female pigs clustered best into 3 groups whereas the male samples fit best into 5 groups as shown in Figure 4. Though the number of clusters by sex was similar between the 2 age groups, animals that grouped together during 115-124 d did not appear to remain in the same group in the later stage.

We investigated the relationship between clusters of the OTUs and 180-217d animals with regard to fat depth measures. Animals and OTUs were clustered into 5 and 20 groups, respectively, as shown in Figure 5. Average estimated breeding values (EBV) was calculated for each animal cluster. The relative abundance of OTUs in groups 3, 5, 6, 8, 13, 16, 17 and 19 appeared to be significantly correlated with fat depth EBV.

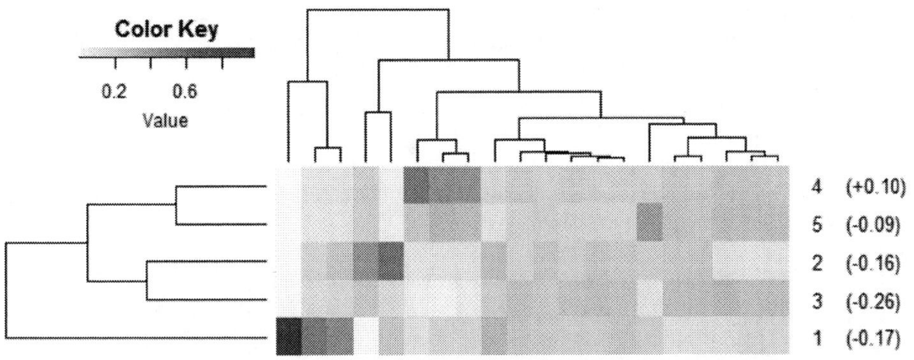

Figure 5. A heatmap of relative abundance of bacterial clusters within pig's fecal microbiome. Five animal clusters, 1 – 5, with animal counts of 10, 153, 180, 174, 92 respectively. Average breeding value of fat depth for animals within each animal cluster is presented in parentheses next to cluster number. The 380 taxonomic units were clustered into 20 groups. Within each OUT cluster, level of redness shows average OTU count relative to other animal clusters.

Microbiome contribution to the overall daily gain variability is reported on tables 1 and 2. For both traits measured (ADGM and ADG15) the gut microbiome absorbed a significant portion of the phenotypic variation, ranging from approximately 3% for microbiome composition at weaning on ADG15 to more than 65% for microbiome composition at week 15 on ADG15. In both cases the largest amount of variance was absorbed when the composition at week 15 was fit. The microbiome effect eroded the most variance from the residual effect and PE, while a significant portion was also absorbed from the animal genetic effect.

Table.1 Proportion of variances estimated for ADGM

	A	A+M $^{(1)}$	A+M $^{(2)}$	A+M $^{(3)}$
AIC	9855.71	9856.75	9842.15	9842.95
PE (%)	11.52	10.47	7.15	6.54
A (%)	13.20	12.65	6.07	7.29
M (%)	N/A	6.26	57.21	46.64
Residual (%)	75.28	70.62	29.57	39.53

N/A: Not available.

AIC: Akaike information criterion.

$^{(1), (2), \text{and } (3)}$, representing microbial data at weaning, 15 weeks of age, and off-test, respectively.

Table.2 Proportion of variances estimated for ADG15

	A	A+M $^{(1)}$	A+M $^{(2)}$	A+M $^{(3)}$
AIC	10217.89	10219.40	10182.36	10215.58
PE (%)	15.36	14.40	7.20	12.88
A (%)	11.11	10.90	3.51	9.60
M (%)	N/A	2.74	66.14	13.68
Residual (%)	73.53	71.96	23.15	63.84

N/A: Not available.

AIC: Akaike information criterion.

$^{(1), (2), \text{and } (3)}$, representing microbial data at weaning, 15 weeks of age, and off-test, respectively.

Point estimates for the heritabilities of the 57 taxonomical families ranged from low to moderately high ranging from less than 5% to almost 50%, according to family and time point. Different patterns of h^2 (from low to high and vice versa) were observed across time for different families possibly reflecting the overall abundance of a particular family across the trial. A plot of the h^2 for a sample of the families fitted is presented in figure 6.

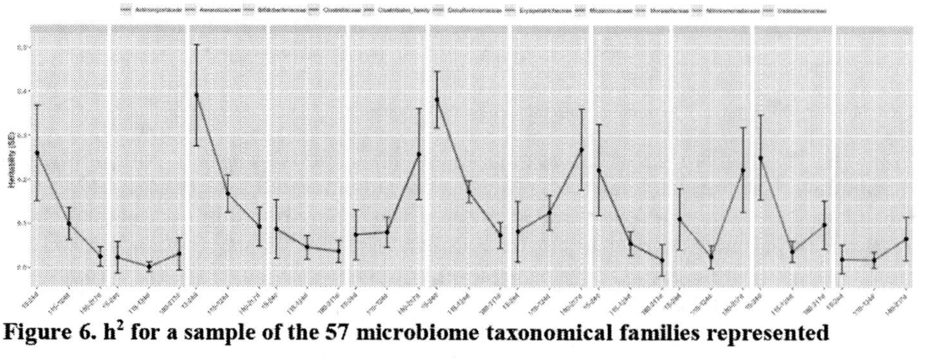

Figure 6. h^2 for a sample of the 57 microbiome taxonomical families represented

REFERENCES

Arthur, J.P.F., Herd, R.M., 2008. Residual feed intake in beef cattle. Revista Brasileira de Zootecnia 37, 269–279. doi:10.1590/S1516-35982008001300031

Cho, I., Blaser, M.J., 2012. The Human Microbiome: at the interface of health and disease. Nat Rev Genet 13, 260–270. doi:10.1038/nrg3182

Colleau, J.J., Tribout, T., 2008. Optimized management of genetic variability in selected pig populations. Journal of Animal Breeding and Genetics 125, 291–300. doi:10.1111/j.1439-0388.2008.00738.x

Escobar, J.S., Klotz, B., Valdes, B.E., Agudelo, G.M., 2014. The gut microbiota of Colombians differs from that of Americans, Europeans and Asians. BMC Microbiol. 14, 311. doi:10.1186/s12866-014-0311-6

Gill, S.R., Pop, M., DeBoy, R.T., Eckburg, P.B., Turnbaugh, P.J., Samuel, B.S., Gordon, J.I., Relman, D.A., Fraser-Liggett, C.M., Nelson, K.E., 2006. Metagenomic Analysis of the Human Distal Gut Microbiome. Science 312, 1355–1359. doi:10.1126/science.1124234

Goodrich, J.K., Waters, J.L., Poole, A.C., Sutter, J.L., Koren, O., Blekhman, R., Beaumont, M., Van Treuren, W., Knight, R., Bell, J.T., Spector, T.D., Clark, A.G., Ley, R.E., 2014a. Human Genetics Shape the Gut Microbiome. Cell 159, 789–799. doi:10.1016/j.cell.2014.09.053

Goodrich, J.K., Waters, J.L., Poole, A.C., Sutter, J.L., Koren, O., Blekhman, R., Beaumont, M., Van Treuren, W., Knight, R., Bell, J.T., Spector, T.D., Clark, A.G., Ley, R.E., 2014b. Human Genetics Shape the Gut Microbiome. Cell 159, 789–799. doi:10.1016/j.cell.2014.09.053

Heinritz, S.N., Mosenthin, R., Weiss, E., 2013. Use of pigs as a potential model for research into dietary modulation of the human gut microbiota. Nutrition Research Reviews 26, 191–209. doi:10.1017/S0954422413000152

Hoque, M.A., Kadowaki, H., Shibata, T., Oikawa, T., Suzuki, K., 2009. Genetic parameters for measures of residual feed intake and growth traits in seven generations of Duroc pigs. Livestock Science 121, 45–49. doi:10.1016/j.livsci.2008.05.016

Hoque, M.A., Kadowaki, H., Shibata, T., Suzuki, K., 2008. Maternal and Direct Genetic Parameters for Production Traits and Maternal Correlations among Production and Feed Efficiency Traits in Duroc Pigs. Asian-Australasian Journal of Animal Sciences 21, 961–966. doi:10.5713/ajas.2008.70641

Howard, J.T., Jiao, S., Tiezzi, F., Huang, Y., Gray, K.A., Maltecca, C., 2015. Genome-wide association study on legendre random regression coefficients for the growth and feed intake trajectory on Duroc Boars. BMC Genetics 16, 59.

Isaacson, R., Kim, H.B., 2012. The intestinal microbiome of the pig. Animal Health Research Reviews 13, 100–109. doi:10.1017/S1466252312000084

Jiao, S., Maltecca, C., Gray, K.A., Cassady, J.P., 2014a. Feed intake, average daily gain, feed efficiency, and real-time ultrasound traits in Duroc pigs: II. Genomewide association. Journal of Animal Science 92, 2846–2860. doi:10.2527/jas.2014-7337

Jiao, S., Maltecca, C., Gray, K.A., Cassady, J.P., 2014b. Feed intake, average daily gain, feed efficiency, and real-time ultrasound traits in Duroc pigs: I. Genetic parameter estimation and accuracy of genomic prediction. Journal of Animal Science 92, 2377–2386. doi:10.2527/jas.2013-7338

Koch, R.M., Swiger, L.A., Chambers, D., Gregory, K.E., 1963. Efficiency of Feed Use in Beef Cattle. Journal of Animal Science 22, 486–494. doi:10.2134/jas1963.222486x

Kohl, K.D., Amaya, J., Passement, C.A., Dearing, M.D., McCue, M.D., 2014. Unique and shared responses of the gut microbiota to prolonged fasting: a comparative study across five classes of vertebrate hosts. FEMS Microbiology Ecology 90, 883–894. doi:10.1111/1574-6941.12442

Ley, R.E., 2015. The Gene-Microbe Link. Nature 518, S7–S7. doi:10.1038/518S7a

Pig & poultry

McGlone, J., Pond, W.G., 2003. Pig Production: Biological Principles and Applications. Cengage Learning.

Odle, J., Lin, X., Jacobi, S.K., Kim, S.W., Stahl, C.H., 2014. The Suckling Piglet as an Agrimedical Model for the Study of Pediatric Nutrition and Metabolism. Annual Review of Animal Biosciences 2, 419–444. doi:10.1146/annurev-animal-022513-114158

Parks, B.W., Nam, E., Org, E., Kostem, E., Norheim, F., Hui, S.T., Pan, C., Civelek, M., Rau, C.D., Bennett, B.J., Mehrabian, M., Ursell, L.K., He, A., Castellani, L.W., Zinker, B., Kirby, M., Drake, T.A., Drevon, C.A., Knight, R., Gargalovic, P., Kirchgessner, T., Eskin, E., Lusis, A.J., 2013. Genetic Control of Obesity and Gut Microbiota Composition in Response to High-Fat, High-Sucrose Diet in Mice. Cell Metabolism 17, 141–152. doi:10.1016/j.cmet.2012.12.007

Pflughoeft, K.J., Versalovic, J., 2012. Human Microbiome in Health and Disease. Annual Review of Pathology: Mechanisms of Disease 7, 99–122. doi:10.1146/annurev-pathol-011811-132421

Zhang, Q., Widmer, G., Tzipori, S., 2013. A pig model of the human gastrointestinal tract. Gut Microbes 4, 193–200. doi:10.4161/gmic.23867

Proc. Assoc. Advmt. Anim. Breed. Genet. 22:229-232

BEHAVIOURAL TRAITS RECORDED IN GILTS AND ASSOCIATIONS WITH REPRODUCTIVE PERFORMANCE AS GROUP-HOUSED SOWS

K.L. Bunter

Animal Genetics and Breeding Unit, University of New England, Armidale, NSW 2351, Australia

SUMMARY

Gilts from two maternal lines were recorded for flight time (FT, N=8854) and scored for the count of lesions resulting from fighting 24 hours after selection and mixing into new groups (N=3238). Anterior (ANT) and posterior lesion counts were scored on a progressive four point scale representing none to multiple lesions (0-3), and aggressive gilts (0/1 scores) were defined by ANT>1. Lesion counts over the whole body were subsequently rescored pre-farrowing (PFBLES). Genetic correlations were estimated between these behavioural traits and average daily gain (ADG), gilt removals without a farrowing event and first parity litter size (TB: total born; NBA: number born alive) and birth weight. All behavioural traits scored for gilts and pregnant sows pre-farrowing were lowly to moderately heritable (h^2<0.15), implying that selection could alter FT or reduce fighting behaviour and hence skin lesions at different time points. However, lesion scores for gilts were not highly correlated with later PBFLES, reflecting changes to individual participation in fighting behaviour over time. Skin lesion traits were also uncorrelated with FT. These traits therefore represent assessment of different behaviours. Flight time was genetically correlated with ADG (rg: 0.24±0.10) only. All behavioural traits were generally not significantly correlated (either genetically or phenotypically) with subsequent reproductive performance, implying a neutral association between behavioural phenotypes and selection criteria in maternal lines. However, low adverse phenotypic correlations between lesion and locomotion scores or gilt removals illustrate a detrimental impact of fighting on welfare. Using management strategies to reduce fighting and developing a better understanding of the genetic basis for long term behaviour remain important for ongoing improvement of welfare and performance of group housed sows.

INTRODUCTION

Relative to stall housing, reintroduction of group housing for sows during gestation has several positive welfare benefits, but also enables negative interactions between sows. In particular, aggression amongst sows within groups can compromise their welfare and reproductive performance (Anil *et al.* 2006; Spoolder *et al.* 2009). However, observing behaviours of individual pigs directly is time consuming and impractical, and therefore an individual's contribution to aggression is frequently unknown. In addition, aggressive behaviours alter as the social hierarchy is established within stable groups (Anil *et al.* 2006), so the timing of observation is important. Behavioural indicators previously studied in commercial growing pigs include flight time (Crump 2004; Hansson *et al.* 2005) and the counting of skin lesions resulting from fighting, with anterior lesions in particular used as an indicator of participating in reciprocal fighting (Turner *et al.* 2006, Turner *et al.* 2009). In this study I investigated the genetic basis of behavioural traits recorded for purebred pedigreed gilts at selection, and their associations with lesion scores recorded again pre-farrowing, along with first parity reproductive performance under group housing during gestation.

MATERIALS AND METHODS

From January 2013 through to December 2014, approximately 10000 gilts from two maternal lines (Large White and Landrace) were recorded at a single site for lifetime average daily gain (ADG, g/day) at 24 weeks of age. Behavioural traits recorded concurrently included flight time and shortly after skin lesion scores resulting from fighting. Flight time (FT, s) was recorded upon release from weigh scales at the end of performance testing (N=8854)(see Crump 2004), while skin lesion counts were scored (N=3238) 24 hours after mixing into new groups <1 week later. Skin lesions resulting from fighting were scored on a progressive four point scale separately for each quarter, as 0: no lesions; 1: 1-5 lesions; 2: 6-10 lesions; and 3:10+ lesions. Scores were summed into anterior (ANT) and posterior (POS) quarters and regrouped (0, 1-2, 3-4, 5-6) into 4 scores for analyses. Gilts with ANT>1 were classified as an aggressive behavioural type (AGRO).

Gilts selected as breeding replacements were subsequently exposed to boars after 28 weeks of age and mated using AI. Gilts removed from the herd without a farrowing event were identified (REM0=0/1; removed=1). Pregnant gilts were housed in small static groups throughout gestation. A subset (N=1929) were re-scored for the count of skin lesions over their whole body upon transfer to the farrowing house (PFBLES) using the same scale as above (0-3), along with locomotion (PFLOCO: 0-3) and condition scores (-1,0,1) representing under-, at target, or over-condition. Reproductive performance traits recorded in the first parity included total born and number born alive (TB and NBA, pigs/litter). A subset of sows had records for average piglet weight at birth (ABWT, kg/piglet). Historical and male sibling data for ADG, all contemporary gilt reproductive data and 4 generations of pedigree were used to estimate genetic parameters.

Parameter estimates were obtained using linear mixed models under an animal model with ASREML software (Gilmour *et al.* 2009). Systematic effects included year-month of recording (24 levels), line (2 levels), and gender (M vs F, for ADG only). An additional random effect for birth litter was fitted when significant (P<0.05) based on a likelihood ratio test. Correlations between traits were estimated using a series of bivariate analyses.

RESULTS AND DISCUSSION

Raw data characteristics are provided in Table 1, along with heritability estimates for each trait. Behavioural traits scored for gilts (FT, ANT, POS and AGRO) and pregnant sows pre-farrowing (PFBLES) were lowly to moderately heritable, at the lower end of the range in heritability estimates reported for FT (Hansson *et al.* 2005) or lesion counts (Turner *et al.* 2009; Desire *et al.* 2015) for growing pigs. Scoring was preferred to lesion counts from the perspective of increasing the number of animals which could be recorded in a commercial setting, but reducing continuous traits to scoring categories can reduce estimates of heritabilities. The relatively lower heritabilities in our study could also indicate that as animals mature the genetic contribution to aggressive behaviours decreases (e.g., through learned behavioural responses to mixing).

Flight time was not significantly correlated, genetically (rg) or phenotypically (rp) with lesion score traits (not shown). This suggests that variation in FT is not associated with aggressive behaviours implied by lesion scores. Flight time was significantly correlated with ADG in this and previous studies (Hansson et al 2005), but the correlations with reproductive outcomes were negligible (Table 2). Growth is genetically uncorrelated with litter size traits (Bunter et al, 2010), supporting this result. Therefore, FT did not seem to yield any behavioural information strongly associated with either welfare or future sow reproductive performance.

With respect to skin lesion scores, relatively few gilts remained unmarked (0 scores) by 24 hours after mixing. Lesions were more common on the anterior than posterior parts of the body, but greatly reduced over the whole body before farrowing (Table 1) (see Bunter and Boardman, 2015). Anterior scores and POS were highly correlated with each other (rg: 0.99±0.05; rp: 0.66±0.05) but not significantly correlated with PFBLES observed approximately six months later (range rg: 0.20 to 0.30±0.23; range rp: 0.01 to 0.02±0.02)(Table 2). The genetic correlation between AGRO and PFBLES was stronger (0.42±0.26), but rp remained negligible. Lesion counts greatly decreased in the time interval between selection and farrowing because gilts were regrouped after mating and subsequently housed in stable groups. Aggressive interactions are known to reduce over time within stable groups (Anil *et al.* 2006). Lesion score traits had low positive genetic correlations (rg: 0.11 to 0.23, P>0.05) with ADG, similar in magnitude to those presented by Desire *et al.* (2015) and rp were negligible.

Table 1. Raw data characteristics including the distribution across scores (Distribution: %×100), along with heritability (h^2), common litter effects (c^2) and the phenotypic variance (σ^2_p). (na: not applicable; ns: P>0.05; PF: pre-farrowing)

Trait	N	Mean (SD)	Distribution -1/0/1/2/3	Parameters h^2_{SE}	c^2_{SE}	σ^2_p
Flight time: FT (s)	8854	1.00 (0.54)	na	$0.07_{0.02}$	ns	0.324
Anterior score: ANT (0-3)	3238	2.82 (1.59)	na/8/38/39/15	$0.14_{0.04}$	$0.07_{0.03}$	2.34
Posterior score: POS (0-3)	3237	2.05 (1.34)	na/16/52/28/4	$0.12_{0.04}$	$0.11_{0.03}$	1.62
Aggressive type: AGRO (0/1)	3238	0.54 (0.50)	na/46/54/na/na	$0.12_{0.03}$	ns	0.235
Av. daily gain: ADG (g/day)	30926	575 (79.4)	na	$0.19_{0.02}$	$0.07_{0.01}$	4935
Gilt removal: REM0 (0/1)	3575	0.25 (0.43)	na/75/25/na/na	$0.10_{0.03}$	$0.06_{0.03}$	0.180
PF lesion score: PFBLES (0-3)	1929	0.92 (0.74)	na/29/54/14/3	$0.10_{0.04}$	ns	0.516
PF locomotion: PFLOCO (0-3)	1945	0.34 (0.58)	na/72/23/5/0	$0.05_{0.04}$	$0.08_{0.04}$	0.333
PF condition: PFCS (-1/0/1)	1950	0.01 (0.38)	7/85/8/na/na	$0.09_{0.04}$	ns	0.139
Total born: TB (pigs/litter)	5097	11.8 (2.94)	na	$0.11_{0.02}$	ns	8.38
Born alive: NBA (pigs/litter)	5097	11.1 (2.86)	na	$0.10_{0.02}$	ns	7.95
Av. birth weight: ABWT (kg/pig)	2154	1.38 (0.22)	na	$0.36_{0.05}$	ns	0.042

Table 2. Genetic (1st row) and phenotypic (2nd row) correlations (SE in subscript) between behavioural traits (FT: flight time; ANT: anterior scores; POS: posterior score; AGRO: aggressive phenotype; PFBLES: PF lesion score) and performance outcomes

Traits	FT	ANT	POS	AGRO	PFBLES
Av. daily gain	$0.24_{0.10}$	$0.16_{0.12}$	$0.22_{0.12}$	$0.11_{0.14}$	$0.23_{0.15}$
	$0.07_{0.01}$	$-0.01_{0.02}$	$-0.01_{0.02}$	$-0.01_{0.02}$	$-0.00_{0.02}$
Gilt removal	$0.18_{0.17}$	$0.21_{0.21}$	$0.19_{0.21}$	$0.26_{0.18}$	$-0.15_{0.25}$
	$0.02_{0.02}$	$0.03_{0.02}$	$0.01_{0.02}$	$0.04_{0.02}$	$0.01_{0.03}$
PF locomotion	$0.42_{0.27}$	$-0.02_{0.32}$	$-0.20_{0.33}$	$0.41_{0.28}$	$0.92_{0.25}$
	$0.04_{0.03}$	$-0.00_{0.02}$	$-0.01_{0.02}$	$-0.01_{0.03}$	$0.10_{0.02}$
PF condition	$0.11_{0.21}$	$0.07_{0.23}$	$0.14_{0.23}$	$-0.25_{0.26}$	$-0.09_{0.27}$
	$-0.00_{0.03}$	$-0.02_{0.02}$	$-0.00_{0.02}$	$-0.05_{0.03}$	$-0.06_{0.02}$
Total born	$-0.14_{0.15}$	$0.15_{0.17}$	$0.10_{0.17}$	$0.10_{0.17}$	$0.01_{0.22}$
	$0.01_{0.02}$	$0.01_{0.02}$	$-0.02_{0.02}$	$0.00_{0.02}$	$-0.00_{0.02}$
Born alive	$-0.11_{0.16}$	$0.11_{0.18}$	$0.04_{0.18}$	$0.04_{0.19}$	$-0.05_{0.22}$
	$0.02_{0.02}$	$0.01_{0.02}$	$-0.02_{0.02}$	$0.00_{0.02}$	$-0.00_{0.02}$
Av. birth weight	$0.09_{0.13}$	$-0.12_{0.15}$	$-0.17_{0.16}$	$-0.01_{0.16}$	$-0.30_{0.20}$
	$0.02_{0.03}$	$-0.04_{0.03}$	$-0.05_{0.03}$	$-0.04_{0.03}$	$0.04_{0.04}$

The most significant phenotypic associations were between AGRO and REM0 and PBFLES with PBLOCO. These particular combinations represent traits measured close together in time. Low positive phenotypic correlations between these traits indicated that fighting of gilts post-selection increased undesirable (forced) removals. This association is not linear, however, because the highest scoring gilts are more likely to be removed (Bunter, 2015). Since correlations represent linear associations, non-linear associations can lower estimates. Similarly, sows which engaged in fighting pre-farrowing showed evidence of compromised locomotion and elevated rates of lameness pre-farrowing (Lumby *et al.* 2015). Genetic correlations between these trait combinations mirrored the direction of phenotypic correlations, but standard errors were large.

Behavioural traits were generally not significantly correlated with reproductive outcomes for first parity sows. The exceptions were low negative phenotypic correlations between POS and ABWT, and between AGRO and ABWT or PFCS, which suggest that gilts engaged in fighting post-mixing were more likely to have poorer condition and lighter piglets at their first farrowing.

The overall lack of significant genetic correlations between the behavioural and other traits resulted from the relatively low magnitude of most estimates combined with large standard errors. Negligible phenotypic correlations also reflect accompanying near zero residual correlations. In combination, these results imply that measures of behavioural traits on gilts will not provide much information on later behaviour, or indirectly on reproductive outcomes of group housed sows. This included skin lesion traits, which directly reflect detrimental interactions between animals. Studies which have reported positive correlations between skin lesion counts repeatedly recorded younger animals over a short time frame without remixing in the interim (eg. Desire *et al.* 2015; Turner *et al.* 2009). Results from this study support the conclusions of Turner *et al.* (2009) that selection against high lesion counts would reduce aggression at mixing. However, while rg tended to be positive between repeated scores, results from this study throws some doubt on interpreting longer term outcomes from selection based on earlier lesion scores. Our results do not support strong genetic associations between the behaviour of finisher gilts and their later scores pre-farrowing or their reproductive performance outcomes, but do support some more immediate consequences from fighting (eg removals). Therefore, management strategies to reduce fighting and understanding genetic contributions to long term behaviour remain important for improving welfare of group housed sows. Overall, genetic correlations between behavioural traits and reproductive outcomes were generally favourable, implying that selection on maternal attributes would be expected to have neutral to favourable effects on the fighting behaviour of gilts.

ACKNOWLEDGEMENTS

Funded under Pork CRC project 1C-107. The author would like to thank contributions by Rivalea Australia Pty Ltd, Cherie Collins, Rebecca Athorn, Kate Boardman and Jemma Lumby.

REFERENCES

Anil L., Anil S.S., Deen J., Baidoo S.K., and Walker R.D. (2006) *Canadian J. Vet. Res.* **70**: 128.

Bunter K.L., Lewis C.R.G., Hermesch S., Smits R., and Luxford B.G. (2010) *Proceedings 9th WCGALP*, Leipzig, Germany

Bunter K.L. (2015) *Anim. Prod. Sci.* **55**: 1509.

Bunter K.L. and Boardman K.M. (2015) *Anim. Prod. Sci.* **55**: 1493.

Crump (2004) *AGBU Pig Genetics Workshop. p107.*

Desire S., Turner S.P., D'Eath R.B., Doeschl-Wilson A.B., Lewis C.R.G., and Roehe R. (2015) *Appl. Anim. Behav. Sci.* **166**: 52.

Desire S., Turner S.P., D'Eath R.B., Doeschl-Wilson A.B., Lewis C.R.G., and Roehe R. (2015) *J. Anim. Sci.* **93**: 3303.

Gilmour A.R., Gogel B.J., Cullis B.R. and Thompson R. (2009) '*ASREML user Guide Release 3.0*' VSN International Ltd, Hemel Hempstead, UK.

Lumby J.C., Bunter K.L., and Wynn P.C. (2015) *Anim. Prod. Sci.* **55**: 1510.

Spoolder H.A.M., Geudeke M.J., van der Peet-Schwering C.M.C., Soede N.M. (2009) *Livestock Sci.* **125**: 1.

Turner S.P., White I.M.S., Brotherstone S., Farnworth M.J., Knap P.W., Penny P., Mendi M., and Lawrence A.B. (2006) *Animal Sci.* **82**: 615.

Turner S.P., Roehe R., D'Eath R.B., Ison S.H., Farish M., Jack M.C., Lundeheim N., Rydhmer L., and Lawrence A.B. (2009) *J. Anim. Sci.* **87**: 3076.

Proc. Assoc. Advmt. Anim. Breed. Genet. 22:233-236

GENOTYPE BY PARITY INTERACTIONS WERE NOT FOUND FOR GROWTH IN AUSTRALIAN PIGS

S. Hermesch[1] and B. G. Luxford[2]

[1]Animal Genetics and Breeding Unit, a joint venture of NSW Department of Primary Industries and University of New England, University of New England, Armidale, NSW 2351, Australia
[2]Rivalea Australia, Corowa NSW, 2646

SUMMARY

Growth is known to be lower for gilt progeny in comparison to progeny from multiparous sows. Information about the effects of dam parity on other performance traits is often limited. The aim of this study was to quantify the effects of dam parity on performance traits and to evaluate genotype by parity interactions for growth rate of pigs. Dam parity had strongest effects on growth. Estimates of growth were 14 to 22 g/day lower in gilt progeny in comparison to progeny from older sows. Feed intake and feed conversion ratio were -0.023 to -0.066 units lower for progeny from older sows in comparison to gilt progeny. Dam parity had no biological importance for backfat and muscle depth. Growth was defined as a different trait for progeny from the first to the fifth parity of dams for analyses of genotype by parity interactions. Heritabilities varied from 0.13 to 0.20 for the three growth traits. Maternal genetic effects were low and slightly higher for progeny from older sows (0.047). Genotype by parity interactions for growth of pigs were not found based on high estimates of genetic correlations between different growth traits (range:0.83 to 0.98) and current selection practices that define growth as the same trait for progeny from different dam parities can be continued based on the results of this study.

INTRODUCTION

Progeny of gilts grow more slowly than progeny from older sows (Standal, 1973). This fact has long been known in pig industries although scientific studies quantifying this effect are sparse. Recently, Hermesch and Li (2013) showed that the reduction in growth rate of pigs from gilt litters varied between herds. Growth rate of gilt progeny was 6.7 to 21.1 g/d lower than progeny growth from third-parity litters. The exact reasons for this variation in the reduction of growth rate of gilt progeny between herds are unknown.

The gap in performance of gilt progeny relative to progeny of older sows may widen if the implications of continued selection are not fully understood in regard to optimal gilt management on farms. Selective breeding continues to focus on improving efficient lean meat growth and reproductive performance of sows. This selection emphasis affects characteristics of sows and genetic improvement of growth and backfat will lead to larger and leaner sows. For example, associations between estimated breeding values and sow characteristics estimated by Hermesch *et al.* (2010) indicate that sows have the genetic potential to be 30 kg heavier every 10 years as a result of genetic improvement of progeny growth of about 100 g/day over 10 years. Downward selection of backfat of 5 mm in progeny implied that sows had the genetic potential to be 7.5 mm leaner. These genetic associations, however, are often not fully expressed due to management of gilts and sows. For example, feed intake may be restricted in gilts and sows reducing the weight of gilts and sows relative to the genetic potential for weight gain in breeding females. Gilts are now considerably heavier and leaner at mating and first farrowing due to selection, and may be less able to support the lean meat growth potential of their progeny. It was the aim of this study to investigate the effects of dam parity on performance of progeny and to estimate genotype by parity interactions for growth.

MATERIALS AND METHODS

Data. Performance of progeny were recorded from 2000 to 2015. These data were combined with information about reproductive performance of sows. Editing procedures focused on good cross-classification of effects and completeness of litter and sow characteristics for progeny performance. Only progeny from the first 6 parities of sows were considered. These conditions were fulfilled for 262,193 pigs in total which were recorded in two locations and included male and female pigs from 6 genetic lines. Pigs were recorded at an average age of 151.90 (± 8.94) days and an average body weight of 91.21 (± 13.05) kg to obtain information about average daily gain (ADG), fat depth (FD) and muscle depth (MD).

A proportion of pigs were tested for daily feed intake (DFI) using electronic feeders. Feed intake records collected from 2003 to 2010 were included in the analyses. Entire-male pigs entered electronic feeders at an average age of 120.60 (± 5.47) days and a body weight of 70.92 (± 8.01) kg. The test period was 35.75 (± 2.54) days long and pigs were fed *ad libitum*. Additional traits available for these pigs were average daily gain prior to test (ADG1) and growth rate during test (ADG2) as well as DFI and feed conversion ratio (FCR). Records exceeding 3 standard deviations from the mean were deleted for all traits.

Analysis. The GLM (SAS 2014) procedure was used to derive the fixed effect model for each trait and to estimate least squares means for the effect of dam parity on performance traits. Dam parity, which had 5 levels because parity 5 and 6 were combined into 1 level, was added as an additional fixed effect to the base model for each trait. The base model included line and contemporary group based on week of birth at each location for all traits. Sex was fitted for ADG, FD and MD only because other traits were only available for entire males. Backfat and MD were adjusted for the weight at recording which was fitted as a linear and quadratic covariate. Weight of pigs at start of test to record DFI was only significant for DFI as a linear and quadratic covariate.

Genotype by parity interactions were evaluated for growth which was defined as a separate trait for progeny from each parity (ADG-P1 to ADG-P5). Variance and covariance components were estimated with ASReml (Gilmour *et al.* 2009) in univariate and bivariate analyses fitting an animal model. Additional random effects fitted in univariate analyses were maternal genetic and permanent environmental effects of dams. For bivariate analyses, only additive genetic and permanent environmental effects of dams were fitted because estimates of maternal genetic effects were low and partially confounded with permanent environmental effects of sows. Further, the residual covariance was fixed at zero because growth traits were recorded on different animals and it was not possible to estimate residual and subsequently phenotypic correlations.

RESULTS AND DISCUSSION

Effects of dam parity. Estimates of growth for gilt progeny were 22 g/day lower in comparison to progeny from second and third parity sows (Table 1). This difference in growth was reduced to 17 and 14 g/day between progeny from gilts versus progeny of fourth and fifth-parity sows. Further analyses showed that growth of gilt progeny in comparison to progeny from the second to third parity was 16 to 18 g/day lower in 2004 and 2009 for ADG, while the difference increased to 26 to 32 g/day in 2013 to 2015 (for details see Hermesch, 2015). In comparison, growth of gilt progeny was 5 to 20 g/day lower than growth of progeny from third-parity sows in the 9 herds investigated by Hermesch and Li (2013).

Early growth is expected to be more strongly affected by characteristics of the dam. Growth prior to test (ADG1) was affected by dam parity, which conversely had no significant effect on growth during the test period (ADG2) (P values, Table 1). Dam parity affected DFI and FCR significantly and gilt progeny had inferior performances in these traits. Progeny from multiparous sows ate -0.025 to -0.066 kg less feed per day than gilt progeny and had a better FCR (difference of -0.015 to -0.058

kg/kg). However, these differences in performance of gilt progeny to progeny from older sows were not observed in a second independent herd analysed by Hermesch (2015). The effects of dam parity on performance traits should be investigated for each population because estimates of dam-parity effects on growth of progeny were variable between herds and over time.

Table 1. Number of observations (*n*), means and standard deviations (SD) as well as predicted differences between first and subsequent parities for performance traits observed in progeny.

Trait	*n*	Mean	SD	Parity 2	Parity 3	Parity 4	Parity 5/6	P value
ADG (g/d)	261,919	600	77	22	22	17	14	< 0.0001
ADG1 (g/d)	7,679	588	62	18	16	13	13	< 0.0001
ADG2 (g/d)	7,679	860	197	1	7	-5	-7	0.41
DFI (kg)	7,537	2.44	0.47	-0.025	-0.044	-0.045	-0.066	0.0003
FCR	7,537	2.06	0.43	-0.013	-0.058	-0.023	-0.050	0.004
BF (mm)	215,066	10.2	2.29	-0.14	-0.13	-0.12	-0.13	< 0.0001
MD (mm)	214,172	43.4	5.87	-0.09	0.24	1.08	0.71	< 0.0001

Abbreviations: ADG: average daily gain, ADG1: ADG until 70 kg prior to feed-intake test, ADG2: ADG during feed-intake test, DFI: daily feed intake, FCR: feed conversion ratio, BF: backfat, MD: muscle depth.

Genetic parameters. Heritability estimates for growth were 0.16 for progeny from 3 parities in comparison to estimates of 0.13 (± 0.01) and 0.20 (± 0.02) for ADG-P3 and ADG-P5 (Table 2). Maternal genetic effects were consistent for ADG-P1 to ADG-P4 and slightly higher for ADG-P5 (0.047 ± 0.013). Common litter effects were higher for ADG-P1 to ADG-P3 (0.09 and 0.10) in comparison to lower estimates of 0.07 and 0.04 for ADG-P4 and ADG-P5. Heritability estimates and common litter effect estimates obtained in this study for different growth traits were within the range of estimates presented by Hermesch and Jones (2012) for overall growth based on subsets of the data used in the current study. Inclusion of maternal genetic effects as an additional random effect decreased heritability estimates by 0.01 for all traits. Meanwhile, the permanent environmental effect of the dam was reduced by the magnitude of estimate of the maternal genetic effect for each growth trait, demonstrating high sampling correlations between these 2 random effects for these growth traits. These changes in variances between models indicate that data structure were not sufficient to disentangle these two maternal effects when traits were defined separately for each dam parity. However, maternal genetic effects are generally low for growth in pigs (e.g. Johnson *et al.* 2002) and estimates of variance components for the 5 growth traits followed expectations.

Table 2. Number of observations (*n*), heritabilities (h^2, with standard errors (se)), maternal genetic (m^2) and permanent environmental effect of dam (c^2) as well as phenotypic variance (Vp) for average daily gain (ADG) of progeny from the first (ADG-P1) to the fifth (ADG-P5) parity of dams.

Trait	*n*	h^2 (se)	m^2 (se)	c^2(se)	Vp
ADG-P1	100,662	0.16 (0.01)	0.024 (0.005)	0.10 (0.005)	4811
ADG-P2	72,298	0.16 (0.01)	0.025 (0.006)	0.09 (0.006)	5022
ADG-P3	44,430	0.13 (0.01)	0.021 (0.008)	0.10 (0.009)	4898
ADG-P4	25,732	0.16 (0.02)	0.029 (0.014)	0.07 (0.014)	4885
ADG-P5	24,320	0.20 (0.02)	0.047 (0.014)	0.04 (0.013)	4826

Estimates of genetic correlations were high, ranging from 0.83 to 0.98 between traits (Table 3). Genetic correlations tended to decrease as the difference in parities increased for definitions of growth traits. Overall, these estimates indicate that a genotype by parity interaction can be ignored in pig breeding programs as is currently the case. No estimates of genotype by parity interactions were found in the literature. A comparable investigation may be the analyses of genotype by sex interactions because sex of pigs is another systematic effect for performance traits. This interaction was investigate by Crump *et al.* (1997), who found no significant genotype by sex interactions. The magnitude of genotype by environment interactions depends on the difference between environments that genotypes experience. Differences in environments provided by dams to progeny in different parities were not large enough for the detection of genotype by parity interactions.

Table 3. Genetic correlations (above diagonal, with standard errors (se)) and correlations due to permanent environmental effect of dam (below diagonal) for average daily gain (ADG) of progeny from the first (ADG-P1) to the fifth (ADG-P5) parity of dams.

Trait	ADG-P1	ADG-P2	ADG-P3	ADG-P4	ADG-P5
ADG-P1		0.94 (0.02)	0.96 (0.03)	0.93 (0.04)	0.83 (0.04)
ADG-P2	0.25 (0.03)		0.98(0.03)	0.97 (0.04)	0.90 (0.04)
ADG-P3	0.26 (0.04)	0.25 (0.04)		n.e.[1]	0.93 (0.05)
ADG-P4	0.24 (0.06)	0.23 (0.06)	n.e.[1]		0.89 (0.05)
ADG-P5	0.26 (0.07)	0.34 (0.06)	0.40 (0.06)	0.31 (0.08)	

[1] Correlations could not be estimated

CONCLUSIONS

Dam parity had the greatest effect on growth. Gilt progeny grew more slowly than progeny from older sows and differences were larger in more recent years. Further, a higher DFI and higher FCR was observed for gilt progeny in the current population. Estimates of dam-parity effects on performance of progeny were variable and this effect should be evaluated on farms to ensure management of gilts and sows is optimal for each genotype. No genotype by parity interactions were found for growth and current selection practices for growth can be continued.

ACKNOWLEDGEMENTS

This study was funded by Australian Pork Limited under project APL2014/458. Data provided by the industry partner is gratefully acknowledged.

REFERENCES

Crump R.E., Haley C.S., Thompson R. and Mercer J. (1997) *J. Anim. Sci.* **65**: 275.

Gilmour A.R., Gogel B.J., Cullis B.R. and Thompson R. (2009) 'ASReml User Guide Release 3.0.' VSN International Ltd, Hemel Hempstead, HP1 1ES, UK.

Hermesch S. (2015) Final Report for Australian Pork Limited, Project 2014/458.

Hermesch S., Jones R.M., Bunter K.L. and Gilbert H. (2010) 9th WCGALP, paper 292

Hermesch S. and Jones R.M. (2012) *Animal* **6**: 1904.

Hermesch S. and Li L. (2013) In 'Manipulating pig production XIV. Proceedings of the 14th Biennial Conference of the Australiasian Pig Science Association', p. 53, editors J.R. Pluske and J.M. Pluske.

Johnson Z.B., Chewning J.J. and Nugent R.A. 3rd (2002) *J. Anim. Sci.* **80**: 1470.

SAS (2014). Base SAS® '9.4 Procedures Guide: Statistical Procedures, Third Edition'. SAS Institute, Cary, NC, USA.

Standal N. (1973) *Acta Agr. Scand.* **23**: 225.

Proc. Assoc. Advmt. Anim. Breed. Genet. 22:237-240

EXTENSION OF ENVIRONMENTAL DESCRIPTORS USED TO ANALYSE SIRE BY ENVIRONMENT INTERACTION FOR GROWTH OF PIGS

S.Z.Y. Guy[1], J. Harper[2], L. Li[3], P.C. Thomson[1,3] and S. Hermesch[3]

[1]School of Life and Environmental Sciences, University of Sydney, NSW, 2006
[2]Rivalea Australia, Corowa NSW, 2646
[3]Animal Genetics and Breeding Unit, a joint venture of NSW Department of Primary Industries and University of New England, University of New England, Armidale, NSW 2351, Australia

SUMMARY

Environmental descriptors based on contemporary group estimates of average daily gain (ADG), backfat (BF), daily feed intake (DFI) and muscle depth (MD) were used to evaluate sire by environment interactions (S×E) for growth rate of pigs. Further, these descriptors were combined using principal component analysis and the first principal component (PC1) was used as an overall environmental descriptor. Use of the environmental descriptors based on MD, BF and DFI did not detect any S×E for growth. However, significant S×E was detected using the environmental descriptor based on ADG and also the overall descriptor based on PC1, where the S×E variance components accounted for 2.1% and 1.8% of the phenotypic variance. While an environmental descriptor that encompasses more traits is expected to capture more environmental variation, use of the environmental descriptor based on ADG alone may be adequate to describe phenotypic variability attributed to S×E for growth.

INTRODUCTION

Environments can be quantified by contemporary group (CG) estimates of performance traits, adjusted for systematic and genetic effects. Environmental descriptors based on CG estimates of multiple production traits have been used in dairy cattle, which have been applied to the evaluation of genotype by environment interaction for fertility traits (Strandberg *et al.*, 2009). In pigs, estimates of CG effects based on number born alive and numbers weaned have been used to quantify disease environments affecting sow reproductive performance (Herrero-Medrano *et al.*, 2015). Meanwhile, use of CG estimates to describe growth rate of the pig has been limited to lifetime average daily gain (ADG) and backfat (BF) (Guy *et al.*, 2015; Li and Hermesch, 2016).

The objective of this study was to extend the traits used to derive environmental descriptors, to also include daily feed intake (DFI) and muscle depth (MD). These environmental descriptors will be used individually, as well as combined into an overall descriptor, to describe variation in the growth performance of sire progeny across different environments, i.e. sire by environment interaction (S×E) for growth. It is hypothesised that use of more traits will capture additional variation in the environment, and hence the ability to detect S×E will improve.

MATERIALS AND METHODS

Data. Pedigree and production records were available from a commercial piggery, located in the Riverina region of NSW, Australia. Only pigs that had records for all traits of interest were included in the study. Feed intake was only recorded from between 2004 and 2010 for entire males from 2 lines. These boars were housed in the normal production environment until 112 days of age on average, then moved to pens equipped with electronic feeders. After an adjustment period of 5-7 days, boars were weighed and classified 'on test'. Only boars with a test age of between 109 and 133 days were included in analysis. The average weight at start of test was 71.3 ± 7.6 kg (mean ± SD). Boars were on test for an average of 36 days. For analysis, DFI was defined as the average amount of feed consumed per day during the testing period (kg/day). Measurements for BF (average of measurements at last rib and base of tail, mm) and MD (between the 3rd and 4th last ribs, mm) were taken at the end of the test period with real-time ultrasound. For inclusion in the analysis, all production traits were restricted to within 4 standard deviations of the raw mean.

Boars that were tested in the same week and year were assumed to be under the same managerial and environmental conditions, and were therefore allocated to the same CG. The minimum size of the CGs was set at 15 pigs, giving a total of 255 CGs. The CG sizes ranged from 16 to 107 pigs, with an average of 30 pigs. There were on average 11 sires represented in each CG.

After data cleaning, there were 7,746 individual records, representing 448 sires and 2,565 dams from 4,245 litters. The average weight at the end of the test period was 102.5 ± 10.7 kg, at an average age of 157 ± 7 days.

Analysis. In the first step of analysis, environmental descriptors were derived from animal models for the 4 production traits using ASReml (Gilmour *et al*., 2009):

$$ADG_{ijklm} = \mu + Line_i + Season_j + Animal_k + Litter_l + CG_m + \varepsilon_{ijklm}$$

$$BF_{ikmn} = \mu + Line_i + \beta_1 EndWeight_n + Animal_k + CG_m + \varepsilon_{ikmn}$$

$$DFI_{ijklmp} = \mu + Line_i + Season_j + \beta_1 StartWeight_p + Animal_k + Litter_l + CG_m + \varepsilon_{ijklmp}$$

$$MD_{ikmn} = \mu + Line_i + \beta_1 EndWeight_n + Animal_k + CG_m + \varepsilon_{ikmn}$$

where μ is the overall mean for the trait of interest. All models contained the fixed effect of the ith Line (2 levels), and random additive genetic effect of the kth animal, random effect of the mth test week-year CG and random residual effect ε, which was unique to each trait. $Litter_l$ as a random effect was significant only for ADG and DFI. Additional fixed effects included $Season_j$ (4 levels) for ADG and DFI, covariate of weight at end of test period ($EndWeight_n$) for BF and MD, and covariate of weight at start of test period ($StartWeight_p$) for DFI.

Estimates of CG effects were extracted from each of these models and combined through principal component analysis using the prcomp() function in R (R Core Team, 2016). Principal component analysis combines variables by producing weighted linear combinations that capture maximum variation. It is therefore dependent on scale, so CG estimates were scaled to a variance of 1. The first principal component (PC1) was used as the overall descriptor. Environments were categorised by partitioning each environmental descriptor into quintiles. Pigs were assigned an environment according to the CG they belonged to, with each pig having an environment based on the CG estimates of the 4 traits, as well as the overall descriptor.

In the second step of analysis, S×E for growth was evaluated using the environments characterised from the 5 environmental descriptors derived in the first step. A separate sire model was used for each descriptor:

$$y_{ijklmnp} = \mu + Line_i + Season_k + Sire_l + S{\times}E_{lm} + Litter_n + CG_p + \varepsilon_{ijklmnp}$$

where y_{ijklmn} is the ADG of the jth progeny of sire l in the mth environment (E). The amount of S×E for growth was quantified by the S×E variance component.

RESULTS AND DISCUSSION

Boars had a mean ADG of 653.8 ± 65.0 g/day, a mean BF measurement of 8.8 ± 1.8 mm, a mean DFI of 2.10 ± 0.37 kg/day and a mean MD of 45.5 ± 5.8 mm.

Estimates of CG effects ranged from -53.5 to 56.6 g/day for ADG, -1.66 to 2.18 mm for BF, -0.46 to 0.49 kg/day for DFI, and -5.04 to 10.49 mm for MD. Pearson's correlations between these CG estimates were all positive and less than 0.15, except for between ADG and DFI (0.39). This suggests these 4 traits capture different aspects of the environment.

Genetic parameter estimates for each trait are presented in Table 1 to assess the fitted models used to derive environmental descriptors. Heritability estimates for the 4 traits align with previous studies (Hermesch, 2008), although they were slightly lower due to the inclusion of CG variance component in the calculation of the phenotypic variance estimate. The estimated common litter effect was lower than expected, which may be due to a low average of 1.8 boars/litter tested.

Table 1. Genetic parameter estimates of average daily gain (ADG) (g/day), backfat (BF) (mm), daily feed intake (DFI) (kg/day) and muscle depth (MD) (mm), using models from which contemporary group estimates were used as environmental descriptors

Trait	$\hat{\sigma}^2_P$ ± SE	$\hat{\sigma}^2_E$ ± SE	$\hat{h}^2$ ± SE	$\hat{c}^2$ ± SE	$\hat{i}^2$ ± SE
ADG	4063.2 ± 87.6	2328.3 ± 82.7	0.22 ± 0.03	0.06 ± 0.01	0.15 ± 0.01
BF	2.34 ± 0.06	1.16 ± 0.05	0.29 ± 0.02	-	0.21 ± 0.02
DFI	0.12 ± 0.003	0.06 ± 0.002	0.22 ± 0.03	0.04 ± 0.01	0.23 ± 0.02
MD	28.8 ± 0.95	13.7 ± 0.50	0.21 ± 0.02	-	0.31 ± 0.02

Abbreviations of estimates: $\hat{\sigma}^2_P$ = phenotypic variance, $\hat{\sigma}^2_E$ = residual variance, $\hat{h}^2$ = heritability, $\hat{c}^2$ = proportion of phenotypic variance attributed to common litter effect, $\hat{i}^2$ = proportion of phenotypic variance attributed to contemporary group effect

The CG estimates based on the 4 traits were combined through principal component analysis. The first principal component (PC1) explained 37.5% of the variation, and the second principal component (PC2) explained 26.1%. For PC1, the greatest emphasis was placed on ADG and DFI, with loadings of 0.60 and 0.64 respectively. The PC1 loading for BF was 0.35 and 0.31 for MD. Meanwhile, PC2 placed the greatest emphasis on the carcass traits, with loadings of 0.59 for BF, 0.65 for MD, -0.40 for ADG and -0.26 for DFI. These loadings suggest associations between the descriptors based on ADG and DFI, and also between the descriptors based on BF and MD.

The environments characterised by the 5 descriptors were used in sire interaction models to evaluate S×E for growth rate, and results are presented in Table 2. Estimates of additive genetic variance using descriptors based on BF and MD were larger but not appreciably different, considering their standard errors. Other variance components, except for the S×E term, remained fairly consistent across models using different environmental descriptors.

Table 2. Genetic parameter estimates for the analysis of sire by environment interaction (S×E) for growth rate, using environmental descriptors based of average daily gain (ADG), backfat (BF), daily feed intake (DFI), muscle depth (MD), and all 4 traits combined using the first principal component (PC1)

Descriptor	$\hat{\sigma}^2_A$ ± SE	$\hat{\sigma}^2_{CG}$ ± SE	$\hat{\sigma}^2_{S\times E}$ ± SE	$\hat{\sigma}^2_P$ ± SE	$\hat{h}^2$ ± SE	$\hat{c}^2$ ± SE
ADG	1022.1 ± 173.9	532.1 ± 65.6	**87.0 ± 37.9**	4060.3 ± 87.5	0.25 ± 0.04	0.10 ± 0.01
BF	1104.9 ± 171.5	586.8 ± 68.3	12.6 ± 28.9	4099.0 ± 90.1	0.27 ± 0.04	0.11 ± 0.01
DFI	1067.5 ± 171.6	584.1 ± 68.3	32.4 ± 30.9	4095.1 ± 89.9	0.26 ± 0.04	0.11 ± 0.01
MD	1123.4 ± 171.5	587.3 ± 68.3	0.28 ± 30.2	4099.2 ± 90.1	0.27 ± 0.04	0.11 ± 0.01
PC1	1009.5 ± 173.9	561.0 ± 67.0	**74.1 ± 36.5**	4079.8 ± 88.8	0.25 ± 0.04	0.10 ± 0.01

Abbreviations of estimates: $\hat{\sigma}^2_A$ = additive genetic variance (calculated as 4 times the sire variance component estimate), $\hat{\sigma}^2_{S\times E}$ = sire by environment interaction variance component. Other abbreviations as explained in Table 1.
Note: Significant S×E in bold

There was no or minimal S×E for growth detected using the environmental descriptor based on MD, BF, and DFI, with the interaction terms accounting for 0.01%, 0.3% and 0.8% of the phenotypic variance, respectively. However, there was significant S×E for growth when using the environmental descriptor based on PC1, which accounted for 1.8% of the phenotypic variance.

Although not substantially different, the environmental descriptor based on ADG accounted for even more phenotypic variance at 2.1%. Therefore, the ability to detect S×E for growth rate was greatest using either the environmental descriptor based on ADG or the overall descriptor.

The trait used to quantify the environment is usually based on the same trait that is being modelled. For example, numbers born alive was used to quantify disease environments, in which sow reproductive performance was assessed using numbers born alive (Herrero-Medrano *et al.*, 2015). This was also the case for the environmental descriptor based on ADG used in this current study. While ADG appears to be the driver of PC1, this overall descriptor may appear to be a more objective measure of the environment as it does not solely depend on the trait being modelled. However, use of PC1 does not appear to capture more variation in the environment to increase the ability to detect S×E, and the descriptor based on ADG alone appears sufficient.

Estimates of heritability for ADG were lower using the animal model in the first step of analysis compared to the sire model estimates in the second step of analysis, although they were not appreciably different when taking standard errors into account. Other variance components were stable across models except for estimates of litter effect and residual variances, which were both larger in the sire model. Higher estimates of litter effect in sire models may be attributed to the dam genetic effect being absorbed by the litter component.

The environmental descriptors were partitioned into quintiles to allow for ~ 1,500 pigs classified in each environment. This resulted in 10-17% of sires with progeny across all 5 environments, and 22-25% with progeny in only 1 environment. The ability to detect S×E is greatest when sires are represented across all environments, which can be achieved if the descriptor is partitioned into fewer environments. However, this needs to be balanced out with the need for sufficient differences between environments in order to detect S×E for growth.

CONCLUSIONS

This paper considers CG estimates of alternative production traits as a practical way to quantify the pig environment. The sire interaction model provides a simple method to evaluate the presence of S×E for selection decisions, where estimated breeding values for sires are available across all environments, as well as for specific environments. While a descriptor that encompasses alternative traits may be a more objective measure of the environment, use of the environmental descriptor based on ADG alone may be sufficient to capture most of the phenotypic variability attributed to S×E.

ACKNOWLEDGEMENTS

This study is funded by a Pork CRC postgraduate scholarship and Pork CRC Project 2B-105. Special thanks to Rivalea for making the data available.

REFERENCES

Gilmour A.R., Gogel B.J., Cullis B.R. and Thompson R. (2009) 'ASReml User Guide Release 3.0' VSN International Ltd, Hemel Hempstead, HP1 1ES, UK.

Guy S.Z.Y., Hermesch S. and Thomson P.C. (2015) *Proc. Assoc. Advmt. Anim. Genet.* **21**: 457.

Hermesch S. (2008) *Anim.* **2**: 1178.

Herrero-Medrano J.M., Mathur P.K., ten Napel J., Rashinidi H., Alexandri P. *et al.* (2015) *J. Anim. Sci.* **93**: 1494.

Li L. and Hermesch S. (2016) *J. Anim. Breed. Genet.* **12**: 336.

R Core Team (2016). *R Foundation for Statistical Computing*, Vienna, Austria.

Strandberg E., Brotherstone S., Wall E. and Coffrey M.P. (2009) *J. Dairy Sci.* **92**: 3437.

Proc. Assoc. Advmt. Anim. Breed. Genet. 22:241-244

ESTIMATES OF EFFECTIVE POPULATION SIZE AND INBREEDING LEVEL FOR THREE AUSTRALIAN PIG BREEDS

O. D'Augustin[1,2], V. Börner[1] and S. Hermesch[1]

[1] Animal Genetics and Breeding Unit (AGBU), a joint venture of NSW Department of Primary Industries and University of New England, Armidale, NSW 2351, Australia
[2] University of Rennes 1, Rennes, France

SUMMARY

Selective breeding may result in higher inbreeding levels which can lead to inbreeding depression and limit future genetic gain. This study quantified inbreeding levels and evaluated effective population sizes for Large White (LW), Landrace (LR) and Duroc (DU) populations in Australia. Pedigree data from 1994 to 2015 representing about 12 generations on average were explored with the software package PopRep by Groeneveld *et al.* (2009) which provides multiple population parameters. Pedigree completeness was highest in 2004 and 2005 when it reached about 95% and 80% in the third and sixth generation. Average inbreeding levels were highest for these years with averages of 0.031, 0.034 and 0.050 in LW, LR and DU, respectively. Two herds joined the across-herd genetic evaluations at that time and pedigree completeness varied from 80 to 90% and from 60 to 70% in the third and sixth generation in subsequent years leading to lower estimates of inbreeding levels. Estimates of effective population size varied from 64 to 98 in LW, from 52 to 108 in LR and from 42 to 61 in DU over time. These estimates of effective population size are imprecise and an underestimate of true effective population sizes given the limited time period considered and the extent of missing pedigree.

INTRODUCTION

Selective breeding may result in higher inbreeding levels leading to inbreeding depression, which is a decrease of the population fitness, because of the accumulation of deleterious recessive alleles (Falconer & Mackay 1996; Ouborg *et al.* 2010). Higher inbreeding levels may also limit future genetic gain because of a lack of genetic variation between individuals. It is therefore important to be able to quantify inbreeding levels in order to estimate population structure and avoid these problems. Moreover, if very few sires (compared to dams) are used for artificial insemination, the effective population size (*Ne*) is lowered, which increases inbreeding. According to Kimura and Crow (1963), "the effective population size is defined as the size of an idealized population that would have the same amount of inbreeding or of gene frequency drift as the population under consideration." The aim of this study was to estimate inbreeding levels and to evaluate effective population size of 3 Australian pig populations.

MATERIALS AND METHODS

Pedigree for the Large White (LW), Landrace (LR) and Duroc (DU) breeds were extracted from the National Pig Improvement Program database (http://npip.une.edu.au). Data were recorded from January 1994 to September 2015 for LW, and from January 1995 to September 2015 for LR and DU (Table 1). The pedigree of these 3 breeds were explored using the software package PopRep by Groeneveld *et al.* (2009).

Pedigree Completeness. The pedigree completeness statistically quantifies the percentage of missing animals over generations. It is important to quantify completeness of pedigree because estimates of inbreeding levels and effective population sizes are affected by this parameter. If there are too many missing animals, the results will be biased because the inbreeding coefficients will be underestimated. This will also affect estimates of effective population size as they depend on the rate of inbreeding per generation. Pedigree completeness was computed with the following formula:

$$I_{d_k} = \frac{1}{d} \sum_{i=1}^{d} a_i$$

where d is the number of generations considered, k represents the paternal or maternal line of the individual I, and a_i is the proportion of known ancestors in the generation i (MacCluer *et al.* 1983). Pedigree completeness ranges from 0 to 1 and increases for later generations as more pedigree information becomes available over time. For the first generation, pedigree completeness is often 100% because pedigree of parents is known.

Table 1: Data structure for the Large White (LW), Landrace (LR) and Duroc (DU) breeds

	Number of animals	Number of males	Number of females	Number of sires	Number of dams	Number of herds	Period of recording
LW	264,296	133,020	131,276	2,287	13,206	11	1994-2015
LR	147,160	74,727	72,433	1,351	6,995	8	1995-2015
DU	53,931	27,511	26,420	643	2,445	8	1995-2015

Effective population size and inbreeding level. PopRep uses 6 different methods to compute inbreeding levels within population, leading in turn to 6 estimates of effective population size (*Ne*). The standard method (Falconer and Mackay, 1996) to calculate Ne is $Ne = 1/(2\Delta F)$, where ΔF is the rate of inbreeding per generation. In order to decide which method is the best to use, PopRep computes 2 side conditions which have to be met for the method to be reliable. These 2 side conditions require positive estimates of Ne for at least 4 years and impose a limit on the variation of estimates of Ne over time. Based on these conditions, the method by Gutiérrez *et al.* (2009) was the only one to be reliable for the 3 breeds based on the 2 side conditions (data not shown). This method provided most consistent estimates of Ne over time because it considered the complete pedigree length.

RESULTS AND DISCUSSION

The amount of pedigree data available affects maximum levels of inbreeding and subsequently effective population size. The current study was based on data from 20 years. Given the average generation interval of 1.8 in LW and LR and 1.7 in DU, these data represent about 12 generations on average. A longer time period of 35 years was considered by Welsh et al. (2010) who reported results for 19 generations equivalent of an average generation interval of 1.8 years. In comparison, Krupa et al. (2015) used pedigree data over 25 years from 1988 to 2013 and the maximum number of generations traced varied from 20 to 25 between breeds. These differences in number of generations should be taken into account when comparing inbreeding levels.

Pedigree Completeness. Pedigree information was complete (100%) for the first generation in 12 (LW), 13 (LR) and 10 (DU) years of the 20 years from 1995 to 2014. In the first year when data were available (1994 for LW and 1995 for LR and DU), pedigree completeness started with 33% for the third generation and then increased linearly for about 6 years as fewer generations were censored (Figure 1). Similarly, pedigree completeness started with 17% for the sixth generation and increased linearly for the following 10 years of pedigree recording. Maximum pedigree completeness plateaued at about 95% in the third generation for DU and LR and decreased to a range of 80 to 90%

in subsequent years. In comparison, maximum pedigree completeness was about 80% in the sixth generation and decreased to a range of 60 to 70% afterwards. The trend in pedigree completeness over time was slightly different for LW in comparison to LR and DU. Pedigree completeness was initially lower and reached its highest value at a later point while pedigree completeness was slightly higher in later years. Animals with unknown pedigree were introduced in 2004 and 2006 when 2 new herds joined the scheme and some importation of unknown animals continued over time. These importations of unknown animals reduced pedigree completeness in the Australian populations. Specific time points for importation of unknown animals into Czech pig breeds were also visible in the trends for pedigree completeness shown by Krupa et al. (2015). These importations occurred in the early 1990s and until 2013, pedigree completeness converged to nearly 100% for all 6 generations in LW and LR populations. Czech DU and Pietrain had further importations over time and in 2013, pedigree completeness varied from 75 to 90% in the third to sixth generation in these 2 breeds. Overall, these trends demonstrate the extent of missing pedigree observed in pig breeding populations.

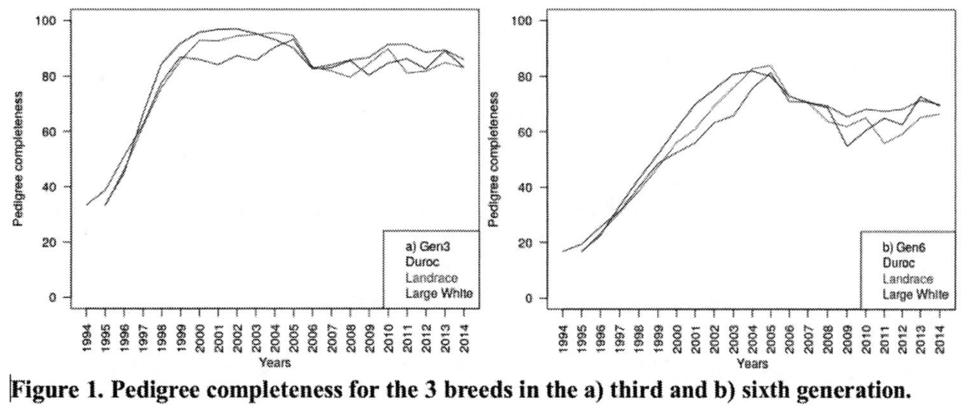

Figure 1. Pedigree completeness for the 3 breeds in the a) third and b) sixth generation.

Effective population size. Pedigree data was not complete for 2015 and population estimates for that year are not shown because the computation of Ne may be incorrect. Similarly, not all herds were represented in the 1994 data for LW and the high Ne estimated for this year was likely an over-estimate (Figure 2a.) From 1995 until 2014, estimates of Ne varied from 64 to 98 in LW, from 52 to 108 in LR and from 42 to 61 in DU. Effective population sizes varied from 30 to 225 across years and breeds in the study by Krupa *et al* (2015). In 2012, population sizes were more stable and varied from 35 for DU to 83 in Pietrain, similar to the range observed in this study.

Inbreeding. Average inbreeding levels of all animals increased until 2004 and 2005 (Figure 2b). For these years average inbreeding levels were highest with averages of 0.031, 0.034 and 0.050 in LW, LR and DU, respectively. Since then, breeds were less stable with herds leaving or joining the recorded populations which is reflected in lower pedigree completeness and more variable estimates of average inbreeding coefficients over time. In comparison, mean inbreeding levels were about 0.045 for Yorkshire, Duroc and Hampshire, about 0.065 for Landrace and about 0.075 for Berkshire after 12 generations in a US study (Welsh *et al.*, 2010). Approximately 12 generations were considered in the current study and mean inbreeding levels continued to increase after 12 generations for the US pig populations. Considering a longer time period and more generations with more complete pedigree information is expected to result in higher estimates of mean inbreeding levels and lower estimates of effective population size.

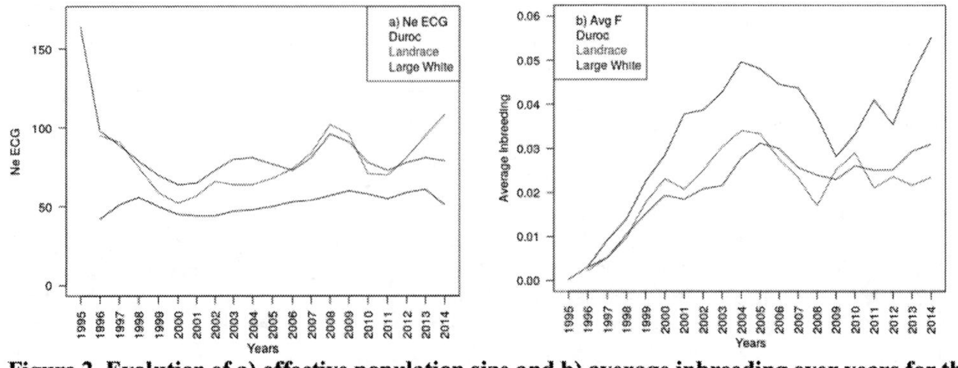

Figure 2. Evolution of a) effective population size and b) average inbreeding over years for the 3 breeds using the complete pedigree to estimate rate of inbreeding per generation.

CONCLUSION

Maximum estimates of inbreeding levels varied from 0.031 to 0.050 for the 3 breeds and effective population sizes varied from 42 to 108 over years for these breeds. These estimates were similar to estimates presented in other studies. However, pedigree was incomplete and estimates of inbreeding levels and subsequently effective population sizes are imprecise and an under-estimate of true population values. Pedigree data is often incomplete in livestock populations and estimates of effective population size based on pedigree information should only be regarded as 'guestimates'. Further research should focus on evaluation of the benefits of using genomic information for estimating inbreeding levels more accurately and the impact of higher inbreeding levels on performance and fitness traits.

ACKNOWLEDGEMENTS

The authors thank the breeders who provided the data.

REFERENCES

Falconer D.S. and MacKay T.F.C. (1996) 'Introduction to Quantitative Genetics' 4th ed. Longman, Harlow.

Groeneveld, E., Van der Westhuizen, B., Maiwashe, A., Voordewind, F. and Ferraz, J.B.S. (2009) *Genet. Mol. Res.* **8**: 1158.

Gutiérrez, J.P., Cervantes, I. and Goyache, F. (2009) *J. Anim. Breed. Genet.* **126**: 327.

Kimura, M. and Crow, J.F. (1963) *Evolution* **17**: 279.

Krupa, E., Źáková, E. and Krupová, Z. (2015) *Asian Australasian J. Anim. Sci.* **28**: 25.

MacCluer, J.W., Boyce, A.J., Dyke, B., Weitkamp, L.R., Pfennig, D.W. and Parsons C.J. (1983) *J. Hered.* **74**: 394.

Ouborg, N.J., Pertoldi, C., Loeschcke, V., Bijlsma, R. (Kuke) and Hedrick, P.W. (2010) *Trends in Genetics* **26**: 177.

Welsh, C.S., Stewart T. S., Schwab C. and Blackburn H.D. (2010) *J. Anim. Sci.* **88**: 1610.

Proc. Assoc. Advmt. Anim. Breed. Genet. 22:245-248

GENETIC PARAMETERS FOR BODY-WEIGHT TRAITS OF A NATIVE POULTRY BREED IN THAILAND

S. Tongsiri[1,2,3], M.G Jeyaruban[1], S. Hermesch[1], J. H. J. van der Werf[2], L. Li[1], and T. Chormai[3]

[1]Animal Genetics and Breeding Unit, a joint venture of NSW Department of Primary Industries and University of New England, Armidale, NSW 2351, Australia
[2]School of Environmental and Rural Science, University of New England, Australia
[3]Department of Livestock Development, Thailand

SUMMARY

The heritabilities and genetic correlations for body-weights measured at day-old (BW1D), and at 4 (BW4), 8 (BW8), 12 (BW12), 16 (BW16), 20 (BW20) and 24 (BW24) weeks of age, and also at first egg (BWFE) of Thai native chickens (Lueng Hang Kao Kabinburi, LHKK) were estimated using Restricted Maximum Likelihood (REML) procedures. Data were from the Kabinburi Livestock Research and Breeding Center, Thailand and the records contained five generations of performance records on 11,588 birds born from 2003 to 2007. Estimates of heritabilities for additive genetic effects of body-weight traits ranged from 0.10 to 0.51. Heritabilities for maternal genetic effects ranged from 0.04 to 0.25, except for BWFE, which had no significant maternal genetic effects. Significant maternal permanent environmental effects were observed for all traits, except for BW24 and BWFE. Estimates of additive genetic correlations between the body-weight traits ranged from 0.25 to 0.99. Estimated heritabilities and genetic correlations between body-weight traits suggest that the growth performance of LHKK chickens can be improved by selection on one or more juvenile weight traits.

INTRODUCTION

Poultry production in Thailand can be categorised as a) intensive poultry production using commercial strains (80%), and b) backyard poultry production using native chickens (20%) (Information and Communication Technology Center 2016). The Thai native chicken (TC) has been very popular among Thai backyard poultry farmers for many centuries. This is because the TC expresses high disease resistance and can be raised with low-quality feed. Moreover, they are a source of household income and protein supply for rural Thai farmers. The TC has had increased importance since the avian influenza outbreak in Thailand in 2004 (Avian Influenza Control Center 2006). During the outbreak, TC was an important source of poultry meat for domestic consumption due to import and export restrictions imposed on poultry products.

TC meat is preferred by Thai consumers due to its unique taste and texture. However, the meat is 2 to 3 times more expensive than commercial broiler meat. This is because the TC meat production does not adequately support domestic consumption and also it's tastier than broiler meat. The TC has low mature body-weight and reaches 1.5 to 2 kg at 7 months of age (Thummabood *et al.* 2000). Lueng Hang Kao Kabinburi (LHKK) is one of the most popular breeds of the TC and therefore, identified by the Department of Livestock Development (DLD), Thailand to improve its meat production.

The heritability and genetic correlations of production traits are important factors in developing any selection scheme. However, knowledge about genetic parameters for native chickens under tropical climate condition is limited. Therefore, the objective of this study was to estimate genetic parameters on growth traits at different ages for LHKK under tropical condition in Thailand.

MATERIALS AND METHODS

Animal and Data. Pure-bred dual-purpose LHKK chickens were housed on a Thai government farm at the Kabinburi Livestock Research and Breeding Center, between 14.0478° North latitude and 101.3725° East longitude in the Eastern region of Thailand. Data were recorded for five generations from 2003 to 2007. Chickens were randomly mated and mainly selected for breed specific plumage characteristics. Seventy males and three hundred and fifty females were maintained each year to produce 21,500 mixed sex chicks. About 4,000 chicks were retained as replacements and the rest (17,500) were issued to farmers. The replacement chicks were grown on deep litter housing from one day-old to 21 days of age at 7.5 chicks per one square meter and fed with diet containing 18% crude protein and 2,900 Kcal ME/Kg of energy. They moved to grower pens at 21 days of age. During the growth period, chicks were allowed to scavenge during daytime and were sheltered at night. At 22 weeks of age, selected chickens were moved to individual battery cages until culling after 1 year of laying. Traits considered in this study were body-weights at day-old (BW1D), and at 4 (BW4), 8 (BW8), 12 (BW12), 16 (BW16), 20 (BW20), and 24 (BW24) weeks of age, and body-weight at first egg (BWFE) mean age of 28 weeks. Records more than 3 standard deviations from the mean of the data were eliminated. The total number of birds in the pedigree were 17,883 and the number of birds with records were 11,588 from 1,461 dams and 486 sires.

Statistical analyses. SAS (SAS Institute Inc., Cary NC USA) was used to calculate descriptive statistic and to identify significant fixed effects. Genetic parameters were estimated via Restricted Maximum Likelihood (REML) using a mixed linear model and WOMBAT software (Meyer 2007). A log likelihood ratio test was used to test significance of random effects and to identify the best model. For all body weight traits, except for BW24 and BWFE, the model was:

$$Y_{ijklm} = y_i + h_j + s_k + a_l + m_m + pe_m + e_{ijklm}$$

Where: Y_{ijklm} is one of the six body-weight traits measured on animal l, in hatch j within year i with sex k, a_l is the random additive genetic effect of animal l, m_m is maternal genetic effect of dam m, pe_m is permanent environmental effect of dam m and e_{ijklm} is the random error associated with this observation. The covariance between additive and maternal effect was assumed to be zero. Only additive genetic and maternal genetic effects were fitted for BW24 and only additive genetic effect was fitted for BWFE. A series of bivariate analyses were used to estimate genetic and phenotypic correlations.

RESULTS AND DISCUSSION

Descriptive statistics. The descriptive statistics for the eight body-weight traits of LHKK chickens measured for five generations are summarized in Table 1. The mean weights observed for the eight traits were similar to the means reported by the Bureau of Animal Husbandry and Genetic Improvement (2016) for four breeds of TC. However, the mean weights of BW8, BW12 and BW16 were slightly heavier (200g) than those of another popular dual-purpose Thai indigenous chicken breed called Pradu Hangdum (Na-Rungsri *et al.* 2007). The higher BWFE of LHKK compared to other native chickens (Sangdaoreung *et al.* 2005) suggests that LHKK chickens could be improved as a dual-purpose chicken.

Genetic parameters. For most weight traits, direct heritability estimates were generally higher than maternal heritability estimates, or the ratio of permanent environmental dam effects variance to total variance (Table 1). Estimates of direct heritability varied from 0.10 (±0.02) to 0.51 (±0.06). The highest heritability was estimated for BWFE and the lowest was for BW1D. Estimated heritabilities for body-weight traits were within the range reported in previous studies. For Pradu Hangdum, Na-Rungsri *et al.* (2007) reported heritabilities of 0.43, 0.46 and 0.39 for BW8, BW12 and BW16, respectively. For BWFE, Boonkum *et al.* (2014) estimated a heritability of 0.51. Estimated maternal heritabilities were low to moderate and ranged from 0.04 (±0.02) to 0.25 (±0.04) for all traits, except for BWFE, for which no significant maternal effects was found. Estimated variance ratio for permanent environmental effects of dams were low to moderate for all traits and ranged from 0.11 (±0.02) to 0.28 (±0.03), except for BW24 and BWFE. Slight increase in permanent environmental effect of dam from BW8 to BW20 might be due to difficulty in portioning maternal and permanent environmental effects with few repeated observations for dams.

Table 1 Descriptive statistic, and estimated heritabilities (±SE) for direct (h^2_d), and maternal genetic effect (h^2_m), the variance ratio for permanent environmental effects (c^2) of dam and phenotypic variance (σ^2_p) for body-weight traits of LHKK chickens

Traits[1]	No. of records	Mean	SD	h^2_d	h^2_m	c^2	σ^2_p
BW1D (g)	11588	30.93	3.38	0.10±0.02	0.25±0.04	0.28±0.03	12.03
BW4 (g)	11201	218.91	56.68	0.28±0.03	0.05±0.02	0.11±0.02	1473.90
BW8 (g)	10807	642.08	138.74	0.33±0.03	0.04±0.02	0.11±0.02	9609.67
BW12 (kg)	9777	1.10	0.21	0.38±0.03	0.05±0.02	0.13±0.02	0.03
BW16 (kg)	8948	1.49	0.31	0.32±0.03	0.08±0.03	0.17±0.02	0.06
BW20 (kg)	7643	1.81	0.41	0.28±0.03	0.05±0.02	0.19±0.03	0.09
BW24 (kg)	6157	2.12	0.47	0.28±0.04	0.11±0.02		0.10
BWFE (kg)	1428	2.05	0.25	0.51±0.06			0.06

[1] BW1D, body-weight at day-old; BW4, BW8, BW12, BW16, BW20 and BW24 are body-weights at 4, 8, 12, 16, and 20 and 24 weeks of age, respectively; BWFE, body-weight at first egg.

Table 2 Direct genetic (a), maternal genetic (m) and permanent environmental of dam (pe) correlations (above diagonal) and phenotypic correlation (below diagonal) between body-weights at day-old (BW1D) and 4 (BW4), 8 (BW8), 12 (BW12), 16 (BW16), 20 (BW20), 24 (BW24) weeks of age, and body-weight at first egg (BWFE) of LHKK chickens

Trait	Effect	BW1D	BW4	BW8	BW12	BW16	BW20	BW24	BWFE
BW1D	a		0.37	0.29	0.27	0.30	0.25	0.35	0.65
	m		0.65	0.80	0.67	0.67	0.71	0.66	
	pe		0.31	0.03	-0.13	-0.17	-0.14		
BW4	a	0.22		0.86	0.74	0.60	0.67	0.52	0.55
	m			0.98	0.97	0.79	0.89	1.00	
	pe			0.81	0.58	0.57	0.59		
BW8	a	0.16	0.68		0.97	0.90	0.89	0.81	0.77
	m				0.97	0.98	0.93	1.00	
	pe				0.90	0.84	0.80		
BW12	a	0.14	0.56	0.79		0.98	0.95	0.92	0.78
	m					1.00	1.00	1.00	
	pe					0.97	0.93		
BW16	a	0.14	0.47	0.70	0.85		0.99	0.97	0.81
	m						0.99	1.00	
	pe						0.99		
BW20	a	0.12	0.47	0.66	0.79	0.90		0.99	0.85
	m							1.00	
BW24	a	0.19	0.39	0.59	0.76	0.85	0.90		0.93
BWFE	a	0.14	0.29	0.48	0.58	0.57	0.64	0.63	

Estimated standard error for genetic and phenotypic correlations varied from 0.01 to 0.18 and 0.00 to 0.02, respectively.

Phenotypic correlations and correlations for additive, maternal genetic and permanent environmental effects of dam between the body weight traits of LHKK chickens are presented in Table 2. The phenotypic correlations between weight traits at different ages varied from 0.12 to 0.90 and were highest for body-weight traits at adjacent age points. The additive genetic correlations among body-weight traits were generally high and positive, except for correlation with BW1D. Genetic correlations varied from 0.25 to 0.99. Moderate genetic correlations were observed between BW1D and other weight traits, except with BWFE. The highest genetic correlations (0.99) was observed between BW20 and BW16 and BW24. Correlations between maternal genetic effects at different ages ranged from 0.65 to 1.00 and the correlations between permanent environmental effects of dam ranged from -0.17 to 0.99.

Genetic correlations between body-weights at different stages of growth are lacking in literature for the native chickens. However, Lwelamira *et al.* (2009) reported high correlations for additive genetic effects (0.60-0.93) and for phenotypic effects (0.54-0.74) for body-weights at 8, 12, 16, and 20 weeks of age of local chickens in Tanzania. Niknafs *et al.* (2012) reported moderate to high genetic correlations between BW1D and BW8 (0.57) and BW12 (0.36) for local chicken in Iran. The high genetic correlation between BW1D and BWFE suggest that selection for heavy mature hen will indirectly increase the weight of day-old chicks. Moreover, high genetic correlations between body-weights at early growth with BWFE suggest that selecting heavier juvenile birds would increase mature weight in LHKK chickens.

Moderate to high heritabilities for direct additive genetic effects on body-weight traits suggests that selection for higher body-weight will increase growth rate and meat production of LHKK. This will improve its value as a dual-propose breed. Furthermore, high genetic correlations between body-weights measured during the growth period and the weight measured at maturity suggest that selecting for higher body-weight between 8 to 24 weeks of age would increase the mature weight of LHKK chickens. However, before implementing selection for growth rate, the correlated responses in egg production of LHKK chickens need to be investigated.

ACKNOWLEDGEMENTS

The authors wish to thank Department of Livestock Development in Thailand for conduction and data collection. We thank Agricultural Research Development Agency (Public Organization), ARDA, of Thailand for financial support given to Ms Siriporn Tongsiri for her PhD study.

REFERENCES

Avian Influenza Control Center (2006) 'Avain Influenza control in Thailand'. Department of Livestock Development, Bangkok.

Boonkum, W., Pobwongsa, I. and Chankitsakul, V. (2014) *Khon Kaen Agr. J. (Suppl.1)* **42**:255.

Bureau of Animal Husbandry and genetic Improvement (2016) 'Productivity of Thai native chickens under free range and organic systems'. Department of Livestock Development, Bangkok.

Information and Communication Technology Center (2016) 'Number of chicken and poultry farmers in Thailand' Department of Livestock Development, Bangkok.

Meyer, K. (2007) *J. Zhej Uni. Sci.* **8**:815.

Lwelamira, J., Kifaro, G. C. and Gwakisa.P. S. (2009) *Trop. Anim. Health. Prod.* **41**:51.

Na-Rungsri, D., Akkahart, K. and Leotaragul, A. (2007) *DLD* **50**:210.

Niknafs, S., Nejati-Javaremi, A., Mehrabani-Yeganeh, H. and Fatemi, S. A. (2012) *Trop. Anim. Health. Prod.* **44**:1437.

Sangdaoreung, S., Tangtaweewipat, S., Cheva-Isarakul, B., Sirisrisawas, B. Punyawee, S. and Na ChaiangMai, P. (2005) *Kasetsart J.* **43**:1.

Thummabood, S., Kliaewkamonthut, B., Thummabood, A. and Morathob, S. (2000) 'Crossbred of Thai native chickens production' Department of Livestock Development, Bangkok.

Proc. Assoc. Advmt. Anim. Breed. Genet. 22:249-252

DEVIATIONS AROUND KINSHIP EXPECTATIONS AT VARIOUS SNP MARKER DENSITIES IN A POPULATION OF BROILER CHICKEN

A. Reverter[1], Y. Li[1], A. George[2], J. Henshall[3], R. Sapp[3], R. Okimoto[3], R. Hawken[3] and S. A. Lehnert[1]

[1]CSIRO Agriculture & Food, Queensland Bioscience Precinct, Queensland 4067, Australia.
[2]CSIRO Data61, Ecoscience Precinct, Queensland 4102, Australia.
[3]Cobb-Vantress Inc., Siloam Springs, Arkansas 72761-1030, USA.

SUMMARY

We compare four low-density SNP panels containing 151, 400, 1,000 (1K) and 3,000 (3K) SNP selected from a higher density chip of 50K SNP in their ability to correctly infer 7 kinship relationships (from self-self to grand-mother – grand-offspring) in 4,217 commercial broiler chicken. Self–self relationships estimated from the diagonal elements of the genomic relationship matrix (GRM) were symmetric and centred at 1.0 regardless of the panel used. However, genomic relationships for other relationships were centred slightly left to the expected value indicating possible genotype or pedigree errors. Relationships estimated using either the 1K or the 3K SNP panels were almost undistinguishable from those estimated using the whole 50K chip. However, the two lowest density panels produced relationships with long-tailed distributions. We conclude that a SNP panel of 1K SNP is a cost-effective tool to estimate relationships among individuals.

INTRODUCTION

The ability to correctly infer relationships among individuals underpins the utility of SNP genotype data. This ability is of particular relevance in the development of low-density panels for the implementation of cost-effective genomic strategies. Judge *et al.* (2016a) have recently explored the optimal use of low-density SNP panels for breed assignment in Angus and Hereford cattle. The authors conclude that at least 300 to 400 SNP are needed to accurately predict breed proportions. Similarly, working with various cattle and sheep populations Strucken *et al.* (2016), concluded that at least 700 SNP are needed to fully exclude false positives in parentage assignments. Other authors have evaluated the use of low-density panels for imputation to higher density in cattle (Ogawa *et al.* 2016), sheep (Ventura *et al.* 2016), pig (Badke *et al.* 2014), and chicken (Wang *et al.* 2013).

Here we present four low-density SNP panels containing 151, 400, 1,000 and 3,000 SNP and compare them with the higher density chip of 50K SNP based on their ability to estimate relationships in a population of 4,217 commercial broiler chicken from 22 overlapping generations.

METHODS

Animal resources and relationships considered. We used a total of 4,217 broiler chicken (3,139 females and 1,078 males) from 22 overlapping generations of a commercial line of Cobb-Vantress Inc. The birds were selected from a larger population to ensure parents and grandparents contained within the sample had genotypes for ~50,000 (50K) SNP from the high-density Avian chip from Illumina Inc.

In total, there were 795 dams with genotypes, 117 sires with genotypes and 133 grand-dams with genotypes. With these, seven types of animal to animal relationships were explored including (1) self – self (N = 4,217); (2) full-sibs (N = 29,599 pair combinations); (3) Father – offspring (N = 2,915 pairs); (4) Mother – offspring (N = 2,708 pairs); (5) Paternal half-sibs (N = 186,716 pairs); (6) Maternal half-sibs (N = 1,560 pairs); and (7) Grand-mother – grand-offspring (N = 5,327).

Using the Method 1 of VanRaden (2008) we built the genomic relationship matrix (GRM) across the 4,217 birds using the 50K chip, as well as with the four low-density SNP panels described next.

Low-density panels. For the formation of the low-density SNP panels, we developed a 6-component fitness function to be optimised that included (1) minor allele frequency (MAF); (2) equidistance to ensure uniform genome coverage; (3) distance to known gene; (4) significance of the association to feed-related phenotypes; (5) pleiotropy test statistics; and (6) connectivity in a co-association network.

We used simulated annealing for the optimisation process. Simulated annealing (Kirkpatrick *et al.* 1983) is a heuristic search algorithm for global optimization, using iterative random movements to approximate optimum solution and has gained popularity in the context of livestock genetics and genomics including studies with cattle (Schierenbeck *et al.* 2011) and poultry (Chapuis *et al.* 2016).

Initially, three SNP densities were considered: 400 SNP, 1,000 (1K) SNPs and 3,000 (3K) SNPs. Importantly, these panels were nested such that the 400 SNP in the small panel were contained in the 1K SNP of the medium panel, and these were themselves contained in the larger 3K panel.

In addition, a smaller panel of only 151 SNP was developed. This panel was made of SNP (1) in the coding region of genes reported to be of relevance in the feed efficiency literature; (2) significant (P< 0.01) in the GWAS for at least one of seven feed-related phenotypes previously undertaken; and (3) Included in the 3K SNP panel.

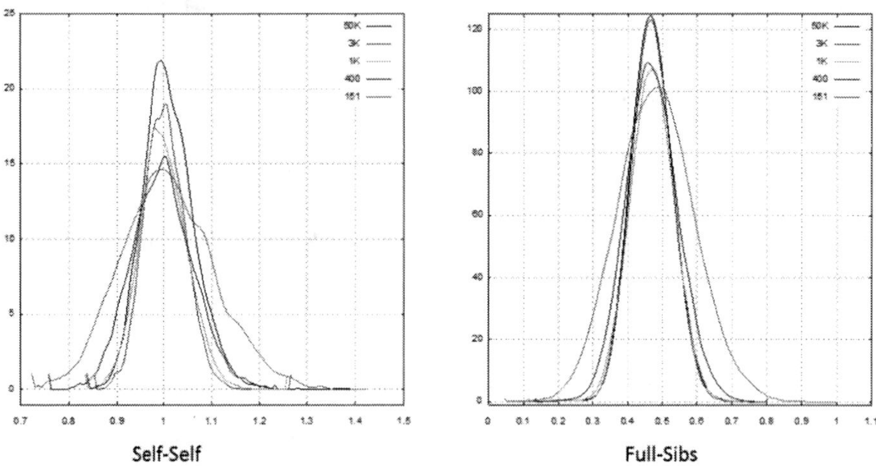

Figure 1. Distribution of genomic relationships for self-self (diagonals elements of the genomic relationship matrix) and full-sibs estimated based on SNP panels of various densities.

RESULTS

Table 1 shows the summary statistics (including mean, standard deviation, minimum and maximum) for genomic relationships estimated using either the high-density 50K SNP chip or the four low-density SNP panels considered is this study and for the seven types of pedigree-based kinships available in our dataset of 4,217 broiler chicken.

Self-self relationships based on the diagonal elements of the GRM were all centred at the expected value of 1. However, the spread was much higher for the panels with only 400 or 151 SNP. Indeed, across all types of relationships considered, the very low density panels of 400 and 151 SNP yielded estimated relationship with higher variation compared to the panels of higher density.

This deviation from expectation is made apparent in Figure 1 for the case of self-self and full-sib relationship in the five SNP panels.

Table 1. Summary statistics for genomic relationships estimated using the high-density 50K SNP chip and four low-density SNP panels for seven types of pedigree-based kinships

Kinship[A]	Panel	No Pairs	Mean	SD	Min.	Max.
SS	50K	4,217	1.009	0.055	0.837	1.404
	3K	4,217	1.001	0.044	0.853	1.267
	1K	4,217	0.999	0.050	0.834	1.259
	400	4,217	0.997	0.066	0.759	1.238
	151	4,217	1.008	0.099	0.723	1.423
FS	50K	29,599	0.469	0.061	0.134	0.769
	3K	29,599	0.470	0.060	0.133	0.749
	1K	29,599	0.470	0.064	0.158	0.739
	400	29,599	0.470	0.075	0.114	0.804
	151	29,599	0.481	0.110	0.045	1.003
FO	50K	2,915	0.467	0.047	0.335	0.727
	3K	2,915	0.469	0.042	0.358	0.719
	1K	2,915	0.469	0.045	0.334	0.710
	400	2,915	0.469	0.056	0.320	0.743
	151	2,915	0.469	0.087	0.187	0.826
MO	50K	2,708	0.466	0.045	0.344	0.736
	3K	2,708	0.468	0.039	0.363	0.727
	1K	2,708	0.472	0.046	0.322	0.731
	400	2,708	0.474	0.057	0.297	0.728
	151	2,708	0.462	0.089	0.168	0.798
PHS	50K	186,716	0.236	0.053	-0.072	0.575
	3K	186,716	0.235	0.054	-0.092	0.566
	1K	186,716	0.232	0.058	-0.109	0.585
	400	186,716	0.231	0.070	-0.205	0.622
	151	186,716	0.241	0.103	-0.214	0.765
MHS	50K	1,560	0.251	0.091	0.082	0.621
	3K	1,560	0.252	0.090	0.071	0.637
	1K	1,560	0.248	0.096	0.045	0.633
	400	1,560	0.250	0.112	-0.014	0.697
	151	1,560	0.254	0.122	-0.154	0.760
GMGO	50K	5,327	0.239	0.059	0.063	0.539
	3K	5,327	0.234	0.058	0.048	0.499
	1K	5,327	0.233	0.062	0.014	0.484
	400	5,327	0.231	0.075	-0.008	0.523
	151	5,327	0.233	0.106	-0.100	0.634

[A]SS = self-self; FS = full sibs; FO = father – offspring; MO = mother – offspring; PHS = paternal half-sibs; MHS = maternal half-sibs; GMGO = grand-mother – grand-offspring.

Notably, the distribution of estimated genomic relationship for full-sibs was not centred at the expected value of 0.5 and instead averaged ~0.47 for all SNP panels considered (Figure 1, left panel). This same anomaly was reported by Lourenco *et al* (2015) and was attributed to both genotype and pedigree errors. Indeed, with the possible exception of genomic relationships estimated for self-self and for maternal half-sibs which was centred at the expected value of 1.0 and 0.25, respectively, all other relationships were centred at a value slightly lower than the expectation. Further research is needed to ascertain whether errors in pedigree and/or genotypes are responsible for this anomaly.

Table 2 presents the correlation between genomic relationships estimated using the 50K SNP chip and the four low-density SNP panels. On average, this correlation decreased from 0.864 when using the 3K SNP panel to 0.465 when using the 151 SNP panel. However, the decrease was not linear, with the smallest being by 10.7% from the 3K to the 1K panel (0.864 to 0.771), and the largest by 27.1% from 400 to 151 SNP panels (0.638 to 0.465).

Table 2. Correlation between genomic relationships estimated using the high-density 50K SNP chip and four low-density SNP panels for seven types of pedigree-based kinships

Panel	Type of Kinship[A]							Average
	SS	FS	FO	MO	PHS	MHS	GMGO	
3K	0.752	0.892	0.834	0.807	0.889	0.966	0.905	0.864
1K	0.607	0.808	0.729	0.699	0.797	0.933	0.828	0.771
400	0.455	0.682	0.595	0.501	0.654	0.869	0.713	0.638
151	0.292	0.484	0.411	0.341	0.470	0.732	0.525	0.465
Average	0.526	0.716	0.642	0.587	0.702	0.875	0.743	

[A]SS = self-self; FS = full sibs; FO = father – offspring; MO = mother – offspring; PHS = paternal half-sibs; MHS = maternal half-sibs; GMGO = grand-mother – grand-offspring.

Averaged across the four low-density panels, self-self relationships (from diagonal elements of the GRM) were the least correlated (r = 0.526) with the ones obtained with the 50K panel, followed by mother-offspring (r = 0.587) and father-offspring (r = 0.642). The highest average correlation was observed for maternal half-sib combinations (r = 0.875).

CONCLUSIONS

In contrast to other livestock species, broiler chicken have large full-sib families implying a large benefit in adopting genomic evaluation compared to pedigree-based evaluation. However, this benefit relies on accurate estimation of relationships among individuals. This accuracy is affected when using low-density panels as a cost-effective alternative to genomic evaluation with a 50K panel. We conclude that a panel of 1,000 SNP can be used to reliably estimate relationships. However, further research is needed to ascertain the potential impact on the breeding goal of a selection line when the SNP in a low-density panels have been selected according to a fitness function that includes the association of SNP to traits in the breeding objective.

REFERENCES

Badke YM, Bates RO, Ernst CW, Fix J, and Steibel JP (2014) *G3 (Bethesda)* **4**: 623.

Chapuis H, Pincent C, and Colleau JJ. (2016) *J. Anim. Breed. Genet.* **133**: 3.

Judge MM, Kelleher MM, Kearney JF, Sleator RD, and Berry DP (2016a) *Animal* **24**: 1.

Kirkpatrick S, Gelatt CD Jr, and Vecchi MP. (1983) *Science* **220**: 671.

Lourenco DA, Fragomeni BO, Tsuruta S, Aguilar I, Zumbach B, et al. (2015) *Genet. Sel. Evol.* **47**: 56.

Ogawa S, Matsuda H, Taniguchi Y, Watanabe T, Takasuga A, et al. (2016) *Anim. Sci. J.* **87**: 3.

Schierenbeck S, Pimentel EC, Tietze M, Körte J, Reents R, et al. (2011) *J. Dairy Sci.* **94**: 6143.

Strucken EM, Lee SH, Lee HK, Song KD, Gibson JP, et al. (2016) *J. Anim. Breed. Genet.* **133**: 13.

VanRaden PM. (2008) *J. Dairy Sci.* **91**: 4414.

Ventura RV, Miller SP, Dodds KG, Auvray B, Lee M, et al. (2016) *Genet. Sel. Evol.* **48**: 71.

Wang C, Habier D, Peiris BL, Wolc A, Kranis A, et al. (2013) *Poult. Sci.* **92**: 1712.

PRECISION MEDICINE: REALISTIC EXPECTATIONS FOR PREDICTION OF RISK TO HUMAN COMPLEX DISEASE

Naomi Wray

The University of Queensland, St. Lucia, Queensland 4072, Australia.

SUMMARY

I will review the progress and prospects of risk prediction for disease in people. Fundamental to this progress is that data sharing is now common-place for both association study summary statistics and of individual-level measures.

GENETIC RISK PREDICTION

The methodology of genetic risk prediction of human disease parallels genetic evaluation in livestock. However, fundamental differences reflect the data structure available for generating and validating predictors, and that the prediction goal is of an uncommon binary phenotype of an individual, rather the mean value of a quantitative trait in the next generation. Genetic predictors of common complex genetic diseases, can never be diagnostically accurate for an individual, but genetic risk stratification could have clinical utility. For example, risk stratification could identify a high-risk class that includes the majority of those who will become affected in their lifetime (high sensitivity) even though the majority of those in this high-risk class will not be affected (poor specificity). Accurate prediction of a phenotype, requires the genetic predictor to be enhanced to include non-genetic risk factors. The genomics era allows the inclusion genomic biomarkers, which could reflect the downstream consequences disease and of non-measured environmental risk factors.

I will review the progress and prospects of risk prediction for disease in people. Fundamental to this progress is that data sharing is now common-place for both association study summary statistics and of individual-level measures. For example, the UK Biobank (Sudlow et al. 2015; www.ukbiobank.ac.uk) study of 500,000 people with deep phenotyping and genome-wide genotype data now presents opportunities for quantitative genetics methods common in livestock to be applied to human data, and provides new opportunities for cross-fertilisation of ideas between disciplines. While disease risk prediction receives much hype in the era of personal or precision medicine it is important to not to oversell what can be realistically achieved.

REFERENCES

Sudlow C, Gallacher J, Allen N, Beral V, Burton P, Danesh J, et al. UK Biobank: An Open Access Resource for Identifying the Causes of a Wide Range of Complex Diseases of Middle and Old Age. PLoS Med. Public Library of Science; 2015;12: e1001779.

Proc. Assoc. Advmt. Anim. Breed. Genet. 22:255-263

BEEF CATTLE GENOMIC SELECTION IN TROPICAL ENVIRONMENTS

L.G. Albuquerque[1,2], G.A. Fernandes Júnior[1], R. Carvalheiro[1,2]

[1]School of Agricultural and Veterinarian Sciences, São Paulo State University (Unesp), Jaboticabal, SP, Brazil
[2]National Council for Scientific and Technological Development (CNPq), Brasília, DF, Brazil

SUMMARY

Compared to dairy cattle, beef cattle genomic selection is in an early stage. Nevertheless, good perspectives and opportunities for its application are foreseen or are already underway. Genomic selection is expected to benefit beef production by allowing identifying genetic superior animals earlier and more accurately as well as to select for traits that are difficult and expensive to measure such as meat quality and feed efficiency, among other applications. We started this paper discussing the importance of beef cattle production in the tropics, than we presented some results from genomic studies and applications of genomic selection in the tropics, using the Nellore breed (*Bos indicus*) as a case study.

BEEF CATTLE PRODUCTION IN THE TROPICS

Tropical regions correspond to the Earth territories situated between the Tropic of Cancer (northern hemisphere) and the Tropic of Capricorn (southern hemisphere), covering countries from Central (25 countries) and South America (10 countries), Africa (47 countries), Asia (15 countries) and Oceania (17 countries), with over 4779 million ha in extent of land, in which, around 40% correspond to forest ecosystems (Chidumayo and Gumbo 2013). These regions are important for the world food production and security (Foley *et al.* 2011) and cattle are a vital source of animal protein (Porto-Neto *et al.* 2014). According to FAO (Food and Agriculture Organization of the United Nations), there are in the world around 1.47 billion head of cattle and about 65 percent are located in tropical areas. The two countries with the largest number of cattle in the world, India (302 million head) and Brazil (219 million head), are situated in the tropics. The number of cattle is also expressive in other tropical regions such as Africa (312 million head), Central America (47 million head) and Oceania (40 million head). Data from the United States Department of Agriculture (USDA) shows that, in 2016, from a total of 60,486 and 9,439 (1,000 Metric tons) carcass weight equivalent worldwide produced and exported, respectively, 17,489 and 5,340 (~29% and ~57%) were produced and exported by Brazil (9,284 and 1,850), India (4,250 and 1,850), Australia (2,075 and 1,385), and Mexico (1,880 and 255).

The tropics are characterized by warm temperatures throughout the year, usually above 18 ^{o}C and seasons are commonly divided in wet and dry. As the beef cattle production in these areas is based on grass-fed, it is common to observe animals gaining body weight during the rainy season and losing or keeping it during the dry season, which usually leads to slaughter of old animals (Millen *et al.* 2011), affecting meat quality. In addition, the natural infestations of ecto and endo- parasites and the high temperature and humidity are challenges that the farmers in the tropics have to deal with (Porto-Neto *et al.* 2014). Because of this, adaptation to tropical environmental conditions is an essential trait to cattle reared in these areas. As a general rule, tropically adapted breeds (Zebu cattle) are more efficient than non-adapted (Taurine cattle) in such tropical environment conditions (Porto-Neto *et al.* 2014), explaining the prevalence of Zebu breeds in the beef production systems in the tropics. In Brazil, for example, about 80% of the cattle have Zebu contribution, mainly of Nellore breed (Carvalheiro 2014).

GENOMIC STUDIES IN THE TROPICS

Genomic selection presents an opportunity for commercial breeders to increase the rates of genetic progress in beef, primarily, through increased accuracy of estimated breeding values on young animals (Johnston et al. 2012). In general, results from several studies support the feasibility of applying genomic selection in tropical regions (Table 1). One of the main advantage of genomic selection is the possibility to accurately select animals early in life, being especially useful for the selection of traits that are difficult or expensive to measure like fertility, disease resistance, methane emissions, feed conversion, and carcass and meat quality (Hayes *et al.* 2013; Carvalheiro 2014). Traditionally, evaluation of these traits in sires requires progeny tests since selection candidates cannot be directly assessed, increasing both costs and generation intervals.

Table 1. Genomic prediction accuracies in tropical beef production

Traits	Prediction accuracy	Reference
Brazilian beef cattle		
Sum of SFA	0.12 to 0.24	
Sum of MUFA	0.07 to 0.13	Chiaia *et al.* (2017)
Sum of PUFA	0.45 to 0.56	
Carcass traits	0.21 to 0.47	Fernandes Júnior *et al.* (2016a)
Feed efficiency	0.06 to 0.58	Silva *et al.* (2016)
Growth, reproductive and visual score	0.17 to 0.74	Neves *et al.* (2014)
Brazilian beef cattle		
Feed efficiency, growth, carcass and meat quality traits	0.13 to 0.48	Bolormaa *et al.* (2013)
Growth, reproductive carcass traits	0.20 to 0.45	Johnston *et al.* (2012)

SFA: saturated fatty acids; MUFA: monounsaturated fatty acids; PUFA: polyunsaturated fatty acids

Genetic markers have also been used in genome-wide association studies (GWAS) in order to identify genomic regions with major effects. Promising quantitative trait loci have already been identified. The QTL that harbors the PLAG1 gene, for example, has been associated with growth and carcass and meat quality traits in both Australian beef cattle and Brazilian Nellore cattle (Fortes *et al.* 2013; Utsunomiya *et al.* 2013; Porto-Neto *et al.* 2014; Fernandes Júnior *et al.* 2016b; Magalhães *et al.* 2016). Genomic studies in the tropics have also been focused on the identification of chromosome regions associated with traits related to sexual precocity. Together, Costa *et al.* (2015) and Regatieri et al. (2017), for example, reported 43 candidate genes for age at first calving, early pregnancy and heifer rebreeding.

Adaptation- and temperament-related QTLs have also been identified. Using GWAS in a crossbred (*taurine x indicine*) cattle population, Porto-Neto *et al.* (2014) identified an extended genetic region centered around the MSRB3 gene on BTA5 affecting several traits related to climatic adaptation of tropical cattle including parasite resistance, yearling weight, body condition score, coat color and penile sheath score. In a Nellore population, Valente *et al.* (2016) reported the existence of nine candidate regions (BTA1 at 73 Mb, BTA2 at 65 Mb, BTA5 at 22 Mb and 119 Mb, BTA9 at 98

Mb, BTA11 at 67 Mb, BTA15 at 16 Mb, BTA17 at 63 Kb, and BTA26 at 47 Mb) affecting animal temperament. According to the last authors, these genomic regions harbor genes such as PARK2, GUCY1A2, CPE and DOCK1 that are, respectively, related to dopaminergic system, memory formation, biosynthesis of peptide hormone and neurotransmitter and brain development. The understanding of genetic control of traits related to adaptation and cattle temperament should contribute to improve the productivity and animal welfare in the tropics.

APPLICATIONS OF GENOMIC SELECTION IN THE TROPICS: NELLORE BREED AS A CASE STUDY

There are different important breeding programs and research groups working on genomic selection applied to beef cattle in the tropics. We will focus on applications of genomic selection for the Nellore breed as a case study because of our research background and due to the importance of this breed for the global beef market (Carvalheiro 2014, USDA 2016). Nellore breeding programs also represent a successful case of partnership between academy and industry. The history behind the establishment and evolvement of the different commercial Nellore breeding programs running independently in Brazil was described by Ferraz and Fries (2004) and Carvalheiro (2014). Currently, these breeding programs jointly control over half a million Nellore cows per year. We will list some genomic selection applications from part of these Nellore breeding programs that are working closely to our research group, so we are more aware of what they are doing. They are CIA de Melhoramento (www.ciademelhoramento.com.br), DeltaGen (www.deltagen.com.br), Nelore Qualitas (www.nelorequalitas.com.br) and PAINT (www.crvlagoa.com.br). We would like to emphasize that there are other important research groups and breeding programs in Brazil also working with genomic selection applied to Nellore and other breeds.

Selection of progeny test candidates

The selection of young sires to be progeny tested in Nellore breeding programs is performed based on selection indexes presenting low to moderate accuracy (~0.5), when genomic information is not used. Under the current breeding scheme, young sires have their semen distributed when they are ~2 years old and have their final proof (based on progeny performance) with ~5 years old. For not presenting highly accurate proofs, these young sires are usually not used intensively until their final proof attests their genetic superiority. As a consequence, the generation interval is increased, constraining the genetic gain.

Genomic selection has increased the accuracy of selection of young sires. For instance, in some breeding programs genotyped young sires have been selected with an average accuracy of 0.75, i.e. 50% higher than the average accuracy of regular proofs. Investments and collaborations among the breeding programs are being done aiming to increase their reference populations and the accuracy of genomic predictions. The target is evaluating young sires with accuracies comparable (>0.85) to proven bulls.

Due to genotyping costs, some breeders perform a first screening based on regular proofs to select the animals to be genotyped, than choose the young sires to be progeny tested based on their genomic enhanced proof. Typically, ten times more candidates are genotyped than the animals to be tested. For example, if a breeding program intents to progeny test 50 young sires in a specific breeding season, the top 500 based on regular proofs are genotyped an their genomic enhanced proofs finally determine those to be selected. Some breeders have already decided to genotype all yearling animals, without pre-screening on regular proofs.

Although this application of genomic selection presents some advantage, increasing the accuracy of selection of young sires would have a limited impact on the genetic gain if they are not used more intensively when they are still young. Fortunately, the increased accuracy of genomic predictions is motivating some breeders to use young sires more intensively.

Intensifying the use of young sires

Historically, Nellore breeders have been using, on average, no more than 30% of young sires to mate their cows (Figure 1a). With few exceptions, generation interval of sires is generally around (or even greater than) seven years. In general, breeders are more comfortable in using proven bulls. Moreover, the trade-off between accuracy and generation interval makes it difficult to technically convince farmers to use young sires more intensively, e.g. the ratio between accuracy and generation interval for young sires (0.5/4.0=0.125) is similar to that for proven bulls (0.9/7.0=0.128). Genomic selection is changing this pattern. As genotyped young sires are presenting higher accurate proofs (compared to young sires without genomic information), some breeders are intensifying their use and, as a consequence, obtaining higher response to selection (Figure 1b).

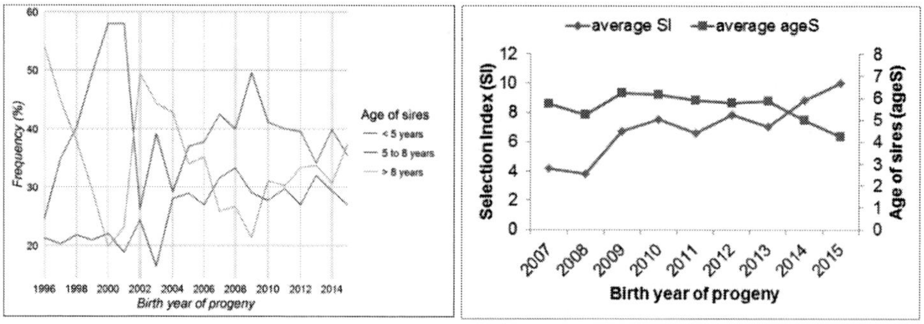

Figure 1. (a: left) Frequency (%) of progeny by age class of sires and year of birth, for Aliança Nelore dataset (~100,000 calves/year); and (b: right) Genetic trend and average age of sires at Jacarezinho farm (~10,000 calves/year).

As the reference population gets better (larger and more representative of the population) and allows obtaining more accurate proofs, genomic selection cancels the trade-off between accuracy and generation interval. It has been predicted that in the near future the seedstock Nellore cows will be mated only with young sires, which is a dramatic 'change of paradigm' on breeder's behavior. Assuming an average accuracy of 0.8 for the genomic enhanced proofs of young sires, this strategy would result in a ratio between accuracy and generation interval equal to 0.2 (0.8/4), a substantial increase compared to the ratio from the scheme without genomics. Indeed, dairy cattle breeders, especially Holstein breeders, had already witnessed this change in their breeding scheme after the advent of genomic selection (Van Eenennaam *et al.* 2014).

Selection of donors

According to the Brazilian Society of Embryo Technology, Brazil has been producing, through *in vitro* fertilization (IVF), more than half a million embryos per year, being, approximately, half of that from beef cattle breeds. In the past, embryo technology was mainly used in Brazil by "elite" herds focused on producing "show type" animals. Lots of embryos used to be produced from cows without any genetic proof and raised in artificial environments. Fortunately, this pattern has changed partially because of the drastic development of IVF, which is becoming more reliable and feasible, but also due to the position conquered by the breeding programs that nowadays lead the genetic market as seedstock providers, position which used to be occupied by "show type" animals.

Previously to genomics, a typical technical recommendation for breeders was to select, as donors, the top cows with reasonably accurate proofs, what generally resulted in selecting old cows. Genomic selection has allowed intensifying the use of young cows or heifers as donors, for increasing the accuracy of their genetic proofs. This strategy, of producing more progeny from genetically superior young animals through the synergistic adoption of genomic selection and reproduction technologies, is predicted to promote substantial increase in genetic gain compared to more conventional breeding schemes (Carvalheiro 2014). Caution should be made to certify that the heifers and young cows have superior and reasonably accurate genomic proofs for maternal and reproduction traits to be selected as donors, in order to produce replacement heifers.

Genotyping of embryos

As previously mentioned, IVF and embryo production have been used in large scale by some farms in Brazil (>1,000 embryos implanted/farm/year, with pregnancy rates around 40%). This reproduction technology provides an outstanding opportunity for increasing the genetic progress if sires and donors are properly chosen and if a reasonably good pregnancy rate of implanted embryos is attained. The genetic progress could be even higher if the genetic merit of embryos were predicted more accurately (using genomic information for example) before they were implanted. Genomic predictions of biopsied and genotyped embryos are already being obtained for dairy cattle (Saadi *et al.* 2014).

A recent study showed the feasibility of genotyping Nellore biopsied embryos and obtaining their proofs more accurately (Carvalheiro *et al.* 2017). Farmers can use this information, for example, to decide which embryos to implant, as their genomic proofs may substantially deviate from what is expected based on parents average. Another application would be to implant the embryos using a customized approach, matching the genomic profile of the embryo with the customers' needs. For example, farmers more focused on producing high quality beef, could decide to implant just embryos presenting good genomic predictions for marbling and tenderness.

Screening young sires in commercial herds

It is estimated that Brazilian commercial herds have roughly 35 million Nellore cows to be mated under natural mating (NM). If we assume one bull per 25 cows under NM and an annual replacement rate of bulls of 20%, the commercial herds need around 280,000 young replacement bulls per year. Nellore breeding programs running in Brazil jointly control about 500,000 cows, and produce around 40,000 young top bulls per year. So, breeding programs produce less than 15% of the young bulls demanded by the commercial herds, i.e. most of the commercial Nellore cows under NM are being mated with non-proven bulls.

In theory, genomic selection has the potential to change this scenario, as the commercial herds have now a tool to predict the genetic merit of bulls without a traditional proof. This application of genomic selection is technically questionable since there is evidence of the presence of substructures (based on genomic kinship) among Nellore subpopulations (Utsunomiya *et al.* 2013). In this case, the prediction equation developed using data from one subpopulation (e.g. breeding program) will not necessarily work properly in another subpopulation (e.g. commercial herd), particularly if these subpopulations are unrelated and if the developed prediction equation is more influenced by relatedness and co-segregation than by linkage disequilibrium between markers and QTL (Sun *et al.* 2016). This technical issue is even more relevant if we consider that Nellore presents lower level of linkage disequilibrium between markers at short distances compared to taurine breeds (Pérez O'Brien *et al.* 2014).

However, there are some commercial farms that already started using genomic predictions for screening young sires from their own herds. A typical use is being performed by large operation commercial farms (>10,000 cows) that, due to logistical and labor constrain, do not control their herd in a breeding program but have some genetic links with seedstock herds for using their genetic material (bull, semen, etc.). It is believed that these commercial farms could replace part of the bulls used under NM with their own produced young sires. An example is illustrated in Figure 2 where a large commercial farm pre-screened over 2,000 yearling contemporary males based on phenotypic appraisal, chose 272 to be genotyped and selected 69 to be used as replacement, based on their genomic proof (Index>5). A principal component analysis of the genomic relationship matrix revealed that the 272 genotyped animals were within the same cluster of the reference population used to calculate their genomic proofs, suggesting that the accuracy of their genomic predictions (0.38-0.58) were not overestimated.

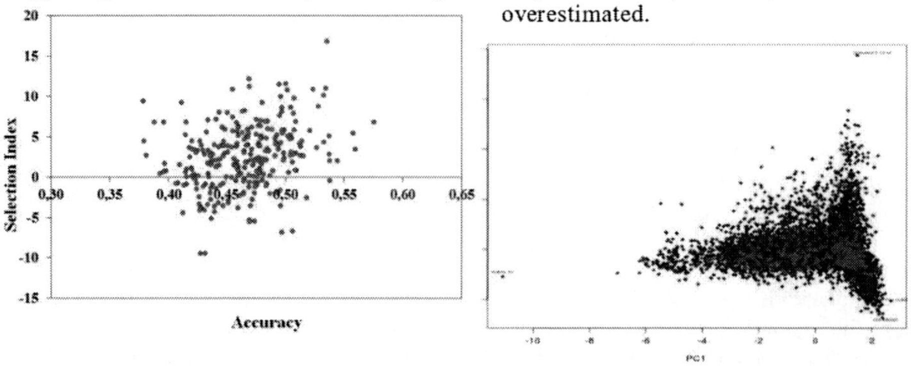

Figure 2. (a: left) Index and accuracy of genomic proofs of young sires from a commercial herd; and (b: right) Principal component analysis plot (x-axis: PC1; y-axis: PC2) based on genomic relationship matrix (blue=reference population, red=selection candidates).

Increasing selection intensity for reproduction traits

The Nellore breeding programs in Brazil use different strategies to select for reproduction traits. The most common are independent phenotypic culling, discarding heifers and cows that are not pregnant at the end of the breeding season, and accounting for reproduction traits in the selection index. In general, the selection indexes adopted give more emphasis to growth and carcass traits than to reproduction traits. This fact is often explained by the low heritability and low accuracy of genetic proofs for reproduction traits.

Motivated by the increase in accuracy obtained with genomic predictions, some breeders are given more weight to reproduction traits on their selection indexes. There are also some programs that are replacing, in their selection indexes, EPDs of indirect traits (e.g. scrotal circumference) by EPDs of traits directly associated with reproduction (e.g. age at first calving or heifer pregnancy). This strategy is expected to promote a substantial increase in genetic gain compared to the conventional strategies (without genomics), given a good prediction equation for reproduction traits.

Selection for expensive and difficult to measure traits

The opportunity to better select for expensive and difficult to measure traits figures amongst the most important applications of genomic selection. A representative example in beef cattle is the selection for meat quality traits. Without genomics, selection for this type of traits is constrained by its cost-effectiveness as it requires huge investments on phenotyping and on progeny testing, resulting in limited genetic gain due to either low accuracy of genetic proofs or long generation intervals.

Despite presenting good adaptation to tropical conditions and an extraordinary capacity to convert (low quality) pasture on meat production, Nellore cattle tends to present lower quality beef (in terms of tenderness, for example) compared to some Taurine breeds. This helps to explain the huge investments and efforts that Brazilian research groups and breeding programs are doing to establish reference populations and develop prediction equations for meat quality traits. Important studies are revealing the existence of genetic variation and the feasibility of applying genomic selection for these traits (Aboujaoude *et al.* 2016; Feitosa *et al.* 2016; Fernandes Júnior et al.

2016a; Gordo et al. 2016; Chiaia *et al.* 2017), attaining prediction accuracies of about 0.4 for some relevant traits (e.g. tenderness) that previously to genomics were not evaluated. Motivated by the results of these studies some breeders are establishing a consortium to develop strategies that would warranty the improvement and maintenance of the prediction equations for carcass and meat quality traits. Efforts and investments are also been made to develop prediction equations for traits related to feed efficiency (Silva *et al.* 2016) and, more recently, methane emission.

Genomic predictions accounting for GxE

Genotypic information has allowed not only obtaining more accurate genetic proofs in different environments but also identifying young animals with less sensitivity to environmental variation (not published). This will help breeders to better explore genotype by environment interaction, which is commonly an important source of phenotypic variation in tropical environments (Cardoso and Tempelman 2012; Chiaia *et al.* 2015; Santana *et al.* 2015). Breeders are now able to select young sires to produce under specific conditions without the necessity to progeny test them in different environments.

Other applications

Many other uses of genotypic information in Nellore breeding programs are emerging. For instance, GWAS are being performed on morphologic and functional traits such as testicular hypoplasia and skin depigmentation, and important candidate major genes are being identified for these traits (H.H.R. Neves, personal communication). If their effects are confirmed, this information could be used on mating plans or on developing genetic tests, aiming to reduce the economic losses caused by the incidence of these problems. Another example of application is the use of genotypic information to perform genomic control of inbreeding (Sonesson *et al.* 2012). As multiple-sire mating is a regular practice in some farms, due to the large number of cows under natural mating, estimates of inbreeding based on genotypic information are expected to be more reliable than those based on incomplete or erroneous pedigree, allowing controlling inbreeding more effectively.

CONCLUDING REMARKS

Genetic improvement has an important role in increasing efficiency and competitiveness of beef cattle production in the tropics. There are several challenges and opportunities to genetically improve more effectively beef cattle herds in the tropics and genomic selection has shown to be a key tool to increase genetic progress of economically relevant traits. Some applications of genomic selection in Nellore cattle from Brazil were listed but more will certainly come or are already been applied by other breeding programs, breeds and countries. Individually, genomic selection applications may have a moderate impact on the breeding programs but considered together they are expected to significantly improve the genetic progress, profitability and sustainability of beef production in the tropics.

REFERENCES

Aboujaoude C., Pereira A.S.C., Feitosa F.L.B., Lemos M.V.A. et al. (2016) *Anim. Produc. Sci.*, http://www.publish.csiro.au/AN/AN16107

Bolormaa S., Pryce J.E., Kemper K., Savin K. et al. (2013) *J. Anim. Sci.* **91**: 3088.

Cardoso F.F., Tempelman R.J. (2012) *J Anim Sci* **90**:2130.

Carvalheiro R. (2014). In Proc. 10th WCGALP. CD-ROM.

Carvalheiro R., Neves H.H.R., Bresolin T. Arnold D.R., et al. (2017) *Proceedings* AAABG 2017.

Chiaia, H.L.J; Lemos, M.V.A.; Venturini, G.C.; Aboujaoude, C. et al. (2015) J.Anim.Sci. **93**:1503. Chiaia H., Peripoli E., Silva R., Aboujaoude C. et al. (2017) *Meat Sci.* **128**: 60.

Chidumayo E.N. and Gumbo D.J. (2013) *Energy Sust. Develop.* **17**: 86.

Costa R.B., Camargo G.M.F., Diaz I.D.P.S., Irano N. et al. (2015) Genet. Sel. Evol. **47**: 67. Feitosa F.L.B., Olivieri B.F., Aboujaoude C., Pereira A.S.C. et al. (2016) *J. Appl. Genet.* **58**:123.

Ferraz, J.B.S., Fries, L.A. (2004). In Proc 8th WCGALP. http://www.cabi.org/cabdirect/FullTextPDF/2006/20063169265.pdf.

Fernandes Júnior G.A., Rosa G.J.M., Valente B.D., Carvalheiro R. et al. (2016a) *Genet. Sel. Evol.* **48**: 7.

Fernandes Júnior G.A., Costa R.B., Camargo G.M.F., Carvalheiro R. et al. (2016b) *J. Anim. Sci.* **1**: 4087.

Foley J.A., Ramankutty N., Brauman K.A., Cassidy E.S. *et al.* (2011) *Nature* **478**: 337.

Fortes M.R.S., Kemper K., Sasazaki S., Reverter A. et al. (2013) *Anim. Genet.* **44**: 636.

Gordo D.G.M., Espigolan R., Tonussi R.L., Fernandes Junior G.A. et al. (2016) *J. Anim. Sci.* **94**:1821.

Hayes B.J., Lewin, H.A. and Goddard M.E. (2013) *Trends Genet.* **29**: 4.
Johnston D.J., Tier B. and Graser H.U. (2012) *Anim. Prod. Sci.* **52**: 100.
Magalhães A.F.B., Camargo G.M.F., Fernandes Júnior G.A., Gordo D.G.M. et al. (2016) *Plos One* 11: e0157845.
Millen D.D., Pacheco R.D.L., Meyer P.M., Rodrigues P.H.M. et al. (2011) *Anim. Front.* **1**: 2.
Neves H.R., Carvalheiro R., O`brien A.M.P., Utsunomiya Y.T. et al. (2014) *Genet. Sel. Evol.* **46**: 17.
Pérez O'Brien A.M., Mészáros G., Utsunomiya Y.T., Sonstegard T.S., *et al.* (2014). *Livest. Sci.* **166**:121.
Porto-Neto L.R., Reverter A., Prayaga K.C., Chan E.K.F. et al. (2014) *PLOS ONE* 9(11): e113284.
Regatieri I., Boligon A.A., Costa R.B., Souza F.R.P. et al. (2017) *Anim. Reprod. Sci.* **177**: 88.
Saadi H.A.S., Vigneault C., Sargolzaei M., Gagné D., et al. (2014) *BMC Genomics* **15**:889.
Santana Jr M.L., Eler J.P., Bignardi A.B., Menendez-Buxadera A., et al. (2015) *J Applied Genetics* **56**:219.
Silva R.M.O., Fragomeni B.O., Lourenço D.A.L., Magalhães A.F.B. et al. (2016) *J. Anim. Sci.* **94**: 3613.
Sonesson A., Woolliams J.A., Meuwissen T.H.E. (2012). *Genet Sel Evol* **44**:27.
Sun X., Fernando R., Dekkers J. (2016). *Genet Sel Evol* **48**:77.
United States Department of Agriculture (USDA) (2016) Livestock and poultry: world markets and trade. Foreign Agricultural Service/USDA; Office of Global Analysis.
Utsunomiya Y.T., Carmo A.S., Carvalheiro R., Neves H.H.R. et al. (2013) *BMC Genet.* **14**: 52.
Valente T.S., Baldi F., Sant´anna A.C., Albuquerque L.G. and Costa M.J.R.P. (2016) *PLOS ONE*
11: e0156956.
Van Eenennaam A.L., Weigel K.A., Young A.E., Cleveland M.A., Dekkers J.C.M. (2014) *Annual Rev. Anim. Biosc.* **2**: 105.

ACKNOWLEDGMENTS
We would like to acknowledge the breeding programs CIA de Melhoramento, DeltaGen, Nelore Qualitas and PAINT for their support in our research studies, and the funding agencies CNPq and FAPESP #2009/16118-5, for financial support.

Proc. Assoc. Advmt. Anim. Breed. Genet. 22:265-268

A MULTI-TRAIT APPROACH TO INCORPORATING FOREIGN PHENOTYPES AND GENOTYPES IN GENOMIC PREDICTIONS TO INCREASE ACCURACY AND REDUCE BIAS

B.J.Hayes[1,2], G. Nieuwhof[1,3], and M. HaileMariam[1]

[1]AgriBio, Department of Economic Development, Jobs, Transport and Resources, Victoria, Australia
[2]Queensland Alliance for Agriculture and Food Innovation, Centre for Animal Science, University of Queensland, Queensland, Australia
[3]Datagene,Agribio,Bundoora,Australia

SUMMARY

We investigated improvements in reliability of genomic estimated breeding values (GEBV) for key dairy traits as a result of including a large number of genotypes of dairy bulls, with North American daughter performance, in the Australian genomic reference set. Two strategies for incorporating the North American information into Australian genomic evaluations were compared, a single trait approach, where the phenotypes used were de-regressed Interbull proofs (DRP), and a multi-trait approach, where North American performance and Australian performance (DYD of bulls based on Australian daughters) were treated as two, potentially correlated, traits. The two strategies were compared by assessing the correlation of GEBV and DYD from Australian daughters in a set of validation bulls, for milk, fat and protein, somatic cell count, survival, daughter fertility, stature and overall type. Including genotypes of bulls with North American daughter performance in Australian genomic evaluations improved the correlation of GEBV and DYD in the validation bulls for all traits, by between 3% and 7% for production, and up to 15% for fertility and survival. The single trait approach resulted in bias (GEBV over predicting DYD) for some traits including survival, somatic cell count and overall type, while the multi-trait approach gave unbiased GEBV for these traits.

INTRODUCTION

Reliability of genomic estimated breeding values (GEBV) for animals without a phenotype of their own or for daughters (eg young unproven bulls) is a function of the heritability of the trait, the proportion of genetic variance explained by the markers, the genetic diversity of the population, and the number of animals in the reference population where SNP effects are estimated (Daetwyler et al., 2008; Goddard, 2009). The reliability of GEBV is also a function of how closely related young genomic bulls are to the reference population. One way of improving reliabilities of GEBV would be to expand the reference set to include bulls with only overseas daughter information, including those that are sires or grandsires of young genomic bulls used in Australia. This requires a method that appropriately accounts for genotype by environment interaction between Australia and the other countries.

Here we investigate improvements in reliability of Australian genomic breeding values (ABVg) for key dairy traits that can be achieved for young, unproven bulls as a result of including a large number of genotypes of bulls with North American daughter performance into the Australian genomic reference set. Two strategies for incorporating the North American information into Australian genomic evaluations were compared,

1) a single trait approach, where the phenotypes used were de-regressed Interbull multiple across country evaluation (MACE) proofs, and

2) a multi-trait approach, where North American performance (daughter yield deviation, DYD, of bulls based on North American daughters) and Australian performance (DYD of bulls

based on Australian daughters) were treated as two, potentially correlated, traits.

The two strategies were compared by assessing the correlation of genomic estimated breeding values (GEBV) and DYD from Australian daughters) in a set of validation bulls (born in or after 2008). The increase in this correlation relative to a single trait approach with only the current Australian reference set was evaluated. The regression of DYD on GEBV was also evaluated, to determine if there was any bias (i.e. if the GEBV systematically over estimated or underestimated the proofs of top ranking bulls when the bulls had daughters). Traits investigated were production (milk, fat and protein kg), somatic cell count, survival, daughter fertility, stature and overall type (standardised traits with a mean of 100 and standard deviation of 5).

MATERIALS AND METHODS

Genotypes for 18,377 North American registered bulls with daughter records were extracted from the Northern American Cooperative Dairy DNA Repository (CDDR) database, and 13,072 bulls and cows from the ADHIS (Australian dairy herd improvement scheme) database. A set of 36,655 SNP common to the Australian evaluation and present in the North American genotypes was identified. Any missing genotypes were imputed using Beagle 3.2 (Browning and Browning 2009). The traits investigated were milk yield, fat yield, protein yield, somatic cell count (SCC), survival (longevity), stature, overall type (overall conformation score was the corresponding Interbull trait) and fertility (daughter pregnancy rate).

North American phenotypes were daughter yield deviations from the US, for milk yield, fat yield, protein yield, somatic cell count (SCC) and survival. For fertility, the North American PTA was de-regressed as suggested by Van Raden (pers comm). For type traits, de-regressed breeding values were used (de-regression removed the pedigree contribution of the EBV), where bulls had at least 50 daughters scored for the trait, using the procedure of Liu (2009). Australian phenotypes were daughter trait deviations (DTDs) for all traits.

The data were split based on year of birth into reference and validation sets. Bulls (either North American or Australian) born before 2008 were included in the reference set, used to calculate SNP effects, and bulls born in or after 2008 were used in the validation set. There were 275 bulls in the validation set, and only Australian daughter information was used in the validation

Three models were fitted to the data 1) Single trait model, Australian information only (de-regressed MACE proofs as phenotypes), 2) Single trait model, Australian and North American information (de-regressed MACE proofs as phenotypes) 3), Multi-trait model, using daughter trait deviations for bulls with daughters in Australia, and DYD for bulls with North American daughters (as described above). Where DYD were not available, de-regressed proofs were used.

The multi-trait model was $\begin{bmatrix} y_1 \\ y_2 \end{bmatrix} = \begin{bmatrix} I_1 & 0 \\ 0 & I_2 \end{bmatrix} \begin{bmatrix} \mu_1 \\ \mu_2 \end{bmatrix} + \begin{bmatrix} Z_1 & 0 \\ 0 & Z_2 \end{bmatrix} \begin{bmatrix} g_1 \\ g_2 \end{bmatrix} + \begin{bmatrix} e_1 \\ e_2 \end{bmatrix}$ where $\mathbf{y_1}$ and $\mathbf{y_2}$ are the vector of response variables (i.e. trait 1 is the DTD of Australian bulls and trait 2 are the DYD of bulls with North American daughters), $\mathbf{I_1}$ and $\mathbf{I_2}$ are identity matrices, $\boldsymbol{\mu_1}$ and $\boldsymbol{\mu_2}$ is the vector of intercepts of DTD and DYD, $\mathbf{Z_1}$ **and** $\mathbf{Z_2}$ are the design matrices that relate genomic breeding values with the individuals, $\mathbf{g_1}$ and $\mathbf{g_2}$ is the vector of genomic breeding values for DTD and DYD, and $\mathbf{e_1}$ and $\mathbf{e_2}$ are vectors of random residuals for DTD and DYD. It was assumed that $\begin{bmatrix} g_1 \\ g_2 \end{bmatrix}$ $\sim N(0, G \otimes T)$, where $\mathbf{T} = \begin{bmatrix} \sigma^2_{g1} & \sigma_{g12} \\ \sigma_{g12} & \sigma^2_{g2} \end{bmatrix}$, the variance-covariance matrix of DTD and DYD, and $\begin{bmatrix} e_1 \\ e_2 \end{bmatrix} \sim N(\mathbf{0}, \boldsymbol{I} \otimes \boldsymbol{R}), where\ \boldsymbol{R} = \begin{bmatrix} \boldsymbol{R}_{11} & \mathbf{0} \\ \mathbf{0} & \boldsymbol{R}_{22} \end{bmatrix}$, the residual variance-covariance matrix of DTD and DYD, with weights on phenotypes for bulls and cows according to Garrick et al. (2009), and **G** is the genomic relationship matrix. ASReml (Gilmour et al., 2009) was used to estimate variance components, including genetic correlations between performance in North American and Australia.

The three different models were used to predict GEBV for the validation bulls, born 2008 and later. The following statistics were assessed: correlation between DTD and GEBV for bulls in the validation set, and the slope of the regression (b) of DTD on GEBV for validation bulls.

Note that to (considerably) simplify implementation in routine evaluations, if a bull has Australian daughters, only his Australian information is used. Bulls are included for the second country trait (eg North America) only if they have daughters in that second country and not in Australia. This means it is not necessary to consider residual correlations among the countries. A second step to simplify implementation was to pre-correct records in each country for the mean and sex effect. Then the solutions to the multiple trait model are (Ignoring fixed effects, and with t the elements of the inverted T matrix, eg t^{11} is the element in the first row and column of T^{-1}):

$$\begin{bmatrix}\hat{g}_1\\ \hat{g}_2\end{bmatrix} = \begin{bmatrix} Z_1'R^{11}Z_1 + G^{-1}t^{11} & G^{-1}t^{12} \\ G^{-1}t^{12} & Z_2'R^{22}Z_2 + G^{-1}t^{22}\end{bmatrix}\begin{bmatrix}Z_1'R^{11}y_1\\ Z_2'R^{22}y_2\end{bmatrix}$$

RESULTS AND DISCUSSION

When de-regressed MACE proofs were used as the phenotype in a single trait analysis, the correlations r(GEBV,DTD) for production were relatively high, and improved by up to 7% (fat) with the addition of the North American data (Figure 1A).

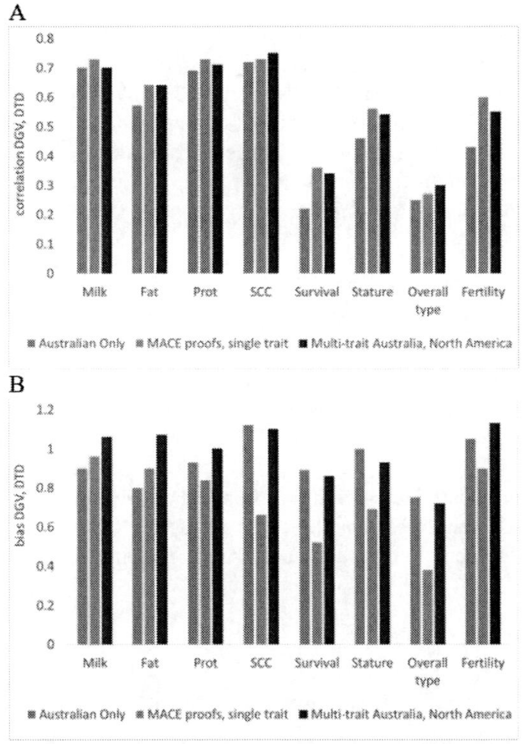

Figure 1. A. Correlation of genomic estimated breeding values (GEBV) for eight dairy traits, using three genomic prediction models. B. Regression of daughter yield deviation on GEBV for eight dairy traits using three genomic prediction models.

The multi-trait approach did result in regression coefficients (slopes) of DTD on GEBV in validation bulls closer to one for some traits (Figure 1B). Particularly survival, somatic cell count and overall type were closer to one with the multi-trait approach.

As a result of running the multi-trait model, genetic correlations between North America and Australia were estimated for all traits considered. These were slightly higher than, but close to the Interbull reported correlations, Table 1. This is interesting, as the multi-trait model uses genomic information only, while the Interbull correlations are based on pedigree.

Table 1. Genetic correlations between Australia and the US, estimated either from the multi-trait genomic model, or from pedigree (Interbull reported correlations).

Trait	Multi-trait genomic estimate	Interbull*
Milk	0.81	0.77
Fat	0.81	0.76
Prot	0.75	0.75
SCC	0.73	0.77
Survival	0.75	0.69
Stature	0.95	0.89
Overall type	0.72	0.64

*http://www.interbull.org/index

Using either a multi-trait approach or a single trait approach (using de-regressed MACE proofs) to add North American daughter performance information to the reference population for calculating GEBV resulted in an increase in r(DTD,GEBV) for a set of validation bulls. The multi-trait approach resulted in slightly less bias (slope of DTD on GEBV for validation bulls) for some traits, and is therefore the preferred approach for these traits. Estimates of genetic correlations between North America and Australia derived from the genomic information were similar to, but slightly higher than, the published Interbull correlations.

ACKNOWLEDGMENTS

The authors are grateful to the CDDR, George Wiggans, Jay Weiker Gordon Doak, Kent Weigel, Jaques Chesnais, Daniel Abernethy and Matthew Shaffer for facilitating this research project.

REFERENCES

Browning, B. L., and S. R. Browning. 2009. A unified approach to genotype imputation and haplotype phase inference for large data sets of trios and unrelated individuals. Am. J. Hum. Genet. **84**:210–223.

Garrick, D. J., J. F. Taylor, and R. L. Fernando, 2009. Deregressing estimated breeding values and weighting information for genomic regression analyses. Genet. Sel. Evol. **41**:55.

Liu, Z. 2009. Deregressing MACE proofs for genomic evaluations. Proteje meeting, Brussels, Belgium.

Proc. Assoc. Advmt. Anim. Breed. Genet. 22:269-272

SOME ALTERNATIVE COMPUTATIONAL STRATEGIES FOR SINGLE-STEP NATIONAL GENOMIC EVALUATIONS

D.J. Garrick[1,2,3], H. Cheng[3], B.L. Golden[2] and R.L. Fernando[3]

[1]Institute of Veterinary, Animal & Biomedical Sciences, Massey University, Hamilton, NZ
[2]ThetaSolutions LLC, Atascadero, California, USA
[3]Department of Animal Science, Iowa State University, Ames, Iowa, USA

SUMMARY

Single-step genomic evaluations combine pedigree and phenotypic information on genotyped and non-genotyped individuals. Such an evaluation can be undertaken using a so-called breeding value model that fits the breeding values of the genotyped and non-genotyped animals (e.g. single-step GBLUP) or using an equivalent so-called marker effects model that directly fits the marker effects. The single-step marker-effects models allow alternatives such as BayesA and mixture models such as BayesB, BayesC or BayesR to be fitted in the context of the single-step analysis. This paper reviews alternative formulations of these equivalent models. The marker-effects formulations of the models are practical options for national genomic evaluations. The most efficient algorithm among those available depends upon the number of marker loci and the numbers of genotyped and non-genotyped animals.

INTRODUCTION

The classical model equation for genetic evaluation using best linear unbiased prediction (BLUP) describes the phenotypes for one or more traits in terms of fixed effects, random additive breeding values, and residual effects that capture that part of the phenotype that cannot be explained by the fixed effects or breeding values (Henderson, 1973). Estimation of breeding values by fitting the mixed linear model typically assumes the pedigree-based additive relationship matrix describes the variance-covariance among breeding values (Henderson, 1973). Henderson (1974) suggested the model equation might be rearranged for computational advantage as explicitly demonstrated in Henderson (1985). That concept was exploited by Quaas and Pollak (1980) in their derivation of the multiple-trait reduced animal model which allowed an animal model to be fitted with little more effort than that for fitting the sire-maternal grandsire models that were commonly used at that time. Nejati-Javeremi *et al.* (1997) showed how to compute a genomic relationship matrix and suggested that be used in place of the additive relationship matrix, a model now known as GBLUP. Meuwissen *et al.* (2001) proposed several models that explicitly fitted haplotype effects rather than breeding values. Those methods varied according to whether the variance ratio for haplotype effects was a known constant (BLUP), an unknown haplotype specific variable (BayesA, BayesB), and whether or not some haplotypes were assumed to have zero effect (BayesB). The breeding value model and BLUP marker effects models were shown to be equivalent (e.g. Stranden and Garrick, 1997). Expanding GBLUP to a more general setting with a model that appropriately accounts for a pedigree including genotyped and non-genotyped animals in a single step was introduced by Legarra *et al.* (2009). That single-step GBLUP (ss-GBLUP) model represented a major advance, and is computationally attractive when there are many more markers than genotyped animals, and all markers are weighted equally to form the genomic relationship matrix. Two marker-effects models are reviewed here which are equivalent to ss-GBLUP and practical for national evaluation. Both allow the model for marker effects to be extended when variance ratios are marker specific and unknown (like BayesA), or follow more general mixture models (BayesB, or BayesR of Erbe *et al.* 2012).

EQUIVALENT MODELS FOR JOINT USE OF GENOTYPED AND NON-GENOTYPED ANIMALS

Single-step GBLUP. Defining a vector of phenotypic records as $\boldsymbol{y}_i$, incidence matrices of fixed effects and breeding values as $\boldsymbol{X}_i$ and $\boldsymbol{Z}_i$, vectors of unknown fixed effects ($\boldsymbol{b}$), random effects ($\boldsymbol{u}_i$) and residuals $\boldsymbol{e}_i$, with the subscript i denoting g=genotyped or n=non-genotyped animals, the model equation can be written as

$$\begin{bmatrix}\boldsymbol{y}_n\\ \boldsymbol{y}_g\end{bmatrix}=\begin{bmatrix}\boldsymbol{X}_n\\ \boldsymbol{X}_g\end{bmatrix}\boldsymbol{b}+\begin{bmatrix}\boldsymbol{Z}_n & \boldsymbol{0}\\ \boldsymbol{0} & \boldsymbol{Z}_g\end{bmatrix}\begin{bmatrix}\boldsymbol{u}_n\\ \boldsymbol{u}_g\end{bmatrix}+\begin{bmatrix}\boldsymbol{e}_n\\ \boldsymbol{e}_g\end{bmatrix}, \text{ with } var\begin{bmatrix}\boldsymbol{e}_n\\ \boldsymbol{e}_g\end{bmatrix}=\begin{bmatrix}\boldsymbol{R}_n & \boldsymbol{0}\\ \boldsymbol{0} & \boldsymbol{R}_g\end{bmatrix},$$

and following Legarra *et al.* (2009) with the genetic variance being σ_u^2,

$$\boldsymbol{H}=\frac{1}{\sigma_u^2}var\begin{bmatrix}\boldsymbol{u}_n\\ \boldsymbol{u}_g\end{bmatrix}=\begin{bmatrix}\boldsymbol{A}_{nn}+\boldsymbol{A}_{ng}\boldsymbol{A}_{gg}^{-1}(\boldsymbol{G}-\boldsymbol{A}_{gg})\boldsymbol{A}_{gg}^{-1}\boldsymbol{A}_{gn} & \boldsymbol{A}_{ng}\boldsymbol{A}_{gg}^{-1}\boldsymbol{G}\\ \boldsymbol{G}\boldsymbol{A}_{gg}^{-1}\boldsymbol{A}_{gn} & \boldsymbol{G}\end{bmatrix},$$

which is somewhat formidable. However, Aguilar *et al.* (2010) showed that, for full-rank $\boldsymbol{G}$,

$$\boldsymbol{H}^{-1}=\begin{bmatrix}\boldsymbol{A}^{nn} & \boldsymbol{A}^{gn}\\ \boldsymbol{A}^{ng} & \boldsymbol{A}^{gg}+\left(\boldsymbol{G}^{-1}-\boldsymbol{A}_{gg}^{-1}\right)\end{bmatrix},$$

which allows existing software used to obtain breeding values in national evaluations using PCG iteration (e.g. Tsuruta *et al.* 2001) to be relatively trivially modified by including an extra step to compute matrix-vector products for the difference matrix $\left(\boldsymbol{G}^{-1}-\boldsymbol{A}_{gg}^{-1}\right)$. This ss-GBLUP approach was computationally appealing in the early days of genomic prediction, when there were fewer than 40,000 animals genotyped. As the number of genotyped animals increased, the effort to form the dense difference matrix and compute its matrix-vector products increase rapidly. Various strategies to avoid that effort have been devised and implemented, including computing matrix-vector products in parts as $\left(\boldsymbol{G}^{-1}-\boldsymbol{A}_{gg}^{-1}\right)\boldsymbol{x}=\boldsymbol{G}^{-1}\boldsymbol{x}-\boldsymbol{A}_{gg}^{-1}\boldsymbol{x}$. Using properties of partitioned matrix inverses allows efficient computation of the product $\boldsymbol{A}_{gg}^{-1}\boldsymbol{x}$ without ever forming $\boldsymbol{A}_{gg}^{-1}$ (Masuda *et al.* 2017). An approximation known as APY (Misztal *et al.* 2014) has been used to compute the matrix product $\boldsymbol{G}^{-1}\boldsymbol{x}$. That approximation can in some cases give identical values as for $\boldsymbol{G}^{-1}\boldsymbol{x}$ computed directly, but the lower bounds for APY in general circumstances have not been established.

Single-step GBLUP with marker effects. There are several practical alternatives for fitting the single-step model that do not require $\boldsymbol{G}^{-1}$, nor even require $\boldsymbol{G}$ to be full rank, and these equivalent models have the additional advantage that they can accommodate various priors for marker effects, allowing single-step models for marker effects akin to BayesA, BayesB and BayesR that cannot be fitted using ss-GBLUP.

Liu *et al.* (2014) rearranged the model to include equations for the marker effects, $\boldsymbol{\alpha}$, in addition to the breeding values of genotyped and non-genotyped individuals. An advantage of that representation is that it does not require the matrix $\boldsymbol{G}$, nor its inverse. However, it involves the inverse of the matrix $\boldsymbol{A}_{gg}$, which is dense. A computational strategy was proposed to avoid computing the inverse, but it requires solving a dense system of equations of order equal to the number of non-genotyped animals, and such solution is required every round of PCG or for every Gibbs sample if a model with Bayesian priors for marker effects is to be fitted. We will not consider that representation further.

Hybrid model. Fernando *et al.* (2014) wrote $\boldsymbol{u}_g=\boldsymbol{M}_g\boldsymbol{\alpha}$ as in Meuwissen *et al.* (2001) where $\boldsymbol{M}_g$ are marker covariates observed on genotyped animals, and partitioned $\boldsymbol{u}_n$ into two components, that part of the breeding values of non-genotyped animals that can be explained by the breeding values of genotyped relatives, and an independent part (imputation error, $\boldsymbol{\epsilon}$) not explained by those relatives. That is, $\boldsymbol{u}_n=\boldsymbol{M}_n\boldsymbol{\alpha}+\boldsymbol{\epsilon}$, where non-genotyped marker covariates are "imputed" using best linear prediction as $\boldsymbol{M}_n=\boldsymbol{A}_{ng}\boldsymbol{A}_{gg}^{-1}$ which can be obtained efficiently by directly solving the sparse set of equations $\boldsymbol{A}^{nn}\boldsymbol{M}_n=-\boldsymbol{A}^{ng}\boldsymbol{M}_g$ and is easily done in parallel. The resulting "hybrid" model equation is therefore written as

$$\begin{bmatrix} y_n \\ y_g \end{bmatrix} = \begin{bmatrix} X_n \\ X_g \end{bmatrix} b + \begin{bmatrix} Z_n M_n & 0 \\ 0 & Z_g M_g \end{bmatrix} \alpha + \begin{bmatrix} Z_n \\ 0 \end{bmatrix} \epsilon + \begin{bmatrix} e_n \\ e_g \end{bmatrix},$$

which is solved by fitting mixed model equations that explicitly include effects for $\boldsymbol{\alpha}$ and $\boldsymbol{\epsilon}$. Defining the variance of the vector of marker effects as $\boldsymbol{I}\sigma_\alpha^2$, the inverse variance-covariance matrix for these effects required to form the mixed model equations are

$$var^{-1}\begin{bmatrix} \alpha \\ \epsilon \end{bmatrix} = \begin{bmatrix} I\frac{1}{\sigma_\alpha^2} & 0 \\ 0 & A^{nn}\frac{1}{\sigma_u^2} \end{bmatrix}.$$

The calculations involving M_n which appears in the off-diagonal of the mixed model equations become formidable when that dense matrix is large, as is the case when there are millions of non-genotyped animals and a large number of markers, but that effort can be reduced when the number of genotyped animals is much less than the number of non-genotyped animals by exploiting the identity $\boldsymbol{A}^{nn}\boldsymbol{M}_n = -\boldsymbol{A}^{ng}\boldsymbol{M}_g$ and storing only $\boldsymbol{M}_g$ as detailed in Fernando *et al.* (2016a). Implementing that approach requires repeated solving of sparse equations of the form $\boldsymbol{A}^{nn}\boldsymbol{x} = \boldsymbol{q}$ within each PCG iteration. This effort is akin to that required to implement the approach of Masuda *et al.* (2017) in ss-GBLUP. If the variance components σ_α^2 or σ_u^2 are assumed not to be known, and/or if mixture priors are to be used for marker effects, this hybrid model can be readily fitted using single-site Gibbs sampling, a model that does not have an equivalent ss-GBLUP form.

Super hybrid model. A further equivalent model involving marker effects can be derived as in Fernando *et al.* (2016b). In circumstances where the number of genotyped animals may be large, perhaps millions, it can be efficiently implemented for national evaluation, especially if there are more genotyped than non-genotyped animals. The model equation is written as

$$\begin{bmatrix} y_n \\ y_g \end{bmatrix} = \begin{bmatrix} X_n \\ X_g \end{bmatrix} b + \begin{bmatrix} 0 \\ Z_g M_g \end{bmatrix} \alpha + \begin{bmatrix} Z_n \\ 0 \end{bmatrix} u_n + \begin{bmatrix} e_n \\ e_g \end{bmatrix},$$

which is solved by fitting a mixed model involving $\boldsymbol{\alpha}$, as in the hybrid model, along with $\boldsymbol{u}_n$ as in ss-GBLUP. We refer to this model here as the super-hybrid model. The inverse variance-covariance matrix for the fitted effects is given by

$$var^{-1}\begin{bmatrix} \alpha \\ u_n \end{bmatrix} = \begin{bmatrix} I\frac{1}{\sigma_\alpha^2} + M_n{}'A^{nn}M_n\frac{1}{\sigma_u^2} & M_g{}'A^{gn}\frac{1}{\sigma_u^2} \\ A^{ng}M_g\frac{1}{\sigma_u^2} & A^{nn}\frac{1}{\sigma_u^2} \end{bmatrix},$$

which only involves the matrix of imputed marker genotypes $\boldsymbol{M}_n$ in a quadratic form on the diagonal, and that symmetric matrix has order equal to the number of markers which in national evaluations can nowadays be an order of magnitude less than the number of genotyped individuals. Comparison of the computing effort in this super-hybrid model relative to the hybrid model for a national cattle evaluation dataset is in Fernando *et al.* (2016b).

All of these equivalent models, namely ss-GBLUP which fits breeding values for non-genotyped and genotyped animals, the ss-GBLUP model with breeding values and marker effects, the hybrid model which fits marker effects and imputation residuals for non-genotyped animals, and the super-hybrid model which fits marker effects and breeding values for non-genotyped animals, can be extended to more complex forms of models. These include those that fit additional polygenic effects not captured by markers, those that fit maternal genetic and maternal permanent environmental effects, and those accommodating multiple traits, those with repeated measures, those including random regression polynomials, those with heterogeneous variances, in addition to breed, heterosis and group effects as required in multi-breed analyses.

DISCUSSION

The two marker effects models reviewed here are equivalent to ss-GBLUP when the genomic relationship matrix is full rank and the variance parameters are known. These marker-effects models may require greater computational effort than ss-GBLUP when the number of genotyped animals is small. The relative effort for the hybrid model that fits marker effects and imputation residuals for non-genotyped animals, compared to the super-hybrid model that fits marker effects and breeding values for non-genotyped animals, varies according to the number of markers and numbers of genotyped and non-genotyped animals. For analyses involving millions of genotyped animals, one or other or both of the marker effects models will be more efficient than ss-GBLUP. Implemented in a Gibbs sampler, these models can readily accommodate alternative priors including those representing mixture distributions, which in some situations leads to higher accuracy of prediction than ss-GBLUP (Lee et al. 2017). Furthermore, using Gibbs sampling will provide samples from the relevant posterior distributions which can be used to provide estimates of the prediction error variances and prediction error covariances, as well as the posterior means that represent the estimates of the breeding values. Both of these marker effects models have been prototyped in multi-breed multiple-trait national evaluations including maternal effects. The super-hybrid model is currently being implemented in the Pan-American Cattle Evaluation (PACE) run by ABRI for Hereford cattle, and in the North American multi-breed analysis run by International Genetic Solutions (IGS) which is the largest North American evaluation in terms of pedigree size. The super-hybrid model is also being used by global companies for pig, chicken and dairy cattle evaluation.

CONFLICT OF INTEREST

DJG and BLG are co-founders of Theta Solutions LLC, a company that licenses BOLT software which is capable of fitting all the models described in this paper.

REFERENCES

Aguilar I., Misztal I., Johnson D.L., Legarra A., Tsuruta S. and Lawlor T. (2010) *J. Dairy Sci.* **93**:743.

Erbe M., Hayes B.J., Matukumalli L.K., Goswami S., Bowman P.J., Reich C.M., Mason B.A. and Goddard M.E. (2012) *J. Dairy Sci.* **95**:4114.

Fernando R.L., Cheng H., Golden B.L. and Garrick D.J. (2016a) *Genet. Sel. Evol.* **48**:96.

Fernando R.L., Cheng H. and Garrick D.J. (2016b) *Genet. Sel. Evol.* **48**:80.

Fernando R.L., Dekkers J.C.M. and Garrick D.J. (2014) *Genet. Sel. Evol.* **46**:50.

Henderson C.R. (1973) In *Proc of the Animal Breeding and Genetics Seminar in Honor of J.L. Lush. ASAS and ADSA Champaign, IL*

Henderson C.R. (1974) *J. Dairy Sci.* **57**:963.

Henderson C.R. (1985) *J. Dairy Sci.* **68**:2267.

Lee J., Cheng H., Garrick D., Golden B., Dekkers J., Park K., Lee D. and Fernando R. (2017) *Genet. Sel. Evol.* **49**:2.

Legarra A., Aguilar I. and Misztal I. (2009) *J. Dairy Sci.* **92**:4656.

Liu Z., Goddard M.E., Reinhardt F., and Reents R. (2014) *J. Dairy Sci.* **97**:5833.

Masuda Y., Misztal I., Legarra A., Tsuruta S., Laurenco D.A.L., Fragomeni B.O. and Aguilar I. (2017) *J. Animal Sci.* **95**:49.

Meuwissen T.H.E., Hayes B.J. and Goddard M.E. (2001) Genetics **157**:1819.

Misztal I., Legarra A. and Aguilar I. (2014)) *J. Dairy Sci.* **97**:3943.

Nejati-Javaremi A., Smith C. and Gibson J.P. (1997) *J. Animal Sci.* **75**:1738.

Quaas R.L., and Pollak E.J. (1980) *J. Animal Sci.* **51**:1277.

Stranden I. and Garrick D.J. (1997) *J. Dairy Sci.* **92**:2971.

Tsuruta S., Misztal I. and Stranden I. (2001) *J. Animal Sci.* **79**:1166.

Proc. Assoc. Advmt. Anim. Breed. Genet. 22:273-276

A NOVEL NUMERICAL METHOD TO QUANTIFY THE CONTRIBUTION OF GENES TO THE POPULATION STRUCTURE

P. Kasarapu[1], L. R. Porto-Neto[1], M. R. S. Fortes[2], S. A. Lehnert[1], M. A. Mudadu[3], L. Coutinho[4], L. Regitano[5], A. George[6] and A. Reverter[1]

[1] CSIRO Agriculture and Food, Queensland Bioscience Precinct, Brisbane, Queensland, Australia
[2] School of Chemistry and Molecular Biosciences, The University of Queensland, Australia
[3] Embrapa Agricultural Informatics, Av. Andre Tosello, 209 Campinas, Sao Paulo, Brazil
[4] University of Sao Paulo, Brazil
[5] Embrapa Southeast Livestock, Rodovia Washington Luiz, Km 234, Sao Carlos, Brazil
[6] CSIRO, DATA61, Ecosciences Precinct Brisbane, Queensland, Australia

SUMMARY

Principal component analysis (PCA) using genome-wide single nucleotide polymorphism (SNP) genotype data is traditionally used to determine distinct groups in a population. We present a novel numerical approach to quantify the importance of each gene to the emerging clusters as informed by PCA. Our method is based on modelling the coefficients (SNP weights) of the first principal component using mixtures of Normal distributions. We applied our approach to three distinct datasets of cattle, chicken, and sheep. We were able to identify subsets of genes in the cattle and chicken genomes that are likely to be important determinants for understanding the phenotypic differences among various disparate livestock populations.

INTRODUCTION

The utility of PCA in discriminating individuals according to breed differences has been well documented in the literature. The Bovine HapMap Consortium used PCA as a central method in the elucidation of genetic structure across biologically diverse breeds (Gibbs *et al.* 2009). Other studies relied on PCA to measure the genetic divergence between indicine and taurine cattle (Bertolini *et al.* 2015). PCA also informs machine learning based classification methods to predict the individual ancestry of cattle (Bertolini *et al.* 2015).

In our study, we used PCA as a starting point to identify a set of genes that have the discriminatory power to identify the lineage of a particular population. We build a model based on the contributions of the SNP to the first principal component (PC1). These are the coefficients of the PC1 which are produced as part of the PCA. The empirical distribution of the SNP reveals distinct modes. We used the output from PCA as a first step to project the data on to the maximum variable direction and used statistical machine learning based mixture modelling to quantify the contribution of genes to the respective lineages.

MATERIALS AND METHODS

Animals and genotypes. We tested our method on three datasets - cattle, chicken, and sheep.

Cattle: We used data from 18,363 animals and 19 breeds belonging to the Beef CRC (http://www.beefcrc.com) and Nelore data from Mudadu *et al.* 2016. The cattle belong to a spectrum of lineages ranging from pure *Bos indicus* (BI; N=5,536 cattle) to pure *Bos taurus* (BT; N=7,589). Additionally, we have 5,238 cattle that are crossbred or tropically adapted composites which are classified as *Bos taurus – Bos indicus* (BTI) breeds. The original data had genotypes for 729,068 SNP. We considered SNP located in autosomal chromosomes and mapped within 1Kb of a known gene to capture SNP associated with protein-coding regions. We further pre-processed the data so that we retained those genes that have at least the median number of 6 SNP to ensure that the genes are minimally represented. The final dataset contained 246,864 SNP in 8,631 genes.

Chicken: The data were from 988 chickens from 4 commercial lines of broilers (Hudson *et al.* 2015), denoted as Lines A (N = 204), B (N = 244), C (N = 254), and D (N = 286). Lines A and B have been generated to select females, whereas lines C and D are to select males. The data had genotypes for 51,713 SNP. After removing monomorphic SNP and retaining those within 20 Kb of a gene (Reyer *et al.* 2015), we considered 36,395 SNP located in 12,642 genes.

Sheep: We used data from the Sheep Hapmap project (http://www.sheephapmap.org/) including 1,222 animals distributed across 9 regions and genotypes for 49,034 SNP. We considered SNP that were not monomorphic and those within 30 Kb of a known protein coding gene (Miller *et al.* 2011), which resulted in 26,077 SNP spanning 12,737 genes.

Principal Component Analysis and Gene contribution to lineage. We used PLINK (Chang *et al.* 2015) to perform the principal component analysis (PCA) and considered only the PC1 as it explains the maximum variability in the data and extracted the weights of each SNP to that component. We used mixture modelling to quantify the contribution of the genes to the lineages.

Mixture modelling is a statistical method to construct a probability distribution by combining the effects due to several component probability distributions. We considered the Normal component distributions to model the probability of the SNP weights in PC1, which were best modelled using two component distributions. Formally, a two-component mixture is defined as

$$\Pr(x) = \underbrace{w\,\mathcal{N}(x;\mu_1,\sigma_1)}_{p_1} + \underbrace{(1-w)\,\mathcal{N}(x;\mu_2,\sigma_2)}_{p_2}$$

where x corresponds to the data (SNP weights in PC1), $\Pr(x)$ is the probability distribution of the mixture, w is the weight of the first component in the mixture, μ_1, μ_2 and σ_1, σ_2 denote the means and the standard deviations of the two Normal ($\mathcal{N}$) components, respectively. As part of statistical inference, the mixture parameters, that is, $w, \mu_1, \mu_2, \sigma_1, \sigma_2$ were estimated using the EMMIX software (McLachlan *et al.* 1999). After estimating the mixture parameters, the contribution of each SNP to each components is given by its *posterior probability*, that is,

$$m_1 = \frac{p_1}{p_1 + p_2} \text{ and } m_2 = \frac{p_2}{p_1 + p_2}$$

where m_1 and m_2 are the posterior probabilities of the given SNP to belong to the first and second component, respectively. The values p_1 and p_2 constitute the two parts of $\Pr(x)$. Note that $m_1 + m_2 = 1$ which implies that for a given gene, m_1 and m_2 correspond to the contributions (memberships) of that gene to the two components of the mixture. As an example, to estimate a gene's contribution to the indicine content in bovine genome, we average the posterior probabilities (m_1 values) of the corresponding SNP in its coding region. A gene contributes to both the indicine and taurine components of the bovine genome. The value m_1 denotes the amount of contribution (as a percentage) to the indicine lineage. We infer that the left mode corresponds to *Bos indicus* because the animals with negative SNP weights are Nelore/Brahman cattle.

RESULTS AND DISCUSSION

The PCA of the cattle and chicken datasets reveals distinct clusters based on their respective lineages; the *Bos indicus, Bos taurus* and *Bos taurus – Bos indicus* breeds are separately clustered (Figure 1a). The PC1 and PC2 explain 21.8% and 2.3% variation in the data respectively. Similarly, for the chicken dataset, we observe Lines A and B distinctly clustered whereas Lines C and D are overlapping (Figure 1b). Mixture modelling of the SNP weights along PC1, (Figure 1c,d) resulted in distinct modes corresponding to the indicine and taurine components of the bovine genome. The estimate of the mixing proportion is $w = 0.31$ establishing an effective membership of 31% *Bos indicus* and 69% *Bos taurus* genes for this particular population. For the chicken data, PC1 and PC2 explain 22% and 3.6% variation in the data, respectively. The value $w = 0.49$ implies an almost equal number of genes contributing to male and female lines.

Further, 64 and 718 genes have a contribution of at least 95% to the indicine and taurine components, respectively. In the chicken genome, there are 1,072 and 1,386 genes with least 95% contribution. The study of these candidate genes can aid our understanding of ancestry-related differences in gene expression and susceptibility of a given lineage to exhibit a certain phenotype.

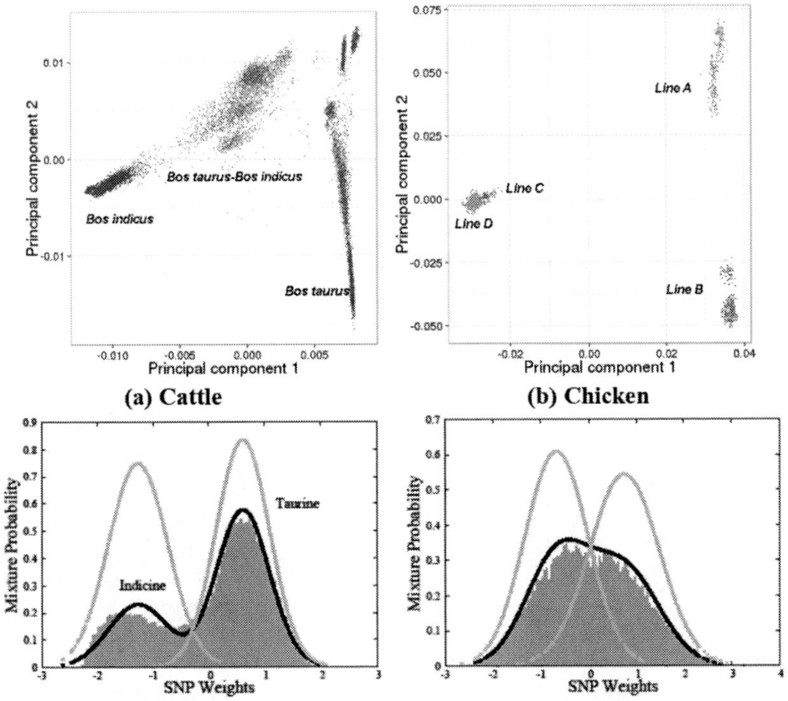

(a) Cattle (b) Chicken

(c) Cattle (d) Chicken

Figure 1. (a)-(b) PCA of SNP genotypes resulting in distinct clusters of animals based on their lineages. (c)-(d) Mixture modelling of SNP weights along PC1. Red indicates the actual distribution of SNP weights, grey curves are the individual Normal distributions, and black curve is the mixture model obtained by combining the two Normal distributions based on the mixing proportions.

The PCA of the sheep data revealed a cluster with sheep (Outgroup, Wildsheep) widely scattered and having negative PC1 values (Figure 2a). On removing these outliers, we note a star-shaped cluster (Figure 2b) with PC1 and PC2 accounting for 4.5% and 2.2% variation, respectively. Mixture modelling shows three distinct modes for the full data (Figure 4c), whereas there is a clear unimodal distribution for the filtered data (Figure 4d). This finding highlights the importance of pre-processing the data prior to our analysis. Kijas *et al.* (2012) suggests the absence of distinct lineages and strong historic mixing, in agreement with our observation of a unimodal distribution.

CONCLUSIONS

Our method based on the mixture modelling of SNP weights captures the genes responsible for the underlying population structure and potentially serves to establish a relationship between the evolutionary structure and phenotypic variation in livestock populations.

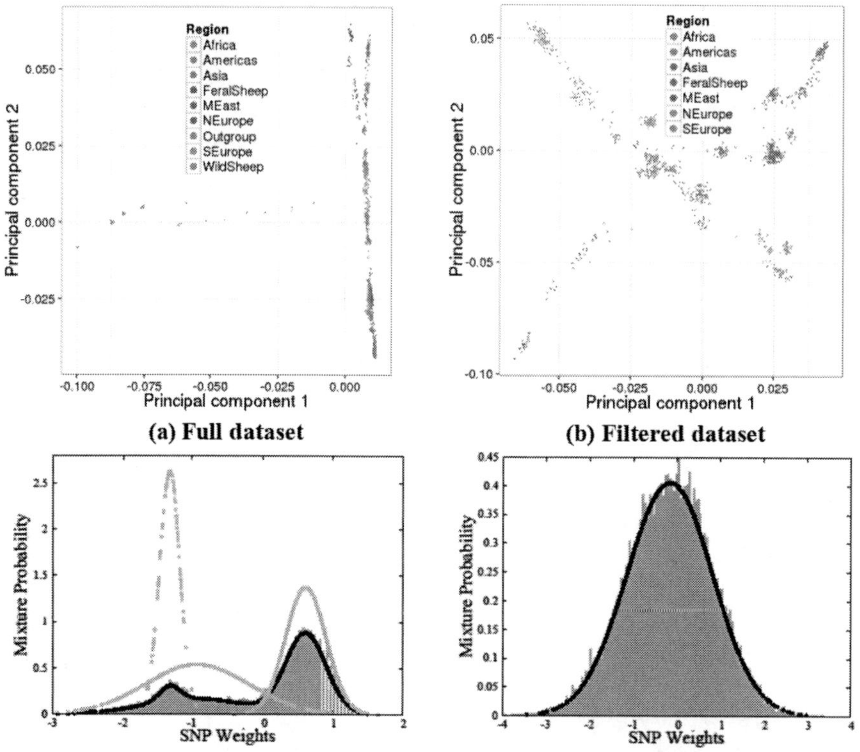

(a) Full dataset (b) Filtered dataset

(c) Full dataset (d) Filtered dataset

Figure 3. PCA and mixture modelling of SNP weights for the Sheep Hapmap data. The full and filtered datasets consist of 1,222 and 1,105 animals respectively.

ACKNOWLEDGEMENT

We acknowledge the support of Cobb-Vantress Inc. in allowing us to use their chicken data.

REFERENCES

Bertolini F., Galimberti G., Calo D.G., Schiavo G., Matassino D., *et al.* (2015) *Journal of Animal Breeding and Genetics,* **132**: 346.

Chang C.C., Chow C.C., Tellier L.C., Vattikuti S., Purcell S.M., *et al.* (2015) *GigaScience,* **4**.

Gibbs R.A., Taylor J.F., Tassell C.P.V., Barendse W., Eversoie K.A., *et al.* (2009) *Science,* **324**: 528.

Hudson N. J., Hawken R., Sapp R., Reverter A. (2015) *Proc. AAABG,* **21**: 153.

Kijas J.W., Lenstra J.A., Hayes B., other members of International Sheep Genomics Consortium (2012) *PLOS Biology*, **10**: e1001258.

McLachlan G.J., Peel D., Basford K.E., Adams P. (1999) *Journal of Statistical Software*, **4**: 1.

Miller J.M., Poissant J., Kijas J.W., Coltman D.W., International Sheep Genomics Consortium (2011) *Molecular Ecology Resources*, **11**: 314.

Mudadu M.A., Porto-Neto L.R., Mokry F.B, Tizioto P.C., Oliveira P.S.N, *et al.* (2016) *BMC Genomics*, **17**: 235.

Reyer H., Hawken R., Murani E., Ponsuksili S., Wimmers K. (2015) *Scientific Reports*, **5**: 16387.

Proc. Assoc. Advmt. Anim. Breed. Genet. 22:277-280

COMMITTED MATINGS UNDER MATE SELECTION

B.P. Kinghorn

School or Environmental and Rural Science, University of New England, Armidale, Australia

SUMMARY

As part of implementing a breeding program we aim to manage the numbers of parents selected, and the relationships between them, in order to manage genetic diversity. However, this ignores the fact that some breeding candidates may recently have been widely used, leaving embryos and juveniles in the system that are not yet in the list of candidates, but which are likely to make future breeding contributions. Use of such candidates should be inhibited somewhat to avoid their over-representation in the program as a whole, leading to increased loss of diversity.

A method to account for these prior contributions when implementing the breeding program is introduced, and indeed this makes notable impact to reduce loss of genetic diversity. This is a more correct way to implement the breeding program, however the alternative of using parameters that drive the program in a more conservative (high diversity) direction had similar impact both on the rate of increase in inbreeding coefficient and on genetic response.

Whatever method is used, it is advisable to test policy and parameters using simulation, to more confidently target an appropriate rate of increase in coancestry and inbreeding, in competition with genetic gains and other issues.

INTRODUCTION

Truncation selection on estimated breeding value is predicted to maximise genetic merit in the offspring generation. However, this is generally not the best strategy for maximising genetic gain in the longer term, as lack of attention to genetic diversity will generally lead to reduced opportunity for genetic gains in later generations.

Optimal Contributions Selection (OCS) provides a solution to this (eg. Meuwissen, 1997). We aim to optimise the contribution of each individual to the breeding population, giving rewards not only for high genetic merit, but also for low relationship to the rest of the individuals selected.

For this, we need to consider the relationship between each selection candidate and all other selection candidates of both sexes, in a numerator relationship matrix among candidates (A) or a genomic equivalent, or a hybrid of these. From the selection decisions made we can calculate x, the vector of relative genetic contributions from each candidate, summing to ½ for each sex. The mean parental coancestry is $x'Ax/2$, and this is the measure of the mean relationship that we aim to keep low in order to keep genetic diversity high. For this we want many small contributions in x. However, for high genetic gain ($x'G$, where G is the vector of EBVs or index values) we want to focus genetic contributions on the most meritorious animals, giving relatively few large contributions in x. Overall, we aim to find the vector x that maximises $x'G + \lambda x'Ax/2$, where λ is a negative weighting that determines the relative emphasis on genetic diversity. Here we focus on mean parental coancestry, $x'Ax/2$.

In scenarios where individuals can only be used as a parent at one mating round, we do not need to consider relationship with ancestors in the pedigree, because the prior contributions of these ancestors are accounted for in the pedigree of the current candidates. In all other scenarios, any prior contributions in previous matings are accommodated if the progeny resulting from these matings are either permanently culled or included in x and A as current candidates. However, if these progeny are not culled and are too young to be candidates – they are juveniles, or even embryos – then their potential future contributions need to be accommodated in some way.

For example, if a bull that is a current candidate has recently been widely used, and has a large number of juvenile and/or embryo progeny in the population, then there should be some inhibition on his further widespread use, and yet these progeny have no impact in $x'Ax/2$ as they are not represented in either x or A. The same argument can be made eg. for a first-use bull whose brother or other relative has juveniles or embryos in the herd.

This issue has been handled in the past by adding these juveniles and embryos in the list of candidates, but restricting their matings to other juveniles and embryos in "virtual matings" that are not of course invoked (Kinghorn et al, 2008). This means that these animals and their relationships with real candidates are accommodated in calculation of parental coancestry. However, somewhat arbitrary decisions have to be made about factors such as the maximum numbers of matings to be allocated to them, and even how many embryos to generate for each pregnant female.

An exception here may be when genomic and/or other information on juveniles is available, whereby these can be included as candidates in the main data file, with juvenile male x juvenile female Grouping set up such that juvenile mating results can be used for selection purposes, but without actual mating, as they are not sexually mature. In this case, Committed Matings for adult selection candidates need only cover matings that have not yet resulted in birth of progeny entered in the main data file as juvenile candidates.

MATERIALS AND METHODS

This paper presents a more appropriate method of handling this issue of contributions from juveniles and embryos that are not represented in the current candidates. Rather than add these animals to the list of candidates, the matings that gave rise to them are added to the list of matings to be made. This means that their parents are allocated the status of candidates in the datafile, whether or not they are candidates for the current mating round, and the matings that gave rise to them are made a fixed part of the solution as "Committed Matings".

This has been implemented in the program Matesel, with detailed instructions at http://matesel.une.edu.au. In addition to the main data file, the user supplies a list of "Committed Matings" that must be included in the solution. The user can choose to include these previous matings in the declared constraints on usage of candidates at the current mating round – otherwise the program makes appropriate adjustments so that only current constraints need to be declared.

To test this method, Matesel was used to make all selection decisions in a 20-year breeding program simulated using the program PopSim, developed from the version available within Genup (http://bkinghor.une.edu.au/genup.htm). PopSim has recently been used in a similar manner by Cowling et al (2016), who give some detailed description.

A breeding population typical of *Bos Indicus* cattle was simulated, the key features being 100 females mated each year, bulls and cows first mated to drop progeny at 3 years of age, culling for age at 6 and 10 years for bulls and cows, mating group sizes constrained between 20 and 40 females, random mate allocation, and BLUP EBVs leading to a multi-trait index. Bulls and cows can be culled by Matesel at intermediate ages for reasons related to genetic merit and impact on diversity.

In the **Control** treatments, no account was taken of the matings made by candidates in previous years, whereas in the **CM** (Committed Matings) treatments, previous matings that had not yet resulted in progeny of breeding age were accommodated as described above.

The Target Degree mode of balancing genetic gain and genetic diversity was used in Matesel, with 0 degrees relating to full emphasis on genetic gain and 90 degrees relating to full emphasis on genetic diversity. Treatment Control25 used 25 target degrees for a moderate outcome, and Control32.5 used 32.5 degrees for a more conservative outcome (more diversity at the cost of lower short-term genetic gain). Treatments CM25 and CM17.5 used the Committed Matings approach with lower target degrees, for reasons that become apparent.

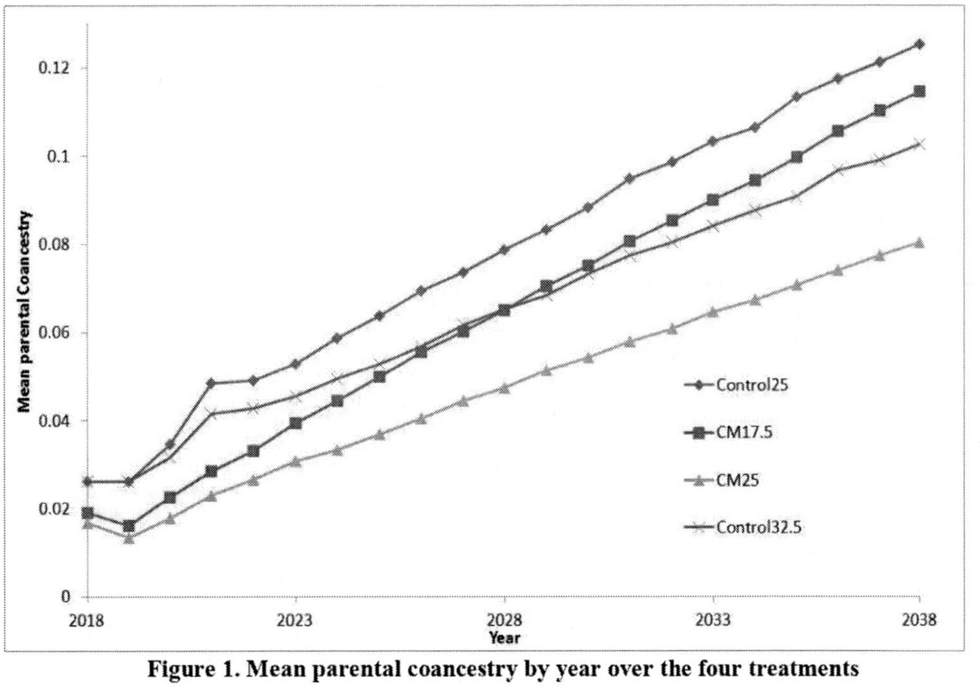

Figure 1. Mean parental coancestry by year over the four treatments

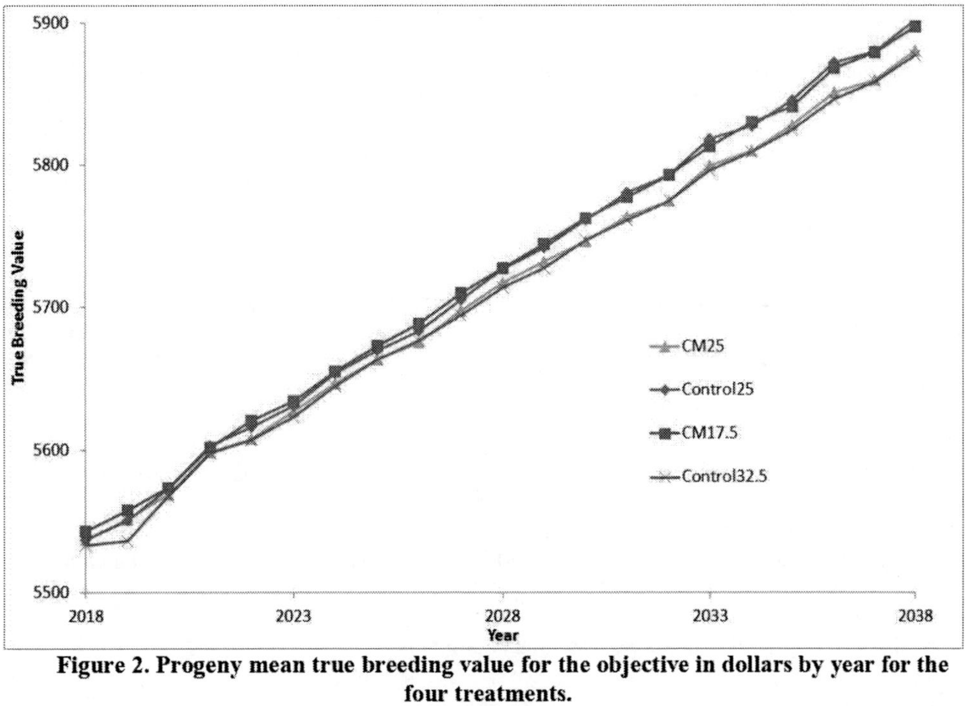

Figure 2. Progeny mean true breeding value for the objective in dollars by year for the four treatments.

RESULTS AND DISCUSSION

Figure 1 shows mean parental coancestry and figure 2 shows true breeding value in progeny for the breeding objective, over the 20 years of breeding. All results are averages over 50 replicates.

Notice that the Control programs immediately show a higher level of coancestry at the start of the programs. This is to be expected, as these treatments involve fewer candidates and hence a lower apparent effective population size. The CM results are more correct, as they include essentially a full generation of individuals, from embryo to older parents, whereas the Control treatments exclude consideration of embryos and juveniles.

However, what is more important, both here and in real breeding programs, is the *rate of change* in coancestry. In particular, CM25 increases nearly 30% more slowly than Control25, showing the impact of accommodating the prior contributions involved.

Figure 1 shows that this slower rate of increase in coancestry in CM25 is approximately matched by a more conservation control policy of 32.5 Target Degrees. Moreover, the faster rate of increase in coancestry for Control25 is approximately matched by a more aggressive CM policy of 17.5 Target Degrees.

Figure 2 shows that the genetic responses are higher and approximately equal for the treatments with higher rates of increase in coancestry (Control25 and CM17.5), and likewise, responses are lower and approximately equal for the treatments with lower rates of increase in coancestry (Control32.5 and CM25). In the longer term we can expect genetic progress in the more conservative programs to benefit from increased conservation of genetic diversity.

A simple conclusion is that, despite the use of the Committed Matings method being the correct thing to do, it is possible to not do this, but to compensate by adopting an appropriately more conservative policy, with a higher value for Target Degrees (or constraint to a lower mean parental coancestry if using that approach to manage coancestry). This is perhaps understandable, as a more conservative policy operates not only on current matings, but also on all previous matings, and it seems that the impact on overall optimal contributions is appropriately balanced.

What is "an appropriately more conservative policy"? The actual level of coancestry is probably not so important – this depends critically on the amount and quality of pedigree and/or genomic information available, as well as on prior breeding policies. We should ***predict*** the rate of change in coancestry over time, and yet this is very difficult to do for a real (multi-faceted) breeding program. However, we can ***discover*** the rate of change in coancestry using simulation, as in this paper.

The recommendation here is to discover what the rate of change in coancestry is as a function of parameters/policy chosen, and to choose parameters accordingly. Under random mating, the rate of change in inbreeding coefficient is expected to be equal to the rate of change in coancestry, so choose a rate of change that fits your attitude to ΔF, in competition with genetic gain and other issues.

REFERENCES

Cowling W.A., Li L., Siddique K.H.M., Henryon M., Berg P., Banks R.G., Kinghorn B.P. (2016) *J Exp Bot 2016 erw406*. doi: 10.1093/jxb/erw406

Kinghorn B. P., Banks R., Gondro C., Kremer V. D., Meszaros S. A., Newman S., Shepherd R. K., Vagg R. D. and van der Werf J. H. J. (2008) *In: Adaptation and fitness in animal populations. J. H. J. van der Werf, R. Frankham, H. -U. Graser and C. Gondro. Berlin, Springer*. Pages 191-200.

Meuwissen T.H.E. (1997) *J Anim. Sci.* **75**: 934-940.

Proc. Assoc. Advmt. Anim. Breed. Genet. 22:281-284

USING MACHINE LEARNING METHODS TO IDENTIFY SUBSETS OF SNP FOR GENOMIC PREDICTION

B. Li[1,2], A. George[3], A. Reverter[1] and Y. Li[1]

[1] CSIRO Agriculture & Food, 306 Carmody Road, St Lucia, QLD, Australia
[2] School of Computer Science and Technology, Shandong Technology and Business University, Yan Tai, Shandong, P. R. China
[3] CSIRO Data61, Dutton Park, QLD, Australia

SUMMARY

Machine learning methods have gained popularity dealing with high dimensionality, highly correlated structure, or "large P, small N" genomic data problems. The methods have been shown to be efficient in GWAS and candidate gene identification. However, the utility of methods in identifying a subset of single nucleotide polymorphism (SNP) for genomic prediction of breeding values has not been explored before. In this study, using 40,184 SNP genotypes and the live weight phenotypes from 1,097 Brahman cattle, we examined the power of two machine learning methods, Random Forests and Gradient Boosting Machine, in the identification of top 1,000 or 3,000 SNP and using them for building a genomic relationship matrix (GRM) for genomic prediction of breeding values. Our results clearly show that using the subsets of SNP identified by the two methods resulted in the improvement both in the heritability estimate and the genomic prediction accuracy.

INTRODUCTION

Machine learning methods have gained popularity dealing with high dimensionality, highly correlated structure, or "large P, small N" problems arising from large genomic data analyses. Two of these methods, Random Forests (RF; Breiman, 2001) and Gradient Boosting Machine (GBM; Friedman, 2001), have been shown to outperform the conventional GWAS methods in association mapping and genomic-wide prediction of estimated breeding values (GEBV) (Chen and Ishwaran 2012; Lukbe *et al.* 2013; González-Recio *et al.* 2014; Waldmann 2016). However, the utility of these methods in identifying a subset of SNP to estimate GEBV has not been evaluated before. In this study, we examined the efficiency of RF and GBM for the identification of a subset of markers and tested these small panels using a GEBV approach.

MATERIAL AND METHODS

Data. We used a SNP dataset consisting of 40,184 SNP genotypes from 1,097 Brahman cattle from the Legacy Database of the CRC for Beef Genetic Technologies (www.beefcrc.com). The animals varying from 373 to 509 days old came from 57 contemporary groups and were measured for live weight (the average being 308.64 kg (± 38.85) with the range from 180 to 430 kg). A quality check of the marker data resulted in the removal of 2,102 SNP having MAF <0.01 or with missing genotypes due to full genotype requirement by the machine learning methods. A total of 38,082 SNP were used for the final analysis. Since machine learning methods are non-parametric approaches, they cannot directly fit fixed effects in the model to account for environmental effects. Therefore prior to any analysis, the phenotypic values were adjusted for the fixed effects of the contemporary group and age. The residuals from the linear model of analysis of variance were used as phenotype for the evaluation of the machine learning methods.

Machine leaning methods – RF and GBM. Details of the RF method can be found in Breiman (2001). In brief, RF uses a bootstrapping method to randomly select a subset of animals as the training dataset (default being two third of the total number of animals), and a subset of SNP (default being a squared root of total number of SNP) to form a decision tree that splits the sampled individuals into two subsamples with different weight range values. The remaining individuals (one third) are then used as the validation dataset to determine the prediction error of the SNP tree on the phenotypes. The process repeats until a large number of decision trees are forming a forest (the parameter Ntree determines the size of a forest). Each individual decision tree building excise is independent to each other (with replacement). An individual SNP variable importance value (VIM) is determined by averaging the prediction error values of the SNP across all forest trees. GBM also generates multiple random samples to form trees, but subsequent samples always rely on the outcomes from the previous samples. It builds the trees iteratively by adding all "weak learners" – small trees with only a few SNP splits that predict the phenotypes with high bias but low variance (Lubke *et al.* 2013). Therefore, GBM reduces the prediction error by reducing bias through adding all the outcomes from a large number of models. Each method has its own parameter for measuring a SNP variable importance value (VIM). In RF, it is the %IncMSE (% increasing in mean squared error), while in GBM it is the Relative Influence - maximal cumulated estimated improvement in MSE. In both methods, the higher the VIM value, the more important the SNP is. The R libraries randomForest and gbm (https://www.rstudio.com/) were used for the analysis. The parameter Ntree was set as 2,000, the default values were used for RF and the learning rate of 0.1 for GBM.

Identification of top SNP and Gene Ontology (GO) Enrichment Analysis. Based on the ranked SNP VIM values from RF and GBM, the top 1,000 and 3,000 SNP were selected. The sets of genes near the top SNP or all the SNP with positive VIM values were examined for biological processes using the *Bos taurus* Reference from the PANTHER program (http://www.pantherdb.org/).

Construction of additive genomic relationship matrices using top SNPs for estimating genetic variances and genomic prediction of phenotypes. The additive genomic relationship matrix (GRM) was constructed using either 1,000 or 3,000 top SNPs from all animals, following the same method as in our chicken study (Li *et al.* 2016). An additive genomic model, fitting the GRM as random effect and the contemporary group and age as fixed effects, was then applied to estimate the genetic variance explained by each subset of top SNP (1,000 or 3,000). A random five-fold cross-validation scheme was used, i.e. randomly splitting 1097 animals into 5 equal-size groups and each group (20% of the population) was in turn assigned with missing phenotypic values and used as the validation set. The prediction accuracy was calculated as the correlation between the GEBVs of the animals with no phenotypic values and the true phenotypes of the animals adjusted for fixed effects. The program Qxpak v5.02 (Perez-Enciso and Misztal 2011) was used for the analyses.

RESULTS AND DISCUSSION

Profiles of SNP VIM values from RF and GBM. Figure 1 shows the distribution of the ranked SNP VIM values in RF and GBM. It can be seen that the majority of the SNP had very small or zero VIM values in RF and GBM. Of 38,082 SNP, 18,453 (48.5%) and 16,600 (43.6%) SNP were identified with the positive VIM values in RF and GBM, respectively. Between the two methods, there were 8,797 SNP in common. In RF, we also found a total of 6,660 SNP (17.5%) with negative VIM values, corresponding to the lower end of the distribution (Figure 1, RF graph). These negative values indicate that these SNP were problematic and should not be included in a prediction model. The reason was that the new prediction models using randomly permuted SNP positions on the decision trees had a much smaller mean squared error value (MSE) than the initial prediction model, hence a negative %InMSE value.

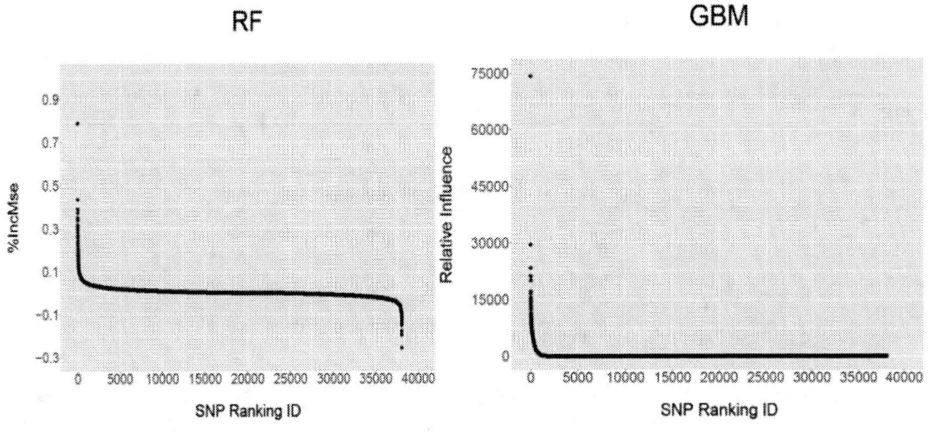

Figure 1. The distribution of ranked SNP variable importance values from RF (%IncMSE) and GBM (Relative Influence).

Gene enrichment analysis for SNP with positive VIM values from RF or GBM. When the sets of genes that were closest to the top 3,000 SNP or all the SNP with the positive VIM values were examined, we found that the top 3,000 SNP were primarily involved in the development, system development, visual perception, nervous system development and cellular activity ($p < 0.0001$). The evidence was much stronger for the genes near all the SNP with positive VIM, involving the growth pathways of development process (RF, P=1.54E-07; GBM: P= 2.09E-08) and system development (RF: P = 5.38E-07; GBM: P = 2.05E-07).

Both RF and GBM identified the same SNP with highest VIM value. It was ARS-BFGL-NGS-1712 mapped to gene BMPER (BMP binding Endothelial Regulator) on BTA4. A literature search found that BMPER played vital roles in adipocyte differentiation, fat development and energy balance in human and mouse (Zhao et al. 2015). The SNP was a very good candidate for selecting for increased body weight and rump length in cattle breeding (Zhao et al. 2015).

Table 1. Estimates of genetic variance and heritability (h^2) for live weight using different subsets of top ranking SNP identified by RF and GBM with additive genomic model

Method	No of Markers	Genetic Variance	Residual Variance	h^2
RF	1,000	332.60	256.78	0.565
	3,000	373.64	233.56	0.616
GBM	1,000	402.99	204.22	0.664
	3,000	417.05	184.08	0.694
All SNP	38,082	391.29	313.25	0.555

Estimates of genetic variance and heritability (h^2). Table 1 shows the REML estimates of genetic variance and h^2 for a subset of 1,000 or 3,000 top SNP identified by RF or GBM. Equivalent analysis using all 38,082 SNP are also listed in Table 1. It can be seen that there was a significant improvement in the h^2 estimate when the top 3,000 SNP from either RF or GBM were used in an additive genomic model. GBM performed particularly well in both 1,000 or 3,000 SNP cases, where the genetic variance estimates were higher than using all 38,082 SNP. Both RF and GBM captured complex SNP-SNP interactions, hence, resulted in an increased genetic variance.

Table 2. Prediction accuracy of GEBV for live weight using the top 1,000 or 3,000 SNP identified by RF and GBM methods

Methods	R1*	R2	R3	R4	R5	Average
RF1000	0.362	0.449	0.422	0.528	0.477	0.448
RF3000	0.321	0.408	0.421	0.443	0.440	0.407
Average	0.353	0.441	0.404	0.482	0.461	0.428
GBM1000	0.429	0.474	0.546	0.518	0.551	0.504
GBM3000	0.433	0.460	0.469	0.548	0.541	0.490
Average	0.418	0.463	0.476	0.501	0.510	0.474
All SNP	0.134	0.200	0.209	0.275	0.228	0.209

* Randomly selected 20% animals without phenotypic values

Accuracy of GEBV. Table 2 shows the accuracy of GEBV with a subset of SNP markers using an additive genomic model and a random split five-fold cross-validation scheme. In comparison to the additive model with all available SNP, surprisingly, the average prediction accuracy from either top 1,000 or 3,000 SNP outperformed the whole SNP panel, regardless the sources of the SNP chosen from RF or GBM. The prediction accuracy values from RF and GBM were double the amount of those of all SNP, ranged from 0.41–0.45 in RF and 0.43–0.50 in GBM.

Applications of large-scale SNP panels for genomic selection programs have a mixed success in livestock species (Waldmann 2016). While in the dairy cattle industry the genomic prediction of phenotypic values for production traits has achieved high success, the accuracy of GEBVs in beef cattle has been low (Waldmann 2016). We know from large number of GWAS and genomic prediction studies that the majority of SNP had little or no effects on phenotypes at all. This raises the question whether there is a benefit to use only small panel of SNP for genomic prediction? Our results here indicate that the machine learning methods, especially GBM, are efficient methods in identifying a subset of SNP with direct link to the candidate genes affecting the growth trait. It is possible to build a low density SNP panel for a genomic selection program.

In this study, we only examined a phenotype of moderate heritability in beef cattle. Further studies, using systematic approaches, are needed to validate the efficiency of machine learning methods in building low density SNP panels for different species or populations, optimal subset of SNPs and a range of phenotypes with different heritability values.

ACKOWLEDGEMENTS

We would like to acknowledge the financial supports for B. Li from the High Education Science and Technology Planning Program of Shandong Provincial Education Dept. (J16LN14), and Shandong Provincial Science and Technology Development Program (China) (2014GGX101044).

REFERENECES

Breiman L. (2001) *Machine Learning*. **45**: 5.

Chen X. and Ishwaran H. (2012) *Genomics*. **99**: 323.

Conzalez-Ricio O., Rosa G.J.M., Gianola D. (2014) *Livest. Sci.* **166**: 217.

Friedman, J. (2001). *Ann. Stat.* 29: 1189.

Li Y., Hawken R., Sapp R., George A., Lehnert S.A., Henshall J.M. and Reverter A. (2016) *Poult. Sci.* **0**:1.

Lubke G.H., Laurin C., Walters R., Eriksson N., Hysi P., Spector T.D., Montgomery G.W., Martin N.G., Medland S.E. and Boomsma D.I. (2013) *J Data Mining Genomics Proteomics* **4**:143.

Waldmann, P. 2016. *Genet Sel Evol*. **48**:42.

Perez-Enciso M. and Misztal I. 2011. *BMC Bioinformatics*.**12**:202.

Proc. Assoc. Advmt. Anim. Breed. Genet. 22:285-288

STRATEGIES TO USE WHOLE GENOME SEQUENCE DATA FOR GENOMIC PREDICTION IN DAIRY CATTLE

I. van den Berg[1], I.M. MacLeod[2], P.J. Bowman[2,3], T. Wang[2] and M.E. Goddard[1,2]

[1] Faculty of Veterinary & Agricultural Science, University of Melbourne, Victoria, Australia
[2] AgriBio, Department of Economic Development, Jobs, Transport & Resources, Victoria, Australia
[3] Biosciences Research Centre, La Trobe University, Victoria, Australia

SUMMARY

Sequence data may potentially increase prediction accuracy compared to medium or high density (HD) SNP markers, by containing causative mutations directly rather than relying on linkage disequilibrium between markers and causative mutations. Besides causative mutations, sequence data contains a much larger number of variants that have no effect on the analysed trait. A Bayesian variable selection model could be used to assign large effects only to the causative mutations. In practice, however, analysing millions of sequence variants is computationally challenging. Therefore, we tested an approach to split up the analysis per chromosome, correcting for all other chromosomes using HD estimates, in a simulation study, using a faster, hybrid version of the Bayes R variable selection model. While directly computing breeding values based on effects estimated per chromosome resulted in a reduced accuracy, reanalysing all variants that were selected per chromosome resulted in a similar accuracy to analysing all variants simultaneously, especially when HD variants were included.

INTRODUCTION

Sequence data can potentially increase prediction accuracy compared to medium or high density SNP markers, by containing causative mutations directly rather than relying on linkage disequilibrium (LD) between markers and causative mutations. However, the majority of sequence variants have no effect and can introduce noise into the prediction. In theory, Bayesian variable selection models could assign larger effects to the causative mutations, and zero effects to the rest. In practice, correctly estimating the effects of millions of variants in high LD with each other is computationally challenging, and the results of genomic prediction using sequence data so far have been variable. Using a Bayesian variable selection model to analyse all variants simultaneously, van Binsbergen *et al.* (2015) reported a slightly lower accuracy with full sequence data than with high density (HD) genotypes. Other approaches, using various methods to preselect variants, show sometimes an increase in accuracy (Brøndum *et al.* 2015; Macleod *et al.* 2016; van den Berg *et al.* 2016), while others found no increase in accuracy but increased bias (Calus *et al.*, 2016; Veerkamp *et al.*, 2016).

Our objective was to find an approach to approximate genomic prediction of whole genome sequence data with a Bayesian variable selection model. To parallelise the analysis, we tested analysing chromosomes separately after correcting the phenotypes for all other chromosomes using HD estimates. Results obtained per chromosome were either directly used to compute breeding values, or used to preselect variants for subsequent analysis with all chromosomes together. The analysis with all selected variants was performed either with or without the HD variants. A dataset with a limited number of realised imputed sequence variants and simulated phenotypes was used, to enable comparison with prediction using all sequence variants at once.

MATERIALS AND METHODS

The dataset used was the AUS-Sim simulated dataset described in more detail by Macleod *et al.* (2016). The dataset contained realised genotypes for 3,047 Holstein bulls, 4,942 Holstein cows, 770 Jersey bulls, 1,553 Jersey cows, 869 Red Holstein bulls, 741 Australian Red cows and 114 Australian Red bulls. The data was split up in a reference population containing all Holstein and Jersey individuals, and a validation population containing all Australian Red and Red Holstein individuals. Pedigree information for all individuals was obtained from the Australian Dairy Herd Improvement Scheme (ADHIS) and Interbull.

Two sets of genotypes were used, the HD set containing genotypes for 600,641 SNP on the Illumina BovineHD beadChip, and the SEQ set, containing 994,019 imputed sequence variants selected based on their function annotations. The HD genotypes were either obtained by direct genotyping, or imputation from the Illumina BovineSNP50 chip. The SEQ set contained 45,026 non-synonymous coding variants, 578,734 variants within 5 Kb upstream and downstream of genes, or in three/five prime untranslated genic regions, and 370,259 variants on the HD chip.

Quantitative trait loci (QTL) were simulated by randomly sampling 4,000 variants from all SEQ variants. QTL effects were sampled from three normal distributions with a mean of zero and variances of $0.0001\,\sigma_g^2$, $0.001\,\sigma_g^2$ and $0.01\,\sigma_g^2$ for 3,485 small, 500 medium and 15 large QTL, respectively, where σ_g^2 is the additive genetic variance. Subsequently, the true breeding value (TBV) of individual j was computed as $TBV_j = \sum_{i=1}^{4000} x_{ij} a_i$, where x_{ij} is the standardised genotype of individual j for QTL i, and a_i the additive effect of QTL i. An environmental effect was sampled from a normal distribution and added to the TBV to obtain a phenotype with a heritability of 0.6. A Holstein breed effect was sampled from $N(10,1)$ and added to the TBV for all Holstein individuals.

Genomic prediction was done using the hybrid version of the Bayes R mixture model described by Wang *et al.* (2016). This assumes that variant effects were drawn from four distributions with $N(0,0\sigma_g^2)$, $N(0,0.0001\sigma_g^2)$, $N(0,0.001\sigma_g^2)$ and $N(0,0.01\sigma_g^2)$. The hybrid model first uses an Expectation-Maximization (EM) model to estimate variant effects, the proportion of variants assigned to each of the four distributions, fixed effects (breed and sex), polygenic effects and residual variance. Subsequently, the converged estimates from the EM module were used as starting values for a Monte Carlo Markov Chain (MCMC) module that was run for 10,000 iterations. The analysis either included all variants for the full MCMC chain, or dropped a proportion of variants directly after the EM part, after 200 MCMC iterations, or after 10,000 MCMC iterations based on their probability to be included in any of the non-zero distributions. After some variants were dropped from the model another 10,000 iterations of the MCMC chain were performed. The mixing proportions at the moment of dropping were added to the prior of the mixing proportions for the remaining analysis. The analysis was done with either all variants together (FULL), split up per chromosome (CHR), with variants selected per chromosome but rerun with all chromosomes together (KEPT), and KEPT but including the HD variants (KEPT+HD). For FULL and CHR, the prior for the mixing proportions was [1,1,1,1], while for the KEPT and KEPT+HD, the posterior estimate of the mixing proportions obtained by FULL was used. Accuracies were calculated as the correlation between TBVs and GEBVs, and bias was calculated as the regression of TBVs on GEBVs.

RESULTS AND DISCUSSION

The accuracy of all scenarios using sequence data was higher than that using the HD genotypes, as shown in Table 1. For all scenarios, the accuracy for Red Holstein was larger than that for Australian Red. This was expected, as Red Holstein individuals are closer related to the Holstein population in the reference population than the Australian Red individuals. The accuracy obtained with the hybrid, analysing all sequence variants simultaneously, corresponded with the accuracy Macleod *et al.* (2016) obtained with the same dataset, using Bayes R. This confirms that, in agreement with Wang *et al.* (2016), the hybrid is a good, faster alternative for Bayes R. Table 1 shows the results of dropping some of the variants after 10,000 MCMC iterations and then running 10,000 additional iterations. When all variants were analysed together, dropping up to 90% of the variants had minimal effect on the accuracy. However, as shown in Figure 1, when variants were dropped immediately after the EM or after only 200 MCMC iterations the reduction in accuracy increased when more variants were dropped, especially for Australian Red. Dropping variants after 10,000 MCMC iterations does, however, increase the computing time because 10,000 additional iterations were run after dropping some variants. Therefore, these results show that while this strategy can be used to select variants associated with a trait, it does not help to reduce the computing time.

Table 1. Accuracy and bias of genomic prediction.

			Accuracy		Bias	
Data	**Analysis**	**DropProp**	**AusRed**	**RedHol**	**AusRed**	**RedHol**
HD	FULL	0.0	0.45	0.64	0.83	0.99
SEQ	FULL	0.0	0.60	0.66	1.07	0.97
		0.7	0.59	0.66	1.03	0.96
		0.9	0.59	0.66	1.01	0.96
SEQ	CHR	0.0	0.56	0.65	0.97	0.93
		0.7	0.56	0.65	0.97	0.93
		0.9	0.56	0.65	0.96	0.93
SEQ	KEPT	0.7	0.59	0.65	1.03	0.94
		0.9	0.59	0.65	1.01	0.93
SEQ+HD	KEPT+HD	0.7	0.60	0.66	1.04	0.94
		0.9	0.60	0.67	1.02	0.95

HD = high density genotypes, SEQ = sequence variants, FULL = all chromosomes in a single analysis, CHR = separate analysis for each chromosome, KEPT = variants selected by CHR reanalysed together, KEPT + HD = same as KEPT but including HD genotypes; dropProp = proportion of variants that are dropped after 10,000 MCMC iterations, ausRed = Australian Red, redHol = red Holstein

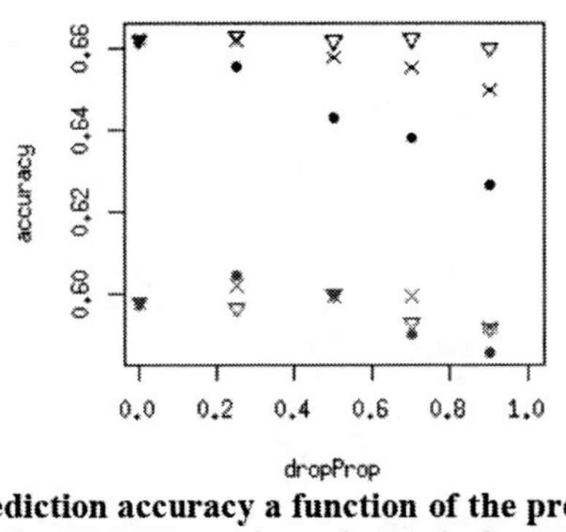

Figure 1. Reduction in prediction accuracy a function of the proportion of dropped variants. Circles = variants dropped after EM, X = variants dropped after 200 MCMC iterations, triangle = variants dropped after 10,000 MCMC iterations, black = Red Holstein, red = Australian Red.

Splitting up the analysis per chromosome resulted in a large reduction in elapsed time required to complete the analysis (between 1.9 and 4.5 hours per chromosome, instead of 55 hours when all chromosomes were analysed together), but also reduced the accuracy. Using the HD variants to correct for the rest of the genome assumes independence between effects on chromosomes, while in reality, there could be LD across chromosomes, and the sum of small effects on different chromosomes can contribute to a polygenic effect.

Selecting variants one chromosome at a time and then analysing them all together resulted in an accuracy almost equal to analysing all sequence variants simultaneously (the KEPT row in Table 1). There was, however, still a slight reduction in accuracy compared to the analysis where no variants were dropped. The vast majority of variants that were dropped would have ended up in the distributions with zero or very small effects. Therefore, they may contribute to a polygenic effect rather than be linked to specific QTL. In the analysis including both the variants selected per chromosome, as well as HD variants, the accuracy increased slightly and was equal to that obtained using all sequence variants.

Even though the results in this simulation are rather positive, in reality, the advantage of sequence data is likely to be smaller. For example, in the simulation, it was assumed that all QTL are segregating across breeds and have the same effects across breeds. In reality, only a proportion of variants segregates across breeds (Raven *et al.* 2014), and it is likely that their effects are not exactly the same, for example due to differences in minor allele frequencies (MAF). Another factor that could reduce the advantage of sequence data over high or medium density is imputation accuracy. Most sequences are obtained by imputation rather than direct sequencing, and this introduces errors in the genotypes. Furthermore, sequences used in this simulation were preselected based on functional annotations, strongly reducing the number of variants. Reducing the number of variants made it possible to compare analysing all variants simultaneously with strategies to split up the analyses. When these strategies are applied to datasets containing millions of variants, the large number of variants may induce problems to accurately estimate effects simultaneously, especially for variants that are in high LD with each other.

Our analyses show that preselecting sequence variants with a Bayesian variable selection model per chromosome and subsequently using those variants for genomic prediction, preferably combined with genome wide makers, could be an alternative to analysing full sequence data directly.

ACKNOWLEDGEMENTS

This research was supported by the Center for Genomic Selection in Animals and Plants (GenSAP) funded by The Danish Council for Strategic Research.

REFERENCES

Brøndum R.F., Su G., Janss L., Sahana G., Guldbrandtsen B., Boichard D. and Lund M.S. (2015) *J. Dairy Sci.* **98**:4107.

Calus M.P.L., Bouwman A.C., Schrooten C. and Veerkamp R.F. (2016). *Genet. Sel. Evol.* **48**:49.

Macleod I.M., Bowman P.J., Vander Jagt C.J., Haile-Mariam M., Kemper K.E., Chamberlain A.J., Schrooten C., Hayes B.J. and Goddard M.E. (2016) *Genet. Sel. Evol.* **17**:144.

Raven, L-A., Cocks B.G. and Hayes B.J. (2014) *BMC Genomics* **15**:62.

van Binsbergen R., Calus M.P.L., Bink M.C.A.M., van Eeuwijk F.A., Schrooten C. and Veerkamp R.F. (2015) *Genet. Sel. Evol.* **47**:71.

van den Berg I., Boichard D. and Lund M.S. (2016) *Genet. Sel. Evol.* **48**:83.

Veerkamp R.F., Bouwman A.C., Schrooten C. and Calus M.P.L. (2016) *Genet. Sel. Evol.* **48**:95.

Wang T., Phoebe Chen Y-P., Bowman P.J., Goddard M.E. and Hayes B.J. (2016) *BMC Genomics* **17**:744.

Proc. Assoc. Advmt. Anim. Breed. Genet. 22:289-292

A PIPELINE FOR THE ANALYSIS OF MULTI-OMICS DATA

Loan To Nguyen[1,2], Marina R. S. Fortes[1] and Antonio Reverter[3]

[1]School of Chemistry and Molecular Biosciences, The University of Queensland, Australia,
[2] Faculty of Biotechnology, Vietnam National University of Agriculture, Vietnam,
[3] CSIRO Agriculture & Food, Queensland Bioscience Precinct, Queensland 4067, Australia

SUMMARY

We describe an analytical pipeline to exploit the results from RNA sequencing (RNA-Seq) experiments combining a series of processes from data normalization to network inference. The pipeline makes use of numerical approaches aimed at identifying key regulators via the regulatory impact factor (Reverter *et al.* 2010) metrics. It also employs the partial correlation and an information theory (Reverter and Chan 2008) for the identification of significant edges in the construction of gene co-expression networks. Key nodes in the network include differentially expressed genes, transcription factors, tissue specific genes as well as genes harboring SNPs found to be associated with the phenotype(s) of interest. The pipeline has already been successfully employed in two beef cattle studies, dealing with the onset of puberty and feed efficiency. In the present paper, we describe a pipeline to analyze RNA-Seq data, focus on relevant genes, generate gene co-expression networks and identify emerging clusters within the network to provide new insight about the subject matter under scrutiny.

INTRODUCTION

Gene expression is the process which transferring the information of the gene into the production of a functional product. Genes may be expressed at specific tissue or only at certain physiological state in the animal life cycle. By measuring the abundance of gene products (RNA transcripts) in a tissue at a specific physiological state, the gene expression rate can be evaluated. Using gene expression analysis to identify candidate genes and biomarkers could ultimately enhance the accuracies of genomic prediction for key traits.

RNA sequencing (RNA-Seq) is a next-generation sequencing technique developed in 2008 for the analysis of gene expression across the entire transcriptome (Mortazavi *et al.* 2008; Wang *et al.* 2009). RNA-Seq was first applied in model organisms including Arabidopsis (Lister *et al.* 2008), yeast (Nagalakshmi *et al.* 2008) and mouse (Mortazavi *et al.* 2008), but has rapidly increased its popularity to a number of other organisms including human (Sultan *et al.* 2008) and bovine (Huang and Khatib 2010). RNA-seq is high-throughput and the analysis of large-scale datasets has a wide range of applications, however, every RNA-seq experimental scenario may have different optimal methods for analyses. New approaches are currently being developed (Han *et al.* 2015). Here we provide a step-by-step recipe on how to use the pipeline to analyze RNA-Seq data, focus on relevant genes, generate gene co-expression networks and identify emerging clusters within the network to provide insight about the subject matter under scrutiny. Without entering in detailed numerical intricacies (published elsewhere and cited herein), we discuss the essential principles of the analytical methods of each step in the process.

METHODS

In what follows, we provide a step-by-step recipe on how to exploit RNA-Seq data in order to identify differential expressed genes, key regulatory genes and generate gene co-expression network, in combination with algorithms such as RIF (Reverter *et al.* 2010) and PCIT (Reverter and Chan 2008). Figure 1 provides a schematic of the flow chart for this analytical pipeline.

Generally, the pipeline used for the analysis of multi-omics data requires a series of four steps as follows:

Step 1 – RNA-Seq Experimental Resource. In order to infer differentially expressed genes and gene co-expression networks in our multi-omics pipeline, the following resources are required: 1) the RNA-Seq data comprising at least two experimental conditions; 2) the experiment data conducting at least in two tissues. In the puberty example, the two experimental conditions would be the pre- or post-puberty stages; while the reproductive tissues of interest could include hypothalamus, pituitary, ovaries and uterus, as well as tissues related to the onset of puberty such as liver, fat and muscle. Other experimental setting could include healthy versus disease states, various breeds and/or various time points as conditions.

Step 2 – Normalization via Mixed-Model Equations. The ability of mixed-models in terms of their power to accommodate covariance structures in various forms is well documented in the animal breeding and genetics literature. Similarly, mixed-models are the ideal tool for the normalization of gene expression data (Reverter *et al.* 2005). Aiming for parsimony the simplest model will contain the library as the only fixed effect, and the interaction effect of gene by animal by condition by tissue and the residual as the only random effects:

$$Y = \text{Library} + \text{Gene} + \text{Gene*Animal*Condition*Tissue} + \text{Error}$$

The solution of the Gene*Animal*Condition*Tissue (GACT) interaction are used as the normalized mean expression (NME) of each gene in each animal and tissue. However, combinations of lower order gene interactions, such as Gene*Animal, Gene*Condition and Gene*Tissue are also possible. Additionally, the GACT solutions for all the animals from the same condition could be averaged to obtain the NME of each gene in each condition and tissue. The NME values will provide the basis for the computation of differential expression and tissue-specificity.

Step 3 – Selection of 'Relevant' Genes. To facilitate the task of generating and analysing the resulting network, only a subset of genes will be used according to the following four categories: differentially expressed (DE) genes, tissue-specific (TS) genes, genes harbouring SNP reported to be associated with a phenotype or phenotypes of relevance, and significant regulators such as transcription factors (TF). Next, we briefly describe the way in which each category is identified.

Differentially expressed (DE) genes. Typically, the contrast of interest will be comprised of the (possibly differential) expression of a given gene in a given tissue across the two (or more) conditions under study. These can be obtained directly from the NME and the statistical inference on the contrast performed based on a number of approaches of which a simple t-test is quite possibly optimal, preferably after correction for multiple testing using either Bonferroni or (preferably) Benjamini and Hochberg methods (both described in (Benjamini and Hochberg 1995)).

Tissue specificity. Similarly, the NME can be used to reveal the expression of each gene in each tissue and then compute the proportion of a gene's total expression in each of the tissues (ie. based on the NME of a gene in a tissue divided by the sum of the NME of the same gene summed across all tissues). This could be done either within or across the two (or more) conditions under study. In doing so, tissue-specific (TS) genes will be identified from those genes whose expression in a given gene is higher than in any other tissue by a particular amount such as fold-based bearing in mind that a gene can be TS for one tissue only. Additionally, using comparative genomics from human studies, we can source the identity of TS genes from the Tissue-specific Gene Expression and Regulation database (TIGER: http://bioinfo.wilmer.jhu.edu/tiger/).

Genes harbouring associated SNP. Today there is a plethora of GWAS in the literature quite possibly studying a condition similar (even identical) to the one in our current study. The results from these studies can be mined to retrieve the genes surveyed in our RNA-Seq study that are reported to harbour SNP associated with a phenotype or condition similar or preferably identical to the one in our current study.

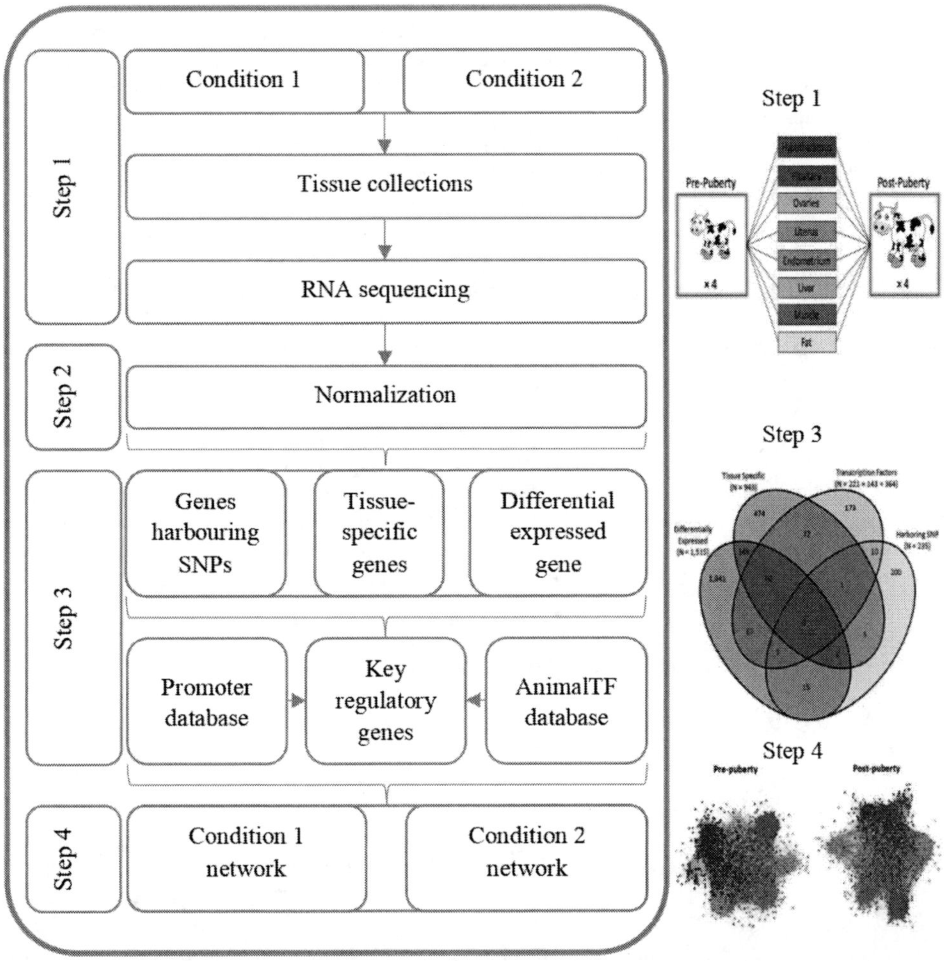

Figure 1. Flow chart of the pipeline for the RNA sequencing analysis (left) and illustrations adapted from Canovas *et al.* (2014) in the context of the onset of puberty in Brangus heifers

Key regulators. In order to identify the regulators (not necessarily TF) present among the genes surveyed in our RNA-Seq study, we mine to the Animal Transcription Factor Database (http://www.bioguo.org/AnimalTFDB/). Among these, we define as significant or "key" regulators those with statistically significant RIF metrics (using DE, TS and SNP harbouring genes as targets) and/or those with binding motif in the promoter region of DE, TS and/or SNP harbouring genes. In more detail, RIF comprises a set of two metrics designing to evaluate the regulatory power of molecules by exploring their differential connectivity to other influential genes (eg. those differentially expressed) in two contrasting conditions of interest (eg. pre- and post-puberty).

Step 4 – Network Inference and Visualisation Analysis.
For the network inference, we use the DE, TS, key TF and SNP harbouring genes as nodes and significant connections are identified using the partial correlation and information theory (PCIT) algorithm either through the original FORTRAN90 source code (Reverter and Chan 2008) or through an R package (Watson-Haigh *et al.* 2010). The PCIT exploits the twin concepts of partial correlation and mutual information. In brief, PCIT ascertain the significance of a given correlation between 2 entities (e.g., genes or network nodes) after accounting for all other genes in the dataset. Importantly, the output from PCIT can be viewed with Cytoscape (Shannon *et al.* 2003), a software program for analysing and visualizing gene co-expression network. In order to characterize network features, many Cytoscape plug-ins are available. Of these plug-ins, we recommend MCODE (Bader and Hogue 2003) to identify highly interconnected gene clusters, and BINGO (Maere *et al.* 2005) to determine which Gene Ontology terms are significantly overrepresented in a set of clustered genes. Hopefully, these clusters may have biological significance within the context of the phenotype under study.
One final process in the analysis of the resulting network is to identify the best trio of TF among those spanning the majority of the network topology. To this end, we search for TF with lots of connections in the network but few in common as these indicate redundancy.

CONCLUSIONS

The biological complexity and the rapid accumulation of publicly data arise the need to develop efficient tools for large-scale multi-dimensional data analysis. We conclude that the proposed analytical pipeline is a useful procedure providing an opportunity screen and identify key regulatory genes as well as generate regulatory networks with predictive power for the phenotype under investigation. Therefore, it may also be a significant tool for integrating different RNA-seq dataset and different levels omics data in order to investigate the complexity of biological subjects.

REFERENCES

Bader G. and Hogue C. (2003). *BMC Bioinformatics.* **4**: 2.

Benjamini Y. and Hochberg Y. (1995). *J Royal Stat Soc Series B (Methodological).* **57**:289.

Canovas A., Reverter A., DeAtley K.L., Ashley R.L., Colgrave M.L. *et al.* (2014) *PLoS One.* **9**: e102551.

Han Y., Gao S., Muegge K., Zhang W. and Zhou B. (2015) *Bioinform Biol Insights.* **9**: 29.

Huang W. and Khatib H. (2010) *BMC Genomics.* **11**: 1.

Lister R., O'Malley R.C., Tonti-Filippini J., Gregory B.D., Berry C.C. *et al.* (2008) *Cell.* **133**: 523.

Maere S., Heymans K. and Kuiper M. (2005). *Bioinformatics.* **21:** 3448.

Mortazavi A., Williams B.A., McCue K., Schaeffer L. and Wold B. (2008) *Nat Methods.* **5**: 621.

Nagalakshmi U., Wang Z., Waern K., Shou C., Raha D. *et al.* (2008) *Science.* **320**: 1344.

Reverter A., Barris W., McWilliam S., Byrne K.A., Wang Y.H. *et al.* (2005) *Bioinformatics.* **21**: 1112.

Reverter A. and Chan E.K.F. (2008) *Bioinformatics.* **24**: 2491.

Reverter A., Hudson N.J., Nagaraj S.H., Perez-Enciso M. and Dalrymple B.P. (2010) *Bioinformatics.* **26**: 896.

Shannon P., Markiel A., Ozier O., Baliga N.S., Wang J.T. *et al.* (2003) *Genome Res.* **13**: 2498.

Sultan M., Schulz M.H., Richard H., Magen A., Klingenhoff A. *et al.* (2008) *Science.* **321**: 956

Wang Z., Gerstein M. and Snyder M. (2009) *Nat Rev Genet.* **10**: 57.

Watson-Haigh N.S., Kadarmideen H.N. and Reverter A. (2010) *Bioinformatics.* **26**: 411.

Proc. Assoc. Advmt. Anim. Breed. Genet. 22:293-300

GENETICS OF BRISKET DISEASE IN BEEF CATTLE: A NOT SO HIGH ALTITUDE PROBLEM

M. G. Thomas[1], G. M. Krafsur[1], T. N. Holt[1], R. M. Enns[1], S. E. Speidel[1], F.B. Garry[1], A. Canovas[2], J. M. Medrano[3], R. D. Brown[4], K.R. Stenmark[4], and J.M. Neary[5]

[1]Colorado State University, Fort Collins, CO 80521, United States
[2]University of Guelph, Ontario, N1G2W1, Canada
[3]University of California, Davis, CA 95616, United States
[4]University of Colorado, Anschutz Medical Campus, Denver, CO 80045, United States
[5]Texas Tech University, Lubbock, TX 79409, United States

SUMMARY

Brisket disease, also termed high altitude disease (HAD), has been observed for a century in beef production systems at altitudes > 1,600 m. This disease is often diagnosed as cattle that died of right heart failure; however, there is limited data collection for this trait. Pulmonary arterial pressure (PAP) is an indicator trait that can be used to assess pulmonary hypertension. Cattle with PAP values ≤ 41 mmHg are considered as low risk of developing HAD, whereas cattle with values ranging from 42 to 49 mmHg are considered moderate and those with values ≥ 50 mmHg are considered high risk. This trait has moderate heritability and the phenotype and EPD are used in yearling bull and heifer selection programs. Multi-omics approaches involving SNP-chip genotypes and RNA-sequence gene expression levels indicate that yearling PAP is very polygenic and influenced by gene-pathways of inflammation, tissue remodeling, and metabolism; therefore, these results suggest that this trait could be improved with genomic selection. Recently, late feedlot death in fattening cattle at altitudes < 1,600 m has been observed with etiology of right and left heart failure. This finding adds complexity to understanding brisket disease and challenges the designing of breeding programs for healthy cattle.

INTRODUCTION

High altitude disease (HAD) has been a problem in beef cattle production systems within the Rocky Mountain region of the United States for approximately a century (Glover and Newsom 1917). This disease is commonly known as "Brisket Disease" due to the swelling of this anatomical region. The swelling is a result of the thoracic cavity and pericardium filling with fluid when an animal experiences the hypoxia-induced physiological cascade that leads to right heart failure. Economic losses associated with morbidity and mortality of cattle in response to this disease have been documented in several reports, despite the incidence being relatively low (< 5%) in native cattle. However, the incidence of this disease can expand to levels of approximately 40% in cattle not adapted to altitudes > 1,600 m (Will and Alexander 1970; Salman et al. 1991; Holt and Callan, 2007). High altitude disease is caused by hypoxia-induced pulmonary hypertension; however, collection of the true phenotype is often impossible in extensive pastures of mountain beef production systems; consequently, cattle are screened for disease susceptibility (i.e., risk) using an indicator trait, mean pulmonary arterial pressure (mPAP). This hemodynamic metric reflects the steady-state resistance to blood flow experienced by the right ventricle and a greater pressure, measured in mmHg, indicates greater pulmonary arterial resistance to flow (Holt and Callan 2007).

Mean PAP is a moderately heritable trait in growing Angus cattle (h^2 of 0.26 to 0.46; Crawford et al. 2016). Most cattle that are measured for this phenotype are yearlings (i.e., ~365 days of age) and British and Continental *Bos taurus* breeds although it should be noted that the neonatal calf is also very sensitive to hypoxia (for review: Stenmark et al. 2013). Since yearling mPAP is the most common phenotype used to determine if cattle are tolerant of hypoxia, seedstock producers in high altitude production systems use this information in sire selection and heifer replacement programs. Expected progeny difference (EPD) for mPAP has been estimated for several breeding programs in the Western United States; however; breed association sponsored EPD for this trait are currently limited due to need for development of standardized data collection programs for this specific trait and due to the requirement that cattle be tested while physically residing at these elevations.

Genomic selection provides opportunity to improve accuracy of EPD and provide breeding values for traits that are difficult to measure, if ample data can be collected for training processes (Garrick and Fernando 2014). Therefore, it is very logical to suggest that genomic selection is a viable approach to ameliorate HAD. This review will describe the efforts of the research team at the Colorado State University Beef Improvement Center to combat HAD.

A more recent occurrence challenging our understanding of brisket disease in mountain production systems is the observation of the pulmonary hypertension and heart failure in fattening feedlot cattle (Neary et al. 2015ab). This is a growing concern to the feedlot segment of the beef industry that exist on what is known as the "Great-Plains" of North America, which are altitudes < 1,600 m (i.e., not so high altitude). This review will also describe what is becoming known as "late feedlot death" as to provide perspective as to the physiological and production challenges that hypoxia is causing beef production systems in the United States.

HIGH ALTITUDE DISEASE AND PULMONARY ARTERIAL PRESSURE

High altitude disease is recognized by the swelling of the brisket in live cattle; however, if the animal perishes, then postmortem evaluation of the organs in the thoracic cavity reveals malformation (i.e., enlargement) of the right ventricle of the heart and several anomalies of the pulmonary artery and lungs. As a consequence of restrictive-remodeling of the pulmonary artery, the liver will develop a unique nutmeg color. The World Health Organization classifies this form of pulmonary artery hypertension in people as Group 3. The phenotype of HAD in cattle can be confused with consequences of bronchiopneumonia or other respiratory infections that may have caused or exacerbated an animal's hypoxic state (Neary et al. 2013). There is no data collection process for the trait of HAD. This is primarily due to limited access to cattle in extensive and rugged pastures in mountainous beef production systems. Data collection from cattle in pastures with this topography has also led us to also study genetics of grazing distribution (Bailey et al. 2015).

Because of the challenges associated with collecting the phenotype of HAD, mPAP is used as an indicator trait to determine if cattle are tolerant of the hypoxic conditions related to increasing altitude. Mean PAP is estimated from the collection of systolic and diastolic PAP measured within the pulmonary artery. These measures are performed on cattle in a squeeze-chute (crush) by a licensed veterinarian that developed these specific skills. To effectively measure this phenotype, it is recommended that the cattle reside at an elevation greater than 1,500 m for at least 3 weeks. It should be noted that this phenotype is also influenced by sources of variation such as breed, gender, pregnancy, age, elevation, concurrent and (or) previous illness, environmental conditions, etc. (Holt and Callan 2007).

In the Rocky Mountain region of the United States, mPAP is annually measured on > 10,000 replacement bulls and heifers. Many beef producers use the raw phenotype and the following categories to cull high risk cattle. Specifically, cattle with PAP values ≤ 41 mmHg are considered low risk of developing HAD, whereas cattle with values ranging from 42 to 49 mmHg are considered moderate and those with values ≥ 50 mmHg are considered high risk (Holt and Callan, 2007). Colorado State University Department of Animal Sciences manages a research facility to study performance and adaptability of Angus cattle in a high elevation beef production system. This research facility is known as the Colorado State University Beef Improvement Center (i.e.,

Rouse Angus Ranch near Riverside Wyoming) This facility a seedstock and cow/calf operation that raises its own replacement females, has grazing lands that range from 2200 to 2800 m in elevation, and supports 430 mother cows and associated animals (e.g., bulls, replacement heifers, steers, etc.). This breeding program uses estrous synchronization and artificial insemination (AI) technologies to coordinate a progeny testing program involving the companies of Genex, American Breeder Services, and Select Sires. Each of these companies contributes bulls that they want to evaluate for tolerance to high altitude. In the past 15 years, > 300 Angus AI sires have been mated in this program. There are several groups of Angus cattle from the Rouse Angus breeding program described in Table 1. The distribution of the mPAP data from this herd appears to have an out of proportion and non-Gaussian distribution of elevated pressures (Figure 1). Therefore, approximately 50.8% of the cattle would be categorized as low risk, 38.1% as moderate risk, and 11.1% as high risk for developing HAD.

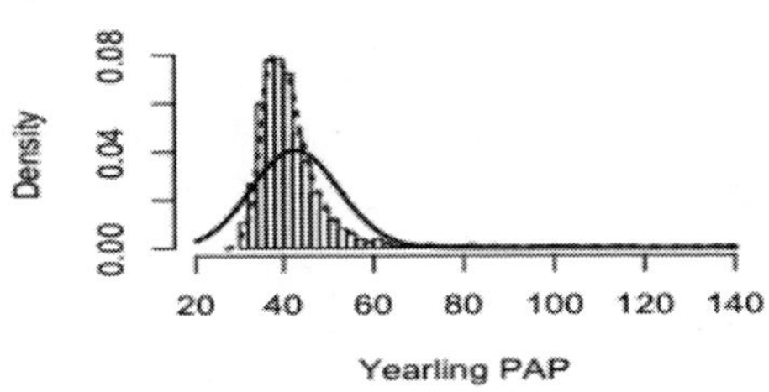

Histogram of mPAP (mmHg, x-axis) in yearling Angus bulls, heifers, and steers (n = 5,659) studied at the Colorado Beef Improvement Center (i.e., Rouse Angus Ranch; elevation 2,200 m; Zeng, 2016). The y-axis is the percentage (density) of the number of animals at each level of the mPAP phenotype.

In addition to the mPAP data from Rouse Angus cattle, Table 1 also presents results from additional groups of cattle and species. Note that grazing crossbred cows sampled at sea-level have PAP values approximately 9 mmHg higher than healthy humans and other large ruminants thought to be adapted to high altitude (i.e., American Bison and Yak)Also, the two groups of yearling bulls and fattening steers, which would typically be gaining 1.5 kg/day of body weight, have substantially higher PAP values than other types of animals. These data provide evidence to suggest that cattle have higher PAP than most animals and these values increase when the animals are fed high-gain diets. It should also be noted that the fattening Angus steers in the first row have PAP values within the range as people suffering from hypertension described in the last row of the data; therefore, providing validation evidence that mPAP is effective indicator trait for HADThe health consequences of mPAP in these fattening steers will be described further in the section titled "Late Feedlot Death".

HERITABILITY AND EXPECTED PROGENY DIFFERENCE

Mean PAP was reported to be a trait of moderate (h^2 = 0.34) to high (h^2 = 0.46) heritability using data from a registered Angus herd in Carbondale Colorado (elevation 1,981 m; Enns et al. 1992; Shirley et al. 2008). Using data from the Rouse Angus herd, which are purebred cattle, but not registered, a moderate (h^2 = 0.26) heritability was estimated using records from 1993 to 2014 (n = 5,776; bulls and heifers; Crawford et al. 2016). Zeng (2016) reported similar results using bull, heifer, and steer data, which also included the 2015 calf crop from this breeding program. A genetic correlation of 0.67 was observed between yearling (365 days of age) and weaning (205 days of age) in mPAP in the Dissertation of Zeng (2016). Slight to moderate, and positive genetic correlations ($\leq$ 0.22) were observed between growth traits and yearling mPAP in these two studies; however, stronger relationships ($\leq$ 0.50) were reported by Shirley et al. (2008), but the PAP values in that study were collected at weaning. Nonetheless, the positive relationships between growth traits and PAP values could be unfavorable for overall animal health with aggressive selection for growth.

Colorado State University's Center for Genetic Evaluation of Livestock has calculated EPD for PAP for several cattle breeding programs located in the western mountainous states of the United States. For the two Angus breeding programs described in the previous paragraph, genetic trend was slight. This is most likely a result of the unique distribution of the numeric trait of PAP (Figure 1), the continuous introduction of new sires through the progeny test program, and (or) a selection program that typically only culls medium and high risk animals. In general, 90% of the animals will have an EPD ranging -4 to +4, with the remaining 10% increasing up to +19 mmHg. The underlying physiology of this trait and its interaction with altitude suggest that there is likely a limit in genetic reduction of PAP; therefore, Zeng (2016) conducted research evaluating genetic correlations among PAP and growth traits, heritability, and EPD using the categorical veterinary risk descriptions of PAP (i.e., low, moderate, and high)These genetic correlations and heritability estimates were calculated using a threshold model that assumed a continuous underlying normal distribution of liabilities. The overall results of this effort were very similar to those obtained using the raw phenotypes. The EPD from the threshold models were converted to a probability scale. These types of EPD are often difficult for the beef producers to interpret; therefore, the decision was made to continue to calculate the EPD using the raw data as the breeders and bull buyers in this region of the world are very familiar with the phenotype of PAP, which is expressed in the units of mmHg.

MULTIOMICS STUDY OF PULMONARY ARTERIAL PRESSURE

Pulmonary arterial pressure data is currently being used for EPD calculations of traits indicative of tolerance to high altitude in cattle. There are no data collection systems for incidence of high altitude disease as collection of "death" data is difficult in these extensive environments where predators and topography may mean mortalities are never observed. To date, 3,000 cattle from the Rouse Angus herd have been high density genotyped. Sixty-six Angus animals have been genotyped with the BovineSNPHD (~777,000 SNP) and the rest with BovineSNP50 (54,0001 SNP). Genome-wide association studies (GWAS) have been performed to identify quantitative trait loci (QTL). Figure 2 presents results from Angus weanling steers (i.e., 6 months of age) described in the studies of Neary et al. (2014). Figure 3 presents the results from yearling Angus cattle described in the studies of Zeng (2016). Cumulatively, these results indicate that PAP is a very polygenic trait. These findings parallel the results from various gene expression studies that extracted RNA from heart, lung, and blood peripheral blood mononuclear cells in cattle exposed to hypoxic conditions and compared to samples from calves in normoxic conditions. Fibroblasts harvested from the pulmonary artery were an important cell type in these studies as their unique-phenotype is indicative of hypoxia-indued arterial remodeling. In these results, multi-gene pathways of inflammation, tissue remodeling, and metabolism were prominent. The latter is a very intriguing result and suggests that hypoxic cells have altered-glycolytic metabolism (i.e., Warburg effect; Stenmark et al. 2013; Newman et al. 2015; Li et al. 2016).

Since PAP EPD uses data collected from yearling cattle, an RNA-Seq study was conducted to obtain gene expression data from steers being grown as contemporaries to the bulls at the Rouse Angus ranch. This study involved identifying High- and Low-PAP individuals as to obtain RNA from muscle (i.e., right and left heart ventricle and longissimus dorsi) and the pulmonary system tissues (pulmonary artery, aorta, and lung). This approach allowed study of transcriptome-gene expression and also provided sequence for SNP. The initial results of this effort suggested that more than 1,000 genes were differentially expressed between high and low PAP groups in the right ventricle (Canovas et al. 2016). Splice variant analyses revealed several hundred differentially expressed genes in RNA from the right ventricle, aorta and pulmonary artery. Pathway and transcription factor (i.e., gene regulatory analyses) also revealed numerous genes involved in inflammation and several other indicators of heart failure. Examples of such gene-pathways were

IL-8/IL-10 signaling, leukocyte extravasation and factors promoting cardiogenesis, coagulation, thrombin and cardiac hypertrophy signaling. There were some responses among the high and low groups in hypoxia inducible factors (HIF); however, this is a very large and complex gene family requiring additional study.

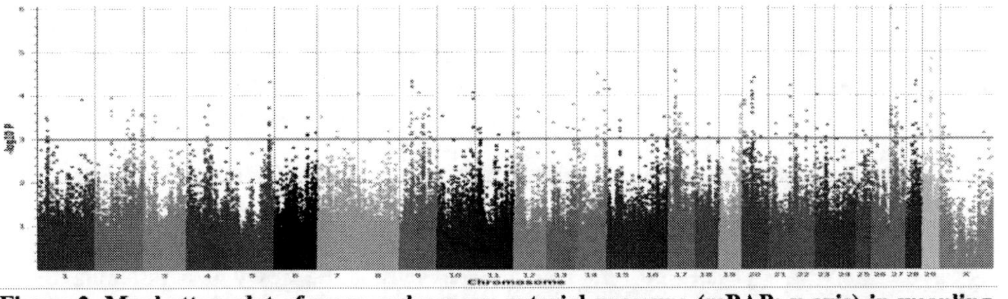

Figure 2. Manhattan plot of mean pulmonary arterial pressure (mPAP; y-axis) in weanling Angus calves (n = 66) at 2200 m. Genotypes were from BovineSNPHD (770,000 SNP) and were used in a single SNP analysis (Neary, 2014).

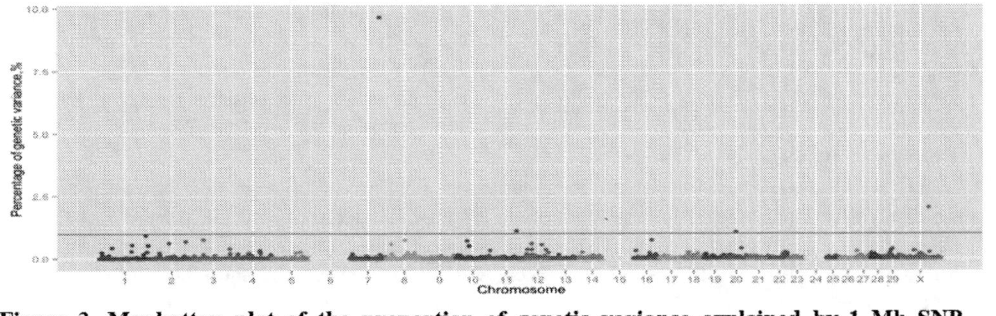

Figure 3. Manhattan plot of the proportion of genetic variance explained by 1 Mb SNP windows for deregressed estimated breeding value of mean pulmonary arterial pressure (mPAP; y-axis) in yearling Angus cattle (n = 2,582) at 2,200 m. Genotypes were from BovineSNP50 (54,001 SNP) and marker effects and associated variances were generated with Bayes B (Zeng, 2016).

LATE FEEDLOT DEATH

Between the years of 2000 and 2012, the incidence of death due to right heart failure doubled in an epidemiologic study of North American feedlot cattle (Neary et al. 2015a). This study involved 1.56 million cattle at 15 feedlots at elevations < 1,300 m. The death rate from congestive heart failure in this study was ~10 steers per 10,000 steers on feed. This is an interesting observation, as also during these span of years, carcasses of finished steers reached their highest weight since the United States Department of Agriculture started reporting these data in 1944. This organization reported average live and carcass weights of steers on January 17, 2017 as 661 and 425 kg, respectively.

Pulmonary artery remodeling and subsequent right heart failure has been thought to be a disease related to hypoxia at altitudes > 1,500 m, so most of the animals affected by this condition were calves within cow/calf operations in mountainous beef production systems; however, veterinary diagnostic laboratory reporting programs have observed an increase in feedlot cattle post-mortem submissions. Many of these cattle came from feedlots of < 1,500 m. The first row of Table 1 presents PAP data collected from fattening feedlot steers. The PAP values are extremely high relative to several other animal groups and somewhat analogous to values in people that are suffering from various cardiopulmonary conditions involving pulmonary hypertension (Moraes et al. 2000; Bossone et al., 2013). Krafsur et al. (2017) are studying feedlot steers that perished from heart failure in the late feeding period. Remodeling of the pulmonary artery and the right heart ventricle was observed in these cattle as would be expected. However, an interesting observation in these tissues was that these steers also contain evidence of left ventricle heart malformations as exhibited in Figure 4. The World Health Organization would categorize this type of pulmonary hypertension as Group 2, which is often associated with obesity. Feedlot steers with United States quality grades of choice and prime are approximately 30% body fat, which is comparable to obese people with body mass indexes ≥ 30 kg/m^2. Late feedlot death is a growing concern and data collection systems are needed to understand if these traits are related to the measures commonly collected on yearling cattle for risk of HAD.

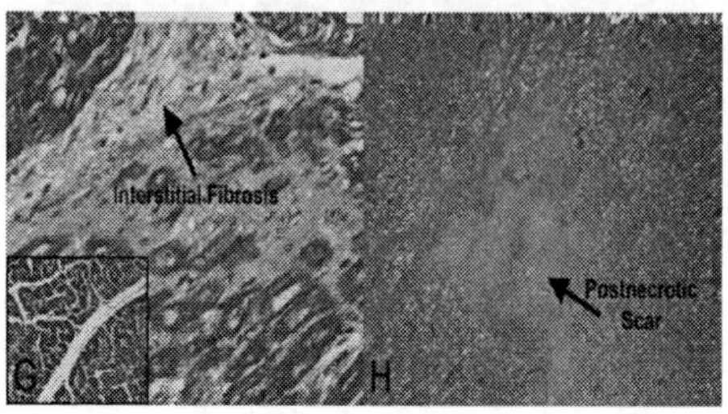

Figure 4. Histopathological images from the left ventricle of the heart of a fattening *Bos taurus* steer from a low elevation (550 m) feedlot. G. Interstitial fibrosis of left ventricle. High magnification (400x); arrow: fibrocollagenous matrix separating and entrapping left ventricular cardiomyocytes. Inset: normal myocardium. **H.** Heart with postnecrotic scar (100x). Arrow: fibro-collagenous matrix replacing cardiomyocytes in the left ventricle.

CONCLUSIONS

High altitude disease is a problem in mountainous beef production systems; however, there are no data collections systems for this phenotype, so an indicator trait, mPAP, is used in selection. This indicator trait from measures of yearling Angus cattle has unique a non-Gaussian distribution, but appears suitable for EPD calculation. Mean PAP is very polygenic and results of multi-omics studies suggest genomic selection approaches could provide opportunity to improve breeding value estimations. Evidences suggest that all age classes of cattle (i.e., neonate, weanling, yearling, and finishing steers) are responsive to hypoxia and additional research is needed to help understand genetic relationships among these age groups and the influences of their growth rates on HAD.

REFERENCES

Anand I.S., Harris E., Ferrari R., Pearce P. and Harris P. (1986) *Thorax* **41**: 696.

Bailey D.W., Lunt S., Lipka A., Thomas M.G., Medrano J.F., Canovas A., Rincon G., Stephenson M.B and Jensyn D. (2015) *Range. Ecol. Mgt.* **68**: 142.

Bailey D.W., Thomas M.G., Holt T.N., Stephenson M.B., Enns R.M. and Speidel S.E. (2016) *Livestock Sci.* **190:** 76.

Bossone E., D'Andrea A., Citro R., Argiento P., Ferrara F., Cittadini A., Rubenfire M. and Naeije R. (2013) *J. Am. Soc. Echocardiogr.* **26**: 1.

Canovas A., Cockrum R., Brown R., Riddle S., Neary J.M., Holt T.N., Medrano J.F., Islas-Trejo A., Enns R.M., Speidel S.E., Cammack K., Stenmark K.R. and Thomas M.G. (2016) *Proc. Int. Soc. Anim. Genet.* p. 78 (abstract P3050), Salt Lake City, UT, July 2016.

Crawford N.F., Thomas M.G., Holt T.N., Speidel S.E., and Enns R.M. (2016) *J. Anim. Sci.* **94**: 4483.

Crawford N.F., Enns R.M., Speidel S.E., LaShell B., Holt T.N. and M.G. Thomas (2017) *Prof. Anim. Scientist* **33**: (in press).

Enns R.M., Brinks J.S., Bourdon R.M. and Field T.G. (1992) *Proc. West. Sect. Am. Soc. Anim. Sci.* **43**: 111.

Garrick D. and Fernando R.L. (2014) in The Genetics of Cattle, 2nd Edition, 475-498, editors D. Garrick and A. Ruvinsky, CAB International, Boston, MA, USA.

Glover G.H. and Newsom I.E. (1917) Brisket Disease: bulletin no. 204 revised and abbreviated. Bull. No. 229, Colorado Agricultural College, Fort Collins.

Holt T.N. and Callan R.J. (2007) *Vet. Clin. North Am. Food Anim. Pract.* **23**: 575.

Krafsur G.M., Brown R.D., Holt T.N., Gould D.H., Garry F., Riddle S., Neary J.M., Enns R.M., Thomas M.G. and Stenmark K.R. (2017) *Am. J. Respir. Crit. Care Med.* **192**: (abstract 8137 accepted Jan. 12, 2017)

Li M., Riddle S., Zhang H., D' Alessandro A., Flockton A., Serkova N.J., Hansen K.C., Moldvan R., McKeon B.A., Frid M., Kumar S., Li H. Liu H., Canovas A., Medrano J.F., Thomas M.G., Ilioska D., Plecita-Hlavata L., Jezek P., Pullamsetti S., Fini M.A., El Kasmi K.C., Zhang Q. and Stenmark K.R. (2016) *Circulation* **134**: 1105.

Moraes D.L., Colucci W.S. and Givertz M.M. (2000) *Circulation* **102**: 1718.

Neary J.M., Gould D.H., Garry F.B., Knight A.P., Dargatz D.A. and Holt T.N. (2013) *J. Vet. Diag. Invest.* **25**: 210.

Neary J.M. (2014) PhD Dissertation, Colorado State University.

Neary, J. M., Wildman B.K., Booker C.W. and Morley P.S. (2015a). *J. Vet. Intern. Med.* **30:** 326.

Neary J.M., Garry F.B., Holt T.N., Thomas M.G. and Enns R.M. (2015b) *J. Anim. Sci.* **93**: 3854.

Newman J.H., Holt T.N., Cogan J.D, Womack B., Phillips J.A., Li C., Kendall Z., Stenmark K.R., M.G. Thomas and Brown R.D. (2015) *Nat. Commun.* **6:** 1.

Salman M.D., King M.E., Odde K.G. and Mortimer R.G. (1991) *J. Am. Vet. Med. Assoc.* **198:** 968.

Shirley, K.L., Beckman D.W. and Garrick D.J. (2008). *J. Anim. Sci.* **86:** 815.

Stenmark K.R., Yeager M.E., El Kasmi K.C., Nozik-Grayck E., Gerasimovskaya E.V., Li M., Riddle S. and Frid M.G. (2013). *Annu. Rev. Physiol.* **75:** 23-47.

Will D.H. and Alexander A.F. (1970). In Bovine Medicine and Surgery, pp. 412-430. Am. Vet. Publ., Wheaton, IL.

Zeng, X. (2016) PhD Dissertation, Colorado State University.

Table 1. Review of mean pulmonary arterial pressure (mPAP) values in cattle and closely related ruminant species and humans.

Animals	Altitude (m)	mPAP (mmHg)	mPAP Range (mmHg)	References
Angus fattening steers; n = 30	2,200/1,400	54.1 ± 2.7	42-143	Krafsur et al. 2017 Neary et al. 2015
Yearling Angus bulls (gain-test); n = 1,397	2,200	45.8 ± 0.3	29-139	Zeng 2016
Yearling bulls of several *Bos taurus* breeds (gain test); n = 2,426	2,200	45.1 ± 0.8	29-145	Crawford et al. 2017
Mature Angus cows; n = 44	2,200-2,800	42.8 ± 0.8	31-55	Bailey et al. 2016
Yearling Angus heifers (grazing); n = 3,489	2,200	41.4 ± 0.2	22-135	Zeng 2016
Yearling Angus steers as grazing stockers; n = 773	2,200	41.1 ± 0.2	27-138	Zeng 2016
Angus-crossed cows (*Bos taurus* and *Bos indicus*), n = 49	0	34 ± 0.5	28-41	Holt, personal communication
Mature American Bison; n = 6	2,200	29.8 ± 0.8	28-34	Holt, personal communication
Mature Himalayan Yak; n = 6	3,700	20.2 ± 1.4	18-21	Anand et al. 1986
Healthy Human (meta-analysis)	≤ 400	25	15-35	Bossone et al. 2013
Human hypertension (meta-analysis)	≤ 150	>35	15-70	Moraes et al. 2000

Research in this report supported with funds from NIH-5-PO1 HL014985-40A1 (Stenmark) and the John E. Rouse Endowments of Colorado State University (Thomas and Enns).

Proc. Assoc. Advmt. Anim. Breed. Genet. 22:301-304

BREEDING FOR LOW METHANE IS ASSOCIATED WITH LONGER DAYS TO CALVING IN FIRST-PARITY ANGUS FEMALES

K.A. Donoghue[1], T.L. Bird-Gardiner[1], P.F Arthur[2] and R.M. Herd[3]

[1] NSW Department of Primary Industries, Agricultural Research Centre, Trangie, NSW 2823
[2] NSW Department of Primary Industries, EMAI, Menangle, NSW 2568
[3] NSW Department of Primary Industries, Beef Industry Centre, Armidale NSW 2351

SUMMARY

Angus heifers (n=497) that had been divergently selected for methane yield (MY) were used to examine the impact of selection for MY on reproductive performance. Joining records for first-parity heifers over 3 different birth years were used to assess pregnancy rate, days to calving, calving rate, weaning rate, weight of the subsequent calf at birth and weaning, and weight of the female at calf weaning. Selection for MY did not have any impact on pregnancy rate, or weight of the subsequent calf at birth or weaning. However, Low MY heifers calved later in the season and were lighter in weight at calf weaning than High MY heifers. Evidence from this study, along with other studies, indicates that selection for Low MY is associated with calving later in the calving season due to a delay in the onset of puberty in Low MY heifers. Further investigations are needed to confirm any association between calving and weaning rates and genetic variation in MY and to gain better understanding of impact of selection for methane yield on lifetime reproductive performance.

INTRODUCTION

The contribution of livestock to worldwide greenhouse gas emissions is significant (14.5%) and primarily from ruminants (Gerber *et al.* 2013). Methane emission traits have been reported as moderately heritable in sheep (Pinares-Patino *et al.* 2011) and beef cattle (Donoghue *et al.* 2016a). While selection for low methane genotypes is now possible, the impact of this selection on female fertility traits is unknown. The objective of this study was to investigate whether differences exist between heifers from divergent methane yield selection lines for reproductive performance traits recorded during their first parity. These traits include pregnancy, calving and weaning rates, days to calving, weight of calf at birth and weaning, and weight of heifer at weaning of first calf.

MATERIALS AND METHODS

The females used in this project were heifers from the methane yield selection lines within two research herds of Angus cattle at the Agricultural Research Centre, Trangie, NSW, Australia. Methane yield (MY) is the amount of methane emitted per day divided by the weight of feed eaten on a dry-matter basis (g CH_4/kg DMI). Details on the creation of the selection lines and measurement of methane emissions can be obtained from Donoghue *et al.* (2016a). Females were born in 2009 (n=197); 2011 (n=177) and 2012 (n=123). All females available each year were joined at an average age of 500 days, except for heifers born in 2009, who were joined at approximately 780 days of age. Heifers were exposed to bulls for 9 weeks, in single sire mating groups at a ratio of 30 females per bull. The females were pregnancy tested by trans-rectal ultrasonography approximately 10 weeks after the end of joining. Only data from the females' first joining season was considered in this study.

Reproductive performance of the heifers was assessed by pregnancy rate (PREG), calving rate (CALV) and weaning rate (WEAN), defined as the percentage of heifers that were pregnant, percentage of heifers that calved, and percentage of heifers that weaned a calf, respectively, out of the total number exposed to a bull. These rates were coded as binomial traits (e.g. 0, non-pregnant; 1, pregnant). Age and weight of the females at joining were also recorded.

Days to calving (DTC) was also studied as one of the reproductive performance traits, and was calculated as the number of days between the start of the joining season and the date of birth of the subsequent calf. Females that failed to calve were assigned a penalty value which was 21 days greater than the last female to calve in their contemporary group. Weight of the calf at birth and weaning, and weight of the female at the weaning of her calf were also recorded.

Data for weight traits were collected throughout the life of the females, including birth (BWT), weaning (WWT), yearling (YWT) and final (FWT) weight, which were measured at birth and at mean (±SD) age of 231 (±25), 422 (±23), and 603 (±79) d, respectively. In addition body composition traits were measured at mean (±SD) age of 585 (±97) d, and comprised ultrasound P8 fat depth (P8), eye muscle area (EMA) and intramuscular fat % (IMF). Within herd MY Estimated Breeding Values (MY EBV) were obtained for all heifers.

Model of analysis. All the traits, except the binomial traits (PREG, CALV and WEAN) were analysed using the Mixed Model procedure (PROC MIXED) of SAS (SAS Institute, Cary, NC). The model used for the analysis of weight (BWT, WWT, YWT and FWT), body composition (P8, EMA and IMF) traits, joining age (JOINAGE), joining weight (JOINWT) and weight of female at weaning of calf (MCWT) included the fixed effects of selection line (Low or High line), herd (Trangie1 or Trangie2) and birth year (2009, 2011 or 2012), the random effect of sire of heifer and the random residual error. Age of the heifer was included as a covariate in the model for all traits except BWT and JOINAGE, while age of dam was included in the model for BWT and WWT. Interactions between fixed effects were included when significant ($P < 0.05$).

For DTC, the model included the fixed effects of selection line and contemporary group, the random effect of sire of heifer and random residual error. Contemporary group for DTC included all females joined to the same sire in the same year. For calf birth (CALFBWT) and weaning (CALFWWT) weight, the model included the fixed effects of selection line and birth year, sex of the calf, the random effects of sire of the calf, sire of the heifer and the random residual error. Age of the heifer was included in the model for both traits, while age of the calf was included in the model for CALFWWT. Interactions between fixed effects were included when significant ($P < 0.05$).

The binomial traits (PREG, CALV and WEAN) were analysed using the Generalised Linear Model (PROC GENMOD) procedure of SAS with a logit link function. The model fitted included the fixed effects of selection line and contemporary group, sire of heifer and the random residual error. Contemporary group was as defined for DTC. Age at calving was also included in the model for CALV. Correlations between reproductive performance traits and weight, body composition and methane traits were examined using the CORR procedure in SAS.

RESULTS AND DISCUSSION

Least-squares means for weight, body composition traits and MY EBVs are presented in Table 1. There was a significant divergence between selection lines in MY EBV, being 0.20g CH_4/kg DMI ($P < 0.0001$; Table 1). There were no significant differences between MY selection line heifers for weight traits or EMA (all $P > 0.05$; Table 1). Bird-Gardiner *et al.* (2016) analysed a larger data set from the same project that included both males and females, and also reported no significant differences between MY selection lines for weight traits. Heifers from the Low MY selection line were significantly fatter at the P8 site ($P < 0.05$) and had significantly higher IMF% ($P < 0.05$) than heifers from the High MY selection line, which was unexpected given that the genetic corrections with MY reported by Donoghue *et al.* (2016a) were not statistically significant.

Table 1 Least-squares means (±SE) of weight and body composition traits and methane yield EBVs of heifers from divergent selection lines for methane yield

Trait	Selection Line		Significance[a]
	Low Methane Yield	High Methane Yield	
Number of heifers	227	270	-
Birth Weight (kg)	33.0 (0.3)	32.9 (0.2)	ns
Weaning Weight (kg)	229.8 (5.3)	232.8 (5.2)	ns
Yearling Weight (kg)	343.3 (2.0)	344.9 (1.8)	ns
Final Weight (kg)	419.3 (2.2)	421.8 (2.1)	ns
P8 fat (mm)	9.9 (0.6)	9.0 (0.6)	*
EMA (cm^2)	59.8 (0.5)	59.5 (0.5)	ns
IMF (%)	5.3 (0.09)	5.1 (0.09)	*
MY EBV[b] (gCH_4/kg DMI)	-0.12 (0.02)	0.08 (0.02)	**

[a] ** Significance at $P \leq 0.0001$; * Significance at $P \leq 0.05$; ns Non-significance at $P > 0.05$
[b] Methane Yield Estimated Breeding Value

Least-squares means for reproductive performance traits are presented in Table 2. There were no significant selection line differences in any of the reproductive performance traits of the heifers ($P > 0.05$; Table 2), except for DTC and MCWT where heifers from the Low MY selection line calved significantly later and were significantly lighter in weight at weaning of their first calf than heifers from the High MY selection line ($P < 0.05$; Table 2).

Table 2 Least-squares means (±SE) for transformed (T) and untransformed (UT) values of reproductive performance traits of heifers from divergent selection lines for methane yield

Trait	Data type	Selection Line		Significance[a]
		Low MY	High MY	
Number of heifers joined	-	227	270	-
Age at joining (days)	-	604.5 (1.3)	601.4 (1.2)	ns
Weight at joining (kg)	-	419.2 (2.6)	420.8 (2.6)	ns
Pregnancy Rate (%)	T	1.95±0.28	2.41±0.31	ns
	UT	86.6	91.1	-
Calving Rate (%)	T	1.80±0.43	2.21±0.45	ns
	UT	80.6	85.6	-
Days to Calving (days)	-	301.4 (1.2)	298.3 (1.1)	*
Calf Birth Weight (kg)	-	31.8 (0.3)	31.5 (0.3)	ns
Weaning Rate (%)	T	1.16±0.21	1.30±0.21	ns
	UT	75.8	78.2	-
Calf Weaning Weight (kg)	-	226.0 (9.5)	229.7 (9.6)	ns
Cow Weight at Weaning (kg)	-	499.3 (5.6)	510.6 (5.3)	*

[a] ** Significance at $P \leq 0.0001$; * Significance at $P \leq 0.05$; ns Non-significance at $P > 0.05$

Phenotypic correlations between selected weight, body composition, methane and reproductive performance traits are reported in Table 3. MY was not significantly correlated with PREG or DTC (Table 3), but was significantly phenotypically correlated with CALV, WEAN, P8 and MCWT. The significant correlation between MY and P8 and MCWT provides further evidence, along with the observed divergence between the selection lines for these traits (Table 2), of a significant association between MY and these two traits. The significant correlations of MY with CALV and WEAN is at odds with the lack of divergence between the selection lines for these traits (Table 2), and thus any conclusion about the strength of these associations must await further analysis of a larger dataset. DTC was significantly (P < 0.0001) correlated with P8 and MCWT (Table 3), with calving later in the season associated with greater P8 fat depth and heavier weight at calf weaning.

Table 3 Phenotypic correlations[a] between selected growth, body composition, methane and reproductive performance traits in heifers from divergent selection lines for methane yield

Trait	Pregnancy Rate	Calving Rate	Days to Calving	Weaning Rate	P8 fat	Cow WT at weaning
Methane Yield	0.006	0.12^{*}	-0.09	0.13^{*}	-0.46^{**}	0.24^{**}
P8 fat	-0.01	-0.07	0.13^{**}	-0.04	-	-0.18^{**}
Cow WT at weaning	-0.47^{**}	-0.47^{**}	0.23^{**}	-0.54^{**}	-0.18^{**}	-
Joining Age	-0.01	-0.09^{*}	0.003	-0.07	0.29^{**}	-0.14^{**}
Joining WT	-0.01	-0.06	0.006	-0.05	0.22^{**}	0.16^{**}

[a] Significantly different from 0 at ** $P \leq 0.0001$, and * $P \leq 0.05$

Selection for MY did not have any impact on pregnancy rate, or weight of the subsequent calf at birth or weaning. However, Low MY females calved later in the calving season and were lighter in weight at calf weaning than High MY females. Calving later in the season could be the result of Low MY females reaching puberty at a later age than High MY females. Donoghue *et al.* (2016b) using a subset of the data in this study reported the regression of MY EBV on age at puberty, with a decrease in MY EBV significantly associated with later age at puberty. A delayed onset of puberty would be expected to be associated with calving later in the season, as observed in Low MY heifers in this study, and a shorter interval between calving and second parity joining, which may explain why Low MY females were lighter in weight at weaning of the first calf.

CONCLUSIONS

It is concluded from this study that selection for Low MY is associated with calving later in the calving season in first-parity heifers. Evidence from a study on a subset of this data would indicate that these results are due to a delay in the onset of puberty in Low MY females. Further investigations are needed to confirm any association between calving and weaning rates and genetic variation in MY, and to determine if any difference in reproductive performance persists through subsequent parities to gain better understanding of impact of selection for methane yield on lifetime reproductive performance.

REFERENCES

Gerber P.J., Steinfeld H., Henderson B., Mottet A., Opio C., Dijkman J., Falcucci A and Dijkman J. (2013). 'Tackling climate change through livestock – A global assessment of emissions and mitigation opportunities'. FAO, Rome.

Bird-Gardiner T.L., Donoghue K.A., Arthur P.F., Herd R.M. and Hegarty R.S. (2015) *Proc. Assoc. Advmt. Anim. Genet.* **21**: 122.

Donoghue K.A., Bird-Gardiner T.L., Arthur P.F., Herd R.M. and Hegarty R.S. (2016a) *J. Anim. Sci.* **94**: 1438.

Donoghue K.A., Wilkins J.F., Bird-Gardiner T.L., Arthur P.F. and Mula D.J. (2016b) *Proc. Aust. Soc. Anim. Prod.* **31**: 1219.

Pinares-Patino C.S., McEwan J.C., Dodds K.G., Cardenas E.A., Hegarty R.S., Koolaard J.P. and Clark H. (2011) *Anim. Feed Sci. Tech.* **166-67**: 210.

Proc. Assoc. Advmt. Anim. Breed. Genet. 22:305-308

GENETIC ASSOCIATION OF YOUNG MALE TRAITS WITH FEMALE REPRODUCTIVE PERFORMANCE IN BRAHMAN AND SANTA GERTRUDIS CATTLE.

M.G. Jeyaruban and D.J. Johnston

Animal Genetics and Breeding Unit[1], University of New England, Armidale, NSW 2351

SUMMARY

This study investigated the potential value of traits measured in young bulls as genetic indicators for female reproductive performance in two tropical beef cattle breeds. Genetic parameters for the male traits percentage of normal sperm (PNS) and scrotal circumference (SC), and their genetic relationships with first (DTC_1) second (DTC_2) and all days to calving measures (maximum of six) (DTC_{all}) were estimated in Brahman (BRAH) and Santa Gertrudis (SANTA) cattle from records submitted for BREEDPLAN evaluation. Estimated heritabilities for PNS and SC were 0.20±0.06 and 0.45±0.02 for BRAH and 0.17±0.05 and 0.43±0.02 for SANTA, respectively. Genetic correlations between PNS and DTC_1, DTC_2 and DTC_{all} were -0.67±0.28, -0.79±0.25 and -0.47±0.22 in BRAH and -0.18±0.21, -0.28±0.27 and -0.20±0.20 in SANTA, respectively. Genetic correlations of SC with DTC_1, DTC_2 and DTC_{all} were -0.26±0.12, -0.25±0.13 and -0.19±0.08 in BRAH and -0.02±0.09, -0.19±0.13 and 0.00±0.09 in SANTA. These results showed that PNS and SC measured at 18 to 24 months of age in young BRAH and SANTA bulls were moderately heritable and their genetic correlations with DTC were in the same direction in both breeds. PNS and SC had higher genetic correlations with early DTC measures compared to DTC measured in older cows, indicating that they may be more related to early reproduction than lifetime reproduction. In addition, PNS was more strongly related with all DTC measures than SC in both breeds, which suggests that PNS may be a better indicator trait than SC for improving female reproduction in tropical breeds in Northern Australia.

INTRODUCTION

Cow and bull reproductive efficiency are important for the productivity and profitability of beef cattle producers in northern Australia. Various measures of female fertility were investigated to improve female reproductive efficiency. Days to calving (DTC) is one such measure and is implemented in the BREEDPLAN genetic evaluation as the key measure of genetic merit for female reproduction. However, low heritability, low intensity of selection and observations relatively late in life, limit the capacity to improve female fertility using DTC measures only. Therefore, identification of male traits, which have high genetic associations with female fertility, could greatly assist the improvement of reproductive performance of beef cattle in Northern Australia. From a previous study (Meyer *et al.* 1991), scrotal circumference (SC) in males is used as an indicator trait for female fertility in BREEDPLAN evaluation. However, Johnston *et al.* (2014) identified percentage of normal sperm (PNS) as being strongly genetically correlated with female reproductive traits in tropical breeds, in a research data set. These findings need to be validated using industry data before PNS could be included in routine BREEDPLAN genetic evaluation. Therefore, the aim of this study was to compare the genetic correlations of PNS and SC with early and lifetime female reproduction traits in Brahman (BRAH) and Santa Gertrudis (SANTA) cattle, to discover whether there are additional or better genetic indicators to improve reproduction efficiency of beef cattle in Northern Australia.

---

[1] AGBU is a joint venture of the NSW Department of Primary Industries and University of New England

MATERIALS AND METHODS

Data for PNS, collected as part of the bull breeding soundness evaluation for young BRAH and SANTA bulls (of Beef CRC and industry), along with the pedigree data and performance data for other traits were obtained from respective breed society databases. The PNS data were measured between 540 to 800 days of age and for bulls with multiple PNS records, only the first record was used. Data for all DTC (DTC_{all}) and SC were submitted by members of BRAH and SANTA societies for their respective BREEDPLAN evaluations. Scrotal circumference (SC) was measured between 300 to 700 days of age. DTC for natural mating was defined as the number of days from the "bull in date" (first day of introducing bulls) until the subsequent calving of the cow (Johnston and Bunter, 1996). DTC_{all} were split into DTC from the first mating (DTC_1) and DTC from the second mating (DTC_2), to describe early reproductive measures. For BRAH, DTC_1 was measured between 650 to 925 days of age and DTC_2, the second measurement for cows with a DTC_1. For SANTA, the DTC_1 was measured between 270 to 640 days of age. All non-calver cows were included by assigning a penalty DTC record as described by Johnston and Bunter (1996).

The number of records and descriptive statistics for all traits are given in Table 1. For BRAH, there were 215, 1884, 1199 and 1199 sires with progeny recorded for PNS, SC, DTC_1 and DTC_{all}, respectively. There were 128 and 816 common sires with progeny recorded for PNS and DTC_1, and SC and DTC_1, respectively. For SANTA, there were 221, 2385, 619 and 619 sires with progeny recorded for PNS, SC, DTC_1 and DTC_{all}, respectively. There were 119 and 560 common sires with progeny recorded for PNS and DTC_1, and SC and DTC_1, respectively.

Genetic variances and variance ratios were estimated by restricted maximum likelihood (REML) using a univariate animal model, and genetic correlations were estimated using a series of bivariate animal model analyses, with three generations of pedigree in WOMBAT (Meyer 2007). Models included contemporary group as a fixed effect, age of measurement as a covariate and the random additive genetic effect of animal for all traits. An additional random common environmental effect of animal was fitted for DTC_{all} to account for repeated records. Contemporary group definitions for DTC and SC were as defined in BREEDPLAN (Graser *et al.* 2005) and for PNS were formed by accounting for herd of origin, year of birth, birth type, previous weight management groups and date of measurement.

RESULTS AND DISCUSSION

Raw means by breed are presented in Table 1. For BRAH, the means for PNS, SC and DTC_{all} were 68.2, 26.2 and 358 and for SANTA, were 73.1, 32.3 and 358, respectively Estimated heritabilities were moderate for PNS and SC in both breeds (Table 1) (0.20 and 0.17 for PNS and 0.45 and 0.43 for SC in BRAH and SANTA, respectively). Estimated heritability for PNS in BRAH was of similar magnitude to the 0.25 reported by Corbet *et al.* (2013) for the Beef CRC study. Estimated heritabilities for SC in BRAH and SANTA were in agreement with the range of estimates reported by Cammack *et al.* (2009). The moderate heritability estimates for PNS and SC suggest that both traits could be improved by selection in BRAH and SANTA. Heritability estimates for all DTC measures in both breeds were low. Estimated heritabilities for DTC_2 were higher than the estimates for DTC_1 in both breeds. Estimated heritabilities for DTC_1 and DTC_2 were lower than the values of 0.22 and 0.20 reported for DTC_1 and DTC_2 of BRAH, respectively, by Johnston *et al.* (2014a) using research data.

Estimated genetic correlations between bull traits and DTC measures are presented in Table 2. Genetic correlations between PNS and DTC were in the same direction for BRAH and SANTA (Table 2). Moderate to strong negative genetic correlations were estimated between PNS and DTC measures and were of larger magnitude for BRAH than SANTA. The difference in the magnitude of the genetic relationship between PNS and all three DTC measures in the two breeds may be attributed to differences in their age at first mating. BRAH heifers were one year older than the

SANTA heifers at first mating, and it is likely that a higher proportion of BRAH than SANTA heifers would be cycling at the time of first mating. Estimated genetic correlations between PNS and DTC_2 were of higher magnitude than the correlations of PNS with DTC_1 and DTC_{all} in both breeds. For BRAH, the genetic correlations between PNS and all three DTC measures were significantly different from zero. Estimated genetic correlations between PNS and DTC_1 and DTC_2 in BRAH were stronger than the values reported by Johnston *et al.* (2014b), which ranged from -0.69 to -0.04.

Table 1. Number of records, descriptive statistics, additive genetic variance (σ_a^2) and heritability (h^2) with se in parenthesis for percentage normal sperm (PNS), scrotal size (SC) and first (DTC_1), second (DTC_2) and all days to calving (DTC_{all}) for Brahman and Santa Gertrudis.

Trait (units)	No. of records	Mean	SD	Min	Max	σ_a^2	h^2
			Brahman				
PNS (%)	2330	68.2	26.0	1.0	99.0	119.7	0.20 (0.06)
SC (cm)	23247	26.2	4.9	13.4	41.5	2.6	0.45 (0.02)
DTC_1 (days)	11800	363.9	58.2	270.0	491.0	145.9	0.07 (0.02)
DTC_2 (days)	3349	370.6	46.6	271.0	490.0	202.4	0.13 (0.04)
DTC_{all} (days)	29552	358.6	51.8	270.0	491.0	111.3	0.06 (0.01)
			Santa Gertrudis				
PNS (%)	2078	73.1	20.2	2.0	99.0	65.7	0.17 (0.05)
SC (cm)	35663	32.3	4.1	20.0	44.0	2.9	0.43 (0.02)
DTC_1 (days)	5794	377.3	60.9	273.0	491.0	311.8	0.13 (0.03)
DTC_2 (days)	2042	360.4	47.7	273.0	491.0	228.4	0.14 (0.05)
DTC_{all} (days)	15104	358.6	53.2	270.0	491.0	156.9	0.08 (0.01)

Genetic correlations between SC and DTC measures for BRAH and SANTA were in the same direction as between PNS and the three DTC traits, except for DTC_{all} in SANTA. However, for both breeds, the magnitude of the genetic correlations with SC was lower than those estimated between PNS and DTC. Genetic correlations of SC with DTC_1 and DTC_2 were within the range of -0.35 to -0.21 reported between SC and both DTC_1 and DTC_2 for BRAH by Johnston *et al.* (2014b).

Table 2. Genetic correlations between PNS and SC with first (DTC_1), second (DTC_2) and all days to calving (DTC_{all}) of Brahman and Santa Gertrudis cattle.

Trait (units)	Genetic correlation		
	DTC_1 (days)	DTC_2 (days)	DTC_{all} (days)
		Brahman	
PNS (%)	-0.67 (0.28)	-0.79 (0.25)	-0.47 (0.22)
SC (cm)	-0.26 (0.10)	-0.25 (0.13)	-0.19 (0.08)
		Santa Gertrudis	
PNS (%)	-0.18 (0.21)	-0.28 (0.27)	-0.20 (0.20)
SC (cm)	-0.02 (0.09)	-0.19 (0.13)	0.00 (0.09)

This study showed that PNS had stronger genetic correlations than SC with all DTC measures in BRAH and SANTA. This suggests that PNS is a better indicator trait than SC to improve female reproduction in these breeds, under Australian conditions. Inclusion of PNS, along with SC, in the genetic evaluation for BRAH and SANTA will improve the accuracy of prediction for DTC. Consistently higher genetic correlations were observed between PNS and DTC_2 than between PNS and DTC_1, which implied that PNS had a higher genetic correlation with lactation anoestrus interval than heifer puberty in DTC_1, in both breeds. These results are consistent with those reported by Johnston *et al.* (2014b). Although a similar pattern was observed for the genetic correlations between PNS and DTC measures in both breeds, the differences in the magnitude of the correlations for BRAH and SANTA were attributed to the physiological stage they were in at the time of recording the DTC measures. For BRAH, the first DTC record was from females mated as 2 year olds, which could have a higher proportion of heifers cycling than SANTA (which could submit results from yearling matings) at the time of measuring DTC_1.

CONCLUSIONS

PNS and SC measured in young BRAH and SANTA bulls were moderately heritable. Selecting bulls for higher PNS and SC at 2 years of age is expected to lead to a correlated response of reduced days to calving in cows in both breeds, with the effect stronger in early matings. Stronger genetic relationships between PNS and all DTC measures in both breeds suggest that PNS is a better indicator trait than SC, in terms of estimating genetic merit, and hence providing potential for improving female reproduction in tropical breeds in Northern Australia. PNS could be included in BREEDPLAN genetic evaluation as indirect selection criteria for improving female reproduction in tropical breeds. Furthermore, difference in the magnitude of the genetic correlations of PNS with DTC_1 and DTC_2 and low to moderate correlation (less than 0.6) between DTC_1 and DTC_2, suggest that DTC records in BREEDPLAN evaluation could be split into DTC_1 and DTC_2 to better describe early female reproduction.

ACKNOWLEDGEMENT

The authors would like to thank the Meat and Livestock Australia (MLA) for their financial support through L.GEN.0174 and the Cooperate Research Centre for Cattle and beef quality, Australian Brahman Breeders Association Limited and Santa Gertrudis Australia for providing data for this study.

REFERENCES

Cammack K.M., Thomas M.G., and Enns R.M. (2009) *Prof. Anim. Sci.* **25:** 517.

Corbet N.J., Burns B.M., Johnston D.J., Wolcott M.L., Corbet D.H., Venus B.K., McGowan M.R. and Holroyd R.G. (2013) *Anim. Pro. Sci.* **53:** 101.

Forni S. and Albuquerque L.G. (2005) *J. Anim. Sci.* **83:** 1511.

Graser H-U., Tier B., Johnston D.J. and Barwick S.A. (2005) *Aust. J.Exp.Agric.* **45**:913.

Johnston D.J. and Bunter K.L. (1996) *Livest. Pro. Sci.* **45:** 13.

Johnston D.J., Barwick S.A., Fordyce G., Holroyd R.G., Williams P.J., Corbet N.J. and Grant T (2014a) *Anim. Pro. Sci.* **54:** 1.

Johnston D.J., Corbet N.J., Barwick S.A., Wolcott M.L. and Holroyd R.G. (2014b) *Anim. Pro. Sci.* **54:** 74.

Meyer K., Hammond K., Mackinnon M.J., and Parnell P.F. (1991) *J. Anim. Sci.* 69**:** 3533.

Meyer K. (2007) *J. Zhej. Uni.* **8:** 815.

Proc. Assoc. Advmt. Anim. Breed. Genet. 22:309-312

WEIGHTING FACTORS FOR GENOMIC INFORMATION USED IN SINGLE-STEP GENOMIC SELECTION IN AUSTRALIAN BEEF

Y. D. Zhang, A. Swan, D.J. Johnston and C. J. Girard

Animal Genetics and Breeding Unit*, UNE, Armidale, NSW

ABSTRACT

Single-step genomic evaluation utilises all phenotypes, pedigree and genotypes and could significantly enhance beef cattle genetic evaluation. An appropriate weighting factor for genomic and pedigree information is required to predict single-step estimated breeding values (EBVs). This study assessed the optimal weighing factor lambda (λ, ranging between 0 and 1 for none to 100% weighing on genomic information) for a series of beef traits using an empirical approach. The optimal value of λ was identified from the maximum accuracies of genomic predictions by internal cross-validation. The estimated genomic accuracies for Brahman cattle ranged from 0.23 to 0.70 for traits with adequate numbers of genotypes and phenotypes. The accuracy of genomic predictions generally increased as the λ weighting factor increased for a range of traits and typically approached an asymptote towards the optimal λ. For traits with adequate numbers of records, the optimal λ values ranged from 0.4 to 0.8.

INTRODUCTION

Application of genomic selection in livestock enables more accurate selection of animals at younger ages, and for hard to measure and sex-limited traits. Ultimately, the use of genomic selection can increase genetic gain. Best Linear Unbiased Prediction (BLUP) is a traditional and reliable tool to estimate breeding values and it has served animal breeders well. Genomic BLUP (GBLUP) works in the similar way to BLUP, but substitutes the pedigree based relationship matrix A with the genomic relationship matrix *G*. The recently developed single step genomic BLUP (ssGBLUP) by Legarra *et al.* (2009) and Christensen and Lund (2010) makes use of genotypes, all phenotypes and pedigree information, aiming to streamline the application and enhance the accuracies of EBV. The variance matrix of EBVs for ssGBLUP combines *A* and *G*, and the inverse matrix (H^{-1}) required to solve the ssGBLUP equations has a simple form as shown by Aguilar et al. (2010): $H^{-1} = A^{-1} + \begin{bmatrix} 0 & 0 \\ 0 & G^{-1} - A_{22}^{-1} \end{bmatrix}$. An appropriate weighting of pedigree and genomic information when constructing *G* is required because SNP marker panels do not explain all of the additive genetic variation (e.g. Goddard *et al.* 2011). A modified genomic relationship matrix is typically used, as $G = \lambda G_m + (1 - \lambda)A_{22}$, where λ is the fraction of the additive genetic variance explained by markers, ranging between 0 and 1. This study assessed the optimal weighing factor λ using an empirical approach.

MATERIALS AND METHOD

Data. Phenotypes, pedigree and genotypes for this study were from the BREEDPLAN database for Brahman cattle. Traits in this analysis included growth (5 traits), ultrasonic scanning body composition (6 traits), carcase characteristics (6 traits), flight time, scrotal circumference and days to calving (DTC). Table 1 summarises the pedigree, records, number of genotypes available for each trait. DTC was measured repeatedly. On average, every animal had 2.5 DTC records and 4.6 for genotyped animals. In total there were 7166 animals genotyped and included in the H^{-1} matrix.

*AGBU is a joint venture of The NSW Department of Primary Industries and UNE.

Statistical models. Single step GBLUP (ssGBLUP) analyses were performed in Wombat (Meyer 2007), using all available records for each of these traits. The data were obtained from the BREEDPLAN database, with phenotypes pre-adjusted for all fixed effects but contemporary group. The model fitted contemporary group as the sole fixed effect, and the additive genetic breeding value as a random effect. In addition, for birth weight, weaning weight and yearling weight maternal genetic effects were also fitted, and DTC was analysed using a repeatability model. Analyses were performed with the H^{-1} matrix calculated for a range of λ values between 0 and 1 with an increment of 0.1. The optimal value of λ was identified by the highest accuracies of ssGBLUP EBV via five-fold cross-validation. Animals with both genotypes and phenotypes were split into five groups, based on half-sib family structure, with no progeny within half-sib families allocated to more than one group. In each of the five analyses, four groups were used as the genomic reference to predict EBVs of the fifth (test) group. Phenotypes for animals in the test group were omitted from the training data, but their pedigree and genotype data were included in the H^{-1} matrix in order to obtain their EBVs. This cross-validation was performed for the range of λ values from 0 to 1.0.

Table 1. Summary of data for each trait, numbers of animals (N) and number of sires (Sires) for phenotypes and genotypes, and average size of test set in 5-fold cross-validation (ncv)

Trait	Phenotypes		Genotypes		
	N	Sire	N	Sire	ncv
Birth weight (BWD)	19567	1145	2072	219	357
Weaning weight: 200 day (WWD)	198250	5249	5677	653	796
Yearling weight: 400 day (YWD)	101415	4382	4607	510	701
Final weight: 600 day (FWD)	102490	4370	4295	506	647
Mature cow weight (MCW)	8433	930	1241	155	203
Heifer scan eye muscle area (HEA)	10562	714	1814	84	341
Bull scan eye muscle area (BEA)	10852	1013	1459	140	260
Bull scan rib fat (BRF)	9921	963	839	121	141
Bull scan p8 fat depth (BP8)	10128	971	854	127	141
Carcase weight (CWT)	2982	178	933	89	171
Carcase P8 depth (CP8)	2675	146	911	89	167
Carcase rib fat (CRF)	2569	146	859	88	156
Carcase intramuscular fat (CIM)	2703	154	926	89	170
Shear Force (SHF)	2584	146	898	89	163
Flight Time (FLT)	7756	280	1195	81	227
Scrotal size (SS)	27709	2049	1686	261	263
Days to calving (DTC)	18178	1349	1130	139	178
Pfizer MBV Tenderness (MPT)	6909	1158	1920	173	342

The accuracies of genomic predictions were calculated as the correlation between EBVs and adjusted phenotypes, scaled by the square root of the heritability of the trait, which was estimated using all records and pedigree. The means of the five scaled correlation coefficients are presented as the accuracy. For repeated records (DTC), adjusted phenotypes were calculated as the average residual from a repeatability model fitting contemporary group, then weighted according to Garrick *et al.* (2009). The heritability used to calculate the accuracy for DTC was also adjusted according to $h^2_{adj} = h^2/(t + \frac{1-t}{n})$, where t is the repeatability and n is the average records per animal within each test set. EBVs for three λ values (0, 0.5 and 1.00) were compared in five classes where animals were phenotyped or genotyped or both.

RESULTS AND DISCUSSION

Results are summarised in Table 2. There was a wide variation in the value of λ at the highest accuracies from 0.1 to 1.0. The highest accuracies of EBV ranged from 0.15 for CIM to 0.70 for SS. For traits with reasonable number of records (BWD, WWD, YWD, FWD, MCW, HEA, BEA, SS, FLT), the λ values ranged from 0.4 to 0.8, and the corresponding accuracies of EBV ranged from 0.23 to 0.70. The λ value at the highest accuracy for CRF (0.1) differed markedly from most traits, possible due to the quality of phenotypes for this trait (carcase might be trimmed prior to measurement). As the maximum was approached, accuracy was relatively insensitive over a large range in λ values. This was observed in most traits as the response surface generally approached an asymptote.

Table 2. Results of Brahman cross-validation tests for a range values of λ, with estimated heritability (h^2), maximum accuracy (r_max), λ_max (λ at r_max), and range in λ where accuracy varied by -0.01 around r_max (λ_low to λ_high)

Trait	h^2	λ_max	r_max	λ_low	λ_high
BWD	0.45	0.60	0.53	0.30	1.00
WWD	0.32	0.40	0.45	0.20	0.70
YWD	0.38	0.60	0.33	0.30	0.90
FWD	0.43	0.70	0.53	0.30	1.00
MCW	0.60	0.80	0.40	0.50	1.00
HEA	0.30	0.90	0.23	0.50	1.00
BEA	0.29	0.50	0.37	0.30	0.70
BRF	0.28	0.90	0.26	0.60	1.00
BP8	0.42	0.70	0.28	0.40	1.00
CWT	0.51	0.30	0.47	0.20	0.60
CP8	0.30	1.00	0.27	0.50	1.00
CRF	0.26	0.10	0.24	0.00	0.30
CIM	0.25	0.40	0.15	0.10	0.80
SHF	0.27	0.80	0.41	0.40	1.00
FLT	0.28	0.50	0.51	0.20	0.80
SS	0.43	0.70	0.70	0.40	1.00
DTC	0.05	0.80	0.34	0.60	1.00
MPT	0.72	1.00	0.50	0.70	1.00

Table 3 shows impacts of three values of λ on EBVs for CP8, DTC, MCW and WWD. The variation in EBVs increased from λ=0 to 0.5 for phenotyped or phenotyped but not genotyped classes of animals, but less in moving from λ=0.5 to 1.0. In contrast, the variation in EBVs increased with from λ=0 to 1.0 for genotyped and genotyped but not phenotyped classes of animals, and so for most traits in both genotyped and phenotyped class. Correlations of EBVs across three values of λ were consistently high for P and P-G classes. Understandably, this was due to the impact of direct phenotypic information. For G or G-P animals, EBVs were predicted through a combination of pedigree and genomic relationships, and correlations between λ=0.5 and λ=1.0 were always very high (0.93 to 0.97); lower correlations were observed for EBVs between λ=0 and λ=0.5 (0.88 to 0.95). The correlation the λ=0 and λ=1.0 further decreased (0.70 to 0.85).

Table 3. Comparison of EBV for animals in phenotyped (P), phenotyped but not genotyped (P-G), genotyped (G), genotyped but not phenotyped (G-P), and both phenotyped and genotyped (P+G) classes over three values of λ (0, 0.5 and 1.0). EBV standard deviations for three values of λ (0 = sd0, 0.5 = sd50, 1.0 = sd100), and correlations between EBVs (e.g. r50_100 for correlation of EBV between λ values 0.5 and 1.0)

Trait	Group	N	sd0	sd50	sd100	r0_50	r0_100	r50_100
CP8	P	2675	1.30	1.41	1.36	0.99	0.97	0.99
CP8	P-G	1764	1.30	1.40	1.33	1.00	0.99	1.00
CP8	P+G	911	1.30	1.44	1.42	0.98	0.94	0.99
CP8	G	7166	0.74	0.87	0.98	0.91	0.77	0.96
CP8	G-P	6255	0.61	0.75	0.89	0.87	0.70	0.96
DTC	P	18178	4.52	4.77	4.82	0.99	0.97	0.99
DTC	P-G	17048	4.50	4.73	4.76	0.99	0.97	0.99
DTC	P+G	1130	4.86	5.38	5.72	0.97	0.90	0.98
DTC	G	7166	3.31	3.92	4.51	0.91	0.79	0.97
DTC	G-P	6036	2.93	3.58	4.24	0.89	0.75	0.97
MCW	P	8433	26.96	27.46	26.71	1.00	0.99	1.00
MCW	P-G	7192	26.88	27.24	26.45	1.00	1.00	1.00
MCW	P+G	1241	27.41	28.66	28.08	0.99	0.96	0.99
MCW	G	7166	17.07	19.28	21.15	0.92	0.81	0.97
MCW	G-P	5925	13.96	16.61	19.32	0.88	0.73	0.96
WWD	P	198250	8.81	8.92	8.80	1.00	0.99	1.00
WWD	P-G	192573	8.85	8.95	8.81	1.00	1.00	1.00
WWD	P+G	5677	7.51	8.09	8.56	0.95	0.84	0.96
WWD	G	7166	7.40	8.01	8.49	0.94	0.83	0.96
WWD	G-P	1489	6.69	7.33	7.91	0.92	0.79	0.95

In view of these results, a value of λ=0.5 has been adopted in preliminary ssGBLUP analyses for BREEDPLAN, but this may change as more experience is gained with the method. In future, a high weighting factor, for example, λ=0.7, could be considered, as shown in Table 2, a high λ will be beneficial to most traits, but adversely affect WWD, CWT and CIM. The current data structure for the Brahman Breedplan analysis may have impacts on findings. Further study using data from other breeds, e.g. Angus, is required to validate these results.

REFERENCES

Aguilar I., Misztal I., Johnson D.L., Legarra A., Tsuruta S., *et al.* (2010) *J. of Dairy Sci.* **93**:743.
Christensen O.F. and Lund M.S. (2010) *Genet. Sel. Evol.* **42**: 2.
Legarra A., Aguilar I. and Misztal I. (2009) *J. of Dairy Sci.* **92**: 4656.
Meyer K. (2007) *J. of Zhejiang Uni. Sci. B* **8**: 815.
Goddard M., Hayes B. and Meuwissen T. (2011) *J. of Anim. Breed. Genet.* **128**: 409.
Garrick D.J., Taylor J.F. and Fernando R.L (2009). *Genet. Sel. Evol.* **41**:55.

Proc. Assoc. Advmt. Anim. Breed. Genet. 22:313-316

GENETIC AND PHENOTYPIC CHARACTERIZATION OF MSA INDEX AND ITS ASSOCIATION WITH CARCASE AND MEAT QUALITY TRAITS IN ANGUS AND BRAHMAN CATTLE.

M.G. Jeyaruban, D.J. Johnston and B.J. Walmsley

Animal Genetics and Breeding Unit[1], University of New England, Armidale, NSW 2351

SUMMARY

MSA Index (MSA_I) predicts overall eating quality of a carcase from predictions of tenderness, juiciness, flavour and overall liking based on extensive consumer taste panel studies. Prices paid by processors for MSA graded meat are consistently higher than non-MSA graded cattle. There is significant industry pressure to determine the level of genetic control for MSA_I. This study used data from Angus (ANGS) and Brahman (BRAH) BIN cattle to estimate genetic parameters for MSA_I and their genetic and phenotypic relationships with key carcase and meat quality traits. Heritabilities for MSA_I were 0.50±0.09 and 0.49±0.16 in ANGS and BRAH cattle, respectively. In ANGS, genetic correlations between MSA_I and carcase weight (CWT), carcase rump fat (CP8), ossification score (OSS), MSA marble score (MSA_M), shear force (SF) and meat colour L (Col_L) were 0.41±0.12, 0.02±0.15, -0.22±0.18, 0.96±0.02, -0.33±0.19 and 0.40±0.16, respectively and in BRAH, were 0.12±0.21, 0.19±0.24, -0.51±0.20, 0.94±0.05, -0.18±0.25 and 0.46±0.21. Similar trends were observed in the correlations between MSA_I and carcase and meat quality traits in both breeds. This study showed that MSA_I has a very high genetic association with MSA_M and, to a lesser extent, OSS in both breeds. Selecting for higher MSA_M and lower OSS will genetically improve MSA_I of carcases of ANGS and BRAH cattle, and the addition of MSA_I as an additional trait in the evaluation would contribute almost no additional information about meat quality or value.

INTRODUCTION

Eating quality is important if the beef industry in Australia is to remain competitive in the world and domestic markets. Eating quality refers to the compositional quality and the palatability of meat, and in the 1990s, consistency of beef eating quality was identified as a key issue for marketing Australian beef (Bindon 2001). Beef consumers were unsure of how to identify beef of acceptable quality and this led to a decline in domestic beef consumption (Polkinghorne *et al.* 2008). Furthermore, concerns with the fat content of beef, and associated health implications, and a decline in understanding of beef cuts, cooking methods and an inability to predict quality from product appearance also impacted on demand for beef products on the domestic market. Meat and Livestock Australia (MLA) developed the Meat Standards Australia (MSA) grading system to provide consumers with assurance of eating quality (Watson *et al.* 2008). MSA eating quality scores are a combination of consumer assessed tenderness, juiciness, flavour and overall liking of meat products. Initially, the grading system assigned an eating quality score to specific muscle portions cooked by defined methods (Watson *et al.* 2008). Subsequently, a single number and standard national measure called the MSA Index (MSA_I) was developed to predict overall eating quality of a whole carcase (Thompson 2014).

The MSA model predicts the eating quality of 39 cuts in a carcase using measurements collected by accredited MSA graders (MSA Index, Meat and Livestock Australia, 2014). MSA_I is a weighted average of these scores for the 39 MSA cuts for the most common corresponding cooking method, ranging from 30 to 80 and expressed to 2 decimal places, to represent the

[1] AGBU is a joint venture of the NSW Department of Primary Industries and University of New England

predicted eating quality of a whole carcase. The MSA_I is independent of any processing inputs and is calculated using only attributes influenced by pre-slaughter factors. It is a benchmark which can be used across all processors, geographic regions and over time, and reflects the impact on eating quality of management, environmental and genetic differences between cattle at the point of slaughter (MSA Index, Meat and Livestock Australia, 2014). Currently, over 40 processors are grading MSA beef, with prices received for MSA graded g cattle being consistently higher than non-MSA cattle (Southern Beef Technology services, 2015). There has been a recent request from industry to examine genetic and non-genetic influences on MSA_I and genetic relationships between MSA_I and key carcase and meat quality traits. This study aimed to estimate these in temperate and tropical breeds.

MATERIALS AND METHODS

Carcase and meat quality data used for this study were recorded as part of the Angus (ANGS) Sire Benchmarking Program and Brahman (BRAH) beef information nucleus (BIN) project up to June 2016. ANGS steers were grain fed with a ration containing an energy level of 12MJ/kg for 300 days and Brahman steers were finished on pasture. ANGS and BRAH steers were killed at mean age of 794 and 963 days, respectively. Hot carcase weight and hot P8 fat depth were recorded before the carcase entered the chiller. Carcases were dressed according to AUSMEAT standard specifications (AUS-MEAT 2005) while limiting the subcutaneous trimming to not influence the fat depth measurement at the P8 or 12/13th rib measurements sites. Carcase weight (CWT), rump P8 fat (CP8), MSA marble scores (MSA_M), and ossification scores (OSS) were measured by MSA certified graders. Samples were collected from the *Longissimus dorsi* and transported to a laboratory where meat colour was recorded as Minolta l (COL_L) and shear force (SF) was measured as described by Perry et al. (2001). MSA_I was calculated as the weighted average of the predicted eating quality scores using the empirical modelling described by Thompson (2014). Inputs to this calculation included hormone growth promotant status, milk fed vealer status, sale yard status, sex, *Bos indicus* content, hump height, CWT, OSS, 12/13th rib fat and MSA_M. The number of records and descriptive statistics for all traits are given in Table 1. The number of sires with progeny recorded differed across the six traits, ranging from 83 to 123 for ANGS and 72 to 80 for BRAH.

For each trait, records that were more than three standard deviations from the mean were removed as outliers. A univariate linear animal model was used to estimate genetic parameters for carcase and meat quality traits in both breeds:

$$Y_{ik} = cg_i + \beta_1 \text{ age/weight}_k + \beta_2 (\text{age/weight}_k)^2 + a_k + e_{ik}$$

Where Y_{ik} is the trait of interest of animal k in a fixed contemporary group i (cg_i), age/weight is age or weight of animal k, β_1, and β_2 are regression coefficients for linear and quadratic effects of age or weight of animal (included in models based on significant level), a_k is the random additive genetic effect of animal k and e_{ijk} is the random error associated with this prediction. Contemporary groups for all traits were defined based on protocols for carcase traits in the BREEDPLAN evaluation (Graser *et al.* 2005). Genetic variances, variance ratios and genetic correlations were estimated by restricted maximum likelihood (REML) using univariate and bivariate animal models, with three generations of pedigree, using WOMBAT (Meyer 2007).

RESULTS AND DISCUSSION

Raw means by trait and breed are presented in Table 1. Design of this study will not allow direct comparison of breeds. This is because BRAH and ANGS were subjected to different finishing regimes, slaughtered at different abattoirs, at different ages and graded by different graders at different quartering sites.

Table 1. Number of records and the descriptive statistics for carcase and meat quality data

Trait	Angus[1]					Brahman				
	No.	Mean	SD	Min	Max	No.	Mean	SD	Min	Max
Carcase traits										
CWT (kg)	1394	460.1	37.1	334.0	568.0	898	314.4	24.1	227.0	382.0
CP8 (mm)	1383	23.1	6.1	10.0	41.0	891	12.0	4.0	2.0	24.0
OSS (score)	1383	154.8	110.0	110.0	200.0	894	138.6	13.7	110.0	190.0
Meat quality traits										
MSA_I (score)	1349	65.1	1.7	60.0	69.9	629	53.7	1.7	49.5	58.5
MSA_M (score)	1382	515.2	115.6	160.0	880.0	886	267.2	67.5	120.0	490.0
SF (kg)	737	3.8	0.6	2.0	5.7	881	4.5	0.8	2.8	7.5
Col_L (score)	1384	42.7	2.8	35.7	50.3	891	38.3	2.6	30.5	46.2

[1] Design of this study will not allow direct comparison of breeds.

Estimated heritabilities for MSA_I and genetic and phenotypic correlations with carcase and meat quality traits are given in Table 2. Heritabilities were similar for MSA_I in both breeds (0.50 in ANGS and 0.49 in BRAH). Heritabilities for carcase and meat quality traits were moderate to high for both breeds. For ANGS, estimated heritability for MSA_M agreed with the value of 0.48 reported by Barwick *et al.* (2009) for ANGS crosses. However, estimates for CWT and CP8 in ANGS were higher than the values reported for temperate breeds by Reverter *et al.* (2003) and the estimates for meat quality traits were also higher than the values reported for temperate breeds by Johnston *et al.* (2003). Except for OSS, heritability estimates for carcase and meat quality traits in BRAH were higher than the estimates reported by Wolcott *et al.* (2009). Both ANGS and BRAH steers used in this study were killed at a higher age than those in the previous studies and this led to higher means and variations (SD) for carcase and meat quality traits in the two breeds.

The genetic correlation between MSA_I and carcase and meat quality traits was variable in sign and magnitude, but was in the same direction for both breeds. For ANGS, MSA_I had a moderately positive genetic correlation with CWT (0.41), a low or no correlation with CP8 (0.02) and a moderately negative correlation with OSS (-0.22). For meat quality traits in ANGS, MSA_I had a highly positive genetic correlation with MSA_M (0.96), a moderately positive genetic correlation with Col_L (0.40) and a moderately negative correlation with SF (-0.33).

Table 2. Heritabilities (h^2) and genetic correlations (r_g) between MSA Index and carcase and meat quality traits (standard error in parenthesis)

Type	Traits						
	MSA_I	CWT	CP8	OSS	MSA_M	SF	COL_L
				Angus			
h^2	0.50 (0.09)	0.66 (0.10)	0.48 (0.19)	0.22 (0.07)	0.48 (0.09)	0.43 (0.14)	0.31 (0.08)
r_g		0.41 (0.12)	0.02 (0.15)	-0.22 (0.18)	0.96 (0.02)	-0.33 (0.19)	0.40 (0.16)
				Brahman			
h^2	0.49 (0.16)	0.59 (0.14)	0.36 (0.11)	0.36 (0.11)	0.37 (0.11)	0.38 (0.12)	0.44 (0.12)
r_g		0.12 (0.21)	0.19 (0.24)	-0.51 (0.20)	0.94 (0.05)	-0.18 (0.25)	0.46 (0.21)

In BRAH, MSA_I had a low positive genetic correlation with CWT (0.12), a low positive correlation with CP8 (0.19) and a moderately negative correlation with OSS (-0.51). For meat quality traits, MSA_I had a positive genetic correlation with MSA_M (0.94), a moderately positive genetic correlation with Col_L (0.46) and a moderately negative correlation with SF (-0.18).

The high positive genetic correlations between MSA_I and MSA_M, along with moderately negative correlations with OSS in both breeds, suggests that selection primarily for higher marbling and lower OSS will improve the MSA_I of carcases in ANGS and BRAH cattle. Currently, marble scores are included in the BREEDPLAN evaluation of these two breeds and, therefore, the genetic evaluation of meat quality is being adequately addressed in the BREEDPLAN evaluation of these two breeds.

CONCLUSIONS

MSA-Index measured in Angus and Brahman steers were moderately heritable and had very similar genetic correlations with carcase and meat quality traits in both breeds. The very high genetic correlations with MSA_M and moderate to high negative genetic correlations with OSS indicate that improving those two traits will improve the MSA_I in both breeds. Marble Score included in BREEDPLAN evaluation is, therefore, expected to underpin the genetic of meat quality as assessed by MSA_I in Angus and Brahman cattle. Further evaluation of OSS, with more data, is required before being included in the BREEDPLAN evaluation.

ACKNOWLEDGEMENT

The authors would like to thank the Meat and Livestock Australia (MLA) for their financial support for research (B.BFG.0050) and the Angus Society of Australia and Australian Brahman Breeders' Association for providing data for this study through their BIN programs.

REFERENCES

AUS-MEAT (2005) 'Handbook of Australian meat' 7th ed. AUS-MEAT Ltd, Tingalpa, DC Qld.

Barwick S.A., Johnston D.J., Wolcott M.L., Wilkins J.F. and Mckiernan W.A. (2009) *Proc. Assoc. Advmt. Anim. Breed. Genet.* **18**:484.

Bindon B.M. (2001) *Aust. J.Exp.Agric.* **41**: 843.

Graser H-U., Tier B., Johnston D.J. and Barwick S.A. (2005) *Aust. J.Exp.Agric.* **45**:913.

Johnston D.J., Reverter A., Ferguson D.M., Thompson J.M. and Burrow H.M. (2003) *Aust, J. Agri. Res.* **54**: 135.

Meat and Livestock Australia (2014) 'MSA Index'. Available at http://solutionstofeedback.mla.com.au/Cattle/Meat-Standards-Australia-Grading/MSA-Index (accessed 27 September 2016).

Meyer K. (2007) *J. Zhej. Uni.* **8**: 815.

Perry D., Thompson J.M., Hwang I.H., Butchers A. and Egan A.F. (2001) *Aust. J.Exp.Agric.* **41**: 981.

Polkinghorne R., Thompson J.M., Watson R., Gee A. and Porter M. (2008) *Aust. J.Exp.Agric.* **48**: 1351.

Reverter A., Johnston D.J., Perry D., Goddard M.E. and Burrow H.M. (2003) *Aust. J. Agri. Res.* **54**: 119.

Southern Beef Technology services and Tropical Beef Technology services (2015) 'Breeding for improved MSA compliance and increased MSA Index value'. Winter 2015 Update. Available at https://issuu.com/sbtstbts/docs/sbts_tbts_2015_winter_news_update_1 (accessed 15 November 2016).

Thompson J. (2014) 'Bone out for MSA quality index'. Meat and Livestock Australia NSW.

Watson R., Polkinghorne R. and Thompson J.M. (2008) *Aust. J.Exp.Agric.* **48**: 1368.

Wolcott M.L., Johnston D.J., Barwick S.A., Iker, C.L., Thompson J.M. and Burrow H.M. (2009) *Anim. Prod. Sci.* **49**: 383.

Proc. Assoc. Advmt. Anim. Breed. Genet. 22:317-320

DEVELOPMENT OF THE BEEF GENOMIC PIPELINE FOR BREEDPLAN SINGLE STEP EVALUATION

N.K. Connors, J. Cook, C. Girard, B. Tier, K. Gore, D.J. Johnston, and M.H. Ferdosi

Animal Genetics and Breeding Unit*, University of New England, New South Wales, Australia

SUMMARY

Single step genomic BLUP (SS-GBLUP) for BREEDPLAN beef cattle evaluations is currently being tested for implementation across a number of breeds. A genomic data pipeline has been developed to enable efficient analysis of the industry-recorded SNP genotypes for incorporation in SS-GBLUP analyses. Complex data collection, along with format and/or naming convention inconsistencies challenges efficient data processing. This pipeline includes quality control of variable formatted data, and imputation of genotypes, for building the genomic relationship matrix required for implementation into single step evaluation.

INTRODUCTION

Genomic information from high density SNP panels has been incorporated into the Australian beef cattle genetic evaluation system, BREEDPLAN, since 2011, by "blending" EBVs from the standard analysis with direct genomic values (DGVs) from independent genomic prediction analyses using selection index theory. The ultimate goal has been to include all available information including pedigrees, phenotypes, and genotypes in a single analysis, known as single step genomic BLUP (Legarra *et al.* 2014).

One of the major practical challenges for including genomic information in genetic evaluations has been the development of scalable data-management systems (Swan *et al.* 2012) which can handle the increasing number of genotypes with increasing density of SNPs (Johnston *et al.* 2012). Quality control of the data becomes increasingly important, as inclusion of genotypes raises questions with regards to existing pedigree and potential breed. This paper describes the data pipeline developed for incorporating genomic information into SS-GBLUP analyses for BREEDPLAN, from on-farm DNA collection, through data quality control and building the genomic relationship matrix (GRM), to implementation within single step evaluation.

INDUSTRY DATA STRUCTURE

The genomic pipeline from sampling DNA on-farm to genomic evaluation is the most complex data recording process involved in genetic evaluation, and is regularly subject to errors. Samples are often handled by several people at different points in the pipeline, genotyping can be carried out by a number of different research and commercial entities using a variety of platforms, and ensuring data consistency has proved difficult.

Currently, Australian beef cattle genetic evaluations are organised individually by breed societies using databases that are maintained by the Agricultural Business Research Institute (ABRI) in most cases and using BREEDPLAN evaluation software licenced to ABRI (Graser *et al.* 2005), apart from Angus Australia who maintain their own database. At the time of writing, the role of the Animal Genetics and Breeding Unit (AGBU) within the single step genomic pipeline is to collate genotypes from various breed societies and construct a GRM which is used in

* AGBU is a joint venture of NSW Department of Primary Industries and University of New England

single step evaluations conducted routinely by ABRI. In future it is intended for the pipeline to be incorporated into the recording and processing at breed societies and ABRI for routine SS-GBLUP evaluation.

GENOMIC DATA PROCESSING

The genomic pipeline begins upon receiving raw genotypes from a genotyping lab (Figure 1). The DNA sample must be assigned to an animal ID, usually provided by the breed society, either by name, society ID, or BREEDPLAN database number. This process has significant issues with regards to mismatching of samples to animals, particularly with historic data. Often issues with animal identities (e.g. additions/changes to suffix/prefix, duplicate names, etc.) has meant DNA samples have been attributed to the wrong animal. Thus far this has been a major hurdle in the roll-out of SS-GBLUP, as animals with simple identity changes/errors, which in turn lead to pedigree errors, will be rejected from the GRM downstream. Ensuring consistent sample identification is critical but not always successful.

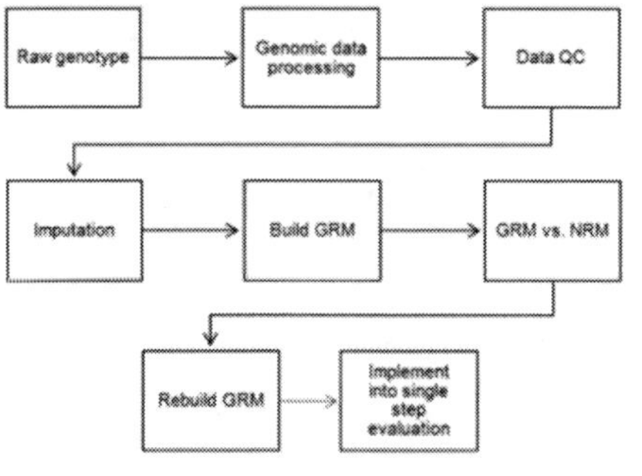

Figure 1. Genomic pipeline flow chart.

DATA QUALITY CONTROL

For quality assurance purposes, raw genotypes should be provided with GenCall (GC) scores for each SNP and a SNP map file to ensure consistency across SNP panels. The SNP maps may be used for imputation and checking recombination events, and allow the genotypes to be readily converted to a consensus 150K wide format genotype. The 150K formats allow consistency across all genotypes regardless of panels/chips, and enable high-throughput data management and quality control.

With a consistent format across all genotypes, the data undergoes a quality control (QC) analysis, with filters including average GC score, missing SNPs, SNPs with low GC scores, and allele frequencies. Animals are removed from the dataset used to construct the GRM based on the following criteria:

- Less than 79% calls with a GC score > 0.6
- More than 20% missing SNPs on the observed panel
- Average GC score less than 0.6
- Sire or dam younger than 550 days (based on recorded pedigree and date of birth)
- More than 50% SNPs heterozygous
- Minor and major allele frequencies are higher than 80% or lower than 20%
- Inconsistency between assigned sex and genotype determined sex

In each case, where a genotype fails due to poor data quality, the sample/animal is flagged with the breed society and/or lab for either re-genotyping the sample or re-sampling if possible.

A 4K subset of SNPs consistent across all panels is used as a further check for the animal's suitability for the GRM, checking for breed composition, parentage/pedigree, and duplicate genotypes (greater than 90% similarity). Currently the GRM is built for purebred animals only, and as such only animals with a minimum 80% of a single breed proportion (Boerner 2017) are included. At this point, any obvious pedigree errors will be identified and either corrected or the animals will be removed from the dataset. Animals failing to meet the required criteria of the data QC will be rejected from the GRM dataset, and provided a diagnostic code describing the cause of rejection. An example of the number of genotypes removed from a GRM dataset after quality control filters are applied is shown in Table 1.

IMPUTATION

In some instances, multiple genotypes of half-sib families with the same sire are available, enabling the sire's genotype to be imputed. Previous studies have shown that the imputation accuracy depends on the SNP density and the number of half-sibs for that sire (Ferdosi *et al.* 2014). Although un-genotyped sires with small half-sib families can be imputed, the imputed genotype will contain considerable amount of missing markers and the accuracy of imputation will be low. For inclusion in the GRM dataset, half-sib families larger than 11 individuals were considered for sire imputation, with imputation and haplotyping methods similar to those implemented in the "hsphase" algorithm (Ferdosi *et al.* 2014). The phased offspring are retained in a haplotype library for FImpute (Sargolzaei *et al.* 2014). SNP loci with more than 80% missing genotypes across animals are removed, and the missing SNPs are imputed using the haplotype library and the corrected pedigree.

Table 1. Number of genotypes removed from a GRM dataset after quality control process

Quality control filter	Number of genotypes
Total	12169
Less than 79% SNPs with GC score above 0.6	167
More than 20% SNPs missing	4
Average GC score less than 0.6	8
Extreme major/minor allele frequencies (>80% and <20%)	15
Breed proportion less than 80%	489
Duplicate genotype and sample id - multiple platforms	730
Duplicate genotype - different sample id	21
Duplicate sample id - different genotype	7
Inconsistent sex (pedigree vs genotype)	82
Incorrect sire or dam	282

GENOMIC VS PEDIGREE RELATIONSHIP QUALITY CONTROL

The GRM is built using VanRaden's method 1 (VanRaden 2008). With the inclusion of genomic information, previously unidentified relationships are discovered. These relationships may simply be previously unknown or not recorded, or may be an artefact of inbreeding within the population. Regardless of the reason, the additional information provided by the GRM to identify relationships not seen in the NRM will increase the accuracy of EBVs.

However, there will also be discrepancies between genomic and pedigree relationships, most likely due to incorrect recording; even well recorded herds have a fraction of their calves (3-5%) with incorrect pedigree (Johnston *et al.* 2012). It is possible that the recorded sire of an animal appears 'disproven' using genomic information, in which case there are a number of possible scenarios. The recorded pedigree may be incorrect, or the genotype sample may be of the wrong animal (e.g. sampling mix up, sample identity error, etc.). The issue is knowing which scenario is correct. There are a number of actions possible with the information available:

- Ignore the genotype and continue with the pedigree relationship (i.e. genotype wrong)
- Use the genotype to fix the pedigree relationship (i.e. pedigree wrong)
- Remove animal (i.e. uncertain whether pedigree or genotype is correct).

If the genomic relationship is ignored, a new genotype and/or sample should be requested. If the pedigree is corrected based on the genotype, this correction must be performed at the breed society level. It is possible that additional genotyping may change the GRM over time, as more half-sib relationships become available and new pedigree discrepancies will appear, or animals may be re-genotyped. In some instances, duplicate genotypes will occur, whereby the sample ID are the same, and the genotypes different; or the genotypes are the same, but the sample IDs are different. In these instances, it is difficult to identify which is correct, and as such both genotypes are unrecoverable. Table 1 provides an example of the number of genotypes removed from a GRM dataset after identifying duplicate samples and pedigree errors.

There are a number of assumptions in the building of the GRM with respect to using an unselected base population with little inbreeding, which can affect the genomic relationships (VanRaden 2008). Thus the issue of genomic and pedigree relationship discrepancies remains contentious, as the 'correct' action is not always obvious.

CONCLUSIONS

Increasing use of high density genomic information has the potential to improve the accuracy of genetic evaluations, and rates of genetic gain in the beef industry. This must be supported with efficient data pipelines which automate the quality control and analysis of genotypic data for inclusion into routine genetic evaluations. The genomic pipeline described here aims to do this, though difficulties arise due to complex data recording processes, multiple sample/data handling points, multiple laboratories, commercial entities and breed societies. Carefully structured and consistent data handling among the various participants will enable a smooth transition to SS-GBLUP, providing a repeatable, traceable, and auditable process, which is documented to ensure the highest quality and to identify changes over time for the Australian beef industry.

ACKNOWLEDGEMENTS

This research includes data generated by the Beef CRC. This research is supported by Meat and Livestock Australia (MLA) project B.BFG.0050. and L.GEN.0174.

REFERENCES

Boerner V. (2017) *Proc. Assoc. Advmt. Anim. Breed. Genet.* **22:** "(in press)".

Ferdosi M.H., Kinghorn B.P., van der Werf J.H., Lee S.H., Gondro C. (2014) *BMC Bioinformatics* **15:** 172.

Graser H.-U., Tier B., Johnston D.J., Barwick S.A. (2005) *Aust. J. Exp. Agr.* **45:** 913.

Johnston D.J., Tier B., Graser H.-U. (2012) *Anim. Prod. Sci.* **52:** 106.

Legarra A., Christensen O.F., Aguilar I., Misztal I. (2014) *Livest. Sci.* **166:** 54.

Sargolzaei M., Chesnais J.P., Schenkel F.S. (2014) *BMC Genomics* **15:** 478.

Swan A.A., Johnston D.J., Brown D.J., Tier B., Graser H.-U. (2012) *Anim. Prod. Sci.* **52:** 126.

VanRaden P.M. (2008) *J. Dairy Sci.* **91:** 4414.

Proc. Assoc. Advmt. Anim. Breed. Genet. 22:321-324

EFFECTS OF IRISH BEEF INDEXES AND BREEDING PROGRAMS ON GREENHOUSE GAS EMISSIONS

C. Quinton[1], F. Hely[1], T. Byrne[1], P. Amer[1], A. Cromie[2]

[1] AbacusBio Ltd., Dunedin, New Zealand
[2] Irish Cattle Breeding Federation, Ireland

SUMMARY

Effects of Irish beef Maternal Replacement Index and Beef Data and Genomics Program on system-wide greenhouse gas (GHG) emissions intensity were predicted. Expected index selection responses of increased offspring feed intake, decreased carcass weight and conformation, and increased carcass fat were predicted to increase system GHG intensity. These were offset by expected decreases in offspring mortality, cow and heifer live weights, calving interval, and age at first calving, and increased cow survival that were predicted to reduce system GHG intensity. Summed over responses in all traits, system GHG intensity was predicted to be reduced by 0.0088603 kg CO_2e/kg meat/breeding cow/year/€ index. Genomic selection and AI strategies were predicted to improve genetic progress and reduce total CO_2e by 5-10% after 20 years.

INTRODUCTION

Beef cattle genetic improvement programs have a key role in reducing global greenhouse gas (GHG) emissions. Genetic gains in livestock productivity and efficiency can reduce GHG emissions when expressed on a per-animal or intensity basis (i.e. emissions per unit of product) (Wall *et al.* 2010; Capper 2011; Hayes *et al.* 2013), and these changes from selection are permanent and cumulative over generations. Recognizing this, the Irish government has launched the Beef Data and Genomics Program (BDGP) as a major initiative to accelerate genetic progress for beef maternal efficiency traits and reduce GHG emissions. The potential system-wide impacts of trait genetic changes on GHG emissions therefore need to be quantified.

The objectives of this study were to quantify the effects of each trait in this index on system GHG emissions intensity, and predict the overall effects of genetic change from index selection and BDGP strategies on system GHG emissions intensity.

MATERIALS AND METHODS

The ICBF beef Maternal Replacement Index is an economic index containing offspring and cow production, carcass, reproduction and survival traits. System-wide emissions intensity (EI) per breeding cow per year was calculated as the sum of all system emissions (Σe) divided by the sum of all meat produced in the system (Σm), as follows:

$$EI = \frac{\sum e}{\sum m} = \frac{(o \times e_{offspring}) + (r \times e_{replace}) + (e_{cow})}{(o \times m_{offspring}) + (m_{cow})}$$

where o=number of slaughtered offspring per breeding cow per year, r=number of replacements reared per breeding cow per year, $e_{offspring}$=gross emissions (kg CO_2e) per slaughtered offspring over its lifetime, $e_{replace}$=gross emissions per replacement over her rearing period, e_{cow}=gross emissions per breeding cow per year, $m_{offspring}$=meat output (kg meat) per slaughtered offspring, and m_{cow}=meat output per breeding cow per year. These factors were considered as functions of the index genetic traits g.

The change in EI per change in each index trait was calculated as the partial derivative of EI with respect to each trait g (dEI/dg), as follows:

$$\frac{dEI}{dg} \approx \frac{1}{\sum m}\left\{\begin{aligned}&\left[\beta_{e.offspring}(g)\times o\right]+\left[\beta_{e.replace}(g)\times r\right]+\beta_{e.cow}(g)-\left[\beta_{m.offspring}(g)\times o\frac{\sum e}{\sum m}\right]\\&-\left[\beta_{m.cow}(g)\times\frac{1}{\sum m}\right]+\left[\beta_{r}(g)\times e_{replace}\right]+\left[\beta_{o}(g)\times\left(e_{offspring}-m_{offspring}\frac{\sum e}{\sum m}\right)\right]\end{aligned}\right\}$$

where for each trait, $\beta_{e.offspring}(g)$, $\beta_{e.replace}(g)$, and $\beta_{e.cow}(g)$ are traits effects on gross emissions per slaughtered offspring over its lifetime, per replacement reared from birth until it becomes a breeding cow, and per breeding cow per year, respectively; $\beta_{m.offspring}(g)$ and $\beta_{m.cow}(g)$ are trait effects on meat produced per slaughtered offspring, and per breeding cow per year, respectively; $\beta_r(g)$ and $\beta_o(g)$ are trait effects on number of replacement heifers required and number of offspring reared, per breeding cow per year, respectively; and other variables as previously defined.

For the current age-constant slaughter system, numbers of animals were o=0.6 and r=0.2. Gross emissions were calculated from feed intake, assuming feed intake of offspring, replacements, and cows of 3970.6 kg DM, 3522.4 kg DM, and 2874.6 kg DM, respectively, and conversion of 0.583 kg CO_2e/kg DM (Fennessy *et al.* 2015). Therefore, $e_{offspring}$=2314.9 kg CO_2e, $e_{replace}$=2053.6 kg CO_2e, and e_{cow}=1675.9 kg CO_2e, per breeding cow per year. Meat outputs were $m_{offspring}$=234.61 kg meat and m_{cow}=35.09 kg meat, per breeding cow per year. Therefore, Σe=3475.5 kg CO_2e, Σm=175.9 kg meat, and EI=160.5 kg CO_2e/kg meat, per breeding cow per year.

Estimated trait effects $\beta(g)$ are shown in Table 1. Effects on gross emissions were based on how trait changes affect feed intake, with conversion of 0.583 kg CO_2e/kg DM. Offspring emissions were affected by feed intake. Replacement emissions were affected by heifer live weight based on additional feed required of 9.406 kg DM/kg LW. Cow emissions were affected by cow live weight based on additional feed required of 3.197 kg DM/kg LW, and age at first calving based on additional feed required of 5.432 kg DM/d delay until calving. Meat produced was affected by carcass weight, conformation and fat (Drennan *et al.* 2009). Number of replacement heifers required was affected by cow survival. Number of offspring reared was affected by offspring mortality and calving interval. Other index traits were assumed to have no influence on equation terms.

Trait responses to index selection (trait unit/€ index value; Table 1) were predicted from linear regressions of individual bulls' ICBF proofs for each trait to their Maternal Replacement Index value. Trait-wise yearly responses in emissions intensity from index selection were calculated from trait dEI/dg, multiplied by the trait number of discounted genetic expressions (Table 1; Amer *et al.* 2001) and the predicted trait responses to index selection. Values were summed over all traits to obtain total response in EI per unit of index genetic gain.

RESULTS AND DISCUSSION

Effects of index traits on system emissions intensity. Offspring feed intake, mortality, carcass fat, cow and heifer live weights, calving interval, and age at first calving had numerically positive relationships with system EI, while offspring carcass weight, carcass conformation, and cow survival had negative relationships with system EI (Table 1). These values could potentially be used as weightings in an index to evaluate individual bulls for GHG emissions intensity. However, an emissions-only index would not consider trait economics. A more practical index for would combine economics of production from the Replacement Index with the GHG emissions intensity changes and consider potential trade-offs between direct farm profit improvement and GHG intensity.

Expected responses to index selection. Expected responses of increased offspring feed intake, decreased offspring carcass weight and conformation, increased carcass fat, and decreased cow carcass weight were predicted to increase system GHG intensity (Table 1). These were offset by expected decreases in offspring mortality, cow and heifer live weights, calving interval, and age at first calving, and increased cow survival that were predicted to reduce system GHG intensity (Table 1). Cow live weight, calving interval and survival had the greatest effects on system GHG intensity, while other traits had comparatively minor effects. Summed over responses in all traits, system GHG intensity was predicted to be reduced 0.0088603 kg CO_2e/kg meat/breeding cow/year/€ index.

Table 1. Maternal Replacement Index trait specific effects, effects on system emissions intensity, discounted genetic expressions (DGE) per year, and predicted trait unit and emissions intensity responses to index selection

Trait (unit)	Specific effects in model (change/trait unit)	Effect on EI (kg CO_2e/kg meat/trait unit)	DGE	Trait response (trait unit/€ index)	EI response (kg CO_2e/kg meat/€ index)
Feed intake (kg DM/d)	$\beta_{e.offspring}$(FI)=0.583 kg CO_2e	0.0020	0.54	0.0005	0.0000005
Mortality (%)	β_o(M)=-0.01 offspring	0.1320	1.10	-0.0023	-0.0003297
Carcass weight (kg)	$\beta_{m.offspring}$(CW)=0.686 kg meat	-0.0463	0.54	-0.0205	0.0005131
Carcass conformation (score)	$\beta_{m.offspring}$(CC)=4.072 kg meat	-0.2746	0.54	-0.0017	0.0002507
Carcass fat (score)	$\beta_{m.offspring}$(CF)=-2.982 kg meat	0.2011	0.54	0.0013	0.0001455
Cow live weight (kg)	$\beta_{e.cow}$(CLW)=1.864 kg CO_2e	0.0106	2.204	-0.1147	-0.0026804
Heifer live weight (kg)	$\beta_{e.replace}$(HLW)=5.484 kg CO_2e	0.0062	0.614	-0.1147	-0.0004393
Calving interval (d)	$\beta_{e.cow}$(CI)=-1.232 kg CO_2e, β_o(CI)=-0.0027 offspring	0.0292	2.204	-0.0283	-0.0018198
Age at first calving (d)	$\beta_{e.cow}$(AFC)=3.167 kg CO_2e	0.0180	0.614	-0.0454	-0.0005025
Cow survival (%)	β_r(S)=-0.008 heifers	-0.0940	2.204	0.0193	-0.0039989
Cow carcass weight (kg)	$\beta_{m.cow}$(CCW)=0.6 kg meat	-0.00002	0.288	-0.0777	0.0000004

These findings are consistent with studies that have found GHG emissions benefits arising from productivity and efficiency gains over time (Wall *et al.* 2010; Capper 2011; Hayes *et al.* 2013). Generally, increasing growth rate and numbers of animals in a system will increase overall feed intake and resultant gross GHG produced by the system. However, genetic and management improvements have also increased system-wide efficiency, meaning that more product is made per unit feed input. This comes from more efficient feed utilization on an individual animal basis, plus improved reproductive and survival rates that mean each breeding animal and associated replacements can produce more output-generating animals.

Genetic change from selection generates permanent and cumulative effects on traits, and therefore system-wide reductions in GHG achieved through selection will continue over generations. In previous studies (Hely *et al.* 2016; Hely and Amer 2016), genetic trends for the Irish BDGP were predicted for three scenarios: 1) current selection with Replacement Index; 2) genomic selection, increasing use of top progeny-tested maternal AI bulls to 30% in pedigree herds and 20% in commercial suckler herds; and 3) genomic selection with use of elite AI sires increased to 50% in pedigree herds and 30% in commercial herds (Table 2). Applying these predicted trends to the estimated reduction of 0.0088603 kg CO_2e/kg meat/breeding cow/year/€ Replacement Index, and maintaining a fixed population size of 800,000 breeding cows, annual GHG emissions can be reduced up to 9.5% which corresponds to a total reduction of 3335 kt CO_2e after 20 years (Table 2).

Therefore, Irish beef genetic improvement initiatives are predicted to have important outcomes of reducing GHG emissions.

Table 2. Predictions of genetic trends in Maternal Replacement Index from 3 scenarios, and corresponding percent annual and total GHG emissions reductions after 5 and 20 years with constant population size

Scenario	Index trend (€/y)	5y annual GHG	5y GHG (kt CO_2e)	20y annual GHG	20y GHG (kt CO_2e)
1) Current selection	+1.67	-0.4%	-34	-1.5%	-481
2) Genomic selection + increased use of elite AI sires	+5	-1.9%	-229	-5.4%	-1952
3) Genomic selection + maximum use of elite AI sires	+9	-3.1%	-350	-9.5%	-3335

ACKNOWLEDGEMENTS
This work was funded by the European Union Rural Development Program, and the Irish Department of Agriculture, Food and the Marine.

REFERENCES

Amer P.R., Simm G., Keane M.G., Diskin M.G. and Wickham B.W. (2001) *Livest. Prod. Sci.* **67**: 223.

Capper J.L. (2011) *J. Anim. Sci.* **89**: 4249.

Drennan M.J., McGee M., Conroy S.B., Keane M.G., Kenny D.A. *et al.* (2009) Report RMIS No. 5369, Beef Production Series No. 89. Teagasc, Ireland.

Fennessy P., Byrne T. and Proctor L. (2015) Report B.CCH.6133. Meat and Livestock Australia Ltd., North Sydney, NSW, Australia.

Hayes B.J., Lewin H.A., and Goddard M.E. (2013) *Trends Genet.* **29**: 206.

Hely F. and Amer P. (2016). The industry structures required to maximise genetic gains in the Irish beef industry. Retrieved on 16 Dec 2016 from https://www.icbf.com/wp/wp-content/uploads/2016/06/Benefits-of-GIBB-and-Gene-Ireland-Report-March-2016.docx

Hely F., Amer P., Quinton C., Byrne T. and Cromie A. (2016) Book of Abstracts of the 67th Annual Meeting of the European Association for Animal Production, 29 Aug. – 2 Sept. 2016, Belfast, UK, p. 193.

Wall E., Simm G. and Moran D. (2010) *Animal* **4**: 366.

Proc. Assoc. Advmt. Anim. Breed. Genet. 22:325-328

INVESTIGATING VARIATION IN THE TEST LENGTH REQUIRED TO ESTIMATE THE TRAIT OF RESIDUAL ENERGY INTAKE IN GROWING MATERNAL LAMBS

P.L. Johnson, J. Wing, K. Knowler and P. Johnstone

AgResearch Invermay, Mosgiel, New Zealand

SUMMARY

Residual energy (or feed) intake (REI) is one measure of feed efficiency, and is an estimate of whether an animal is consuming more or less energy for its biological outcomes than predicted. To date, little research has been conducted in sheep, and a multi-year trial is underway to generate data for New Zealand maternal breeds, firstly targeting growing lambs. Data required to estimate REI includes daily feed intake and live weight information from which growth rate can be calculated. A key to data collection for the trait of REI is to determine the test length required to accurately estimate REI. A dataset was available on approximately 600 growing maternal breed lambs from 3 cohorts that were measured for daily feed intake of lucerne pellets for a period of 42 days (after 14 days adjustment), with live weight measured twice weekly during the test period. The full dataset was subsetted to simulate reduced test lengths and environmental variance was calculated for each cohort-data subset. Additionally the correlation between the REI estimates from the reduced length datasets and the full dataset was also estimated. The results suggested that the variance of all traits stabilised within 21 to 28 days, and a correlation of greater than 0.90 existed between the estimates made on the data collected in 21 days versus the complete 42 day dataset. These results suggest that the environmental variances stabilises quicker in lamb studies than in beef studies which require a minimum of 56 to 70 days' worth of data.

INTRODUCTION

Residual energy (or feed) intake (REI) is one measure of feed efficiency, and is an estimate of whether an animal is consuming more or less energy for its biological outcomes than predicted. Less research has been conducted on the trait of REI in sheep (compared with other production species), however, a series of studies are now being undertaken which are seeking to investigate the phenotypic and genetic variability of feed efficiency in sheep.

As has been the case in all other species, one of the keys to generating feed efficiency data is the development of an optimum test period in which feed intake, live weight and liveweight gain (the key variables in the REI model) are to be measured. This needs to be a balanced decision as the cost of data collection is high, but equally too short ~~of~~ a measurement period will result in poor parameter estimation. In cattle the traits of feed intake and liveweight gain have been shown to require different test periods to obtain accurate (minimised variance) estimates (Archer *et al.* 1997).

The paper investigates the optimum test period for young ewes, using a 42 day test period dataset collected on approximately 600 9-month old growing New Zealand maternal ewes.

MATERIALS AND METHODS

Animals and data. The animals used in this study are the first cohorts of a multi-year trial investigating the trait of REI in New Zealand maternal sheep breeds. Details of the animals and traits measured during the test period are in Johnson *et al.* (2016) with the addition of animals from Greenhouse Gas selection lines (Elmes *et al.* 2014). Briefly, 3 cohorts of 200 9-month old growing ewes of composite New Zealand maternal genetics were housed in an indoor feed intake facility in mobs of 40 and given *ad libitum* access to lucerne pellets via automated feeders which recorded individual feeding events per animal through the use of electronic identification tags. The adjustment period was 14 days and the test period was 42 days. The live weight (LWT) of the animals was measured twice weekly in the morning, un-fasted.

Analyses. Residual energy intake was estimated as described in detail by Johnson *et al.* (2016) using the model first described by Koch *et al.* (1963). Briefly, REI is the residual value of a regression model where energy intake is the dependent variable with mid-test metabolic live weight ($LWT^{0.75}$), and daily liveweight gain (average daily gain: ADG) fitted as independent variables.

In order to investigate the impact on environmental variance of reduced test lengths datasets based on cumulative days' worth of data were generated. For daily energy intake, 42 datasets were generated for each of the 3 cohorts including all data collected up to and including the day represented by the dataset. Specifically, dataset 1 only contained daily energy intake (DEI) data collected on day 1 of the trial, with Dataset 2 containing data collected on days 1 and 2 of the trial. Dataset 42, the final dataset, contained data from all of the days within the test period. Given LWT was only measured twice weekly only 13 of the datasets included additional live weight data. For the datasets that contained additional LWT data, ADG and mid-test metabolic live weight were re-calculated and REI re-estimated. A summary of the data for the full 42 day dataset is in Table 1. The traits of DEI, LWT, ADG and REI from each cohort-subset of data were individually analysed in GenStat Version 13 (Payne *et al.* 2009) using a REML model and the estimate of error (environmental) variance reported. All 42 datasets within a cohort were analysed for the trait of DEI, however, only datasets containing additional LWT data were analysed for the remainder of the traits. The environmental variances from each analysis were collated for each trait and plotted per cohort against day of trial to demonstrate the change in environmental variance within increasing amounts of data contributing to the trait estimation.

Correlation coefficients were estimated for each cohort for REI contrasting the full 42-day dataset with the sequential datasets. The correlation coefficients were plotted against day of trial to observe the change in correlation with increasing amounts of data contributing to the estimation of REI.

Table 1. Summary statistics (mean ± std (range)) for 3 cohorts of 9-month old ewe lambs measured to estimate residual energy intake, with data collected over the full 42 day test period.

	Cohort 1	Cohort 2a	Cohort 2b
Mid-test period metabolic midweight ($LWT^{0.75}$)	19.2 ± 1.8 (14.1 – 25.7)	21.2 ± 1.6 (16.2 – 26.3)	22.2 ± 1.8 (17.0 – 27.6)
Daily energy intake (MJ ME/day)	21.6 ± 2.9 (12.7 – 29.3)	23.3 ± 3.0 (15.5 –31.38)	29.7 ± 3.2 (18.6 –36.8)
Average daily gain (g/day)	314 ± 50 (190 – 478)	317 ± 58 (178 – 503)	380 ± 64 (205 – 702)
Residual energy intake (MJ ME/day)	0.0 ± 1.0 1.0 (-2.8 – 3.4)	0.0 ± 1.0 (-2.5 – 2.8)	0.0 ± 1.0 (-4.1 – 2.5)

RESULTS AND DISCUSSION

Optimising the test length to estimate residual feed intake values is important in determining the potential throughput of animals tested, and also the cost of generating the trait data if it is to be implemented into breeding programmes. The changes in environmental variance estimates with increasing test length are presented in Figure 1 for 3 cohorts of growing maternal 9-month old ewes. For all traits the environmental trait variances stabilise with 21 to 28 days of measurements. The only trait to show a small consistent trend of an increase in variance is live weight, however, this reflects differences in growth rate between individuals which results in further divergence between animals across the time period.

The phenotypic correlation between REI estimated for the different test lengths with the estimate from the full 42 day test length are in Figure 2 for the 3 cohorts. The correlation between the full dataset was greater than 0.9 with just 21 days' data, and greater than 0.95 with 32 days' data.

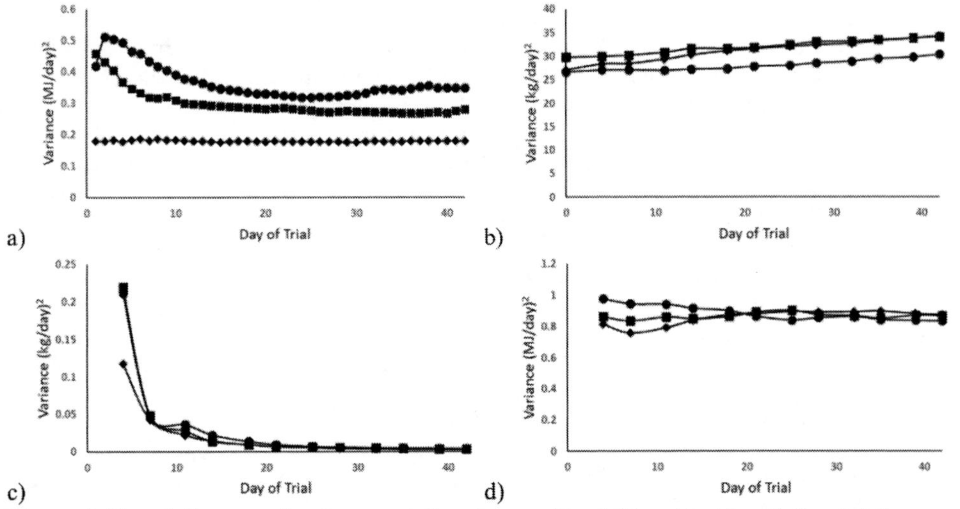

Figure 1. Cumulative error (environmental) variance with additional test length for a) daily energy intake, b) live weight, c) average daily gain and d) residual energy intake for three cohorts (◆,1; ■,2; ●,3) of New Zealand maternal ewe lambs measured for daily intake through an indoor facility capturing daily feed intake data with live weight measured twice weekly.

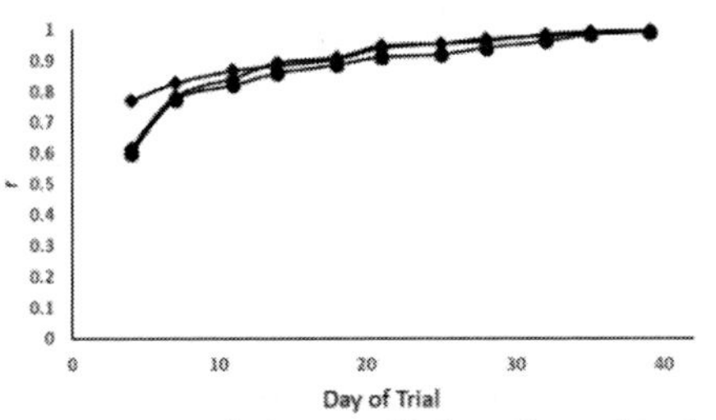

Figure 2. Correlation coefficient for residual energy intake calculated within increasing test length compared to the full 42-day test length for three cohorts (◆,1; ■,2; ●,3) of New Zealand maternal ewe lambs measured for daily intake through an indoor facility capturing daily feed intake data.

In beef studies, it has been concluded that the length of time required for the variance of feed intake to stabilise is less than is required to obtain stable variances for the growth rate of the animals being measured (Archer *et al.* 1997; Wang *et al.* 2014). However, further research has demonstrated that the frequency with which the live weight measurements are made also influences the test length required to obtain stable growth rate data (Archer *et al.* 1999; Kearney *et al.* 2004). The conclusion from these studies was that feed intake in cattle can accurately be estimated with 35 days data, with growth estimated accurately with 56 days using daily automated weighing, but up to 70 days if only weighed fortnightly.

There is less published literature investigating test length in sheep. In the study of Cockrum *et al.* (2013) only variance estimates for REI were reported, and did not observe the same level of stabilisation within the short time frames observed in this study. However, the proportion of variation in RFI explained by live weight and growth rate in their study was considerably lower than that reported for Cohort 1 by Johnson *et al.* (2016), which suggests that overall their feed intakes were influenced by other factors not accounted for in their models which could have contributed to increased variability. Unpublished results from a study in Merinos support the findings of this study, in that their feed intakes stabilised by 3 weeks (*B. Pagagoni pers. comm.*).

All production traits are subject to environmental variance, as they are not strictly under genetic control. The results from this study support that the trial design, including the feed offered and the facility developed do not result in a large amount of ongoing environmental variability, and as such allow phenotypic estimates of REI to be obtained within a relatively short time frame when compared to the cattle equivalents.

Given the ultimate aim of the genetic selection for REI, as reported by Archer *et al.* (1997), there is a further need to consider the genetic correlations for different test lengths. Such an analysis will be conducted once further cohorts are collected. Based on the results of Archer *et al.* (1997) it is likely the genetic correlations will be high for at least an equivalent if not shorter time period than is required to obtain high phenotypic correlations. At the time of publication, the feed intake facility used in this study is being re-located. Further analysis on subsequent cohorts measured in the new facility will be required to confirm the findings of this paper. If it is validated, the current test length for growing maternal lambs has the potential to be reduced.

ACKNOWLEDGEMENTS

This is a Beef + Lamb New Zealand Genetics project funded by the Ministry for Business, Innovation and Employment and Beef + Lamb Zealand. The animals sourced from the Woodlands Progeny Test and the Greenhouse Gas Selection lines were funded by the New Zealand Agricultural Greenhouse Gas Research Centre and Pastoral Greenhouse Gas Research Consortia, who are partners in this project for extended measurements not reported herein. AgResearch capital expenditure funding supported the development and production of the automated feeders.

REFERENCES

Archer, J.A., Arthur, P.F., Herd, R.M., Parnell, P.F. and Pitchford, W.S. (1997) *J. Anim. Sci.* **75:** 2024.

Archer, J.A., Arthur, P.F., Herd, R.M., Richardson, E.C. and Burton, D.A. (1999) *Proc. Assoc. Advmt. Anim. Breed. Genet.* **13:**247.

Cockrum, R.R., Stobart, R.H., Lake, S.L. and Cammack, K.M. (2013) *Small Rumin. Res.* **113:**313.

Elmes, S.N., Bain, W.E., Greer, G.J., Hickey, S.M., Young, E.A., Pickering, N.K., Rowe, S.J., Knowler, K.J., Pinares-Patino, C.S. and McEwan, J.C. (2014) *Proc N.Z. Soc. Anim. Prod.* **74:** 142.

Johnson, P.L., Miller, S.P. and Knowler, K.J. (2016) *Proc N.Z. Soc. Anim. Prod.* **76:** 34.

Kearney, G.A., Knee, B.W., Graham, J.F. and Knott, S.A. (2004) *Aust. J. Exp. Agric.* **44:**411.

Koch, R.M., Siger, L.A., Chambers, D. and Gregory, K.E. (1963) *J. Ani. Sci.* **22:** 486.

Payne, R.W., Murray, D.A., Harding, S.A., Baird, D.B. and Soutar, D.M. (2009) *GenStat for Windows 12th Edition. Introduction.* Hemel Hempstead, UK: VSN International.

Wang, Z., Nkrumah, J.D., Li, C., Basarab, J.A., Goonewardene, L.A., Okine, E.K., Crews Jr, D.H. and Moore, S.S. (2014) *J. Anim. Sci.* **84:** 2289.

Proc. Assoc. Advmt. Anim. Breed. Genet. 22:329-332

INHERITANCE OF WRINKLE IN MERINO SHEEP AT DIFFERENT AGES

J.C. Greeff[1], L.J.E. Karlsson[2] and A.C. Schlink[3]

[1]Department of Agriculture and Food WA, South Perth WA 6151
[2]RMB 314, Bridgetown WA 6255
[3]University of Western Australia, Crawley, WA

SUMMARY

The inheritance of wrinkle on the neck and on the tail was investigated in a medium micron Merino flock in a Mediterranean environment from birth to hogget shearing. It was moderate to highly heritable (0.32 to 0.74) at all ages but it was not genetically the same trait at the different ages. Log transforming the data did not generally increase the heritability of the trait. Body wrinkle at birth was genetically moderately to lowly correlated with neck and skin wrinkle at later ages. The best time to score tail wrinkle in a Mediterranean environment is between yearling and hogget age as this is when the heritability and variation of the trait was the highest.

INTRODUCTION

High breech or tail skin wrinkle is one of the most important predisposing factors for breech strike in Merino sheep (Greeff *et al.* 2014; Smith *et al.* 2009). Morley *et al.* (1952), Brown *et al.* (2010), Greeff *et al.* (2014), Hatcher and Preston (2015) and Smith et al. (2009) showed that skin wrinkle is a heritable trait and Greeff *et al.* (2014) showed that neck wrinkle at yearling age was the best indicator trait to select for reduced wrinkle. Scholtz *et al.* (2010) reported that the correlations between neck wrinkle, body wrinkle and breech wrinkle scored at hogget age were generally high ($r_g > 0.89$), implying that selecting on any trait will result in a correlated response in the other. However, the Mediterranean regions of Australia suffer from large fluctuations in rainfall from hot, dry summers to cold, wet winters. This results in dramatic changes in the feed supply during the year, often requiring animals to be supplementary fed during the dry times of the year. Schlink *et al.* (2000) and Herselman and King (1993) showed that large changes in skin wrinkles and skin weight can occur during the year in Merino sheep on high and low planes of nutrition. Adding the effect of different times of crutching and shearing, growers need to consider a range of environmental and management factors under their specific production system, in selecting the best indicator trait for reduced skin wrinkle at different ages.

MATERIAL AND METHODS

The Australian Wool Innovation Breech strike flock of the Department of Agriculture and Food Western Australia was used to investigate the inheritance of neck and tail wrinkle during the year up to hogget age, as well as the genetic and phenotypic relationships between neck and tail wrinkle from birth to hogget shearing.

The flock consisted of 3623 lambs born from 1674 dams that were mated to 80 rams over a four-year period (2010 to 2013). The lambs were scored for body wrinkle at birth and for neck wrinkle and tail wrinkle at marking, weaning, early post-weaning, post-weaning, yearling and hogget shearing (16 months), using the visual sheep scoring system (AWI, 2008). As the animals in the flock were relatively plain, and previous studies (Greeff, submitted) have shown that tail wrinkle was a better indicator trait to differentiate between sheep than breech wrinkle, tail wrinkle was used in this study. Skin wrinkle was scored from 1 (plain) to 5 (high wrinkle count), and half scores were allocated where appropriate.

Lambs were born in July/August, and marked in late August. They were weaned at approximately 100 days of age. The lambs were shorn approximately 5 weeks after weaning, and crutched at 12 months of age in July the following year. They were shorn with 12 months wool in early December when the sheep were approximately 18 months of age.

STATISTICAL ANALYSIS

ASREML (Gilmour *et al.* 2009) was used to analyse the data. A series of univariate analysis were carried out on each of the skin traits (body wrinkle at birth, neck and tail wrinkle at marking, weaning, early post-weaning, post-weaning, yearling, hogget and at post hogget shearing) to obtain estimates of the heritability of skin wrinkle at different ages on the neck and tail. Year of birth (2010-2013), age of the dam (2-7 years), birth status (single or multiple) and sex of the lamb were fitted as fixed effects. All two-way interactions were initially fitted. Statistically non-significant factors were dropped from the final model. This was followed by a bivariate analysis to obtain genetic and phenotypic (co)variances to estimate the phenotypic and genetic correlations between skin wrinkle on the neck and on the tail at different ages. Two analyses were carried out on each trait: one on the raw trait scores and one where the trait had been log transformed (log(wrinkle+10)) to normalise the data as it was skewed distributed in all cases, with low numbers of animals in the higher wrinkle categories.

RESULTS AND DISCUSSION

Table 1 shows that the highest expression of wrinkle was recorded for body wrinkle at birth. This is contrary to eastern state results which show that the highest expression of skin wrinkle was at weaning (Dun and Eastoe, 1970). In this study, the average neck and tail wrinkle then reduced by more than one unit score at marking, weaning, early and post weaning. This trend can be explained as it is difficult to score wrinkle on sheep with long wool. However, at yearling and at hogget age, tail wrinkle increased and was higher than neck wrinkle. This was probably due to the fact that tail wrinkle was scored post shearing in order to be able to better differentiate between animals. In addition, the scoring system may have also contributed as half scores were also used to capture as much of the differences as possible.

Table 1. Average skin wrinkle scores (± SD) and the number of records in different ages

	Birth	Marking	Wean	Early post weaning	Post weaning	Yearling	Hogget
Body wrinkle	2.86						
SD	1.04						
Neck wrinkle		1.73	1.68	1.85	1.63	1.26	1.43
SD		0.55	0.52	0.50	0.37	0.58	0.53
Tail wrinkle		1.39	1.28	1.47	1.02	2.11	2.12
SD		0.42	0.45	0.41	0.09	0.95	0.94
No. of records	3623	2782	3585	1944	3501	1709	1680

Body wrinkle at birth showed the highest amount of phenotypic variation, followed by tail wrinkle at yearling age. The lowest amount of phenotypic variation was found for post weaning wrinkle score. The heritability estimates of neck (Table 2) and tail wrinkle (Table 3) in general agreed with previously published results (Morley, 1952). However, it differed at different ages. The highest estimate on the raw data was at yearling age (0.74) which decreased to 0.37 when log transformed. Transformation of wrinkle sometimes increased the heritability estimate but in other cases it decreased it. It appears that the accuracy of selection is higher for untransformed skin wrinkle score data in this flock.

The phenotypic correlations between the neck wrinkle traits varied from 0.27 to 0.65. Traits closer together had higher correlations than traits further apart. However, neck wrinkle at yearling age appears to be less strongly correlated with the neck wrinkle traits scored at other times.

The genetic correlation between neck wrinkle, and between tail wrinkle at different ages, were higher than the phenotypic correlations. Tail wrinkle and neck wrinkle were genetically highly correlated when scored at the same age ($r_g > 0.085$) (estimates not shown). However, the genetic correlations between body wrinkle at birth, and neck or with tail wrinkle was generally low. Body wrinkle at birth was most correlated with neck wrinkle ($r_g = 0.54$) and tail wrinkle ($r_g = 0.58$). This indicates that body wrinkle at birth is not genetically the same trait as neck or skin wrinkle later in life.

Table 2. Phenotypic variances (Vp), heritability (on diagonal), phenotypic correlations (above diagonal), and genetic correlations (below diagonal) of body wrinkle at birth (BBDWR) and neck (NKWR) at marking (M), weaning (W), early post weaning (E), post weaning (P), yearling (Y) and hogget (H) shearing

	BBDWR	MNKWR	WNKWR	ENKWR	PNKWR	YNKWR	HNKWR
Raw wrinkle scores							
Vp	1.01	0.31	0.27	0.25	0.14	0.33	0.28
BBDWR	**0.53**	0.37	0.26	0.24	0.14	0.17	0.19
MNKWR	0.54	**0.66**	0.48	0.51	0.30	0.27	0.40
WNKWR	0.36	0.77	**0.42**	0.54	0.39	0.42	0.38
ENKWR	0.34	0.78	0.77	**0.55**	0.47	0.39	0.47
PNKWR	0.20	0.59	0.64	0.78	**0.37**	0.35	0.39
YNKWR	0.25	0.55	0.69	0.71	0.64	**0.32**	0.65
HNKWR	0.33	0.64	0.68	0.82	0.74	0.90	**0.47**
SE range	0.04-0.07	0.02-0.08	0.02-0.07	0.02-0.08	0.02-0.08	0.02-0.05	0.02-0.02
Log transformed wrinkle score (+10)							
Vp	0.006	0.002	0.002	0.002	0.001	0.002	0.002
BBDWR	**0.53**	0.36	0.26	0.24	0.14	0.16	0.19
MNKWR	0.52	**0.62**	0.50	0.47	0.31	0.27	0.39
WNKWR	0.39	0.79	**0.38**	0.61	0.39	0.48	0.37
ENKWR	0.37	0.80	0.82	**0.48**	0.49	0.39	0.46
PNKWR	0.21	0.61	0.64	0.83	**0.39**	0.36	0.37
YNKWR	0.23	0.45	0.77	0.59	0.62	**0.37**	0.66
HNKWR	0.36	0.67	0.66	0.85	0.77	0.92	**0.43**
SE range	0.04-0.10	0.02-0.08	0.02-0.07	0.02-0.10	0.02-0.06	0.02-0.07	0.02-0.05

CONCLUSIONS

Neck and tail wrinkle were moderately to highly heritable traits and log transforming the data did not appear to improve the accuracy of selection. Body wrinkle at birth and neck and tail wrinkle at marking, was not strongly genetically correlated to skin wrinkle at later ages. This study shows that the best time to measure skin wrinkle on the tail or neck was after crutching at yearling age or after shearing at hogget age in a Mediterranean environment as this is when the heritability and phenotypic variation was highest. This time also coincided with the fly season in this environment.

Table 3. Phenotypic variances (Vp), heritability (on diagonal), phenotypic correlations (above diagonal), and genetic correlations (below diagonal) of body wrinkle at birth (BBDWR) and tail wrinkle (TAWR) at marking (M), weaning (W), early post weaning (E), post weaning (P), yearling (Y) and hogget (H) shearing

	BBDWR	MTAWR	WTAWR	ETAWR	YTAWR	HTAWR
Raw wrinkle scores						
Vp	1.01	0.16	0.19	0.16	0.52	0.17
BBDWR	**0.54**	0.39	0.19	0.25	0.16	0.16
MTAWR	0.58	**0.56**	0.40	0.48	0.26	0.33
WTAWR	0.28	0.52	**0.43**	0.51	0.35	0.21
ETAWR	0.42	0.67	0.80	**0.46**	0.55	0.48
YTAWR	0.37	0.58	0.71	0.87	**0.74**	0.66
HTAWR	0.27	0.55	0.40	0.79	0.90	**0.47**
SE range	0.05-0.08	0.02-0.07	0.02-0.07	0.02-0.06	0.02-0.05	0.02-0.05
Log transformed wrinkle score (+10)						
Vp	0.006	0.001	0.001	0.001	0.002	0.001
BBDWR	**0.54**	0.37	0.19	0.24	0.15	0.17
MTAWR	0.56	**0.58**	0.41	0.43	0.23	0.35
WTAWR	0.31	0.55	**0.35**	*	0.39	0.21
ETAWR	0.38	0.65	*	**0.42**	*	*
YTAWR	0.35	0.47	0.72	*	**0.37**	0.66
HTAWR	0.28	0.59	0.41	*	1.01	**0.48**
SE range	0.04-0.10	0.02-0.10	0.02-0.08	0.02-0.07	0.03-0.06	0.02-0.06

* Not estimable

ACKNOWLEDGEMENT

The researchers acknowledge the generous financial support provided by Australian woolgrowers and the Commonwealth through Australian Wool Innovation Limited. The authors also thank the Research station staff for managing the flock and Nicola Stanwyck for the collection of the data.

REFERENCES

AWI (2008) Visual sheep scoring system. www.sheepgenetics.org.au.

Brown, D.J., Swan A.A. and Gill J.S. (2010) *Anim. Prod. Sci.* **50**: 1060.

Dun R.B. and Eastoe R.D. (1970) Science and the Merino breeders. NSW Dept of Agriculture.

Greeff J.C., Karlsson L.J.E. and Schlink A,C, (2014) *Anim. Prod. Sci.* **54:** 125.

Herselman M.J. and King B.R. (1993) *Karoo Agric.* **5**: 3.

Hatcher S., Preston J.W.V. (2015) *Small Rumin. Res.* **130**: 36.

Morley F.H.W. (1952). NSW Dep of Agriculture Science Bulletin No 73.

Schlink A.C., Sanders M. and Hollis D.E. (2000) *Asian Aust. J. Anim. Sci.* **13**. Suppl. A: 253.

Scholtz A.J., Cloete S.W.P., van Wyk J.B., Misztal I., du Toit E. and van der Linde T.C. de K. (2010). *Anim. Prod. Sci.* **50**: 210.

Smith J.L., Brewer H.G. and Dyall T. (2009) *Proc. Assoc. Advmt. Anim. Breed. Genet.* **18**: 334.

Proc. Assoc. Advmt. Anim. Breed. Genet. 22:333-336

GENOTYPING STRATEGIES OF SELECTION CANDIDATES IN SHEEP BREEDING PROGRAMS

T. Granleese[1,2], S.A. Clark[1,2], J.H.J. van der Werf[1,2]

[1] School of Environmental and Rural Science, University of New England, Armidale, Australia, 2351

[2] Cooperative Research Centre for Sheep Industry Innovation, Armidale, 2351

SUMMARY

Genotyping strategies for both ram and ewe selection candidates were investigated to maximise the benefit of genomic selection while minimising genotyping costs. Through stochastic simulation we investigated both early and late-stage genomic selection of rams using a selection index that contained an early in life measurement (post-weaning weight) and a hard to measure trait (intra-muscular fat) that was not measured on selection candidates. We also simulated genotyping strategies for female selection candidates in breeding programs using natural mating, multiple ovulation and embryo transfer (MOET) or juvenile *in vitro* fertilisation and embryo transfer (JIVET). Our results showed that genomic selection of rams lifted genetic gain by 40%. Genomic testing the top 20% of ram selection candidates achieved 80% of the maximum benefit using late-stage genomic selection, while testing 47% of the top ranked rams implementing early-stage genomic selection was required to achieve 80% of the benefit. Genetic gain lifted by a maximum of 15-65% for genomic testing in (only) ewe selection candidates. To achieve 80% of the maximum benefit of genomic selection, 65%, 35% and 45% of ewe selection candidates required genomic testing each year for natural, MOET and JIVET breeding programs, respectively. Genotyping ram selection candidates provided the best value for money.

INTRODUCTION

Limited research has been published about strategies to genotype selection candidates while minimising costs. Van der Werf *et al.* (2014) and Horton *et al.* (2015) proposed that genotyping 20% of ram selection candidates will return 80% of the maximum potential benefit (i.e. compared with testing all rams) and assuming 2-stage genomic selection. Van der Werf *et al.* (2014) also discussed multiple trait indexes with unfavourable correlations between early measured traits and late-in-life- or hard-to-measure traits and agreed with Sise and Amer (2009) that more candidates would need to be tested by genomic methods compared with simple single-trait indices. There is little published data investigating genotyping methods of the optimisation of ewe selection candidates, particularly when using female reproductive technologies.

This paper aims to investigate genotyping strategies for early-stage and late-stage selection for ram selection candidates using a multi-trait index with a hard-to-measure trait to maximise genetic progress while minimising testing costs. This paper also investigates genotyping strategies in ewes.

MATERIALS AND METHODS

Stochastic simulation was used to model closed breeding schemes for 500 sheep. For each scenario we generated a base population of unrelated animals, and subsequently established a 15-year breeding program with overlapping generations. We simulated an early-in-life measured trait (post-weaning weight - PWT) and a trait that did not get measured (intra-muscular fat - IMF). Heritabilities, genetic and phenotypic (co)variances for the parameters were used from Swan *et al.* (2015). Each year, individual animals had breeding values estimated (EBV) via pedigrees based on multi-trait Best Linear Unbiased Prediction (BLUP) using ASReml software (Gilmour et al. 2009).

For each breeding program and index the impact of genomic selection (GS), assuming all animals had genomic information available at birth, was assessed. The cost of GS was not accounted for in this study. Genomic information was modeled following the method of Dekkers (2007) which simulates a genomic breeding value as a correlated trait with a heritability of 0.999 and a correlation r to the measured trait, where r is the accuracy of the genomic breeding value for each trait. The accuracy of the genomic test varied for each trait (Swan *et al.* 2015).

Genotyping strategies

Truncation selection was used in all breeding programs (i.e. the highest ranked rams were randomly mated with the highest ranked ewes). Number of candidates genotyped in each breeding program ranged from zero (control) to 100% (maximum benefit). No ewe or ram selection candidates were genotyped in Scenarios 1-2 and Scenarios 3-5, respectively.

Early-stage selection juvenile rams (Scenario 1)

Juvenile rams were eligible to be genotyped prior to any phenotypic measurements or BLUP breeding values. Juvenile rams were sorted from highest ranked to lowest ranked based on parent average breeding values. No ewe selection candidates were genotyped in this scenario.

Late-stage selection mature rams (Scenario 2)

Ram selection candidates were eligible to be genotyped at genetic evaluation that included a measurement of PWT and a BLUP calculation had been made. These rams were then sorted from highest lowest based on index breeding values. No ewe selection candidates were genotyped in this scenario.

Late-stage selection mature ewes for natural mating (Scenario 3)

Ewe selection candidates prior to their first year of mating were eligible to be genotyped after the initial measurement of PWT and a BLUP calculation had been made. These ewes were then sorted similar to Scenario 2. In the natural mating scenario 500 ewes were selected with the probability of one progeny born per mating.

Late-stage selection mature ewes for MOET matings (Scenario 4)

Ewe selection candidates prior to their first year of mating were eligible to be genotyped after the initial measurement of PWT and a BLUP calculation had been made. These ewes were then sorted similar to Scenario 2. In the MOET mating scenario 125 ewes were selected with the probability of zero to eight progeny born per mating with an average of 4 similar to Granleese *et al.'s* (2016) method.

Early-stage selection juvenile ewes for JIVET matings (Scenario 5)

Juvenile ewes were eligible to be genotyped prior to any phenotypic measurements or BLUP breeding values similar to Scenario 1. Because the generation interval for dams in JIVET sheep breeding programs are as little as 6 months, Scenario 5 required 2 rounds of mating a year. In the JIVET mating scenario 64 ewes were selected annually (or 32 each mating round) with the probability of zero to sixteen progeny born per mating, with an average of 8, similar to Granleese *et al.'s* (2016) method.

RESULTS AND DISCUSSION

When there was no genotyping in any strategy MOET breeding programs yielded the highest genetic gain (Figure 1a). However as genotyping increased in selection candidates, JIVET breeding programs yielded the highest genetic gain (Figure 1a). This corresponds to Granleese *et al.'s* (2016) results. However, genotyping male selection candidates resulted in the most cost-effective way to increase genetic gain in a breeding program when compared to the cost of female reproductive technologies.

The maximum increase due to genomic selection in rams lifted genetic gain by 40% (Figure 1b). Genomic testing of the top 20% of ram selection candidates achieved 80% of the maximum benefit using late-stage genomic selection, while 47% required testing in early-stage genomic testing

(Figure 1c). This demonstrates the importance of using initial measurements and screening on breeding values that use selection indexes that have late-in-life or hard to measure traits. Genetic gain lifted by a maximum of 15-65% for genomic testing in ewe selection candidates (Figure 1b). To achieve 80% of the maximum benefit of genomic selection, 65%, 35% and 45% of ewe selection candidates required genomic testing each year for natural, MOET and JIVET breeding programs, respectively (Figure 1c).

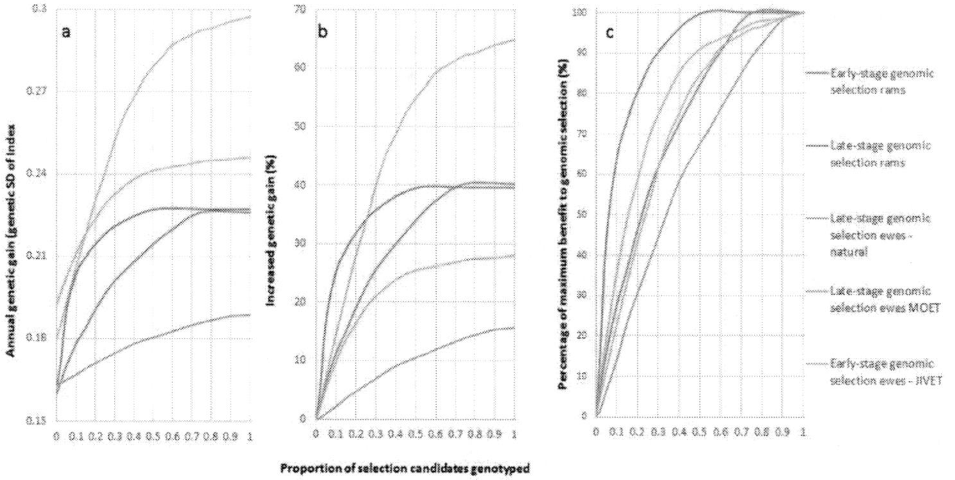

Figure 1: All x-axis are presented in proportion of selection candidates genotyped. a) Genetic gain in genetic standard deviations of the breeding objective; b) Increased genetic gain from genotyping proportions of selection candidates (note that zero % is zero genotyping); c) Percentage of maximum benefit of genotyping selection candidates (note that 100% genotyping is 100% of the benefit)

With the additional cost of producing lambs using female reproductive technologies (Granleese *et al.* 2017), genomic selection of ram selection provides the most favourable cost-benefit. Van der Werf *et al.* (2014) raised the idea of genotyping proportions of male selection candidates to receive the majority of the potential benefit. Our study demonstrates similar outcomes and reinforces that recording initial information prior to genotyping is crucial to achieving the "20-80" rule in ram selection candidates. Furthermore, important rules can be learned for using genomic selection in female selection candidates. It seems uneconomical to genotype all female selection candidates in natural mating or artificial insemination programs, particularly while genotyping costs are still relatively high. It would also be rare to find any sheep flocks in Australia that have their entire nucleus drop born to reproductive technologies as in our scenarios 4 and 5. However, many sheep studs in Australia have a proportion of their lambs born via reproductive technologies. This study and previous studies demonstrate the strong synergies between the two. Therefore, from this study we can use lessons to apply practically. For example if a breeder wanted to perform 10 MOETs or JIVETS on ewes, to get 80% of the maximum genomic selection benefit they should genotype 35 or 45 selection candidates, respectively.

CONCLUSIONS

Genotyping strategies in sheep breeding programs are necessary to reduce cost. This study provides evidence that late-stage genomic selection is far more efficient than early-stage genotyping methods particularly when there is late-in-life or hard-to-measure traits in the breeding objective.

We demonstrate that strong synergies exist between genomic selection and female reproductive technologies and show that genotyping efficiencies exist too with females when using reproductive technologies.

REFERENCES

Dekkers J.C.M. (2007) *J. Anim. Br. Gen.* **124**: 331.

Gilmour A.R., Gogel B.J., Cullis B.R., Welham S.J., Thompson R. (2009) ASReml User Guide Release 3.0 VSN International Ltd, Hemel Hempstead, HP1 1ES, UK.

Granleese T., Clark S.A., Swan A.A., van der Werf J.H.J. (2016) *Anim. Prod. Sci.*, published online.

Granleese T., Clark S.A., Kinghorn B.P., van der Werf J.H.J. (2017) *J. Anim. Breed. Gen.*, submitted.

Sise J.A., Amer P.R. (2009) *Proc. Assoc. Advmt. Anim. Breed. Gen.*, Barossa Valley, Australia, **18**: 29.

Swan A.A., Pleasants T., Pethick D. (2015) *Proc. Assoc. Advmt. Anim. Breed. Gen.*, Lorne, Australia, **21**: 29.

Van der Werf J.H.J., Banks R.G., Clark S.A., Lee S.J., Daetwyler H.D., Hayes B.J., Swan A.A. (2014) *Proc. 10th World Cong. Gen. Appl. Livest. Prod.*, Vancouver, Canada, **10**: 352.

Proc. Assoc. Advmt. Anim. Breed. Genet. 22:337-339

MARKETING AND DELIVERY OF RAMSELECT WORKSHOPS IN SOUTH AUSTRALIA: A COORDINATED APPROACH

A.L. Collins[1], M. Cousins[2], B.L. Ashton[3], S. Milne[4], D. Milne[4], B. Hancock[5], P. Schulz[6], L. Hogan[7] and F.D. Brien[8]

[1] Rural Solutions SA, Primary Industries and Regions SA (PIRSA), Port Augusta, South Australia
[2] Cousins Merino Service, Burra, South Australia
[3] Sheep Consultancy Service Pty Ltd, Port Lincoln, South Australia
[4]Richmond Hill Agribusiness, Branxholme, Victoria, Australia
[5] Rural Solutions SA, PIRSA & Sheep CRC, University of Adelaide, Roseworthy South Australia
[6]Schulz Livestock, Field, South Australia
[7]CRC for Sheep Industry Innovation, Armidale, New South Wales, Australia
[8]University of Adelaide, Roseworthy, South Australia

SUMMARY

Achieving increased adoption of objective genetic information has been a long-running issue in the Australian sheep industry, particularly for wool sheep. Improving the understanding within the commercial ram buying sector of Australian Sheep Breeding Values and how they can enhance the accuracy of genetic selection will lead to improved genetic gain in commercial flocks and increase demand for seedstock producers to adopt objective genetic selection. RamSelect workshops are designed to improve the knowledge and confidence of ram buyers in the use of breeding values to make informed decisions when purchasing rams based on both objective and visual assessment. A coordinated approach to the marketing and delivery of RamSelect workshops has been used in South Australia with great success.

INTRODUCTION

Sheep Genetics (SG) provides a national genetic information and evaluation service for the meat and wool sectors of the Australian sheep industry, delivered as LAMBPLAN, MERINOSELECT and DOHNE. Sheep are ranked according to various production traits using Australian Sheep Breeding Vales (ASBVs). ASBVs provide an estimate of an animal's genetic merit and performance that will be passed onto their progeny. ASBVs enable sheep breeders to select rams based on genetic potential, rather than their visual appeal, which can be affected by environment and management. Selecting animals on their genetic merit for the traits that are important to the business will improve the rate of genetic gain, more reliably meet market specifications and improve overall productivity.

Commercial ram buyers can use ASBVs to objectively compare rams and identify those that best suit their production system and target markets. However, the use of ASBVs by commercial ram buyers has historically not been as high as it could be (particularly in the wool/Merino sector), due mainly to a lack of understanding and thus mistrust of the science, and its application to their business.

The CRC for Sheep Industry Innovation (Sheep CRC) and the NSW Department of Primary Industries, with input from Sheep Genetics, Meat and Livestock Australia, Australian Wool Innovation, the South Australian Research and Development Institute (SARDI) and the private sector, developed the RamSelect workshop with the aim of increasing ram buyer confidence in using ASBVs to select the best value rams for their breeding objective. The workshop provides skills in:

- understanding ASBVs and indexes and how to use them

- understanding the value of selecting rams based on the expected performance of their progeny using ASBVs
- assessing, grading and valuing rams for purchase using both visual assessed traits and ASBVs
- describing a breeding objective
- developing a strategy for buying rams at the next auction or selection day

RamSelect received the Award for Excellence in Innovation at the 2014 CRC Association Awards. It was deemed to have resulted in significant uptake of new research and resulted in major improvements to industry productivity.

APPROACH

In April 2013, Rural Solutions SA, SARDI and the Sheep CRC recognised that a more coordinated approach to delivery of training to commercial sheep producers and their service providers on the commercial application of ASBVs, may increase the uptake and adoption of this proven technology. A coordinator was appointed to work within South Australia. The role was to identify and approach potential workshop hosts, coordinate a network of RamSelect deliverers, create promotion and marketing opportunities, and manage a state wide delivery calendar.

A network of ten service providers from both the government and private sector was established. New deliverers, who had not delivered the workshop previously were mentored and trained through co-delivery of the workshop. Regular phone meetings of the network were held where planning, strategies and experiences were discussed, and these were complimented by regular email contact. Opportunities were identified and delivery coordinated across the state. Potential hosts were identified by the network members and the coordinator was able to contact them personally. The most successful approach in contacting studs as potential hosts was a personal email with information attached about the workshop, with a follow-up phone call.

Workshop hosts were encouraged to actively seek involvement from local stock agents and also to involve local students when appropriate.

The coordinator was able to organise promotional articles which appeared in print in the State rural press, regional newspapers and breed society publications, as well as electronic media. Press coverage of workshops across a range of breeds was also organised. Attendance at key industry events such as the Adelaide Show and SG Regional Forums also enabled promotion of the concept of hosting workshops. Education was also organised with the rural media, to ensure that rural reporters had a sound understanding of ASBVs and how to report them.

RESULTS AND DISCUSSION

During the period April 2013 to October 2015, 34 RamSelect workshops were delivered in South Australia, providing the latest approaches to using ASBVs for ram selection to 364 producers, 62 stock agents and 274 tertiary and secondary students. The most popular time for workshop delivery was in July – August which is in the lead up to most studs' annual sales. This means that workshop participants have an opportunity to apply their learnings in a practical situation very soon after the workshop. The majority of studs who hosted workshops were from the traditional meat breeds, but an increasing number of Merino studs became involved towards the end of this period. Stock agents were actively encouraged to be involved with workshops, and the large number (62) who did participate over this time has led to significant increases in their knowledge and confidence using ASBVs in ram buying decisions.

Evaluation of the RamSelect workshops nationally has demonstrated very positive support from participants:

- 93% indicated they now understood how to make better use of ASBVs
- 87% of participants could see how to improve their ram selection to increase returns
- 87% indicated they would use ASBVs to help select rams in the future
- 74% said they would ask their ram breeder for more information on the rams he is offering
- Participants rated the workshop 8.5 out of 10 for usefulness in assisting with ram selection
- Participants were highly likely to recommend the workshop to others

Many seedstock producers have reported an increased interest in the use of ASBVs by commercial ram buyers when they are making their ram selections. Producers who previously were not interested in looking at ASBV figures prior to purchase, possibly due to mistrust as a result of lack of understanding, now request the figures and actively scrutinise them prior to making ram purchase decisions.

The key success factors with this approach taken in South Australia were:

- A coordinated network of experienced, knowledgeable and credible deliverers, covering all regions of the State, and with a range of experiences enabling effective matching of deliverers with workshop opportunities
- Active networking of the delivery group providing access to extensive producer networks
- The network provided a supportive professional group that enabled sharing of experiences and ideas with workshop delivery
- Two person delivery of the workshops works well for engagement and enabled mentoring of new deliverers
- The dedicated coordinator was able to spend time approaching potential hosts personally, and also attend industry events with the express purpose of identifying potential workshop hosts
- The coordinator was also able to dedicate time to promotional activities

ACKNOWLEDGEMENTS

The Sheep CRC co-funded the delivery of RamSelect workshops with participants paying a fee to attend and hosts providing substantial in-kind contributions. The role of coordinator was funded by a grant from the former Department of Further Education, Employment, Science and Technology in South Australia, administered by SARDI and by the Sheep CRC.

Proc. Assoc. Advmt. Anim. Breed. Genet. 22:341-344

USE OF GENOMIC DATA TO DETERMINE BREED COMPOSITION OF AUSTRALIAN SHEEP

P.M. Gurman, A.A. Swan and V. Boerner

Animal Genetics and Breeding Unit[1], University of New England, Armidale, NSW 2351, Australia

SUMMARY

The Australian sheep industry is characterised by the use of many sheep breeds and regular crossing among them. For the purposes of genetic evaluation, sheep are currently assigned breed proportions based on pedigree. SNP genotypes have been used in many applications to reveal population structure including livestock breeds. In this paper, we investigate the use of SNP genotypes to determine breed structure in Australian sheep breeds using the genotype database assembled for genetic evaluation. Algorithms implemented in two software programs, ADMIXTURE and BreedComp were able to identify sheep breeds and genetic groups within the Merino breed. These results can potentially lead to more accurate identification of breed content, and more accurate predictions of breeding value through improved allocation to genetic groups.

INTRODUCTION

A number of sheep breeds play an important role in the Australian sheep meat and wool industries, with the Merino dominant for wool production, Border Leicester, Coopworth and composite breeds used for maternal performance, and Poll Dorset, Texel, and Suffolk and White Suffolk used as terminal sires for meat production. Crossing among these breeds is common not only at the commercial level, but also in seed-stock flocks where some breeders seek to exploit breed differences. Considerable genetic diversity is present within the Merino breed, such that many flocks are considered to be different genetic groups for the purposes of evaluation. Currently, the evaluation system accounts for breed and within-breed genetic group differences using the Westell-Quaas approach (Westell *et al.* 1988), in which the breed composition of each animal is modelled through the pedigree. With increasing availability of genomic data, the utility of this data in estimating sheep breeds and genetic groups within breeds has been examined. Various authors have investigated the use of genomic data to identify population structures in beef breeds (Sölkner *et al.* 2010; Kuehn *et al.* 2011a; VanRaden *et al.* 2011; Frkonja *et al.* 2012) and sheep breeds (Dodds *et al.* 2013). In this paper, we investigate the use of genomic data to identify breed and within-breed population structures in Australian sheep.

MATERIALS AND METHODS

50K SNP genotypes (as described by Moghaddar *et al.* (2015)) and pedigree-based breed proportions were collated for 31,125 sheep from the reference and industry populations used for genomic evaluation in Australia. These data contained records for straight-bred sheep with 623 Border Leicester, 1,966 Poll Dorsets, 28 Texels, 37 Suffolks, 39 White Suffolks, and 14,440 Merinos, where "straight-bred" is defined here as containing at least 0.95 of that breed proportion from the pedigree. Of the recorded straight-bred Merino sheep, some were recorded as straight-bred of a particular Merino group with 456 as 'ultra-fine', 2,907 as 'fine-medium' and 967 as 'strong'. The genomic relationship matrix (G) was calculated using the method by Yang *et al.* (2010) and a singular value decomposition performed on the G matrix, such that $G = U\Sigma V$. Vectors of the U

[1]AGBU is a joint venture of NSW Department of Primary Industries and the University of New England

matrix were then analysed visually in 'R' (R Core Team 2016) to determine if breeds and genetic groups could be identified.

Analytical tools were used to predict breed and genetic group proportions including supervised ADMIXTURE (Alexander *et al.* 2009) and constrained genomic regression, hereby referred to as BreedComp (Boerner, 2017), which is a constrained (i.e. estimated proportions are less than one) version of the approach by Chiang *et al.* (2010) and Kuehn *et al.* (2011). For both algorithms, a training set of animals was created, comprising animals from straight-bred animal clusters for each breed or genetic group of Merinos based on previous pedigree-based breed proportions. This training set contained seven breeds, 497 Border Leicesters, 1902 Poll Dorsets, 21 Texels, 36 Suffolks, 339 'ultra fine' Merinos, 488 'fine-medium' Merinos and 216 'strong' Merinos, resulting in 3,499 sheep used for training and 27,626 sheep for validation. While a group of straight-bred White Suffolks were recorded, these animals did not appear to be genetically different to sheep recorded as partly White Suffolk. This breed was not included in the training animals, with animals previously assigned to this breed attributed to proportions of the other breeds by the algorithms.

Prediction accuracy was measured using Root Mean Squared Error (RMSE), based on the differences between current pedigree based breed assignments, and those estimated by the predictive algorithms. RMSE values for each breed or genetic group were calculated by $RMSE = \sqrt{(\sum_{i=1}^{n}(\hat{q}_i - q_i)^2)/n}$ where $\hat{q}_i$ is the breed proportion predicted using genomic data for the ith animal, q_i is the breed proportion from the pedigree for the ith animal, and n is the number of animals modelled. RMSE was calculated for each breed individually, as well as an overall value across all breeds. Algorithms producing lower RMSE values were deemed to produce more accurate estimates.

RESULTS AND DISCUSSION

Breeds could be differentiated in plots of the first two vectors of the U matrix (see Figure 1). Distinct clusters of straight-bred Border Leicester, Poll Dorset and Merino animals could be identified in the extremes of these plots from their pedigree based breed assignments and a small cluster of Suffolk animals could be identified in the Suffolk plot. Further, groups of crossbred animals could also be differentiated in between the clusters of straight-bred animals, e.g. a group of ¼ Merino and ¾ Border Leicester can be seen in the lower middle of the their respective plots. This is also true for other clusters of animals in these plots, which can be attributed visually to varying combinations of breeds.

It was also possible to identify genetic groups of Merino animals in the 4th and 5th vectors of the U matrix (see Figure 2), with differentiation of ultra-fine, fine-medium and strong sheep possible. It is evident from these plots that some sheep previously assigned to the fine-medium group may instead belong to the ultra-fine group. This would suggest that historically the fine-medium category has become a default category for sheep that have been hard to group. It can also be seen in Figures 1 and 2 that many animals appear to have been assigned to the wrong breed or genetic group.

The ADMIXTURE and BreedComp algorithms were able to predict breed and genetic group proportions based on the training animals provided to them, with BreedComp appearing to provide slightly more accurate estimates. RMSE values are presented, where possible (see Table 1), with BreedComp producing lower RMSE values for a larger number of breeds than ADMIXTURE. Caution is warranted in interpretation of these values presented here because of errors in the pedigree-based assignments. In addition, some Merino sheep of unknown type have been allocated to a default category.

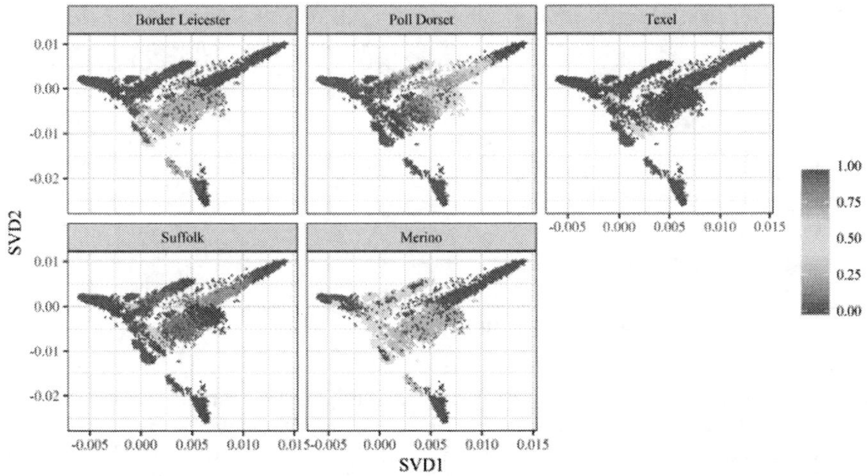

Figure 1: Pedigree based breed proportions of main sheep breeds in each genotyped animal. Red points indicate animals with 100% content of the given breed, and blue points 0%.

BreedComp was able to identify all breeds and genetic groups of sheep included in the training animals. ADMIXTURE did not identify the Texel breed from the training animals, instead identifying a fourth genetic group of merinos. Some sheep were reassigned by both algorithms, for instance, many sheep were reclassified from the fine-medium Merino group to the ultra-fine group. This can be seen in the RMSE values for these categories (see Table 1) which were larger than for the other categories. Importantly for routine application, BreedComp was approximately 45 times faster than ADMIXTURE, with BreedComp running for 4.1 minutes on a single CPU core, while ADMIXTURE ran for 3.1 hours on 28 CPU cores. BreedComp appears to be slightly more accurate and faster than ADMIXTURE for these data.

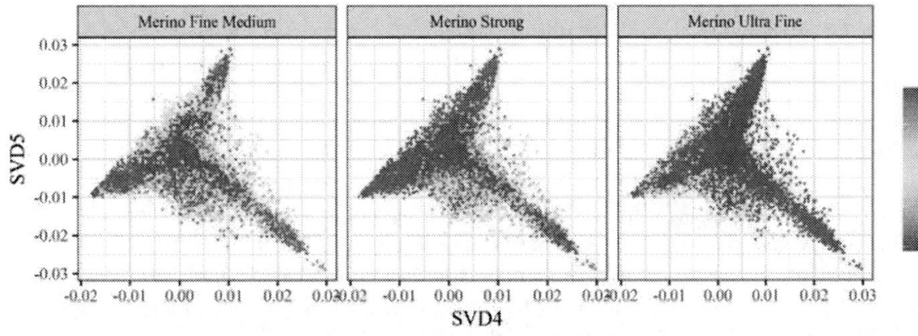

Figure 2: Pedigree based proportions of each genetic group for Merinos. Red points indicate animals with 100% content of the given breed, and blue points 0%.

Table 1: RMSE for breed composition predictions for ADMIXTURE and BreedComp.

Breed	ADMIXTURE[1]	BreedComp
Overall	0.181	0.170
Border Leicester	0.059	0.055
Poll Dorset	0.059	0.095
Texel	-	0.027
Suffolk	0.172	0.072
Merino (Ultra-Fine)	0.242	0.298
Merino (Fine-Medium)	0.254	0.247
Merino (Strong)	0.190	0.190

[1]ADMIXTURE was unable to identify the Texels thus a RMSE value was not calculated.

Another issue with using these algorithms to estimate breed proportions is their inability to estimate breed proportions for breeds lacking genotyped straight-bred animals. For instance, a small cluster of sheep can be identified just to the upper right of the Merinos that are ½ Merino and ½ Dorper. These sheep are currently being assigned by BreedComp as ¾ Merino and ¼ Poll Dorset. Without the inclusion of straight-bred Dorper sheep in the data, these sheep cannot be correctly classified. A small cluster can also be identified that contains a portion of Coopworth.

Application of the BreedComp algorithm would allow for more accurate estimation of breeding values for Australian sheep, especially for animals without pedigree information and only genomic data. Animals previously assigned to the wrong breed or genetic group can also be reassigned, further improving genetic evaluation systems.

REFERENCES

Alexander D.H., Novembre J., Lange K. (2009) *Genome Res.* **19**:1655.

Boerner V. (2017) Proc. Assoc. Adv. Anim. Breed. Genet. **22**: these proceedings.

Chiang C.W.K., Gajdos Z.K.Z., Korn J.M., Kuruvilla F.G., Butler J.L., *et al.* (2010) *PLOS Genet.* **6**:e1000866.

Dodds K.G., Newman S.-A.N., Auvray B., McEwan J.C. (2013) Proc. Assoc. Adv. Anim. Breed. Genet. **20**:274.

Frkonja A., Gredler B., Schnyder U., Curik I., Sölkner J. (2012) *Anim. Genet.* **43**:696.

Kuehn L.A., Keele J.W., Bennett G.L., McDaneld T.G., Smith T.P.L., *et al.* (2011a) *J. Anim. Sci.* **89**:1742.

Moghaddar N., Gore K.P., Daetwyler H.D., Hayes B.J., van der Werf J.H.J. (2015) Genet. Sel. Evol. **47**:97.

Sölkner J., Frkonja A., Raadsma H.W., Jonas E., Thaller G., *et al.* (2010) *Interbull Bull.* 62.

VanRaden P.M., Olson K.M., Wiggans G.R., Cole J.B., Tooker M.E. (2011) *J. Dairy Sci.* **94**:5673.

Westell R.A., Quaas R.L., Van Vleck L.D. (1988) *J. Dairy Sci.* **71**:1310.

Yang J., Benyamin B., McEvoy B.P., Gordon S., Henders A.K., *et al.* (2010) *Nat. Genet.* **42**:565.

R Core Team (2016) R: A Language and Environment for Statistical Computing R Foundation for Statistical Computing, Vienna, Austria. http://www.R-project.org/.

Proc. Assoc. Advmt. Anim. Breed. Genet. 22:345-348

GENETIC CORRELATIONS BETWEEN LAMB SURVIVAL, BIRTH WEIGHT, AND GESTATION LENGTH DIFFER BETWEEN BIRTH TYPES

O.A. Kelly, M.L. Hebart, F.D. Brien and W.S. Pitchford

School of Animal and Veterinary Sciences, University of Adelaide, Roseworthy SA 5371 Australia

SUMMARY

Lamb survival significantly limits the productivity of the Australian sheep industry, with twin and multiple-born lambs suffering greater mortality rates than single-born lambs. Using data from the Sheep CRC Information Nucleus Flocks, correlation estimates for gestation length, birthweight, and lamb survival across different birth types were examined. Gestation length and lamb survival are uncorrelated genetically in both singles and twins (0.02 and 0.04) but have a low genetic correlation of 0.27 for multiple-born lambs. Birthweight was lowly and negatively genetically correlated to singles survival (-0.29), negligibly correlated for twins (-0.15), and lowly positively correlated for multiples (0.37). The results of this study demonstrate that the influence of birthweight on survival varies significantly across different litter sizes, and selecting for birthweight to improve lamb survival would not be beneficial for survival across all birth types.

INTRODUCTION

Lamb survival is a significant problem in Australia, severely limiting the productivity of the sheep flock (Hinch and Brien 2014). Phenotypically, birthweight is one of the largest influences on lamb survival within the first few days of life (Oldham *et al.* 2011). Litter size significantly impacts lamb survival, with twin survival rates reported at 27% below single-born survival (Kleemann and Walker 2005) and is associated with lower birthweights and slower growth rates (Dwyer and Morgan 2006). Gestation length is also known to be shorter for larger litters, leading to lower birthweights and lower lamb viability (Dwyer and Morgan 2006; Li and Brown 2015).

Lamb survival has a very low direct heritability, with estimates generally calculated around 0.01-0.03 and only marginally higher estimates for maternal heritability, averaging 0.05-0.06 (Safari *et al.* 2005; Brien *et al.* 2010), suggesting that direct selection to improve survival would be inefficient. Indirect selection to improve lamb survival may be more effective, however it is vital to account for any agonistic and antagonistic effects on survival as a consequence of correlations with other traits when indirect selection is practiced. The genetic parameters for birthweight are widely varied throughout literature, having anywhere between a low to moderate heritability and often with little to no genetic correlation with lamb survival despite the phenotypic curvilinear relationship (Brien *et al.* 2014). Despite reports of lamb survival and birthweight varying significantly depending on litter size, very few papers have examined this idea in detail (Li and Brown 2015). This paper reports on the analysis of the relationship of birthweight and lamb survival across different litter sizes and discusses whether they should be treated as separate traits depending on birth type.

MATERIALS AND METHODS

Information Nucleus Flock. The data used was from the Sheep CRC Information Nucleus Flock (INF), a collection of records from 2007-2011 over eight locations across Australia with Merino and crossbred ewes that were inseminated to Merino, maternal and terminal breed rams. Further details on design, data collection and management of the INF have been reported by Fogarty *et al.* (2007) and Geenty *et al.* (2014). Gestation length (GL) was treated as a trait of the lamb and calculated from AI dates (conception) and lamb birth dates, with records above 160 days and below 138 days removed due to biological improbability. Lamb survival to three days (S3) was chosen as the primary focus as it accounts for 66% of lamb mortalities to weaning. Birthweight (BWT) was also analysed.

Statistical Analysis. The statistical package ASReml (Gilmour 2015) was used to estimate genetic and phenotypic variance and covariance components, heritability, and correlations between GL, BWT, and S3 by fitting a linear mixed model with restricted maximum likelihood. An animal model was initially attempted, however, due to a lack of convergence, this was unable to test the hypotheses herein and so a sire model was used throughout the study. The dam permanent environmental effect was also removed from the model due to lack of depth within the pedigree.

Model 1. Three univariate analyses were run to estimate variance components for GL, S3 and BWT. Analyses included the fixed effects of type of birth (TOB; singles, twins, multiples=3+), age of dam (AOD; 2-8+), sex (male (M) or female (F)), location (8 sites), genetic groups (dam breed, sire breed) and year (2007-2011), significant two-way interactions (AOD by year, flock by year, flock by year by TOB) and the random terms of sire and dam. The sire variance includes ¼ additive genetic whereas dam includes ¼ additive genetic variance plus maternal genetic and environmental effects. A trivariate analyses was performed to estimate genetic correlations between GL, S3 and BWT.

Model 2. Separate univariate models for singles, twins and multiples were fitted to estimate separate sire and dam variance components for each TOB (singles, twins and multiples) for the three traits, GL, S3 and BWT. The fixed effects and random terms were as outlined in Model 1. Multiple records per litter were randomly removed for gestation and treated as missing as they all had the same gestation length, as per the technique used by Li and Brown (2015), leaving a total of 15,097 gestation length records. A series of bivariate analyses were performed using Model 2 fixed and random effects to estimate the genetic and phenotypic correlation between GL, LS3 and BWT.

RESULTS

Basic statistics and heritability estimates were calculated for the overall traits as outline in Model 1 and for the separated traits as outlined in Model 2 (Table 1). Separating the traits by TOB (Model 2) was a significant improvement over Model 1 (Table 1).

Table 1: Summary statistics and heritability estimates for gestation length (GL), birthweight (BWT) and lamb survival to three days (S3) for the overall trait (Model 1) and separated by type of birth (Model 2) with the Likelihood Ratio Test (LRT) statistic comparing Model 1 to Model 2 (* = significant at the 0.001 level - 32.91 at 12 degrees of freedom)

		Mean	σ	Count	h^2	LTR
GL		149.4 days	2.6 days	15 097	0.53 ± 0.05	
	Single	149.8 days	2.6 days	7 267	0.53 ± 0.05	79.38*
	Twin	149.1 days	2.5 days	6 762	0.54 ± 0.15	
	Multiple	148.6 days	2.5 days	1 068	0.54 ± 0.15	
BWT		4.8 kg	1.1 kg	23 619	0.16 ± 0.02	
	Single	5.5 kg	1.1 kg	7 267	0.21 ± 0.03	386.82*
	Twin	4.6 kg	1.0 kg	13 229	0.12 ± 0.02	
	Multiple	3.8 kg	0.9 kg	3 123	0.16 ± 0.07	
S3		0.88	0.32	23 619	0.02 ± 0.01	
	Single	0.94	0.25	7 267	0.01 ± 0.02	320.88*
	Twin	0.89	0.32	13 229	0.04 ± 0.01	
	Multiple	0.72	0.45	3 123	0.15 ± 0.06	

An initial analysis was performed to calculate the correlation between GL, BWT, and S3 as singular traits in a univariate model (Model 1; Table 2) before the traits were considered separate by TOB (Model 2). Phenotypic correlations were not calculated between overall traits because the model was improved by separating the traits and it was deemed unnecessary.

The between trait genetic correlations for BWT and S3 had an increasing trend, being lowly negative for single lambs, negligibly negative for twins, and lowly positive for multiples (Table 2). The correlation between GL and S3 were negligible for single and twin lambs while there was a low correlation for multiples (0.27), though with a large standard error. The phenotypic correlation between S3 and BWT was greater for larger litters (Table 2), with the phenotypic correlation between GL and S3 being negligible for all litter sizes.

Table 2: Genetic sire correlations between gestation length (GL), birthweight (BWT) and lamb survival to three days (S3) by type of birth (with standard errors)

	Genetic			Phenotypic		
	GL-S3	GL-BWT	BWT-S3	GL-S3	GL-BWT	BWT-S3
Overall	-0.01 (0.11)	0.36 (0.06)	-0.31 (0.13)			
Singles	0.07 (0.30)	0.31 (0.08)	-0.29 (0.42)	-0.02 (0.01)	0.34 (0.01)	0.07 (0.01)
Twins	-0.03 (0.13)	0.41 (0.08)	-0.15 (0.17)	0.04 (0.01)	0.33 (0.01)	0.18 (0.01)
Multiples	0.27 (0.24)	0.53 (0.22)	0.37 (0.29)	0.04 (0.03)	0.29 (0.03)	0.28 (0.02)

DISCUSSION

Birthweight has a complex relationship with lamb survival, as described in literature. The two traits are known to be phenotypically linked, with some reports referring to birthweight as one of the biggest factors influencing the initial survival of the lamb, with the relationship of a negative quadratic nature (Hatcher *et al.* 2009; Celi and Bush 2010; Oldham *et al.* 2011). Hatcher *et al.* (2009) have described the optimum phenotypic birthweight for lamb survival as being different between singles, twins, and multiples, with attempts to select for higher birthweight to improve twin and multiple-born lamb survival potentially resulting in a decrease in single-born lamb survival rates due to dystocia. Considering this, an expected trend was seen in the genetic correlation (Table 2) between birthweight and lamb survival where single and twin-born lambs were lowly negatively correlated, -0.29 and -0.15 respectively, while multiple-born lambs were lowly positively correlated (0.37). The overall genetic relationship between birthweight and survival was lowly negative (-0.31, Table 2) and within the range reported in previous literature (Brien *et al.* 2014); segregation by type of birth gave a clearer indication that lamb survival is a genetically separate trait across birth types (Table 1; Kelly *et al.* 2016). Female pigs, which consistently have large litter sizes, similarly demonstrate a low positive genetic correlation between survival and birthweight (Tabuaciri *et al.* 2010). Phenotypically, the correlation between birthweight and survival increases as litter size increased (Table 3), indicating that birthweight had a more significant influence on survival in larger litter sizes and is likely due to the smaller size of lambs born as multiples. Despite following an expected trend, survival and birthweight were not as highly correlated as expected. Although the likelihood ratio test (Table 1) provides evidence that separating by birth type improves the statistical model, the precision of these correlations are low. Given the low heritability and variation, combined with the lack of precision in correlations estimates, suggests birthweight is unlikely to be a suitable indicator trait for lamb survival.

The genetic correlations between gestation length and birthweight (Table 2) for twins and multiple-born lambs follow the expected trend of consistently moderately positive, as reported in piglets (Rydhmer *et al.* 2008). This aligns with the overall genetic correlation of 0.36 between gestation length and birthweight (Table 2). This suggests that longer gestation length may improve the birthweight of multiple-born lambs and potentially improve their survival, but would also increase birthweight of single and twin lambs and this could be detrimental to their survival. The direct correlation between survival and gestation length for single and twin-born lambs was negligible (Table 2), which differs from earlier results from Li and Brown (2015), who reported low positive correlations between gestation length and lamb survival for single and twin-born lambs.

Birthweight is critical to early lamb survival and is associated with many genetic factors, such as litter size. With the economic push to increase litter size in sheep (Swan 2009), it's vital to understand the interaction between litter size, birthweight, and survival. The results of this study demonstrate that genetically selecting for birthweight to improve lamb survival does not appear to be beneficial for survival across different litter sizes. Despite interesting correlations seen in separating birthweight and lamb survival by birth type and that this is an improvement on the single-trait model (Table 1), these correlations and their precisions are low (Table 2). Furthermore, treating the traits as separate by birth type would be difficult to implement in a practical breeding plan, although economic value for birthweight may change with mean litter size.

REFERENCES

Brien F.D., Cloete S.W.P., Fogarty N.M., Greeff J.C., Hebart M.L., Hiendleder S., Hocking Edwards J.E., Kelly J.M., Kind K.L., Kleemann D.O., Plush K.L. and Miller D.R. (2014) *Anim. Prod. Sci.* **54**: 667.

Celi P. and Bush R. (2010) in 'International Sheep and Wool Handbook', pp. 223-257, editor D.J. Cottle, Nottingham University Press.

Dwyer C.M. and Morgan C.A. (2006) *J. Anim. Sci.* **84**(5): 1093.

Fogarty N.M., Banks R.G., van der Werf J.H.J., Ball A.J. and Gibson J.P. (2007) *Proc. Assoc. Advmt. Anim. Breed. Genet.* **17**: 29.

Geenty K.G., Brien F.D., Hinch G.N., Dobos R.C., Refshauge G., McCaskill M., Ball A.J., Behrendt R., Gore K.P., Savage D.B., Harden S., Hocking-Edwards J.E., Hart K. and van der Werf J.H.J. (2014) *Anim. Prod. Sci.* **54**: 715.

Gilmour A. (2015) ASReml 4, *VSN International Limited.*

Hinch G.N. and Brien F. (2014) *Anim. Prod. Sci.* **54**: 656.

Hatcher S. and Atkins K.D., Safari E. (2009) *J. Anim. Sci.* **87**(9): 2781.

Kelly O.A., Hebart M.L., Brien F.D. and Pitchford W.S. (2016) *Proc. Aust. Sci. Anim. Prod.* **31**: 135.

Kleemann D.O. and Walker S.K. (2005) *Theriogenology* **63**(8): 2075-2088.

Li L. and Brown D.J. (2015) *Anim. Prod. Sci.* **56**(5): 934.

Oldham C.M., Thompson A.N., Ferguson M.B., Gordon D.J., Kearney G.A. and Paganoni B.L. (2011) *Anim. Prod. Sci.* **51**: 776.

Rydhmer L., Lundeheim N. and Canario L. (2008) *Livest. Sci.* **115**(2-3): 287.

Safari E., Atkins K.D., Fogarty N.M. and Gilmour A.R. (2005) *Proc. Assoc. Advmt. Anim. Breed. Genet.* **16**: 28-31.

Swan A. (2009) *ACIAR Proc.* **133**: 170-176.

Tabuaciri P., Bunter K.L. and Graser H.U. (2010) *AGBU Pig Genetics Workshop*: 65-72.

Proc. Assoc. Advmt. Anim. Breed. Genet. 22:349-352

PREDICTION ACCURACIES FOR POLLED AND HORNED MERINO SHEEP USING DIFFERENT GENETIC MODELS

N. Duijvesteijn[1,2], S. Weerasinghe[1] and J.H.J. van der Werf[1,2]

[1]School of Environmental and Rural Science, Univ. of New England, Armidale, NSW, Australia
[2]Cooperative Research Centre for Sheep Industry Innovation, Armidale, Australia 2351

SUMMARY

Previous studies have mapped the responsible locus for the polled phenotype to the 3' region of *RXFP2* at ovine chromosome 10. SNPs to determine whether the insertion is present are neither on the Ovine 50K nor on the OvineHD. In this study we tested different strategies for prediction of the horn phenotype, including single SNP, multiple SNP haplotypes and GBLUP. In total, 4,001 Merino sheep with HD genotype information were used. Prediction accuracies were calculated for each sex separately. Models with the highest prediction accuracies for horned used either single SNPs or 3-SNP haplotypes and also included a polygenic effect estimated based on traditional pedigree relationships. The accuracies of predicting the 'horned' phenotype were 0.338 for females and 0.724 for males. For predicting 'polled' phenotype, the best models were the same but included a genomic relationship matrix. The accuracies were 0.713 for females and 0.618 for males. Results show that prediction accuracy is high using a single SNP, although not unity as the causative mutation is not genotyped, but likely also because females show incomplete penetrance. As long as there is no genotype from a single SNP causative mutation, additional information through pedigree is valuable for the prediction of horned and polled phenotype.

INTRODUCTION

The genetic background of the polled phenotype has long been studied in horned species such as cattle and sheep (Castle 1940, Georges et al. 1993). The causative mutation in sheep has been mapped to chromosome 10 (Johnston et al. 2011). A 1.78-kb insertion in the 3'-untranslated region of *RXFP2* causes the polled phenotype, as described by Wiedemar and Drogemuller (2015). However, this insertion is not completely explaining the phenotype in different sheep breeds (Lühken et al. 2016). The mode of inheritance is complex as expression differs between sexes and there is not yet a single locus model with complete penetrance. Currently the causative mutation is neither on the Illumina Ovine 50K chip nor on the OvineHD 600K chip. SNPs close to the region of insertion are currently used to predict the phenotype. The aim of this study is to test various strategies for predicting horned or polled phenotypes, including single SNP, multiple SNP haplotypes and SNPBLUP.

MATERIALS AND METHODS

Population and phenotypic data. The data consisted of purebred Merino sheep including Dohne Merino and polled Merino. The phenotype recorded was polled, scurs, knobs or horns, which was analysed as polled / non-polled and horned / non-horned. In total, 4,001 sheep were used. Table 1 shows the distribution of polled and horned status between the two sexes.

Table 1. Number of observed phenotypes for male and female Merinos.

Sex	Polled	Non-Polled	Horned	Non-Horned
Female	1325	1123	88	2360
Male	1042	511	481	1072

Genotypes. Of all 4,001 animals in the dataset, 3,708 were genotyped with the Ovine 50K. The remaining animals were genotyped with the Ovine 12K and imputed up to 50K. All 4,001 animals were further imputed up to 600K. All 600K genotyped animals (~2300) were used for imputation, including 445 animals from the data set used in this study. In total 510,175 SNPs passed quality control and 17,280 SNPs were located at OAR10.

Statistical analysis. We applied three methods to predict the phenotype polled or horned status. To select the best single SNP for prediction, we ran a local GWAS for chromosome 10 (OAR10). The single SNP was either fitted solely (base model), or together with a polygenic effect (fitted either by a traditional pedigree or by a genomic relationship matrix).

The second method was using haplotypes. A haplotype was formed using the most significant SNPs from the single SNP GWAS (3, 5 or 10 SNPs). Genotype data was phased using EAGLE. Only haplotypes with a frequency >1% were fitted in the model, and otherwise placed in a bin (sum of all low frequency haplotypes). The number of haplotypes formed from three, five or 10 SNPs, was equal to three, three, and seven, respectively.

The third method was applying a GBLUP analysis using a GRM based on all SNPs from the 600K (Yang et al. 2010) or only those SNPs from OAR10. Additionally, a dominance relationship matrix was added to the model based on the same two sets of SNPs (Zhu et al. 2015). Breeding values from the additive and dominance GRM were summed to get the predicted phenotype.

Mode of inheritance. We compared various models where the mode of inheritance was investigated. The model including a sex-dependent effect for the additive and dominance variance resulted in the best predictions (results not shown). Therefore, whenever possible, this mode of inheritance is used for prediction.

Validation. A fivefold cross-validation was performed. In each replicate, 20% of the data was randomly blinded and the phenotype was predicted. Prediction accuracy was defined as the correlation of the breeding value with the 0/1 phenotype.

RESULTS

The local GWAS for polled / non polled and horned / non horned clearly indicates the known region with highly significant associations around 29.5 Mb (Figure 1). The most significant SNP for polled was OAR10_29546872.1 which differed from the most significant SNP for horned which was OAR10_29458450, although both SNPs are in high LD (r^2=0.985). Those SNPs were used to perform the single SNP analyses.

In Table 2 and 3, the frequencies of the genotypes with phenotype polled and horned by sex is shown. Using frequencies to determine the polled or horned status across the validation gives an average prediction accuracy of 0.71 for horned and 0.63 for polled (base model, Table 4). In Table 4, the results of the different genetic models are shown. The highest accuracy for predicting polled was achieved when using a GRM additional to the single SNP, resulting in a correlation of 0.713 for females and 0.618 for males. The highest accuracy for predicting horned was by using pedigree relationships additional to the single SNP, which resulted in a correlation of 0.338 for females and 0.724 for males. Models where haplotypes were used resulted in similar accuracies as the single SNP approach. Haplotypes formed from 3 and 5 SNPs (hap3 and hap5), gave very similar prediction accuracies, where hap10 had a lower prediction accuracy.

DISCUSSION AND CONCLUSION

The most significant SNP for polled and horned where very close to the causative mutation, (OAR10_29546872.1: 29512572 and OAR10_29458450: 29458450) of which OAR10_29546872.1 has been used by the Sheep CRC (J. van der Werf, pers. comm). Dominik et al. (2012) found a SNP OAR10_29389966_X.1 to be most predictive in Merino sheep. This SNP was also in the top10 of most significant SNPs in our GWAS. The reported SNP by Johnston et al. (2011)

OAR10_29448537.1 did not occur in the top100 SNPs of our GWAS.

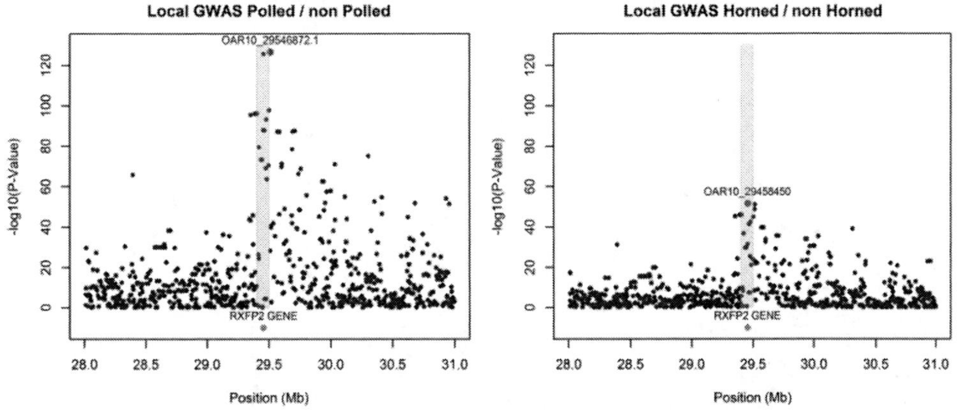

Figure 1. Local GWAS plot for the traits polled and horned of OAR10. The grey rectangle indicates the location of the gene *RXFP2* (29.4-29.5 mb). The most significant SNP is indicated in red.

Table 2. Frequencies of the SNP OAR10_29546872.1 per sex for the phenotype polled and probabilities for being polled, and frequencies of the SNP OAR10_29458450 per sex for the phenotype horned and probabilities for being horned.

Sex	Genotype	Non Polled	Polled	Probability Polled	Non Horned	Horned	Probability Horned
Female	0	1058	174	**0.14**	1151	81	**0.07**
	1	353	811	**0.77**	1047	6	**0.01**
	2	25	138	**0.84**	162	1	**0.01**
Male	0	675	29	**0.04**	229	475	**0.67**
	1	340	385	**0.53**	719	6	**0.01**
	2	27	97	**0.78**	124	0	**0.00**

A model including pedigree information additional to the single SNP or haplotype had a better prediction accuracy compared to using only a single SNP for the prediction of both horned and polled. When the single SNP was not explicitly fitted (local GRM model), the prediction was reduced. For the trait polled and horned highly predictive SNPS close to the known causative mutation should be modelled explicitly. Applying methods which shrink all SNP effects equally like GBLUP will therefore have a lower prediction accuracy in the presence of a large QTL. Mixture models such as Bayes B or C, should perform better.

Clearly prediction accuracy was not close to one, in spite of highly significant SNPs close to a known causative mutation. This indicates that the most significant SNP is not in complete LD with the causative mutation or it does not confer complete penetrance. This is also indicated by the explained variance from the genotypes. For the trait horned, 85% of the phenotypic variance was explained by the single SNP, and 95% of the phenotypic variance when pedigree was also included. For the trait polled, 67% of the phenotypic variance was explained by the single SNP, and 80% of the phenotypic variance when pedigree was also included.

Table 4. Prediction accuracies for horned and polled for the different models for the whole dataset or split by sex with or without fitting a polygenic effect (Ped).

		Correlation Horned			Correlation Polled		
Method	Ped	Average	Female	Male	Average	Female	Male
Single SNP	-	0.711	0.173	0.721	0.630	0.644	0.581
Single SNP	A	0.723	0.338	0.724	0.644	0.665	0.580
Single SNP	GRM	0.719	0.302	0.723	0.686	0.712	0.617
Hap3	A	0.723	0.324	0.726	0.647	0.671	0.579
Hap5	A	0.721	0.324	0.723	0.646	0.673	0.573
Hap10	A	0.673	0.224	0.681	0.632	0.658	0.560
Hap3	GRM	0.722	0.302	0.727	0.687	0.713	0.618
Hap5	GRM	0.721	0.300	0.725	0.676	0.703	0.604
Hap10	GRM	0.696	0.285	0.691	0.670	0.697	0.599
GRM OAR10	GRM	0.391	0.226	0.628	0.617	0.657	0.574
GRM	GRM	0.380[1]	0.273	0.561	0.580[2]	0.620	0.526

[1]Four of the five replicates converged. [2]Only two of the five replicates converged.

Differences between males and females have been described previously (Dolling 1961, Dominik et al. 2012) in Merino sheep. Possibly incomplete penetrance is causing the sporadic horned phenotype in females, and makes prediction more difficult (prediction accuracy 0.338 vs 0.724 for horned in females and males).

Different approaches to validate the different genetic models (e.g. regress back to 0/1 trait by using a threshold on the predicted phenotypes) could clarify results further, and will be investigated additionally.

To conclude, prediction of polled and horned is already successful using a single SNP (~0.7), although not 1 as the causative mutation is not genotyped (on the new 15K Ovine chip it should be present), but likely also because females show incomplete penetrance. Additional information through pedigree is valuable for the prediction of the horned and polled phenotype as long as the causative mutation is not genotyped.

ACKNOWLEGEMENTS

The authors acknowledge the contributions of people from breeders and many CRC participants that contributed to the Sheep CRC Information Nucleus flocks. Also Klint Gore is acknowledged for his help on retrieving data.

REFERENCES

Castle, W. (1940). *J. Hered.* **31**: 486-487.

Dolling, C. (1961). *Crop and Pasture Sci.* **12**: 535-361.

Dominik, S., J. Henshall and B. Hayes (2012). *Anim. Genet.* **43**: 468-470.

Georges, M., R. Drinkwaterz, T. Kingz, et al. (1993). *Nat. Genet.* **4.**

Johnston, S. E., J. McEwan, N. K. Pickering, et al. (2011). *Mol. Ecol.* **20**: 2555-2566.

Lühken, G., S. Krebs, S. Rothammer, et al. (2016). *Genet. Sel. Evol.* **48**: 78.

Wiedemar, N. and C. Drögemüller (2015). *Anim. genet.* **46**: 457-461.

Yang, J., B. Benyamin, B. P. McEvoy, et al. (2010). *Nat. Genet.* **42**: 565-569.

Zhu, Z., A. Bakshi, A. A. Vinkhuyzen, et al. (2015). *Am. J. Hum. Genet.* **96**: 377-385.

Proc. Assoc. Advmt. Anim. Breed. Genet. 22:353-360

CAN GRAZING LIVESTOCK IN DEVELOPING COUNTRIES BENEFIT FROM USE OF GENOMIC SELECTION?

H.M. Burrow[1], M.L. Wolcott[2], A. Maiwashe[3], M.L. Makgahlela[3], B.J. Hayes[4], J.G. Rees[5] and M.J. Bradfield[6]

[1] UNE Business School, University of New England, New South Wales, Australia
[2] Animal Genetics and Breeding Unit, University of New England, New South Wales, Australia
[4] Animal Production Institute, Agricultural Research Council, Private Bag X2, Irene, South Africa
[4] Queensland Alliance for Agriculture and Food Innovation, University of Queensland, Australia
[5] Research and Innovation Systems, Agricultural Research Council, P.O. Box 8783, Pretoria, South Africa
[6] Agricultural Business SA (Pty) Ltd, P.O. Box 543, Irene, South Africa

SUMMARY

Genomic selection is transforming animal and plant breeding across developed countries globally, with economic benefits of billions of dollars annually. Despite huge potential for livestock industries in developing countries to achieve similar transformations, to date there has been very limited use of genomic selection in grazing livestock in those countries. This is attributable to several major challenges. This paper discusses those challenges and proposes options to overcome or reduce them. It also identifies priority areas of research that must be undertaken if grazing livestock in developing countries are to benefit from genomic selection. Achieving genetic gains through genomic selection in smallholder cattle herd in Southern Africa and opportunities to extend the concept to other livestock species, and other developing countries, is also briefly examined.

INTRODUCTION

The world's population is predicted to increase from 7 billion in 2011 to 9 or 10 billion by 2050, with most growth occurring in Africa and Asia (Gerland *et al.* 2014). Incomes of many people in these countries are increasing, and with rising incomes, demand for meat and dairy products is also increasing. The increased demand is predicted to continue through to 2050 (Delgado *et al.* 1999). This presents a major opportunity for livestock industries in tropical and sub-tropical environments where ruminant species consume pastures that have few alternative economical uses.

To capture these opportunities, livestock enterprise and industry efficiency must increase by 2.0-2.5% p.a., equivalent to doubling outputs from constant resources over the next 35 years (Mullen 2012). Due to pressures on agriculture in developed countries and the developing world's requirement for high volumes of low-cost food, much of that increased production must occur in the regions of greatest need i.e. Africa and Asia. This increased demand for food is leading to greater competition for inputs such as land, water, fertilizer, grain and labour, driving up costs of livestock production. Climate change is predicted to add to the challenge (Hughes 2003), requiring livestock that are productive under hotter and drier climates and, in the tropics and sub-tropics, requiring animals which tolerate increased ecto- and endo-parasitic burdens and vector-borne diseases.

To double outputs from constant resources by 2050, farmers need to adopt cost-effective, transformational technologies for use in animals that are well adapted to their production environments. Traditional technologies delivering incremental changes will assist in improving productivity, but use of genomic information in proven genetic improvement methods could accelerate the required productivity improvements by increasing the rates of genetic gain for all economically important traits in livestock.

Genomic selection is the use of genome-wide genetic markers to estimate the genetic merit of individual animals (Meuwissen *et al.* 2001). Genome wide markers are required to capture variation from all the mutations affecting complex traits such as yields and fertility. Recently Meuwissen *et al.* (2016) reported that most economically important traits in livestock are affected by somewhere between 2,000 and 10,000 genes. Single Nucleotide Polymorphisms (SNPs) are the markers most commonly used in genomic selection, owing to low cost of genotyping and ability to genotype tens or hundreds of thousands of SNP in a single assay. Genomic selection therefore estimates the effect of all the SNP on the target trait simultaneously. Genomic selection is now transforming animal and plant breeding across developed countries globally, with enormous economic benefits. However there are no known examples of the successful use of genomic selection in grazing livestock in developing countries. This paper examines the constraints to use of genomic selection in those countries and explores opportunities to overcome them particularly in ruminants (beef and dairy cattle, sheep and goats) grazed at pasture in the tropics and sub-tropics where the greatest opportunities for productivity improvements also exist.

CONSTRAINTS TO USE OF GENOMIC SELECTION IN DEVELOPING COUNTRIES

In this paper we exclude ongoing development of the genomic selection methodology per sé as there are highly-competent research groups around the world undertaking such development. We are confident they will overcome any methodology issues, probably well ahead of development of solutions to other constraints that discourage the use of genomic selection in developing countries. Several of those constraints are also common to extensive livestock production systems in developed countries. The constraints and possible solutions are discussed briefly below.

Lack of phenotypes recorded in accurately-defined contemporary groups. In developed and developing countries, the major constraint to use of genomic (and conventional) selection in livestock is the difficulty and expense of accurately identifying appropriate fixed effects and contemporary groups and measuring the full range of economically important productive and adaptive traits required to achieve a well informed and balanced breeding objective. As discussed by Burrow and Henshall (2014), technology may in future provide a way of measuring animals, but it cannot replace the statistical imperative that, for the measurements to be meaningful, contemporary groups of appropriate structure and sufficient size are required. In extensive production systems and in developing countries, this presents difficulties in both managing and routinely recording large cohort groups. However if the design is inadequate in terms of contemporary group size and structure, the measurements will not provide useful predictions of genetic merit. This is perhaps the greatest constraint for smallholder farmers in developing countries, as often they own only a small number of breeding animals and they generally practice year-round joining.

To overcome this constraint, reference populations which are specifically designed to accurately manage and record animals within contemporary groups and capture data for the traits of interest have been established in some developed countries to exploit the opportunities provided by genomic selection. Examples of these populations in beef cattle are described by Upton *et al.* (2001) for growth, feed efficiency and carcase and beef quality and Burrow *et al.* (2003), Barwick *et al.* (2009) and Johnston *et al.* (2009) for the full range of productive and adaptive traits in the breeding objective. Van der Werf *et al.* (2010) and Swan *et al.* (2012) describe similar populations designed to capture data for a range of productive attributes in meat and wool sheep. A large study in the USA also developed specific populations to record resistance or susceptibility to Bovine Respiratory Disease in both beef and dairy cattle (BRD CAP 2017).

In future, Meuwissen *et al.* (2016) anticipate that accuracy of within-breed genomic selection will be achieved by use of very large within-breed reference populations. Alternately, genomic selection may be applied across-breeds, with accuracy obtained from across-breed reference populations and high-density genomic selection methods focusing on causative genomic regions discovered through programs such as the 1000 bull genome project (Hayes *et al.* 2014). In their opinion, it is highly likely that future applications will increasingly turn towards across-breed genomic selection (Meuwissen *et al.* 2016). This has considerations for application of genomic selection to developing countries as discussed in a later section of this paper.

Pedigrees and relationships. The benefits from knowledge of pedigree are well understood, allowing progeny test or BLUP-based selection in preference to selection on phenotype. In other than an intensive management system though, the costs are significant. Genomics provides a genuine alternative to the labour-intensive practices of single-sire mating and mothering-up, with the proviso that genotyping is cost-effective. The cost of the cheapest DNA assays on the market are now of the same magnitude as the cost of obtaining a tissue sample and of data management, and the cost of moderate-density SNP assays is not much more. With a thousand-SNP panel, parentage assignment is trivial. With a slightly larger (tens of thousands) SNP panel such as those used for genomic selection, it is not necessary to estimate discrete pedigrees at all, with relationships based solely on genomic similarity (Meuwissen *et al.* 2001).

Creating genetic linkages across livestock populations. Establishing large reference populations where expensive or hard-to-measure traits are routinely measured in accurately-defined contemporary groups requires that genetic linkages be created with the seedstock and commercial livestock populations targeted for genetic improvement. The best way of achieving these genetic linkages is through widespread use of artificial insemination (AI), though there are difficulties with AI programs under both extensive production systems and in developing countries. However as has been shown in the beef industry in northern Australia, where beef producers are prepared to put in the effort, successful AI programs are feasible. Another option for consideration where breeding populations are within relatively close proximity is to rotate sires amongst the breeding herds/flocks so genetic linkages are created through natural mating. Very importantly, the concept of genetic linkage changes with genomic information. When genomic information is available, what is needed is for chromosome segments to be represented across herds and environments, not sires or relatives per se. This is one of the advantages of genomic selection: it may be much easier to have chromosome segments, from a common ancestor quite a number of generation ago, represented across herds and environments, rather than progeny of link sires.

Need for consistent trait definitions across livestock populations. When designing resource populations for use in genomic selection programs, consideration must be given to trait definitions, to ensure that animals in multiple populations are recorded for the same trait(s). Alternately the resource populations need to be large enough to allow estimation of genetic correlations with indicator traits, if consistent recording of the same trait(s) cannot be achieved across all populations. Again, estimating these genomic correlations and genotype by environment interactions becomes more straightforward with genomic information, as what is required is observations of the traits/environments on common chromosome segments, rather than sires progeny (Visscher *et al.* 2104; Hayes *et al.* 2016).

Lack of infrastructure and human capacity. Two problems of major significance in developing countries are: a) the lack of infrastructure required to undertake all aspects of a genomic selection program including on-farm management and phenotyping, laboratory testing of animal samples, data capture and storage and lack of computing facilities etc.; and b) lack of human capacity, particularly in areas of technological capability and data analysis and interpretation.

Possibility of Genotype x Environment interactions. Livestock breeders, and particularly those in developing countries in the tropics, need to be cognisant of the possibility of Genotype x Environment (GxE) interactions which could arise from vastly different markets and/or production systems. An earlier review of the literature concluded that GxE interactions were problematic if poorly adapted breeds were compared across temperate and tropical environments, but they were unlikely to be a problem in tropically adapted beef cattle and sheep grazed in either temperate or tropical environments (Burrow 2012). However that review was based on production systems that aimed to optimise female reproductive performance and achieve premium meat market specifications from sale animals. It did not include consideration of vastly different market endpoints e.g. targeting high-value meat markets in developed countries *cf.* production of high volumes of low-cost meat in the beef, sheep and goat industries of many developing countries. A similar example can be found in the dairy industry, where 'high-performing' genetics now used very widely across developed countries globally are often found to be sub-optimal when used in low-input dairy farms in developing countries.

Difficulties of negotiating collaborative research agreements. An important constraint which must be considered during development of any research collaborations is the much greater difficulty and complexity of completing formal research agreements in developing countries (*cf.* those in developed countries). In the authors' experiences in different developing countries, completion of the formal agreements can sometimes take several years longer than anticipated (which also often means the research needs in the draft agreements have substantially changed in the interim period).

OVERCOMING THE CONSTRAINTS: PHENOTYPING AND GENOTYPING USING BEEF CATTLE EXAMPLES FROM SOUTHERN AFRICA

As flagged by Meuwissen *et al.* (2016), improved accuracy of genomic selection will be achieved in future by use of very large within- or (more likely) across-breed reference populations. The within-breed approach has already proved very successful in dairy and other livestock industries in developed countries. However it is not clear how farmers in developing countries could establish such reference populations due to a lack of both funding for phenotyping and genotyping and technical capacity to design and manage the populations within appropriate contemporary groups. Maiwashe and Banga (2013) suggested that in terms of funding, genotyping and phenotyping should be considered as international and national responsibilities respectively. But livestock farmers in developed countries had already adopted a more commercially-oriented 'user-pays' approach. Subsequently though, the South African government (through its Technology Innovation Agency - TIA) initiated a 'Beef Genomics Program' (BGP) in 2014 (and a similar program for dairy cattle in 2016 and potentially also for sheep and goats in future) in conjunction with seedstock breeders in South Africa and other Southern African countries, with the aim of developing within-breed reference populations designed to capture the phenotypes and genotypes necessary for genomic selection.

Beef Genomics Program (BGP) in Southern Africa. Currently, seedstock cattle breeders in Southern Africa use different genetic evaluation services and breeders aligned with the different service providers therefore use different approaches to phenotyping and genotyping. Under TIA funding guidelines, each breed society develops its own strategy with respect to use of genomic information. In 2016, 12 cattle breeds (Afrikaner, Beefmaster, Bonsmara, Boran, Brahman, Charolais, Drakensberger, Hereford, Limousin, Santa Gertrudis, Simbra and Simmental) were actively participating in the BGP with Brangus, Nguni and Tuli anticipated to participate from 2017 (Becker, 2016).

Beef cattle breeds aligned with SA Stud Book (26 breeds comprising 60% of registered beef cattle in South Africa, with the Bonsmara being the predominant breed; Stud Book, 2017) are currently focusing on genotyping primarily Bonsmara, Beefmaster and Drakensberger animals which already have Estimated Breeding Values (EBVs) for growth, feed conversion ratio and reproduction traits. Growth and reproduction phenotypes are recorded on the properties of birth of the animals, while feed efficiency is measured on selected bull calves from each on-farm weaning cohort in central bull testing stations. This approach means the phenotyping costs for existing phenotypes are met directly by the seedstock breeders, whereas the genotyping costs are met by the BGP. As a result, proof-of-concept for genomic selection will be demonstrated with the accuracy of available EBVs being marginally increased in these breeds with the inclusion of genomic information. However phenotypes for other economically important traits such as carcass and meat quality and adaptation are not currently recorded in those breeds, though it is likely they will be included in the anticipated next round of funding in 2018.

The Livestock Registration Foundation (LRF) represents the remaining 40% of registered beef cattle in South Africa, 80% of registered beef cattle in Namibia and 100% of registered beef cattle in Zimbabwe. These breeds have collectively agreed to undertake a 3-year program to collect data in structured contemporary groups using AI or natural mating across designated seedstock herds. All breeds are working at establishing a biobank for storage of DNA samples and genomic reference populations for each participating breed. The LRF breeds are focusing on establishing new phenotypes for traits which are economically important to measure (e.g. feed efficiency, carcass and meat quality and non-traditional measures of reproduction), as well as developing stronger genetic linkages between and within breeds across countries. They also intend to examine the potential for cross-continent genetic evaluations and cross-breed genomic evaluations (Becker 2016).

One genuine opportunity for Southern Africa, to overcome the difficulty of maintaining very large resource populations with accurate phenotypes and matching genotypes for all economically important traits derived from accurately-defined contemporary groups, would be to deliberately create genetic links with existing beef cattle resource populations in developed countries such as Australia. Australian resource populations could include: a) the Beef Information Nucleus (BIN) herds which comprise a nation-wide progeny test program for five cattle breeds, developed collaboratively by seedstock breeders and cattle breed societies in conjunction with Meat and Livestock Australia (MLA; Beef Information Nucleus 2017); and b) the MLA-funded '*Repronomics*®' project which is building on the cattle and extensive phenotypic and genotypic databases from the previous Beef CRC herds (Johnston *et al.* 2009) to specifically develop new female reproductive traits using a combination of female and male reproductive traits, novel molecular genetics approaches and innovative application strategies (Johnston 2016). This type of collaboration may also have the added benefit of addressing and at least partially overcoming the lack of laboratory infrastructure that is a common constraint in developing countries.

If it was possible to achieve cross-country collaborations to allow pooling of phenotypes based on common definitions of traits and matching genotypes derived from genetically-linked and accurately-defined contemporary groups, significant benefits would be created for the cattle industries of all partner countries due to a previously unstated constraint that breeders of tropical beef cattle in Australia and Africa have few alternatives to link with other cattle breeds in a similar way that breeders of temperate breeds have done in developed countries to maximise the numbers of animals, thereby sharing the costs of phenotyping and genotyping more broadly, whilst also significantly increasing the accuracy of genomic selection. However significant new research would be needed to benefit the full range of economically important traits and livestock breeds.

OVERCOMING OTHER CONSTRAINTS

Genetic/genomic linkages across livestock populations. Assuming cross-country collaborations can be negotiated as suggested above, there will be a need at the outset to specifically design the extent of genetic and genomic linkages required across the different resource populations.

Use of consistent trait definitions. Generally 'traditional' phenotypes such as weights and weight gains tend to be consistent across different resource populations. However for most economically important traits there will be a need for further research to either estimate genetic correlations between alternative measures of the traits or to re-define measurements using common definitions to allow valid use of genomic selection across the populations. In beef cattle this would require investigation of alternative measures of male and female reproductive traits, bull traits as indicators of male and female reproduction, alternative measures of carcass and meat quality attributes, feed intake and feed efficiency and cost-effective methods of measuring cattle resistance to environmental stressors such as parasites, diseases and high temperatures and humidity.

Lack of human capacity. As indicated by Maiwashe and Banga (2013), where genomics research has occurred in developing countries, it has tended to focus on 'low-hanging fruits' that are also relatively low-cost e.g. use of SNP data to select against genetic defects, breed characterisation, selection for individual heterozygosity to manage inbreeding, parentage verification and individual SNP associations with phenotypes based on relatively small numbers of animals, with the latter primarily being part of PhD-level training. However the technical capacity required to design and fully implement a genomic selection program is now largely lacking amongst livestock researchers in many/most developing countries, primarily because to date there simply has been no need for such expertise. If it was possible to collaborate with developed countries to implement genomic selection across resource populations as suggested above, then the lack of technical expertise would become a critical deficiency. That deficiency could be overcome either by outsourcing the essential services to a developed country with appropriate expertise (not desirable from a developing country perspective) or implementing intensive training programs, ideally with staff from the developing countries visiting international laboratories to undergo the essential training. A particular need identified in Southern Africa is that training in genomic selection and selection indexes needs to occur across all levels from university through to technical and industry levels.

Technical capacity is also required for the development of some phenotypes such as ovarian and carcass ultrasound scanning, measurement of indicators of male reproductive performance, animal body-condition scoring, measurement of meat quality attributes etc. as well as mentoring and quality assurance training for the intermediaries who will ultimately become responsible for training farmers and farm workers.

PRIORITY RESEARCH AREAS

Assuming it is possible to overcome most of the constraints identified above (as we believe to be the case), several priority research areas would need to be addressed to develop and implement a genomic selection program for breeders of grazing livestock across developing and developed countries. The first priority is to establish large reference populations, with animals measured for the target traits and genotyped in appropriate environments. Then priorities include:

Cross-country genetic/genomic evaluations. Estimates of genetic/genomic relationships would need to be developed for the full range of economically important traits included in the pooled phenotypes and genotypes. This would include examining the scope for combining data across countries for multi-trait genetic/genomic evaluations at a trait by country level to inform an understanding of the differences and similarities of traits not recorded identically in different countries. There would also be a need to construct joint G matrices to inform the capacity for joint single-step analysis. Assuming it is feasible, joint single-step analyses could then be undertaken to deliver prototype cross-country joint evaluations. A logical next-step from joint single-step analyses would be the coordinated use of young sires initially within breeds across countries and possibly through shared use of the MateSel program (MateSel 2017). Protocols for the coordinated use of young sires would be informed by the joint G matrix and joint evaluation results.

Use of sequence data in genetic evaluations. A number of research areas could be undertaken to promote the use of genomic selection in developing countries. Specifically with regard to potential collaborative beef research across Australia and Africa described above, this might include:

- Development of a new low-cost SNP chip for use in *Bos taurus*, *Bos indicus* and tropically adapted *Bos taurus* breeds (the southern African Sanga breeds and the West/East African taurine breeds). This would allow an improvement in the accuracy of genomic EBVs across the range of cattle breeds, composites and their crosses used on both continents;
- Development of computationally efficient genomic evaluation algorithms that utilise whole

genome sequence data suitable for multi-breed and crossbred evaluations. As demonstrated by Kemper *et al.* (2015) and Macleod *et al.* (2016), using 50K SNP genotypes and BLUP methodology does not enable genomic estimated breeding values that work across breeds and in crossbreeds/composites (i.e. do not result in increased accuracy in composites when information from the founder breeds are included). Much higher density of markers (up to whole genome sequence) enables multi-breed predictions where breeds not in the reference set, or with only limited numbers in the reference set, can achieve more accurate genomic evaluations;

- Detection of embryonic lethal and other deleterious mutations in the major breeds used in Africa and Australia based on a haplotype analysis to examine if there are regions in the genome where some haplotypes are never observed in a homozygous state, despite the frequencies of these haplotypes being high enough that multiple animals are expected to be homozygous for the haplotypes.

Multi-breed genomic evaluations. Within breeds, accuracy of genomic selection depends on the number of animals in the reference population and strength of linkage disequilibrium (LD) and family relationships between the reference and selection candidates. Across breeds, factors such as differences in LD, allele frequencies and SNP effects between breeds also impact on the accuracy. Pooling reference populations across breeds appears to be a promising method to increase the size of the reference population, particularly in numerically smaller populations, with the proviso the populations being pooled are not genetically distant (Kizilkaya *et al.* 2010). This component of the research would: i) use deterministic methods to determine the prediction accuracy with smaller numbers of genotyped animals before incurring high costs of large-scale genotyping; and ii) undertake a formal breeding program design to determine the feasibility of pooling data across multi-breed populations across continents. Assuming it is feasible to pool data across breeds, the research would also examine options to implement a single-step, multi-breed genomic evaluation.

Selection indexes and GxE interactions. This would involve two elements: i) definition of production system x target market examples on a breed x country basis to determine the extent of differences/similarities between and across objectives in the different regions; and ii) if appropriate, extension of the selection index modelling to include new traits relevant to the production marketing systems for the particular grazing livestock.

APPLICATION OF RESULTS TO SMALLHOLDER FARMERS IN SOUTHERN AFRICA

Once genomic selection is implemented in southern African commercial herds, it will then be relatively straightforward to transfer the benefits of improved genetic gain to smallholder farmers using bulls and semen from superior sires in schemes similar to the ARC's '*Kaonafatso ya Dikgomo*' (KyD – animal recording; KyD 2017) scheme in South Africa. Currently the KyD scheme is assisting smallholder farmers across all provinces in South Africa to continually improve their cattle production through recording and monitoring productivity and profitability and providing advice on production, animal health and marketing.

OPPORTUNITIES TO EXTEND THE COLLABORATIONS TO OTHER DEVELOPING COUNTRIES AND OTHER LIVESTOCK SPECIES

Assuming results from the multi-breed genomic evaluations mentioned in the research section above indicate feasibility, there is good potential to extend the BGP concept to other African countries such as Kenya and West Africa in partnership with the International Livestock Research Institute. However the cattle breeds commonly used in East and West Africa are not the same as those participating in the BGP across Northern Australia and Southern Africa, so achieving an expansion in tropically adapted beef resource populations to other countries will also depend on the devlopment of the proposed SNP panel for African and other cattle breeds as described above.

Opportunities also exist to expand the concept to other livestock species, including planning already underway to establish sheep resource populations across Australia and South Africa.

ACKNOWLEDGEMENTS

Heather Burrow would like to acknowledge the significant impact that Dr John Vercoe had on her research, her career and her career directions over close to three decades through his always good-humoured ways of challenging, managing, mentoring, cajoling and cheerleading. She is very honoured to dedicate this paper to his memory.

REFERENCES

Barwick S.A., Wolcott M.L., Johnston D.J., Burrow H.M. and Sullivan M. (2009) *Anim. Prod. Sci.* **49**: 351.

Becker J. (2016) www.sastudbook.co.za/images/photos/DrJohnBecker.pdf

Beef Information Nucleus (2017) https://www.mla.com.au/globalassets/mla-corporate/generic/about-mla/beef-information-nucleus.pdf

BRD CAP (2017) Available online at: http://www.brdcomplex.org/

Burrow H.M. (2012) *Animal* **6**: 729.

Burrow H.M. and Henshall J.M. (2014) *Proc. 10th World Cong. Genet. App. Livest. Prod.* Paper 44.

Burrow H.M., Johnston D.J., Barwick S.A., Holroyd, R.G., Barendse W. *et al.* (2003) *Proc. Ass. Adv. Anim. Breed. Genet.* **15**: 359.

Delgado C., Rosegrant M., Steinfeld H., Ehui S. and Courbois C. (1999) *Internat. Food Policy Res. Inst.* **8**: 1.

Gerland P.E., Raftery A.D., Sevcikova H., Li N., Gu D. *et al.* (2014) *Science* **346**: 234.

Hayes B.J., Macleod I.M., Daetwyler H.D., Bowman, P.J., Chamberlain A. *et al.* (2014) *Proc. 10th World Cong. Genet. App. Livest. Prod.* Paper 183.

Hayes B.J., Daetwyler H.D and Goddard M.E. (2016) *Crop Science* **56**: 2251.

Hughes L. (2003) *Aust. Ecology* **28**: 423.

Johnston D.J., Barwick S.A., Corbet N.J., Fordyce G., Holroyd R.G., Williams P.J. and Burrow H.M. (2009) *Anim. Prod. Sci.* **49**: 399.

Johnston D.J. (2016) Banana Stn. Field Day *http://www.brahman.com.au/bin_info_downloads.html*

Kemper K.E., Reich C.M., Bowman P.J., Vander Jagt C.J., Chamberlain A.J., Mason B.A., Hayes B.J. and Goddard M.E. (2015) *Genet. Sel. Evol.* **47**: 29.

Kizilkaya K., Fernando R.L. and Garrick D.J. (2010) *J. Anim. Sci.* **88**: 544.

KyD (2017) https://www.youtube.com/watch?v=Jo13IyYLX38

Maiwashe A. and Banga C.B. (2013) *Proc. South African Soc. Anim. Sci.*, Bloemfontein, 23-26 June.

MateSel (2017) http://breedplan.une.edu.au/index.php?option=com_content&task=view&id=64

Macleod I.M., Bowman P.J., Vander Jagt C.J., Haile-Marian M., Kemper K.E., Chamberlain A.J., Schrooten C., Hayes B.J. and Goddard M.E. (2016) *BMC Genomics* **17**: 144.

Mullen J. (2012) *Aust. Farm Inst. Research Report* pp. 5-24.

Meuwissen T., Hayes B.J. and Goddard M.E. (2001) *Genetics* **157**: 1819.

Meuwissen T., Hayes B.J. and Goddard M.E. (2016) *Animal Frontiers* doi:10.2527/af.2016-0002.

Stud Book (2017) http://www.sastudbook.co.za/dc36/sa-stud-book/breeders'-societies/beef-cattle/beef-cattle-breeders'-societies.html

Swan A.A., Johnston D.J., Brown D.J., Tier B. and Graser H.-U. (2012) *Anim. Prod. Sci.* **52**: 126.

Upton W., Burrow H.M., Dundon A., Robinson D.L. and Farrell E. (2001) *Aust. J. Exp. Ag.* **41**: 943.

Van der Werf J.H.J., Kinghorn B.P. and Banks R.G. (2010) *Anim. Prod. Sci.* **50**: 998.

Visscher P.M., Hemani. G., Vinkhuyzen A.A., Chen G.B., Lee S.H., Wray N.R., Goddard M.E. and Yang J. (2014) *PLoS Genet.* 10(4):e1004269.

Proc. Assoc. Advmt. Anim. Breed. Genet. 22:361-364

BREEDPLAN IN A GENOMICS WORLD – OPPORTUNITIES AND CHALLENGES

H. Nivison

Agricultural Business Research Institute, University of New England, Armidale NSW 2354

SUMMARY

The Agricultural Business Research Institute has been providing estimates of breeding values to cattle breeders for over 40 years. In that time, it has progressed from within herd, to BREEDPLAN, Group BREEDPLAN, and multi-country analyses. For most of that time, the data analysed was phenotypic, with genomic data only included fairly recently via blending.

In April 2017, ABRI released a production BREEDPLAN run to the Australian Brahman Breeders' Association using Single-Step methodology. This full multi-trait analysis, using pedigree, performance and genomic data simultaneously, was a world first for beef cattle.

The incorporation of genomic data into routine Single-Step BREEDPLAN runs opens many opportunities to cattle breeders around the world, including the potential to significantly enhance the accuracy of analyses, allow breeders to make more accurate selection decisions, and therefore increase the rate of genetic gain. However, there remain significant challenges to overcome before these opportunities can be fully realised.

INTRODUCTION

This presentation is intended to examine the practical implications, opportunities and challenges for cattle breeders resulting from the introduction of Single-Step BREEDPLAN analyses. It is not a scientific paper. There are other technical papers available detailing Single-Step BREEDPLAN methodology in these proceedings.

Hugh Nivison has been Managing Director of ABRI since October 2015. He is not a geneticist, coming from a background of sheep and cattle breeding in northern NSW. Hugh holds a Bachelor of Veterinary Science from the University of Sydney, and is an Adjunct Associate Professor at the School of Veterinary Medicine with the University of Queensland. His career has been spent working in agricultural production (primarily livestock) in Australia and overseas.

The Agricultural Business Research Institute (ABRI) is responsible for commercialising the BREEDPLAN suite of software, holding the exclusive licence from the owners of the software, Meat and Livestock Australia (MLA), University of New England (UNE) and New South Wales Department of Primary Industries (NSWDPI). The software is developed by the Animal Genetics and Breeding Unit (AGBU), a joint venture of UNE and NSWDPI, with funding from MLA.

ABRI provides genetic analyses for 84 discrete beef cattle breed associations or clients, many as part of combined multi-country analyses (Trans-Tasman Angus, Pan-American Hereford, Southern-African Brahman for example). ABRI and AGBU are collaborating in two MLA Donor Company funded projects to expand capacity to provide multi-country analyses, and to investigate the technical and data limitations on providing full multi-trait, multi-breed analyses.

DISCUSSION

The practical outcome for any genetic analysis system must be increased rate of genetic gain through more accurate and more timely selection decisions on farm. The key components of that desired outcome are:

1. More accurate predictions
2. Earlier predictions
3. On-farm adoption

OPPORTUNITIES

- **Pedigree**
  - Provided animals have genomic data, Single-Step BREEDPLAN accurately assigns relationships based on the true genetic comparisons of the animals. Whereas relationships have traditionally been described as ½, ¼, 1/8 etc. they can now be described more accurately. Half-sibs can range from 0.16 to 0.34 (vs 0.25) allowing increased accuracy from the BLUP calculation
  - Single-Step BREEDPLAN identifies errors in existing pedigrees that have gone uncorrected previously. Significant improvements in accuracy result from correct parentage assignment, and elimination of previously unknown pedigree errors.
  - Single-Step BREEDPLAN can fill the blanks for some animals where there was no previously recorded pedigree if the parent/s and offspring have genomic information. Some breed societies already require DNA parent verification, but this process can add accuracy for those breeds where this does not currently occur.
  - Accurate assignment of genetic relationships is a key factor in producing accurate BLUP analyses. Single-Step BREEDPLAN allows for greater accuracy in pedigree than was previously available.
- **Hard to Measure (HTM) traits**
  - Many economically important traits are difficult to measure on animals retained as seedstock sires and dams. Carcase and long term fertility traits are obvious examples.
  - Animals that are related genomically, although perhaps not by pedigree, to animals that have the phenotypic records for HTM traits will be able to receive EBVs for those traits if the relationship and accuracy is high enough.
  - Early selection of replacement females using fertility traits generated by Single-Step BREEDPLAN can significantly increase the rate of genetic gain for those traits as opposed to waiting for the animal to generate phenotypic data.
- **Animals with no performance**
  - Animals with no phenotypic records can receive accurate EBVs for a wide variety of traits provided they have a genomic result, and are closely enough related to animals with phenotypic data in the analysis.
  - Dairy heifer selection is an excellent example of this practice, but its application in beef is likely to be less as there is not the dominance of small numbers of sires as in dairy herds.
- **Combining discrete data sets**
  - Data sets that currently have no linkage via pedigree may be used to inform a Single-Step BREEDPLAN analysis via their genomic linkage. Abattoir data combined with BREEDPLAN data holds an exciting prospect of better informing both seedstock and commercial cattle selection systems.
  - Highly accurate carcase EBVs for breeding animals, and highly accurate feed efficiency estimates for feedlot cattle are some of the possibilities.

CHALLENGES

As attractive as the opportunities are for cattle breeders, there remain some disincentives that may prevent uptake of the technology.

- (mis)Understanding
  - There will be some major EBV changes, and some of those will be for well-known and well used sires. While the changes will have justification (pedigree correction for example), some breeders will see this as an example of the analyses being unreliable.
  - Single-Step BREEDPLAN is currently designed to work for pure-bred animals. During the analysis, a Breed percentage is calculated based on an individual's relationship with a reference population representative of the various breeds existing in Australia. Animals less than a threshold (80% for example) are excluded as not being purebreds. Unfortunately, some breeders have seen this as meaning animals less than 100% are not "Pure" and are attempting to use this analysis procedure for political gain.
  - Managing expectations will remain a challenge. Some breeders expect that they will now be able to just pull a tail hair and receive full BREEDPLAN EBVs without any data collection irrespective of how closely they are related to other animals in the analysis.
  - Breed Societies and others charged with administration of the pedigree, performance and genomic databases will have an increased responsibility for ensuring the accuracy of these data sources. Potential errors in pedigree, or breed will need to be investigated, and if confirmed will need to be rectified. Telling a breeder the recorded pedigree for his well-used sire is incorrect will be an uncomfortable role for breed society staff. This is likely to be a short term issue until the various inconsistencies are resolved.
- Cost
  - SNP data is currently expensive to collect when considering the sample collection and testing charges together. As volume increases for laboratory testing, costs can be expected to decrease, but the on-farm cost will remain similar. While cattle prices remain buoyant, producers are likely to embrace the technology. However, if they are forced to prioritise discretionary expenditure in a downturn, they may reconsider their participation in genomics, particularly on a whole-herd scale.
  - There will need to be some consideration given to the differing influence of phenotypic and genomic data in the Single-Step BREEDPLAN analyses, including the option of differential pricing structures based on the value of the contribution from different data sources. Accurate collection of phenotypic data, particularly for HTM trait will need to be encouraged, potentially via financial incentives.
- Reduced phenotypic recording
  - The beef industry in general, and breed societies in particular will need to actively ensure sufficient, accurate and linked phenotypic performance data continues to be collected to enable the Single-Step BREEDPLAN analyses. Systems including Reference populations and BINs can provide this data, but are expensive to operate, and will likely require industry funding as they are usually beyond the financial abilities of individual breeds. Innovative alternative methods of collecting and generating this data in a more cost-effective manner should be investigated.

  - The education of seedstock and commercial breeders on the importance of phenotypic data, and possible financial incentives for collection will need to be a focus for both breed societies and the wider industry.
- No of tested animals
  - Breeds with smaller populations may initially struggle to implement Single-Step BREEDPLAN due to their low numbers of total genomic results. The GBLUP method of Single-Step BREEDPLAN requires the generation of a Genomic Relationship Matrix (GRM) which may have stability issues at low level of results.
- Technical
  - The sheer volume of extra data generated by combining genomic results with existing pedigree and performance datasets will lead to challenges in both storage and transmission. Breed societies and the BREEDPLAN service will need to develop innovative methods for ensuring efficient and cost effective data management during Single-Step BREEDPLAN runs.
  - The technical complexity of combining datasets for multi-country or multi-breed analyses is further complicated by the addition of genomic data. Many of the current BREEDLAN analyses are conducted on a multi-country basis, and further expansion of this service is planned. BREEDPLAN is also exploring opportunities for robust multi-breed, multi-trait analyses and the associated complexity of merging the pedigree, performance and genomic datasets involved.
  - The much greater computational requirements of Single-Step BREEDPLAN could have potentially slowed the speed of analysis down considerably. Innovative development by AGBU has ensured that the full multi-trait Single-Step BREEDPLAN analyses run in a time comparable with existing BREEDPLAN runs. Further enhancements in software and hardware are planned, and will be essential as many BREEDPLAN clients move to more frequent evaluations.

BREEDPLAN in a genomics world (Single-Step BREEDPLAN) will provide cattle breeders with more accurate analyses, leading to more accurate selection decisions, and resulting in increased rates of genetic gain. The challenges involved in achieving this outcome are not insignificant, but the rewards of adoption should ensure the opportunities will be realised.

Proc. Assoc. Advmt. Anim. Breed. Genet. 22:365-368

AN UPDATE ON GENETIC PROGRESS IN THE AUSTRALIAN SHEEP INDUSTRY

A.A. Swan[1], R.G. Banks[1], D.J. Brown[1] and H.R. Chandler[2]

[1] Animal Genetics Breeding Unit*, University of New England, Armidale, Australia
[2] Sheep Genetics, Meat and Livestock Australia, Armidale, Australia

SUMMARY

Substantial genetic progress has been made by the Australian sheep industry in the era of across-flock genetic evaluation which began in the early 1990s. Rates of gain in standard indexes increased throughout the 2000's for maternal breeds and terminal sires, but have plateaued or slightly decreased since 2010. For Merinos, the rate of gain has remained relatively constant over the same period. Average rates of gain for each breed group are currently as high as 94% of "potential" gain for terminals sires, 84% for Coopworth and maternal composite flocks, 49% for Border Leicesters, and up to 47% for Merinos. However, the top 20% of breeders are exceeding potential gain for all breed groups except Border Leicester, and it is the poor performance of the bottom 20% of breeders which leads to lower performance on average for a breed group, particularly for Merinos.

INTRODUCTION

The Australian sheep industry has made substantial and measurable genetic progress since the advent of across-flock genetic evaluation systems in the early 1990's, underpinned by pedigree and performance recording in ram breeding flocks and extensive use of artificial insemination. The effectiveness of selection has varied across different breed groups, with Swan *et al.* (2009) finding that terminal sire breeds were achieving 110% of "potential" gain, maternal breeds up to 79%, and Merinos 33%. In this study we present an updated analysis of genetic progress for the major Australian sheep breed groups.

MATERIALS AND METHODS

The main across-flock genetic evaluations for Australian sheep are currently conducted within three breed groups, maternal breeds, Merinos, and terminal sire breeds (Brown *et al.* 2007). Genetic trends were estimated from analyses published in January 2017, averaging estimated breeding values (ASBVs) and index values by year of birth, starting at 1989, the first year where significant numbers of animals were available, and ending at 2015, the most complete recent cohort. For terminal sires, the breeds selected where Poll Dorset, Dorset, Texel, and White Suffolk, while for Merinos, flocks of Australian Merino or Australian Poll Merino origin were selected, excluding central test sire evaluation flocks. For maternal breeds, Border Leicester (BL) was considered separately to Coopworth and maternal composite flocks (CM), because the former are used in a production system based on crossbred ewes, while the latter are used in self-replacing production systems.

Trends were calculated for six standard indexes, the Maternal Dollar index (MATDOL) for both BL and CM; Dual Purpose Plus (DPP), Merino Production Plus (MPP), and Fibre Production Plus (FPP) for Merinos; and Carcass Plus (CPLUS) and Lamb 2020 (LP2020) for terminal sires. Because these indexes are presented to breeders expressed on different scales, the results were scaled by the standard deviation of each breeding objective.

Rates of gain in indexes were calculated in sliding 10 year windows, by regressing average index value on year of birth. So for example, the rate of gain for the year 2000 was the estimated slope of the regression for years 1991 to 2000.

Rates of gain in indexes were compared to potential rates of gain based on deterministic selection index predictions. The assumptions used in these calculations were full pedigree recording, with the traits typically recorded by breeders in each breed group, including birth weight (BWT), weaning weight (WWT), post-weaning weight (PWT), yearling weight (YWT), adult weight (AWT), eye muscle (EMD), fat (FAT), worm egg count (WEC), fleece weight (CFW), fibre diameter (FD), staple strength (SS), and number of lambs weaned (NLW). A summary of the traits included for each index is shown in Table 1, along with index accuracy based on the traits measured, and potential gain per year. The latter was calculated as the index accuracy multiplied by i/L, where i is the selection intensity and L the generation interval, both calculated from recent data in the evaluation databases for each breed group.

Table 1: index accuracy, potential gain per year (per standard deviation of objective), and traits measured to calculate index accuracy.

Index	Accuracy	Potential gain	Traits measured to calculate index accuracy
MATDOL	0.45	0.199	BWT, WWT, PWT, EMD, WEC, CFW, NLW
DPP	0.30	0.078	YWT, AWT, EMD, CFW, FD, SS
MPP	0.36	0.094	YWT, AWT, CFW, FD, SS
FPP	0.44	0.116	YWT, AWT, CFW, FD, SS, WEC
CPLUS	0.61	0.230	WWT, PWT, EMD, FAT
LP2020	0.60	0.228	BWT, WWT, PWT, EMD, FAT, WEC

Trends were also calculated for individual flocks currently in the evaluation, restricted to those with more than 50 progeny per year with ASBVs, and 7 or more cohorts present out of the most recent 10. There were 38 BL flocks and 20 CM flocks in the maternal analysis, 138 flocks in the Merino analysis, and 274 flocks in the terminal sire analysis. The rate of gain for each flock was calculated between 2006 and 2015, and compared to the potential gain. These results were summarised within the top 20% of flocks and bottom 20% of flocks.

Finally, the contribution of individual traits to index gain over time was calculated by estimating the rate of gain for each trait within the sliding 10 year windows described above, multiplying by the relative economic value, and expressing as a percentage of the total index gain.

RESULTS AND DISCUSSION

Index trends in Figure 1 show that substantial genetic progress has been achieved since 1989. Most is observed for the CPLUS index in terminal sires (approximately 5 Standard Deviations), followed by LP2020. For the MATDOL index, CM have increased by over 3 SD, and BL by 1.5 SD. Merinos have made the least gain, between 0.5 and 1 SD.

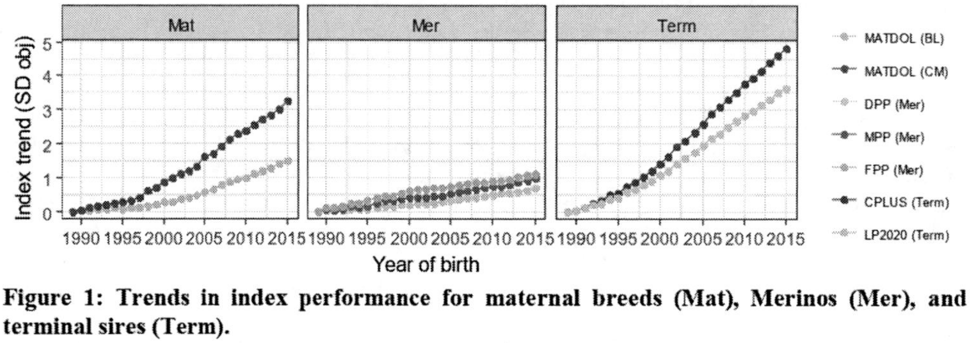

Figure 1: Trends in index performance for maternal breeds (Mat), Merinos (Mer), and terminal sires (Term).

As shown in Table 1, Merino indexes have the lowest index accuracies and potential genetic gain, and this is accounted for in Figure 2. In addition, Figure 2 shows that the rate of gain in terminal sires and the maternal breeds increased substantially through to approximately 2010, but has since plateaued, or perhaps declined slightly in the case of terminal sires. For Merinos, the rate of gain has been slowly increasing for the DPP and MPP indexes from a low point in 2005, while the rate of gain has been declining for the FPP index, which reflects a reduction in emphasis throughout the industry on fibre diameter, the dominant trait in this index.

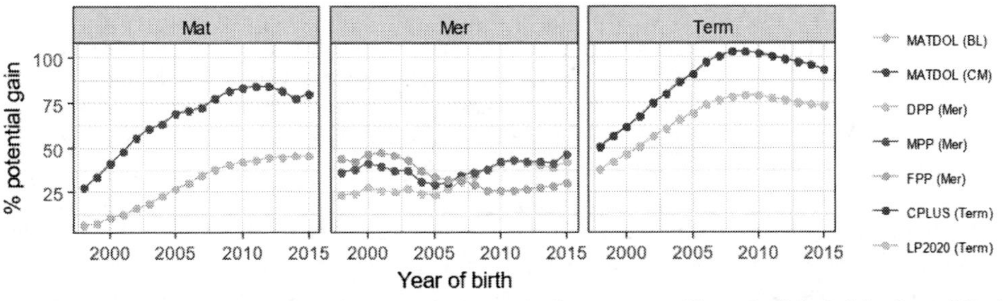

Figure 2: Percentage of potential rate of index gain for maternal breeds (Mat), Merinos (Mer), and terminal sires (Term).

While a plateauing of the rate of gain may not necessarily be an issue, it needs to be noted that the potential rates of gain we have used here have been deliberately set at a conservative level in order to match the recording programs and population structures within the current databases. Exceeding these potential gains is possible, by increasing selection accuracy with better recording programs, optimising breeding programs, and/or utilisation of across-flock and across-breed differences in performance. As shown in Table 2, the top 20% of breeders exceed the potential gain in most cases. In addition, for all cases with the exception of CPLUS in terminal sires (94% of potential gain) and MATDOL in CM (84%), the realised gain is substantially lower than the potential: 49% for MATDOL in BL, and 43, 47, and 37% for DPP, MPP, and FPP in Merinos.

Table 2: Percentage of potential genetic gain for top 20% of flocks, bottom 20% of flocks and mean across flocks.

Percentile	MATDOL (BL)	MATDOL (CM)	DPP (Mer)	MPP (Mer)	FPP (Mer)	CPLUS (Term)	LP2020 (Term)
Top 20%	82	125	102	111	91	134	107
Bottom 20%	19	39	-4	-6	-3	45	35
Mean	49	84	43	47	37	94	73

Possible reasons for not achieving potential rates of gain include firstly the lack of measurement of key traits in the index. This is the case for the LP2020 index in terminal sires, where the measurement of WEC is not common in ram breeding flocks. Secondly, selection may be taking place on traits outside the index, for example, in Merinos there has been a move from horned to poll Merinos over the last decade, and possible selection on breech and other visual traits. In terminal sires, ASBVs for eating quality traits have been available since 2011, and these are negatively correlated to some traits in the CPLUS and LP2020 indexes. An additional factor in maternal breeds is that maternal composites and Coopworths have had greater ability to capitalise on across-breed effects than Border Leicesters. However, taking these points into account, it seems that the biggest issue is simply that selection is ineffective in some flocks. On average, selection in Merinos and BL is much less effective than in CM and terminal sires, but the top 20% of flocks are approaching and exceeding the potential gain in all breeds (Table 2). It is poor performance in the bottom 20% of flocks which is reducing average gain, particularly for Merinos. This means that one strategy to increase industry gain would be to focus on improving poor performing flocks, which could be achieved by better application of relatively simple technology. Sheep Genetics is introducing a program working with individual breeders through the "RAMping Up Genetic Gain" project in collaboration with NSW DPI, AGBU, and the Sheep CRC. Through this program, breeders can assess the quality and quantity of information in their breeding program, determine how effectively it can be used in the evaluation analysis, and consider how well it is then used to make selection decisions. It should also be noted that while the differences between the top and bottom flocks are shown to be large, we have not assessed how the industry impact of different flocks may vary.

The changing contribution of individual traits is shown for the main indexes (those which are achieving the most gain for their respective breeds) in Figure 3. In the early phases of across-flock evaluation, body weight (WWT and PWT) made the biggest contribution to gain for maternal breeds and terminal sires. Since 2005, the contribution of eye muscle depth (EMD) has increased for both breed groups, as well as reproduction (NLW) for the maternal breeds. This is partly due to measurement of these traits becoming more common, and has contributed to increases in the overall rates of gain seen in Figure 2. In the CPLUS index, fat depth (FAT) has a negative relative economic value, because the breeding goal has essentially been to improve lean meat yield. The influence of fat on the index has been declining, partly because of periodic updates to CPLUS to reduce the relative economic value, and because of a positive genetic correlation between FAT and EMD.

In the early phases of evaluation for Merinos, fibre diameter was the dominant trait, but from 2005 the influence of body weight (YWT) began to increase. Note that adult body weight in Merinos (AWT) is making a negative contribution to index gain because although the genetic trend is positive, the relative economic value is negative due to the impact of the trait on ewe flock feed costs. In terms of overall response, higher adult weight is favourably associated with early growth and reproduction. In a third phase from 2010, emphasis on fleece weight (CFW) has substantially increased, and this is related to a small negative contribution to the index from fibre diameter (FD) caused by a small positive trend in the trait.

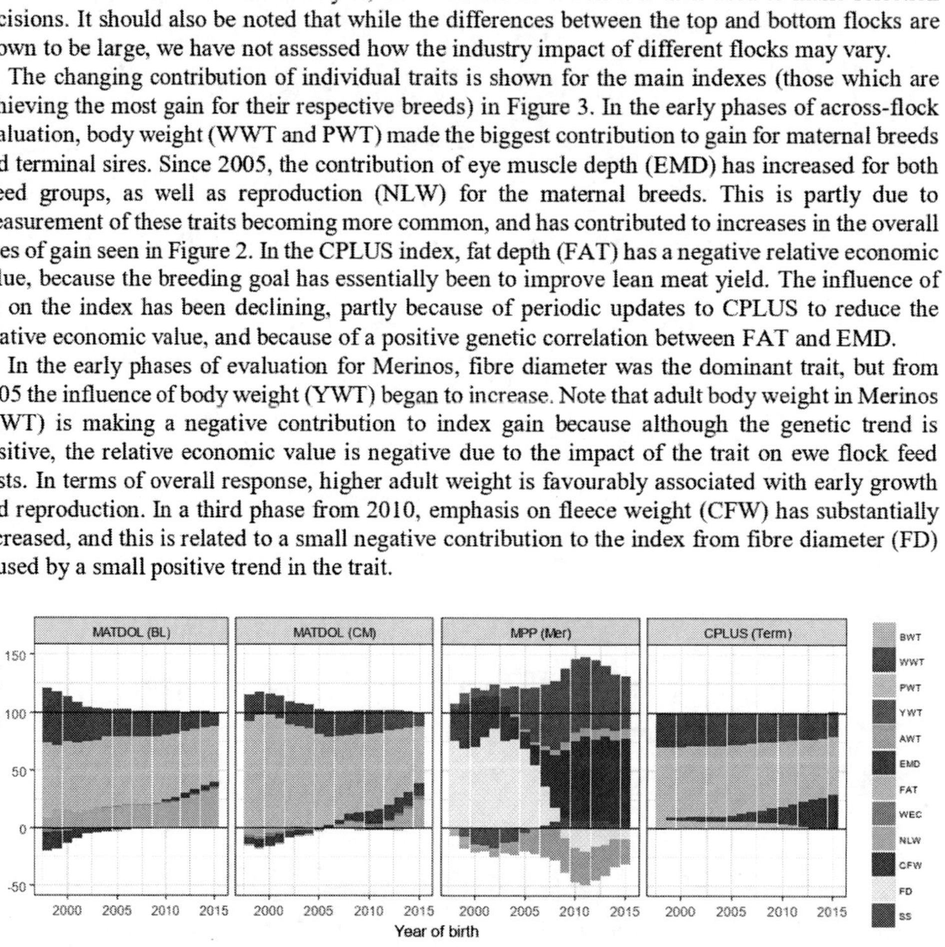

Figure 3: Changes in the contribution of individual traits to gain in selected indexes (Maternal Dollar index (MATDOL) for Border Leicester (BL) and maternal composite flocks (CM), Merino Production Plus (MPP), and Carcass Plus (CPLUS) for terminal sires over time.

REFERENCES

Brown, D.J., Huisman, A.E., Swan, A.A., Graser, H-U., Woolaston, R.R., Ball, A.J., Atkins, K.D., and Banks, R.G. (2007) *Proc. Conf. Assoc. Anim. Breed. Genet.* **17**: 187.

Swan, A.A., Brown, D.J., and Banks, R.G. (2009) *Proc. Conf. Assoc. Anim. Breed Genet.* **18**: 326.

Proc. Assoc. Advmt. Anim. Breed. Genet. 22:369-372

ECONOMIC VALUE OF GENOMIC SELECTION IN A VERTICALLY INTEGRATED BEEF CATTLE PRODUCTION SYSTEM

J.W. Buchanan[1], M.D. MacNeil[2,3], R.C. Raymond[3], A.R. McClain[3], and A.L. Van Eenennaam[1]

[1]University of California, Davis, CA, 95616
[2]Delta G, Miles City, MT 59301 and University of the Free State, P.O. Box 339, Bloemfontein 9300, South Africa
[3]J.R. Simplot Land and Livestock, Grand View, ID, USA

SUMMARY

Two genetic evaluations were carried out for a vertically-integrated beef production operation running a 600 cow Charolais purebred herd providing terminal sires for a 10,000 commercial cow herd to determine the effect of genotyping on accuracy of genetic predictions, rate of genetic gain, and discounted revenue for breeding sires. Genetic evaluation A included phenotypes from purebred and crossbred offspring and a pedigree containing purebred relationships as well as sire-calf relationships for crossbred calves derived from parentage assignment. Genetic evaluation B contained the same information as A, with the addition of 9K genotypes for purebred and crossbred animals. Genotyping resulted in an 11.9% increase in the average accuracy (R_{TI}) of the estimated breeding values (EBV) over parentage assignment alone. Gene flow methodology was used to estimate the cumulative discounted expressions (CDE) resulting from the selection of a genetically superior purebred (PB) and commercial (CM) sire. Additional discounted revenue derived from the increased accuracy due to 9K genotyping in genetic evaluation B was \$465 for a CM sire, and \$10,355 for a retained PB sire. The cumulative net present value (CNPV) over a 20 year planning horizon was \$9,400,910 and \$17,930,183 for scenario A and B, respectively, assuming 25% of the CM progeny were assigned parentage or genotyped annually at a cost of \$15/parentage assignment or \$35/9K genotype. These estimates assume the value from genetic improvement is returned to the enterprise. In this scenario genotyping PB selection candidates and some proportion of CM progeny resulted in a positive return on investment over parentage assignment alone.

INTRODUCTION

Adoption of genomic technology in the beef cattle industry provides an opportunity to accelerate genetic gain and increase income (Meuwissen *et al.* 2013). In a vertically integrated production system there is opportunity to capture additional profit generated from genotyping by implementing a genetic evaluation using some combination of phenotypes, pedigree information, and genotypes (Aguilar *et al.* 2010). Increased genetic gain from genotype information results from an increase in the accuracy of the prediction of genetic merit and the reduction of generation interval through genomic selection on young unproven sires (Todd *et al.* 2014). The objective of this study was to compare the accuracy of genetic evaluations obtained from pedigree relationships derived from parentage with those obtained when using ~9K genotypes. A secondary objective was to calculate the estimated additional economic returns associated with the accelerated genetic gain in both scenarios. Inference was to a genetic evaluation program for a vertically-integrated, two-tiered beef cattle production system producing Charolais terminal sires for 10,000 commercial cows.

MATERIALS AND METHODS

Data used in this study consisted of records from a Charolais purebred herd combined with feedlot and carcass performance records from their crossbred calves finished at a common feeding facility. Historic pedigree information (n=8,361 pedigree records) was available for the purebred herd. During the course of three years, 568 purebred Charolais bulls and 8,776 crossbred calves were genotyped using a combination of the GeneSeek Genomic Profiler LD (26k) v.1.1-4 and HD (76k) SNP arrays, which allowed for the reconstruction of sire-calf pedigree relationships. There were 8,549 SNP markers common to all animals in the evaluation after trimming for call rate ($\geq$0.90) and removing markers on sex chromosomes. Phenotypes collected from purebred Charolais bulls included birth weight (BW), weaning weight (WW), GrowSafe (GrowSafe Systems, Ltd, Airdrie, AB Canada) dry matter intake (DMI), average daily gain (ADG), ultrasound 12TH rib fat depth (URFAT), ultrasound intramuscular fat percentage (UIMF), and ultrasound ribeye area (UREA). Phenotypes collected from crossbred calves finished in the feedlot included WW (collected at feedlot arrival), DMI, 12th rib fat depth (FAT), marbling score (MARB) determined by image analysis (VBG 2000 E+V, Oranienburg, Germany), and carcass ribeye area (REA).

Pedigree-based and genomic EBVs were estimated using the single-step approach to simultaneously evaluate genotyped and non-genotyped animals with pedigree information. Inclusion criteria for purebred sires to be evaluated in this study were a recorded pedigree relationship, a genotype, ultrasound and DMI records, and at least one recorded crossbred calf. For evaluations where crossbred carcass traits were available, ultrasound indicator traits were analyzed in a bivariate animal model (MacNeil *et al.* 2010). Beef Improvement Federation (BIF) EBV accuracies were calculated using standard errors derived from single-step GBLUP according to BIF guidelines (2015).

$$BIF_{acc} = 1 - \sqrt{\frac{Prediction\ error\ variance}{Additive\ genetic\ variance}} \qquad [1]$$

This accuracy was then transformed to an approximation of the correlation between the true and estimated breeding value (R_{TI}):

$$R_{TI} = \sqrt{1 - (1 - BIF_{acc})^2} \qquad [2]$$

The average accuracy in each genetic evaluation was also used to estimate the expected genetic gain in profit per year ($\Delta G_{\$/yr}$) using economic index coefficients for each trait:

$$\Delta G = \sum_{j=1}^{n} \alpha_j\ I\ R_{TI_j}\ \sigma_{A_j} \qquad [3]$$

Where α_j equals the economic value for trait *j* ($/marketed crossbred carcass), *I* equals the selection intensity of PB or CM bulls, R_{TI_j} equals the average accuracy for trait *j*, σ_{A_j} equals the additive genetic standard deviation for trait *j*. Selection intensities were chosen to create a replacement rate that would maintain the current population structure.

The gene flow method of Hill (1974) was utilized to estimate the cumulative discounted expression (CDE) resulting from the selection of a genetically superior CM or PB sire using the population structure, age classes, and selection intensity for a combined 600 cow nucleus herd and 10,000 commercial cow production system as described (Van Eenennaam *et al.* 2011). A discount rate of 5% and a 20 year planning horizon was used to determine the present value resulting from the future expression of production traits after selection decisions have been made. The discounted revenue derived from CM sires was estimated as:

$$\$CM = \frac{\Delta G\ x\ CDE_{CM}\ x\ No.of\ commercial\ cows}{No.\ of\ yearling\ commercial\ bulls\ retained} \qquad [4]$$

Similarly, the discounted revenue derived from PB sires was estimated as:

$$\$PB = \frac{\Delta G\ x\ CDE_{PB}\ x\ No.of\ commercial\ cows}{No.\ of\ PB\ sires\ retained} \qquad [5]$$

In addition, cumulative net present value (CNPV) over a 20 year planning horizon was calculated using the accuracies and resulting rate of genetic gain, along with the internal rate of return (IRR).

RESULTS AND DISCUSSION

Table 1 displays the population structure and breeding system assumptions used in this study.

Table 1. Population structure and system parameters commercial (CM) and purebred (PB).

Parameter	Assumed Value
No. of PB bull calves born each year	231
PB bull:cow ratio	1:25
No. of PB cows	600
No. of PB bulls selected each year	10 (4.3%, $i = 2.11$)
No. of bulls selected as CM bulls	154 (69.7%, $i = 0.50$)
CM bull:cow ratio	1:20
No. of CM cows	10,000
Age structure of PB bulls (2 to 4 yr)	0.41, 0.33, 0.26
Age structure of CM bulls (2 to 5 yr)	0.34, 0.27, 0.22, 0.17
Age structure of cows in CM herd	0.2, 0.18, 0.17, 0.15, 0.12, 0.09, 0.05, 0.03, 0.01

Average accuracy of EBVs for the traits included in the genetic evaluation are shown in Table 2 for the 248 purebred Charolais bulls that were included in this comparison. Genetic evaluation A and B contained the same pedigree and phenotypes, but genotypes were added to genetic evaluation B. The addition of genotypes to the genetic evaluation resulted in an 11.9% increase in the average accuracy over parentage assignment alone.

Table 2. Accuracy (R_{TI}) of Genetic evaluation. Evaluation (A) contained phenotypes from purebred and crossbred animals and a pedigree derived from SNP parentage assignment and (B) with the addition of 9K SNP genotypes

	Accuracy of Genetic Evaluation	
Trait[1]	A	B
D2H	0.474 ±0.008	0.490 ±0.005
DMI	0.569 ±0.018	0.648 ±0.006
FAT	0.628 ±0.008*	0.749 ±0.008*
HCW	0.602 ±0.009	0.721 ±0.005
MARB	0.621 ±0.011*	0.712 ±0.005*
REA	0.666 ±0.008*	0.725 ±0.004*
UFAT	0.597 ±0.003*	0.606 ±0.006*
UIMF	0.661 ±0.002*	0.712 ±0.002*
UREA	0.630 ±0.004*	0.655 ±0.003*
WW	0.699 ±0.006	0.711 ±0.004
YG	0.635 ±0.009	0.745 ±0.004

*Bivariate model with carcass traits evaluated with ultrasound indicator trait.

[1]D2H = days to harvest, FAT = carcass backfat thickness, MARB = camera-based marbling score, UFAT = ultrasound backfat thickness, UIMF = ultrasound intramuscular fat, UREA = ultrasound ribeye area.

The discounted revenue from genetic evaluation is shown in Table 3. Additional discounted revenue derived from accuracy due to genotyping in genetic evaluation B was $465 per CM sire. If 154 CM sires are retained each year (69.7%, $i = 0.50$) as breeding males that produce commercial offspring, then the annual economic return becomes $71,610 on an enterprise basis.

Increased discounted revenue derived from genotyping for PB sires was $10,355 If 10 PB sires are retained as herd sires (4.3%, $i = 2.11$) this value becomes $103,550 on an enterprise basis. The total discounted revenue derived from genotyping is then $175,160 for the enterprise per year. Additionally, the CNPV derived from estimates of genetic gain in $/PB bull in both genetic evaluations was estimated over the first 20 years of selection. Assuming a cost of $15/parentage test and $100 or $45 in phenotyping costs for PB and CM animals, respectively, the initial investment for scenario A was $591,880. Assuming a cost of $35/test for the GGPLD and the same phenotyping costs, the initial investment for scenario B was $778,760 (Table 3).

Table 3. Discounted revenue per year and cumulative net present value (CNPV) for purebred (PB) and commercial (CM) sires in genetic evaluation scenarios A and B.

Genetic Evaluation	Income/expense source	N	Expense	Income	Total
A	CM progeny phenotyping ($100/hd)	1875	$187,500		
	PB bull phenotyping ($100/hd)	231	$23,100		
	Parentage ($15/hd)	2106	$31,590		
	Startup cost		$591,880		
	20 year CNPV				$9,400,910
	20 year internal rate of return				9.2%
B	CM progeny phenotyping ($100/hd)	1875	$187,500		
	PB bull phenotyping ($100/hd)	231	$23,100		
	Genotyping ($35/hd)	2106	$73,710		
	Startup cost		$778,760		
	20 year CNPV				$17,930,183
	20 year internal rate of return				16.3%
Difference in CNPV after 20 yr					$8,529,273

This example assumes 25% of the CM progeny and 100% of the PB males were assigned parentage or genotyped each year of the 20 year period. Scenarios A and B reach breakeven value after 12 and 10 years of selection, respectively. Scenario B also generates approximately $8,529,273 additional cumulative revenue over 20 years. These economic returns may be inflated as they were based on single trait accuracies which may be overestimated as they did not account for information that might be provided from correlated traits.

The CNPV estimate suggests a positive return on investment can be derived from 9K genotyping young PB selection candidates and a portion of the CM progeny in this two-tiered beef cattle production system as compared to a genetic evaluation using a pedigree containing purebred relationships and sire-calf relationships for crossbred calves based on parentage analysis alone.

ACKNOWLEDGEMENTS

This project was supported in part by funds from the USDA Agriculture and Food Research Initiative Competitive Grant no. 2011-68004-30367 from the USDA National Institute of Food and Agriculture, the W. K. Kellogg endowment and the California Agricultural Experiment Station of the University of California–Davis. The authors would like to acknowledge GeneSeek®, a Neogen Corporation Company, for in kind contributions to this project, and both Tom Granleese and Julius van der Werf for providing spreadsheets to calculate gene flow and cumulative net present value.

REFERENCES

Aguilar I., Misztal I., Johnson D.L., Legarra A., et al. (2010) *J. Dairy Sci.* **93**: 743-752.

Beef Improvement Federation Guidelines for Uniform Beef Improvement Programs. (2010) 9th ed. Raleigh, NC: Beef Improvement Federation; 2010. 182 p.

Hill W.G. (1974) *Anim. Prod.* **18**: 117-139.

MacNeil M.D., Nkrumah J.D., Woodward B.W., and Northcutt S.L. (2010) *J. Anim. Sci.* **88**: 517-522.

Meuwissen T., Hayes B., and Goddard M. (2013) *Annu. Rev. Anim. Biosci.* **1**: 221-237.

Todd D. L., Roughsedge T., and Woolliams J.A. (2014) *Animal* **8**: 388-394.

Van Eenennaam A.L., van der Werf J.H., and Goddard M.E. (2011) *J. Anim. Sci.* **89**: 307-320.

Proc. Assoc. Advmt. Anim. Breed. Genet. 22:373-376

THE DISTRIBUTION OF RISK AND REWARD IN EXTENSIVE LIVESTOCK IMPROVEMENT SYSTEMS, THEIR CONSEQUENCES AND POSSIBLE RESPONSES

R.G Banks

Animal Genetics and Breeding Unit (a joint venture of NSW Department of Primary Industries and the University of New England), University of New England, Armidale, NSW 2351, Australia

SUMMARY

A range of players or sectors make investments into genetic improvement in the extensive livestock industries, but overall returns are heavily dependent on decisions made by bull- (and ram-) breeders. They in turn rely on sales of genetic material to cover their own investments and maintain profitability. Some broad-scale characteristics of the investments and of the markets for genetic material are reviewed, leading to the observation of very high uncertainty in those markets. This uncertainty is almost certainly acting as a brake on genetic progress, and some possible approaches to reduce the uncertainty are considered. Such approaches will aim to improve efficiency of the market for genetic material, and will need to be designed to be robust, transparent and simple and cheap to apply.

INTRODUCTION

For any livestock improvement system, return on investment must be satisfactory both to maintain profitability and to fund further improvement. For single, vertically-integrated enterprises, this should be straightforward in terms of accounting and response, and the same conclusion likely applies for breeding enterprises in multi-vendor situations, provided that they have enough scale and market share. For multi-enterprise industries, such as beef cattle and sheep in Australia, things may not be so straightforward. Discussion of opportunities, and research, is often quite sensibly focussed on technical questions, which can ultimately be summarised as how to achieve higher accuracy of selection for a given investment and with possible constraints on inbreeding. Increasingly however, and especially when there is a mix of public, collective and private investment, some focus is on how to improve investment return: in simple terms, how to increase the rate of genetic progress?

In livestock industries such as beef cattle and sheep, investment in genetic improvement can be grouped into two categories: performance recording, and research, development and extension. The returns ultimately accrue as increased value of sales of products and margins distributed in some way through the value chain, or at worst reduced rate of decline in the real value of these things. Who makes the investments, and what returns they receive, could have important effects.

This paper focusses on these distributions of risk and reward, and explores whether they might be affecting rates of genetic progress, and if so, what responses might be considered. This paper focusses on the beef industry, but the limited available evidence suggests similar patterns apply in sheep.

RATES OF PROGRESS ACHIEVED

Previous reports of genetic progress being achieved in the beef and sheep industries indicate that:

a) Averages are consistent with, or higher than, those achieved in other countries (Swan 2009, these proceedings)
b) There is wide variation in rates within and between breeds (Johnston, 2007).

Adoption rates, estimated as proportion of sires entering the market either with BVs themselves, or sired by animals with BVs, are moderate to high.

Together, these observations suggest that the technologies and the forms in which they are offered/provided, enable very satisfactory genetic progress. Given the opportunity costs of genetic progress being slower than potential, it is reasonable to ask what causes the difference.

INVESTMENTS AND RETURNS

Data collection and availability of total investments and returns is limited, but some high level approximations can be used. Table 1 summarises estimates for the beef industry for the decade leading to the end of the Beef CRC.

Table 1: Estimated annual investments in Research, Development and Extension/Adoption and Implementation in the Australian beef industry, 2002-2012

Stage	Estimated Annual Investment ($m)			Funding Source %		
	Cash	In-kind	Total	Breeders	Producer levies	Gov'ts
Strategic research	$7.48	$8.10	$15.58		6%	94%
Applied RDE	$1.00		$1.00		50%	50%
Extension to breeders	$0.50		$0.50	42%		58%
Implementation	$2.50		$2.50	50%		50%
Routine evaluation	$0.60		$0.60	100%		
Data collection in studs	$3.75		$3.75	100%		
	$15.83	**$8.10**	**$23.93**			

The key observation from this summary is that overall investment is substantial, and it is sharply divided between pre-implementation Research, Development and Extension/Adoption, borne predominantly by tax-payers and producers, and implementation, which is borne predominantly by breeders. This raises the question of how returns flow to the different investors.

In the case of estimated returns, there are 3 sources (breed, herd, and within-herd):

- Estimates of returns to sectors, using Equilibrium Displacement modelling (Zhao et al, 2000). These estimate that the distribution of returns from improved productivity, or marketing (which can be treated as a proxy for improved product quality) is approximately 30% to producers, and approximately 67% to domestic and international consumers (Table 2). Note that the form such benefits (to either the consumer, or ultimately the producer and breeder) take when consumers essentially pay the world price may not be clear, and some of the estimated consumer benefits may accrue to local land values, to the benefit of domestic producers.

Table 2: Distribution of Returns (%) by sector R&D investment into improved production or improved demand (Zhao et al, 2000)

Sector	Beef	
	Production Research	Promotion Research
Producers	24-34	20-30
Feedlots	0.1-0.2	0.3
Processors	1	1
Retailers	4	4-7
Domestic Consumers	50-55	50-65
Overseas Consumers	8-9	5-12

- Data from breed societies on bull sales, at the breed and herd level.
    - Breed level: at the time of writing, only a very limited sample of such data has been analysed: average prices for Angus bulls across years, and average prices for Angus studs

within a year. The former analysis suggests a relatively constant relationship between the average price across all bulls sold in a year and the prevailing price for young cattle. The ratio of average bull price in dollars to Eastern Young Cattle indicator in c/kg across the period 1996-20104 is 10.8:1. Using a standard carcase weight and number of lifetime progeny, this ratio equates to 5.4% of total on-farm earnings per bull (and therefore approximately 1.6% of total value chain income per bull).

- Herd level: the regression of herd average bull price in 2014 (adjusted for season and state) on herd average merit for $Index for the 2012 drop, in Angus herds across Australia is = ($ 17.26 x Herd Average Merit) – 1813, with an r-squared of 6.6%. Certainly this is a small sample, but there are two interesting aspects in these numbers: firstly, that the regression – the amount paid to the breeder per $Index point, is close to half the proportion of the value-added per bull received by the commercial producer, meaning that the bull-buyer is on average sharing the "rewards" from genetic improvement with the bull breeder; and secondly, that there is a great deal of variation around the regression, meaning that there is considerable uncertainty for the breeder about how much reward he/she will receive for the genetic improvement generated and offered for sale.
- Data from samples of individual bull sales: Van Eeenannam (pers. comm.) analysed data from individual stud sales for Angus studs for regression of price on index value, within stud, and found regressions ranging from $80-160 per index point, with r-squared values in the range 20-26%. Similarly, analysis of data from 8 Angus studs for the 2016 selling year reveal regressions of sale price on $Index across studs averaging $88 extra per $Index point (range $34 to $134) and r-squared averaging 19% (range 7% to 32%). The data for these sales also show very variable but sometimes strong relationships of price with weight of sale day (not shown).

Some caution is needed in considering the meaning of these results:

- it is likely that buyers of Angus bulls have the most appreciation of BREEDPLAN information, reflecting the focus on the technology and its extension over many years,
- the price data includes stud and herd bull sales, in varying proportions across studs, but this is likely to bias upwards both the across-herd and within-herd regressions of price on merit,
- the studs for which individual regressions have been investigated could be realistically described as technology leaders, and so their clients may not be representative of the average bull-buyer.

IMPLICATIONS

Overall investment into genetic improvement in the beef industry can be categorised into "pre-commercialisation" investment (R&D, extension) and commercialisation or implementation investment. The former creates potential for genetic progress and hence wealth generation, the latter converts potential into reality.

The distribution of the investment into these 2 categories almost perfectly maps to 2 categories of investors: the former to tax-payers and commercial producers acting collectively, and the latter to bull breeders. In general analyses of return on investment in genetic improvement show very favourable long-term and industry- or community-wide returns. However, such outcomes are completely dependent on the behaviour of the breeding sector, or more precisely on the large number of heterogeneous agents who comprise it: – how much they invest, how they invest and how they make selection decisions. Those decisions all ultimately depend on the returns obtained from selling genetic material – and more particularly on both the expected level of return and the uncertainty about that expectation.

The limited analysis of sale results for Angus cattle has two main messages. First, there is a relatively stable relationship between overall average price of bulls and the current price of commercial cattle. It is as if the market estimates and applies some function of steer price in determining the price it is willing to pay for bulls. Second, there is some variation around this overall average, related to herd merit and animal merit within herd. So, the market can be said to be paying for genetic merit, albeit with very wide variation about how much it pays. And further, there is very wide variation around the across- and within-herd regressions, indicating considerable uncertainty for the bull breeder. This has most relevance to genetic improvement in relation to the across-herd regression of average price on merit. The low r-squared implies that obtaining a reliable return on investment in (additional) herd recording and faster genetic progress cannot be guaranteed.

In these circumstances, it is not unreasonable to expect under-investment (the utility value of an investment being its expectation minus a function of its uncertainty, and the uncertainty in this case is clearly high), and potentially some excess caution in decisions relating to breeding direction and selection differential. Both these responses limit the overall returns, for all investors. Accordingly, there would be value in industry and community consideration of strategic mechanisms to reduce the uncertainties. Note that this does not relate to industry and community support for RDE – those investments are to create potential. The mechanisms needed are to in some way reduce the uncertainty and/or improve the returns – which can almost be considered as modifying the implicit license under which the knowledge and tools of genetic evaluation and improvement are made available to the breeding sector, which in turns supplies a service called genetic improvement.

Currently, the rest of industry incentivizes the breeding sector via the market for genetic material, which the data reviewed here suggests to have very low efficiency in its signalling – there is a lot of noise in the market, as reflected by the low and variable r-squared values. One approach to changing the terms of the transaction between breeders and the rest of the value chain, developed in the grains industry of this country, is to collect an end-point royalty on production, and funnel that back to breeding companies. This levy can be used to offset breeding program costs. This approach goes part way towards solving the problem, but the link between returns and genetic merit is still filtered through market perceptions and knowledge informed by a system of variety comparison. This is essentially almost equivalent to funding breeders for their recording costs, although it does not deliver that specificity. This still leaves the returns for better breeding decisions uncertain. It would be worth considering whether other or additional mechanisms could be developed that more directly reward genetic merit, and thus increase the r-squared in relationships between price and merit.

These ideas may seem impractical or anti-market, but in fact groups of interested players or their agents define rules of operation for many markets. Leaving genetic improvement, and the wealth it can generate, solely to an imperfect market has only one merit, and that is that it is easy.

CONCLUSION

Markets for genetic material (beef in this case) are the means by which incentives for genetic improvement are delivered, and hence by which returns on investment generated. These markets appear to be very noisy despite signs of underlying rationality. Mechanisms to reduce that noisiness are worth investigation, as such reduction should improve incentives and hence overall returns.

REFERENCES

Johnston, D.J. (2007) *Proc. Assoc. Advmt. Anim. Breed. Genet.* **17**:8

Swan, A.A., Brown, D.J., and Banks, R.G. (2009) *Proc. Conf. Assoc. Anim. Breed Genet.* **18**: 326.

Swan, A.A., Banks, R.G., Brown, D.J., and Chandler, H.R. (2017) *Proc. Conf. Assoc. Anim. Breed Genet.* **22** *(in press)*

Zhao, X., Mullen, J.D., Griffith, G.R., Griffiths, W.E. and Piggott, R.R. (2000) Economic Research Report No. 4, NSW Agriculture, Orange.

Proc. Assoc. Advmt. Anim. Breed. Genet. 22:377-383

APPLYING NEW GENOMIC TECHNOLOGIES TO ACCELERATE GENETIC IMPROVEMENT IN BEEF AND DAIRY CATTLE

A. Canovas[1], J. Casellas[2], M. Thomas[3] and J. F. Medrano[4]

[1]University of Guelph, Centre for Genetic Improvement of Livestock, Department of Animal Bioscience, Ontario, N1G2W1, Canada. [2]Universitat Autonoma de Barcelona, Bellaterra, Spain. [3]Colorado State University, Fort Collins, CO 80521, United States. [4]University of California, Davis, CA 95616, United States

SUMMARY

In recent years, breeders have combined the use of phenotypic appraisal and the estimation of breeding values (PTA or EPD) to make genetic selection decisions in beef and dairy cattle that have resulted in a steady genetic gain of 2% per year. However, the most extensive application of genomics has occurred in the dairy industry with the estimation of molecular breeding values that has improved selection efficiency to a much higher order of magnitude. Despite a growing molecular and physiological understanding of complex traits, little is known about the genes determining the traits and their precise function, and a significant unexplained source of variation of phenotypes remains in livestock. Within this context, a more complete understanding of the genes and regulatory pathways and networks involved in economically important traits (i.e. fertility and reproduction, feed efficiency, meat quality and carcass traits) in beef and dairy cattle will provide knowledge to help improve genetic selection and reproductive management. Therefore, high throughput -OMICS technology (i.e., transcriptomics, metagenomics, metabolomics, as well as epigenomics amongst several others), will complement these tools and further advance identification of functional genes within a systems biology approach.

INTRODUCTION

The field of genetics and genomics in most of the livestock species has experienced a dramatic technological revolution in the last 5 years. During the earlier 2000's, the emphasis in livestock genetics was in linkage maps with single markers and quantification of some genes using real time qPCR. The first bovine genome assembly was published in 2009 (Elsik et al., 2009), and since that time, the development and use of various whole genome-omics tools has accelerated research in cattle genetics (Reverter et al., 2013; Snelling et al., 2013). After the development of the livestock genome sequences, new technologies such as microarrays were introduced allowing the study of gene expression of the entire transcriptome. With that, the concept of "Genetical Genomics" was developed with the idea of integrating structural and functional genomic data, combining gene expression from microarray technology with genotype data from marker genotypes. Currently, the field of Genetical Genomics or the integration of structural and functional genomic data, has expanded as high throughput tools have become available for genomic analysis such as high-density (HD) genotyping-chips (Illumina, San Diego, CA), whole genome sequencing and genotyping by sequencing and RNA-Sequencing to measure the gene expression in the entire transcriptome (in a more accurate way than microarray technology; Wickramasinghe et al. 2014).

Currently, with all the new available technologies in livestock combined with statistical methodologies, the integration of structural and functional genomics information with other – OMICS into a systems biology approach has allowed development of a better biological understanding of phenotypes. As part of the genomics tool box the HD-genotyping SNP chips such as 50K and 800K, whole genome sequencing technologies are now available in most of the livestock species and have been extensively utilized, such as in dairy cattle, in genetic improvement. As a part of high throughput tools available for genomic analysis, RNA-

Sequencing allows measuring not only gene expression, but also examining genome structure identifying SNP and other structural variation such as insertions, deletions and splice variants (Wickramasinghe et al. 2014; Cánovas et al. 2010). The expectation is that the integration of all these types of genomic data will accelerate the genetic improvement by improving accuracy of selection and reducing the generation interval.

Combining the information from the –OMICS technologies together with metabolic pathways and functional/biological analysis into a Systems biology approach allows the identification of functional SNP increasing the accuracy of selection. Briefly, trancriptomics using RNA-Sequencing technology measures gene expression of the entire transcriptome, as well as structural variation (i.e. SNP, insertions, deletions and splice variants in coding regions). Also, it allows the identification of the differential expressed splice variants affecting genes overlapping significant SNPs in significant QTL regions. Metagenomics, identifies and quantifies the microbial profiles and determine their functionality in specific bovine samples (i.e. rumen fluid) using 16SrRNA based profiling of microbial populations as well as partial protozoa and fungal 18S rRNA (Handerson et al., 2015). Metabolomics, identifies circulating plasma and/or rumen fluid metabolites as potential biomarkers to assess physiological and hormonal states relative to performance by Gas-chromatography-Mass-spectrometry (GC-MS) and NMR analysis. For each analysis, internal standards are included for normalization to allow comparisons between samples. Temporal and treatment changes in metabolites can be determined to be significant after normalization and can be correlated to the microbial profiles and gene expression obtained from RNA-Sequencing. The microbial profiles and/or microbial activities of each animal can be then defined and associated with the phenotype of interest using a multivariate approach.

The particular benefits of new integrated high throughput genomics technologies will be most likely for genetic improvement of those traits that are difficult to measure such as feed efficiency, methane emission and fertility and reproduction traits.

BEEF AND DAIRY CATTLE GENETIC IMPROVEMENT USING GENOMIC TECHNOLOGIES

In general, the beef industry is highly segmented (seedstock, cow-calf, feedlot, processor, retailer and consumer). This segmentation is seen in the way various stakeholders along the supply chain make logical decisions for their own business, yet in the light of the entire chain, these decisions are detrimental to the end product and consumer satisfaction. With little or no market signal from consumer to seedstock, needed changes are not made for the end product to be desired and competitive with other protein choices of the consumer. Poor information flow along the beef supply chain is leading to massive inefficiency as demonstrated in the dramatic decline in high yielding carcasses in the past decade (down from 66% to 41%), Canadian Beef Grading Agency data). This drop resulted from overfeeding in an attempt to increase the level of intramuscular fat (marbling). The past thirty years has seen a dramatic and consistent decline in per capita beef consumption, due in part as a result of inconsistent product, primarily in the important trait of tenderness. Over the past thirty years (and even more drastically the last 10 years), beef consumption per capita in Canada has declined from 27 to 19 Kg while that of chicken has risen from 21 to 33 kg.

The many studies of the beef industry completed in the past 20 years all identified lack of information flow as a serious impediment to the industry. This is more currently demonstrated by the inclusion of connectedness (information flow along and back the supply chain) as one of four target areas of focus in the Canadian Cattlemen's Association's (CCA's) national beef strategy (beefstrategy.com). The utilization of genomics in the beef industry has been limited (compare to dairy industry), due in part to the 1) complicating factor of multiple breeds and correspondence of SNP effects, 2) in general, the genetic evaluations are carried out by the multiple breed associations using different methods to calculate the EPDs, 3) reporting different EPDs for different breeds confusing the data comparisons from different evaluations (Van Eenennaam 2016). International Genetics Solutions (IGS) is trying to leverage a multibreed database that enables the comparison of EPDs across breeds associations (Chianina, Gelbvieh, Limousin, Maine-Anjou, Red Angus, Simmental, Shorthorn, Canadian Simmental, Canadian Gelbvieh, Canadian Limousin, Canadin Angus and Canadian Shorthom).

Genomic approaches offer an opportunity to accelerate the genetic improvement, but genomics requires accurate phenotypes (and genotypes) from genomically linked individual animals. Therefore, the focus is on collecting data to enable generation of GEBV for all animals in a population. Industry breeding strategies will then incorporate these new genetic evaluations to improve production efficiency and to provide a more consistent and competitive product to consumers. Also, beef producers often make breeding decisions without current knowledge of the science, existing systems and tools that they can use on farm to benefit from this research. Taken in isolation, a research finding can be confusing and lack any meaningful connection to the real world of production in which producers make decisions. Thus, producers require more concise, layman's terms presentation of the benefits of implementing change in their businesses, as well as a more clearly defined stepwise approach they can take to implement that change. Therefore, is important to do an effort to translate and transfer the research knowledge associated to genetics and the importance of collecting phenotypes (and genotypes) which is key to affecting change necessary to enhance and accelerate innovation in beef industry.

Using genomics-based approaches to improve beef genetics will result in increased production efficiency, making beef a more attractive protein option than current relative to poultry and pork. Effective use of genomics will also provide the consumer with a more consistently tender product and therefore help to address declining consumption. Therefore, it's important to capture and make use of data in the development of genomically enhanced Expected Progeny Differences (GEPD's) which are simply more accurate indicators of an animal's true genetic merit. Although there are phenotypes (related to meat quality and carcass traits) and genotype data collected, more phenotype and genotype data are required to enable generation of GEPDs to incorporate into beef industry breeding strategies. New sets of data will be crucial to enlarge the reference set of animals with genotype and phenotype, but most important, will provide the researchers with additional information collected at the whole chain scope, which will be unique to improve the quality of prediction equations and at the same time, enhance economical analysis of the desired traits.

GEPDs have a significant role to play in the effective selection for these traits. The selection efficiency with GEPDs will be higher with the knowledge gained from functional genomics studies. Despite a growing molecular and physiological understanding of complex traits such as carcass traits, little is known about the genes determining the traits associated to tenderness and their precise function, and a significant unexplained source of variation of phenotypes remains in beef. Within this context, a more complete understanding of the genes and regulatory pathways and networks involved in economically important traits such as tenderness and carcass traits in beef cattle will provide knowledge to help improve genetic selection. By leveraging -OMICS and systems biology, we can develop more robust approaches for genetic selection that can be incorporated to industry breeding plans for improving meat quality focusing on tenderness and carcass traits in beef cattle.

Large databases containing genomic and phenomic information, together with the data generated using the new genomic technologies (for example, gene expression data from RNA-Sequencing together with the results from GWAS (using the phenotypes and genotypes) could be subjected to genome wide association studies to identify quantitative trait loci (QTLs) for carcass quality which will assist in the ranking of functional variants. These analyses can be used to develop prediction equations for gEPDs which are incorporated into multi-trait value indices that perform well in commercial cattle.

Gains in prediction accuracy achievable by the addition of subsets of functional variants to existing panels can be quantified by imputing the variants into training and testing populations already genotyped. Information about variant function can be incorporated into genomic predictions through the use of the BayesRC method (MacLeod et al. 2014) or weighted genomic best linear unbiased prediction (GBLUP). Another important aspect to successfully exploit the power of applying the new genomic technologies to accelerate the genetic improvement in beef and dairy cattle is the annotation of the bovine genome. Although the first bovine genome assembly in 2009 and the subsequent releases in the last years, still there are several gaps and errors in the annotation of the current bovine genome. In order to improve the annotation and taking advantage of the results generated using the new –OMICS technologies available, there is the international initiative on the Functional Annotation of Animal Genomes (FAANG); a coordinated international action to accelerate Genome to Phenome (see web site for more detail: http://www.faang.org/index). The project is expecting to close the genotype-to-phenotype gap providing new information on genetic variants that explain variation in the target traits that can be used to increase the accuracy of the genomic predictions.

APPLYING GENOMICS TO IMPROVE BULL AND COW FERTILITY AND THEIR ASSOCIATION WITH FEED EFFICIENCY

Transmission ratio distortion (TRD) occurs when one of the two alleles from a heterozygous locus is preferentially transmitted to the progeny. This phenomenon typically causes a departure of the expected Mendelian inheritance ratios in the offspring. TRD has been reported in a broad range of organisms including plants, insects, fish, birds, and mammals. However, little is known of its effects on livestock species. Several biological mechanisms can cause TRD, including the preferential transmission of one of the two alleles carried by a heterozygote parent to the zygote at the time of fertilization, also known as meiotic drive, as well as embryo or fetal failure or differential viability during early neonatal life under a given genotype. Independent of the specific cause, TRD must be viewed as a genetic mechanism that can have important effects impairing fertility or viability in the early developmental stages. In this study the analysis of TRD will be used to discover new regions of the genome and the genes located in those regions that may have an effect on reproduction. Such regions, if they affect male and female fertility through embryonic or early calf mortality, are expected to change the distribution of genotypes among progeny in relation to the genotypes of the parents. Once TRD regions have been identified, their phenotypic effects will be analyzed by genetical genomics, integrating structural and functional genomic data through associations (GWAS) using the new bovine genome assembly ARS-UCD1.0 that is currently being developed with fertility phenotypes or genetic evaluations (e.g., sire conception rate, non-return rate and semen quality traits) and functional studies (gene ontology and biological pathways analysis).

In addition, the association between SNP located in the genes with TRD and feed efficiency traits will also be examined in order to study possible correlation between feed efficiency traits and fertility in young beef and dairy bulls. Although feed costs are a major factor influencing the profitability of beef production, successful reproduction in the cow-calf sector is a primary driver affecting profitability. The number and percentage of cows successfully bred during the breeding season is a major factor influencing the profitability of the cow/calf operation. Bull fertility plays a key role in the success of calf production. Identifying bulls with superior fertility and with superior feed efficiency could significantly impact cow-calf production efficiency. However, it is important to understand the relationships between fertility and feed efficiency to avoid undesirable consequences of selection for a single trait as the beef and dairy cattle industry is beginning to more aggressively select animals for improved feed efficiency.

Genomic regions with TRD will be identified as well as the key regulator genes and functional SNP affecting fertility and reproduction in 9 beef and 1 Dairy cattle breeds and its correlation with feed efficiency in young bulls. In order to identify the causal mutations, further validation of the most promising genes found will be investigated applying the new OMCIS technologies using the two groups of animals with extreme phenotypes (fertility traits). Regarding the studied beef breeds, populations with a ranged from 340 to 75,000 trios of genotypes (genotypes in both parental and offspring generations) will be used to identified the TRD regions in Angus (225,984), Beef Booster (997), Charolais (863), Gelbvieh (927), Hereford (1,077), Limousin (717), Simental (827), Alberta Composite (747) and Guelph Composite (1,596; most of them, Angus x Simmental). Genotyping from the different beef breeds were performed using different genotyping platforms (medium density panels (i.e., 50K SNP), and high density panels (i.e., 700K SNP). In dairy cattle, more than 1,100,000 genotypes from Holstein provided by Canadian Dairy Network (CDN, Guelph, Canada), Semex Alliance and CDCB will be used in the proposed project to enhance statistical power and validate the results from the 9 beef cattle breeds under study.

In relation to reproduction, another example of combining the new genomic technologies is the analysis of puberty in beef cattle by examining the genes and regulatory pathways and networks involved in this complex physiological event. Puberty is the process by which animals mature into an adult capable of sexual reproduction (Dorn and Biro 2011). The process to achieve the puberty is similar in the two bovine sub-species (i.e., Bos indicus and Bos taurus), but occurs at markedly older ages in Bos indicus heifers (Rodrigues et al., 2002; Nogueira, 2004). However, despite a growing molecular and physiological understanding of the reproductive system, knowledge of the precise mechanisms regulating puberty in cattle is limited, and phenotypic identification of animals that undergo puberty at an early age is costly and labor-intensive. Therefore, enhancing our understanding of the genes and regulatory pathways and networks involved in bovine puberty can provide a window to help improve genetic selection and reproductive management in cattle.

Whole genome single nucleotide polymorphism (SNP)-chip and RNA-Sequencing data from the hypothalamus have been used to construct gene networks associated with puberty in cattle (Fortes et al. 2010 and 2011). Results from these approaches allowed postulating that regulatory loci underlying the quantitative trait loci (QTL) associated with heifer fertility traits influence puberty. Livestock production traits are usually complex and involve multiple tissues. Therefore, the transcriptome of five tissues related to reproduction (i.e. hypothalamus, pituitary gland, ovary, uterus, and endometrium) has been characterized as well as tissues known to be relevant to growth and metabolism needed for cattle to achieve puberty (i.e., longissimus dorsi muscle, fat, and liver) in PRE and POST puberty heifers using RNA-Sequencing (Cánovas et al. 2014). In order to exploit the power of complementary -OMICS analyses, PRE and POST puberty co-expression gene networks were constructed by combining the results from RNA-Sequencing, genome-wide association study (GWAS), and bovine transcription factors. As a result of combining the power of the different genomic technologies this reduced the complexity of the large lists of SNP and/or genes identified associated with puberty (Figure 2). Thus, combining the results from RNA-Sequencing and GWAS identified a total of 25 eQTL associated to heifer fertility. Applying the new genomic technologies and integrating the structural and functional genomic data revealed key transcriptional regulators (i.e., PITX2, FOXA1, TSG1D1, DACH2, LHX4, PROP1 and SIX6). As a validation of the approach used combining data from several genomic technologies, six genes captured in the cattle network were concordant with the human network that reported 30 loci for age at menarche using other functional analysis (Elks et al. 2010). Results from multiples sources of -omics data will facilitate the design of breeding strategies to improve fertility in Bos indicus-influenced composite cattle (Cánovas et al. 2014). Multi-tissue omics analyses improve understanding of the number of genes and their complex interactions for puberty in cattle. These results also help discovering genes that contain biologically relevant SNP-genotypes to validate in an independent population that can be used in genetic improvement processes of Bos indicus-influenced composite cattle.

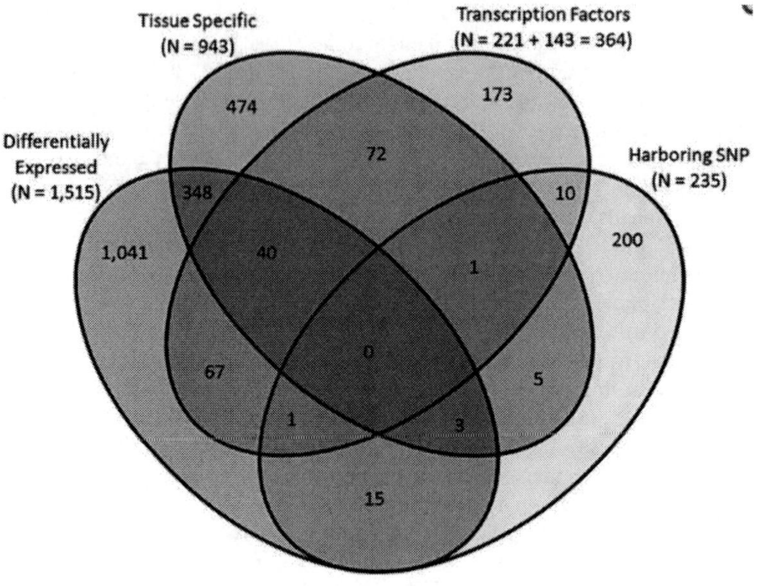

Figure 1. Combining results from different genomic technologies to identify the key regulator genes and/or functional SNP associated with puberty using the list of: 1) Differentially expressed genes between pre- and post-puberty, 2) Genes tissue specific, 3) Transcription factors and 4) Genes harboring SNP observed with GWAS and associated with indicator traits of puberty (age at puberty as measured by the presence of the first corpus luteum, first service conception and heifer pregnancy).

Genetic advancement in the beef industry has been limited in a number of ways including the high cost of determining genetic merit (low selection efficiency). While genomics has proven highly effective in the dairy industry, the challenges faced by the beef sector will require a more sophisticated approach than what is currently available. Utilization of genomics in the beef industry has been low to date, due in part to the complicating factor of multiple breeds and correspondence of SNP effects. The incorporation of new –OMICS technologies will enhance the accuracy of determination of genetic merit and make genomics more applicable and useful to beef breeders. In this way, the new high throughput tools available for genomic analysis will build upon all of the genomics work to date and have the effect of making genomics applicable to an entire sector, which would have a particular benefit to beef production. That paradigm shift will provide beef breeders with a tool that will let them become much more competitive with other protein producing sectors through improved feed efficiency, weight gain in a reduced time span, more consistent product and reduced negative environmental impact.

In beef cattle, the best genomics prediction using SNP markers has an accuracy of only 20%, which is insufficient for impact and adoption. However, as described by Karisa *et al.* (2014), metabolites could explain over 75% of the variation in residual feed intake (RFI) and result in ~80% accuracy of prediction. One of the objectives is to reduce the complexity of the large lists of SNP and/or genes associated with a trait of interest to identify the key regulator genes and functional SNP (and genomic regions) associated to feed efficiency and methane emission by integrating functional genomics with the new –OMICS technologies into a system biology approach for more reliable results that could be implemented in dairy and beef breeding strategies to improve the accuracy of selection.

To have a more complete understanding of the biological knowledge and the genes and mutations affecting feed efficiency, it is also important to study the microbiota profile associated to the genotype of the animal and its interactions affecting the feed efficiency in rumen fluid. A metatranscriptome (using also RNA-Sequencing) approach can used to identify the active microbial phylotypes including bacteria, archaea, eukyrotic (fungi and protozoa) and rumen microbial activities from animals with different RFI ranking. Metatranscriptomic can not only generate taxonomy profiles for active organisms, but also can be used to evaluate the activities of the rumen microbiome. Also, metagenomics to identify and quantify microbial profiles and determine their functional capacity in bovine rumen fluid samples. Total DNA is extracted and partial bacterial and archaeal 16S rRNA gene as well as partial protozoa and fungal 18S rRNA gene is amplified as previously described by Handerson *et al.* (2015) and sequenced using Illumina-Miseq. The purpose of this analysis is to profile the rumen microbiota and rumen microbial metagenomic profiles generated can be correlated with feed efficiency and methane emission phenotypes.

REFERENCES

Cánovas, A., Rincon G., Islas-Trejo A., Wickramasinghe S. and Medrano J.F. (2010) *Mammal. Genome* **21**: 592.

Cánovas, A., A. Reverter, K. DeAtley, R. Ashley, M. Colgrave, et al. (2014) *PLOS One* **9(7)**:e102551.

Dorn, L.D., and F.M. Biro. 2011. Journal of Research *Adolscence* **21**: 180–195.

Elsik, C.G., R.L. Tellam, and K.C. Worley. 2009. *Science* **324**: 522–528.

Elks, C.E., J.R. Perry, P. Sulem, D.I. Chasman, et al. (2010) *Nature Genetics* **42**: 1077–1085.

Fortes, M.R.S., A. Reverter, Y. Zhang, E. Collis, S.H. Nagaraj, et al. (2010) *Proceedings National Academic of Science* **107**: 13642–13647.

Fortes, M.R.S., A. Reverter, S.H. Nagaraj, Y. Zhang, N.N. Jonsson, et al. (2011) *Journal of Animal Science* **89**: 1669–1683.

Henderson, G., F. Cox, S. Ganesh, A. Jonker, et al. (2015) *Scientific Report* **25**: 14567.

MacLeod, H.D., H.D. Daetwyler, P.J. Bowman, A.J. Chamberlian, C.J. Vander Jagt, et al. (2014) *Proceedings, 10th World Congress of Genetics Applied to Livestock Production*.

Nogueira, G.P. (2004) *Animal Reproduction Science* **82–83**: 361–372.

Reverter, A., and M.R.S. Fortes (2013) *Journal of Animal Science* **91**: 530–536.

Rodrigues, H.D., J.E. Kinder, and L.A. Fitzpatrick (2002) *Biology Reproduction* **66**: 603–609.

Snelling, W.M., R.A. Cushman, J.W. Keele, C. Maltecca, M.G. Thomas, et al. (2013) *Journal of Animal Science* **91**: 537–552.

Van Eenennaam, A. (2016) Recent Developments in Genetic Evaluations and Genomic Testing. www.ebeef.org.

Wickramasinghe, S., A. Cánovas, G. Rincon, and J.F. Medrano (2014) *Livestock (Review)* **166**: 206-216.

Proc. Assoc. Advmt. Anim. Breed. Genet. 22:385-388

THE *REPRONOMICS* PROJECT - ENABLING GENETIC IMPROVEMENT IN REPRODUCTION IN NORTHERN AUSTRALIA.

D.J. Johnston[1], T.P Grant[2] T.J. Schatz[3], B.M. Burns[4], G. Fordyce[5], R.E. Lyons[6]

[1]Animal Genetics and Breeding Unit*, University of New England, Armidale, NSW 2351, Australia
[2]Queensland Department of Agriculture and Fisheries, Toowoomba, QLD
[3]NT Department of Primary Industries and Fisheries, GPO Box 3000, Darwin, NT
[4]QLD, Department of Agriculture and Fisheries 25 Yeppoon Road, Parkhurst, QLD
[5]Queensland Alliance for Agriculture and Food Innovation, UQ, Charters Towers, QLD
[6]Animal Genetics Laboratory, University of Queensland, Gatton, QLD

SUMMARY

This paper provides a brief description of a large breeding and genotyping project currently being conducted in northern Australia using three tropically-adapted beef breeds. Intensive recording of early-in-life female reproduction phenotypes using real-time ultrasound on large numbers of females will significantly increase the size of the genomic reference populations for these key traits and breeds. This phenotypic data, along with high density SNP genotypes, will provide crucial data to enable the effective implementation of new BREEDPLAN genomic evaluations currently being developed for the Australian beef industry, and will assist in increasing the accuracy of selection, especially in young bulls. The project design will also allow investigations of *GxE* interactions for female reproduction traits; the potential development of new traits; and methodologies required for the implementation of across-breed genomic evaluations.

INTRODUCTION

Reproduction is a key profit driver in northern Australia and the recent Beef CRC northern reproduction project, and earlier research (e.g. Hetzel *et al.* 1989), showed there is a clear role for genetics in improving commercial weaning rates. Heritability and genetic variance estimates for reproduction traits from the Beef CRC suggest considerable genetic progress is possible in both Brahmans and Tropical Composites. This is particularly the case for component traits of reproduction measured earlier in a cow's reproductive life. Two of these traits identified as important for genetic improvement are heifer age at puberty (Johnston *et al.* 2009) and first-lactation anoestrous interval (Johnston *et al.* 2014a). Male reproductive measures are also heritable (Corbet *et al.* 2013) and can be used with other novel traits as indirect measures of reproduction (Johnston *et al.* 2014b; Barwick *et al.* 2014). Genomics research has also shown that this emerging technology has the potential to add significantly (Hawken *et al.* 2012 and Zhang *et al.* 2014) to our ability to make genetic progress in reproduction (Barwick *et al.* 2014). However to move the research outcomes more rapidly into industry breeding programs requires the collection of considerably more reproductive phenotypes and genotypes to build the size of the genomic reference populations across an increased number of northern beef breeds. This is the aim of a new 5 year project (MLA *B.NBP.0759*) known as the Repronomics^{TM} project that has been running for 3.5 years and has generated significant numbers of calves and has recorded large numbers of females for age at puberty, lactation anoestrous interval, calving and weaning rates, along with many other traits in 3 major northern beef breeds. These records are being combined with DNA SNP genotypes on all project animals, as well as key industry animals, to drive new genomics enhanced BREEDPLAN evaluations. This paper provides a brief description of the experimental design and an update on the recording that is occurring in the project.

---

* AGBU is a joint venture of NSW Agriculture and the University of New England

LOCATION AND BREEDS

The project is utilising the 3 numerically largest tropically adapted beef breeds (*viz*. Brahman, Droughtmaster and Santa Gertrudis) in northern Australia. The research is being conducted on Queensland Department of Agriculture and Fisheries (DAF) research facilities (Brian Pastures, Gayndah; Spyglass, Charters Towers) and the Northern Territory Department of Primary Industry and Fisheries' Douglas Daly Research Farm, Daly River. The project is also utilizing industry seedstock herds located throughout Queensland. The Douglas Daly Brahman herd is part of a large long-term fertility selection experiment (Schatz *et al.* 2010). The Brian Pastures and Spyglass herds consist of pedigree and performance recorded females, and include a proportion of ex-Beef CRC cows used as base females. All herds are fully BREEDPLAN recorded and genetic linkages exist with the Beef CRC project, the Brahman Beef Information Nucleus herds, the Smart Futures Fund Next-Gen project industry herds and other key industry seedstock herds. At Brian Pastures, all 3 breeds are present, whereas at Spyglass there are Brahman and Droughtmasters. At each location, the breeds are managed and recorded together, and this is providing unique data for the future development of across-breed EBVs.

SIRES USED

The sires used in the project include naturally mated bulls purchased by DAF, and in recent years AI sires have been chosen that are currently influential in each of the breeds in terms of the number of offspring generated in the last 5 years. In addition, some emerging young sires have been used in each breed. The sires are chosen particularly if they have limited or no daughters recorded for reproduction in BREEDPLAN. Poll status is considered, but is not the primary selection criterion. The aim is to generate 15-20 daughters from each sire and intensively record them for early reproduction. For Droughtmasters, a selection of older sires has also been used to allow estimation of breed genetic parameters, particularly for female reproduction traits. To-date, the project has generated progeny on 236 sires, with 86 currently having 10 or more daughters, not including calves generated in the latest 2017 born calf crop.

FEMALES GENERATED AND KEY TRAIT RECORDING

Currently the project has generated approximately 4,200 calves from 4 year drops. Breeding is by natural mating for maidens and first-lactation cows. At the Douglas Daly herd, heifers are mated as yearlings, but in the DAF herds the heifers are mated first as 2-year olds. The majority of older cows have been used for AI (2 rounds fixed-time program) to generate progeny on the key industry sires, and although the project has experienced below average seasonal conditions in the first 2 years, the resultant calving rates to AI have been very good, averaging 50%.

The project is using real-time ultrasound and highly skilled ultrasonographers to perform regular ovarian assessments on all females to accurately determine the follicle development, and importantly, the presence of a corpus luteum (CL). Every year the cohort of maiden heifers are regularly scanned to determine age at observed CL (see Figure 1) which is used as a measure of age at puberty for each heifer. All first-lactation cows are also regularly scanned during the mating season to determine their time to return to cycling post-calving. Post-weaning all acyclic females continue to be scanned until a CL is observed. To-date the project has recorded more than 1,500 heifers on the research station herds for age at puberty and about 1,000 first-lactation cows for their anoestrous interval.

All females are regularly recorded for body weight, hip height, body condition score, subcutaneous fat depth, and eye muscle area. At calving, each cow is scored for calving ease, teat and udder score, maternal behaviour and body condition. The data is checked to ensure highest quality and continuously loaded onto a custom-built project database. All BREEDPLAN traits are regularly extracted and sent to ABRI's northern multi-breed research database and made available for use in routine BREEDPLAN evaluations of the 3 breeds.

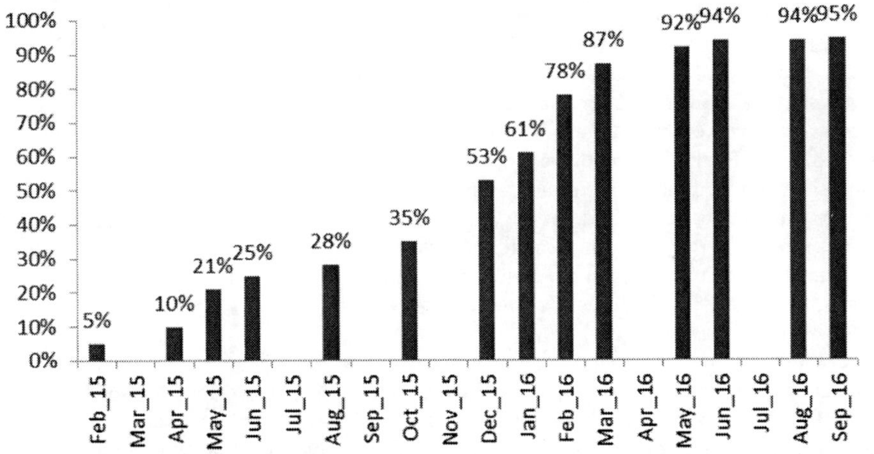

Figure 1. Cumulative frequencies of first observed CL from regular ovarian scanning of the 2014 drop Spyglass heifers weaned mid-2014.

DNA GENOTYPING

To enable the development of genomic selection, all females have been DNA parent verified and genotyped with a 25K SNP chip, and all project sires are genotyped with an 80K *Bos indicus* SNP chip. Large numbers of seedstock animals have also been genotyped, including sires in co-operator industry herds (2-3 herds/breed) and other sires in Brahman and Santa Gertrudis with high accuracy BREEDPLAN days to calving EBVs. As the project progresses, cohorts of young bulls in co-operator seedstock herds will be genotyped, and will provide a demonstration of the benefits in increased accuracy from the project recording. All DNA data is checked and stored on the project database and is available for inclusion in subsequent genomic evaluations.

STEER PROGENY

All calves generated in the project are intensively recorded from birth to weaning - including accurate birth date, birth weight, gestation length (AI calves only), calving ease and survival, weaning weight, flight time, plus several other research measures. The Douglas Daly male calves remain entire, whereas at branding the DAF bull calves are castrated. After weaning, several steer cohorts have entered a northern MLA Donor Company-funded BIN project. The steers are grown-out and recorded for post-weaning performance and subsequent full abattoir carcase and meat quality assessments. The steer recording complements the female recording at the research stations, and completes the suite of key profit driver traits for northern beef production systems.

ENABLING GENOMIC SELECTION

The new single-step procedure represents a seminal change in the evolution of the BREEDPLAN genetic evaluation system. Single-step evaluations (Misztal *et al.* 2009) allow simultaneous use of existing pedigree relationships (for the majority of animals) in conjunction with a genomic relationship matrix (GRM) of genotyped animals from high density SNP profiles (e.g. 20K or 50K). The procedure simplifies, and is equivalent to, the use of genomic information currently using estimated genomic values derived from genomic prediction equations, and allows the evaluation to be continually updated as additional phenotypes and genotypes are added. Currently, the existing pedigree relationship matrix allows differences in phenotypic performance to be transmitted to known relatives, whereas the single-step evaluation will allow genetic differences between individuals to be influenced through their degree of genomic relationship. Therefore animals with large amounts of phenotypic information when genotyped will influence the EBVs and accuracies of any animal that is genomically related.

As the genetic evaluations of the tropical breeds move towards single-step methodologies, the data and research outcomes from this project will be pivotal in driving this new era of genetic evaluation. While the Beef CRC genotyped and phenotyped large numbers of tropically adapted cattle for female and male reproduction traits, it did not include Santa Gertrudis or Droughtmaster breeds, a gap that this project is addressing. The project is also generating phenotypes and genotypes on current industry-relevant genetics and this provides the northern breeding industry with the unique opportunity to implement genomic selection, increasing the accuracy of selection of young bulls, particularly for important female reproduction traits.

CONCLUSIONS

The project is well underway and achieving its targets in generating females and intensive recording of female reproduction. Levels of recording of the project-generated females are increasing and the data is feeding into new BREEDPLAN evaluations that will enable tropical breeds to make genetic change in improving female reproduction rates. The project is building uniquely recorded herds that will allow genetics to be compared across environments, and will be a powerful resource enabling industry herds to be benchmarked for reproduction traits, as well as many other traits including overall genetic merit. Finally, the head-to-head management of breeds will provide the necessary data to generate across-breed genomic EBVs for large numbers of traits.

ACKNOWLEDGEMENTS

The authors acknowledge the MLA funding (*B.NBP.0759*) and the contributions of the collaborator institutions: DAF, NTDPIF, UQ, QAAFI, and ABRI. Thanks also to efforts of station mangers and their staff, and to our skilled team of technicians, ultrasound scanners, AI, DNA lab, data managers, project managers and scientists. We thank the support of the northern beef industry, individual breeders, and the participating Breed Societies.

REFERENCES

Barwick SA, Johnston DJ, Holroyd RG, Walkley JRW, Burrow HM (2013) *Anim. Prod. Sci.* **54**: 97.

Corbet NJ, Burns BM, Johnston DJ, Wolcott ML, Corbet DH, Venus BK, Li Y, McGowan MR, Holroyd RG (2013) *Anim. Prod. Sci.* **53**: 101.

Hawken RJ, Zhang YD , Fortes MRS, Collis E, Barris WC, Corbet NJ, Williams PJ, Fordyce G, Holroyd RG, Walkley JRW, Barendse W, Johnston DJ, Prayaga KC, Tier B, Reverter A, Lehnert SA (2012) *J. Anim. Sci.* **90**: 1398.

Hetzel DJS, Mackinnon MJ, Dixon R, Entwistle KW (1989) *Anim. Prod.* **49**: 73.

Johnston DJ, Barwick SA, Corbet NJ, Fordyce G, Holroyd RG, Williams PJ, Burrow HM (2009) *Anim. Prod. Sci.* **49**: 399.

Johnston DJ, Barwick SA, Fordyce G, Holroyd RG, Williams PJ, Corbet NJ (2014a) *Anim. Prod. Sci.* **54**: 1.

Johnston DJ, Corbet NJ, Barwick SA, Wolcott ML, Holroyd RG (2014b) *Anim. Prod. Sci.* **54**: 74.

Misztal I, Legarra A, Aguilar I (2009) *J. Dairy Sci.* **92**: 4648.

Schatz TJ, Jayawardhana GA, Golding R, Hearnden MN (2010) *Anim. Prod. Sci.* **50**: 345.

Zhang YD, Johnston DJ, Bolormaa S, Hawken RJ, Tier B (2014) *Anim. Prod. Sci.* **54**: 16.

Proc. Assoc. Advmt. Anim. Breed. Genet. 22:389-392

MYTHBUSTERS: NON-ADDITIVE GENETIC EFFECTS HAD NEGLIGIBLE IMPACT ON WAGYU CARCASS QUALITY

R.A. McEwin[1], M.L. Hebart[1], S. de Bruin[2] and W.S. Pitchford[1]

[1]School of Animal and Veterinary Sciences, University of Adelaide, Roseworthy Campus SA 5371, Australia
[2]Mayura Station, Millicent, SA 5280, Australia

SUMMARY

Anecdotal evidence suggests that mating specific sire and dam lines will produce superior offspring compared to those of equivalent expected merit but in random combinations. BREEDPLAN allows for genetic selection of sires and dams to advance carcass trait expression in progeny by utilising, mostly, additive genetic relationships. This program does not intentionally account for non-additive genetic effects in the genotype. The influence of non-additive genetic effects in Wagyu, and specifically, their effect on marbling performance has not been widely investigated. Non-additive genetic effects had negligible impact.

INTRODUCTION

Development of a genetic evaluation program in Australia started as the National Beef Recording Scheme (NBRS) in the late 1970's and became BREEDPLAN in 1985 (Graser and Hammond 1985). The purpose of BREEDPLAN is to quantitatively evaluate an individual's genetic merits before they are selected as breeding stock on a breed by breed basis. Genetic variation can be partitioned into two components, additive and non-additive variance (Falconer 1981). BREEDPLAN produces, for each trait analysed, an estimated breeding value (EBV) which is a representation of the additive genetic component of the individual's genotype (Tempelman and Burnside 1989). BREEDPLAN does not intentionally account for non-additive genetic variance, such as dominance or epistatic interactions, although some may be absorbed by a fitted sire x herd interaction effect. This is not reported to breeders, but improves the estimation of the reported EBVs (Graser *et al.* 2005). The program has evolved from the initial two-trait analysis with a sire-maternal grandsire model, to a multi-trait animal model analysis system (Quaas and Pollak 1980), including growth, reproduction and carcass traits. One such carcass trait is marbling performance which is defined as the accumulation of triacylglycerol in muscle tissue occurring primarily within adipocytes located between muscle fibre bundles (Harper and Pethick 2004).

Anecdotal evidence from breeders suggests that matings of specific superior sire and dam lines will produce superior performing offspring compared to other matings of equivalent expected merit but in random combinations. In Wagyu these so called 'superior offspring' would be those that produce the most highly marbled carcasses. This performance above expectation has been attributed to non-additive genetic components in the genotype (Seath and Lush 1940; Tempelman and Burnside 1989) and among breeders is referred to as "Nicking". Few studies have investigated the importance of nicking to marbling performance and even fewer have done so using Wagyu data. The aim of this project is to provide breeders with information as to whether non-additive genetic effects should be included in the genetic evaluation of animals or whether the sole focus should remain on additive genetic effects.

MATERIALS AND METHODS

Data. Data for this study was collected and provided by a single Full-blood Wagyu herd located in the Lower South East agricultural region near Millicent, South Australia. All cattle are born, raised and feedlot finished for 300 days on property. Calves were weaned as a group of similar age and kept within their respective weaning groups, following the same ration program, until slaughter at 2-2.5 years of age. Data supplied was for animals slaughtered between June 2010 and July 2016 and included pedigree information for all animals slaughtered. The dataset consisted of records for 1357 animals comprising 33 and 994 sires and dams respectively, where all sires included had greater than or equal to 10 progeny each. Pedigree went as deep as 5 generations.

Traits of interest. Raw AUS-MEAT chiller assessment records on hot standard carcass weight (HSCW, kg), marble score (MS, score 0-12), loin eye muscle area (EMA, cm^2) and fat depth at the rump P8 site (P8, mm) were utilised for analysis (n=1357). An indicative carcass value (VALUE, $) was calculated based on approximate current values as HSCWx(4+MS) where a carcass with MS=0 was estimated to receive $4/kg and every increase in MS achieved an additional $1/kg.

The average carcass weight was 412 kg with 16 mm of P8 fat, 95 cm^2 EMA, a marble score of 8 and worth $4945. Marbling and P8 were highly variable with a coefficient of variation (CV) of 23% and 37% respectively followed by VALUE (CV 18%), whereas HSCW and EMA were far less variable (CV 9-10%). All traits were normally distributed with the exception of P8 fat that was transformed by taking the square-root of the record prior to analysis.

Model Development and Statistical Analysis. Data were analysed with a general linear mixed model using ASReml 3.0 (Gilmour *et al.* 2009). Model 1 was developed beginning with a sire model and increasing in complexity to include an animal model with random terms fitted. All models included fixed effects of management group (in this case Kill-date was used to account for calving/weaning group), year and season effects (having a correlation of 0.95 and 0.99 with date of birth and feedlot induction date respectively), age of dam (2-10+ years) and sex (heifers and steers). Models for traits other than HSCW included HSCW as a covariate. All significant ($P<0.05$) two way interactions among these fixed effects were included. Inbreeding coefficients were calculated based on all available pedigree information (mean 5% and ranging from 0-26%) and were fitted in Model 3. Models varied in the random terms included.

Model 1. The base "animal" model, which served as a baseline, included the random term of animal to account for the additive genetic effects inherited by the calf.

Model 2. Model 1 plus the maternal additive genetic effect to account for the genes inherited by the dam that affect performance of offspring (e.g. milk production).

Model 3. Model 2 plus non-additive effects modelled as inbreeding and sire by maternal grandsire (SxMGS) interaction were fitted together where;

- Inbreeding was fitted as a random covariate to test the importance of dominance genetic effects contributing to Nicking;
- SxMGS accounts for epistatic genetic effects, inherited from two sire lines, associated with Nicking.

To formally test the importance of the non-additive genetic effects, the likelihood ratio test was used to determine which random terms were significant ($P<0.05$).

RESULTS AND DISCUSSION

Carcass weight, EMA, MS and VALUE were all moderately heritable (0.24, 0.33, 0.39 and 0.36 respectively, Table 1) while P8 was lowly heritable (0.19). The non-additive genetic effect of nicking as estimated from the sire by maternal grand-sire effect was minimal, accounting for up to 0.6% of the phenotypic variance for the muscling traits (HSCW and EMA) and zero variance for the fat traits (P8 and MS). Additionally there was no evidence to suggest that inbreeding accounted for any of the phenotypic variation across traits (Table 1).

Table 1: Chi Squared (χ^2 probability) test of significance (Model 2 vs 3) and proportion (%) of phenotypic variance accounted for by random terms (Model 3) as well as the direct heritability (h_a^2) and accompanying standard error (SE) for each trait of interest (Model 2)

Trait	Additive[1]	Inbreeding[1]	SxMGS[1]	Dam additive[1]	χ^2 Prob	h_a^2	SE of h^2
HSCW (kg)	24.0	0.0	0.5	6.9	0.95	0.24	0.11
P8 (mm)*	19.5	0.0	0.0	3.9	1.00	0.19	0.09
MS (0-12)*	37.5	0.0	0.0	2.1	1.00	0.39	0.12
EMA (cm^2)*	32.5	0.0	0.6	0.0	0.99	0.33	0.09
VALUE ($)	30.9	0.0	0.0	6.0	1.00	0.36	0.11

* indicates trait adjusted for HSCW

[1] random terms fitted in Model 3: Additive; Additive genetic effects, Inbreeding; Inbreeding coefficient partitioning dominance, SxMGS; Sire by maternal grandsire interaction effect partitioning epistatic interactions, Dam additive; partitioning additive genetic effects inherited by the dam.

As inbreeding depression is resultant of increasing homozygosity and hence the loss of dominance effects (reduction in genetic variation), the SxMGS term then would be partitioning any epistatic effects while inbreeding partitioned dominance. Model 3 was not significantly different from Model 2 indicating that the additional random terms, accounting for non-additive genetic components, did not offer a significant statistical improvement on the additive model. However, despite Model 3 not statistically improving the model, it was able to estimate a variance component attributed to nicking for the muscling traits which suggests that these values should not be completely discounted.

Few studies in beef cattle exist to compare these results with and even fewer that include Wagyu data. However, many authors have commented on the influence of non-additive genetic effects on dairy production. Johnson *et al.* (1940) investigated the effect of nicking in Jersey cows and concluded that while their study had not shown nicking to influence pounds of milk butterfat yielded, there was not enough evidence to suggest that nicking does not exist amongst Jersey matings. Seath and Lush (1940) reported similar inconclusive results regarding milk production and butterfat percentage while Tempelman and Burnside (1989) reported that dominance effects, which nicking has been partly ascribed to, were important for milk fat yield. It is clear the previous dairy studies have attributed nicking to dominance effects while the present study has partitioned nicking as epistatic effects. It is possible that the pedigree herein is not sufficiently deep enough to account for all the dominance variation through fitting inbreeding coefficient as a random effect and hence some dominance variation has been absorbed into the SxMGS term.

In a genome wide association scan fitting additive and dominance effects of single SNPs, Bolormaa *et al.* (2015) found that significant ($P<0.001$) dominance effects occur for IMF% accounting for 10% of the phenotypic variance in the trait. This suggests that dominance effects would have an impact on marble score however that was not the result in the present study. The results of Bolormaa *et al.* (2015) were similar with the study herein, in that dominance effects accounted for 0% of the phenotypic variance for P8 fat depth however the author stated that the inclusion of dominance effects did not improve the accuracy of predicting genetic value of individuals.

It could be argued that nicking effects may be present when multi-trait selection is considered, as is currently practiced in industry since multiple traits affect profitability. To investigate this, a multi trait index (VALUE) was created encompassing the estimated dollar value of HSCW and

MS which are the most important economic traits to Wagyu producers. There was no-evidence to suggest that nicking had any effect on VALUE (Table 1).

The results herein demonstrated that non-additive genetic effects (dominance and epistasis) have a negligible impact on carcass traits. This is consistent with a study by Hill *et al.* (2008) that found additive genetic effects have the greatest influence by far, accounting for often 50 to 100% of the total genetic variance for complex traits. Hence the inclusion of non-additive genetic variances in the estimation of individual genetic merit is not likely to result in any potential reward, except for perhaps slight increases in estimation accuracy, and therefore additive genetic variance should continue to have sole focus in Wagyu breeding programs.

ACKNOWLEDGEMENTS

The authors acknowledge the contributions provided by Mayura Station, with a big thanks going to all the members of the team, especially Scott de Bruin, Lee Humphries and Mark Oliver.

REFERENCES

Bolormaa S., Pryce J.E., Zhang Y., Reverter A., Barendse W., Hayes B.J., Goddard M.E. (2015) *Genet. Sel. Evol.* **47**: 26.

Falconer D. (1981) 'Introduction to quantitative genetics' 2nd ed. John Wiley & Sons Inc

Gilmour A.R., Gogel B., Cullis B., Thompson R., Butler D. (2009) ASReml user guide release 3.0. *VSN International Ltd*

Graser H., Hammond K. (1985) *Crop and Past. Sci.* **36**: 527

Graser H., Tier B., Johnston D., Barwick S. (2005) *Anim. Prod. Sci.* **45**: 913

Harper G., Pethick D. (2004) *Anim. Prod. Sci.* **44**: 653

Hill W.G., Goddard M.E., Visscher P.M. (2008) *PLoS Genet* **4**: e1000008

Johnson L.A., Bartlett J.W., Copeland L. (1940) *J. Dairy Sci* **23**: 709

Quaas R.L., Pollak E. (1980) *J. Ani Sci* **51**: 1277

Seath D.M., Lush J.L. (1940) *J. Dairy Sci* **23**: 103

Tempelman R., Burnside E. (1989) http://cgil.uoguelph.ca/pub/articles/nicking.html 02 Feb 2016

Proc. Assoc. Advmt. Anim. Breed. Genet. 22:393-396

FACTORS INFLUENCING GESTATION LENGTH IN TROPICALLY ADAPTED BEEF CATTLE BREEDS IN NORTHERN AUSTRALIA

D.J. Johnston[1] and T.P. Grant[2]

[1]Animal Genetics and Breeding Unit*, University of New England, Armidale, NSW
[2]Queensland Department of Agriculture and Fisheries, Toowoomba, QLD

SUMMARY

Records from artificial insemination programs and accurate birth date recording on 812 calves were used to compute gestation length (GL) for 3 tropically adapted beef breeds in northern Australia. Calves were a subset of those generated across 2 years and 2 locations as part of an ongoing beef genetics project in northern Australia. Analyses revealed few fixed effects were significantly influencing the trait, however calf sex was highly significant in all three breeds with males having 2.1, 3.0 and 3.7 days longer GL than females in Droughtmaster, Santa Gertrudis and Brahman, respectively. Large sire differences (up to 13 days) were also observed within each breed, and indicate a large degree of genetic control on the trait. These results have implications for breeding program design, parentage assignment and the development of a genetic evaluation for this trait, both within and across-breeds.

INTRODUCTION

Gestation length (GL) is a trait that can be easily generated from AI mating records, and is an important component of calving ease (Jeyaruban *et al.* 2016) and may be associated with increased calf losses from low birth weight calves in tropical breeds (Wolcott *et al.* 2016). However, few estimates exist of the heritability and associated fixed effects for GL in tropical beef breeds. The distribution of GL and magnitude of sire differences will have implications for management of AI programs and assigning of parentage. Therefore this study aimed to investigate factors influencing GL and estimate their size of effects in 3 tropically-adapted beef breeds.

MATERIALS AND METHODS

Animals. The animals used in this study were a subset of those from 2 research herds that are currently involved in a large genetics project in northern Australia (MLA *B.NBP.0759*) – known as the *Repronomics*™ project. In brief, the project aims to enable increased accuracy of genomic selection by collecting high quality female reproduction phenotypes and other economically important traits on influential sires in each breed. The phenotypic records will be combined with high density genotyping to drive new single-step genetic evaluation methods which are being developed for Australian beef industry (Johnston *et al.* 2017).

The research herds involved include Spyglass Research Facility (SPY) located 120km NW Charters Towers, QLD, comprising Brahman (BM) and Droughtmaster (DM) herds. The second location is the Brian Pastures Research Facility (BP), Gayndah, QLD, and includes BM and DM herds as well as a Santa Gertrudis herd. The calves used in this study were generated at each location using 2 rounds of fixed-time AI over 2 years. All cows were inspected daily over the calving season to establish accurate date of birth for each calf and sire parentage was determined by DNA verification.

For all breeds, common sires were used across years, and for DM and BM several sires were used across locations. Numbers of records by location, breed and across-location link sires are presented in Table 1. The smaller herd sizes at BP restricted the number of sires used compared to SPY, however

---

* AGBU is a joint venture of NSW Agriculture and the University of New England

all sires used at BP in BM and DM were also used at SPY (plus additional sires). At both locations all matings were as purebreds, however at BP the cow herd included Beef CRC tropical composite cows.

Table 1. Data description for gestation length (days) records in Brahman, Droughtmaster and Santa Gertrudis cattle at each location

Breed	Total N	Location	N	Mean (d)	Std	range	N sires	N progeny by link sires
Brahman	377	Brian Pastures	103	291.7	5.5	277-310	12	309
		Spyglass	274	292.3	5.5	278-309	18*	
Droughtmaster	337	Brian Pastures	52	290.8	5.5	278-303	8	200
		Spyglass	285	288.3	5.5	272-312	15*	
Santa Gertrudis	98	Brian Pastures	98	284.9	5.6	271-298	5	
Total	812						38	

* includes all sires used at Brian Pastures

Trait definition. Gestation length was computed as the number of days between the successful AI date and subsequent date of birth. Data edits included removing records from multiple births and one record was removed as a suspected premature birth GL=266 and only 22kg birth weight. All records were confirmed to be the result of AI mating by DNA sire verification. Two records were removed because the breed of sire was incorrect (i.e. AI straw error). Due to differences in cow age structures in the herds and breeds, cow age was grouped into 3 classes based on cow year of birth. At BP, cow genotype and cow age were confounded and were therefore fitted as a combined effect. Descriptive statistics of the raw data by breed and location are presented in Table 1.

Statistical methods. Analyses were performed for each breed separately using REML procedures in SAS (SAS Institute Inc. Cary, NC, USA). Gestation length (in days) was included in the mixed model as the dependent variable and the initial models included terms for year (year 1, year 2), cow lactation status at AI (wet, dry), sex of calf (male, female), cow age class (old, medium, young), and all first order interactions. For DM and BM analyses also included terms for location (BP, SPY), and importantly, a term for sire x location. For all analyses a term for cow was included as a random effect. Non-significant terms ($P > 0.05$) were sequentially removed to yield the final model for each breed. Least squares means were computed for significant fixed effects using the LSMEANS procedure in SAS.

RESULTS

a) Brahman. A total of 377 records were available for Brahmans from a total of 18 sires with an adjusted mean of 291.7 days. The final model included significant terms for sire, calf sex and location. Sire x location was not significant indicating that although SPY had longer GL there was no evidence of re-ranking of the sires.

Least squares means for gestation length in Brahmans are plotted in Figure 1. Calves from SPY had longer GL than BP, and male calves were 3.7 days longer gestation than females. Large differences also existed between the Brahman sires.

b) Droughtmaster. A total of 337 records were available for Droughtmaster from 15 sires with an adjusted mean of 288.4 days. The final model included significant terms for sire and calf sex. Location and sire x location effects were not significant. Least squares means are plotted in Figure 2 and show male calves were 2.1 days longer gestation than females. Large differences also existed between the sires with a range of almost 2 weeks.

c) Santa Gertrudis. A total of 98 records were analysed for Santa Gertrudis from 5 sires with an adjusted mean of 284.5 days. The final model included sire and calf sex. Male calves were 3.0 days longer GL than females (see Figure 3).

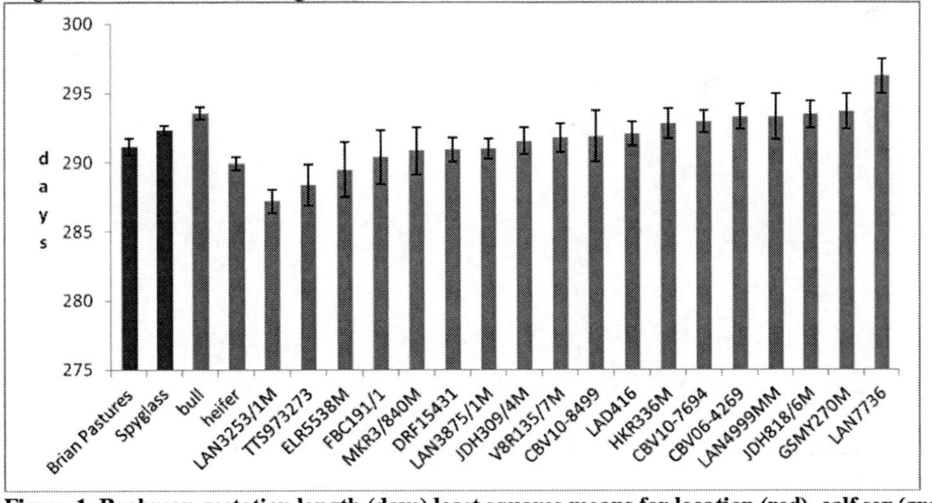

Figure 1. Brahman gestation length (days) least squares means for location (red), calf sex (green) and sires (blue).

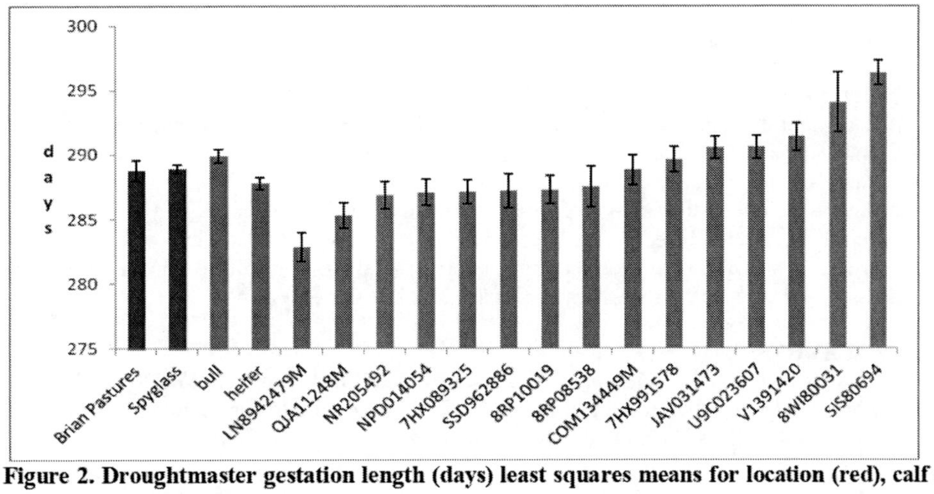

Figure 2. Droughtmaster gestation length (days) least squares means for location (red), calf sex (green) and sires (blue).

DISCUSSION

Although the results and least squares means are not directly comparable across breeds in the current analyses, the identified significant fixed effects and their magnitude of effect were consistent. The large sire differences were not surprising given the high heritability estimates of the trait in temperate beef breed (see Jeyaruban *et al.* 2016), however few estimates exist for tropical beef breeds.

Calf sex had a consistent and large effect in all breeds, with male calves having significantly longer GL than females. Plasse *et al.* (1968) reported a 1.9 day sex difference in purebred Brahmans.

Consistent with the results from the current study, few significant fixed effects were reported by Corbet *et al.* (1997) and Plasse *et al.* (1968). The extended GL of these tropically adapted breeds (especially Brahman) compared to temperate beef breeds is consistent with a review of Chenoweth (1994) and may indicate a mechanism of adaptation, possibly to high ambient temperatures. It could be due to *Bos Indicus* cattle having less uterine capacity as an evolutionary development to reduce internal heat production, and thus requires longer GL to produce a viable calf.

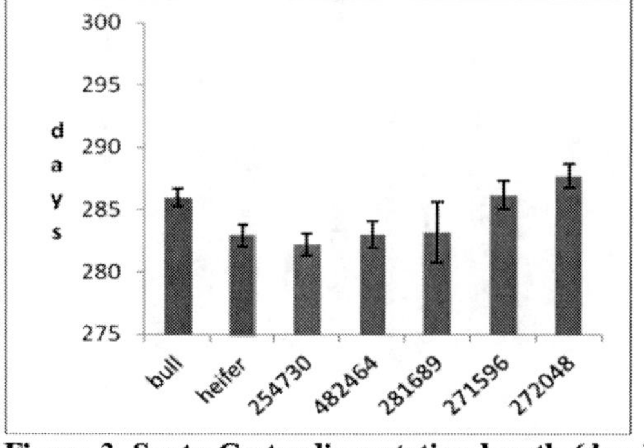

Figure 3. Santa Gertrudis gestation length (days) least squares means for calf sex (green) and sires (blue).

There was no evidence of *sire x location* interactions, however Brahmans had slightly longer GL at SPY compared to BP. This effect may been due to the slight difference in the sample of sires used or in the genetics of the cow herd, but may also be related to the hotter environment at SPY compared to BP, and may be further evidence of heat adaptation impacting GL.

CONCLUSIONS

This study confirmed the longer GL in these tropical breeds compared to temperate beef breeds and has ramifications for the design of breeding programs and parentage assignment. It also highlights the importance of modelling gender in genetic evaluations of the trait. In the future, these data will allow the estimation of genetic parameters (direct and maternal additive variance, and heritability) and this will contribute to each breed's genetic evaluation for GL, as well as to the development of an across-breed EBV for GL. Future work is also possible relating GL to its effects on calf birth weight and calf losses, and possible consequences on female reproductive performance.

ACKNOWLEDGEMENTS

This work was supported by Meat and Livestock Australia (MLA) project *B.NBP.0759*. Thanks to DAF and their Research station management and staff, and especially to our dedicated technicians at the 2 locations. Lastly, thanks to our highly skilled AI team, especially Mr Greg Fawcett.

REFERENCES

Chenoweth P.J. (1994). *Aust. Vet. J.* **71**:422-426.

Corbet N.J., Miller R.G., Fraser P.J., Lloyd A.G. and Burrow H.M. (1997). *Assoc. Adv. Anim. Breed. Genet.* **12**: 300.

Jeyaruban M.G., Johnston, D.J., Tier B. and Graser H-U. (2016). *Anim. Prod. Sci.* **56**:927.

Johnston D.J., Grant, T.P., Schatz, T.J., Burns, B.M., Fordyce, G. and Lyons R.E. (2017). *Assoc. Adv. Anim. Breed. Genet.* **22**: *(this edition)*.

Plasse D, Warnick, A.C., Reese R.E. and Koger M. (1968). *J. Anim. Sci.* **27**:101.

Wolcott M.L., Johnston, D.J., Tier, B. and Zhang Y. (2016). *Final report B.NBP.0761*. MLA, Sydney

Proc. Assoc. Advmt. Anim. Breed. Genet. 22:397-400

BREED COMPOSITION EFFECTS AND GENETIC PARAMETERS FOR PRODUCTIVITY OF TROPICAL BEEF CATTLE

L.R. Porto-Neto, R. McCulloch, B.E. Harrison, R.J. Bunch, W. Barendse, A. Reverter

CSIRO Agriculture & Food, 306 Carmody Rd., St Lucia, Queensland 4067, Australia

SUMMARY

The ability of beef cattle to remain productive in tropical environments is largely determined by heat tolerance and tick resistance. In Australia, crossbreeding and composite breeding with *Bos indicus* cattle have been used to introduce these traits into *Bos taurus* breed backgrounds. We examined SNP genotyping data and phenotypes for six production traits from a Tropical Composite herd in Central Queensland to test whether the *Bos indicus* ancestry (Indicus) percentage was able to explain the performance of beef cattle in tropical Australia. Tick count and coat length were significantly linked to Indicus percentage, while live weight, body condition, rectal temperature and scrotal circumference were not influenced by the Indicus content of the genome.

INTRODUCTION

Tropical Composites and the Brahman breed are the main types of cattle across Northern Australia. The term "Tropical Composite" generally refers, to minimally stable crossbred herds of European ancestry (*Bos taurus*) with varying degree of crosses with *Bos indicus* influenced cattle (Porto-Neto *et al.*, 2014). These types of cattle have been chosen for their adaptation or tolerance to challenges in tropical environments (Barwick *et al.*, 2009; Prayaga *et al.*, 2009).

The complex ancestry of the cattle *per se* does not limit the potential genetic improvement of those herds, but when combined with extensive herd management practices, and lack of relationship (pedigree) information, the intricacy of breed ancestry does add another layer of complexity to derive accurate estimates of genetic merit. In recent years, advanced analytical tools have been developed to better explore genotypes derived from single nucleotide polymorphisms (SNP) arrays. These new methods facilitate broader adoption of genomic technology as it, for instance, gives an alternative approach around the lack of pedigree, and allows the estimation of breed ancestry.

Here we analysed a Tropical Composite herd genotyped for around 50,000 SNP (Harrison *et al.*, 2012), with multiple observations for six phenotypes. We used the molecular data to study the population structure, estimate heritabilities, genomic correlations, breed composition, and to test the effect of ancestry on observed phenotypes.

MATERIALS AND METHODS

We targeted a Tropical Composite commercial population (TXX, n = 877) from central Queensland (Harrison *et al.*, 2012), and, for some analyses, used a set of animals as genotypic references representing ancestral breeds of the targeted population. These included Charolais (CHA, n = 90), Senepol (SEN, n = 69), Belmont Red (BEL, n = 73), and Brahman (BRM, n = 90) (Barwick *et al.*, 2009; Porto-Neto *et al.*, 2013). Most animals were genotyped using the BovineSNP50 v1 (Illumina Inc., San Diego, CA), the remaining animals were genotyped using more recent platforms (e.g. BovineSNP50 v2 or BovineHD). Standard quality control was applied to genotypes. If the analyses required samples that were genotyped using different arrays, only SNP that were in common across all platforms were kept for analyses.

Animals were phenotyped between 2 to 4 times across two years. Most animals were phenotyped around yearling age. The observed phenotypes were transformed tick counts (Tick, log2 of average tick counts), coat score (Coat, average coat score), condition score (Cond, average body condition score), rectal temperature (Temp, average rectal temperature), scrotal circumference (SC, single observation) and live weight (WT, average weight).

To assist the visualization of the populations substructure we ran principal components analysis (PCA) using PLINK 1.9 (Chang *et al.*, 2015), and estimated the ancestral proporti ons of *Bos taurus* and *Bos indicus* using Admixture software (Alexander *et al.*, 2009), as previously described (Porto-Neto *et al.*, 2014). A hexa-variate analysis was run to estimate heritabilities and genomic correlations between phenotypes fitting a precomputed genomic relationship matrix (VanRaden, 2008) in Qxpak v5 (Perez-Enciso and Misztal, 2011). Finally, we tested the effect of ancestral proportions on the observed phenotypes using SAS (SAS Inst., Cary, NC). The statistical model included the covariates of estimated percentage of *Bos indicus* and age at observation for each measurement, and the fixed effects of sex, breed type, management group, and operator (tick counts). For WT, after some exploratory analyses, an additional interaction between percent *Bos indicus* and sex was fitted.

RESULTS AND DISCUSSION

After quality control, around 760 animals (TXX) with genotypes (n = 49,573) and phenotypes were available for analysis, apart from SC which had a reduced number of observations (Table 1). There was large variation in phenotypes within the analysed population.

Table 1. Summary statistics of observed phenotypes

Phenotype	N	Mean	Std. Dev.	Min.	Max.
Tick	758	3.58	1.76	0.00	6.89
Coat	760	3.67	0.92	1.00	8.00
Cond	760	6.33	1.38	3.00	10.00
Temp	757	39.64	0.49	38.30	41.80
SC	248	28.41	3.70	17.50	38.00
WT	760	294.39	73.79	126.50	562.00
Indicus	761	24.48	7.64	5.02	60.45

Transformed tick counts (Tick), Coat score (Coat), Condition score (Cond), Rectal temperature (Temp), Scrotal circumference (SC), Weight (WT), and SNP-based percent estimate of *Bos indicus* ancestry (Indicus).

To explore and visualize potential population substructures we ran PCA analysis using additional cattle samples representing the breeds used during the formation of the targeted population (Figure 1). The majority of the animals appear to have varying proportions of three main ancestral breeds, Charolais, Senepol and Belmont Red with only a minor component of BRM. Moreover, a number of individuals seem to represent pure ancestral breeds, as they cluster together with the reference pure-breed clusters, these were seen with CHA, SEN and BEL clusters.

Using a genomic relationship matrix in the hexa-variate analyses we estimated the heritabilities and genomic correlation between phenotypes (Table 2). Heritability estimates varied from 0.391 (Temp) to 0.492 (Coat). It is worth noting that a known major gene variant affecting coat type that is derived from SEN cattle (Littlejohn *et al.*, 2014) segregates in this population and greatly influences the coat type. For Tick, the heritability (0.466) agrees with previous analyses using the same and one other population (Harrison *et al.*, 2012; Prayaga *et al.*, 2009), but is higher than other reports (Porto Neto *et al.*, 2011). For the other phenotypes, most estimates agreed with those previously reported for another sample of Tropical Composite cattle, maybe with the exception of Temp (higher here), and WT (lower here) (Porto-Neto *et al.*, 2014; Prayaga *et al.*, 2009).

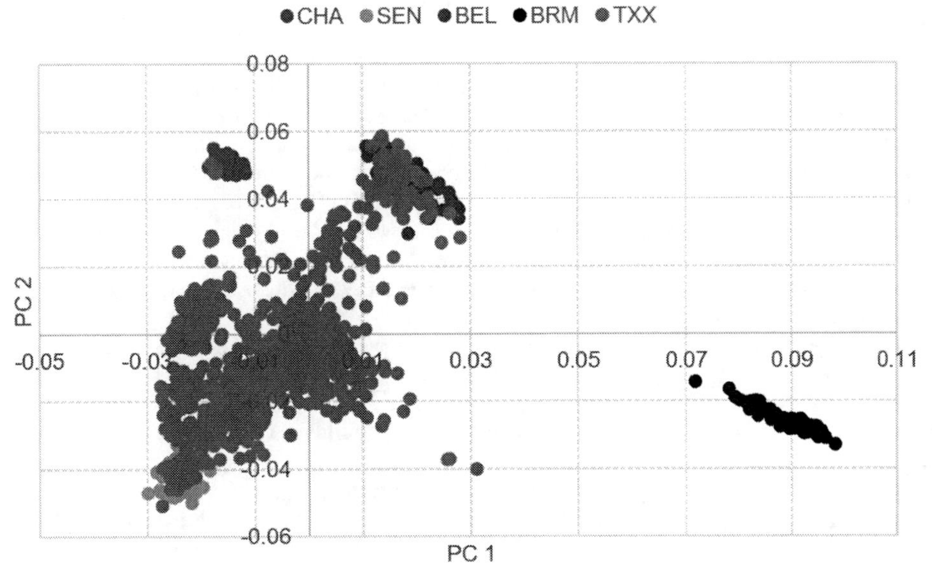

Figure 1. Principal Components Analyses using SNP genome-wide distributed (n = 33,620) including Charolais (CHA), Senepol (SEN), Belmont Red (BEL), Brahman (BRM) and Tropical Composite (TXX).

The estimated genomic correlations, in general, were not strong. Nevertheless, we detected positive correlations varying from 0.402 to 0.612 between Cond, SC and WT, which are different from some previously reported negative genetic and phenotypic correlations between Cond and WT (Porto-Neto *et al.*, 2014). The positive correlation between Tick and Coat (0.207) has the same direction of effect, but differs in magnitude from that previously reported (0.49) in a mixed breed analysis (Prayaga *et al.*, 2009).

Table 2. Estimated heritabilities (diagonal), and genomic correlations (off-diagonal) derived from hexa-variate analyses for observed traits

	Tick	Coat	Cond	Temp	SC	WT
Tick	**0.466**	0.207	-0.140	0.037	-0.143	-0.073
Coat	0.207	**0.492**	-0.066	0.065	-0.052	-0.105
Cond	-0.140	-0.066	**0.346**	0.061	0.402	0.612
Temp	0.037	0.065	0.061	**0.391**	-0.115	-0.014
SC	-0.143	-0.052	0.402	-0.115	**0.395**	0.516
WT	-0.073	-0.105	0.612	-0.014	0.516	**0.429**

Transformed tick counts (Tick), Coat score (Coat), Condition score (Cond), Rectal temperature (Temp), Scrotal circumference (SC) and Weight (WT).

The average percentage of *Bos indicus* ancestry (Table 1) of the population was low (24.48, ranging from 5.02 to 60.45, with only 47 animals > 35.0), confirming the suggestive conclusion from the PCA analyses (Figure 1) of little influence of Brahman cattle within this herd. The effect of Indicus percentage was significant only for Tick (-0.053 ± 0.010; p-value < 0.0001) and Coat (-0.024 ± 0.005; p-value < 0.0001). Both trait values were negatively affected by Indicus percentage, supporting the common knowledge that higher the Indicus content is, the lower is the tick load and the shorter the coat length. We confirm the effect on Coat, but could not detect the previously observed effect of Indicus on Temp, Cond and WT (Porto-Neto et al., 2014). This could perhaps be explained by the much lower Brahman influence within the tested herd, and the additional complexity in dealing with different breed compositions and levels of crossbreeding. An alternative model where the Indicus covariate was nested within breed type resulted in the main effect of breed becoming non-significant except for WT (p-value < 0.05). In addition to the already observed significant effect of Indicus for Tick and Coat, this alternative model estimated significant Indicus effect for some breeds in SC and WT.

The moderate to high heritabilities of all six traits measured in this herd clearly advocate for a genetic approach to tropical beef productivity. The need for Bos Indicus-derived adaptation genes is particularly evident for the tick count and coat length traits, confirming the rationale for using Brahman cattle for their tick resistance and heat tolerance traits in Australian beef breeding. More detailed genomics studies may be able to identify the exact alleles which confer the Bos indicus-derived tick resistance and heat tolerance traits in breeding programs. This will enable the maintenance of tropically adapted *Bos taurus* beef breeds with minimal Brahman influence.

ACKNOWLEDGEMENTS

The authors are thankful to Andres Legarra for providing software to efficiently compute the genomic relationship matrix and its inverse, Miguel Perez-Enciso for providing a customised version of the Qxpak5 software. The Beef CRC and Bovine Hapmap for providing genotypes of the reference breeds. This project was partly funded by Pfizer Animal Health (Zoetis).

REFERENCES

Alexander D.H., Novembre J., and Lange K. (2009) *Genome Res.* **19**: 1655.

Barwick S.A., Johnston D.J., Burrow H.M., Holroyd R.G., Fordyce G., Wolcott M.L., Sim W.D., and Sullivan M.T. (2009) *Anim. Prod. Sci.* **49**: 367.

Chang C.C., Chow C.C., Tellier L., Vattikuti S., Purcell S.M., and Lee J.J. (2015) *Gigascience* **4**.

Harrison B.E., Bunch R.J., McCulloch R., Williams P., Sim W., Corbet N.J., and Barendse W. (2012) *Anim. Prod. Sci.* **52**: 890.

Littlejohn M.D., Henty K.M., Tiplady K., Johnson T., Harland C., Lopdell T., Sherlock R.G., Li W.B., Lukefahr S.D., Shanks B.C., Garrick D.J., Snell R.G., Spelman R.J., and Davis S.R. (2014) *Nature Communications* **5**.

Perez-Enciso M., and Misztal I. (2011) *BMC Bioinformatics* **12**: 202.

Porto-Neto L.R., Reverter A., Prayaga K.C., Chan E.K., Johnston D.J., Hawken R.J., Fordyce G., Garcia J.F., Sonstegard T.S., Bolormaa S., Goddard M.E., Burrow H.M., Henshall J.M., Lehnert S.A., and Barendse W. (2014) *PLoS One* **9**: e113284.

Porto-Neto L.R., Sonstegard T.S., Liu G.E., Bickhart D.M., Silva M.V.B., Machado M.A., Utsunomiya Y.T., Garcia J.F., Gondro C., and Van Tassell C.P. (2013) *BMC Genomics* **14**: 876.

Porto Neto L.R., Jonsson N.N., D'Occhio M.J., and Barendse W. (2011) *Vet Parasitol* **180**: 165.

Prayaga K.C., Corbet N.J., Johnston D.J., Wolcott M.L., Fordyce G., and Burrow H.M. (2009) *Anim. Prod. Sci.* **49**: 413.

VanRaden P.M. (2008) *J. Dairy Sci.* **91**: 4414.

Proc. Assoc. Advmt. Anim. Breed. Genet. 22:401-404

DO PLAINER BODIED MERINO EWES HAVE HIGHER LIFETIME REPRODUCTION RATES?

S. Hatcher[1] and S.I. Mortimer[2]

[1] NSW DPI, Orange Agricultural Institute, Forest Road, Orange, New South Wales, Australia
[2] NSW DPI, Trangie Agricultural Research Centre, Trangie, New South Wales, Australia

SUMMARY

Significant phenotypic relationships between lifetime lamb survival and lifetime net reproduction rate (NRR) with neck and body wrinkle score were identified in 2 of 3 Merino resource flocks. For both traits, the relationships favoured plainer ewes. Commercial producers culling their wrinkliest ewes to reduce the risk of flystrike are unlikely to negatively impact the lifetime reproductive performance of their flocks.

INTRODUCTION

The lifetime productivity of Merino ewes in terms of their wool production, fleece characteristics, parasite resistance and reproduction is becoming increasingly important to Merino producers. This has been driven by the changing relative value of wool to meat production and continued decline in the terms of trade for agricultural commodities. An ever-increasing range of technologies and data management systems now allow either cohorts of animals within a flock to be selected and managed according to the average performance of the group; or individual animals selected and managed according to their performance relative to other individuals in the flock (Atkins *et al.* 2006). Lee *et al.* (2009a) identified potential gains to be made in lifetime NRR by retaining high performing ewes beyond the normal culling age and removing ewes with low reproduction from the breeding flock. Retaining the top 25% or 50% of older ewes for an additional 1-2 lambing opportunities based on pregnancy scanning information combined with udder examination at marking, together with removing poor performers (twice dry ewes) early in life, can improve production and profit in a Merino flock (Lee *et al.* 2014).

Increased public awareness of the animal welfare aspects of surgical mulesing (Greeff *et al.* 2014) has resulted in Merino producers reducing the degree of wrinkling, particularly since 1999 (Brown *et al.* 2010), through selection for plainer bodied animals. In addition, various within flock selection strategies such as selecting replacement breeding ewes with low wrinkle scores, culling individuals with high wrinkle scores and mate allocation (i.e. mating plain ewes with the plainest Merino rams) are now being advocated as a means for commercial producers to both reduce the incidence of flystrike and reliance on mulesing in their flocks (Richards and Atkins 2010). The impact of such phenotypic selection of ewes on their reproductive performance is unknown. Significant phenotypic variation exists for wrinkle scores (Hatcher and Preston 2015) as well as both annual (Safari *et al.* 2007) and lifetime reproduction traits (Lee *et al.* 2009b). While previous studies have reported phenotypic relationships between wrinkle score and annual reproduction events (Crook 1992; Turner and Young 1969), no studies have examined the relationship between wrinkle score and lifetime reproduction traits in Merino sheep. This study reports on the phenotypic relationship between neck and body wrinkle with lifetime NRR and its components.

MATERIALS AND METHODS

Lifetime reproduction data (from 2 - 6 years of age) were collated from three Merino genetic resource flocks (D-Flock, C-Flock and QPLU$) run at the Trangie Agricultural Research Centre. Lambing and weaning performance of the ewes in each of these flocks were routinely recorded. An outline of each flock and its management is provided by Lee *et al.*(2009a). Data were available for 3,300 D-flock ewes (born 1975 - 1983), 1,411 C-flock ewes (born 1984 - 1993) and 5,393 QPLU$ ewes (born 1992 - 2002). Neck and body wrinkle scores were assessed at weaning time in the three flocks using either a 1 - 9 scoring system (D- and C-Flocks, with 1 being low wrinkle score) or a 1 to 5 score (AWI Ltd and MLA Ltd 2013, QPLU$ flock). The 1 - 9 scoring system aligns with the 1-5 system with scores 1, 3, 5, 7, and 9 corresponding to scores 1 to 5 respectively and 2, 4, 6, & 8 the ½ scores in between.

This preliminary study was based on phenotypic information of the number of times each ewe was joined, the number of parities (lambing events), the total number of lambs born and the number of lambs weaned from 2 to 6 years of age. From these data lifetime fertility (no. times lambed/no. times joined), fecundity (no. lambs born/no. times joined), lamb survival (no. lambs weaned/no. lambs born) and NRR (no. lambs weaned/no. times joined) were calculated for each ewe as was the average pre-joining liveweight (kg). ASReml (Gilmour *et al.* 2009) was used to fit the effects of genotype (bloodline within the D-Flock, animals having the same proportion of genes derived from each of the bloodlines within the C-flock, and selection line within-strain within the QPLU$ flock), year of birth and wrinkle score (neck and body wrinkle scores were analysed for each lifetime reproduction trait in separate models). The significance of differences between the ASReml predicted means for neck and body wrinkle score was determined using T-tests. These were based on the least significant difference calculated from the standard error of the difference for each lifetime reproduction trait within each flock and the degrees of freedom.

RESULTS AND DISCUSSION

For the D- and C-Flocks the distribution of wrinkle scores was skewed towards plainer animals (Figure 1), the average wrinkle score for each flock being 2.2 and 2.3 for neck wrinkle and 1.9 and 2.0 for body wrinkle. For these two flocks ewes with score 5 for neck and body wrinkle represented less than 0.4% of the flock. The distribution of wrinkle scores for the QPLU$ flock was less skewed with average wrinkle scores of 3.1 and 2.4 for neck and body wrinkle. In the QPLU$ flock score 5 animals represented 6% and 1.4% of all ewes for neck and body wrinkle respectively.

Figure 1. Distribution of neck and body wrinkle scores in three Merino resource flocks

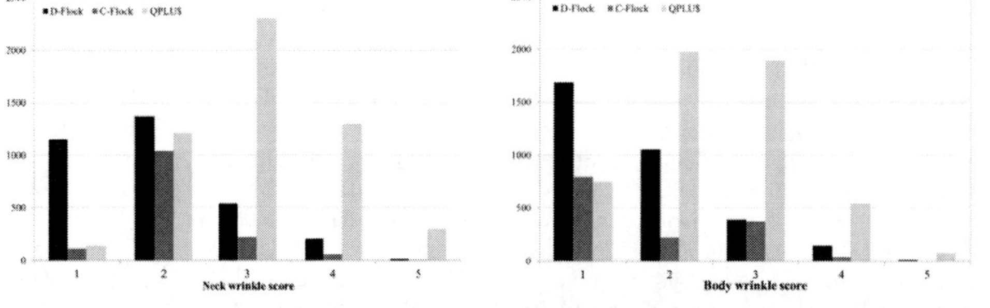

Wrinkle score, neck or body, was not a significant source of variation in either lifetime fertility or fecundity in any of the three Merino resource flocks (Table 1a and 1b). However, while wrinkle score (neck or body) was not a significant source of variation in lifetime lamb survival for the D- or C-Flocks, a significant relationship was evident in the QPLU$ flock ($P<0.001$). For each wrinkle trait, lifetime lamb survival was highest for the plainer bodied ewes (i.e. those with wrinkle scores 1, 2 or 3) compared with the wrinklier ewes (i.e. score 4 or 5) (Table 1a and 1b). For body wrinkle, the relationship was more distinct with score 1 ewes having the highest lifetime lamb survival and lifetime lamb survival significantly decreasing with each increase in wrinkle score (Table 1b).

Table 1. Lifetime NRR, its components (fertility, fecundity and lamb survival) adjusted for genotype and year of birth effects and average pre-joining liveweight from 2-6 years of age, for a) neck and b) body wrinkle score of Merino ewes of three different Merino resource flocks, together with the standard error of the difference (s.e.d.)

a)

Flock	Flock mean	Neck Wrinkle Score 1	2	3	4	5	s.e.d.
		Fertility (no. of times lambing/no. of times joined)					
D-Flock	0.70	0.693	0.672	0.672	0.673	0.771	0.049
C-Flock	0.82	0.844	0.807	0.798	0.849	0.767	0.093
QPLU$	1.30	1.267	1.281	1.285	1.300	1.292	0.031
		Fecundity (no. of lambs born/ no. of times joined)					
D-Flock	1.30	1.308	1.265	1.251	1.251	1.386	0.086
C-Flock	1.35	1.355	1.331	1.323	1.340	1.172	0.149
QPLU$	1.39	1.398	1.372	1.380	1.390	1.373	0.038
		Lamb survival (no. of lambs weaned/no. lambs born)					
D-Flock	0.68	0.679	0.664	0.650	0.650	0.700	0.058
C-Flock	0.74	0.793	0.740	0.739	0.741	0.791	0.107
QPLU$	0.68	0.670a	0.692b	0.685b	0.644c	0.606d	0.028
		Net reproduction rate (no. of lambs weaned/ no. of times joined)					
D-Flock	0.71	0.700	0.668	0.658	0.623	0.699	0.074
C-Flock	0.89	0.970	0.869	0.860	0.920	0.805	0.164
QPLU$	0.97	0.958a	0.973a	0.968a	0.906b	0.834c	0.044
		Average pre-joining liveweight (2-6 years) (kg)					
D-Flock	45.8	45.3a	45.6a	46.5b	46.0c	47.7d	0.86
C-Flock	45.6	47.7	48.8	48.7	49.7	51.3	1.97
QPLU$	55.0	54.1a	55.0b	55.2c	55.1d	56.0e	0.22

b)

Flock	Flock mean	Body Wrinkle Score 1	2	3	4	5	s.e.d.
		Fertility (no. of times lambing/no. times joined)					
D-Flock	0.70	0.694	0.671	0.659	0.686	0.596	0.056
C-Flock	0.82	0.811	0.816	0.800	0.820	0.974	0.093
QPLU$	1.30	1.259	1.285	1.300	1.286	1.344	0.035
		Fecundity (no. of lambs born/ no. of times joined)					
D-Flock	1.30	1.300	1.247	1.256	1.291	1.340	0.097
C-Flock	1.35	1.321	1.343	1.346	1.325	1.216	0.150
QPLU$	1.39	1.357	1.377	1.397	1.366	1.455	0.043
		Lamb survival (no. of lambs weaned/no. lambs born)					
D-Flock	0.68	0.677	0.656	0.638	0.687	0.663	0.065
C-Flock	0.74	0.749	0.739	0.738	0.722	0.744	0.107
QPLU$	0.68	0.704a	0.687b	0.665c	0.611d	0.553e	0.031
		Net reproduction rate (no. of lambs weaned/ no. of times joined)					
D-Flock	0.71	0.699a	0.654b	0.632b	0.672ab	0.502c	0.084
C-Flock	0.89	0.880	0.868	0.878	0.841	0.943	0.164
QPLU$	0.97	0.966ab	0.971a	0.946b	0.851c	0.812c	0.049
		Average pre-joining liveweight (2-6 years) (kg)					
D-Flock	45.8	45.4	45.8	46.3	46.3	45.2	0.97
C-Flock	49.1	48.4	48.9	49.3	49.8	52.3	1.99
QPLU$	55.0	55.3	55.2	55.2	55.1	57.3	0.63

Within each flock, values followed by different letters are significantly different at P = 0.05.

There was no significant relationship between neck wrinkle score and lifetime NRR for either the D- or C-Flocks, however this was significant for the QPLU$ flock ($P<0.001$). For that flock, the lifetime NRR was highest for those ewes with less neck wrinkle (Table 1a). For body wrinkle there was a significant relationship with lifetime NRR in both the D-Flock ($P=0.038$) and the QPLU$ flock ($P<0.001$) but not the C-Flock (Table 1b). Previous research based on annual reproduction events found high wrinkle scores were related to poorer reproductive outcomes (Turner and Young 1969), with more wrinklier ewes weaning half as many lambs during their lifetime compared to plainer bodied ewes (Dun 1964).

These significant differences in lifetime lamb survival and lifetime NRR appear to be unrelated to the average pre-joining liveweight over the lifetime of the ewes. While both neck and body wrinkle score were significant sources of variation in pre-joining liveweight for the D-Flock ($P=0.005$ and $P=0.044$ respectively) and neck wrinkle a significant source of variation in the QPLU$ flock ($P=0.022$), the wrinklier animals tended to have the highest liveweight (Table 1 a and b). Lee *et al.* (2009a) found that pre-joining liveweight was generally poorly correlated with lifetime NRR and that those ewes with the highest lifetime NRR tended to have slightly lower liveweight immediately before joining.

This preliminary study has identified a phenotypic relationship between wrinkle score and lifetime NRR rate and its lifetime lamb survival component. Commercial producers who choose to cull the most wrinkly ewes (i.e. score 4 or 5) and retain those ewes with lower wrinkle scores to reduce flystrike risk are unlikely to have any detrimental impact on the lifetime NRR of their flock. Further work is required to estimate phenotypic and genetic correlations between wrinkle score and lifetime NRR (and its components) in order to determine whether implementing a selection program for fewer wrinkles will impact on the lifetime reproductive performance of the flock in future generations.

REFERENCES

Atkins K. D., Richards J. S., and Semple S. J. (2006) In '8th World Congress on Genetics Applied to Livestock Production'. Belo Horizonte, MG Brazil pp. 05-01.

AWI Ltd and MLA Ltd (2013) Visual Sheep Scores - Researcher Version. Version 2. AWI and MLA, Sydney.

Brown D. J., Swan A. A., and H. U. Graser (2010) In '9th World Congress on Genetics Applied to Livestock Production'. Leipzig, Garmany pp. 0924.

Crook B. J. (1992) Direct and correlated responses to selection for skin fold score in Merino sheep. PhD Thesis The University of New South Wales, Sydney, Australia.

Dun R. B. (1964) Aust. J. Exp. Agric. Anim. Husb. 4: 376-385.

Gilmour A. R., Gogel B. J., Cullis B. R., and Thompson R. (2009) ASReml User Guide Release 3.0. VSN International Ltd, Hemel Hempstead, HP1 1ES, UK.

Greeff J. C., Karlsson L. J. E. and Schlink A. C (2014) Anim. Prod. Sci. 54: 125-140.

Hatcher S. and Preston J. W. V. (2015) Sml Rumin. Res. 130: 36-46.

Lee G. J., Atkins K. D. and Sladek M. A. (2009a) Anim. Prod. Sci. 49: 624-629.

Lee G. J., Atkins K. D. and Sladek M. A. (2009b) Proc. Assoc. Advmt Anim. Breed Genet.18: 382 - 385.

Lee G. J., Sladek M. A., Hatcher S. and Richards. J. S. (2014) Anim. Prod. Sci. 54: 1631-1634.

Richards J. S., and Atkins K. D. (2010) Anim. Prod. Sci. 50: 1053-1059.

Safari E., Fogarty N. M., Gilmour A. R., Atkins K. D., Mortimer S. I., Swan A. A., Brien F. D. Greeff J. C. and van der Werf J. H. J. (2007) Aust. J. Agric. Res. 58: 177-184.

Turner, H. N., and S. S. Y. Young. (1969) Quantitative genetics in sheep breeding. Macmillan of Australia, Victoria.

Proc. Assoc. Advmt. Anim. Breed. Genet. 22:405-408

OPTIMISING THE PROPORTION OF SELECTION CANDIDATES MEASURED FOR METHANE EMISSIONS IN A BEEF CATTLE BREEDING OBJECTIVE THAT INCLUDES FEED INTAKE

D.J. Cottle

School of Environmental and Rural Science, University of New England, Armidale, Australia

SUMMARY

Reducing daily feed intake (DFI) via selection for lower daily methane production (DMP) has the potential to be more cost effective than direct selection for DFI. DMP has a high heritability and high genetic correlation to DFI. The optimal proportion of randomly selected young male and female cattle in which to measure DMP was determined by modelling the measurement costs and response to selection of Angus cattle using the Angus breeding index (ABI) augmented with DMP and DFI in a combined breeding objective (BO), but without DFI being measured. Assuming a 20 year planning horizon, it was not profitable to measure any candidates for DMP. The highest breakeven DMP test cost ($41.80/head) occurred when 38% of males and no females had DMP measures. The selection response for DFI only became negative when at least 52% of males had DMP estimates.

INTRODUCTION

Methane emissions from livestock are receiving increased attention (Cole et al., 2016). Reduction in daily methane production (DMP) can be achieved via direct or indirect selection, e.g. via daily feed intake (DFI), as DFI is a highly correlated trait (Cottle, 2011; Jones et al., 2011). DMP and DFI are both very difficult and expensive to measure in pasture based systems. Robinson and Oddy (2016) suggested that when it is not practical or cost effective to measure DFI, DMP can be used as a proxy for feed eaten. Even at the highest plausible cost of methane emissions they found that the economic benefits from improved feed efficiency when measuring DMP were greater than those from reducing methane emissions.

Key questions to answer in a breeding program are: i) how much can beef producers afford to invest in DMP measurement?; ii) what is the breakeven price (BE) for individual test cost to obtain a positive net present value (NPV)?; iii) what proportions of candidate males or females in the herd should breeders measure?; and iv) what is the predicted impact on DFI of any optimal DMP measurement program? The main aim was therefore to determine the optimal proportions of male and female selection candidates to measure for DMP in a one stage selection program aimed at increasing overall index value. These proportions were determined by modelling the selection costs and responses of Angus cattle selected on the Angus Breeding Index (ABI) with DMP and DFI also included in the combined breeding objective (BO), but with DMP, not DFI, being measured in a random sample of the selection candidates.

MATERIALS AND METHODS

Selection index theory was used (Hazel, 1943). A random proportion (M) of selection candidates were measured for DMP that had an enhanced index with a higher accuracy and a larger standard deviation (SD). Let the total number of selection candidates be N. Selection is across these two cohorts, with M.N and (1-M).N candidates per cohort. Assuming random measurement of DMP, both cohorts will have the same genetic mean, and the SD of the index values within cohort j is $\sigma I_j = rIH_j.\sigma H$, where rIH_j is the accuracy of index j, and σH is the SD of the BO. Each of the three traits (Angus Breeding Index (ABI), DMP and DFI) is represented in the BO and selection on EBV was modelled as based on a single phenotype with heritability equal to the EBV reliability, with phenotypes available on the selection candidate for either trait 1 (ABI) or for trait 1 and trait 2 (DMP). Key parameters used in the indexes are summarized in Table 1. The REVs of trait 2 and 3 are negative (DMP and DFI have a cost), reflecting a typical example of unfavourable correlations. Typically, traits 2 and 3 will be selected in the non-desired direction when only using the ABI, i.e. animals will produce more methane and eat more.

Response to selection was predicted using the distributional properties of the mixture of distributions of animals; those with the ABI only, and those with the enhanced index that also includes DMP (Cottle and van der Werf, 2017). The proportion M of males or females randomly measured for DMP was varied by 1% increments to determine the genetic and economic responses for each value of M. A self-replacing herd of 300 breeding cows was assumed with a 90% calving percentage, annual 5% culling/death rate, with 5% of the male candidates and 42% of female candidates selected for replacement to maintain herd numbers.

Table 1. The key parameter values assumed in the 3-trait model. Trait 2 (T2) and trait 3 (T3) relative economic values (REV) are calculated on a yearly basis to be on the same scale as the trait 1 (T1) genetic standard deviation (GenSD).

Parameters	Trait 1 (Angus Breeding Index: $)	Trait 2 (DMP: kgCO_2e)	Trait32 (DFI: kg DM)
Accuracy of EBV (h)	0.50	0.55	0.60
REV ($/GenSD)	1.0	-3.65	-18.25
GenSD	44.28	0.80	1.92
Correlations	Genetic	Phenotypic	Residual
T1: T2	0.3	0.16	0.1
T1: T3	0.5	0.22	0.1
T2: T3	0.8	0.46	0.3

Notes:
T1: GenSD advised by Dr. Peter Parnell, Angus Australia CEO.
T2: GenSD = phenotypic SD of 42% of 138 g methane/d (Cottle, 2016b) * 25 (greenhouse warming potential of methane) = 1.45 kgCO_2e * accuracy = 0.80. EV = net price of $10/tonne CO_2e (Cottle et al., 2016) = $0.01/ kg$CO_2$e * 365 days = -$3.65/kg CO_2e/year.
T3: GenSD = phenotypic SD of 42% of 7.5kg DM/d (from Minson and McDonald, 1987 and Cottle, 2016b) * accuracy = 1.92. EV = 5c/kg DM * 365 days = -$18.25/kg DM/year, a small increase on feed cost assumed by Cottle et al. (2011) and Robinson and Oddy (2016).

A discounted cost benefit analysis of strategies with and without DMP measurement was based on the increased benefit from the additional genetic gain versus the additional cost of measuring DMP over a time horizon of 20 years with DMP estimates only occurring in the first 10 years and the first genetic benefit from DMP estimates realised in year 2 (Cottle and van der Werf, 2017). Economic assessment was based on estimated combined BO gain, traits' genetic gain, NPV of the cumulative BO ($) gain over 20 years and breakeven (BE) DMP test cost.

RESULTS AND DISCUSSION

An example comparison of the male and female population distributions with either ABI index alone or extended ABI/DMP index with an arbitrary 70% measured for DMP is given in Figure 1. The annual genetic responses of males or females in the combined BO (all 3 traits), ABI, DMP and DFI with different proportions of males or females measured for DMP are shown in Figure 2.

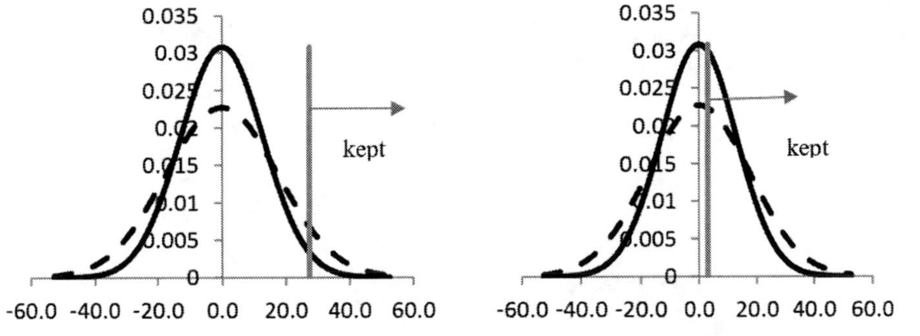

Figure 1. Proportion of candidates versus index value. The Angus Breeding Index (solid line) and extended index (dashed line) with 70% of animals measured for daily methane production (DMP). The proportion of kept males (left) with DMP measurements is higher than for females (right).

The total combined BO value increased by 36% as the proportion of cattle with DMP estimates increased from 0 to 100%, while the responses for ABI, DMP and DFI all became lower, which is in the desired direction for DMP and DFI. It is therefore best to have DMP measurements for all candidates when the cost of measurement is disregarded. However the current estimated cost of measuring DMP was high ($54.64/head, R. Hegarty, pers. comm.), which resulted in it being unprofitable to measure any candidates for DMP.

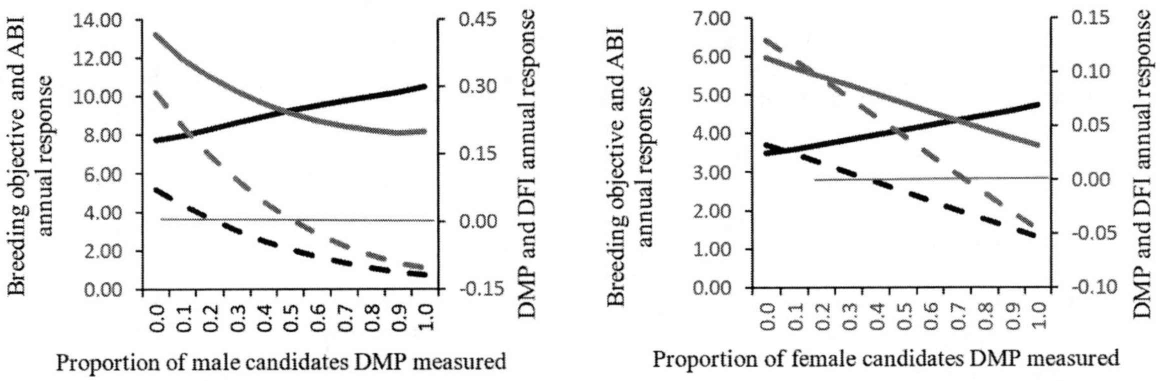

Figure 2. Annual response (per head per year) in males (left) and females (right) in combined breeding objective (BO: $, black, solid line), Angus Breeding Index (ABI: $, grey, solid), daily methane production (DMP: kg CO_2e/d, black, dashed), and daily feed intake (DFI: kg/d, grey, dashed) versus the proportion of candidates measured for DMP. Average generation length (3.4 years) was used, so the total response is the average of the male and female responses.

The highest BE ($41.80 per head) for the DMP test occurred when 38% of males and no females had DMP estimates (Figure 3). At $41.80 additional gains equal costs but DMP and DFI would be lower than when no candidates have DMP measures (Figure 2). Thus the economic situation (NPV) would be no better at this BE with DMP measurement but the environment would be improved from lower methane emissions.

A reduction of DMP from male selection only occurred when at least 23% of males had DMP measures or when 38% of females had DMP measures from female selection. A reduction of DFI from male selection only occurred when at least 52% of males had DMP measures or when 73% of females had DMP measures from female selection (Figure 2: trait intersection with zero line).

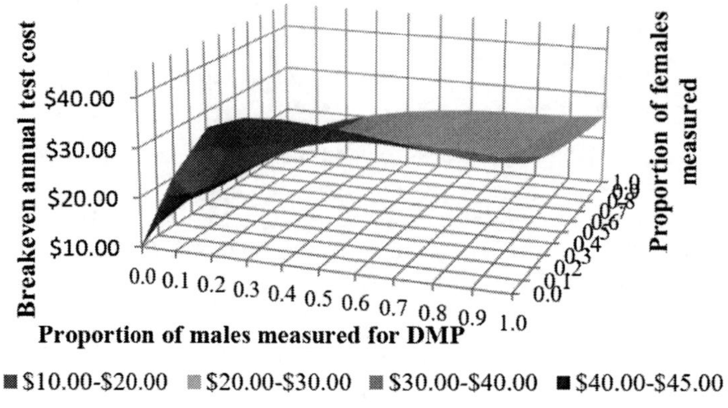

Figure 3. Response in breakeven cost ($/head) to variations in the proportions of males and females measured for daily methane production (DMP) when discounted gains over 20 years with 10 years of measurement were calculated.

Robinson and Oddy (2016) also explored incorporating DMP measurements in BOs which included DFI for cattle, where slaughter weight rather than an industry index was modelled as the first trait. However, only the estimated genetic gains per head for a single round of selection with a selection intensity of 1 were calculated. They therefore didn't study profit, only relative gain, so the optimum proportion of animals to measure for DMP, taking into account costs, was not calculated. They also found that the greatest benefit of including DMP in the BO was as a proxy for DFI. Two stage selection for DMP is difficult if animals choose themselves whether to visit the DMP measurement device and ABI values may not be known at the time of DMP measurement.

ACKNOWLEGEMENTS

Thank you to Tom Gubbins, Te Mania, Robert Wyld, Sapien Technology and Graeme Bremner, UNE for their assistance with data generation and Julius van der Werf, UNE for analysis discussion.

REFERENCES

Cole N.A., Radcliff S., DeVries T. J., Rotz A., Ely D. G. and Cardoso F. (2016) *J. Anim. Sci.* **94**: 3137.

Cottle D.J. (2011) *Proc. Aust. Assoc. Anim. Breed.* **19**: 423.

Cottle D.J, Eckard R., Bray S. and Sullivan M. (2016) *Anim. Prod. Sci.* **56**: 385.

Cottle D.J. and van der Werf J.H.J. (2017) *J. Anim. Sci.*, in press, doi:10.2527/jas2016.1177.

Hazel L.N. (1943) *Genetics* **28**: 476.

Jones F.M., Phillips F.A., Naylor T. and Mercer N.B. (2011) *Anim. Feed Sci. Tech.* **166-167**: 302.

Robinson D.L. and Oddy V.H. (2016) *J. Anim. Sci.* **94**: 3624.

Proc. Assoc. Advmt. Anim. Breed. Genet. 22:409-412

ECONOMIC VALUE OF SELECTION FOR RESIDUAL FEED INTAKE IN THE NEW ZEALAND SHEEP INDUSTRY

J.A.Archer, L.E. Proctor[1], T.J. Byrne[1]

[1] AbacusBio Ltd., Dunedin, New Zealand

SUMMARY

An analysis of the economic benefits and costs of incorporating selection to improve residual feed intake (RFI) in New Zealand's sheep industry was performed. The outcomes indicated that genomic selection for RFI, calibrated on either direct measurement of intake or measurement of greenhouse gases in Portable Accumulation Chambers (PAC) has a positive return when it is assumed that genotyping of candidates in ram breeding flocks is occurring for other traits. A comparison with a hypothetical phenotypic indicator criteria correlated with RFI suggests that further R&D effort is best directed at improving the accuracy of genomic selection in preference to a search for practical indicator measurements.

INTRODUCTION

Feed costs represent a significant component of farm operating costs in New Zealand sheep farming. As such, any reduction in feed costs, while maintaining productivity, may provide an opportunity to increase sheep farming profitability. Genetic improvement is a long-term and sustainable approach to increasing the productivity and profitability of animals, and represents a tool that can be used to improve feed efficiency. This report aims to evaluate the potential benefits from including Residual Feed Intake (RFI) as a criterion in New Zealand sheep breeding programs, and examines 3 alternatives for implementation, being:

1. Genomic selection for RFI calibrated on direct measurement of RFI.
2. Genomic selection for RFI calibrated on indirect measurement of RFI (greenhouse gas production in portable accumulation chambers (PAC)).
3. Phenotype-based selection for RFI based on a hypothetical indicator trait.

MATERIALS AND METHODS

The analysis used standard selection index theory to predict response to selection. A breeding objective was formulated based on models which describe a typical NZ maternal sheep operation. The objective traits included in a standard index include number of lambs born (NLB, Economic weight = 2231c), weaning weight (direct (WWT = 136c) and maternal (WWTM)), carcase weight (CWT = 374c), ewe weight (EWT = -119c), lamb fleece weight (LFW = 261c), hogget fleece weight (HFW = 113c), ewe fleece weight (EFW = 327c), lamb survival (direct (SUR = 8378c) and maternal (SURM)). Three traits were added to this base index, being residual feed intake on growing lambs (RFIl = -112c), replacement hoggets (RFIh = -329), and ewes (RFIe = -495). Genetic correlations between RFI traits were 0.9 (RFIl, RFIh), 0.6 (RFIl,RFIe) and 0.65 (RFIh,RFIe).

A standard group of selection criteria traits formed the base model containing traits commonly used within the Sheep Improvement Ltd genetic evaluation system for NZ sheep industry. These traits included number of lambs born, weaning weight (direct and maternal), carcase weight, hogget fleece weight, survival (direct and maternal), ewe weight, and liveweight at 8 months.

Genomic selection was modeled by including a trait to represent each genomic prediction. The heritability of each genomic trait was set as 0.95, and the accuracy of the prediction was incorporated as a genetic correlation between the genomic prediction and the corresponding profit trait. Where other traits were correlated with the corresponding objective trait, the genomic trait was also assumed to be correlated, with the correlation calculated as the accuracy multiplied by the relevant correlation between phenotype traits. Genomic predictors for WWT, WWTM, LW8, LFW, FW12, EFW and NLB were modelled with accuracies of 0.60, 0.45, 0.57, 0.53, 0.32, 0.35, 0.57 respectively. These accuracies were based off the current NZ genomic prediction accuracies (M.Lee, pers. Comm.) weighted by breed representation with the NZ sheep industry.

Accuracy of potential genomic predictions for RFI calibrated from direct measurement, and RFI and Methane Yield calibrated on PAC measurements were modelled based on the equation of Daetwyler et al. (2013). Calibrations were assumed to be against measurements on hoggets. The number of records used in the equation were calculated based on an assumed number of animals measured per year (400 for RFI directly measured and 1400 for PAC measurements) multiplied by the generation interval (3 years) and 5 generations. Effective population size was set at 500 (J. McEwan, Pers Comm) and genome length was 30 Morgans, giving a Me value of 6,279. Sensitivity analysis of assumed accuracies were undertaken (results not shown). Correlations between the calibrated traits and the objective trait RFIh are given in Table 3, and were multiplied by the accuracy of genomic prediction to give an overall correlation between the genomic prediction and RFIh. Correlations were extended to RFIl and RFIe by multiplying by the relevant correlation with RFIh. The alternative phenotypic predictor trait for rfi was modelled with a heritability of 0.25 and correlations with RFI objective traits of 0.3. These parameters were considered to be realistic, given that very few candidates for strong physiological indicators of RFI have been discovered in 20 years of significant research on this trait. Table 2 summarises the scenarios modelled.

Table 1. Accuracy of genomic predictions, calculated using equation of Daetwyler et al (2013).

	RFI	PAC_RFI	PAC_CH4
Genomic accuracy	0.47	0.50	0.62
N records	6,000	21,000	21,000
Heritability	0.30	0.10	0.19
No. measured per year	400	1400	1400
Generation interval	3	3	3
No. Generations	5	5	5
Correlation with RFIh	1.00	0.57	-0.25

Table 2. Scenarios with different information available. All scenarios included the base phenotypic measurements included as described in the text.

Scenario name	Genomics information[1]	Description
Base	N/A	Represents current recording and selection practices (no information available on RFI).
Base+G	growth, reproduction, wool	Represents current recording and selection practices but in which genomic test results are available on selection candidates in industry breeder flocks for a suite of traits (growth, reproduction, and wool), excluding RFI.
Base+G_R	growth, reproduction, wool, RFI	Genomic test results are available on selection candidates in industry breeder flocks for a young animal RFI trait – RFI genomic predictions are calibrated on individual feed intake measures.
Base+G_P	growth, reproduction, wool, PAC_RFI, PAC_CH4	Genomic test results are available on selection candidates in industry breeder flocks for a young animal RFI/CH4 traits calibrated on individual PAC measures.
Base+G_RP	growth, reproduction, wool, RFIy, PAC_RFI, PAC_CH4	Genomic test results are available on selection candidates in industry breeder flocks for a young animal RFI trait – RFI genomic predictions are calibrated on both individual feed intake measures and PAC measures.

Base+N	N/A	Represents current recording and selection practices but in which a new phenotypic selection criterion (predictor of feed intake) is available on selection candidates in industry breeder flocks.
Base+NG	growth, reproduction, wool	Represents current recording and selection practices plus genomic test results available on selection candidates in industry breeder flocks for a suite of traits, excluding RFI, and a new selection criterion (predictor of feed intake) is measured on selection candidates in industry breeder flocks.

Calculation of costs and benefits.

Costs were calculated as marginal costs over the cost of recording base phenotypic traits in ram breeding flocks. The cost of genomic selection was calculated based on $30 per ram tested. Additional costs of including RFI (direct measurement) into the genomic selection was calculated as $120,000 per year ($300 per lamb tested), or PAC measurements were $70,000 per year ($50 per lamb tested). No additional costs for the implementation of genomic selection were included, as it was assumed that genomic testing was adopted based on the non-RFI traits. The cost of generating industry reference flocks which RFI is measured on was also not included, as these flocks were assumed to be generated for other purposes (e.g. within the current industry progeny test flocks). Cost of the phenotypic indicator trait was $10 per ram.

Benefits were generated based on the response to selection (in cents per ewe per year). This was then used as an input to a model to portray the flow of rams from breeder flocks and therefore the flow of replacement daughters, genetically improved for RFI, into the national flock (i.e. the number of replacements sired by rams from flocks where RFI was integrated into selection decisions) over time based on a standard flock age structure. The analysis assumes that the first performance recorded offspring will be born with an estimated breeding value/ genomic breeding value for RFI in 2020, the first rams genetically improved for RFI will mate ewes in the commercial flock in 2022, and the first daughters arising from sires genetically improved for RFI will enter the commercial sheep flock as replacements in 2024. There is therefore a two-year lag from generation of genetically improved rams to use in the commercial flock, and a further two-year lag until the daughters of those genetically improved rams enter the commercial flock as replacements. An adoption profile was also included, starting with 20% of rams sourced as being improved for RFI, and increasing by 5% every year to a maximum of 70%. The economic value was calculated as the benefits arising from 10 years of selection, accumulated over 20 years. A discount rate of 7% was used.

RESULTS AND DISCUSSION

Total economic responses across all objective traits are shown in Table 3 along with their relativities to the base and base plus genomics scenarios. Adding genomics to the base scenario led to a 12% increase in gain. Including RFI into the genomic predictions gave an additional 2-6% economic response, depending on the calibration used. Response in RFI traits was greatest in scenarios where genomic calibrations were based on direct measurement of RFI, as might be expected given the assumed correlations of calibration traits with the objective traits. Measurement of a phenotypic predictor of RFI in ram breeding flocks produced approximately 2% additional economic response.

A comparison between costs and benefits shows that the case for inclusion of genomics (excluding RFI) in industry breeding programmes is compelling. Given this, the additional cost of generating specific calibrations for RFI as a hard to measure trait is relatively small compared to the additional benefits gained, and so this analysis supports this model as an implementation pathway for inclusion of RFI into industry breeding programmes. Cost:benefit ration might be improved by implementing two-stage selection and reducing costs by only testing the top proportion of rams. Sensitivity results on the accuracy of the genomic predictions (not shown) indicated that there is significant upside to more accurate genomic predictions for RFI. Thus effort into increasing the accuracy is warranted, and under a scenario where one-step genomic evaluation is implemented this would mean: 1) maximising the genetic relatedness between the reference population and selection candidates; 2) increasing the number of animals in the reference population; 3) ensuring the calibration phenotype is the most accurate possible (ie. increasing the heritability of the phenotype eg via appropriate test duration); or a combination of these factors.

The inclusion of a phenotypic indicator trait measured on all selection candidates was modelled on a reasonably optimistic scenario. However, while inclusion of RFI into the breeding programme via this mechanism generated positive returns, the benefits were small relative to the genomic selection scenarios while the costs were significantly larger. Thus to generate a better return than genomic selection, a phenotypic indicator would have to have a combination of parameters which here better than those used in this study. Given the practical considerations around implementation, the requirement for an additional measure to be adopted (vs no additional measurement in ram breeding flocks once genotyping is adopted), the unknown correlated responses in other traits to selection (this analysis assumed no correlated traits), and the R&D risk around identifying such a predictor, the outcomes suggest that further R&D investment would be best directed to improving genomic selection rather than a search for phenotypic predictors.

Table 3. Response to selection (cents/ewe/year), industry benefit and cost ($M over benefit/cost horizons) by RFI genetic improvement scenario.

Scenario	Base	Base+G	Base+G_R	Base+G_P	Base+G_RP	Base+N	Base+NG
Total Response	161.64	173.64	181.40	176.53	183.81	165.22	176.91
Relative to Base		12.0	19.8	14.9	22.2	3.6	15.3
Relative to Base + G			7.8	2.9	10.2	NA	3.3
Total benefit	947.0	993.2	1023.1	1004.4	1032.4	960.8	1005.8
Relative to Base		46.2	76.2	57.4	85.5	13.8	58.9
Relative to Base+G			29.9	11.2	39.2	NA	12.6
Cost Relative to Base – 100% of rams tested		13.3	14.2	13.8	14.7	4.4	17.8
Cost attributable to RFI – 100% of rams tested[1]			0.9	0.5	1.4	4.4	4.4

[1] For Base+G_R, Base+G_P, and Base+G_RP scenarios, the costs attributable to RFI are independent of the percentage of rams tested, whereas for Base+N/G, the costs scale up proportionally; 100% of rams are assumed to be phenotyped for the new selection criteria in the Base+N/G scenarios.

REFERENCES

Daetwyler, H.D., CALUS, M.P.L., PONG-WONG, R., DE LOS CAMPOS, G., Hickey, J.M. 2013 *Genetics* 193:347-365.

Proc. Assoc. Advmt. Anim. Breed. Genet. 22:413-416

NONLINEAR ECONOMIC VALUE FOR NUMBER OF LAMBS BORN IN NEW ZEALAND SHEEP INDEXES

C. Quinton[1], T. Byrne[1], P. Amer[1]

[1] AbacusBio Ltd., Dunedin, New Zealand

SUMMARY

A new non-linear economic value for number of lambs born (NLB) was developed for implementation in the New Zealand (NZ) SIL selection indexes. The function consists of a quadratic relationship between NLB EBV and dollar value at below-optimum commercial NLB levels, and a flat value above the optimum. This caps the reproduction dollar values of individuals with above-optimum NLB EBV that may be over-valued with a linear economic value. When incorporated into the NZ Maternal Worth index, the non-linear reproduction economic value mitigates the risk of very prolific genetics driving individuals' total index.

INTRODUCTION

A typical linear selection index used in genetic evaluation programs is calculated as $I_{linear} = \sum(EBV_i \times b_i)$, where for each trait i in the index, individuals' EBV_i (trait unit) are multiplied by a constant weight b_i. In this way, diverse traits including growth, reproduction, product yield and quality are incorporated to a single overall estimate of an individual's total genetic merit. In the NZ national sheep genetic evaluation, the New Zealand Maternal Worth (NZMW) index includes traits for reproduction (DPR), survival (DPS), growth (DPG), adult size (DPA) and wool (DPW). Each of these traits are weighted in indexes by linear economic values of \$/trait unit (Byrne *et al.*, 2012)

However, there is concern that the current linear economic valuation for reproduction, defined as number of lambs born (NLB), risks overweighting this trait within the NZMW for highly prolific commercial flocks. The current economic weight b_{NLB}=2231 cents/lamb was derived from national population mean NLB of approximately 1.5 lambs. But due to the wide range of breeds and farm conditions across the country, many flocks experience much higher NLB. While increasing NLB in less prolific animals will increase revenue per ewe, at high NLB ewes may not be able to rear all lambs and feed and labour required to rear additional lambs reduces profit per lamb. Therefore, in practice profit per lamb per ewe decreases with increasing NLB and there is an optimum NLB above which production becomes unprofitable. This can be defined within the NZMW total merit index by replacing the current linear reproduction valuation DPR=($EBV_{NLB} \times b_{NLB}$) by a non-linear function that describes the relationship between EBV_{NLB} and profit per ewe for a typical NZ commercial sheep farm. A previous investigation (Martin-Collado *et al.*, 2016) determined that a non-linear then flat function is the most efficient approach to value NLB within the context of a multi-trait selection index.

The objective of this study was to develop a non-linear then flat index weighting for NLB to be applied within NZMW indexes, and evaluate its effect on ram rankings for reproduction and total merit.

MATERIALS AND METHODS

Non-linear function for NLB value. First, the relationship between EBV_{NLB} and commercial phenotype was defined. This is necessary because economic values must reflect what occurs at the commercial level, but breeder flocks are approximately two generations ahead of commercial flocks, and managed differently. Commercial phenotypes were predicted as $y_{NLB}=EBV_{NLB}+1.43$, based on average difference in NLB phenotypes between breeders and commercial producers due to management, genetic lag between breeder and commercial flocks, and NLB genetic trend.

The non-linear function was defined in two parts: a quadratic function $v_q = a + by_{NLB} + cy_{NLB}^2$ describing the relationship between NLB and farm profit up to the optimum NLB phenotype, then a flat constant value at and above the optimum NLB. Coefficients b and c were defined from the existing model for deriving linear economic values for commercial NLB in the NZ sheep industry (Byrne *et al.* 2012). In the range of NLB values of 1.5 to 2.0, the relationship between NLB and marginal economic value (MEV, cents/lamb) can be described by a linear function $MEV = -5899NLB + 10904$ (Figure 1a). The optimal NLB occurs at MEV=0 where y_{NLB}=1.848 lambs. This relationship reflects the first derivative of the quadratic function describing the relationship between NLB and farm profit per ewe.

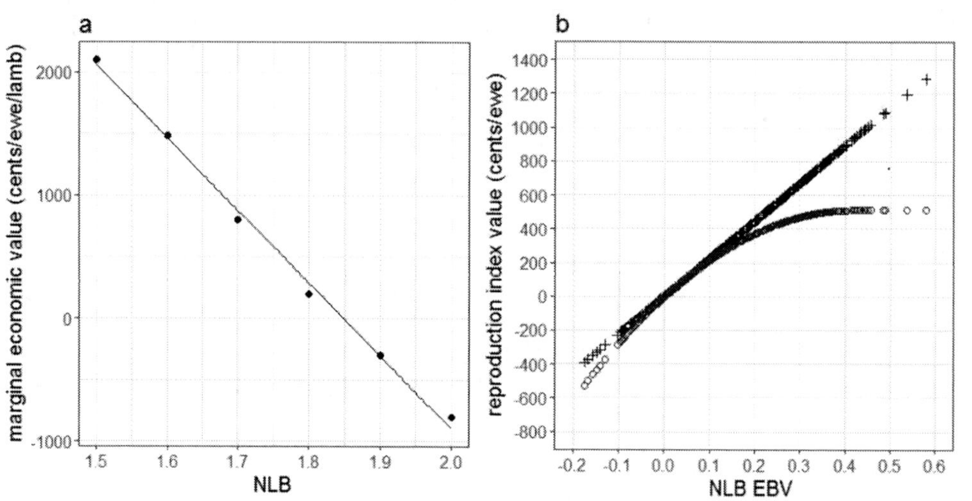

Figure 1. (a) Relationship of marginal economic value with population mean NLB. (b) Relationship of individual proven rams' EBV$_{NLB}$ with profitability according to linear economic value b_{NLB}=2231 cents/ewe/lamb (crosses), or non-linear quadratic then flat function (circles).

Coefficients of the implicit quadratic function were calculated with simple algebraic integration of MEV, so that b=10904 and c=-5899/2=-2949.5. The constant a was calculated to make mean non-linear values similar to the current mean linear values achieved with DPR=($EBV_{NLB} \times b_{NLB}$). If the current national mean EBV$_{NLB}$=0.043 lambs, then the current mean value is (0.043×2231)=95.3 cents. Setting the quadratic equation to obtain this value for a predicted commercial mean NLB phenotype y_{NLB}=1.473 lambs, a=-9566.7. The optimum of this equation occurs at y_{NLB}=1.848 where value=511 cents. The function was then modified to give flat pattern of response as predicted commercial NLB exceeds the optimum y_{NLB}. The final non-linear reproduction value (DPRnl) function was therefore:

$$DPRnl = \begin{cases} -9566.7 + 10904y_{NLB} - 2949.5y_{NLB}^2, \text{if } y_{NLB} < 1.848 \\ 511, \text{if } y_{NLB} \geq 1.848 \end{cases}$$

With this function, the economic weighting applied to NLB for an individual depends on its EBV_{NLB}. Figure 1b illustrates the difference in individual reproduction profitability values calculated with linear DPR vs. the non-linear DPRnl.

Comparison of linear and non-linear values. The effects of non-linear economic value of NLB were evaluated by comparing ram index values and rankings calculated with current linear DPR and NZMW with those calculated with DPRnl. National evaluations for rams born in 2010-2014 were extracted from the SIL database in September 2016. Individuals' predicted commercial phenotype y_{NLB}, reproduction index values DPR and DPRnl, and total merit values NZMW and NZMWnl were calculated. NZMW was the sum of DPR plus DPG, DPA, DPW, and DPS subindexes; NZMWnl was the sum of DPRnl, DPG, DPA, DPW and DPS. Comparisons were done within breed for young rams defined as those born in 2014, and proven rams defined as those with EBV_{NLB} accuracy≥60. Values for DPR vs. DPRnl, and for NZMW vs. NZMWnl were compared with Pearson and Spearman rank correlations.

RESULTS AND DISCUSSION

Results presented here are for one breed only, but are representative of major breed groups tested. Proven rams' DPR and DPRnl (Figure 1b) illustrate how for individuals with near-average EBV_{NLB}, reproduction index values were similar, but above this values diverged. At the highest EBV_{NLB}=0.58, reproduction value dropped from DPR=1291 cents/ewe to DPRnl=511 cents/ewe (the capped maximum value). The non-linear NLB economic value reduced young and proven rams' average reproduction value by 88 and 109 cents/ewe, respectively; these reductions carried through in average total index values (Table 1).

Table 1. Ram group mean values for NLB EBV and predicted commercial phenotype (y_{NLB}), linear and non-linear reproduction values (DPR, DPRnl), growth (DPG), adult weight (DPA), survival (DPS), wool (DPW), and total merit with linear and non-linear reproduction values (NZMW, NZMWnl)

	Young (N=10921)	Proven (N=381)	Top Proven by NZMW (N=100)	Top Proven by NZMWnl(N=100)
EBV_{NLB} (lambs)	0.18	0.17	0.26	0.21
y_{NLB} (lambs)	1.61	1.6	1.69	1.64
DPR (cents/ewe)	401	378	585	478
DPRnl (cents/ewe)	313	269	390	341
DPG (cents/ewe)	1122	1244	1541	1556
DPA (cents/ewe)	-321	-352	-251	-221
DPS (cents/ewe)	207	192	315	349
DPW (cents/ewe)	173	167	245	246
NZMW (cents/ewe)	1582	1629	2435	2408
NZMWnl (cents/ewe)	1493	1520	2240	2271
Corr. DPR-DPRnl	0.962	0.942	0.901	0.928
Corr. NZMW-NZMWnl	0.988	0.979	0.886	0.886
Rank corr. DPR-DPRnl	1	1	0.999	1
Rank corr. NZMW-NZMWnl	0.984	0.975	0.862	0.862

Pearson and Spearman correlations between DPR and DPRnl, and between NZMW and NZMWnl were very high (r>0.94; Table 1). However, these values are an incomplete view of ranking changes. The flat maximum value in the non-linear function reduced variance and skewed distribution of reproduction values (Figure 2a). For average rams, there is little change in value going from linear to non-linear economic value. However, as rams' EBV_{NLB} increases, there is greater loss in index value (Figures 1b, 2b). Rams that were high-ranking with linear NZMW mainly due to extreme high EBV_{NLB} dropped in rank. For the top 100 proven rams as ranked by linear NZMW, the correlations with non-linear values are weaker (Table 1). Effects on other index traits can be seen with means from the top 100 proven rams as ranked by NZMW or NZMWnl (Table 1). Top rams according to NZMWnl, had higher values for growth, adult weight and survival, compared to top rams according to the linear index (Table 1).

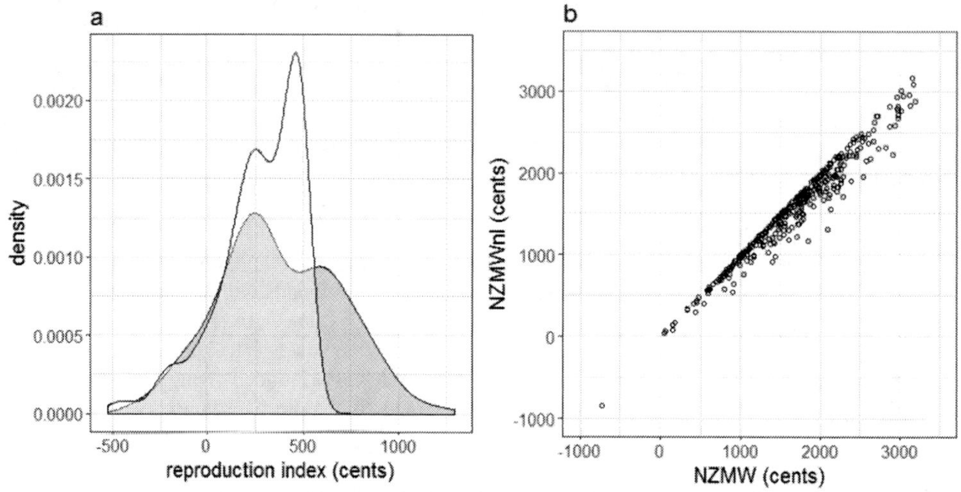

Figure 2. (a) Distributions of linear DPR (dark grey) and nonlinear DPRnl (white) reproduction index values for proven rams group. (b) Relationship of individual proven rams' total linear NZMW with NZMWnl that includes nonlinear NLB value.

The results of this study show that there is a practical way to implement a non-linear function for NLB. The non-linear function tempers the risk of the more prolific genetics badly overshooting optimum NLB. Implementation of this non-linear index function is expected to reduce population-wide selection response for NLB, but increase in response for growth traits and ewe weight.

REFERENCES

Byrne T.J., Ludemann C.I., Amer P.R. and Young M.J. (2012) *Livest. Sci.* **144**: 20.
Martin-Collado, D., Byrne, T.J., Visser, B. and Amer, P.R. (2016) *J. Anim. Breed. Genet.* **133**: 476.

Proc. Assoc. Advmt. Anim. Breed. Genet. 22:417-420

EFFECT OF GxE ON RESPONSES TO SELECTION IN RECORDED MULTI-TIER SHEEP BREEDING SCHEMES

B.F.S. Santos[1, 2], J.H.J. van der Werf[2,3], J.P. Gibson[2] and P.R. Amer[1]

[1]AbacusBio Limited, PO Box 5585, Dunedin 9058, New Zealand
[2]School of Environmental & Rural Science, University of New England, Armidale NSW, Australia
[3]Cooperative Research Centre for Sheep Industry Innovation, Armidale NSW, Australia

SUMMARY

In multi-tiered breeding schemes, the top tier is frequently maintained under different conditions to that of the commercial tier, which may lead to a genotype by environment interaction (GxE) that affects performance in target environments. Genomic selection might be useful for selection of candidates that perform well across environments in integrated breeding schemes. The results of this study demonstrate that there might be benefits from inclusion of phenotypic and genomic data from lower tiers into an integrated reference population, for situations when there is concern about GxE between the Nucleus tier and commercial populations.

INTRODUCTION

Integrated breeding schemes are frequently structured in multiple tiers. The nucleus tier pursues rapid rate of genetic gain and overall mean performance, and supplies grandparent stock to multiplier flocks or herds which provide the commercial tier with sound breeding animals as cheaply as possible (Bichard, 1977). Frequently, these tiers are maintained in different environmental conditions and for various reasons, it is common for the environment in the nucleus to be rather different from the conditions in which commercial animals are kept (James, 2009).

Falconer (1952) considered the performance of improved genotypes might be different under a less favourable environment, leading to a genotype by environment interaction (GxE). To mitigate the potential detrimental effects of impaired environment and genetic merit, breeders should target selection of robust animals capable of performing well in challenging conditions. Integrated breeding schemes could be optimised to allow for the existence of GxE among tiers, consequently increasing productivity of the commercial livestock.

DNA technologies, can contribute to better integration of multi-tier breeding schemes. A better estimation of GxE was observed with genomic selection (GS), when compared to conventional selection methods (Mulder, 2016; Silva *et al.* 2014). However, estimates of genomic breeding values based only in nucleus records might be sub-optimal predictors of direct response in the commercial environment. Genomic information and genomic relationships among individuals improve the accuracy of prediction of breeding values in an optimized reference population (Clark *et al.* 2012), and its design determines the selection response achieved in the target subpopulation. Integrated breeding schemes should source information from multiple tiers and environments (Nirea & Meuwissen, 2016).

Our hypothesis is that phenotypes and genotypes recorded on specific multiplier and commercial individuals will increase genetic progress of integrated breeding schemes and minimize the potentially detrimental effects of GxE. The objective of this study was to compare selection strategies and their effectiveness under different levels of GxE, in Australian fine-wool commercial sheep operations that exploit multi-tier breeding structures.

MATERIALS AND METHODS

A selection index model was used to compare scenarios representing different genetic correlations between nucleus, multiplier and commercial environments and various amounts of recording in the different levels. We compared selection differentials for rams and ewes from the different tiers and the percentage reduction in superiority with increased levels of GxE.

Breeding scheme. The breeding scheme was based on an actual fine-wool commercial operation which maintains a multiplier flock that produces rams to mate commercial ewes, all in the same farm. The multiplier tier is composed of commercial ewes historically selected as better performers, based on a phenotypic index, and mated to elite outside nucleus rams by artificial insemination (AI). In this simulation, either pedigree or genomic selection were assumed in the multiplier tier.

Selection index model. This study applied selection index theory to quantify responses to selection based on a pre-determined multiple-trait breeding objective. The definition of the aggregate breeding value of selection candidates, across tiers, was calculated as the sum of the products of economic weights (ew_j) of the j traits composing the breeding objective and their respective breeding values (ebv_j), computed by $H = \sum_1^n (ew_j \cdot ebv_j)$, as described by Hazel *et al.* (1994). Response to selection was calculated for each of the component traits of the breeding objective, computed as the product of the response in index value and the respective regression coefficients of traits on the index, assuming information sources from appropriate selection candidates. Genetic parameters and trait economic weights required for the calculation of the regression coefficients were supplied by Sheep Genetics. Estimated breeding values were based on phenotypes recorded in the Nucleus or in the Multiplier/Commercial tiers, plus genomic predictions with estimates accounting for or not accounting for GxE.

Selection intensities were determined for each pathway (Table 1), accounting for the dilution of intensities when selecting commercial and multiplier candidates due to having two-stage selection, and only a proportion of animals are recorded and/or genomically tested. We also account for differences in intensities when selecting nucleus rams for the multiplier, and multiplier rams used to follow up AI in the multiplier tier itself.

Table 1. Selection intensity in two-stages in different categories and tiers.

Born in	Used at	Sex	Intensity
Nucleus	Nucleus	Males	2.56
Nucleus	Multiplier	Males	2.35
Multiplier	Multiplier	Males	2.35
Multiplier	Commercial	Males	0.80
Multiplier	Multiplier	Females	1.09

Genotype by environment interaction. We modelled GxE by defining a new set of breeding goal traits (indexed as j') expressed in the multiplier and commercial tier environment, which are genetically different from but correlated with the equivalent traits expressed in the nucleus environment (j). The correlations (r_{GxE}) represented the level of GxE, and ranged from 1.00 to 0. 50. The correlation for goal traits with other traits (i) was calculated as $r_{i,j'} = r_{j,i} \cdot r_{GxE}$, where $r_{j,i}$ is the correlation between traits j and i within an environment. The calculations assumed equal phenotypic variances and heritability of traits expressed in the nucleus and in lower tiers. We also compared the potential effect of "unknown" GxE incidence by directly multiplying the direct response of each goal trait without GxE (i.e. r_{GXE}=1), by the trait economic value and this product further multiplied by the respective correlation representing the levels of GxE.

RESULTS AND DISCUSSION

Selection differentials. As GxE increased, trait selection responses decreased. The superiority of nucleus candidates for goal traits was reduced most by the increasing levels of GxE, -\$0.43 and -\$0.19 per 0.05 increase in GxE from the value of \$10.59 and \$4.53, for males and females respectively, in the absence of GxE (Figure 1a). Selection differentials were larger when GS was used (Figure 1a), compared to pedigree selection (Figure 1b). This might be caused by the additional genomic information which makes GS more accurate. With genomic selection, the level of GxE had a small effect on selection differentials in multiplier tier males (Figure 2a) and females (-\$0.04 and -\$0.03 per 0.05 increase in GxE, respectively) and also in replacement commercial females (-\$0.07). The loss of selection differential remained high however, with pedigree selection (Figure 2b). When GxE was present, but not accounted for in genetic evaluation and selection (i.e. GxE unknown), its detrimental effects were larger, -\$0.48 per 0.05 increase in GxE for nucleus males. In this case (pedigree selection), selection differentials in all tiers were reduced by increased levels of GxE.

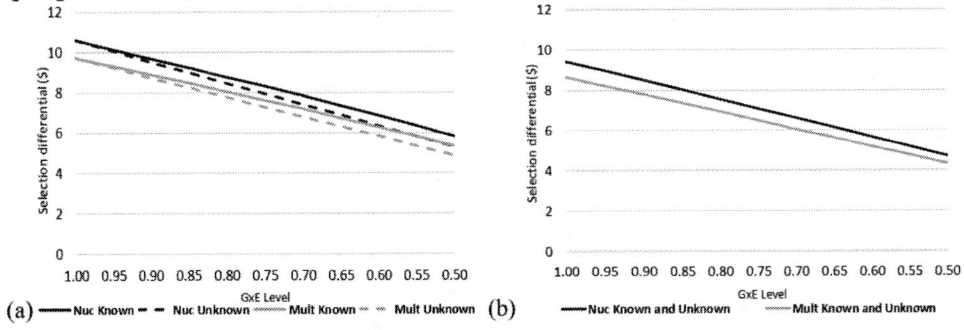

Figure 1- Superiority (in breeding objective terms) of nucleus born rams selected for use in the nucleus (Nuc) and multiplier (Mult) at different levels of known and unknown GxE, based on genomic (a) and pedigree selection (b).

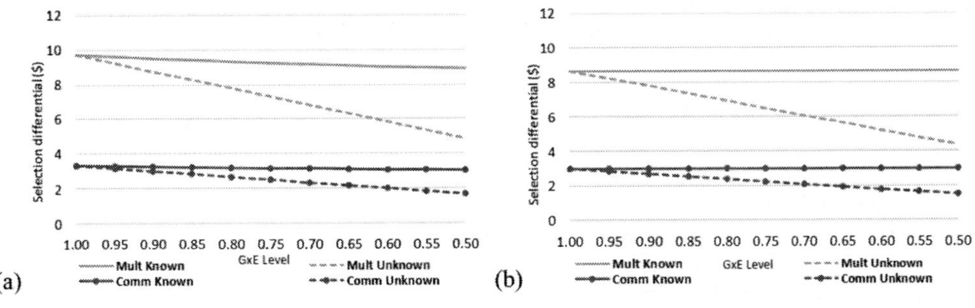

Figure 2- Superiority of multiplier born rams used in the multiplier (Mult) and commercial tier (Comm) at different levels of known and unknown GxE, based on genomic (a) and pedigree selection (b).

Reduction in selection differentials. The reduction in selection differentials when expressed as a percent reduction relative to the no GxE situation was linearly related to the increase in the value of r_{GxE} (Table 2). Comparing across tiers the reduction was greatest for the nucleus, and was slightly bigger in pedigree selection when compared to GS, -5% versus -4% per 0.05 increase in GxE, respectively. Selection differentials with pedigree selection in the multiplier and commercial tiers were unaffected because information sources are recorded in the target environment. GS resulted in small reductions in selection differentials due to GxE in the nucleus. However, in the multiplier and commercial tiers the GxE reduces the value of genomic information relative to the performance records, which are measured in the target environment.

According to Nirea & Meuwissen (2016), a genetic correlation <1 between environments is the result of a combined action of multiple environment challenges. Estimating genomic breeding values from a reference population that includes records from multiple environments might be an alternative way to address GxE in a breeding program, and this should be explored. The results of this study (Table 2) demonstrate that there would be more value from inclusion of phenotypic and genomic data from lower tiers into an integrated reference population as the genetic correlation between the Nucleus tier and commercial populations increases. In Australia, commercial farmers often source rams from nearby environments, similar to their own. While this should minimise problems of GxE, it limits the pool of selection candidates. Environment specific reference populations boosted in size and commercial relevance by training data, i.e. phenotypes and genotypes, from multiple tiers could allow more accurate and appropriate sourcing of high merit rams from other production regions and environments. In this case, genomic selection could be used to provide wider scope for identifying elite individuals from other regions and reduce the detrimental impacts of GxE on realised genetic progress.

Table 2. Percent reduction in selection superiority for both rams and ewes across different tiers at different levels of known GxE, based on genomic (GS) and pedigree selection.

GXE Level	Nucleus		Multiplier		Commercial	
	GS	Pedigree	GS	Pedigree	GS	Pedigree
1.00	-	-	-	-	-	-
0.90	-8.23	-10.05	-2.18	0.00	-2.18	0.00
0.80	-16.95	-20.11	-4.12	0.00	-4.12	0.00
0.70	-25.91	-29.89	-5.81	0.00	-5.81	0.00
0.60	-35.35	-39.95	-7.26	0.00	-7.26	0.00
0.50	-45.04	-50.00	-8.47	0.00	-8.47	0.00

ACKNOWLEDGEMENTS

The authors acknowledge the Cooperative Research Centre for Sheep Industry innovation for funding this study.

REFERENCES

Bichard M. (1977). *Livestock Production Science*, **4**, 245–254.

Clark S.A., Hickey J.M., Daetwyler H.D., van der Werf, J.H.J. (2012). *Genetics, Selection, Evolution : GSE*, **44**(1), 4.

Falconer D.S. (1952). *The American Naturalist*, **86**(830), 293–298.

Hazel L.N., Dickerson, G.E., Freeman, E. (1994). *Journal of Dairy Science*, **77**(10), 3236–51.

Mulder H.A. (2016). *Frontiers in Genetics*, **7**(October), 1–11.

Nirea K.G., Meuwissen T.H.E. (2016). *Journal of Animal Breeding and Genetics*, 1–10.

Silva F.F., Mulder H.A., Knol E.F., Lopes P.S., Guimarães, S.E.F., Lopes P.S., Mathur P.K.,Viana J.M.S., Bastiaansen J.W.M. (2014). *Journal of Animal Science*, **92**(9), 3825–3834.

Proc. Assoc. Advmt. Anim. Breed. Genet. 22:421-424

A SIMPLE METHOD FOR EVALUATING THE GENOTYPE QUALITY OF THE SIRE X CHROMOSOME USING HALF-SIB FAMILIES

M.H. Ferdosi[1], D. Johnston[1], H. A. Al-Mamun[2] and C. Gondro[2]

[1] Animal Genetics and Breeding Unit, University of New England, Australia
[2] School of Environmental and Rural Science, University of New England, Australia

SUMMARY

Recent studies have shown that the bovine X chromosome contains more than a thousand genes, some of which may be economically important. However, since males are heterogametic for the X chromosome, it has, to date, largely been ignored for genomic prediction and its information content has not been fully explored. The genotyping quality of the X chromosome is the first question that must be addressed. In this study, we suggest a simple method to impute the X chromosome of the sire using a half-sib family in order to check genotyping accuracy. The results showed that the suggested method allows for a robust imputation of the X chromosome in ungenotyped sires and is useful for the routine quality control of the genomic data.

INTRODUCTION

Chromosome X contains more than a thousand genes and it is the second largest chromosome in the bovine genome (Su *et al.* 2014). In most genomic prediction applications, the X chromosome is ignored as it requires different algorithms and methods to become useful (Sargolzaei *et al.* 2014; Su *et al.* 2014). Recent studies have shown that there are some genes in the X chromosome that may be economically important (Richardson 2016). The first step in genomic prediction is to evaluate the quality of genotyping. Therefore, it is important to check the genotyping quality of the X chromosome before any further analyses. Previous studies (Ferdosi *et al.* 2014a; Ferdosi *et al.* 2014b) have shown that the sire imputation accuracy from half-sib family genotype data is very high and that the imputed sire can be used to measure genotyping quality. However, we require a different method of sire imputation for the X chromosome.

The sex chromosomes in bovine males consist of an X and Y chromosome with a small region of homology at the telomere called the pseudo-autosomal region (Das *et al.* 2009). Thus, the sire X chromosome can be treated as a mostly haploid chromosome with a small diploid region. The haploid region should not have any heterozygosity and this fact can be used to identify the cut-off between the haploid and diploid regions. In addition, it can be used to identify the animals' gender, i.e. the males should not have any heterozygosity in this region except for genotyping errors and is therefore another way of checking the quality of genotypes. Once the pseudo-autosomal region has been identified, the same method used to impute the autosomal regions of sires (Ferdosi *et al.* 2014a) can be used to impute their genotypes in the pseudo-autosomal region. In this study, we discuss a very simple method to impute the remaining haploid region and illustrate its use for evaluating the genotyping quality of the X chromosome.

METHODS

Genotype data. Female offspring receive the X chromosome from their sire and only half-sib families that included at least one female were used for the sire imputation. The dataset included 8453 Angus (379 half-sib - HS), 4710 Brahman (323 HS), 53 Droughtmaster (9 HS), 1550 Hereford (37 HS), 527 Santa Gertrudis (131 HS), 1325 Wagyu (40HS) and 3411 Hanwoo. The Hanwoo half-sibs were only males and not used for sire imputation. They were only used to identify genotyping errors (heterozygous SNPs) on the X chromosome. The number of female half-sibs in each family was between 1 and 103. The genotyping density was varied (20k, 50k, 80k and 800k) but 150k SNP that covered the 20k, 50k and 80k panels was used as a consensus panel.

Identification of the pseudo-autosomal region. 1218 male individuals with 800k markers were used to identify the pseudo-autosomal region. This was the only region in the X chromosome where, aside from genotyping errors, there can be heterozygous sites. Therefore, the region at the end of X chromosome with a clear heterozygous block was identified as the pseudo-autosomal region in the X chromosome.

Sire imputation of the haploid and pseudo-autosomal regions. The haploid region of X chromosome was imputed by using the homozygous loci that were not in the pseudo-autosomal region in the female offspring. These homozygous loci can be used to infer the sire allele directly since it only had one X chromosome. This makes sire imputation very simple and errors in imputation were only due to genotyping errors. To resolve these genotyping inconsistencies, the rounded average of the homozygous regions in the female offspring were recorded as the sire's allele; i.e. it was sufficient to simply identify the most common homozygous sites in the female offspring. This function is available in the new version of hsphase (Ferdosi *et al.* 2014b). A similar method to hsphase (Ferdosi *et al.* 2014b) but based on the log-likelihood was used to impute the pseudo-autosomal regions of the sires. Finally, the Sire imputation accuracy was calculated as the number of correctly imputed markers that must be common with genotyped X chromosome divided by the total number of markers that were available for both imputed and real sire genotypes.

RESULTS AND DISCUSSION

The pseudo-autosomal region based on the appearance of heterozygous sites was located at the end of chromosome X around position 86.2 Mb in assembly Btau4.6.1. This position is in agreement with the region previously reported by (Das *et al.* 2009). When we aligned chromosome Y with chromosome X using BLAST, we failed to find the expected very large contiguous matching block between the two chromosomes. This could be due to the quality of the assembly of the X and Y chromosomes (Tellam *et al.* 2009). We noticed a lot of missing nucleotides in both of the chromosomes; however, the largest matching block (about 10 kb) on chromosome X was still found around the 86.3 MB region.

Figure 1 shows the boxplot of proportion of heterozygous sites in the haploid region of X chromosome in males. The results showed that the majority of individuals had less than one percent genotyping errors.

The sire imputation accuracy of the X chromosome (haploid region) for 6 cattle breeds is shown in figure 2. Generally, as the number of half-sibs in the families increased, the sire imputation accuracy increased but even small family sizes have high accuracy of imputation. The accuracy was not dependent on the SNP panel of the sire nor the breed. However, the number of SNPs that can be imputed varied according to breed and panel and the number of half-sibs in a family is more important (Figure 3). The 800k and 80k panels were suitable for genotyping the X chromosome in the Brahman and Santa Gertrudis breeds which have an indicine background (Figure 3).

The sire imputation accuracy in the pseudo-autosomal region was 0.93±0.07. This accuracy was lower than for the autosomal chromosomes reported in (Ferdosi *et al.* 2016). In that report, sire imputation accuracy in the autosomal chromosomes was more than 94% for the half-sib families with more than 7 individuals. The small number of markers and mapping errors in the diploid region may be the reason for the lower accuracy of sire imputation. However, in the haploid region only genotyping errors can decrease the sire imputation accuracy as the order of markers is not relevant.

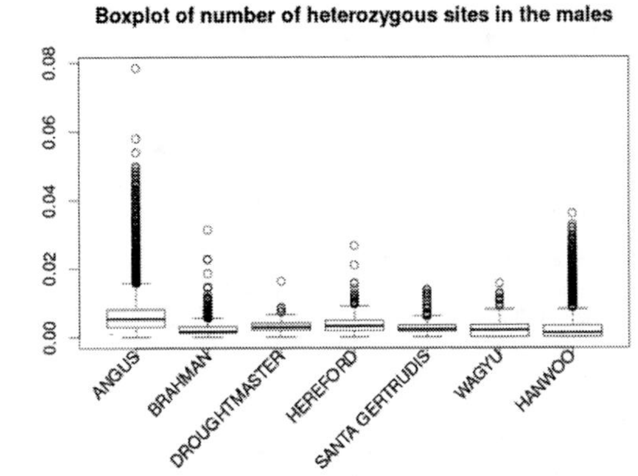

Figure 1. The proportion of number of heterozygous markers in the haploid region of the X chromosome in male animals

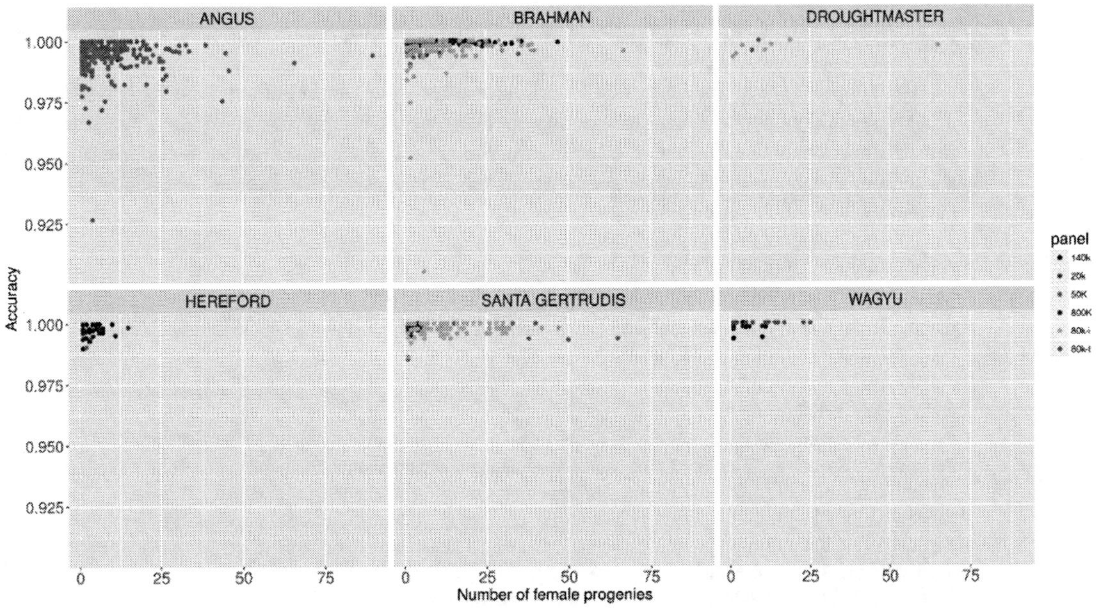

Figure 2. Sire imputation accuracy for six breeds using different panels – haploid region (i: indicus, t: taurus)

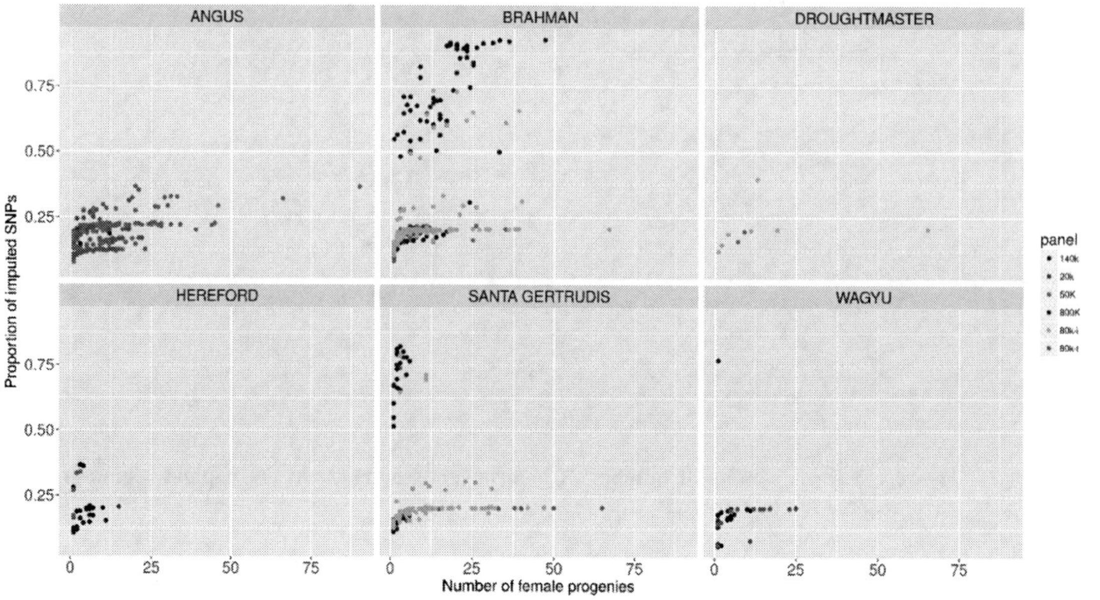

Figure 3. The proportion of SNP that can be imputed – haploid region (i: indicus, t: taurus)

The simple method detailed in this work allows for robust imputation of the X chromosome in ungenotyped sires and is useful for routine checking of the quality of genotyping. We expect that future extensions to genomic prediction methodology will make better use of the information in the sex chromosomes and this work provides an easy framework for routine imputation of the X chromosome.

ACKNOWLEDGEMENTS

MHF and DJ were supported by Meat and Livestock Australia. HAM and CG were supported by a grant from the Next-Generation BioGreen 21 Program PJ01134906 and PJ012611, Rural Development Administration, Republic of Korea and Australian Research Council (DP130100542).

REFERENCES

Das P.J., Chowdhary B.P., Raudsepp T. (2009) *Cytogenetic and genome research* **126:** 139.

Ferdosi M., Börner V., Tier B. (2016) Identifying recombination events and haplotypes in beef cattle from half-sib families. In 'International Conference on Quantitative Genetics. Madison, Wisconsin, USA'.

Ferdosi M.H., Kinghorn B.P., van der Werf J.H., Gondro C. (2014a) *Genet Sel Evol* **46:** 11.

Ferdosi M.H., Kinghorn B.P., van der Werf J.H., Lee S.H., Gondro C. (2014b) *Bmc Bioinformatics* **15:** 172.

Richardson L.A. (2016) *Plos Biol* **14:**

Sargolzaei M., Chesnais J.P., Schenkel F.S. (2014) *Bmc Genomics* **15:**

Su G.S., Guldbrandtsen B., Aamand G.P., Stranden I., Lund M.S. (2014) *Genet Sel Evol* **46:**

Tellam R.L., Lemay D.G., Van Tassell C.P., Lewin H.A., Worley K.C., Elsik C.G. (2009) *BMC Genomics* **10:** 193.

Proc. Assoc. Advmt. Anim. Breed. Genet. 22:425-428

AGREEMENT AMONG GWAS RESULTS FROM DIFFERENT STATISTICAL METHODS AS A STRATEGY TO INCREASE THE POWER OF QTL DETECTION

T.P. Melo[1], B.F. Garcia Neto[1], M.R.S. Fortes[2], L.G. Albuquerque[1,4], R. Carvalheiro[1,4]

[1] School of Agricultural and Veterinarian Sciences, São Paulo State University (Unesp), Jaboticabal, SP, Brazil
[2] School of Chemistry and Molecular Biosciences, University of Queensland, Queensland, Australia
[3] Queensland Alliance for Agriculture and Food Innovation (QAAFI) Institute, University of Queensland, Queensland, Australia
[4] National Council for Scientific and Technological Development (CNPq), Brasília, DF, Brazil

SUMMARY

The power of true positive associations in GWAS for traits affected by many QTL is generally low. This and other unfavorable scenarios pose a problem for the detection of true QTLs, which may lead to false positive associations. The aim of this study was to evaluate if combining the results of different statistical methods may increase the power to detect QTL. We simulated a polygenic trait, with known QTL positions. GWAS was performed using the WssGBLUP and BayesC methods, in a total of 8 different analyses, varying the assumptions of the SNP effects and the phenotypic data used. The results showed that as the number of analyses that a window was detected as important increased, so did the probability of that window containing a true QTL. Windows identified in 7 or 8 analyses were able to detect just some (60.5%) of the true QTL. Windows detected in at least 5 analyses captured 96% of the true QTL, but included some false positives (10.8%). Further studies are recommended, simulating traits with different genetic architectures, under different population structures, to evaluate the reproducibility of the present results.

INTRODUCTION

QTL detection remains a challenge in animal breeding, especially for lowly heritable complex polygenic traits. Under this scenario, Genome Wide Association Studies (GWAS) may present low power or high number of false positives, depending on the significance threshold adopted. Many statistical methods to perform GWAS are available (Meuwissen *et al.* 2001; Habier *et al.* 2011; Wang *et al.* 2012, and others), however their efficiency will depend on several factors such as the genetic architecture of the trait and the modeling assumptions related to the markers effects. Furthermore, other factors such as the linkage disequilibrium and the amount of phenotypic and genotypic information available may also affect the ability of QTL detection (Melo *et al.* 2016).

When a genome region is detected as important by many statistical methods, the evidence that this region harbours a true QTL is supposedly increased (Legarra *et al.* 2015). The aim of this study was to evaluate if the number of statistical methods for which a region is considered to be significant is associated with the power of QTL detection, for a simulated lowly heritable complex trait.

MATERIAL AND METHODS

Simulation. QMSim software (Sargolzae & Schenkel 2013) was used to simulate a trait with heritability and phenotypic variance equal to 0.14 and 1, respectively. A historical population, with constant size of 1,000 animals (500 males: 500 females), was simulated for 1,000 generations. The population size was then decreased until it reached 200 animals (100 females), over another 2,020 historical generations, producing a bottleneck effect and, as a consequence, genetic drift and linkage disequilibrium. The 200 animals from the last generation of the historical population were selected as the founders of an expansion population, simulated over 6 generations. In this expansion process, the number of females grew exponentially and each dam had five offspring in each generation, totaling 16,000 animals (8,000 females) at the end of the expansion process. A total of 240 males and 6,000 females from the last expansion population were randomly selected to be the founders of the selection population. The selection was performed over another 15 generations, using a replacement rate of 20% for males and females, based on estimated breeding values. Phenotypic information of the females (≈45,000) from all generations of the selection population and 2,000 randomly selected genotypes from females of the last three generations were used to perform the GWAS. This small proportion of genotyped animals was chosen to mimic a common situation. The simulated genome had a length of 2,333 cM, 735,293 markers and 7,000 QTLs. The average number of markers and QTLs per chromosome was 16,782 and 158, respectively, randomly distributed over 29 autosomes. It was assumed that QTLs explain 100 % of genetic variance. QTL allele effects were sampled from a gamma distribution with a shape parameter of 0.4, and the phenotypes were generated summing the effects of 1,000 randomly selected segregating QTLs to an error term sampled from a normal distribution with zero mean and variance of 0.86. Ten replicates of the simulation process were performed. More details about the simulation are available in Melo *et al.* (2016).

Statistical methods. Two statistical methods were used to perform the GWAS, namely weighted single-step GBLUP (WssGBLUP; Wang *et al.* 2012) and BayesC (Habier *et al.* 2011). The model adopted for WssGBLUP was: $\boldsymbol{y}=\boldsymbol{1}\mu+\boldsymbol{Z_a a}+\boldsymbol{e}$, where $\boldsymbol{y}$ is the vector of phenotypes, μ is the overall mean, $\boldsymbol{a}$ is the vector of additive genetic effects, $\boldsymbol{1}$ is a vector of ones, $\boldsymbol{Z_a}$ is an incidence matrix relating the phenotypes to $\boldsymbol{a}$, and $\boldsymbol{e}$ is the vector of residuals. The covariance between $\boldsymbol{a}$ and $\boldsymbol{e}$ was assumed to be zero and their variances were considered to be $H\sigma_a^2$ and $I\sigma_e^2$, respectively, where σ_a^2 and σ_e^2 are the direct additive and residual variance, respectively, $\boldsymbol{H}$ is the matrix which combines pedigree and genomic information (Aguilar *et al.* 2010), and $\boldsymbol{I}$ is an identity matrix. The SNP effects ($\hat{\boldsymbol{u}}$) were calculated as in Stranden & Garrick (2009): $\hat{u}=DP'[PDP']^{-1}a_g$, where $\boldsymbol{D}$ is a diagonal matrix that contains the weights for the SNPs, $\boldsymbol{P}$ is a matrix relating genotypes of each locus (coded as 0, 1 or 2 according to the number of copies of allele B) and $\boldsymbol{a_g}$ is a vector with the estimated breeding values of genotyped animals. $\boldsymbol{D}$, $\hat{\boldsymbol{a}}$ and $\hat{\boldsymbol{u}}$ were iteratively recomputed over three iterations. In the first iteration (w1), the diagonal elements of $\boldsymbol{D}$ (d_i) were assumed to be 1 (i.e., the same weight for all markers). For the subsequent iterations (w2 and w3), d_i was calculated as: $d_i=\hat{u}_i 2p_i(1-p_i)$, where $\hat{u}_i$ is the allele substitution effect of the ith marker, estimated from the previous iteration, and p_i is the allele frequency of the second allele of the ith marker. The WssGBLUP was adopted using two sets of data, one considering all available phenotypic information (SI; n=45,000) and another considering phenotypes just from genotyped animals (SII; n=2,000). The three different weights for the SNPs (w1 to w3) and the two sets of data (SI and SII) resulted in six different solutions for the SNP effects obtained under the WssGBLUP method. BayesC was applied under the model: $\boldsymbol{y}=\boldsymbol{1}\mu+\sum_{i=1}^{n}\boldsymbol{g}_i\boldsymbol{b}_i\delta_i+\boldsymbol{e}$, where $\boldsymbol{y}$, $\boldsymbol{1}$, μ and $\boldsymbol{e}$ are as previously described, $\boldsymbol{g_i}$ is the vector with the genotype of the animals for the ith SNP, $\boldsymbol{b_i}$ is the vector containing the allele substitution effect of the ith SNP and δ_i is an indicator variable (0, 1), with parameter π, where π is the fraction of SNPs not included in the model. Two π values were used, 0.99 or 0.999. The genotypes were coded as AA = 0, AB =1 and BB = 2. In summary, a total of 8 analyses were performed: WssGBLUP SI and SII (w1, w2 and w3), and BayesC (π=0.99 and π=0.999). The GWAS results were compared based on the proportion of variance explained by SNPs within consecutive 1Mb windows. For each analysis, the top 20 marker windows, which explained the greatest proportion of genetic variance, were identified and their locations were contrasted with the true QTL position. A true QTL was considered to be mapped when a top marker window was located no more than 1 Mb from a true QTL that explained at least 1% of the genetic additive variance.

RESULTS AND DISCUSSION

The simulation process resulted on average in 16.7 (±2.8) QTLs explaining 1% or more of the genetic variance. Together, the true QTLs explained on average 29.7% (±4.9) of the genetic variance, with the most important QTL explaining on average 5.1% (±2.4). The different analyses presented poor ability to map the QTLs. Individually, they were able to identify between 5.4% (WssGBLUP; SII; w3) and 17.4% (WssGBLUP; SI; w2) of the true QTLs. The power of QTL mapping increased when a window was detected as significant by different analyses (Figure 1). The percentage of true associations increased along with the number of analyses, reaching a maximum of 100% (i.e. 0% of false positives) when a window was identified as important by 7 or 8 analyses. Although presenting just true associations, windows identified in 7 or 8 analyses were able to detect just part (60.5%) of the true QTL, since some QTL were not mapped by 7 or 8 analyses, however this percentage is still high compared with the worse scenario (1.7%) in which a window was detected just by 1 analysis. The maximum percentage of true QTLs identified was observed when a window was considered as important in 5 analyses, where 96% of the true QTL were identified. This scenario (5 analyses) presented, however, 10.8% of false positive associations (Figure 1).

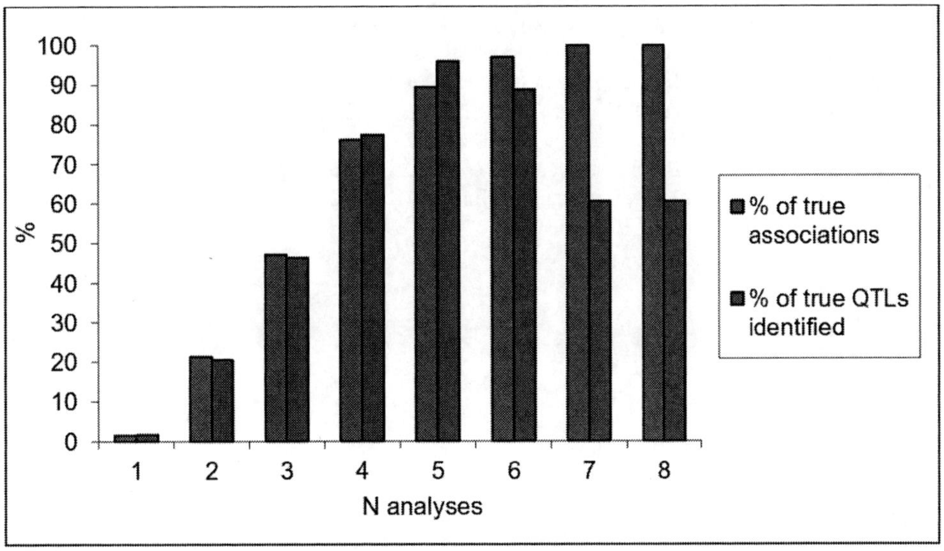

Figure 1. Percentage of true associations and of true QTL detected according to the number of analyses in which a marker window was identified as important

Our results are in accordance with Legarra *et al.* (2015), who recommended using different methods to map QTL more efficiently, arguing that no method is markedly more powerful, being dependent on the genetic architecture of the trait. Van den Berg *et al.* (2013), assessed through simulation the power of BayesC and BayesCπ to detect QTL, and also observed poor ability to detect QTL for lowly heritable complex traits. Unfortunately, the authors did not test if the agreement between results of the different methods/analyses increased the power of QTL detection.

Although our simulation study did not cover all factors affecting the QTL detection in real complex traits, the results provide evidence that the agreement among results from different statistical GWAS methods may be a feasible strategy to map QTL more precisely, especially for lowly heritable polygenic traits. Further studies may investigate the optimal number and combination of statistical methods, under different scenarios of heritability, number of genotyped animals family structure, effective population size, genetic architecture and considering other definitions of true QTLs, which would result in improved power of QTL detection.

In conclusion, our simulation approach demonstrated that agreement among GWAS results from different statistical methods can be used as a strategy to increase the power of QTL detection. This is a promising approach in the context that genomic selection can benefit from identification of true QTL (Pérez-Enciso *et al.* 2015). Our future proposition is to apply these methods to field data collected on beef cattle farms, targeting complex traits.

REFERENCES

Aguilar, I., Misztal, I., Johnson, D.L., Legarra, A., Tsuruta, S. and Lawlor, T.J. (2010) *J. Dairy Sci.* **93**: 743.

Habier D., Fernando R.L., Kizilkaya K., Garrick D.J. (2011) *BMC Bioinformatics* **12**: 1.

Legarra A., Croiseau P., Sanchez M.P., Teyssèdre S., Sallé G., Allais S., Fritz S., Moreno C.R., Ricard A., Elsen J-M. (2015) *Genet Sel Evol.* **47**: 1.

Meuwissen T.H.E., Hayes B.J., Goddard M.E. (2001) *Genetics* **157**: 1819.

Melo T.P., Takada L., Baldi F., Oliveira H.N., Dias M.M., Neves H.H.R., Schenkel F.S., Albuquerque L.G., Carvalheiro R. (2016) *BMC Genetics* **17**: 1.

Pérez-Enciso M., Rincón J.C., Legarra A. (2015) *Genet Sel Evol.* **47**: 43.

Sargolzaei, M. and Schenkel, F.S. (2013) QMSim: User's Guide. p.77.

Stranden I., Garrick D.J. (2009) *J Dairy Sci.* **92**: 2971.

Van den Berg I., Fritz S., Boichard D. (2013) *Genet Sel Evol.* **45**: 1.

Wang H., Misztal I., Aguilar I., Legarra A., Muir W.M. (2012) *Genet Res.* **94**: 73.

Proc. Assoc. Advmt. Anim. Breed. Genet. 22:429-432

A LOOK AT COMPUTATIONS FOR MULTIVARIATE SINGLE-STEP GENOMIC EVALUATIONS FITTING THE 'HYBRID MODEL'

Karin Meyer

Animal Genetics and Breeding Unit*, University of New England, Armidale NSW 2351, Australia

SUMMARY

Computational requirements for single step genomic evaluation fitting a hybrid between breeding value and marker effects models are examined for a simulated example. It is demonstrated that such a model can accommodate large numbers of genotyped animals – readily allowing exploitation of large in-core memory and parallel processing capabilities available with modern hardware – and that a principal component parameterization for multivariate analyses of numerous traits is advantageous.

INTRODUCTION

Genomic evaluation jointly considering genotyped and non-genotyped animals in a so-called single-step (SS) analysis has become routine procedure for many genetic evaluation schemes. Most implementations invoke a formulation which 'simply' replaces the pedigree based relationship matrix in the standard 'breeding value' (BV) model with its counterpart incorporating genomic information; see Legarra *et al.* (2014) for a review. Recently, Fernando *et al.* (2014, 2016) proposed an alternative which does not require construction or inversion of a genomic relationship matrix: the 'hybrid model' (HM) combines a BV model for non-genotyped animals with a 'marker effects' model for genotyped individuals to represent additive genetic effects. Describing strategies for efficient computations, the authors emphasized the scope of the HM to exploit the parallel processing capacities of modern hardware. This paper presents a first look at computational demands of multivariate genomic evaluation under the HM, including an evaluation of a parameterisation to principal components.

THE HYBRID MODEL

Consider records for q traits and let subscripts '1' and '2' denote terms pertaining to n_1 non-genotyped and n_2 genotyped individuals, respectively. Let $\mathbf{I}_q$ denote an identity matrix of size q. Ordering genetic effects by individuals or markers within traits, the multivariate HM model is

$$\begin{aligned} \mathbf{y}_1 &= \mathbf{X}_1\mathbf{b} + \mathbf{Z}_1\mathbf{u}_1 + \mathbf{e}_1 = \mathbf{X}_1\mathbf{b} + \mathbf{Z}_1\left[\left(\mathbf{I}_q \otimes \mathbf{M}_1\right)\boldsymbol{\alpha} + \boldsymbol{\epsilon}\right] + \mathbf{e}_1 \\ \mathbf{y}_2 &= \mathbf{X}_2\mathbf{b} + \mathbf{Z}_2\left(\mathbf{I}_q \otimes \mathbf{M}_2\right)\boldsymbol{\alpha} + \mathbf{e}_2 \end{aligned} \tag{1}$$

with $\mathbf{y}_i$, $\mathbf{b}$, $\mathbf{u}_i$, $\boldsymbol{\alpha}$ and $\mathbf{e}_i$ the vectors of records, fixed effects, breeding values, marker effects and residuals, $\mathbf{X}_i$ and $\mathbf{Z}_i$ the corresponding incidence matrices, and $\mathbf{M}_i$ the matrices of marker counts, appropriately centered and scaled. Marker counts for non-genotyped individuals need to be imputed. This can be done by regression using pedigree information, $\mathbf{M}_1 = \mathbf{A}_{12}\mathbf{A}_{22}^{-1}\mathbf{M}_2$, with $\mathbf{A}_{ij}$ the ij–th submatrix of the numerator relationship matrix and $\boldsymbol{\epsilon}$ accounting for imputation errors (Fernando *et al.* 2014). As formulated, the HM implies that breeding values for genotyped individuals are explained entirely by the markers fitted, but (1) is readily expanded to include additional polygenic effects if this does not hold or any other, additional random effects. Assuming $\text{Var}(\mathbf{u}_1) = \boldsymbol{\Sigma}_G \otimes \mathbf{A}_{11}$, $\text{Var}(\boldsymbol{\epsilon}) \approx \boldsymbol{\Sigma}_G \otimes (\mathbf{A}^{11})^{-1}$ and $\text{Var}(\boldsymbol{\alpha}) = \boldsymbol{\Sigma}_\alpha \otimes \mathbf{D}$, mixed model equations (MME) pertaining to (1) are

$$\begin{bmatrix} \mathbf{X}'\mathbf{R}^{-1}\mathbf{X} & \mathbf{X}_1'\mathbf{R}_1^{-1}\mathbf{Z}_1 & \mathbf{X}_2'\mathbf{R}_2^{-1}\mathbf{Z}_2\left(\mathbf{I}_q \otimes \mathbf{M}_2\right) \\ \mathbf{Z}_1'\mathbf{R}_1^{-1}\mathbf{X}_1 & \mathbf{Z}_1'\mathbf{R}_1^{-1}\mathbf{Z}_1 + \boldsymbol{\Sigma}_G^{-1} \otimes \mathbf{A}^{11} & \boldsymbol{\Sigma}_G^{-1} \otimes \mathbf{A}^{12}\mathbf{M}_2 \\ \left(\mathbf{I}_q \otimes \mathbf{M}_2'\right)\mathbf{Z}_2'\mathbf{R}_2^{-1}\mathbf{X}_2 & \boldsymbol{\Sigma}_G^{-1} \otimes \mathbf{M}_2'\mathbf{A}^{21} & \begin{matrix}\left(\mathbf{I}_q \otimes \mathbf{M}_2'\right)\mathbf{Z}_2'\mathbf{R}_2^{-1}\mathbf{Z}_2\left(\mathbf{I}_q \otimes \mathbf{M}_2\right) \\ +\boldsymbol{\Sigma}_G^{-1} \otimes \mathbf{M}_1'\mathbf{A}^{11}\mathbf{M}_1 + \boldsymbol{\Sigma}_\alpha^{-1} \otimes \mathbf{D}^{-1}\end{matrix} \end{bmatrix} \begin{bmatrix} \hat{\mathbf{b}} \\ \hat{\mathbf{u}}_1 \\ \hat{\boldsymbol{\alpha}} \end{bmatrix} = \begin{bmatrix} \mathbf{X}'\mathbf{R}^{-1}\mathbf{y} \\ \mathbf{Z}_1'\mathbf{R}_1^{-1}\mathbf{y}_1 \\ \left(\mathbf{I}_q \otimes \mathbf{M}_2'\right)\mathbf{Z}_2'\mathbf{R}_2^{-1}\mathbf{y}_2 \end{bmatrix} \tag{2}$$

*AGBU is a joint venture of NSW Department of Primary Industries and the University of New England

PC formulation. A parameterisation to principal components (PC) is obtained by replacing $\mathbf{Z}_i$ with $\mathbf{Z}_i^\star = \mathbf{Z}_i(\mathbf{Q}\otimes\mathbf{I})$, $\mathbf{u}_1$ with $\mathbf{u}_1^\star = (\mathbf{Q}^{-1}\otimes\mathbf{I})\mathbf{u}_1$ and α with $\alpha^\star = (\mathbf{Q}^{-1}\otimes\mathbf{I})\alpha$. A suitable choice is $\mathbf{Q} = \mathbf{E}\Lambda^{1/2}\mathbf{T}$, the 'factor matrix' obtained from the eigen-decomposition of $\Sigma_G = \mathbf{E}\Lambda\mathbf{E}'$ with orthogonal rotation $\mathbf{T}$ to lower triangular form (Meyer *et al.* 2015). Truncating $\mathbf{Q}$ to $r \leq q$ columns, this replaces Σ_G^{-1} in (2) with $\mathbf{I}_r$ and Σ_α^{-1} with $\mathbf{Q}'\Sigma_\alpha^{-1}\mathbf{Q}$, where $\mathbf{I}_r$ denotes an identity matrix of size r.

Computational strategies. Consider the iterative solution of (2) using a preconditioned conjugate gradient (PCG) algorithm. This requires the product of the coefficient matrix in the MME, $\mathbf{C}$, with a vector, $\mathbf{r}$, in each iterate, $\mathbf{Cr} = \mathbf{q}$. Generally, $\mathbf{C}$ is too large to be stored in core. Partition $\mathbf{C}$, $\mathbf{r}$ and $\mathbf{q}$ according to the three types of effects fitted, dropping the subscript '1' from $\mathbf{u}_1$ in the following

$$\mathbf{C} = \begin{bmatrix} \mathbf{C}_{bb} & \mathbf{C}_{bu} & \mathbf{C}_{b\alpha} \\ \mathbf{C}_{ub} & \mathbf{C}_{uu} & \mathbf{C}_{u\alpha} \\ \mathbf{C}_{\alpha b} & \mathbf{C}_{\alpha u} & \mathbf{C}_{\alpha\alpha} \end{bmatrix} \qquad \mathbf{r} = \begin{bmatrix} \mathbf{r}_b \\ \mathbf{r}_u \\ \mathbf{r}_\alpha \end{bmatrix} \quad \text{and} \quad \mathbf{q} = \begin{bmatrix} \mathbf{q}_b \\ \mathbf{q}_u \\ \mathbf{q}_\alpha \end{bmatrix} \tag{3}$$

Submatrices of $\mathbf{C}$ corresponding to $\mathbf{b}$ and $\mathbf{u}$ are the same as in the BV model, i.e. the respective parts of $\mathbf{Cr}$ can be evaluated using sparse matrix multiplication or, for large problems, standard 'iteration on data' techniques. Fernando *et al.* (2016) considered a scenario where the number of markers (m) is relatively small – in the tens rather than hundreds of thousands – so that both $\mathbf{C}_{\alpha\alpha}$ and $\mathbf{M}_2$ of size $n_2 \times m$ could be stored in core. For more than a few traits, however, the former can be too large and even evaluating its distinct $q(q+1)/2$ submatrices (for pairs of traits) once and loading them from out-of-core storage in each PCG iterate may be impracticable. Yet, $\mathbf{M}_1'\mathbf{A}^{11}\mathbf{M}_1$ of size $m \times m$ may be held in core. Fernando *et al.* (2016) described how to impute columns of $\mathbf{M}_1$ for individual markers and how to obtain this product efficiently without the need to store $\mathbf{M}_1$. The authors further emphasized evaluation of partial products, required in solving the MME, in steps. For instance, $\mathbf{C}_{u\alpha}\mathbf{r}_\alpha = (\Sigma_G^{-1}\otimes\mathbf{A}^{12}\mathbf{M}_2)\mathbf{r}_\alpha$ can be separated into dense matrix × sub-vector products for trait i, $\mathbf{M}_2\mathbf{r}_{\alpha,i} = \mathbf{t}_i$, followed by sparse products $\mathbf{A}^{12}\mathbf{t}_i$, and finally pre-multiplication of the resulting vector for all traits with $\Sigma_G^{-1}\otimes\mathbf{I}_{n_1}$. This eliminates the need to store the large, dense matrix $\mathbf{M}_2'\mathbf{A}^{21}$ of size $m \times n_1$. Moreover, evaluating a product of form $\mathbf{SWt}$ (for variable $\mathbf{t}$) in steps as $\mathbf{S}(\mathbf{Wt})$ can be more efficient than $(\mathbf{SW})\mathbf{t}$ (Strandén and Lidauer 1999). Similar decompositions can be employed for the remaining products, exploiting that $(\mathbf{I}_q\otimes\mathbf{M}_2)\mathbf{r}_\alpha$ occurs multiple times and that multiplication with $\mathbf{I}_q\otimes\mathbf{M}_2'$ can be applied to the sum of partial vectors, i.e both computationally intensive products are only needed once per PCG iterate. Detailed steps are summarised in Figure 1.

APPLICATION

To evaluate performance of the HM, data for $q = 16$ traits recorded on 1.5 million animals in 3 generations with $m = 20000$ markers were simulated using `AlphaSim` (Faux *et al.* 2016). Genetic and environmental correlations among traits were assumed to be 0.6 and 0.3 throughout, while heritabilities for odd and even numbered traits were set to 0.5 and 0.2, respectively. Analyses fitted 37,500 randomly assigned fixed contemporary groups and restricted the marker information utilised to randomly selected animals in generation 3, ranging from $n_2 = 0$ to 500000.

Marker counts were centered using frequencies estimated from the data. $\mathbf{M}_1$ was imputed by solving $\mathbf{A}^{11}\mathbf{M}_1 = -\mathbf{A}^{12}\mathbf{M}_2$ (Fernando *et al.* 2016) using sparse matrix factorisation of $\mathbf{A}^{11}$ after reordering to minimise fill-in and triangular solves for blocks of $s = 40$, 100 or 200 markers at a time. Iterative solutions of the MME were obtained using a PCG algorithm with diagonal preconditioner and convergence criterion of 10^{-7}, setting genetic and residual covariances to values reported by the simulation program, assuming $\Sigma_\alpha = \frac{1}{m}\Sigma_G$ and $\mathbf{D} = \mathbf{I}_m$. Analyses used the standard multivariate parameterisation shown above (MV16) or parameterised to r principal component (PCr) for $r = 16$, 14 and 12. Computations were carried out under Linux on a shared machine with 512GB of RAM and

1 $\begin{bmatrix}\mathbf{q}_b\\\mathbf{q}_u\end{bmatrix} := \begin{bmatrix}\mathbf{C}_{bb} & \mathbf{C}_{bu}\\\mathbf{C}_{ub} & \mathbf{Z}_1'\mathbf{R}_1^{-1}\mathbf{Z}_1\end{bmatrix}\begin{bmatrix}\mathbf{r}_b\\\mathbf{r}_u\end{bmatrix}$	6 $\mathbf{q}_u := \mathbf{q}_u + \left(\boldsymbol{\Sigma}_G^{-1}\otimes\mathbf{I}_{n_1}\right)\mathbf{w}$	11 $\mathbf{t} := \mathbf{Z}_2'\mathbf{R}_2^{-1}\mathbf{Z}_2\mathbf{t}$
2 $\mathbf{t} := \left(\mathbf{I}_q\otimes\mathbf{M}_2\right)\mathbf{r}_\alpha$	7 $\mathbf{w} := \left(\mathbf{I}_q\otimes\mathbf{M}_1'\mathbf{A}^{11}\mathbf{M}_1\right)\mathbf{r}_\alpha$	12 $\mathbf{t} := \mathbf{t} + \mathbf{Z}_2'\mathbf{R}_2^{-1}\mathbf{X}_2\mathbf{r}_b$
3 $\mathbf{q}_b := \mathbf{q}_b + \mathbf{X}_2'\mathbf{R}_2^{-1}\mathbf{Z}_2\mathbf{t}$	8 $\mathbf{q}_\alpha := \left(\boldsymbol{\Sigma}_G^{-1}\otimes\mathbf{I}_m\right)\mathbf{w}$	13 $\mathbf{w} := \left(\mathbf{I}_q\otimes\mathbf{A}^{21}\right)\mathbf{r}_u$
4 $\mathbf{w} := \left(\mathbf{I}_q\otimes\mathbf{A}^{12}\right)\mathbf{t}$	9 $\mathbf{w} := \left(\mathbf{I}_q\otimes\mathbf{D}^{-1}\right)\mathbf{r}_\alpha$	14 $\mathbf{t} := \mathbf{t} + \left(\boldsymbol{\Sigma}_G^{-1}\otimes\mathbf{I}_{n_2}\right)\mathbf{w}$
5 $\mathbf{w} := \mathbf{w} + \left(\mathbf{I}_q\otimes\mathbf{A}^{11}\right)\mathbf{r}_u$	10 $\mathbf{q}_\alpha := \mathbf{q}_\alpha + \left(\boldsymbol{\Sigma}_\alpha^{-1}\otimes\mathbf{I}_m\right)\mathbf{w}$	15 $\mathbf{q}_\alpha := \mathbf{q}_\alpha + \left(\mathbf{I}_q\otimes\mathbf{M}_2'\right)\mathbf{t}$

Figure 1. Steps to compute of Cr = q without storing sub-matrices $\mathbf{C}_{\alpha,\alpha}$, $\mathbf{C}_{\alpha\cdot}$ or $\mathbf{C}_{\cdot\alpha}$ in core.

28 Intel Xeon CPU E5-2697 cores, rated at 2.6Ghz, with a cache size of 35MB. Calculations were performed exploiting BLAS and sparse BLAS routines and the parallel direct sparse solver PARDISO to impute $\mathbf{M}_1$, all loaded from the multi-threaded Intel Math Kernel Library 11.3.2. In addition, OpenMP directives were employed to parallelise selected operations, using up to 28 threads. Sparse matrices $\mathbf{C}_{bb}$, $\mathbf{C}_{bu}$, $\mathbf{C}_{ub}$, $\mathbf{Z}_1'\mathbf{R}_1^{-1}\mathbf{Z}_1$, $\mathbf{X}_2'\mathbf{R}_2^{-1}\mathbf{Z}_2$, $\mathbf{Z}_2'\mathbf{R}_2^{-1}\mathbf{Z}_2$, $\mathbf{A}^{11}$, $\mathbf{A}^{12}$ and $\mathbf{A}^{21}$ were held in core, using compressed matrix storage. Whilst $\mathbf{C}_{bu} = \mathbf{C}_{ub}'$ and $\mathbf{A}^{12} = \mathbf{A}^{21\prime}$, holding the additional transposed copies allowed better use of BLAS routines for parallel computations at little increase in RAM required.

RESULTS

Computational requirements to determine $\mathbf{M}_1'\mathbf{A}^{11}\mathbf{M}_1$ and to solve the MME for increasing numbers of genotyped animals are summarised in Table 1, together with selected characteristics of the MME. All times shown are elapsed (wall) times for the specific task, excluding set-up steps.

Most memory (RAM) used was for in-core storage of $\mathbf{M}_2$, of size $n_2 \times m$, while holding $\mathbf{M}_1'\mathbf{A}^{11}\mathbf{M}_1$ for m = 20000 in core required just under 3GB (full-stored). Imputation of $\mathbf{M}_1$ together with calculation of $\mathbf{M}_1'\mathbf{A}^{11}\mathbf{M}_1$ required less than half an hour, with some advantage for larger numbers of markers processed at once and some increase in time required with growing n_2. In comparison, building and inverting the genomic relationship matrix for the same markers required 6, 43, 149 and 381 minutes for $n_2 = 50,100,150$ and 200K, respectively.

Not surprisingly, including genomic information in genetic evaluation increases computational demands to solve the MME by orders of magnitude, the more so the greater the proportion of genotyped individuals. Not only are there substantially more operations per iterate, but for the HM with many non-zero off-diagonal elements in $\mathbf{C}_{\alpha\alpha}$ solutions converged slowly resulting in many more PCG iterates to be performed. Employing a simple, diagonal preconditioner in the PCG algorithm, parameterising to genetic principal components reduces the number of iterates and thus time required substantially. This is due to 'de-correlating' genetic effects for different traits and thus can be less effective when genetic correlations between traits are weak or when more sophisticated conditioning schemes are applied (Meyer 2016). If the number of PCs fitted can be reduced, both RAM and computing time required are decreased further, mainly due to a reduction in the number of equations and therefore the number of operations per iterate. In comparison, a corresponding SS analysis fitting a BV model for n_2 = 50K and MV16 required only 290 iterates and 10 minutes for the solution phase.

DISCUSSION

In practice, genetic evaluation models are more complex than considered here and additional random effects – especially genetic groups – are likely to increase the resources required for HM analyses markedly. Nevertheless, results illustrate that multivariate evaluation under the HM is feasible for numerous traits and many thousands of genotyped animals, especially if aided by a parameterisation to genetic principal components. Computational demands are proportional to the number of markers considered, i.e. may necessitate research efforts to identify appropriate subsets.

Table 1. Computational requirements for genomic evaluation via the hybrid model

		No. of genotyped animals, n_2 (K)								
		0	25	50	75	100	150	200	300	500
Genotyped animals (%)		0	1.7	3.3	5.0	6.7	10.0	13.3	20.0	33.3
No. of equations (M)		24.6	24.5	24.1	23.7	23.3	22.5	21.7	20.1	16.9
RAM to store $\mathbf{M}_2$ (GB)		–	3.7	7.4	11.2	14.9	22.4	29.8	44.7	74.5
Imputation of $\mathbf{M}_1$ and calculation of $\mathbf{M}_1'\mathbf{A}^{11}\mathbf{M}_1$										
RAM (GB)	40[a]	–	26	30	33	37	45	52	67	96
	100	–	29	33	36	40	47	54	69	98
	200	–	33	37	40	44	51	58	73	102
Time (min)	40	–	23	24	24	24	25	26	27	29
	100	–	17	17	17	19	20	21	21	21
	200	–	15	15	15	16	16	16	19	19
Solution of mixed model equations										
RAM (GB)	MV16	35	42	45	48	52	59	66	81	108
	PC16	27	34	37	41	45	52	59	74	102
	PC14	25	32	36	39	43	50	58	73	101
	PC12	23	30	34	38	41	49	56	71	99
No. of iterates	MV16	350	1249	1395	1449	1457	1534	1707	1997	2302
	PC16	167	597	623	638	629	654	668	668	678
	PC14	165	584	624	612	624	653	654	659	680
	PC12	163	572	618	624	635	646	654	653	671
Time (min)	MV16	4	23	33	42	51	66	84	129	237
	PC16	2	15	18	22	29	34	44	68	100
	PC14	2	10	14	18	21	28	35	47	81
	PC12	1	9	13	16	19	24	30	41	68

[a]Block size for imputation of markers

For beef cattle, Saatchi and Garrick (2016) proposed a reduced panel comprising about 2,300 markers and reported predictive performance of more than 80% of that for a full 50K panel. For our example, analyses fitting a BV model converged more quickly, presumably in part due to inclusion of non-important markers in the HM analysis. The HM can be implemented exploiting efficient, off-the shelf linear algebra routines. It appears best suited to analyses with large numbers of genotyped animals where a sparse approximation of the inverse of the genomic relationship matrix is not desirable.

ACKNOWLEDGEMENTS

Work was supported by Meat and Livestock Australia grant L.GEN.1704.

REFERENCES

Faux A.M., Gorjanc G., Gaynor R.C., Battagin M., Edwards S.M., Wilson D.L., Hearne S.J., Gonen S. and Hickey J.M. (2016) *Plant Genome* **9**:1.

Fernando R.L., Cheng H., Golden B.L. and Garrick D.J. (2016) *Genet. Sel. Evol.* **48**:96.

Fernando R.L., Dekkers J.C.M. and Garrick D.J. (2014) *Genet. Sel. Evol.* **46**:50.

Legarra A., Christensen O.F., Aguilar I. and Misztal I. (2014) *Livest. Sci.* **166**:54.

Meyer K. (2016) *J. Anim. Sci.* **94**:4530.

Meyer K., Swan A. and Tier B. (2015) *J. Anim. Sci.* **93**:4624.

Saatchi M. and Garrick D.J. (2016) In *Plant and Animal Genome XXIV Conference*. San Diego, CA, January 9–13, 2016. Abstr.

Strandén I. and Lidauer M. (1999) *J. Dairy Sci.* **82**:2779.

Proc. Assoc. Advmt. Anim. Breed. Genet. 22:433-436

EARLY PREDICTION OF IMPORTANT ADULT WOOL TRAITS

S. Shahinfar[1], and G. Hinch[1], J. Van Der Werf[1], L. Kahn[1]
[1] University of New England, Department of Animal Science, Armidale, NSW, Australia

SUMMARY

Both production and quality play important roles in determining the wool income received by Australian sheep producers. Enabling accurate and early prediction of wool production and quality for individual and groups of sheep can provide useful information assisting on-farm management decision-making. Robustness and high performance of modern prediction methods, namely Machine Learning (ML) algorithms, make them suitable for this purpose. In this research, flock specific environmental data and phenotypic information of yearling lambs were combined to identify the most effective algorithm to predict adult Greasy Fleece weight (aGFW), adult Clean fleece Weight (aCFW), and adult Fibre Diameter (aFD). Those algorithms were evaluated in terms of prediction error and correlation between predicted and actual phenotype in a test dataset.

Multiple linear regression (MLR), Multilayer perceptron (MLP), Model Tree (MT) and Bagging (BG) were used to carry out these predictions and their performance were compared. The MLP method had the poorest performance in all three traits versus, MLR, MT, and BG had very similar performance with BG being superior in all three traits and prediction criteria, with correlation coefficients of 0.93, 0.90, 0.95 and Relative Absolute Error (RAE) of 0.34, 0.41, 0.31 for predicting aGFW, aCFW, aFD respectively.

INTRODUCTION

Phenotypic prediction of wool production of adult sheep based on their early records as yearlings has great management value for sheep producers, allowing them to base their culling decision on an accurate future prediction of wool production for each individual sheep. It is clear that beside genetics, many environmental factors and management practices contribute directly or indirectly in quality and quantity of wool, and predictions need to account for these effects.

Various authors have identified some of the more important factors that affect wool production. For example, Masters et al., (1998) demonstrated that initial liveweight, liveweight change, and supplement choice all have effect on wool growth and staple strength in weaner sheep. Ferguson et al., (2011) reported that liveweight at joining, and liveweight change during pregnancy and lactation acted to regulate wool production of Merino ewes. They used linear prediction models based on a REML approach for predicting CFW, FD, and SS from their data. The authors did not test the model on independent test data thus preventing a generalised understanding. To our knowledge prediction models for wool production of adult sheep based on their yearling records that combine genetic, environment and management effects do not exist.

The objective of this study was to develop and compare the performance of different ML algorithms to predict adult wool production using weather, pasture, animal health and various measurements of related phenotypes, and some related traits. Finally, the best performing model would be selected for further fine tuning and development in the form of a decision support system for industry.

MATERIALS AND METHODS

Data. Data collected over a period of more than 6 years as part of the Sheep CRC Information Nucleus Flocks were used in this study (van der Werf et al., 2010). After editing, the data set contained 7,501 records of animals that had yearling and at least one measurement of their adult GFW; 5,962 record for aCFW and 5,917 record for aFD. Data that were included as phenotypic measurements included, conformation characteristics of sheep that are related to wool production such as Body Wrinkle, health related features such as worm egg count (WEC), and pregnancy status of sheep at yearling. Weather information from each site where the flocks were managed was obtained from the Bureau of Meteorology (BOM). Addition of flock specific in the variable set can be considered as fixed effects in linear mixed model to capture the whole management and perhaps micro-climate effect that might exist in the flock. Also pasture data included predictions of pasture dry weight and digestibility of herbage mass obtained from a simulation model developed by Johnson et al. (2003) were used.

Machine Learning Algorithms. In order to find the best prediction model for practical use, the standard approach is to try a short list of appropriate predictive methods on the data of interest and then pick the best performing method and fine-tune it for use as the predictor tool. In this paper we are comparing a tree based method (MT), a gradient based method (MLP) and an ensemble method (BG) and compare them with the most common statistical method of prediction, Multiple Linear Regression.

a) **Multilayer Perceptron:** is a feedforward artificial neural network that takes a vector of real valued input and calculates a linear combination of these inputs into a set of appropriate outputs. It is well-suited for cases in which the instance space is noisy, complex and intercorrelated (Mitchell, 1997).

b) **Model Tree:** is a type of decision tree developed for numeric prediction. A process similar to decision trees divide and conquer approach is used to partition the multidimensional prediction space of the problem and exploit the partitions (Quinlan, 1992). Values for test instances will be predicted by a linear model stored in each leaf. The MT has been used in prediction of retention pay-off in dairy cattle (Shahinfar et al., 2014). MT often provides accurate and transparent prediction of complex systems with nonlinear and intercorrelated variables.

c) **Bagging:** which stands for bootstrap aggregation, (BG), is an ensemble method in which multiple versions of a predictor will be generated on bootstrap samples of training data to finally drive an aggregated predictor. When predicting numeric values, final prediction is an average over predicted values of all models (Breiman, 1994, and Breiman, 1996). In this study we used Bagging of MT.

Variable Selection Method. In Machine learning practices, it is tempting to include as many variables as possible to the model. Although in theory, having more features should increase the discriminative power of any ML algorithm, nevertheless, in practice often adding irrelevant features can distract the learning algorithm and defect the prediction performance as well as increase the time needed for learning and prediction phase. Full model in this research were consist of 190, 189, and 192 variable for predicting aGFW, aCFW, and aFD respectively.

Greedy hill climbing search in forward manner was used to select a small effective subset of attributes for each trait of interest. Then the same training process was carried out with selected subset of attributes and results were compared (Table 1). The reduced models were as below:

aGFW= Sex + yYLD + yGFW + yBDWR + ytMin_6 + yDryWtAv_9
aCFW = Sex + yCFW + yGFW + yBDWR + yPregScan + yrainAv_3 + ytMin_6 + yDryWtAv_12
aFD = Sex + SireBreed + yOFDA_SpinFine + yOFDA_FDSD + yOFDA_FD + AgeDiff + yFACE + yPregScan + yCS + yDryWtAv3 + yDigA8

where "a" in prefix indicates adult time and, "y" prefix indicates yearling time. BDWR = Body Wrinkle score. tMin_6 = average of minimum temperature in the 6 months prior to first shearing. DryWtAv_9= Dry weight average per hectare in the 9 months prior to first shearing. rainAv_3= Average of Rain fall in the 3 months prior to first shearing. DryWtAv_12= Dry weight average per hectare in the 12 months prior to first shearing. AgeDiff = number of days between first and second shearing. Face= Face Cover Score. CS= Body Condition Score. DryWtAv3 = Dry weight average per hectare in next 3 month after first shearing. DigAv8= Average of Digestibility of pasture in the 8th month after first shearing.

Model Evaluation. To evaluate each Model's performance in 10-fold cross validation framework, three accuracy measurements were considered, Correlation Coefficient between actual and predicted value, Root mean Squared Error (RMSE) and Relative Absolute Error (RAE). Correlation and RMSE are very well known and standard measurements for any prediction method. RAE was used in this research for two main reasons. First, it measures absolute error, which is not affected by outliers. Second, it considers the relative magnitude of the error compared with the predicted value.

$$RAE = \sum_{i=1}^{n} \frac{|\,p_i - a_i|}{|\,a_i - \bar{a}\,|}$$

Where p_i is predicted value; a_i is actual value; and $\bar{a}$ is the prediction by an arbitrary predictor, in this case the average of actual values (Witten and Frank, 2005).

RESULTS AND DISCUSSION

MT and BG always had the best performance in both cases of full (FM) and reduced model (RM) in all performance measurement criteria (Table 1). The superior accuracy of BG is due to ensemble power in which several predictor models will be aggregated to generate a high performance predictor. The superiority of BG over MT was not statistically significant and one could choose MT over BG for practical purposes, of which three are proposed herein. Firstly, the running time on MT is much less demanding than BG. Secondly, despite the black box nature of BG being ambiguous and hard to explain for users, MT is very transparent and intuitive. Thirdly, as a practical rule of thumb in ML, once a single model shows a high prediction performance, ensemble methods will not add much of accuracy. Surprisingly MLP had the poorest performance among all four methods. Perhaps in our case MLP needed much more investigation and fine tunning to deliver a reasonable performance.

In order to assess accuracy and generality of the machine learning of choice, a user should not rely on a simple comparison between two single run or even two 10-fold crossvalidation run. The problem would arise in cases that some algorithems have very close performance and some have certain advantages on others in practice. Repeated 10-fold cross validations were performed on the same partition of data for all four algorithems in use, and Tukey multiple comparison of means were performed on the mean of accuracy criteria. The results is shown in Table 1 using alphabedical superscripts. As multiple means comparison indicated, in most cases there was no significant difference between BG and MT while MLR and MLP were often associated with poorer performance in comparison. As a conclusion our method of choice was MT for early prediction of adult wool traits.

Table 1. Results of 10-fold cross validation for full and reduced models for aGFW, aCFW and aFD with multiple mean comparisons indicated in superscripts.

		Correlation		RMSE		RAE	
Trait	Method	FM*	RM*	FM	RM	FM	RM
aGFW*	MLR^{B**}	0.91^{b}	0.81^{b}	0.73^{b}	1.03^{b}	0.38^{b}	0.54^{b}
	MLP^{C}	0.87^{c}	0.81^{b}	1.03^{c}	1.21^{c}	0.58^{c}	0.68^{c}
	MT^{A}	0.93^{a}	0.92^{a}	0.66^{a}	0.70^{a}	0.35^{a}	0.37^{a}
	BG^{A}	0.93^{a}	0.92^{a}	0.64^{a}	0.69^{a}	0.34^{a}	0.36^{a}
aCFW*	MLR^{BC}	0.88^{c}	0.78^{c}	0.59^{b}	0.78^{c}	0.46^{b}	0.60^{b}
	MLP^{CD}	0.83^{d}	0.78^{c}	0.81^{c}	0.90^{d}	0.65^{c}	0.72^{c}
	MT^{AB}	0.89^{b}	0.87^{b}	0.56^{a}	0.61^{b}	0.43^{a}	0.46^{a}
	BG^{A}	0.90^{a}	0.89^{a}	0.53^{a}	0.57^{a}	0.41^{a}	0.44^{a}
aFD*	MLR^{B}	0.93^{b}	0.91^{c}	1.31^{b}	1.55^{b}	0.32^{a}	0.39^{b}
	MLP^{C}	0.88^{c}	0.88^{d}	2.00^{c}	2.06^{c}	0.54^{b}	0.56^{c}
	MT^{A}	0.94^{a}	0.93^{b}	1.26^{ab}	1.36^{a}	0.31^{a}	0.33^{a}
	BG^{A}	0.95^{a}	0.94^{a}	1.23^{a}	1.29^{a}	0.31^{a}	0.32^{a}

aGFW= Adult Greasy Fleece weight, aCFW= Adult Clean Fleece Weight, aDF= adult Fibre Diameter, *FM= Full Model, RM= Reduced Model. Correlation= correlation between actual and predicted value in test set.
** Alphabedic superscript in method column shows overall method's mean comparison.

REFERENCES

Breiman, L. 1994. Bagging predictors. Technical report. Department of Statistics, University of California Berkeley, CA.

Breiman, L. 1996. Bagging predictors. Mach. Learn. 24:123–140.

Witten, I. H., and E. Frank. 2005. Data Mining, 2nd ed. Elsevier, San Francisco, CA.

Johnson IR, Lodge GM, White RE (2003). The sustainable grazing systems pasture model: description, philosophy and application to the SGS National Experiment. Animal Production Science 43(8), 711-728

Masters DG, Mata G, Liu S. M, and Peterson AD, 1998, Innfuence of liveweight, liveweight change, and diet on wool growth, staple strength, and fibre diameter in young sheep. Aust. J. Agric. Res., 49, 269-277

Ferguson, M B, Thompson AN, Gordon DJ, Hyder MW, Kearney GA, Oldham CM, and Paganoni BL, 2011, The wool production and reproduction of Merino ewes can be predicted from changes in liveweight during pregnancy and lactation. Animal Production Science, 51, 763–775

Mitchell, T. M. 1997. Artificial neural networks. Pages 81–85 in Machine Learning. McGraw-Hill International Edition, New York, NY.

Van der Werf, J. H. J, Kinghorn, B. P, Bank, R. G, 2010, Design and role of an information nucleus in sheep breeding programs. Anim. Prod. Sci. 50: 998-1003.

Proc. Assoc. Advmt. Anim. Breed. Genet. 22:437-440

SNP-PANEL DESIGN FOR DAIRY PROPORTION ESTIMATION AND PARENTAGE TESTING

E.M. Strucken[1], C. Esquivelzeta-Rabell[2], H.A. Al-Mamun[1], C. Gondro[1], O.A. Mwai[3] and J.P. Gibson[1]

[1] School of Environmental and Rural Science, University of New England, NSW, Australia
[2] Pic Improvement Company (PIC), Genetic Services, Hendersonville, USA
[3] International Livestock Research Institute (ILRI), Nairobi, Kenya

SUMMARY

The selection of small numbers of SNPs to analyse population features is an important task in the livestock industry. Populations differ in their genetic architecture, which often requires the selection of population specific SNPs. Different tasks, such as breed proportion prediction or parentage testing, also require specific panels. We tested which selection methods are best for breed proportion estimation and parentage testing in a crossbred dairy population from East Africa. We selected SNPs from a 735k SNP panel (Illumina) based on several methods: **a)** high minor allele frequencies; **b)** high allele frequency differences between ancestral populations; **c)** at random; **d)** with a differential evolution algorithm. Estimates of breed proportions in the subsets were tested against *true* breed proportions based on all 770k SNP obtained from ADMIXTURE. Parentage assignments was based on opposing homozygotes. Panels selected for largest allele frequency differences in ancestral populations gave best results for breed proportion predictions and panels selected for highest minor allele frequency gave best parentage resolution.

INTRODUCTION

The selection of small numbers of SNPs to carry out a variety of genomic test is at the forefront of the livestock industry. Challenges for small SNP panels are the accurate prediction of breed proportions and the assignment of parentages. Knowledge about breed proportions is important to the livestock sector for quality trait marks (*e.g.* Wagyu) but also for breeding decisions, especially in crossbred population. Whilst pure breeds such as Holstein, Jersey, or Wagyu are mostly used in industrialised settings, crossbreds find their application in developing countries where one animal must fulfil multiple purposes (*i.e.* milk and meat). To improve crossbreds, their breed proportion must be determined to choose the best breed or animal for mating. Similarly, assigning parentages is important in the livestock industry, as the pedigree determines factors such as inbreeding, breeding value estimation, or a tracking of agricultural goods. Again, in industrialized settings, record keeping of pedigrees is common practice whilst in developing countries accurate pedigrees are often missing.

Both breed proportion prediction and parentage assignment can be carried out on the basis of genomic information. In theory, however, to accurately predict breed proportions in a crossbred animal, a prior knowledge based on trading history and breeding preferences is required to determine the most likely ancestral breeds. The genomic information of these ancestral breeds is then traced within the crossbred animals. To distinguish the different genomic footprints of the ancestral breeds, it is favourable if the ancestral breeds are genomically different from each other. This should lead to a large allele frequency range of selected markers in the crossbred population.

For parentage assignment, most tests rely on the likelihood that a parent-offspring pair shows the same genotype. Simpler tests only consider homozygous genotypes, especially if only one parent is known. Opposing homozygotes describe the occurrence of a parent displaying one homozygous genotype whilst the offspring displays the other homozygous genotype (Hayes 2011). The more opposing homozygotes are found between two animals, the less likely it is that they are a parent-offspring pair. The highest likelihood, according to Hardy-Weinberg, to observe opposing homozygotes in a population is given for markers with high minor allele frequency. Thus, both breed proportion prediction and parentage assignment depend on different qualities of SNPs.

In this study, we used different selection methods to choose small panels of SNPs (100 to 1500 SNPs) from a 735k panel to determine breed proportions and parentages in a crossbred dairy population of East Africa. Based on the crossbreeding history in Kenya and Uganda (Rege and Tawan 1999; Hanotte *et al.* 2000), an African *Bos taurus* and a *Bos indicus* reference breed as well as 5 European dairy breeds were chosen to determine breed proportions.

MATERIALS AND METHODS

A total of 1,933 crossbred dairy cows from Kenya and Uganda and local indigenous breeds of Ankole (n=43), Nganda (n=17), and Small East African Zebu (Zebu; n=58) were sampled (Dairy Genetics East Africa, DGEA1, project). Additionally, genotypic datasets for N'Dama (as the reference African *Bos taurus* breed; n=20), Nelore (as the reference *Bos indicus* breed; n=20), Guernsey (n=20), Holstein (n=20), and Jersey (n=20) were sourced from the International Bovine HapMap consortium. Further, British Friesian (n=25) from the SRUC in Scotland and Canadian Ayrshire (n=20) from the Canadian Dairy Network (CDN) were used as reference breeds.

All animals were genotyped with the 770k BovineHD Beadchip array (Illumina Inc., San Diego, CA, USA). Genotypes were filtered using *SNPQC* (Gondro *et al.* 2014) with a sample-wise call rate of 90%; a median GC score <0.6; and a GC score<0.6 in at least 10% of the samples. Only markers contained on the 29 autosomal chromosomes were included in the analysis. The cleaned population datasets were merged and included 735,293 SNPs. Markers that were excluded due to quality control criteria in one breed but not in another were set to NA in the breed for which they were excluded.

True breed proportions of the crossbreds were estimated using the full quality controlled data in the ADMIXTURE 1.23 program (Alexander *et al.* 2009). The analysis was supervised with N'Dama, Nelore, Ayrshire, Friesians, Guernseys, Holstein, and Jerseys as assumed ancestral populations.

The pedigree of the crossbreds was reconstructed based on presence or absence of opposing homozygotes (Hayes 2011) and contained 171 cows with 189 offspring, of which 15 cows had two and one cow had three offspring. Parentage testing was based on opposing homozygotes and panel resolution determined based on the separation value (Strucken *et al.* 2014).

Subsets of SNPs ranging from 100 to 1,500 markers were selected based on **a)** highest minor allele frequency in the crossbreds, **b)** absolute allele frequency difference between the ancestral breeds (European dairy breeds vs. a combination of Nelore and N'Dama), at **c)** random (results were averaged across 10 random samples), and **d)** a differential evolution algorithm (Gondro *et al.* 2013, Esquivelzeta *et al.* 2015). Accuracies of dairy proportion prediction were assessed with the coefficient of determination (r^2) between the *true* proportions and the estimated proportions from the subsets. Parentage assignment was assessed with the *separation value* which is based on opposing homozygotes (Strucken *et al.* 2014, 2016).

RESULTS AND DISCUSSION

Allele frequencies showed relatively large interquartile ranges for all selection methods (0.35-0.65) apart for highest MAF (Figure 1). Allele frequencies of SNP subsets were assumed to play a major role for their performance in breed proportion prediction and parentage assignment. Markers with largest allele frequency differences between ancestral breeds should be able to distinguish breed proportions in crossbred animals. Therefore, allele frequencies were expected to show a larger variation in the crossbreds. Markers with a high minor allele frequency, i.e. both alleles occur equally often, have the highest probability to show opposing homozygotes between two unrelated individuals, therefore should work best for parentage assignment.

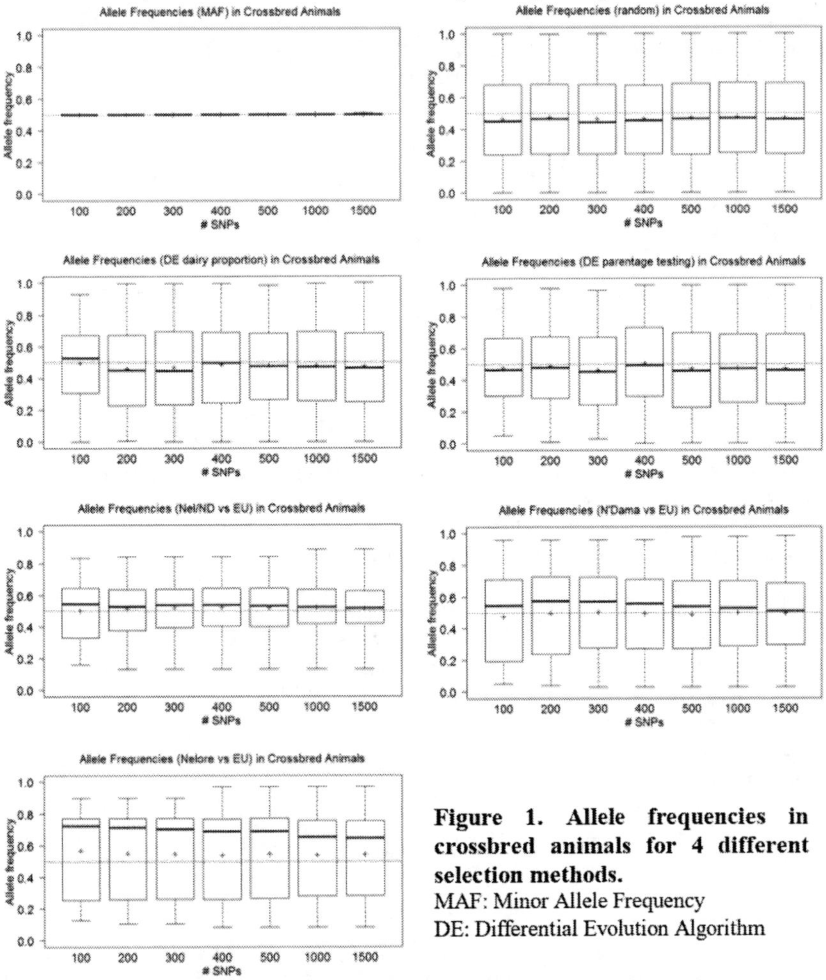

Figure 1. Allele frequencies in crossbred animals for 4 different selection methods.
MAF: Minor Allele Frequency
DE: Differential Evolution Algorithm

Individual breed proportion estimates of the ancestral breeds proved to be highly variable depending on the number of assumed ancestral breeds. Therefore, we used the total proportion of European dairy breeds as a more reliable contrast to the African N'Dama and indicine Nelore. Dairy breed proportions of the crossbred animals were on average 0.7 (SD 0.21).

The various panels predicted total dairy breed proportions with an r^2 of 0.694-0.950 (SE 0.005-0.013) for the smallest subsets of 100 markers (Figure 2a). The best results for all panel sizes was achieved with SNPs selected for largest absolute allele frequency difference between the ancestral breeds.

Lowest numbers of opposing homozygotes were found for panels selected for high minor allele frequency (Figure 2b), thus should perform best for parentage assignments. With 100 markers, however, none of the selection methods resulted in a panel that was able to assign all parentages correctly, as this requires a separation value >0.

All panels that were selected based on the Kenyan and Ugandan crossbred animals were validated in independent crossbred populations of Ethiopia and Tanzania (N=545, N=462). The selection methods rank similarly in the validation populations with the panels selected for largest allele frequency differences between Nelore/N'Dama and the EU dairy breeds performing best for breed proportion prediction and the panels selected for highest minor allele frequency in the Kenyan and Ugandan crossbreds performing best for parentage assignment.

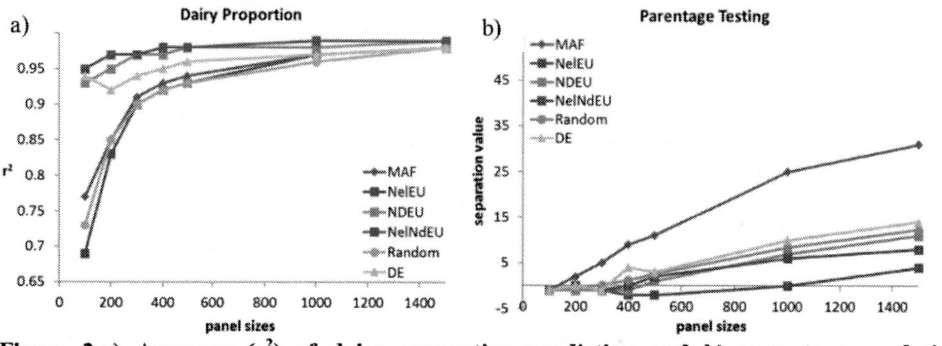

Figure 2.a) Accuracy (r^2) of dairy proportion prediction and b) parentage resolution (separation value) of SNP subsets from 4 different selection methods.
MAF: Minor Allele Frequency; NelEU: Nelore vs EU; NDEU: N'Dama vs. EU; NelNdEU: combined Nelore and N'Dama vs EU; DE: Differential Evolution Algorithm

A combination of panels performing best for breed proportion prediction and parentage assignment performed poorer than the individual panels with same number of SNPs. Further, it showed that breed proportion prediction mainly depends on allele frequencies, i.e. the difference between allele frequencies in ancestral breeds, and the ability to assign parentages is mainly limited by the number of markers (Strucken *et al.* 2016).

ACKNOWLEDGEMENT

We acknowledge the Bill and Melinda Gates foundation. HAM and CG acknowledge the Next-Generation BioGreen 21 Program (PJ01134906), Rural Development Administration, Republic of Korea, and the Australian Research Council (DP130100542).

REFERENCES

Alexander D.H., Novembre J. and Lange K. (2009) *Genome Res.* **19**, 1655.

Esquivelzeta-Rabell C., Al-Mamun H.A., Lee S.H., Lee H.K., Song K.D. and Gondro C. (2015) *Association for the Advancement of Animal Breeding and Genetics Conference (AAABG)*, Lorne, Australia.

Gondro C., Porto-Neto L.R. and Lee S.H. (2014) *Anim. Genet.* **45**, 758-61.

Gondro C. and Kwan P. (2013) In '*Bioinformatics: Concepts, Methodologies, Tools, and Applications*', pp. 105, IGI Global.

Hayes B.J. (2011) *J. Dairy Sci.* **94**, 2114.

Hanotte O., Tawah C.L., Bradley D.G., Okomo M., Verjee Y., Ochieng J. & Rege J.E. (2000) *Mol. Ecol.* **9**, 387.

Rege J.E.O. & Tawan C.L. (1999) *Anim. Genet. Resources Information Bulletin* **26**, 1.

Strucken E.M., Gudex B., Ferdosi M.H., Lee H.K., Song K.D., Gibson J.P., Kelly M., Piper E.K., Porto-Neto L.R., Lee S.H. and Gondro C. (2014) *Anim. Genet.* **45**, 572.

Strucken E.M., Lee S.H., Lee H.K., Song K.D., Gibson J.P. & Gondro C. (2016) *J. Anim. Breed Genet.* **133**, 13.

Proc. Assoc. Advmt. Anim. Breed. Genet. 22:441-444

DEFINING ENVIRONMENTAL STRESS CONDITIONS THAT PRODUCE DIFFERENTIAL SURVIVAL IN BLACK TIGER SHRIMP, *PENAEUS MONODON*

A.R Foote[1,2], C.N Stratford[1,2], E. Giardina[1,3], D.R Jerry[1,2] and N.W Wade[1,4]
[1]ARC Research Hub for Advanced Prawn Breeding, James Cook University, Queensland, Australia.
[2]Centre for Sustainable Tropical Fisheries and Aquaculture, College of Science and Engineering, James Cook University, Queensland, Australia.
[3]Seafarms Group Ltd, Ella Bay Road, Flying Fish Point, Queensland 4860, Australia.
[4]Aquaculture Program, CSIRO Agriculture and Food, Queensland Bioscience Precinct, St. Lucia, QLD, 4067, Australia.

SUMMARY

Breeding for shrimp lines that are resilient to sudden environmental stressors requires accurate definition and measurement of stress traits, along with the additive genetic control of the trait. In a breeding program for *Penaeus monodon,* families that exhibit increased tolerance to acute salinity and ammonia changes are targets for selection; however, there is currently no information on how to best challenge shrimp to induce differential survival and measurement of environmental stress tolerance. This study developed challenge testing methodologies for acute salinity and ammonia tolerance in postlarval *P. monodon.* Results showed differential survival among postlarvae at a range of dose levels. By applying the stress methods developed in a commercial scenario to differentiate between individual families, rankings and selection of more resilient lines could be incorporated in selective breeding programs. Further, the potential to apply these measures as a stress resistance marker in a commercial scenario will be evaluated following the determination of trait heritability and by correlating larval stress performance rankings with grow-out performance.

INTRODUCTION

Domestication and selective breeding programs have led to significant production gains in shrimp farming globally. Whilst selection programs have focused on growth and disease as primary traits to improve, cultured shrimp are also exposed to a range of environmental stressors throughout their production cycle that can affect productivity and survival, either though inducing mortality events, or by decreased growth. Therefore, the ability to identify family lines and select for increased tolerance to sudden environmental stress events may be a desirable trait for future breeding programs. Currently there is no data on how best to conduct challenge tests for environmental stress tolerance in shrimp and whether such traits exhibit significant additive genetic variance (i.e. is heritable). This study focused on developing methods of testing the resilience of postlarvae (PL) to acute salinity and ammonia stress in a way that can be quickly and easily applied in a commercial hatchery prior to stocking into ponds. Following the challenge of sufficient numbers of families, the genetic and genomic basis of this resilience could be determined and may allow heritability to be determined and a genetic marker for stress tolerance to be developed and incorporated into selective breeding programs.

MATERIALS AND METHODS

Penaeus monodon broodstock were sourced from Northern Territory coastal waters and progeny spawned in a commercial hatchery at Flying Fish Point, Queensland. Broodstock maturation and spawning followed routine commercial procedures, with multiple females spawned in communal spawning tanks and the progeny of multiple families reared in communal greenwater rearing tanks. At PL stage 15, 10,000 PL were transported via air freight to Bribie Island Research Centre (BIRC), Woorim, Queensland for experimental testing. At BIRC, PL were stocked at a density of 2500 PL in 5 tonne fibreglass tanks, held indoors with a 12 h day/night photoperiod, receiving 4L/m-1 of filtered 29 ±0.5°C seawater and fed on a diet of commercial flake and pellet.

A range of salinity and ammonia dose rates were tested at a range of PL ages from 15 to 42 (Table 1). Salinity doses were achieved by mixing required volumes of freshwater with undiluted seawater, termed 100% seawater. A calibrated YSI probe was used to ensure 100% seawater salinity levels were consistent over time (38±0.2 ppt). The required volume of 30% ammonia solution AR, NH_4OH (Chem-Supply) was added to the treatments to achieve the required dose, which was also confirmed with a titration kit (API).

Table 1. Salinity and ammonia treatment dose levels at each PL age.

PL age	Salinity rate (% of raw seawater)	PL age	Ammonia rate (mg-1)
15	0, 5, 10, 15, 20, 100	15	11; 21; 53; 214
25	0, 5, 10, 15, 100	22	32; 43; 53; 64
28	0, 5, 10, 15, 100	27	21; 27; 32; 43; 53
32	0, 5, 10, 15, 100	28	16; 21; 27
		34	11; 16; 21; 27; 32

Each treatment was performed in 9 L containers containing 2 L of water. The water temperature of both rearing tanks and experimental containers was 29 ±0.5°C. Treatment container temperature was maintained by placing the containers in a temperature controlled water bath. Treatment salinity or ammonia water parameters (Table 1) were set up prior to commencement of PL stress treatment in batches that were then distributed to each 2 L replicate to ensure consistency among replicates.

Approximately 30 ±10 PL were added to each control and treatment container without prior counting, to reduce handling stress. The total number of animals per treatment was calculated at the subsequent hourly time points when data was collected on the number of live and dead animals. The zero time post treatment was as the animals entered the water, any animals identified as dead in the first 30 s were considered dead prior to entering the experiment, i.e. dead in the rearing tanks, and were removed from the container and excluded from any analyses. An assessment of whether animals were alive or dead was made at 30 min intervals for the first 2 h of treatment then at hour intervals for the remainder of the 5 h experiment. Motile versus non-motile animals were separated by gently swirling the water; motile, live animals would swim and non-motile animals would settle in the centre of the container. The mortality of non-motile animals was then confirmed by gently disturbing them with forceps, if no movement was observed they were considered dead and were subsequently counted and removed from the container. The three control treatments were treated in the same manner with the same level of physical disturbance.

The length of dead PL were measured on 1 mm grid paper for each time point and then the survivors were measured at the end of the experiment.

RESULTS AND DISCUSSION

Suitable parameters for stress testing would be those that showed differential survival of postlarvae (PL), with the ultimate objective of differentiating families that are more or less tolerant to environmental stress. For the practical application of this test in a commercial environment, the ideal mortality curve would have a rate near 50% at 3 h, and a flattening out of mortality beyond. For example the salinity dose of 10% and 15% salinity at PL 32 (Figure 1A), was considered an optimal level (Table 2). While an ammonia rate between 21mg^{-1} and 27 mg^{-1} was recommended at PL 34 (Figure 1B & Table 2). While the range of dose rates and ages have not previously been tested, a dose rate of 20mg^{-1} on PL5 *P. monodon* has reportedly resulted in 53-55% mortality after 72 h (Pan *et al.* 2003).

The results of this study identified initial ranges that achieved the objective of determining suitable stress testing parameters with both salinity stress and ammonia stress for various PL ages (Table 2).

Table 2. Salinity and ammonia stress mortality at 3 h and recommendations on optimal parameters

PL age	Salinity dose level (%seawater)	Mortality (%) at 3 h	Optimal parameter recommendation	PL age	Ammonia dose level (mg^{-1})	Mortality (%) at 3 h	Optimal parameter recommendation
15	0	100	>20%	15	11	2	>21<53
	5	100			21	3	
	10	100			53	100	
	15	100			214	100	
	20	84					
25	0	100	≥15%	22	32	49	<32
	5	98			43	94	
	10	90			53	100	
	15	75			64	100	
28	5	94	10% or 15%	27	21	43	21
	10	62			27	96	
	15	39			32	100	
					43	99	
					53	100	
32	5	94	10% or 15%	28	16	1	16>21
	10	66			21	82	
	15	53			27	97	
	20	20					
				34	11	0	21>27
45	15	4			16	0	
	20	0			21	19	
					27	85	
					32	95	

T-tests showed a significant difference in mortality between all salinity dose levels at 3 h post-treatment ($P<0.05$), except for between 10% and 15% salinity dose levels ($P>0.05$). Significant differences in mortality was also observed between all ammonia dose levels at 3 h ($P<0.05$), except between 32 mg^{-1} and 16 mg^{-1} dose levels ($P>0.05$). This study determined that PL were more tolerant to lower salinity levels at later ages (Figure 2). T-tests revealed significant differences ($P<0.005$) between all ages at the 3 h time point, with the exception of PL 25 that did not differ significantly between PL 15 or PL 32 ($P>0.5$). At PL 15 a salinity ratio of 20% seawater: 80% freshwater resulted in 84% mortality at 3 h post-treatment, while at PL 32 the same salinity resulted in just 20% mortality at 3 h (Figure 3).

Previous studies have found that larger postlarval shrimp have a greater tolerance to salinity stress, often linked with the development of gills (Chong-Roles *et al.* 2014); however, the current study found that mortality was strongly linked to age rather than size (Figures 3 & 4).

This study found that there was no significant effect between PL size and mortality time for salinity or ammonia (ANOVA $P>0.05$); this was only tested at PL 25. The lack of relationship between mortality and size indicated that PL mortality was not simply removing fast or slow growing PL. Furthermore, environmental rearing effects among the PL were minimised as the batch of PL tested underwent communal spawning and rearing where they were subjected to the same environmental conditions. Therefore, this study supports the hypothesis that there may be a significant genetic effect on resilience to environmental stress and subsequently further research should be directed into the genetic, genomic or physiological influences on resilience.

The genetic basis for this resilience as well as heritably may be elucidated in future studies by applying the methods developed in this study to discriminate between resistant and susceptible families. Genomic markers may also be developed that could then be utilised in a selective breeding program to establish more resilient lines, thereby leading to improved survival and production yields in shrimp farming.

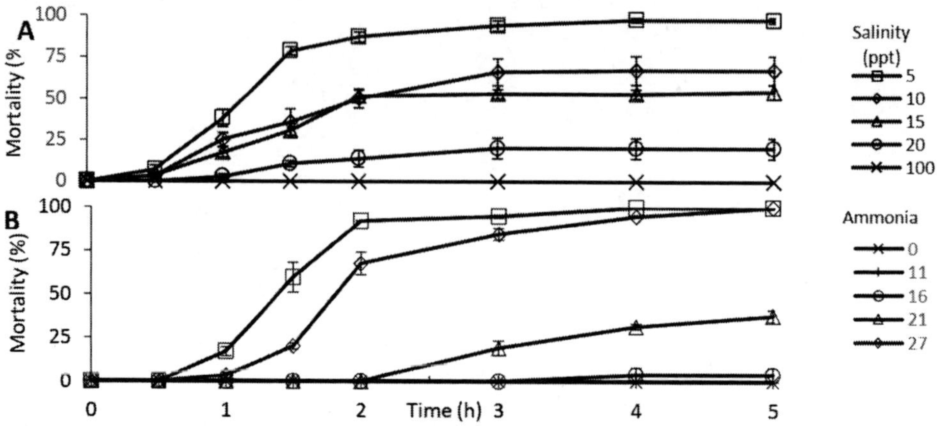

Figure 1. Example of *Penaeus monodon* mortality curves for salinity (A) and ammonia (B) challenge at postlarval ages PL 32 (A) PL 34 (B) (SEM bars).

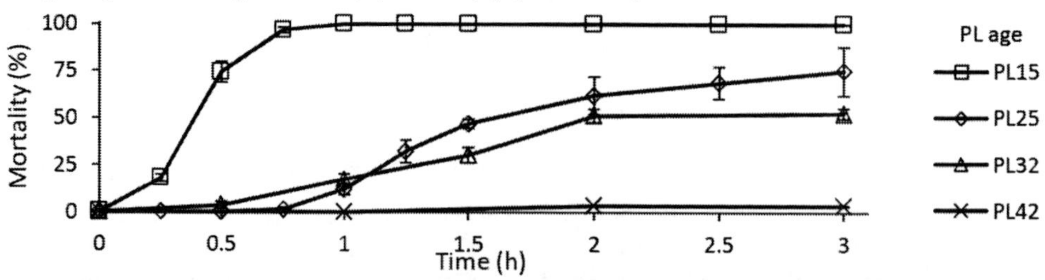

Figure 2. Effect of postlarval age on survival to 15% salinity for *Penaeus monodon* (SEM bars).

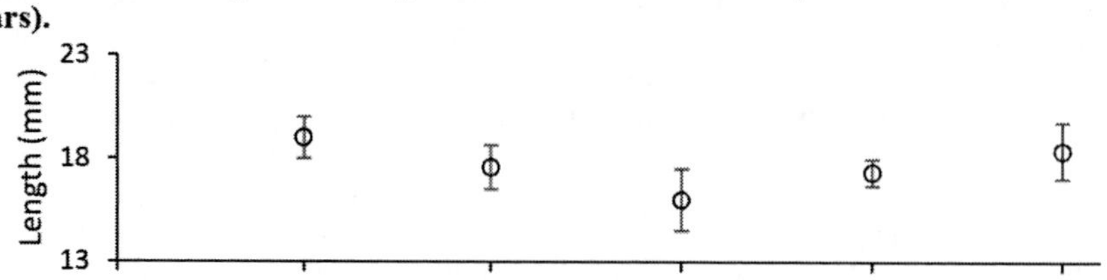

Figure 3. Length of dead postlarve following 16 mg^{-1} of ammonia at PL 25 (SEM bars).

REFERENCES

Pan C., Chien Y., Hunter B. (2003) *Exp. Mar. Biol. Ecol.* **297**: 107.

Chong-Roles J., Charmantier, G., Boulo V., *et al.* (2014) *Aqua.* **422-423**: 267.

Proc. Assoc. Advmt. Anim. Breed. Genet. 22:445-448

RECONSTRUCTION OF UNKNOWN DONOR GENOME FROM CHIMERIC PEARL SAC TISSUE AND HOST GENOTYPES IN *PINCTADA MAXIMA*

M.S. Khatkar[1,2], D.B. Jones[3], P.B. Toole[3,4], H.W. Raadsma[1,2], D.R. Jerry[2,3] and K.R. Zenger[2,3]

[1] Sydney School of Veterinary Science, Faculty of Science, The University of Sydney, Camden, NSW, Australia

[2]ARC Hub for Advanced Prawn Breeding, James Cook University, Townsville QLD, Australia

[3] Centre for Sustainable Tropical Fisheries and Aquaculture, College of Science and Engineering, James Cook University, Queensland, Australia

[4] Atlas South Sea Pearls Ltd., Western Australia, Australia

SUMMARY

Commercial pearl production involves two oysters, the seeded individual (host), and a sacrificed oyster (donor) from which a piece of tissue is utilised during the seeding process. During commercial seeding, it is often difficult to keep track of individual pairing information of host and donor oysters. Here we describe a method for reconstructing donor genotypes from host genotypes and allele frequencies generated from chimeric pearl sac tissue (a mixture of host and donor tissues) at pearl harvest. Using simulation of expected genotype frequencies and genotypes from true samples, we demonstrate that donor genotypes can be constructed with high accuracy. Best results were observed when the error rate of allele frequencies is low, SNPs have a high minor allelic frequency, and when the proportion of the donor tissue in the pearl sac tissue is greater than 20 percent.

INTRODUCTION

Commercial pearl production involves two oysters, the seeded individual (host), and a sacrificed oyster (donor) which produces a piece of tissue (also termed the 'saibo' tissue). This saibo tissue is implanted into the host during the seeding process (Figure 1). Both donor and host oyster tissues are known to be actively expressed during the production of a pearl (Arnaud-Haond et al., 2007), and the quality of pearls produced are influenced by the genomes of both the host and donor oysters (Jerry *et al.*, 2012), (Tayale *et al.*, 2012). Hence, for selective breeding and dissecting the contribution of host and donor oysters, and their interaction on various pearl quality traits, the identification and recording of both animals is critical. In many commercial pearling farms, the tracking of donor oysters is not maintained routinely due to additional management complexity and expense (Jerry *et al.*, 2012) (Jones *et al.*, 2014). However, the identification and validation of donor oysters during pearl harvest at the end of a four year pearl production cycle is integral for conducting genetic studies and making breeding decisions. Under the assumption of the donor genome being present in the pearl sac tissue at harvest, it should be possible to differentiate reciprocal host and donor genomes (and reconstruct genotypes) by using either the known host or donor genotype and pearl-sac allele frequencies.

Here we present a method for reconstructing donor genotypes from host genotypes and allele frequencies from pearl sac tissue (a mixture of host and donor tissues). Using simulations, we demonstrate the effect of genotyping error and variable proportions of host and donor tissue observed within pearl sac samples on the accuracy of reconstructing unknown donor genotypes.

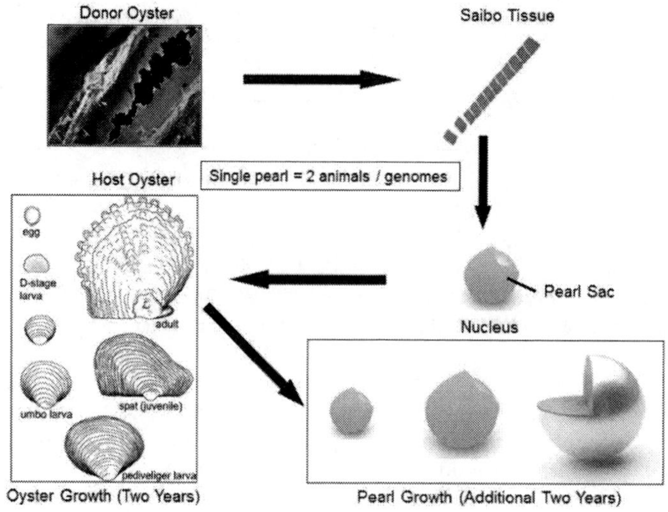

Figure 1. Commercial pearl production via seeding.

MATERIAL AND METHODS

Method of reconstruction of donor genotypes from allele frequency of pearl sac tissue and genotype of host. Let **H** be a vector of genotypes of n SNPs from a host coded as 0 for one homozygotes (AA), 1 for heterozygotes (AB), and 2 for other homozygotes (BB), and **P** a vector between 0 and 1 representing allele frequency (frequency of B allele) of n SNPs of pearl sac tissue. Then reconstruction of the donor genotypes from the allele frequency of the pearl sac tissue and genotypes of the host was done in three steps. First the proportion of host tissue in the pearl sac tissue was estimated as the regression coefficient of host genotypes on allele frequencies of pearl sac tissue from a linear regression model i.e. $\mathbf{2P}=b\mathbf{H}+e$, then allele frequencies of the donor was estimated by $\mathbf{D} = (\mathbf{2P}-b\mathbf{H})/(1-b)$. Finally, the donor genotypes were reconstructed by classifying the frequency estimates **D** of the donor into the nearest genotypic class 0 (AA), 1 (BB) and 2 (BB). The classification error of donor genotypes was calculated from a confusion matrix between the estimated genotypes and the actual genotypes of the donor. A total 1,000 randomly selected pairs of animals were evaluated for each scenario.

Simulation. The working and utility of the above procedure was tested by sampling and simulation. The genotypes of a host genome were generated on 935 SNPs by randomly sampling the genotypes of one animal from a real genotypic dataset on pearl oyster (*Pinctada maxima*) recently generated by genotyping 329 animals with a high-density DArTseq Diversity Arrays SNP panel (Kilian *et al.*, 2012). Similarly genotypes of one donor were generated by sampling another animal from the same dataset. These 935 SNPs had a minimum of 0.3 minor allelic frequency (MAF) in the panel of genotyped animals. Lower thresholds for MAF were also explored. The pearl sac tissue was assumed as a mixture of host and donor tissues in various proportion. The allele frequency for the pearl sac tissue was generated by mixing the donor and the host genotypes in ten different proportions by assuming 5, 15, 25, 35, 45, 55, 65, 75, 85, 95% of host tissue and the remaining respective proportion of the donor tissue. In addition, to accommodate some laboratory error in estimating the allele frequencies, a continuous uniform distribution for error rate in an interval of -0.06 and 0.06 was generated and added to the allele frequencies computed above (while keeping the resultant allele frequencies within a bound of 0 and 1). Other uniform distributions with higher error rates, and normal distributions *N(0, σ2)* with four different error distributions (σ = 0.01, 0.02, 0.05, 0.10) were also evaluated.

RESULTS AND DISCUSSION

The accuracy of reconstructed donor genotypes from pearl sac tissue consisting of different proportion of host and donor tissue, each computed from 1,000 host and donor pairs, are presented in Table 1. These results are based on a uniform error distribution with an interval of -0.06 and 0.06. (Figure 1). The estimated proportion of host genome was close to the actual proportion for most of the scenarios except the scenario 1 and 2 where proportion of the host tissue was small. Nevertheless, the median mis-classification rate for donor genotypes was zero except for the last two scenarios where the proportion of host tissue in the pearl sac was very high (> 85 %). Other laboratory error rates in the estimate of allele frequency and using SNPs with lower MAF were also explored. However, the higher error rates in the estimation of allele frequency resulted in higher misclassification rate of donor genotypes. Similarly using SNPs with lower MAF thresholds also resulted in higher misclassification rate (results not shown).

Table 1. Accuracy of reconstructed donor genotypes from Pearl sac consisted of different proportion of host and donor tissue.

Proportion of host in pearl sac tissue	Median estimated host proportion	QR estimated host proportion	Median correlation between estimated and actual genotype	QR correlation between estimated and actual genotype	Median mis-classification rate (%)	QR mis-classification rate (%)
0.05	0.10	0.02-0.20	1	0.98-1	0	0-0
0.15	0.19	0.12-0.29	1	0.98-1	0	0-0
0.25	0.29	0.22-0.38	1	0.98-1	0	0-0
0.35	0.38	0.33-0.46	1	0.98-1	0	0-0
0.45	0.48	0.43-0.54	1	0.98-1	0	0-0
0.55	0.57	0.53-0.62	0.99	0.98-1	0	0-0
0.65	0.66	0.63-0.70	0.99	0.98-0.99	0	0-0
0.75	0.76	0.74-0.79	0.98	0.97-0.99	0	0-0
0.85	0.85	0.84-0.87	0.96	0.95-0.96	0	0-2.8
0.95	0.95	0.94-0.95	0.74	0.71-0.75	40.3	37.9-43.0

QR is 0.1 and 0.9 quantile based on 1,000 host and donor pairs.

The error rates in constructing the donor genotypes of SNPs with low MAF were higher, and hence we recommend the use of high MAF SNPs. Pedigree information can also be used to correct some of the incorrectly constructed genotypes and genotypes of low MAF SNPs of the donors. These corrected genotypes can then finally be used for GWAS and genomic selection. Genetic relationship between donor and host tissue, especially close relationship such as full-sibs, may also affect the accuracy of reconstruction of donor genotypes and warrants further investigations.

Overall these results suggested that donor genotypes can be constructed with a very high certainty if the error rate in the estimation of allele frequency is low, high MAF SNPs are used, and the proportion of the host tissue in the pearl sac tissue is not very high.

To validate this method further we are now preparing synthetic pools by mixing DNA of pairs of animals in different proportions. In addition we will sample a number of trio samples consisting of host, donor and pearl sac tissues. The procedure of taking samples of pearl sac tissue may also affect the proportion of host and donor genome; a few different procedures of sampling such as slicing the interior pearl sac tissue with a scalpel and using a sterile dental swap to collect cells without an incision on the pearl sac will be evaluated to minimise the proportion of host tissue in pearl sac tissue. These samples will be genotyped with low-density DArTseq Diversity Arrays SNP panel (Kilian *et al.*, 2012). The analysis of these samples will provide estimate of donor tissue in the pearl sac, estimation of laboratory error rate in the allelic frequency of pearl sac, and finally ability to reconstruct donor genotypes.

CONCLUSION

The method presented here provides a way to reconstruct the genotypes of the donor from allelic frequency data on pearl sac tissue and genotypes of the host. The results also suggested that some error in the estimation of allele frequency can be tolerated. However, when a very high proportion of the host tissue is present in the pearl sac tissue, it difficult to reconstruct donor genomes and introduced high error in the estimated genotypes of donor.

REFRENCES

Arnaud-Haond, S., Goyard, E., Vonau, V., Herbaut, C., Prou, J., & Saulnier, D. (2007). Mar Biotechnol (NY), 9(1), 113-116.

Jerry, D. R., Kvingedal, R., Lind, C. E., Evans, B. S., Taylor, J. J. U., & Safari, A. E. (2012). Aquaculture, 338–341, 66-71.

Jones, D. B., Jerry, D. R., Khatkar, M. S., Moser, G., Raadsma, H. W., Taylor, J. J., & Zenger, K. R. (2014). Aquaculture, 434, 476-485.

Kilian, A., P. Wenzl, E. Huttner, J. Carling, L. Xia, H. Blois, V. Caig, K. Heller-Uszynska, D. Jaccoud, C. Hopper, M. Aschenbrenner-Kilian, M. Evers, K. Peng, C. Cayla, P. Hok and G. Uszynski (2012).. Methods Mol Biol, 888, 67-89.

Tayale, A., Gueguen, Y., Treguier, C., Le Grand, J., Cochennec-Laureau, N., Montagnani, C., & Ky, C.-L. (2012). Aquatic Living Resources, 25(3), 269-280.

Proc. Assoc. Advmt. Anim. Breed. Genet. 22:449-452

COMPARISON OF DIFFERENT BREEDING DESIGN OPTIONS FOR LONG TERM GENETIC GAIN AND DIVERSITY IN AQUACULTURE SPECIES

M.S. Khatkar[1,2], G.J. Coman[1,4], P.C. Thomson[1,3] and H.W. Raadsma[1,2]

[1]ARC Research Hub for Advanced Prawn Breeding
[2]Sydney School of Veterinary Science, Faculty of Science, The University of Sydney, Camden, NSW, Australia.
[3]School of Life and Environmental Sciences, Faculty of Science, The University of Sydney, Camden, NSW, Australia.
[4]Aquaculture Program, CSIRO Agriculture and Food, Queensland Bioscience Precinct, St. Lucia, QLD, 4067, Australia.

SUMMARY

Using simulation, we compared the effect of different numbers of families and skewed distribution of family size on long-term genetic gain and inbreeding in aquaculture species. In particular we focused on *P. monodon* specific input parameters and communal rearing of families, and showed that large number of families in a communal breeding scheme are required for increased genetic gain and diversity in addition to mitigating the effect of unequal family contributions. We present a two-stage cost effective scenario implementing combining truncation selection and genomic selection, and showed that 1,000-2,000 animals in the first stage are required for long-term genetic improvement.

INTRODUCTION

The application of genetic markers and genomic selection (GS) in aquaculture is becoming attractive in particular for selection of 'difficult to measure traits' and traits which cannot be directly measured on candidates under selection, and to capture within- and between-family genetic variation. With decreasing cost of sequencing and genotyping the development of SNP panels and application of genomic resources can be rapidly deployed for almost any species, yet limited information is available on the results of practical implementation of GS and have mainly been restricted to the use of simulated data. Such studies demonstrated that GS in aquaculture breeding programmes can increase the accuracy of selection and genetic gains, both in production (continuous) and diseases (dichotomous) traits (Sonesson and Meuwissen, 2009; Nielsen *et al.*, 2011; Lillehammer *et al.,* 2013).

The number of families reared in aquaculture breeding programs is generally limited by the resources especially if the families are produced and reared separately, and this can have a profound impact on inbreeding and long-term genetic gains. More families can be managed if bred and reared communally. However, because of mass spawning the contribution of different families is unequal (Harris *et al.*, 2016), and this can distort selection efficiency and hence genetic gain and inbreeding.

Using simulation, we explored the effect of number of families and unequal distribution of family size on the overall inbreeding and genetic gain on a growth-like trait and examined a cost effective scenario implementing a two-stage selection scheme by combining truncation selection and genomic selection.

MATERIALS AND METHODS

Simulated datasets were generated by QMSim simulation software (Sargolzaei & Schenkel, 2009) by specifying the following input parameters. The initial founder generation was the last of

1,000 historic generations, containing 400 random mating individuals each, equal to the effective population size of wild *P. monodon*. From this founder population, 1, 5, 50,100 or 200 males and 1, 5, 50, 100 or 200 females were used for breeding, with one male mated to one female, each mating producing 200 offspring in each generation. The genetic map of 40 chromosomes each 50 cM long was specified. For each chromosome, 120 biallelic markers and 30 biallelic QTLs were simulated. Mutation rate for both markers and QTLs were set to 2.5E-5 per generation. All the scenarios were simulated with heritability = 0.30 such that heritability due to QTLs = 0.2 and the remaining one third due to polygenic effects. The heritability used was comparable to that of body weight at harvest in shrimp (Sui *et al.*, 2016). The phenotypic variance was set to 1 with mean equal to 0 in the base population. Ten replications for each of the following scenarios were explored:

Scenario A: Single family of 200 progeny produced from the mating of one male with one female in each generation.

Scenarios B-E: Scenario B was generated with five families of 200 progeny each produced from the mating of one male with one female in each generation. Similarly, scenarios C, D and E were generated with 50, 100 and 200 families, respectively.

Scenario F: 100 families with each family producing different numbers of progeny. The numbers of progeny per family were simulated with a discrete probability distribution of $0.3\ (n = 1) + 0.3\ (n = 5) + 0.2\ (n = 50) + 0.1\ (n = 700) + 0.1\ (n = 900) = 1$, where n is the number of progeny. In addition a two-stage selection scheme was implemented. In the first stage different numbers of animals (200, 300, 600, 1000, 2000 or all) were selected randomly from the top 25% of all the animals in the pond, tagged and genotyped. In the stage 2, 100 males and 100 females were selected from the tagged animals based on genomic EBV assuming a selection accuracy of 0.6. For this scenario the phenotypic variance was set to 36 with a mean of 30 in the base population.

Selection Method: The selection of parents to produce specific number of families for the next generation was based on the EBVs, which as noted above had an accuracy of 0.6. Rate of genetic progress was calculated as the change in mean breeding values across generations. Estimates of inbreeding coefficient and mean breeding values were computed for each generation using pedigree information, and compared across ten generations for the scenarios described above.

RESULTS AND DISCUSSION

A comparison of mean inbreeding coefficient and mean phenotypic value after ten generations of selection is presented in Figure 1 which shows that a smaller number of families under selection resulted in increased inbreeding, in particular for single family selection inbreeding was approaching 90% after 10 generations of selection. Inbreeding coefficients based on 50 or more families were generally low (< 10%), and the difference in inbreeding between scenarios with 100 and 200 families was negligible. Rates of genetic gain were highest for using 100 families or more, whereas using single families or low number of families (< 5) resulted in the lowest rate of genetic gain.

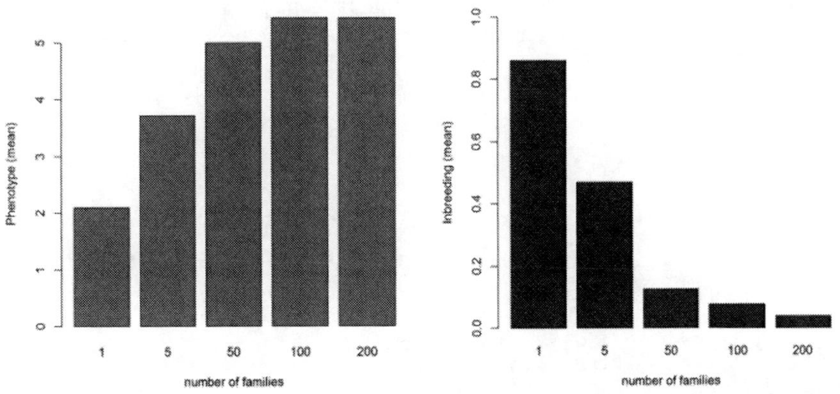

Figure 1. The effect of number of families on mean phenotypic values (A) and inbreeding coefficient (B) after 10 generations of selection (pooled across 10 replicates). Each family contributed an equal number of progeny (*n* = 200). (Scenario A-E).

Estimates of inbreeding coefficient and mean breeding values across ten generation and ten replicates for scenario D are presented in Figure 2. The replicates show a consistent increase in mean inbreeding and breeding value over generations. The differences between replicates were larger for scenarios with smaller numbers of families (results not shown).

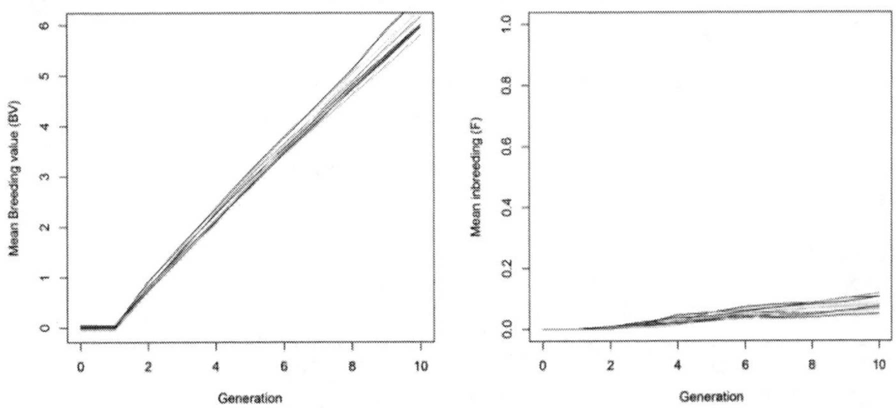

Figure 2. Mean breeding values (A) and inbreeding coefficient (B) over 10 generations of selection for scenario D. Ten replicates are shown by lines with different colours.

Stage-wise selection: Scenario F presents a more practical situation where, due to mass spawning and differential survival, the contribution of different females (families) are unequal in communal breeding and rearing, with some families contributing a large proportion of the progeny *(Harris et* *al.*, 2016). In addition, to reduce the cost of genotyping, we implemented a two-stage selection by combining truncation selection based on phenotype (selecting randomly from the top 25%) in the first stage and selection based on EBV with moderate accuracy of 0.6 in the second stage. Figure 3 shows that selecting 1000-2000 animals in stage one with truncation selection provides most of the genetic gain possible with genotyping all the animals. Inbreeding increased only slightly with larger number of animals selected in the stage 1.

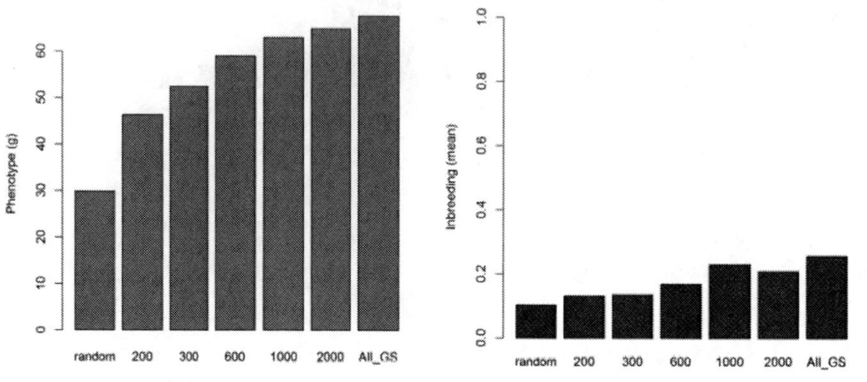

Figure 3. The effect of different number of animals (*x*-axis) in the first stage of two-stage selection scheme (Scenario F). "random" means selecting 200 randomly, "All_GS" selecting all in stage one.

Compared to the scenarios with equal family contribution (presented in Figure 1), the effect of unequal family contribution was more pronounced when the smaller number of families were simulated (results not shown). There was a pronounced increase in inbreeding when the families' contributions are unequal. This is largely due to loss or low representation of families in the subsequent generations.

REFERENCES

Harris, L., Sellars, M., & Perez, F. (2016). Aquaculture International, 24(1), 273-279.

Lillehammer, M., Meuwissen, T. H., & Sonesson, A. K. (2013). Genet Sel Evol, 45, 39.

Nielsen, H. M., Sonesson, A. K., & Meuwissen, T. H. (2011). J Anim Sci, 89(3), 630-638

Sargolzaei, M., & Schenkel, F. S. (2009). Bioinformatics, 25(5), 680-681.

Sonesson, A. K., & Meuwissen, T. H. E. (2009). Genetics Selection Evolution, 41.

Sui, J., S. Luan, K. Luo, X. Meng, X. Lu, B. Cao, W. Li, Z. Chai, N. Liu, S. Xu and J. Kong (2016). Aquaculture Research, 47(9), 2795-2803.

Proc. Assoc. Advmt. Anim. Breed. Genet. 22:453-456

EVALUATION OF POOLED WHOLE GENOME SEQUENCING (POOL-SEQ) TO RECOVER KNOWN GWAS SIGNALS (GENE EFFECTS)

A. Mohamed, L. Porto-Neto, A. Reverter and J. Kijas

CSIRO Agriculture and Food, Queensland Bioscience Precinct, St Lucia, QLD 4067, Australia

SUMMARY

Whole-genome sequencing (WGS) of pools of individuals (Pool-Seq) provides a cost-effective method for genome-wide association studies (GWAS), and offers an alternative to sequencing of individuals that remains cost prohibitive. Pool-Seq is being increasingly used in population genomic studies in both model and non-model organisms. In this paper, the ability of Pool-Seq to recover known GWAS signals was evaluated. Existing GWAS data for 2,112 animals with 729K SNPs were obtained and pooled to simulate data obtained from a pooled WGS approach. Traditional GWAS results was compared with the absolute allele frequency difference (dAF) metric suitable for use with Pool-Seq data. Specifically, we tested the ability of dAF scans to recover known GWAS signals for two different traits with large and moderate gene effects. Pools of different sizes (50, 100 and 200 individuals per pool) were also compared. The results showed the ability of the absolute allele frequency difference (dAF) approach to recover known GWAS peaks obtained by traditional SNP association and recommended the use of a pool size of 100 individuals for DNA pooling.

INTRODUCTION

Recent advances in next generation sequencing (NGS) technologies have tremendously changed genetic research by increasing the number of known molecular markers in both model and non-model organisms such as: single nucleotide polymorphisms (SNPs) (Ellegren 2014). Despite these technical advances, genotyping large numbers of individuals with thousands of SNPs remains costly for large genome-wide association studies (GWAS). In this context, determination of allele frequencies from whole genome sequencing of pooled DNA samples has been suggested as a cost-effective alternative to individual genotyping (Sham *et al.* 2002). Many studies have successfully adopted this approach by comparing allele frequencies between cases and controls in both model and non-model organisms. For example, Abraham *et al.* (2008) performed a genome-wide (case-control) association study to understand Alzheimer's disease in human through the use of DNA pooling and highly significant association with late-onset Alzheimer's disease (LOAD) was observed at the *APOE* locus. To test for loci selected during domestication in chicken, Rubin *et al.* (2010) compared domesticated species to a wild population and identified one domestication-specific adaptation in the thyroid-stimulating hormone receptor (*TSHR*) gene. Pool genome-wide association study (Pool-GWAS) was also used to examine female abdominal pigmentation in *Drosophila melanogaster.* Candidate single-nucleotide polymorphisms (SNPs) near the pigmentation genes *tan* and *bric-à-brac 1* were identified when the allele frequencies in pools of light and dark females were compared (Bastide *et al.* 2013). Moreover, in Atlantic salmon Pool-Seq was used to investigate age at maturation in both wild and domesticated salmon where Ayllon *et al.* (2015) performed a genome wide association study using a pool sequencing approach (20 individuals per pool) of male salmon returning to rivers as sexually mature and revealed that 138 SNPs were significantly associated with sea age at puberty, 74 (48%) of these significant SNPs were located in a region on chromosome 25. More recently, Pool-Seq approach has been successfully deployed to identify genes for the timing of reproduction in Atlantic herring (Martinez Barrio *et al.* 2016).

In this paper, we used existing cattle SNP chip data obtained from individual animals and the associated GWAS results (Porto-Neto *et al.* 2014), to evaluate the power of the pool-seq approach. Using absolute allele frequency difference (dAF) , the ability to recover known GWAS signals was assessed after varying i) number of individuals per pool and ii) trait architecture. Outcomes of this analysis will assist in the design of experiments that seek to use pool-Seq as an alternative to traditional GWAS methodologies.

MATERIALS AND METHODS

Porto-Neto *et al.* (2014) performed a genome-wide association study using 2,112 Brahman cattle with 729,068 SNP genotypes per individual and analysed ten traits related to tropical conditions. Data were retrieved and re-analysed for two different traits, Coat Colour (colour) and rectal temperature (temperature). The first of these was selected to represent traits with large gene effects (colour), while the second exhibits genes of moderate effects. Plink software was used to make a subset of the data for 100 (top and bottom 50), 200 (top and bottom 100) and 400 (top and bottom 200) individuals from the 2,112 animals (representing pool sizes of 50, 100 and 200, respectively) using the --make-bed and --keep functions. Those individuals were assigned into two phenotypes for the GWAS case/control test and two clusters for the delta allele frequency test.

For traditional SNP association approach, an association (GWAS case/control scenario) test was implemented in Plink using the --assoc and --pheno functions. P values of all SNPs were obtained and –logP values were visualised as Manhattan plot generated in the R statistical computing environment.

For absolute allele frequency difference (dAF) approach, allele frequencies were calculated in Plink using the --freq and --within functions. Differences in allele frequencies were calculated for each SNP in the 2 clusters. A Manhattan plot of the absolute values of dAF of all SNPs was generated in R.

A significance threshold ($-\log P \geq 5$) was applied to filter the SNPs and the corresponding absolute values of delta AF of those significant SNPs were retrieved. This threshold was chosen in order to capture enough data for valid comparison and was used previously in GWAS analysis (for example) Cui *et al.* (2016). In order to compare the two approaches, -logP and delta AF values for 1) all significant SNPs and 2) SNPs under peaks were plotted in genomic order. Also simple linear regression was applied and R^2 values were obtained to test the correlation of the results obtained from both approaches for each trait in each pool size used.

RESULTS AND CONCLUSION

Traditional SNP association (GWAS case/control) identified SNPs significantly associated with the two traits under investigation for each of the pool sizes used. Strong GWAS signal(s) were identified in chromosomes 6, 7 and 13 for colour. On the other hand, multiple peaks in many chromosomes were identified for temperature (Figure 1). These findings are consistent with the results in the Porto-Neto *et al.* study.

Absolute allele frequency difference (dAF) results were obtained for the two traits and were compared with the GWAS results. The same GWAS signals were recovered using dAF in each of the two traits for each pool (Figure 1). For example, dAF values of the 32 significant SNPs of the major GWAS peak chromosome 13 in the trait colour showed the same trend as their corresponding –log P values (Figure 2).

Linear regression was used to test the correlation of the results obtained from both approaches for each trait in each pool size. Pool size of 50 showed the least R^2 values in each of the two traits, while there were very small increases in R^2 values (0.02 and 0.01 in colour and temperature, respectively) from pool size 100 to 200 (Table 1). For each trait, number of significant SNPs increased by increasing the pool size, with the pool size of 50 yielding the smallest number of significant SNPs (Table 1).

Table 1. A summary table of the results of comparing SNP association and delta allele frequency approaches for two traits in the Brahman cattle using 3 different pool sizes.

Cattle trait	Pool size	No. of sig. SNPs (-log P ≥ 5)	R^2 value
Colour	50	95	0.04
	100	638	0.7
	200	4,349	0.72
Temperature	50	147	0.00007
	100	951	0.74
	200	1,323	0.75

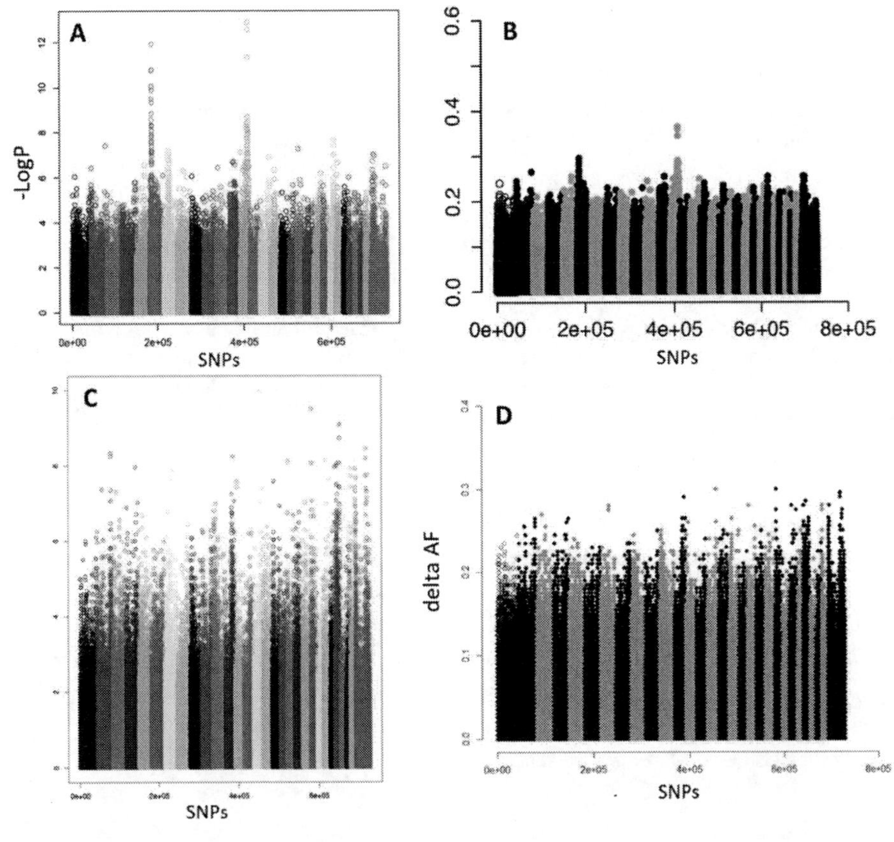

Figure 1. Comparison between GWAS SNP association and absolute allele frequency difference (dAF) approaches using a pool size of 100 (for example) revealed the ability of dAF to recover the same GWAS signals. A and B are Manhattan plots of –log P and absolute values of dAF values, respectively for colour while C and D are Manhattan plots of –log P and absolute values of dAF values, respectively for temperature.

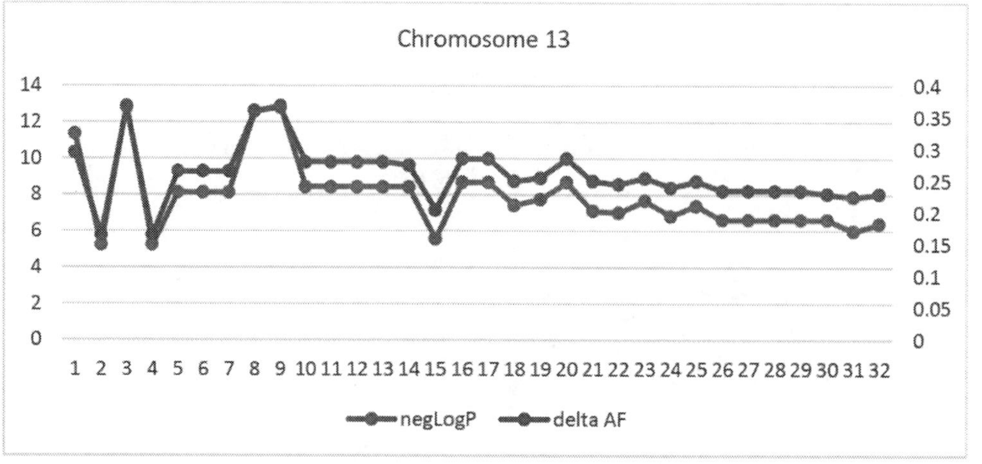

Figure 2. Zoom-in on the 32 significant SNPs (-log P ≥ 5) on a GWAS peak (chromosome 13) for the trait colour showing absolute delta allele frequency values (red) following the same trend as the –log P values (blue).

In conclusion, the absolute allele frequency difference (dAF) approach recovered the same GWAS signals obtained by traditional SNP association approach, for all the two traits under investigation. However, comparing the results from three different pool sizes suggested the use of pool size of 100 individuals for DNA pooling. These results confirm that, for traits controlled by a small number of major genes, the pool-Seq approach is likely to have the power to identify associations using the dAF metric. This opens the possibility to collect samples from only the phenotypic extremes within a population, before searching for associated genomic regions using a simple analytical approach and a modest research budget.

REFERENCES

Abraham R., Moskvina V., Sims R., *et al.* (2008). *BMC Med. Genomics.***1**:44.
Ayllon F., Kjærner-Semb E., Furmanek T., *et al.* (2015). *PLoS Genet.* **11**:11.
Bastide H., Betancourt A., Nolte V., *et al.* (2013). *PLoS Genet.* **9**:6.
Cui, Z., Luo, J., Qi, C., *et al.* (2016). *BMC Genomics*. **17**:946.
Ellegren, H. (2014). *Trends Ecol. Evol.* **29**, 51–63.
Martinez Barrio, A., Lamichhaney, S., Fan, G., *et al.* (2016). *Elife* **5**, e12081.
Porto-Neto L.R., Reverter A., Prayaga K.C., *et al.* (2014). *PLoS ONE*. **9**:11.
Rubin, C. J., Zody M C., Eriksson J. *et al.* (2010). *Nature* **464**, 587–591.
Sham P., Bader J.S., Craig I., *et al.* (2002). **3**: 862-871.

Proc. Assoc. Advmt. Anim. Breed. Genet. 22:457-460

PREPUTIAL EVERSION IN YOUNG, TROPICALLY ADAPTED BULLS IS A USEFUL GENETIC INDICATOR TRAIT FOR IMPROVING FEMALE REPRODUCTION

M.L. Wolcott and D.J. Johnston

Animal Genetics and Breeding Unit*, University of New England, Armidale, NSW 2351, Australia

SUMMARY

Beef CRC research showed that a measure of preputial eversion (PEV: an estimate of the length of preputial mucosa (mm) exposed while a bull stands freely) in tropically adapted bulls at 18 months of age was heritable, and had significant genetic associations with female reproduction performance. The current study examined the sensitivity of genetic parameters for PEV to age by expanding the analysis to include measures at 12 and 24 months. For Brahman bulls, the incidence of non-zero PEV increased with age, from 45 to 59 to 71% at 12, 18 and 24 months respectively. For Tropical Composite bulls, the incidence of PEV was lower and less influenced by age (27 – 31%). Heritabilities for PEV at 12 and 24 months were comparable to those previously reported at 18 months (h^2 = 0.23 to 0.34). These results confirm that if breeders of tropically adapted beef cattle wished to apply selection to improve PEV, this could be undertaken successfully. Genetic correlations of PEV with female age at puberty, lactation anoestrous and lifetime annual weaning rate showed that if PEV were to be exploited as a genetic indicator for female reproductive performance, measures at 18 and 24 months would be more useful than those collected at 12 months of age. Compared to female reproduction traits, preputial eversion is easy to measure and this study suggests it would be a useful trait to add to the genetic evaluation for tropical breeds.

INTRODUCTION

Research reported by Corbet *et al.* (2013) showed that a trait scoring the 'length (in mm) of exposed preputial mucosa' in 18 month old bulls (PEV18) was heritable in both Brahman (BRAH) and Tropical Composite (TCOMP) (h^2 = 0.30 and 0.23 respectively). Johnston *et al.* (2014*b*) showed that lower PEV18 displayed significant genetic relationships with lower age at puberty, and higher lifetime annual weaning rates in BRAH (r_g = 0.33 and -0.71 respectively). In TCOMP, lower PEV18 was genetically associated with lower lactation anoestrus interval, and higher lifetime annual weaning rates (r_g = 0.52 and -0.88). As part of that research preputial eversion at 12 (PEV12) and 24 (PEV24) months of age were also measured. Given the clear potential of PEV18 as an indirect genetic indicator of female reproductive performance, this study aimed to examine the genetics of preputial eversion at 12 and 24 months in BRAH and TCOMP bulls, to estimate the genetic relationships between repeated measures of the trait, and to determine their genetic relationships with key female reproduction traits.

MATERIALS AND METHODS

Bull management and trait definition. The bulls evaluated for this study were from the Beef CRC Northern Breeding Project, and comprised the Brahmans (BRAH) and Tropical Composites (TCOMP) described by Burns *et al.* (2013). That publication also provided a thorough description of the experimental design and animal management. Briefly, bulls bred on 5 co-operating properties in northern Australia from 2004 - 2010 (n = 2742) were transported to Brigalow Research Station (170km southwest of Rockhampton) at weaning for evaluation of male reproductive traits. A smaller cohort were born and raised at the Belmont Research station (n = 1321) and remained there for the duration of the evaluation period (to 24 months of age).

* AGBU is a joint venture of NSW Department of Primary Industries and the University of New England

Burns *et al.* (2013), defined preputial eversion as 'an estimate of the length (mm) of preputial mucosa everted while the bull stands freely', and the trait was scored at 12, 18 and 24 months of age (PEV12, PEV18 and PEV24). For each genotype and age, the proportion of bulls with a non-zero PEV result was calculated to describe the change in the expression of the trait with age.

Female management and reproduction traits. Breeding and management of heifers up to their first annual mating was described by Barwick *et al.* (2009), and Johnston *et al.* (2009) described ultrasound scanning of females to identify age at first corpus luteum (CL), which defined age at puberty (AP). Females were first mated at an average age of 25 months, to calve as 3 year olds. At the start of the second annual mating period, ultrasound scanning to identify the presence of a CL commenced for lactating cows to detect the onset of cycling and allow the calculation of lactation anoestrous interval (LAI: the days from the start of the second annual mating period to the identification of a CL by ultrasound scanning) (Johnston *et al.* 2014*a*). Cows remained in the project until the weaning of calves from their sixth annual mating unless they failed to successfully wean a calf in consecutive years. For all females, lifetime annual weaning rate (LAWR) was calculated as the total number of calves weaned from the first, and up to the sixth mating, divided by the number of annual matings to which the animals were exposed (Johnston *et al.* 2014*a*).

Fixed effect modelling and genetic parameter estimation. Fixed effect models for PEV12 and PEV24 were built as described by Corbet *et al.* (2013) for PEV18. Initial modelling for preputial eversion traits tested the fixed effects of year (2004–10), birth location (six herds), birth month (Sept. to Jan.), post-weaning location (Brigalow or Belmont), dam age (3–9 years), dam previous lactation status (wet or dry), dam management group and all first order interactions. In TCOMP, sire and dam group were fitted to account for the average additive differences between the composite groups and any heterotic effects among combinations of sire and dam groups. Modelling was carried out using the PROC MIXED in SAS (SAS Institute Inc., Cary, NC, USA), with sire fitted as random. Final models were determined by sequentially dropping non-significant terms ($P > 0.05$). Following the methods described by Corbet *et al.* (2013), variance components for PEV12 and PEV24 were estimated in ASReml (Gilmour *et al.* 2009), with animal fitted as random and relationships between animals described using a three generation pedigree. Genetic correlations between preputial eversion traits at different ages, and with female reproduction traits were estimated in bivariate analyses using ASReml. When estimating genetic correlations between male and female traits, the data was edited to remove bull records of dam-offspring pairs where the bull was the resultant progeny of the female trait analysed (Johnston *et al.* 2014*b*).

RESULTS AND DISCUSSION

Preputial eversion data and variance components. Table 1 presents summary statistics, variance components and heritabilities for PEV12 and PEV24 in BRAH and TCOMP bulls, with the already published results for PEV18 (Corbet *et al.* 2013). Results showed that preputial eversion in BRAH tended to increase with measurement age, while the trait was relatively constant from 12 to 24 months in TCOMP. Despite this, in both genotypes, additive genetic variance represented a reasonably constant proportion of the phenotypic, with heritabilities ranging from 0.23 to 0.34. These results support the conclusion of Corbet *et al.* (2013) that preputial eversion could be improved by selection, and suggest that measurements collected in bulls at 24 months of age would be as effective a basis for selection as those collected earlier in life. Evaluating the trait at 24 months would be useful in BRAH, as the incidence of non-zero results increased from 45 to 59 to 71 percent at 12, 18 and 24 months. For TCOMP, the proportion of non-zero preputial eversion scores showed less variation with age (30, 27 and 31% at 12, 18 or 24 months), suggesting that age at measurement would be less important for bulls of this genotype.

Table 1. Number of observations (N), mean, and standard deviation (s.d.), additive (σ^2_a) and phenotypic (σ^2_p) variance, heritabilities (h^2), and its standard error (s.e.) for preputial eversion (mm) in Brahman and Tropical Composite bulls at 12 (PEV12), 18 (PEV18) and 24 (PEV24) months of age.

Genotype	Trait (mm)	N	Mean	s.d.	σ^2_a	σ^2_p	h^2	s.e.
Brahman	PEV12	1357	11	21	69	240	0.29	0.08
	PEV18*	1438	18	16	126	419	0.30	0.08
	PEV24	1430	26	25	182	627	0.29	0.07
Tropical	PEV12	1939	11	22	128	480	0.27	0.07
Composite	PEV18*	2104	10	21	100	429	0.23	0.06
	PEV24	2081	12	25	211	623	0.34	0.06

* Results previously reported by Corbet *et al.* (2013).

Genetic correlations between preputial eversion measured at 12, 18 and 24 months of age. Table 2 shows that genetic correlations between preputial eversion measured at 12, 18 and 24 months of age in BRAH and TCOMP bulls were consistently high ($r_g > 0.8$). The weakest genetic relationship was between preputial eversion at 12 and 24 months in BRAH ($r_g = 0.82$) which is likely to reflect the changing incidence of non-zero results for the trait with age, and provides additional support for the trait being evaluated later for bulls of that genotype. As Brahman bulls tend to be marketed as 2 year olds, the opportunity will be there to evaluate preputial eversion in large contemporary groups, at 18 – 24 months of age, and prior to bull sales and their first mating.

Table 2. Genetic correlations (r_g), and their standard errors (s.e.) between preputial eversion measured at 12, 18 and 24 months of age in Brahman and Tropical Composite bulls.

Preputial eversion (mm)		Brahman		Tropical Composite	
Trait 1	Trait 2	r_g	s.e.	r_g	s.e.
PEV12	PEV18	0.96	0.07	0.98	0.03
PEV12	PEV24	0.82	0.11	0.98	0.03
PEV18	PEV24	0.95	0.07	0.92	0.04

Genetic relationships of preputial eversion measured at 12, 18 and 24 months of age with key female reproduction traits. Table 3 presents the genetic correlations of preputial eversion measured at 12, 18 and 24 months in BRAH and TCOMP bulls, with key female reproduction traits. These suggest that lower preputial eversion was genetically associated with lower age at puberty and higher lifetime annual weaning rates for bulls of both genotypes. These results are consistent with those reported by Johnston *et al.* (2014*b*) for PEV18, (also presented in Table 3). For BRAH, genetic correlations of PEV12 with female traits were of lower magnitude than those at 18 and 24 months, suggesting that if the trait were to be exploited as a genetic indicator of female reproduction, measurements at 18 to 24 months would be more effective. Genetic relationships of preputial eversion with LAI were strongest at 18 months for TCOMP and at 24 months for BRAH (r_g = 0.52 and 0.44 respectively). For both genotypes, measurements of preputial eversion at 18 months of age (r_g = -0.71 and -0.88 for BRAH and TCOMP) displayed the strongest genetic relationships with LAWR. Standard errors were sufficiently high for these however, (due to low heritability of the female trait) that these differences were not statistically significant to those observed for PEV24 (r_g = -0.46 and -0.62 for BRAH and TCOMP).

Table 3. Genetic correlations (r_g) and standard errors (s.e.) of preputial eversion at 12, 18 and 24 months of age (PEV12, PEV18 and PEV24) in Brahman and Tropical Composite bulls with female age at puberty (AP), lactation anoestrous interval (LAI) and lifetime annual weaning rate (LAWR) (units of measurement in parenthesis).

Preputial Eversion (mm)	Female Reproduction	Brahman r_g	Brahman s.e.	Tropical Composite r_g	Tropical Composite s.e.
PEV12	AP (days)	0.15	0.15	0.23	0.16
	LAI (days)	-0.09	0.19	0.51	0.19
	LAWR (%)	-0.23	0.28	-0.61	0.26
PEV18	AP (days)	0.33*	0.13	-0.05*	0.16
	LAI (days)	0.13*	0.16	0.52*	0.25
	LAWR (%)	-0.71*	0.27	-0.88*	0.33
PEV24	AP (days)	0.29	0.14	0.34	0.16
	LAI (days)	0.44	0.18	0.26	0.19
	LAWR (%)	-0.46	0.27	-0.62	0.27

* Results for PEV18 previously reported by Johnston *et al.* (2014*b*).

CONCLUSIONS

This study has shown that some level of preputial eversion was evident in 45% the BRAH bulls at 12 months of age, and that this increased to 71% by 24 months. If the condition is seen as unfavourable by breeders of Brahman cattle, these results show that opportunities exist to apply selection to improve the trait. This study has also confirmed the efficacy of preputial eversion measured in Brahman and Tropical Composite bulls as a genetic indicator for female age at puberty, lactation anoestrous interval and lifetime annual weaning rates. In Brahman, there was evidence that measurements of the trait at 12 months may be less effective as an indirect descriptor of female reproduction than those collected at 18 - 24 months. Compared to female reproduction traits, preputial eversion is easy to measure and these results suggests it would be a useful trait to add to the genetic evaluation for tropical breeds.

REFERENCES

Barwick S.A., Johnston D.J., Burrow H.M., Holroyd R.G., Fordyce G., *et al.* (2009) *Anim. Prod. Sci.* **49**: 367.

Barwick S.A., Johnston D.J., Holroyd R.G., Walkley J.R.W. and Burrow H.M. (2014) *Anim. Prod. Sci.* **54**: 97.

Burns B.M., Corbet N.J., Corbet D.H., Li Y., Crisp J.M., Venus B.K., *et al.* (2013) *Anim. Prod. Sci.* **53**: 87.

Corbet N.J., Burns B.M., Johnston D.J., Wolcott M.L., Corbet D.H., *et al.* (2013) *Anim. Prod. Sci.* **53**: 101.

Gilmour A.R., Gogel B.J., Cullis B.R. and Thompson R. (2009) 'ASReml user guide. Release 3.0.' VSN International: Hemel Hempstead, UK.

Johnston D.J., Barwick S.A., Corbet N.J., Fordyce G., Holroyd R.G., *et al.* (2009) *Anim. Prod. Sci.* **49**: 399.

Johnston D.J., Barwick S.A., Fordyce G., Holroyd R.G., Williams P.J., *et al.* (2014*a*) *Anim. Prod. Sci.* **54**: 1.

Johnston, D.J., Corbet N.J., Barwick S.A., Wolcott M.L. and Holroyd R.G. (2014*b*) *Anim. Prod. Sci.* **54**: 74.

Proc. Assoc. Advmt. Anim. Breed. Genet. 22:461-464

GENETIC AND PHENOTYPIC PARAMETERS FOR FEED EFFICIENCY TRAITS IN AUSTRALIAN ANGUS BEEF CATTLE

J.A. Torres-Vazquez, J.H.J. van der Werf and S.A. Clark

School of Environmental and Rural Science, University of New England, Armidale, Australia, 2351

SUMMARY

Feed intake represents a major cost to all animal production systems. Increasing the efficiency in which animals turn this feed into product can be a major goal in many animal breeding programs. Consequently, Australian beef producers have been measuring many traits associated with feed efficiency in an attempt to increase the accuracy of selection, precision of genetic parameters estimates and ultimately increase the amount of genetic gain achieved. The objective of this study was to estimate genetic and phenotypic parameters for traits associated with feed efficiency from records on 1614 Angus Steers from the Australian Beef Information Nucleus (BIN). Traits analysed included Average Daily Weight Gain (ADG), Metabolic Mid-Weight (MMWT), Daily Feed Intake (FI), Feed Conversion Ratio (FCR) and Residual (or Net) Feed Intake (RFI). Parameters were estimated using bivariate animal models in ASReml. Heritability estimates ranged from 0.12 ±0.06 for FCR to 0.49 ±0.09 for FI. High genetic correlations were estimated between FI and RFI (0.83±0.05) and FI and ADG (0.81±0.08). Significant genetic correlation also existed between ADG and MMWT (0.65±0.12) and between MMWT and FI (0.68±0.08). Heritability estimates show that there would be a favourable response to selection for the efficiency traits in this population. The positive and unfavourable genetic correlation between ADG and RFI, suggest that improving RFI would result in lower ADG. Given this, further studies are required to investigate genetic associations between efficiency traits and other economically important traits, in addition to examine new ways of utilizing feed efficiency information in breeding programs.

INTRODUCTION

Feed intake represents a major input cost in almost all animal production systems (Archer 1999). The efficiency of converting this feed into useable animal products, commonly referred to as feed efficiency, is becoming a common breeding objective. In order to include feed efficiency traits in the breeding goal, genetic parameters are needed for accurate and unbiased prediction of breeding values, as well as to develop selection indices, and to predict selection responses (Hofer, 1998). Several authors have documented significant genetic variation for feed efficiency traits, however, genetic parameters can vary can vary depending on each population.

The objective of the present study was to estimate genetic and phenotypic parameters for Average Daily Weight Gain (ADG), Metabolic Mid-Weight (MMWT), Daily Feed Intake (FI), Feed Conversion Ratio (FCR) and Residual Feed Intake (RFI) from data collected in the Australian Angus Beef Information Nucleus (BIN) (also known as the Angus Sire Benchmarking Program).

MATERIALS AND METHODS

The phenotypic data examined in this study included live weights, and FI measures from 1823 Angus Steers collected from 2013 to 2016 at Tullimba Feedlot. On entry to the feedlot, the animals in this study ranged from 500-600 days of age and weighed approximately 450-500 kg. Initially animals were conditioned for 21 days and fed for an additional 70 days over which time all data was collected. All animals were weighed 6 times over the 70-day test period (fortnightly). ADG was calculated as the regression of weight on time (days), while MMWT was obtained as the mid-point raised to the 0.73 power (Arthur *et al.*, 2001; Berry and Crowley, 2013).

The original information was edited to guarantee the quality of the data to be analysed. Duplicated records and incomplete information were eliminated. FCR was obtained as feed intake divided by the ADG, while RFI was estimated by regressing FI on ADG and MMWT (Arthur and Herd, 2008; Berry and Crowley, 2013).

Fixed effects of mean and Contemporary group (CG) (obtained from BREEDPLAN, which included trial, property management group and feedlot pen) were fitted for all analyses. CG of less of 10 animals were removed from the analysis. In the present study, estimates of both maternal genetic and maternal permanent environmental components for RFI were 0 therefore not included in the model (data do not show).

The final data file consisted of 1614 Angus steers with complete information for ADG, MMWT, FI, RFC and RFI. The pedigree file included an historical file with 21,439 animals with 3,908 sires and 11,610 dams.

Data were analysed using AIREML methodology implemented in the software ASReml (Gilmour *et al.*, 2009). Starting values for bivariate analysis were estimated using single trait animal models, and bivariate animal models were used to estimate genetic and phenotypic correlations.

For each trait, the following bivariate animal model was used:

$$[1]\; y = Xb + Zu + e$$

where **y** is the vector of the phenotypes for the traits; **X** is a matrix relating animals to fixed effects; **b** is vector of fixed effects for the traits analysed; **Z** is a matrix relating animal effects to the data; **u** is a vector which contains animal random effects; and **e** is a vector of residual effects for the analysed traits. Furthermore, expectations and variance matrices of random vectors are described as:

$$E = \begin{bmatrix} y \\ u \\ e \end{bmatrix} = \begin{bmatrix} Xb \\ 0 \\ 0 \end{bmatrix};\; V = \begin{bmatrix} u \\ e \end{bmatrix} = \begin{bmatrix} A \otimes G_0 & 0 \\ 0 & I \otimes R_0 \end{bmatrix}$$

Where $\boldsymbol{G_0}$, and $\boldsymbol{R_0}$ denote 2x2 matrices containing additive genetic and residual covariance components, respectively; **A** is the numerator relationship matrix; **I** is the identity matrix; and ⊗ is the Kronecker product.

RESULTS AND DISCUSSION

Descriptive statistics for the studied traits are presented in Table 1. Heritability estimates, together with genetic and phenotypic correlations using bivariate animal models are presented in Table 2.

Table 1. Descriptive statistics for Average Daily Gain (kg d), Metabolic Mid-Weight (kg), Feed Intake (kg), Feed Conversion Ratio (kg), and Residual Feed Intake (kg d).

Trait	Number	Mean	Min	Max	SD
Average Daily Gain	1614	1.61	0.44	2.90	0.33
Metabolic Mid-Weight	1614	104.31	85.62	135.18	7.35

Feed Intake	1614	14.89	6.78	22.63	2.11
Feed Conversion Ratio	1614	9.55	4.09	32.55	2.09
Residual Feed Intake	1614	0.01	-7.09	5.18	1.70

Table 2. Heritabilities (on diagonal), genetic (above diagonal), and phenotypic correlations (below diagonal) for Average Daily Weight Gain (ADG), Metabolic Mid-Weight (MMWT), Feed Intake (FI), Feed Conversion Ratio (FCR) and Residual Feed Intake (RFI).

	ADG	**MMWT**	**FI**	**FCR**	**RFI**
ADG	0.31 ± 0.07	0.65 ± 0.12	0.81 ± 0.08	-0.61 ± 0.15	0.42 ± 0.17
MMWT	0.35 ± 0.02	0.46± 0.09	0.68 ± 0.08	-0.22 ± 0.23	0.21 ± 0.16
FI	0.46 ± 0.02	0.56 ± 0.02	0.49 ± 0.09	-0.13 ± 0.23	0.83 ± 0.05
FCR	-0.76 ± 0.01	0.00 ± 0.03	0.11 ± 0.03	0.12± 0.06	0.24 ± 0.23
RFI	-0.09 ± 0.03	0.09 ± 0.03	0.78 ± 0.01	0.52 ± 0.02	0.30 ± 0.09

Heritability estimates indicate, for ADG, MMWT, FI and RFI, that a moderate amount of additive genetic variation exists ranging from 0.30 to 0.49. The heritability estimates for ADG, FI and RFI were very close to the pooled heritably estimates published by Berry and Crowley (2013) using a meta-analysis on feed efficiency traits. In this meta-analysis up to 39 scientific publications were analysed, and they reported a polled heritably for ADG, FI and RFI as 0.31±0.014, 0.40±0.012, and 0.33±0.013, respectively. The lowest heritability was for FCR and was associated with a high standard error (0.12 ± 0.06), and although this estimate was lower than the pooled heritability estimate of 0.23±0.013, published by Berry and Crowley (2013), was within the range of the published values (0.09 to 0.46) for different beef cattle populations. In addition to this, FCR is a ratio trait and it has been documented several problems related to predict the genetic change in subsequent generations that can be avoided by using linear index traits such as RFI that increases selection responses (Gunsett, 1984; Arthur *et al.*, 2001). The heritability estimate in this work for MMWT (of 0.46 ±0.09) was slightly higher compared to the estimated provided by Arthur *et al.* (2001) of 0.40±0.02 in Angus cattle of Australia.

The genetic correlations between ADG and FI, between ADG and FCR, between FI and RFI, and between FCR and RFI, were in agreement with the average genetic correlations of 0.76±0.09, -0.57±0.16, 0.82±0.05, and 0.35±0.22, respectively, using a meta-analysis of genetic parameters reported by several authors in beef cattle populations (Berry and Crowley, 2013). In a similar way, genetic correlations between MMWT and ADG, between MMWT and FI are in agreement with the estimated values, published by Arthur *et al.* (2001), of 0.53±0.07 and 0.65±0.03, respectively. Different to the genetic correlation, reported in this work, between FCR and FI (-0.13±0.23), the genetic correlation between RFI and FI was positive (0.83±0.05); in addition, only the genetic correlation of FCR was negative and favourable correlated with ADG, meaning that improving FCR was associated with greater ADG.

The phenotypic correlations between ADG and MMWT, ADG and FI, MMWT and FI, FI and FCR, FI and RFI, and between FCR and RFI were positive. These results are in agreement with the phenotypic correlations reported by Arthur *et al.* (2001) of 0.26, 0.41, 0.63, 0.23, 0.72 and 0.53, respectively. Arthur *et al.* (2001) also reported high negative phenotypic correlation between ADG and FCR of -0.77, similar to the estimate obtained in the present work. Since RFI was obtained by regression, it is expected to be phenotypically independent of the ADG, and MMWT (Arthur *et al.*, 2001). However, in this study a non-zero phenotypic correlation between RFI and ADG and MMWT was observed. A possible explanation for this is that RFI was estimated from many different trials. Once RFI was fitted in a full mixed model some of this variability is corrected for by contemporary group estimates and a non-zero correlation remains. All the phenotypic correlation involving ADG, FI, FCR and RFI were within range estimated by different authors in beef cattle populations (Berry and Crowley, 2013).

In conclusion, all heritability estimates in the current work were within the range of estimates obtained in several beef cattle populations which in most cases involved small numbers of animals. A moderate heritability was found for FCR with high standard errors, suggesting that RFI will represent a better option for improving response to selection. The results from this study suggest that is some situations RFI is neither phenotypically or genetically independent of ADG or MMWT. Given this, additional research is required to investigate other ways to select for feed efficiency and, due to the lack of consistency across several studies and considering that the analysis of feed efficiency traits alone provides little information. Additional research is required to investigate genetic associations between efficiency traits and other economically important traits.

REFERENCES

Archer J. A., Richardson E. C., Herd R. M., and Arthur. P. F. (1999) *Aust. J. Agric. Res.* **50**: 147-161.

Arthur P.F., Archer J. A., Johnston D. J., Herd R. M., Richardson E. C. and Parnel P. E. (2001) *J Anim Sci*. 79(11), 2805-2811

Arthur, J. P. F., and Herd R. M. (2008) *Revista Brasileira de Zootecnia.* **37**: 269-279.

Berry D. P., Crowley J. J. (2013). *J Anim Sci.* **91**: 1594-1613.

Gilmour A. R., Gogel B., Cullis B., Thompson R., and Butler D. (2009). VSN International Ltd, Hemel Hempstead, UK.

Gunsett F. C. (1984) *J Anim Sci.* **59**: 1185-1193.

Hofer, A. 1998. *J Anim Breed Genet.* **115**: 247-265.

Proc. Assoc. Advmt. Anim. Breed. Genet. 22:465-468

EPINETR: A FORWARD-TIME SIMULATOR FOR EPISTATIC NETWORK MODELLING IN R

Dion C. Detterer[1,2], Paul Kwan[1] and Cedric Gondro[1,2]

[1] School of Science and Technology, University of New England, Armidale NSW 2351, Australia
[2] The Centre for Genetic Analysis and Applications, University of New England, Armidale NSW 2351, Australia

SUMMARY

The problem of the missing heritability hinders our understanding of the relationship between genetic markers and complex quantitative traits, in turn limiting informed selection of mates for animal breeding purposes. To this end, we have developed epinetr, a software package for R designed to facilitate the investigation of the possible contribution of gene interaction networks to the missing heritability.

INTRODUCTION

Since the advent of the genome-wide association study (GWAS) in 2005 (Haines *et al.* 2005; Vissler *et al.* 2012), thousands of genetic variants have been identified which contribute to complex traits in either livestock (Tenghe *et. al.* 2016) or humans (Li *et al.* 2016), with an application for livestock being a genetically-informed artificial selection for desirable traits. However, a gap emerged between current heritability estimates for these traits and the contribution of the identified variants: the so-called "missing heritability" problem (Manolio *et al.* 2009; Zuk *et al.* 2014). Several explanations were put forth to explain this disparity (Manolio *et al.* 2009; Eichler *et al.* 2010); among these, the effect of epistasis (i.e. gene-gene interaction) on heritability estimates is an explanation that has attracted considerable attention (Huang 2012; Zuk *et al.* 2012; Bloom *et al.* 2013). Simulations are currently the most viable approach to test epistatic models and how they affect our estimates of additive genetic variance (Hoban *et al.* 2012).

There is thus a need in animal breeding for flexible simulators that can accommodate a wide variety of randomly-generated and user-generated epistatic models while still providing parameters to control other factors. As an aid to further research on the genetic architecture of epistasis, a need also exists for a network-based approach to epistatic modelling in simulators. To this end, we have developed epinetr, a package for the statistical environment R, soon to be submitted to CRAN: epinetr is a forward-time simulator designed specifically for the study of high-order epistatic networks and how they impact estimates of genetic parameters and selection decisions of complex quantitative traits.

This paper first gives an overview of the design decisions behind epinetr, it then discusses the epinetr simulator itself, the features and parameters within the simulator and its ability to handle complex epistatic networks.

DESIGN CONSIDERATIONS

The two broad categories of population genetics simulators form a simple dichotomy: simulators that work forwards-in-time and those that work backwards-in-time (Hoban *et al.* 2012). As can be inferred from the nomenclature, forwards-in-time (or forward-time) simulators start with a population and work forwards to track individuals and pedigrees via selection, recombination and mutation across generations; on the other hand, backwards-in-time (or coalescent) simulators work backwards to infer genetic histories. Forwards-in-time simulators demand more computational resources than backwards-in-time simulators simply due to the level of granularity required (i.e. per-individual simulation); at present, forwards-in-time simulators include Easypop (Balloux 2001),

GenomePop (Carvajal-Rodríguez 2008) and FREGENE (Chadeau-Hyam 2008), none of which include mention of epistatic modelling capabilities in the associated literature. Both simuPOP (Peng and Kimmel 2005) and quantiNemo (Neuenschwander 2008) are forwards-in-time simulators that do allow for statistical epistatic modelling; the same is true for the more recent simulator SELAM (Corbett-Detig and Jones 2016).

Backwards-in-time simulators such as SNPsim (Posada and Wiuf 2003), SIMCOAL2 (Laval and Excoffier 2004), GENOME (Liang *et al.* 2007) and MaCS (Chen *et al.* 2009) are typically more computationally efficient than forwards-in-time simulators, but there is a trade-off: they are not as suited to modelling complexity or natural or artificial selection (Hoban *et al.* 2012). This limits their application to the study of epistatic impact on selection for complex traits.

Existing outside this dichotomy is EpiSIM (Shang 2013), which allows for the simulation of simple 2-way interactions.

The choice was made to build a forward-time simulator, as this allowed for the use of complex selection scenarios. As a further consideration, there is evidence to suggest that epistatic networks exhibit a small world or scale-free structure (Tyler *et al.* 2009; Mackay 2014). While this appears to be a fruitful avenue to pursue, a more general point emerges: the actual network structure may be the key to understanding the underlying mechanics of epistasis, including the relationship between genes and phenotypes. For this reason, epinetr includes the ability to both automatically generate random and scale-free epistatic networks or alternatively input user-defined epistatic networks that can be generated by an external model based on previous knowledge (or a hypothesis) of the underlying architecture of a trait.

In a nutshell, the epinetr package is designed as a tool to investigate potential epistatic sources of missing heritability using network models.

PACKAGE FEATURES

The epinetr package is written for the R statistical software environment, allowing for complex analysis to take place in the same environment as the actual simulation. It includes a set of classes that enable users to perform common operations both before and after the simulation with simple commands, as well as provisions for specifying a large set of population parameters.

Typically, there are 5 broad steps in the workflow:

1. Define population parameters and construct the initial population
2. Attach additive effects to the population
3. Attach an epistatic network to the population and visualise the network
4. Run a forward-time simulation of the population and plot the simulation run

Parameters are specified using a simple parameter file. Below we give an overview of the parameter options available.

Population size, given at initialisation, is fixed throughout the simulation run. However, because litter size is specified by a user-defined probability mass function, some generations may be smaller than the fixed population size. For this reason, another pair of parameters controlling the maximum lifespans of sires and dams may be violated.

Allele frequencies can be inferred from a haplotype file or specified directly, thus allowing for "sideways simulation" (by first using a coalescent simulator to arrive at the allele frequencies); alternatively, haplotypes can be used directly as the initial population.

Both broad- and narrow-sense heritability can be specified, controlling the contributions of additive, epistatic and environmental effects to the overall variance of the trait being studied.

Selection is performed either randomly or via linear ranking; the mutation rate is a single number while recombination probabilities can be optionally specified, thus allowing for the simulation of hotspots. Separate truncation rates for sires and dams can also be specified, as can an initial burn-in period of random selection.

A chromosomal map for the single nucleotide polymorphisms (SNP) is required, with the user determining which SNP are used for quantitative trait loci (QTL) in the epistatic network; alternatively, the user can specify the number of QTL which are then selected from the SNP at random.

The number of times a sire can mate during a single generation can be specified.

Once a population is generated using the above parameters, additive effects across all SNP can then be attached. Effect sizes (i.e. the absolute value of the coefficients) are determined by the restrictions of the population parameters; however, they can be sampled from any distribution specified by the user, including user-defined functions.

Epistatic modelling. By specifying an incidence matrix (with each row representing a QTL and each column representing an interaction between QTL), the user can determine the structure of the epistatic network; alternatively, the system can generate a random or scale-free network for the population with a single command. In either case, the orders of interaction included in the network are specified by the user and limited only by the number of QTL in the population; in addition, scale-free networks can be given a minimum number of interactions per QTL.

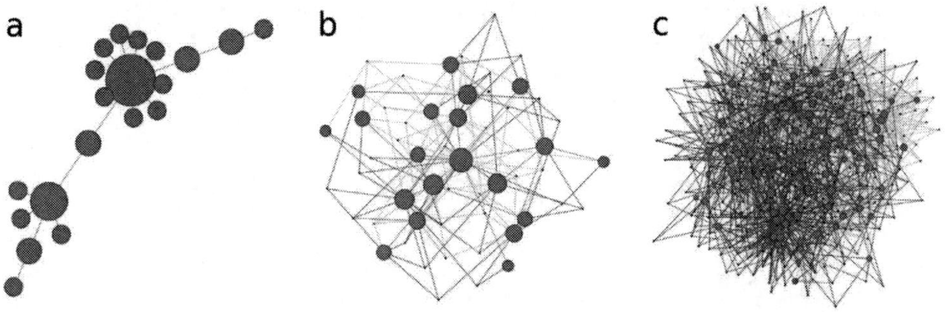

Figure 1. Three unique scale-free epistatic networks generated automatically from within epinetr: a) a 20-QTL network comprised of 2-way interactions; b) a 20-QTL network comprised of 2-, 3- and 4-way interactions; and c) a 100-QTL network comprised of 2-, 3-, 4- and 5-way interactions.

The network structure can be easily visualised using a simple plot command. Figure 1 depicts three potential epistatic scale-free networks generated automatically and visualised from within epinetr.

The result of a simulation run is a set of files giving allele frequencies and pedigrees for each individual in each generation, as well as haplotypes for each individual in the final generation (or, optionally, each generation). Most importantly, the additive, epistatic and environmental contribution to each individual's phenotype is given as an output. Finally, the mean, maximum and minimum phenotypic values within the population across generations can also be easily plotted using a single command.

CONCLUSION

epinetr is an R package designed to facilitate the modelling and analysis of epistatic networks and their effects on estimates of genetic parameters and selection decisions within populations, filling an important niche in population genetics simulation. It is hoped that it will be a valuable tool to better understand how different models of genetic architecture, particularly epistasis, relate to the problem of missing heritability.

ACKNOWLEDGEMENT
This project was supported by a grant from the Next-Generation BioGreen 21 Program (No. PJ01134906), Rural Development Administration, Republic of Korea and Australian Research Council (DP130100542).

REFERENCES

Balloux F. (2001). *J. Hered.* **92**: 301.

Bloom J.S., Ehrenreich I.M., Loo W.T., Lite T.L.V. and Kruglyak L. (2013) *Nature* **494**: 234.

Carvajal-Rodríguez A. (2008) *BMC Bioinformatics* **9**: 223.

Chadeau-Hyam M., Hoggart C.J., O'Reilly P.F., Whittaker J.C., De Iorio M. and Balding D.J. (2008) *BMC Bioinformatics* **9**: 364.

Chen G.K., Marjoram P. and Wall J.D. (2009). *Genome Res.* **19**: 136.

Corbett-Detig R. and Jones, M. (2016) *Bioinformatics* **32**: 3025.

Eichler E.E., Flint J., Gibson G., Kong A., Leal S.M., Moore J.H. and Nadeau J.H. (2010) *Nature Rev. Genet.* **11**: 446.

Haines J.L., Hauser M.A., Schmidt S., Scott W.K., Olson L.M, Gallins P., Spencer K.L., Kwan S.Y., Noureddine M., Gilbert J.R., Schnetz-Boutaud M., Agarwal A., Postel E.A. and Pericak-Vance M.A. (2005) *Science* **308**: 419.

Hoban S., Bertorelle G. and Gaggiotti O.E. (2012) *Nature Rev. Genet.* **13**: 110.

Huang W., Richards S., Carbone M.A., Zhu D., Anholt R.R., Ayroles J.F., Duncana L., Jordana K.W., Lawrencea F., Magwirea M.M., Warner C.B., Blankenburg K., Han Y., Javaid M., Jayaseelan J., Jhangiani S.N., Muzny D., Ongeri F., Perales L., Wu Y., Zhang Y., Zou X., Stone E.A., Gibbs R.A. and Mackay T.F.C. (2012) *Proc. Natl. Sci. USA* **109**: 15553.

Laval G. and Excoffier L. (2004) *Bioinformatics* **20**: 2485.

Li M.J., Liu Z., Wang P., Wong M.P., Nelson M.R., Kocher J.P.A., Yeager M., Sham P.C., Chanock S.J., Xia Z. and Wang, J. (2016) *Nucleic Acids Res.* **44**: D869.

Liang L., Zöllner S. and Abecasis G.R. (2007) *Bioinformatics*, **23**: 1565.

Mackay, T.F. (2014) *Nature Rev. Genet.* **15**: 22.

Manolio T.A., Collins F.S., Cox N.J., Goldstein D.B., Hindorff, L.A., Hunter D.J., McCarthy M.I., Ramos E.M., Cardon L.R., Chakravarti A., Cho J.H., Guttmacher A.H., Kong A., Kruglyak L., Mardis E., Rotimi C.N., Slatkin M., Valle D., Whittemore A.S., Boehnke M., Clark A.G., Eichler E.E., Gibson G., Haines J.L., Mackay T.F.C., McCarroll S.A. and Visscher P.M. (2009) *Nature* **461**: 747.

Neuenschwander S., Guillaume F. and Goudet, J. (2008) *Bioinformatics* **24**: 1552.

Peng B. and Kimmel M. (2005) *Bioinformatics* **21**: 3686.

Posada D. and Wiuf, C. (2003) *Bioinformatics* **19**: 289.

Shang J., Zhang J., Lei X., Zhao W. and Dong Y. (2013) *Genes Genomics* **35**: 305.

Tenghe A.M.M., Bouwman A.C., Berglund B., Strandberg E., de Koning D.J. and Veerkamp R. F. (2016) *J. Dairy Sci.* **99**: 5470.

Tyler A.L., Asselbergs F.W., Williams S.M. and Moore, J. H. (2009) *Bioessays* **31**: 220.

Visscher P.M., Brown M.A., McCarthy M.I. and Yang J. (2012) *Am. J. Hum. Genet.* **90**: 7.

Zuk O., Hechter E., Sunyaev S.R. and Lander E.S. (2012) *Proc. Natl. Sci. USA* **109**: 1193.

Zuk O., Schaffner S.F., Samocha K., Do R., Hechter E., Kathiresan S., Daly M.J., Neale B.M., Sunyaev S.R. and Lander, E.S. (2014*) Proc. Natl. Sci. USA* **111**: E455.

Proc. Assoc. Advmt. Anim. Breed. Genet. 22:469-472

METHODOLOGY FOR QUANTIFICATION OF CIRCULATING CELL-FREE MICRORNA FROM BOVINE PLASMA FOR ANALYSIS OF MEAT QUALITY TRAITS

K.J. Kochan[1,2], S.E. Forman[1], A.E. Hillhouse[2], H.R. Cross[1], and P.K. Riggs[1]

[1]Department of Animal Science, Texas A&M University, College Station, TX, USA
[2]Texas A&M Institute for Genome Sciences and Society, College Station, TX, USA

SUMMARY

MicroRNAs (miRNAs) function as important genetic regulators during growth, development, and cellular processes. Circulating miRNAs, cell free or contained within exosomes, have been detected in all bodily fluids, including plasma. These excreted molecules are not well-understood and may reflect a physiological phenotype, or exert a regulatory function. To develop a means for examining the impact of circulating miRNA function on beef cattle phenotypes, we assessed parameters important for isolation of miRNAs from plasma and subsequent high-throughput expression analysis on a real-time quantitative RT-PCR microfluidics platform. . These methods will facilitate economical expression analysis of circulating microRNAs and their potential association with health status and carcass and meat quality traits in livestock.

INTRODUCTION

MicroRNAs are small (~17-27nt) RNA molecules that function as molecular rheostats to regulate gene expression as part of many physiological processes. These molecules regulate the function of entire networks of genes, increasing the complexity of genetic mechanisms (Jeffries *et al.* 2010). Several miRNAs are known to regulate skeletal muscle phenotypes, and their differential expression within muscle may reflect response to exercise or even variation in activity (Dawes *et al.* 2015, Margolis *et al.* 2016). Recent studies show that circulating miRNAs may reflect health status or response to diet (Ioannidis and Donadeu 2016; Muroya *et al.* 2016). . Thus, we expect that circulating miRNAs can be applied to beef cattle production as informative diagnostic tools for interrogating mechanisms important for muscle growth and fat deposition.

Our overall objective is to develop genomic tools that are directly applicable for food animal research. Our specific interest is to accurately measure and evaluate miRNA expression in peripheral blood and apply this tool for improvement of meat quality. During plasma isolation, lysis of erythrocytes (Kirschner *et al.* 2011, 2013), leukocytes (Al-Soud and Radstrom 2001) or activated thrombocytes (Osman and Falker 2011) can release non-target miRNAs, or iron from hemoglobin or lactoferrin into the plasma, potentially altering expression profiles or inhibiting RT-PCR. In this paper we address critical parameters for experimental handling and processing of plasma from steers on feed for isolation of intact cell-free miRNAs, free of contaminating RNA transcripts or agents inhibitory to downstream analysis. This approach is a necessary first step to enable quantitative analysis of these miRNAs, and their potential relationship to carcass traits.

MATERIALS AND METHODS

Animals, sample collection and processing. Animal handling and sample collection procedures were approved by the Texas A&M Animal Care and Use Committee (AUP #2008-234). . Blood was collected by venipuncture (Vacuette® 18 ga x 1.5 inch needles, Greiner Bio-One North America, Monroe, NC) into 10 ml BD Vacutainer® tubes containing K_2-EDTA (PN 366643, BD Diagnostics, Franklin Lakes, NJ). To assess the effect of handling, an initial pilot study was conducted. Four tubes of blood were collected from a single jugular puncture of 4 Angus steers. . All tubes were gently inverted 10 times immediately after collection. One tube from each set was shaken vigorously to mimic improper handling. Tubes were held at 4°C until processed. Blood in "control" tubes and shaken tubes were processed within 2h of collection. The remaining 2 tubes from each steer were held and processed at 30h and 52h post-collection, respectively.

In a second study, a single tube of blood was collected from each of 88 cross-bred steers every 28d over a period of 7 months, beginning when the steers were approximately 12 months of age. Samples were kept cool and transported to the laboratory within 6 hours. Tubes of blood were centrifuged at 1300 x *g* for 10 min. Plasma was removed with care to avoid disturbing the buffy coat layer, transferred to clear, 1.5 ml tubes, and stored as 500 µl aliquots at -80°C.

Assessment and quantification of hemolysis. Hemolysis was assessed visually and quantified by spectrophotometry, according to the scale in Figure 1. This scale was adapted from a clinical laboratory protocol (http://bit.ly/2kcSm3H). Spectrophotometry was conducted as described by Kirschner *et al.* (2011, 2013), on a Bio-Rad 680 Microplate Reader (Hercules, CA).

CLEAR | SLIGHT | SOME | MODERATE | SEVERE |

Figure 1. Representation of varying levels of hemolysis in bovine blood, ranked left to right as Clear, Slight hemolysis, Some hemolysis, Moderate hemolysis, or Severe hemolysis.

MicroRNA isolation and qRT-PCR. Procedures were conducted according to manufacturer's recommendations (Exiqon, Inc., Woburn, MA). MicroRNA was extracted from 500 µl plasma with the miRCURY™ RNA Isolation Kit-Biofluids (Exiqon). Plasma miRNA was quantified by fluorometric analysis (Qubit® microRNA Assay Kit, Qubit® 2.0 fluorometer, ThermoFisher). Extracted miRNA was reverse-transcribed (RT) into cDNA with the Universal cDNA Synthesis Kit II (Exiqon). Quantitative RT-PCR reactions (10 µl) contained 1x ExiLENT SYBR® Green master mix (Exiqon), 1x ROX (Life Technologies, Carlsbad, CA,), 1 µl primer mix and 2 µl diluted template cDNA. Primers for hsa-miR23a-3p and hsa-miR-451A (PN 204772 and 204734) were used for initial assessment of hemolysis. Amplification was carried out in the ABI 7900HT thermal cycler in 9600-emulation mode (Applied Biosystems, Inc., Foster City, CA).

Microfluidic qRT-PCR. A custom panel of 96 microRNA assays was created (Pick & Mix Panel for Fluidigm, PN 203899, Exiqon), to contain specific miRNAs expected to be present in plasma, or have relevance for muscle growth, fat deposition, and meat quality. Reverse transcription, specific target preamplification (STA) and microfluidic qPCR reactions were performed according to Exiqon's Fluidigm-BioMark recommended protocol (http://bit.ly/2ke3TUF). Plasma RNA input for RT was increased to 8 µl. Samples from severely hemolyzed plasma were excluded from further processing. As a positive control, cDNA prepared from miRNA extracted from *longissimus* muscle was included on the array, with several negative control samples. Amplification was carried out in 96.96 Dynamic Array™ Integrated Fluidic Circuits (IFC) for Gene Expression (PN BMK-M-96.96, Fluidigm, San Francisco, CA) on a Biomark HD thermal cycler. Raw amplification data were analyzed with Sequence Detection System software (SDS v.2.2.2, Applied Biosystems) or Real-Time PCR Analysis software for Biomark (Fluidigm), as appropriate.

RESULTS

Blood collection and handling procedures influence risk of hemolysis. Visual assessment of the pilot study samples confirmed that hemolysis increased if samples were handled roughly or initial processing was delayed. Shaking produced slight and moderate hemolysis and increased holding time greatly increased hemolysis compared to the control samples. In the main study, few samples were moderately to severely hemolyzed when processed within 6 hours of collection. In cases where blood flow into the collection tube was slow, or venipuncture required more than one attempt, moderate to severe hemolysis was likely (11 of 13 records).

Effect of hemolysis on amplification of circulating microRNA. For 7 steers, plasma samples were collected over the course of the study that covered nearly the full range of hemolysis (slight to severe). Free hemoglobin was quantified by spectrophotometry in this subset of 28 samples (Figure 2A). Based on the spectrophotometric analysis, samples scored for hemolysis in the range from clear to moderate on our scale were expected to be acceptable for expression analysis. Blondal *et al.* (2013) reported that hemolysis could also be evaluated by qRT-PCR analysis of miRNAs *miR-23a* and *miR-451*. They found expression of *miR-23a* to be relatively stable in the cell-free fraction. Because *miR-451* is enriched in erythrocytes, an increase in the ratio of miR-*451* to *miR-23a* concentration would indicate hemolysis. Expression of *miR-451* and *miR-23a* in this set of samples was evaluated by quantitative RT-PCR. The difference in threshold cycle quantity (delta C_t) was calculated as described by Blondal *et al.* (2013) and expression measured by qRT-PCR was reduced in the severely hemolyzed group (Figure 2B). Based on these results severely hemolyzed samples were omitted from the microfluidic qPCR experiment.

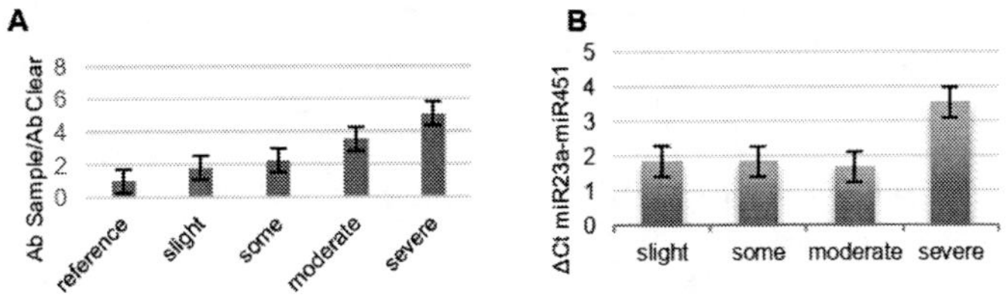

Figure 2. (A) Free hemoglobin in plasma, measured by spectral absorbance at 414 nm and expressed as a ratio of absorbance compared to a clear reference sample. (B) ΔC_t (miR-23a-3p – miR-451a). Severely hemolyzed samples exhibited a 1.8-fold increase in contaminating miRNA abundance compared to those scored as slight ($p<0.05$).

Microfluidic qPCR. Amplification was visually assessed on heat maps for each IFC, and UniSP2 and UniSp6 spike in controls were verified. Seventy-six miRNA assays amplified in at least 1 sample type; 59 amplified cDNA from both plasma and skeletal muscle, 7 amplified only plasma, and 10 amplified only muscle. Poor signals or amplification failure was observed for 17 miRNA assays. MicroRNAs are not abundant in plasma and recovery can be somewhat variable (Brunet-Vega *et al.*, 2015). Modification of the protocol to increase the input volume resulted in amplification signal sufficient for quantification on the Fluidigm IFC platform. While expression was expected to vary across the 240 experimental samples extracted from plasma samples, amplification signal was detected for all 68 assays. Expression failed across all assays for only 2 samples. In 17% of the samples, amplification signal was weak and expression was detected for less than a third of the assays. However, hemolysis status (clear to moderate) was not correlated with general amplification robustness.

DISCUSSION

Proper handling reduces likelihood of plasma hemolysis. Clinical recommendations for human blood collection suggest use of a 21 ga needle, filling the tube to proper capacity, gentle inversion, maintaining ambient temperature between collection and processing, and timely processing (within 2-4 hours) to separate the plasma from the cells (Tuck *et al.* 2009). We found that use of 18 ga needles for cattle, with gentle handling of blood tubes and storage to protect from high heat was sufficient for suitable plasma collection for microRNA isolation. Storage on ice was avoided to prevent activation of platelets, release of miRNA, and contamination of the cell-free fraction (Osman and Falker 2011). Our results indicate prompt processing may be the most important factor in handling, and that plasma should be processed within 6 hours of collection. .

Severe hemolysis may impact data negatively. Several studies have reported methods of quantifying hemolysis and the effect of hemolysis on downstream analysis (Blondal *et al.* 2013; Kirschner *et al.* 2011, 2013). We found that severe hemolysis did alter RT-PCR amplification. By restricting sample quality to hemolysis scored no greater than "moderate," we found that expression measurement via the Fluidigm platform was not inhibited by low level hemolysis.

Significance. This is, to our knowledge, the first description of methodology for use of the Fludigm microfluidics platform expression analysis of circulating microRNAs in biofluids from cattle. Adaptation of a protocol for analysis of microRNA from plasma (Exiqon) resulted in sensitivity sufficient for this platform. This method provides an economical tool to enable PCR-based high-throughput expression analysis of microRNAs. These results are a first step toward systematic evaluation of circulating microRNAs that may play important regulatory roles on growth and development. While not a discovery technique such as RNA-sequencing, this type of approach provides a relatively simple and more economical method for analysis of low-abundance targets over time. We anticipate that investigation of expression patterns of circulating miRNA in relation to carcass and meat quality traits may result in key insights for regulation of muscle growth and fat deposition. Importantly, we expect this approach will provide a new tool for improving meat quality and other desirable phenotypic traits.

ACKNOWLEDGMENTS

We thank Maisie Llewellyn and Dustin Therrien for technical assistance, and Barbara R. Gould and Ken Taylor at Exiqon, Inc., for technical advice and enthusiastic support. Financial support was provided, in part, by the Beef Competitiveness Exceptional Item funding from the Texas legislature.

REFERENCES

Al-Soud W.A. and Radstrom P. (2001) *J. Clin. Microbiol.* **39:** 485.

Blondal T., Jensby Nielsen S., Baker A., Andreasen D., Mouritzen P., *et al.* (2013) *Methods* **59:** S1.

Brunet-Vega, A., Pericay C, Quilez M.E., Ramirez-Lazaro M.J., Calvet X., *et al.* (2015) *Anal. Bioch.* **488**:28-35.

Dawes M., Kochan K.J., Riggs P.K., and Lightfoot J.T. (2015) *Physiol Res* 3:e12469.

Ioannidis J. and Donadeu F.X. (2016) *BMC Genomics* **17**: 184.

Jeffries C.D., Fried H.M., Perkins D.O. (2010) *Int. J. .Biochem. Cell Biol.* **42**:1236.

Kirschner M.B., Kao S.C., Edelman J.J., Armstrong NJ, Vallely M.P., *et al.* (2011) *PLoS ONE* 6: e24145.

Kirschner M.B., Edelman J.J., Kao S.C., Vallely M.P., van Zandwijk N. *et al.* (2013) *Front. Genet.* **4:** 94.

Margolis L.M., Lessard S.J., Ezzyat Y., Fielding R.A., and Rivas D.A. (2016) *J. Gerontol. A. Biol. Sci. Med. Sci.* glw243. doi: 10.1093/gerona/glw243

Muroya S., Shibata M., Hayashi M., Oe M. and Ojima K. (2016) *PLoS ONE* 11: e0162496.

Osman A. and Falker K. (2011) *Platelets* **22:** 433.

Tuck M.K., Chan D.W., Chia D., Godwin A.K., Grizzle W.E., *et al.*, (2009) *J. Proteome Res.* 8: 113.

Proc. Assoc. Advmt. Anim. Breed. Genet. 22:473-476

BREED VARIATION IN TONGUE COLOUR OF DAIRY AND BEEF-CROSS-DAIRY CALVES

L.W. Coleman, H.T. Blair, N. Lopez-Villalobos, P.J. Back, R.E. Hickson

Institute of Animal, Veterinary and Biomedical Sciences, Massey University, Private Bag 11222, Palmerston North 4442, New Zealand.

SUMMARY

Both Angus-cross and Holstein-Friesian-cross-Jersey cattle may have a completely black coat colour, making it difficult to identify breed of newborn calves when Angus bulls are used in New Zealand dairy herds. Holstein-Friesian cattle possess a white spotting gene causing non-pigmentation in coat colour and a pink coloured tongue, whereas Angus cattle have black tongues. The objective of this experiment was to identify whether tongue colour could be a useful predictor of breed in Angus-cross-dairy and dairy-breed calves. Tongue colour of 476 calves soon after birth was classified as being pink, black or pink and black spotted. The conditional probability of a calf with a black tongue being Angus-cross was 0.95 and dairy breed 0.05. The conditional probability of a calf with a pink tongue being dairy breed was 0.85 and Angus-cross 0.15. Culling calves solely on having a black coloured tongue would correctly cull 73% of Angus-cross calves, and retain 90% of dairy-breed calves. Culling calves on possessing a black or pink and black tongue would correctly cull 96% of Angus-cross calves, but also cull 38% of dairy-breed calves. Breed identification on tongue colour alone is insufficient to correctly identify the breed of calves from a New Zealand dairy herd.

INTRODUCTION

If a dairy farmer uses a beef bull to increase value of surplus calves, it can be difficult for the farmer to identify the breed of calves born, so as to retain only dairy-breed calves as replacements. Of beef bulls used in the New Zealand dairy herd, the main breed is Hereford (DairyNZ 2016), in part because the resulting calf will have a white face, making the beef-cross calf easy to identify. When an Angus bull is used, the resulting calves are usually completely black, and they look similar to Holstein-Friesian-cross-Jersey and some Holstein-Friesian calves. Other phenotypic factors might be useful to identify these calves.

The New Zealand dairy herd is comprised predominantly of Holstein-Friesian (33.5%), Jersey (10.1%) and Holstein-Friesian-Jersey crossbreed (47.2%) (DairyNZ 2016). In New Zealand, dairy cows, a straight-bred cow is defined as having ≥14/16 of any one breed's genetics (DairyNZ 2016). Therefore, a cow classified as Holstein-Friesian may have up to 2/16 Jersey genetics.

Previous authors have reported using colour of coat markings, ears and noses in cattle and sheep to identify different genotypes (Pitt 1920, Dry 1924, Ibsen 1933, Dry 1936, Bogart & Ibsen 1937). Coat and tongue colour of straight-bred cattle have been previously investigated (Ibsen 1933). Straight-bred Angus cattle have a completely black coat, with black skin and a black tongue (Ibsen 1933). Straight-bred Holstein-Friesian cattle have a black coat with white patches on the body, white legs below the knee and a pink tongue (Ibsen 1933). Ibsen (1933) proposed that the pink tongue was a result of the animal being homozygous for a recessive white-spotting gene. Straight-bred Jersey cattle are a diluted shade of red with blackened hairs, and black pigmentation on the skin, nose and tongue. The black pigmentation is the result of a dominant black spotting gene (Ibsen 1933). In the same paper, Ibsen proposed that some Jersey cattle also carry the Holstein-Friesian white-spotting gene. Based on Ibsen's theories, an Angus-cross-Holstein-Friesian or Angus-cross-Jersey animal would be expected to be completely black with a black tongue, and black pigmentation on the skin. A Holstein-Friesian-Jersey-cross animal could have a coat colour ranging from the Holstein-Friesian type black with white patches, to a black coated animal with little to no white patches. The tongue colour of a Holstein-Friesian-Jersey-cross would be expected to be pink, pink with black spotting or completely black due to the black spotting gene. The objective of this study was to identify whether tongue colour could be used to identify the breed of calves born from a dairy herd mated to both Angus and dairy bulls, for the purpose of keeping the dairy calves, and culling the calves sired by an Angus bull.

MATERIALS AND METHODS

The experiment was conducted at Limestone Downs dairy farm in Port Waikato New Zealand, with approval from the Massey University Ethics Committee.

Four hundred and seventy-five calves born on the farm during the calving period in 2016 were used in this experiment. Calves were born to cows in a predominantly Holstein-Friesian and Holstein-Friesian-cross-Jersey herd, and sired by an Angus, Jersey or a Holstein-Friesian bull. Calves were identified to dams by the farmer at the time of calf removal, within 24 hours of birth. Calves were identified as being an Angus-cross (n=347), Jersey-cross (n=80), or Holstein-Friesian-cross (n=48) by visual assessment of the animal and through mating records for the probable dam. Visual assessment of the animal consisted of looking at coat colour, physical shape and checking if the calf was polled. The calves suspected of being Angus or Jersey-cross were sampled for DNA parentage analysis (Zoetis, Dunedin, New Zealand). Breed was DNA verified for the Angus-cross and Jersey-cross calves by assigning parentage to a bull of the respective breed.

Tongue colour was recorded at the time of visual assessment. Colour was assessed by opening the mouth of the calf and looking at the top of the tongue. Colour was recorded as being either completely pink, completely black or having a combination of black and pink patches (spotted). No attempt was made to quantify the proportion of pink and black for calves with spotted tongues.

The probability of a calf being a particular breed based on the colour of its tongue was calculated using conditional probability. For example, the probability of a calf being Angus-sired given it had a black tongue was calculated using the equation for conditional probability:

$$P(Angus|black\ tongue) = \frac{P(Angus \cap black\ tongue)}{P(black\ tongue)}$$

where $P(Angus \cap black\ tongue) = \frac{number\ of\ Angus\ calves\ with\ black\ tongue}{total\ number\ of\ calves}$

and $P(black\ tongue) = \frac{number\ of\ calves\ with\ black\ tongue}{total\ number\ of\ calves}$.

RESULTS

Seventy-three percent of Angus-cross calves possessed a black coloured tongue, while only 10% of dairy calves had a black tongue. Holstein-Friesian-cross calves had a lower occurrence of black tongues than Jersey-cross calves (6% and 13% respectively, Table 1). A calf with a black tongue was highly likely (95%) to be Angus and only 1% or 4% likely to be Holstein-Friesian- or Jersey-cross respectively (Table 1).

Table 1: Proportion (%) of calves within each sire breed with each tongue colour, and the conditional probability of a calf possessing a specific coloured tongue being each breed. Dairy breed comprised of Jersey and Holstein-Friesian sired calves (individual breeds in italics)

Sire Breed	n	Black Tongue (B)		Spotted tongue (S)		Pink Tongue (P)	
		Proportion of breed	P (breed | B tongue)	Proportion of breed	P (breed | S tongue)	Proportion of breed	P (breed | P tongue)
Angus	347	73	0.95	23	0.70	4	0.15
Dairy	128	10	0.05	27	0.30	63	0.85
Jersey	*80*	*13*	*0.04*	*34*	*0.23*	*54*	*0.46*
Holstein-Friesian	*48*	*6*	*0.01*	*17*	*0.07*	*77*	*0.39*

A pink coloured tongue was the most common colour for dairy calves (63%) with Holstein-Friesian-cross calves having a 77% incidence and Jersey-cross having a 54% incidence of pink tongues (Table 1). Pink was the least common tongue colour in Angus-cross calves (4%, Table 1). A calf with a pink tongue was most likely to be Holstein-Friesian- (39%) or Jersey-cross (46%), but still had a 0.15 probability of being Angus-cross (Table 1).

A spotted tongue was more common in Jersey-cross (34%) and Angus-cross (23%) calves than Holstein-Friesian-cross (17%) calves (Table 1). Calves with tongues showing pink and black spotting had a greater probability of being an Angus-cross (0.7) than a dairy-cross (0.3) breed (Table 1). The calves with a spotted tongue had a greater probability of being a Jersey-cross (0.23) than a Holstein-Friesian-cross (0.07) calf (Table 1).

The number of calves with spotted tongues from all three breeds raises a question of whether spotted tongue calves should be culled or kept when imposing tongue colour as a culling criterion. If Holstein-Friesian-cross and Angus-cross calves were to be identified solely on tongue colour, and any calf with a black tongue was culled, this experiment indicates that 94% of Holstein-Friesian-cross calves would be correctly retained, however, 27% of Angus-cross calves would also be retained as replacement dairy calves (Table 2). If tongue colour were to be used to identify calves as dairy (Holstein-Friesian- and Jersey-cross) or Angus-cross, culling all calves with a black tongue would unnecessarily cull 10% of dairy breed calves (Table 2).

If calves were to be culled if they had a black, or a spotted tongue, this study suggests 96% of Angus-cross calves would be correctly identified, along with 23% of Holstein-Friesian-cross calves unnecessarily culled (Table 2). When identifying calves as Angus-cross or dairy, the percentage of dairy breed calves unnecessarily culled would be 38% (Table 2).

Table 2: Comparison of the chance of keeping or culling a calf on tongue colour given the sire-breed. Sensitivity is the proportion of correctly kept Holstein-Friesian or dairy (Holstein-Friesian and Jersey) calves. Specificity is the proportion of correctly culled Angus calves

	Culling on black tongue only		Culling on black or spotted tongue	
	Holstein-Friesian	Angus	Holstein-Friesian	Angus
Keep	0.94	0.27	0.77	0.04
Cull	0.06	0.73	0.23	0.96
	Dairy Breed	Angus	Dairy Breed	Angus
Keep	0.90	0.27	0.63	0.04
Cull	0.10	0.73	0.38	0.96

DISCUSSION AND CONCLUSION

Angus-cross calves are more likely to have black tongues and less likely to have pink tongues than dairy calves, however, the high incidence of calves in all breeds with a spotted tongue means that either keeping or culling those with a spotted tongue resulted in a large proportion of falsely identified calves.

The cows in the dairy herd used to produce the calves varied in proportion of Holstein-Friesian and Jersey genetics. There are a small number of cows in the herd which have breeds other than Holstein-Friesian and Jersey in their pedigree, and not all cows have a fully recorded pedigree. Consequently, the full pedigree of each calf cannot be identified, and it is likely that differing proportions of Holstein-Friesian and Jersey genes contribute to the different tongue colours observed.

In reference to Ibsen (1933), there is clearly a relationship to Angus having a black colouring gene and possessing a black tongue. However, 23% of Angus calves had a spotted tongue, which may indicate that tongue colour is more affected by the white-spotting gene from the Holstein-Friesian genetics, than the coat colour is. There were 4% of Angus-cross calves possessing a pink tongue, which may be a result of extreme white-spotting, as the tongues were assessed *in vivo* it cannot be said with certainty that there were no black spots deep in the mouth.

There was one Angus bull used in the experiment, of which 5 of its 10 progeny had a pink tongue, therefore, it is likely that this bull was bred up, and carrying a recessive gene for white spotting. Ibsen (1933) hypothesised that the black-spotting gene causing black tongues in Jersey cattle is dominant over the Holstein-Friesian white-spotting. While this may be true of the 5 white points (four feet and forehead) typical of a Holstein-Friesian being black pigmented in the crossbreed, the theory does not hold up with pigmentation of the tongue. The results from this study suggest that inheritance of tongue colour is more complicated than suggested by Ibsen (1933).

The results of this study suggest that tongue colour may provide useful clues for breed identification because black tongue calves were highly likely to be Angus-cross and pink tongue likely to be dairy calves. It was not infallible however, and the occurrence of spotted tongues raised an issue of whether to keep or cull the calf, as spotted tongues were no more likely to be Angus-cross than dairy. Although it is not reliable as a sole indicator, tongue colour could be combined with other visual assessments to help inform cull/keep decisions.

ACKNOWLEDGEMENTS

This experiment was funded by Beef+Lamb NZ Genetics. The authors would like to thank Angus New Zealand and Limestone Downs farm for their contributions to the project, and also to Geoff Purchas and Joanna Gillingham for their assistance recording tongue colour.

REFERENCES

Bogart, R. & Ibsen, H. L. (1937) *Journal of Genetics*. **35**:31

Dairynz (2016) New Zealand Dairy Statistics 2015-16 https://www.dairynz.co.nz/media/5416078/nz-dairy-statistics-2015-16.pdf

Dry, F. W. (1924) *Journal of Genetics*. **14**:203

Dry, F. W. (1936) *Journal of Genetics*. **33**:123

Ibsen, H. L. (1993) *Genetics*.**18**:441

Pitt, F. (1920) *Genetics*. **9**:281

Proc. Assoc. Advmt. Anim. Breed. Genet. 22:477-480

NO EVIDENCE FOR GENES WITH LARGE EFFECT ON TWINNING IN A BEEF HERD WITH UNUSUALLY HIGH FECUNDITY

L. Cummins[1], C.M. Reich[2,3] and B.J.Hayes[2,3]

[1] "Ivanhoe", 559 Bulart Bridge Road, Cavendish, Vic. 3314 Australia.
[2]AgriBio, Agriculture Research, DEDJTR, , Victoria, Australia
[3]Queensland Alliance for Agriculture and Food Innovation, Centre for Animal Science, University of Queensland, Queensland, Australia

SUMMARY

Fertility is a major driver of productivity and profitability in many livestock production systems. This has inspired a search for mutations with large effects on fecundity. In sheep, at least seven such mutations have been reported. In cattle, mutations with large effects on these traits seem much rarer. In this study, the hypothesis that there are mutations of large effect for fecundity (specifically twinning rate) segregating in a population of cattle with an unusually high frequency of twining was tested. Sixty seven cows in the population, with two years of records of number of calves born per pregnancy, were genotyped with 632K genome wide SNP. In a genome wide association study, no evidence was found for mutations of very large effect on twinning rate (the study had 50% power to detect a mutation explaining 30% of the variance explained at $P<5x10^{-8}$). However, the substantial increases in twinning rate over time as a result of selection achieved in the Ivanhoe herd demonstrates that improvement in this trait is possible, and this might be accelerated by genomic selection.

INTRODUCTION

Fertility is a major driver of productivity and profitability in many livestock production systems. This has inspired investigation into the genetic architecture of fertility and it's component traits – if mutations of large effect are found, and these mutations do not have other deleterious effects (such as on survival), these mutations might be increased in frequency to improve fertility of the population. In sheep, at least seven mutations with large effects on fecundity (number of offspring per dam) have been reported (Davis 2004). For example, the high fecundity of Booroola merino sheep, results from a mutation (FecB) in the bone morphogenetic protein receptor 1B (BMPR-1B) gene (Wilson et al., 2001; Souza et al., 2001, Mulsant et al., 2001) (*Bb* versus *bb* effect of +7 lambs, Bindon 1984). The high fecundity of Inverdale Romney sheep is due to a mutation (FecXI) in the bone morphogenetic protein 15 (BMP15) gene (Galloway et al., 2000) (*iI* versus *ii* effect of 0.6 lambs). In cattle, there is only a single report of a mutation with a large effect on ovulation rate, and on the rate of twins and triplets (Kirkpatrick and Morris 2015). Although some QTL regions with small effects were identified in Norwegian Red and Holstein cattle (Meuwissen et al. 2002, Bierman et al. 2010). The paucity of studies reporting mutations of large effect might be a result of the very low frequency of twins in most cattle populations. This in turn may mean mutations of large effects are at very low frequency and therefore hard to identify. For example, the mutation reported by Kirpatrick and Morris (2015) appears to occur only within a single sire family.

In this study, the hypothesis that there are mutations of large effect for fecundity (specifically twinning rate) in a population of cattle with an unusually high frequency of twining was tested.

MATERIALS AND METHODS

The cattle population was located at "Ivanhoe", Cavendish, Victoria. The herd was established by importing US Meat Animal Research Center (USMARC) Nebraska, Twinner genetics in 2004.

The USMARC Twinner line has been selected for increased twinning rate for over 30 years and had a calving rate of 1.56 per parturition in 2004 (Echternkamp et al 2007).

Sixty seven cows from the herd were selected for genotyping, with the criteria of at least two pregnancies in two years and a range of litter sizes from one in both years to twins in both years. The 67 cows were genotyped with the Bovine HD Array (Illumina, San Diego). Quality control included use of the Illumina GenCall score, genotype calls with <0.6 were excluded, and SNP with multiple or missing map positions were excluded. 632003 SNP remained after quality control as described by (Erbe et al. 2012).

Phenotypes were for each cow's number of calves born alive (2014) and number of foetuses in late pregnancy (2015). For each cow, the two numbers were averaged to get the phenotype that was analysed. A previous study demonstrated a broadly similar trait had a repeatability across years of 0.30 (Gregory et al. 1990). The distribution of the phenotype is given in Table 1.

Table 1. Distribution of number of singles, twins and triplets for sixty seven genotyped cows across two years in the Ivanhoe herd.

Calvings	Number of cows	Phenotype
Singles both years	26	1
Single one year and twins one year	27	1.5
Twins both years	12	2
Twins one year and triplets one year	1	2.5
Triplets one year*	1	3

*Only calved in 2015

A genome wide association study was conducted fitting each SNP in turn, simultaneously with the genomic relationship matrix among the cows to control for population structure, using EMMAX (Kang et al. 2014), with the model $\boldsymbol{y} = \mu + \boldsymbol{x}b + \boldsymbol{Zu}$, where **y** is a vector of the phenotypes, μ is the mean, **x** is vector of SNP genotypes, with genotypes coded as 0 (homozygote first allele), 1 (heterozygote) or 2 (homozygote alternate allele), -2*p, where p is the frequency of the first allele, b is the effect of the SNP, **Z** is matrix allocating records to animals, and u is a vector of polygenic breeding values, where $\mathbf{u} \sim N(0, \mathbf{G}\sigma_g^2)$, with **G** the genomic relationship matrix among animals constructed as described by VanRaden (2008). The heritability of the phenotype was estimated in the same analysis.

RESULTS AND DISCUSSION

The heritability of the fecundity phenotype was 0.12±0.24. This study is too small to estimate heritability, as evidenced by the large standard error. It is perhaps encouraging that the heritability is not zero, and our estimate was close to other estimates (eg 0.06, Gregory et al. 1990).

There was no evidence for mutations of large effect, Figure 1. No SNP had P-values lower than the significance threshold corrected for multiple testing (of 630K SNP), which was $5\text{x}10^{-8}$.

The major limitation of this study is clearly the small number of cows genotyped. This means the study only had the power to detect mutations of very large effect. The study had 50% power to detect a mutation explaining 30% of the variance at $P<5\text{x}10^{-8}$. There are actually some examples of mutations of large effect detected in even smaller cohorts, including a mutation resulting in 4 horns in sheep (Ockert et al. 2016), and mutations associated with ridge back phenotypes in dogs (Salmon Hillbertz NHC, et al. 2007). Kirkpatrick and Morris (2015) detected a mutation with a large effect on bovine ovulation rate in only 131 animals. It can only concluded, from these results, there is no evidence for mutations of really large effect on twinning segregating in the high twinning rate population studied here. Mutations of more modest effect are not ruled out by this study, though it will require much larger numbers of genotyped and phenotyped cows to detect these.

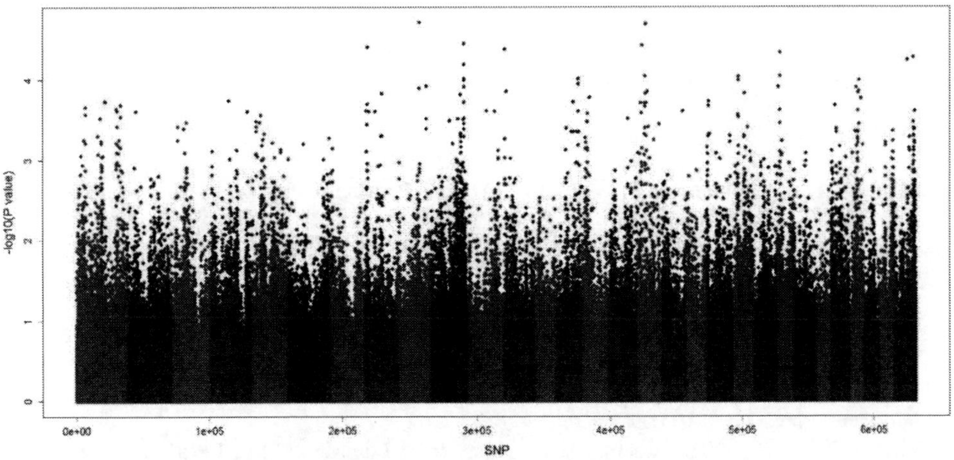

Figure 1. Genome wide association study for number of calves born per calving, with 632003 SNP. SNPs are grouped by chromosome, odd numbered in red and even in blue.

There are at least two implications can be drawn from these results. One is that increasing fecundity in cattle is unlikely to be as simple as introgressing a single mutation – the trait appears to be polygenic at least in this population - so improvement will require phenotypic selection, selection with EBVs for twinning rate or genomic selection. Another interesting implication is that these findings do not point to any obvious targets for genome editing (for example with CRISPR/CAS9), if the aim of the editing was to improve fertility.

The USMARC Twinner selection program, from which the population used in this study was derived, made remarkable progress in increasing ovulation rate and twinning rate, demonstrating the traits have a genetic component, and they are now a unique line of cattle. They have also been demonstrated to have suitable growth and carcass characteristics for beef production in American systems (Gregory et al 1996). So the absence of a major effect mutation obviously does not preclude improvement for twinning rate, and outcrosses of USMARC Twinners with other cattle are likely to produce moderate increases in twinning rate, which could be further increased by suitable selection programs. It has been previously demonstrated that twinning could lead to large increases in production and should be manageable in the temperate climatic zones with smaller farms and better pastures (Cummins et al 1992 and Cummins et al 1994). The major issue is the requirement for a moderately increased supervision input at calving time, which can be assisted by ultrasound scanning of foetal numbers. At Ivanhoe, the USMARC Twinners have been managed on a commercial basis since 2004 and over the 10 years, the twinning rate is 24% per cow mated (Cummins et al 2015). The weaning weight (at about 8 months of age) of twin born calves is about 80% of the weaning weight of single born calves. Within this herd, being born a twin did not reduce the pregnancy rate in 15 month old heifers (Cummins and Cummins 2016).

In conclusion, given the polygenic nature of twining rate in this and most other cattle populations, genomic selection or EBV selection is the best strategy, if increased twinning rates are desired and the production system is suitable.

REFERENCES

Erbe M., Hayes B.J., Matukumalli L.K., Goswami S., Bowman P.J., Reich C.M., Mason B.A. and Goddard M.E. (2012) *J. Dairy Sci.* **95**: 4114.

Salmon Hillbertz NHC, et al. (2007) Nature Genetics 39: 1318.

Cummins L. J. (1992) *Proc. Aust. Soc. Anim. Prod.* **19**: 438-447.

Cummins L. J. (1994) *Proc. Aust. Soc. Anim. Prod.* **20**:27-36

Cummins L.J. & Cummins E.S. (2016) . Animal Production in Australia, *Proc 31st biennial conference.* Pp. 1140

Gregory K.E., Echternkamp S.E. & Cundiff L.V. (1996) **74**: 1223-1233

Cummins L.J., Cummins E. S. & McLeod I.K. (2015). *Proc 4th AVA/NZVA Pan Pacific Veterinary Conference*, Brisbane; 361-365

Meuwissen, T.H., Karlsen, A., Lien, S., Olsaker, I. & Goddard, M.E. (2002) Genetics. **161**:373-9.

Kirkpatrick, B.W. & Morris, C.A. (2015) *Plos One* **10**:e0129025

Davis, G.H. (2004) *Anim Reprod Sci.* **82**:247-53.

VanRaden, P. M. (2008). J. Dairy Sci.

Wilson, T., Wu, Xi-Yang, Juengel, J.L., Ross, I.K., Lumsden, J.M., Lord, E.A., Dodds, K.G., Walling, G.A., McEwan, J.C., O'Connell, A.R., McNatty, K.P., Montgomery, G.W. (2001). *Biol. Reprod.* **64**: 1225–1235.

Souza, C.J., MacDougall, C., Campbell, B.K., McNeilly, A.S., Baird, D.T. (2001). *J. Endocrinol.* **169**: R1–R6.

Mulsant, P., Lecerf, F., Fabre, S., Schibler, L., Monget, P., Lanneluc, I., Pisselet, C., Riquet, J., Monniaux, D., Callebaut, I., Cribiu, E., Thimonier, J., Teyssier, J., Bodin, L., Cognie, Y., Elsen, J.M. (2001) *Proc. Natl Acad. Sci. USA* **98**: 5104–5109.

Bindon 1984

Galloway, S.M., McNatty, K.P., Cambridge, L.M., Laitinen, M.P.E., Juengel, J.L., Jokiranta, T.S., McLaren, R.J.,Luiro, K., Dodds, K.G., Montgomery, G.W., Beattie, A.E., Davis, G.H., Ritvos, O. (2000) Nat. Genet. **25**: 279–283.

Kirkpatrick et al 2012

Kang, H.M., Sul, J.H., Service, S.K., Zaitlen, N.A., Kong, S.Y., Freimer, N.B., Sabatti, C., Eskin, E. 2010. *Nat Genet.* **42**:348-54.

Bierman, C.D., Kim, E., Shi, X.W., Weigel, K., Berger, P., Kirkpatrick, B.W. (2010) **41**:406-16.

Gregory, K.E., Echternkamp, S.E., Dickerson, G.E., Cundiff, L.V., Koch, R.M. & VanVleck, L.D. (1990) *J Anim Sci* **68**; 1867-1876.

Proc. Assoc. Advmt. Anim. Breed. Genet. 22:481-484

GENETIC AND PHENOTYPIC CORRELATIONS BETWEEN VARIOUS GROWTH AND CARCASS TRAITS WITH PRIMAL CUT MEAT YIELD TRAITS IN HANWOO CATTLE

H.A. Al-Mamun[1], B. H. Park[2], M. N. Park[2], S. Kim[2] and C. Gondro[1]

[1] School of Environmental and Rural Science, University of New England, Australia
[2] Animal Breeding & Genetics Division, National Institute of Animal Science, Republic of Korea

SUMMARY

Genetic parameters associated with growth, carcass traits and primal cut yields of 1,098 Korean Hanwoo cattle were investigated using medium density (50k) SNP data. The growth and carcass traits considered in the present study included body weights at different ages (6, 12, 18 and 24 months), cold carcass weight (CWT), eye muscle area (EMA), back fat thickness (BFT) and marbling score (MS). Primal-cut yield (percent of carcass weight; composed of both unique and composite meat cuts from the forequarters and hindquarters) included the yields of chuck (CHK), shoulder (SLD), brisket and flank (BAF), ribs (RIB), tenderloin (TLN), striploin (STLN), sirloin (SLN), top round (TRND), round (RND), fore- and hind-shins (FHS), total primal cut (TPC) and Meat %. Heritabilities for traits were all moderate to high, ranging from 0.24 for WT6m to 0.71 for RND. All of the genetic correlations were positive to varying degrees except those between EMA and BFT, and MS and BFT. Genetic correlations among the primal cuts ranged from 0.46±13 (CHK and RIB) to 0.98±13 (TRND and FHS). Except for RIB (-0.29±13), all the other primal cuts had moderate positive genetic correlations with meat yield percentage.

INTRODUCTION

Hanwoo is the most important cattle in Korea and its history traces back 5,000 years. In Korea, it is highly sought after by consumers at premium prices. Both quality and yield determine the carcass' grade and, consequently, its price. Marbling is the major qualitative factor that determines the carcass' grade and drives most of the profit in the Korean beef industry. However, other important traits have received less attention such as yearling weight which influences both meat quality and quantity, and the yields of the primal cuts that command premium prices. Differences in price exist between different primal cuts and large variation in yield of the primals within each grade has been reported (Moon *et al.* 2003). This variation affects the accuracy of the estimates of grading and consequently there is significant averaging in the payment system. Thus, grading based on carcass weights may not reflect the differences within the carcass primal cuts.

Selection for weight of primal cuts requires genetic parameters for these traits as well as other traits that may be used as selection criteria. Few studies have reported genetic parameters for Hanwoo carcass traits and no report to date has used genomic data to estimate these parameters. In this study, we summarize estimates of genetic parameters for several traits including weight at different ages (6, 12, 18 and 24 months), back fat thickness, eye muscle area, marbling score, carcass weight and different primal-cut weights of Korean Hanwoo cattle using medium density SNP data.

MATERIALS AND METHODS

Animals and Traits. The present study analysed the records of 1,092 Hanwoo males raised under the Korean National Hanwoo Cattle Improvement System from 1997 to 2013. Summary data of different traits are shown in Table 1.

Statistical Analysis. Heritability of each trait was estimated using a univariate model in MTG2 software (Lee and van der Werf 2016). As multi-trait (3 x 3 and more) analyses failed to converge, a series of bivariate analyses using MTG2 were used to calculate the genetic correlations between the traits. Relationship among the animals were accounted for using a genomic relationship matrix (GRM) obtained from SNP data and fitted as a random effect in the model. Phenotypic correlations were calculated as the Pearson correlations between the residuals of the phenotypes after removing the fixed effects using a liner model in R.

RESULTS AND DISCUSSION

Summary statistics for the traits are presented in Table 1. Heritabilities for traits were all moderate to high, ranging from 0.24 for WT6m to 0.71 for RND. Standard errors for the heritabilities ranged between 0.07 and 0.08. Our estimates of heritability differ from the estimates reported by (Choi *et al.* 2015). This may be due to differences in the models used for analysis and sample size. In our analysis, heritabilities were estimated using genomic data (GRM as random effect) whereas Choi *et al.* used pedigree information to estimate the heritabilities. Our heritability estimate for WT12m (0.29±0.07) was slightly lower than literature values ranging from 0.33 to 0.55 (Koots *et al.* 1994; Kemp *et al.* 2002). Our calculated CWT and EMA heritabilities were higher than reported in other literature (Kemp *et al.* 2002; Choi *et al.* 2015). The BFT heritability (0.48±0.08) of the present study was moderate and consistent with the reported BFT by (Reverter *et al.* 2000; Choi *et al.* 2015). In American Shorthorn cattle, (Pariacote *et al.* 1998) reported a similar value (0.46) for carcass fat thickness. Our estimate of MS heritability (0.56±0.08) was slightly higher than those reported in earlier studies ranging from 0.48 to 0.54.

Table 1. Phenotypic mean, standard deviation and heritability with SE

Trait	Mean	SD	h^2 (±SE)
WT6m	169.07	31.08	0.24±0.07
WT12m	320.91	41.27	0.29±0.07
WT18m	483.93	52.08	0.39±0.08
WT24m	634.86	67.66	0.48±0.08
CWT	362.33	41.14	0.56±0.08
EMA	81.28	8.72	0.49±0.07
BFT	8.48	3.3	0.48±0.08
MS	3.38	1.56	0.56±0.08
CHK	12.94	3.71	0.34±0.07
SLD	22.84	2.84	0.62±0.07
BAF	27.92	4.95	0.38±0.08
RIB	55.68	7.59	0.41±0.08
TLN	5.8	0.79	0.49±0.08
STLN	34.8	4.55	0.51±0.08
SLN	7.46	1.08	0.50±0.08
TRND	19.52	2.31	0.70±0.07
RND	31.87	3.75	0.71±0.07
FHS	14.46	2.61	0.32±0.08
TPC	233.28	26.15	0.58±0.08
Meat %	64.46	2.72	0.43±0.07

Significant variation in heritabilities were observed for the different primal cuts. For example, CHK and FHS had low heritability (0.34 and 0.32), whereas TRND and RND had higher heritability (0.70 and 0.71). However, all three loin weights had very similar heritabilities ranging from 0.49 (TLN) to 0.51(STLN). It is difficult to directly compare these heritability estimates with literature values since the definition of primal cuts differ between studies. Nevertheless, in Irish cattle, (Pabiou *et al.* 2009) reported higher heritability for CHK, BAF and SLN, and lower heritability for STLN and RND, and similar heritability for RIB. In terms of meat percentage yield in the present study, our result was in line with the figure (0.42 to 0.47) reported by (Koots *et al.* 1994; Gregory *et al.* 1995). The estimates of heritability for the primal cuts indicate that direct selection may exert a notable influence on traits and that such selection may be accurate because the h^2 estimates of primal cuts are favourable. Table 2 presents the genetic and phenotypic (residuals) correlation between different weight and carcass traits. All of the genetic correlations are positive to varying degrees except those between EMA and BFT; and MS and BFT, which were statistically not different from zero. This may indicate that the traits are independent and genetically distinct. The genetic correlation between weight at different ages and CWT are very high ranging between 0.6 and 0.97, indicating that these traits are probably controlled by similar genes and selection for increased weight is very likely to increase carcass weight. However, selection for carcass traits does not necessarily translate into high meat yield percentage as indicated by the nearly zero correlation (-0.08±0.13) between CWT and Meat % traits. Genetic correlation between EMA and the weight traits are medium and with each of BFT and MS are low to medium.

Table 2. Genetic correlation (above diagonal) and phenotypic correlation (below diagonal) with SE between weight and carcass traits

Trait	WT6m	WT12m	WT18m	WT24m	CWT	EMA	BFT	MS
WT6m	-	0.86±0.05	0.72±0.09	0.63±0.11	0.6±0.11	0.4±0.15	0.15±0.16	0.27±0.15
WT12m	0.86	-	0.94±0.03	0.86±0.04	0.81±0.05	0.51±0.12	0.17±0.15	0.27±0.14
WT18m	0.7	0.89	-	0.97±0.02	0.92±0.02	0.58±0.1	0.28±0.13	0.28±0.13
WT24m	0.61	0.81	0.91	-	0.96±0.01	0.6±0.09	0.23±0.12	0.31±0.12
CWT	0.57	0.77	0.88	0.95	-	0.64±0.08	0.27±0.11	0.26±0.11
EMA	0.28	0.4	0.49	0.54	0.6	-	-0.09±0.13	0.33±0.11
BFT	0.3	0.3	0.28	0.28	0.31	0.05	-	-0.11±0.13
MS	0.14	0.13	0.09	0.08	0.12	0.18	0.1	-

Table 3. Genetic correlations with SE between weight at different ages, carcass traits and primal-cut yields

Trait	WT6m	WT12m	WT18m	WT24m	CWT	EMA	BFT	MS
CHK	0.37±0.18	0.61±0.14	0.61±0.12	0.59±0.11	0.67±0.09	0.58±0.11	-0.16±0.15	0.05±0.14
SLD	0.45±0.13	0.68±0.09	0.73±0.06	0.76±0.05	0.82±0.04	0.70±0.07	-0.04±0.12	0.10±0.11
BAF	0.44±0.16	0.61±0.12	0.73±0.09	0.85±0.06	0.86±0.06	0.60±0.10	0.00±0.15	0.15±0.14
RIB	0.5±0.13	0.71±0.08	0.85±0.05	0.94±0.03	0.96±0.02	0.57±0.10	0.22±0.13	0.44±0.12
TLN	0.59±0.13	0.8±0.08	0.76±0.07	0.76±0.06	0.76±0.06	0.60±0.09	-0.04±0.13	0.18±0.12
STLN	0.52±0.13	0.68±0.08	0.76±0.06	0.83±0.04	0.87±0.03	0.81±0.05	0.03±0.13	0.43±0.11
SLN	0.45±0.14	0.73±0.09	0.77±0.07	0.77±0.06	0.80±0.05	0.85±0.05	0.07±0.13	0.33±0.11
TRND	0.4±0.13	0.62±0.09	0.75±0.06	0.81±0.05	0.86±0.04	0.77±0.06	-0.01±0.11	0.11±0.11
RND	0.45±0.12	0.67±0.08	0.76±0.06	0.8±0.05	0.85±0.03	0.70±0.06	-0.05±0.11	0.20±0.11
FHS	0.4±0.18	0.7±0.12	0.74±0.1	0.78±0.08	0.89±0.06	0.83±0.10	0.16±0.15	0.29±0.14
TPC	0.51±0.12	0.75±0.07	0.85±0.04	0.91±0.02	0.96±0.01	0.76±0.06	0.05±0.12	0.28±0.11
Meat%	-0.21±0.16	-0.12±0.16	-0.15±0.14	-0.1±0.13	-0.08±0.13	0.47±0.12	-0.72±0.08	0.07±0.13

Table 4. Phenotypic correlations between weight at different ages, carcass traits and primal-cut yields

Trait	CHK	SLD	BAF	RIB	TLN	STLN	SLN	TRND	RND	FHS	TPC	Meat %
WT6m	0.23	0.35	0.34	0.52	0.39	0.47	0.41	0.36	0.4	0.26	0.49	-0.22
WT12m	0.35	0.57	0.49	0.68	0.55	0.65	0.57	0.55	0.6	0.44	0.7	-0.19
WT18m	0.44	0.68	0.56	0.75	0.61	0.76	0.65	0.65	0.7	0.54	0.81	-0.17
WT24m	0.47	0.74	0.63	0.82	0.65	0.82	0.69	0.71	0.76	0.58	0.89	-0.16
CWT	0.53	0.8	0.65	0.86	0.69	0.88	0.74	0.76	0.81	0.61	0.94	-0.15
EMA	0.47	0.58	0.43	0.46	0.54	0.69	0.68	0.61	0.6	0.42	0.66	0.2
BFT	-0.02	0.01	0.06	0.36	0.05	0.11	0.1	0.01	0.04	0.06	0.15	-0.48
MS	-0.05	-0.04	-0.12	0.23	0.09	0.2	0.13	-0.05	0.01	0	0.07	-0.13

In fact, BFT has very low correlation with the rest of the traits. The low correlations between CWT and each of BFT and MS were also reported in (Choi *et al.* 2015) 0.12 and 0.25 respectively, and in (Kemp *et al.* 2002) 0.17 and 0.27 respectively. Phenotypic correlations are also following very similar trend of genetic correlations but generally lower compared to the genetic correlations.

Genetic and phenotypic correlations between weight and carcass traits with different primal cuts are given in Table 3 and Table 4 respectively. There is a clear trend of increase in genetic and phenotypic correlations between weights and different primal cuts as the cattle became older. The WT12m exhibited relatively stronger correlation with TLN and the CWT exhibited stronger correlation with STLN compared to other loin cuts. Choi *et al.* (2015) reported that CWT was more associated with the forequarters and WT12m was more associated with the hindquarters. However, in the present study we did not find any significant differences in association between forequarter and hindquarter cuts with either of the CWT and WT12m traits. BFT has very little or no genetic and phenotypic correlations with the primal cuts. EMA has moderate to high genetic and phenotypic correlations with the primal cuts. MS has low to moderate genetic correlations despite showing very low or no phenotypic correlations.

The correlations among the primal cuts are positive to varying degree (data not shown). Genetic correlations among the primal cuts ranged from 0.46±13 (CHK and RIB) to 0.98±13 (TRND and FHS). Except RIB (-0.29±13), all other primal cuts had moderate positive genetic correlation with Meat %. This indicates that, selection for RIB yield may have a small decreasing effect on total meat yield. Phenotypic correlations ranged from 0.19 (CHK and BAF) to 0.91(TRND and FHS).

The objective of this study was to estimate genetic parameters for different weight, carcass and primal cut weights in Korean Hanwoo cattle and to determine their correlations using medium-density SNP data. Our present study was limited by the small amount of available carcass data. Together with the recently obtained estimates, further analysis of a larger carcass data set should allow better prediction of outcomes and enhance ongoing genomic evaluation of Korean Hanwoo cattle. We believe that our results will aid in decision making when carcass traits are to be selected to optimize primal-cut yields.

ACKNOWLEDGEMENT

This project was supported by a grant from the Next-Generation BioGreen 21 Program PJ01134906, Rural Development Administration, Republic of Korea and Australian Research Council (DP130100542).

REFERENCES

Choi T.J., Alam M., Cho C.I., Lee J.G., Park B., et al. (2015) *J. Anim. Sci.* **93**, 1511 - 21.

Gregory K.E., Cundiff L.V. and Koch R.M. (1995) *J. Anim. Sci.* **73**, 1920-6.

Kemp D.J., Herring W.O. and Kaiser C.J. (2002) *J Anim Sci* **80**, 1489-96.

Koots K.R., J.P.Gibson, Smith C. and Wilton J.W. (1994) *Anim Breed* **62**, 308-9.

Lee S.H. and van der Werf J.H. (2016) *Bioinformatics* **32**, 1420-2.

Moon S., Hwang H., Jin K., Lee G., Joo T. et al. (2003) *Asian-Aust J. Anim Sci* **16**, 1049-54.

Pabiou T., Fikse F., Nasholm A., Cromie R., Drennan M.J., wt al. (2009) *J Anim Sci* **87**, 3865-76.

Pariacote F., Vleck L.D.V. and Hunsley R.E. (1998) *J. Anim. Sci.* **76**, 2584–8.

Reverter A., Johnston J., Graser U., Wolcott L. and Upton H. (2000) *J. Anim. Sci.* **78**, 1786–95.

Proc. Assoc. Advmt. Anim. Breed. Genet. 22:485-488

LINKING COMMERCIAL CARCASE DATA TO STUD HERDS: THE POWER OF GENOMICS

L.R. Porto-Neto[1], S. Harburg[2], R.J. Bunch[1], R. Lyons[3], S.A. Lehnert[1], A. Reverter[1]

[1] CSIRO Agriculture & Food, 306 Carmody Rd., St Lucia, Queensland 4067, Australia
[2] North Australian Pastoral Company (NAPCO), Brisbane, Australia
[3] The University of Queensland, School of Veterinary Sciences, Gatton, Australia

SUMMARY

The prediction of breeding values for beef carcase traits using traditional genetic methods requires that the slaughtered cattle have a full pedigree, linking them back to the stud sector of their respective breed, additionally to accurate contemporary group structure and no selective harvesting. Due to these requirements, progress towards EBV for meat quality and other carcase traits in the beef industry has been slow, especially in Northern Australia. Here, an example of how genomic information can be used to cost-effectively feed information from the meat processing sector back to the breeding sector, using pooled DNA genotyping and placing the genotypes of pooled commercial animals and selection candidates at the stud in a hybrid genomic relationship matrix for estimating breeding values. This approach could be used to quickly and cost-effectively build reference populations for commercial performance traits in many livestock breeding applications.

INTRODUCTION

The use of data from commercial cattle to inform genetic improvement is very limited, mainly because the genetic links between the commercial herds and their stud ancestral herds are poorly known. Additionally, the cost of building large genotyped reference populations for the implementation of modern genomic technologies is still prohibitive in most cases.

An alternative approach has been proposed to draw on performance measures that are routinely acquired in commercial populations, and to cost-effectively use DNA information to link these measures to the animals available for selection in the breeding sector (Bell *et al.* 2017; Reverter *et al.* 2016).

Large volumes of carcase data are routinely recorded in the Australian beef industry, but do not usually contribute to the evaluation of genetic merit in the corresponding stud herds. This study exemplifies a cost-effective method for linking carcase data from un-pedigreed commercial animals to the stud sector.

MATERIALS AND METHODS

Animals and measurements. Tail hair samples were collected from 620 commercial Tropical Composite heifers, across two cohorts of similar size. They were feedlot finished (days on feed ~ 249 or 114) and slaughtered under the Meat Standard Australia recording system, which returned more than ten carcase attributes, including weight, MSA marbling score, pH, and hump height. Here we focused on Hump height (Hump) and MSA Index (MSA), which is an index that combines several factors recorded at the processing plant, weighted by its estimated effect, mainly, on meat eating quality (https://www.mla.com.au/marketing-beef-and-lamb/meat-standards-australia/). Additionally, DNA samples from 100 stud sires were included in the analysis. These sires were part of the herd selection program that could directly benefit from the information of the commercial cattle. Under this selection program, these sires will be grandfathers of future commercial cattle.

Pooling DNA and genotyping. Within each cohort, hair samples from each heifer were pooled in groups of five or four individuals according to their carcase phenotype, either MSA or Hump observations. We assembled 101 pools according to MSA, and 40 pools according to Hump (only pools from the extreme values). Each hair pool then had DNA-extracted and genotyped as a single sample using the single nucleotide polymorphism (SNP) genotyping platform GGPHD 150K for beef (Neogen Genomics, Lincoln, USA; http://genomics.neogen.com/pdf/AG151_GGP_TS.pdf). The individual stud sires were genotyped for ~ 50,000 SNP (Zoetis, Kalamazoo, USA). The pooled genotypic data was translated into allele calls using the B-Allele Frequency metric output from GenomeStudio software (Illumina Inc, San Diego, USA). After standard quality control on genotypes, SNP that were in common to both platforms (n = 43,807) were kept for further analyses.

Statistical analyses. Descriptive statistics, phenotypic correlations, and linear models were run in SAS (SAS Inst., Cary, NC). The combined pooled and individual genotypes were used to build a hybrid genomic relationship matrix (hybrid-GRM, Reverter *et al.* (2016)). The hybrid-GRM was then used in an additive genomic model to derive genomic Best Linear Unbiased Prediction (gBLUP) values using Qxpak v5 (Perez-Enciso and Misztal 2011). The statistical model included only heifer cohort as a fixed effect, since season of birth, on-farm management group, time on feed, killing day and killing facility were all confounded within cohort.

RESULTS

Using data from 620 carcases we were able to recover the expected MSA Index effect sizes (http://www.redpoll.org.au/documents/June15_BreedingforMSAComplianceandIndex-1.pdf) for the main attributes known to significantly impact MSA Index, which were carcase weight (0.015 ± 0.002, $p < 0.0001$), Hump (-0.048 ± 0.002, $p < 0.0001$), ossification (-0.069 ± 0.002, $p < 0.0001$), and MSA marbling score (0.016 ± 0.000, $p < 0.0001$). The recovery of known effects gives confidence that this is a representative sample for this Tropical Composite breed and its MSA Index docs not deviate from the expectation.

To make the DNA pooling approach more effective, smaller management groups were removed from the sample, so data from 501 carcases remained for further analyses. We focused on two traits, MSA and Hump, both traits have distinct, overlapping distributions for each cohort (Figure 1). The strong effect of cohort in this sample was mainly attributed to different age at slaughter and feed length.

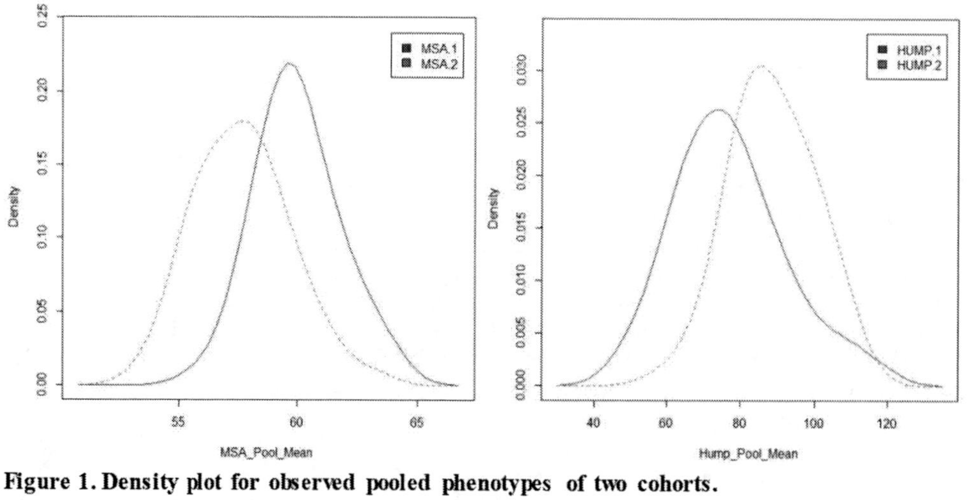

Figure 1. Density plot for observed pooled phenotypes of two cohorts.

For MSA we used data from all carcases (n = 501) as a reference population, split into 101 pooled genotypes, but for Hump we collected only genotypes for pools with extremes values (40 pools, 10 pools of high and low value, within each of the two cohorts of animals). The hybrid-GRM depicted a low relationship between stud sires and all pools. This is not surprising due to the distant relationship between the stud and this generation if the commercial herds. Nevertheless, a variable degree of relationship was detected and can be explored to derive phenotype estimates.

The distribution of gEBV of MSA and Hump for the 100 stud sires is shown in Figure 2A. The MSA gEBV approximates a normal distribution, while the Hump distribution is broader and heavy-tailed. This quite possibly reflects the fact that only the extremes of the phenotypic distribution formed the reference population for Hump, and for MSA the whole distribution of the phenotype was sampled. The negative correlation between Hump and MSA observed in the carcase data, remained present at the level of sire gEBV (Figure 2B). This approach was effective in identifying the potential top and bottom performing sires from the stud herd based on commercial herd performance data.

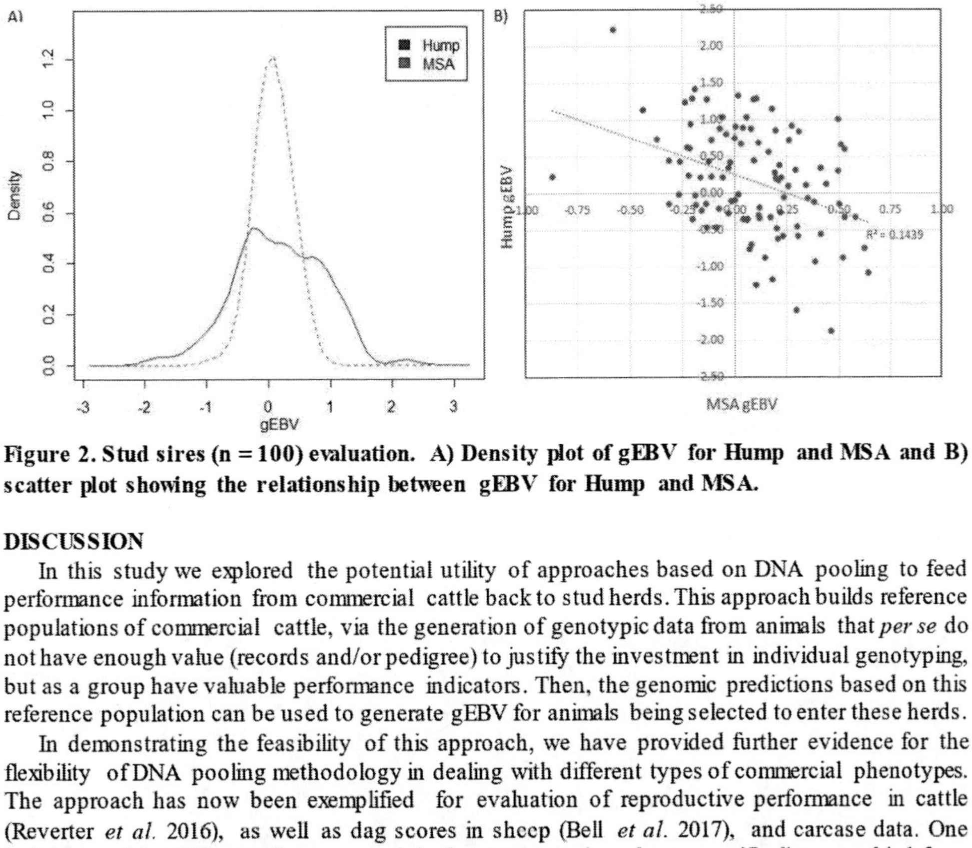

Figure 2. Stud sires (n = 100) evaluation. A) Density plot of gEBV for Hump and MSA and B) scatter plot showing the relationship between gEBV for Hump and MSA.

DISCUSSION

In this study we explored the potential utility of approaches based on DNA pooling to feed performance information from commercial cattle back to stud herds. This approach builds reference populations of commercial cattle, via the generation of genotypic data from animals that *per se* do not have enough value (records and/or pedigree) to justify the investment in individual genotyping, but as a group have valuable performance indicators. Then, the genomic predictions based on this reference population can be used to generate gEBV for animals being selected to enter these herds.

In demonstrating the feasibility of this approach, we have provided further evidence for the flexibility of DNA pooling methodology in dealing with different types of commercial phenotypes. The approach has now been exemplified for evaluation of reproductive performance in cattle (Reverter *et al.* 2016), as well as dag scores in sheep (Bell *et al.* 2017), and carcase data. One limitation of the DNA pooling approach is that genotyped pools are specifically assembled for a given phenotype, so if there are two phenotypes of interest, the pooling process will have to be done twice. If multiple traits of interest have been measured in the reference population and/or genotyping becomes more affordable due to technology developments, the cost savings offered by DNA pooling will be less significant, and individual genotypes may be the better option. In the case of commercial cattle, often no more than one or two key phenotypes are acquired during routine management or monitoring, which means that these approaches are still relevant.

Similar to other methods that estimate breeding values, DNA pooling approaches would also benefit from large reference populations. The larger the reference population is, and the genetically closer to the animals to be tested they are, in general, the more accurate the estimates will be (VanRaden *et al.* 2009). If an equivalent procedure is used to continue collection of commercial phenotypes across different years, the reference population could grow over time, potentially improving the estimates based on it.

ACKNOWLEDGEMENTS

The authors are thankful to the staff of the Animal Genetics Lab (AGL) at School of Veterinary Sciences (UQ), especially Deanne Waine, for the work on pooling DNA samples. Authors are also thankful to Andres Legarra for providing software to efficiently compute the genetic relationship matrix and its inverse, Miguel Perez-Enciso for providing updated customised version of the Qxpak5 software. This project was partly funded by Accelerate Partnership Program from the Queensland Government.

REFERENCES

Bell A.M., Henshall J.M., Porto Neto L.R., Dominik S., McCulloch R., Kijas J., and Lehnert S.A. (2017) *Genet. Sel. Evol.* **49**: 28.

Perez-Enciso M., and Misztal I. (2011) *BMC Bioinformatics* **12**: 202.

Reverter A., Porto-Neto L.R., Fortes M.R.S., McCulloch R., Lyons R.E., Moore S., Nicol D., Henshall J.M., and Lehnert S.A. (2016) *J. Anim. Sci.* **94**: 4096.

VanRaden P.M., Van Tassell C.P., Wiggans G.R., Sonstegard T.S., Schnabel R.D., Taylor J.F., and Schenkel F.S. (2009) *J. Dairy Sci.* **92**: 16.

Proc. Assoc. Advmt. Anim. Breed. Genet. 22:489-492

GENOTYPING OF NELLORE BIOPSIED EMBRYOS

R. Carvalheiro[1,2], H.H.R. Neves[3], T. Bresolin[1], D.R. Arnold[4], A.C. Basso[4] and L.G. Albuquerque[1,2]

[1]School of Agricultural and Veterinarian Sciences, São Paulo State University (Unesp), Jaboticabal, SP, Brazil
[2]National Council for Scientific and Technological Development (CNPq), Brasília, DF, Brazil
[3] GenSys S/S, Porto Alegre, RS, Brazil
[4] In Vitro Brasil S/A, Mogi Mirim, SP, Brazil

SUMMARY

Genomic selection of embryos can boost genetic progress of beef cattle breeding programs by allowing the intensity of selection to increase and the generation interval to be reduced. This strategy depends on a protocol for the biopsy of embryos and DNA amplification, ensuring enough DNA for genotyping, without compromising the embryo's viability. In the present study, the quality of the genotypes of Nellore biopsied embryos was assessed based on genotyping call rate, Mendelian inconsistencies and allele dropouts. The results showed that the genotypes were of a good quality, suggesting feasibility of obtaining genomic prediction of Nellore embryos.

INTRODUCTION

Embryo transfer provides an outstanding opportunity for intensifying the production of genetically superior animals, given that donors and sires are properly selected. The technique of producing embryos of cattle through *in vitro* fertilization (IVF) has been evolving substantially and is becoming more reliable and accessible. For instance, according to the Brazilian Society of Embryo Technology, Brazil has been producing more than half a million embryos per year, through IVF.

Investments in IVF could be optimized if the genetic merit of the embryos could be predicted more accurately. Even if the donors and sires are properly selected, the embryo's genetic merit may substantially deviate from what is expected based on parent average, because Mendelian sampling is responsible for the genetic difference among full-sibs, and accounts for half of the additive genetic variability (Falconer and Mackay 1996). Genomic selection (Meuwissen *et al.* 2001) allows predicting more accurately the genetic merit of embryos, given that they are genotyped for a reasonable number of markers and that a good prediction equation is available. To genotype the embryos, a proper protocol to extract DNA must be developed, without compromising embryo's viability.

Pre-amplification of the DNA extracted from the embryos is required to provide enough DNA for the currently available genotyping platforms. The amplification process usually leads to reduced genome coverage which in turn results in some genotyping errors as, for example, allele dropout at heterozygous loci (Lauri *et al.* 2013). These errors could ultimately compromise the genomic prediction and the feasibility of performing genomic selection on embryos. An alternative to correct part of the genotyping errors is to also genotype the parents of the embryos and use this information to fix inconsistencies, followed by imputation to predict missing genotypes (Saadi *et al.* 2014).

In the present study, the feasibility of obtaining genomic prediction of Nellore biopsied embryos by evaluating the quality of their observed and imputed genotypes is assessed. Genomic predictions and their corresponding accuracies were also computed to envisage the potential benefit of applying genomic selection in Nellore embryos.

MATERIALS AND METHODS

Nellore embryos were produced from ovum pick-up from 28 donors and IVF using semen of two sires. These donors and sires are from a single beef cattle farm (Agropecuária Jacarezinho) which participates in the DeltaGen breeding program (www.deltagen.com.br). A total of 93 embryos were biopsied and genotyped. The biopsy of embryos and DNA extraction were performed according to a protocol developed by In Vitro Brasil S/A (www.invitrobrasil.com.br). The extracted DNA was amplified using commercial kits based on multiple displacement amplification (REPLI-g, Qiagen, Mississauga, ON, Canada). The Illumina Bovine 50K v2 chip (Illumina, San Diego, CA, USA) was used to genotype the embryos, donors and sires. The biopsied embryos were implanted into Nellore recipient cows and presented a pregnancy rate (31%) similar to the rate presented by a control group (24%), suggesting that the DNA extraction did not reduce the embryo's viability.

The software FImpute v2.2 (Sargolzaei *et al.* 2014) was used as in Saadi *et al.* (2014) to check for Mendelian inconsistencies, to fix some genotyping errors and to impute missing genotypes. As the parents of the embryos were also genotyped, family information was initially used by FImpute as the main source of information for fixing the inconsistencies. Afterwards, the fixed 50K genotypes were imputed to HD genotypes (Illumina Bovine HD chip), using family and population information. Finally, the embryos had their direct genomic values (DGV) calculated based on their imputed genotypes and on the prediction equation of DeltaGen breeding program. The reference population for imputation and genomic prediction used in this study has approximately eight thousand animals. The DGVs and their accuracies were calculated using the software GEBV (Sargolzaei *et al.* 2013). The analyses were performed using 34,900 SNPs from the 50K chip and 615,397 SNPs from the HD chip, comprising those SNPs which passed quality control of routine genomic evaluation.

The quality of the genotypes of embryos was mainly assessed by the comparison between observed and imputed genotypes. A better assessment will be performed after the resultant calves are born and genotyped, so the comparison will be made among the embryo-calf pairs.

RESULTS AND DISCUSSION

The average call rate of the embryos' 50K genotypes was equal to 0.93, ranging from 0.75 to 0.98. Seventy four embryos (80%) seemed to present a reasonably good quality of their genotypes (call rate≥0.90). Embryos' genotypes with lower call rates also presented lower levels of heterozygosity (Figure 1), suggesting low quality of the inferred genotypes and the occurrence of allele dropouts at heterozygous loci. Seventeen embryos, most of which with low call rate, showed parentage conflicts based on the original 50K genotype, showing more than 319 Mendelian inconsistencies when their genotypes were contrasted with those from their parents. The poor genotype quality (call rate<0.90) of some embryos were likely caused by the low amount of extracted DNA and the amplification process.

All the parentage conflicts were no longer observed using the imputed ("fixed") genotypes. The FImpute software corrected the Mendelian inconsistencies and imputed all missing genotypes, i.e. the call rate of the imputed genotypes was equal to 1, for both chips (50K and HD). Considering just the SNPs from the 50K chip (for a proper comparison) the level of heterozygosity of the imputed genotypes was, generally, greater than those of the observed genotypes, especially for the embryos originally presenting low call rates. This result indicates that FImpute was able to correct at least part of the allele dropouts. However, even after imputation, there was evidence of underestimation of heterozygosity for the embryos with original genotypes exhibiting low call rates (Figure 1).

Besides fixing the allele dropouts, FImpute uses family and population information to also correct some homozygote SNPs which were miscalled as heterozygotes or as the opposite homozygote. Figure 2 shows that embryos with lower original call rates presented a higher percentage of genotypes corrected due to dropouts and total error rate. In general, the dropouts were responsible for the greatest proportion of changes. The maximum percentage per sample of homozygote SNPs miscalled as heterozygotes or as the opposite homozygote was equal to 0.4% and 0.5%, respectively, whereas it was equal to 1.9% for the dropout. The maximum total change per sample was equal to 2.6% (Figure 2).

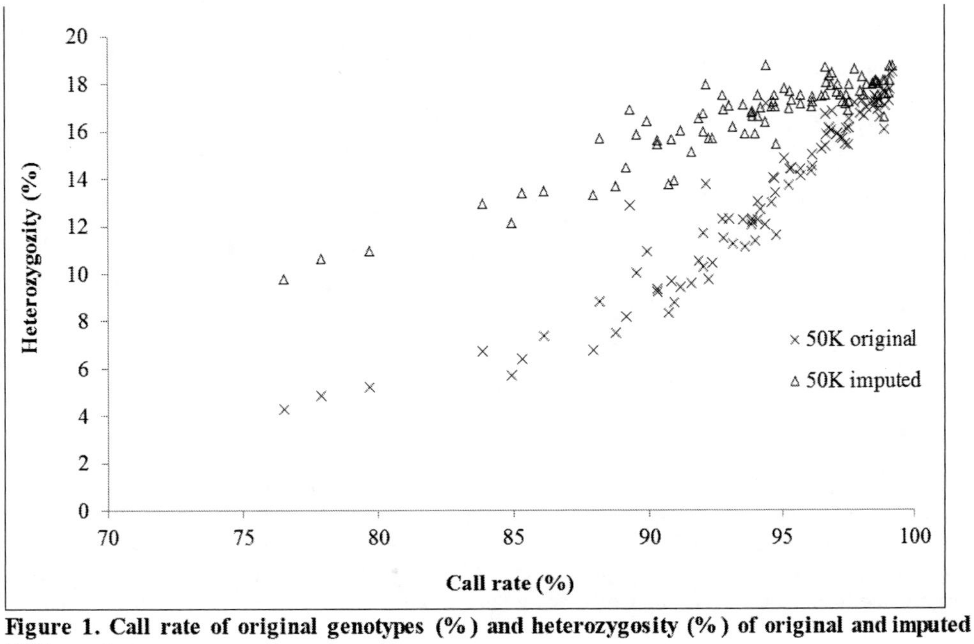

Figure 1. Call rate of original genotypes (%) and heterozygosity (%) of original and imputed 50K genotypes.

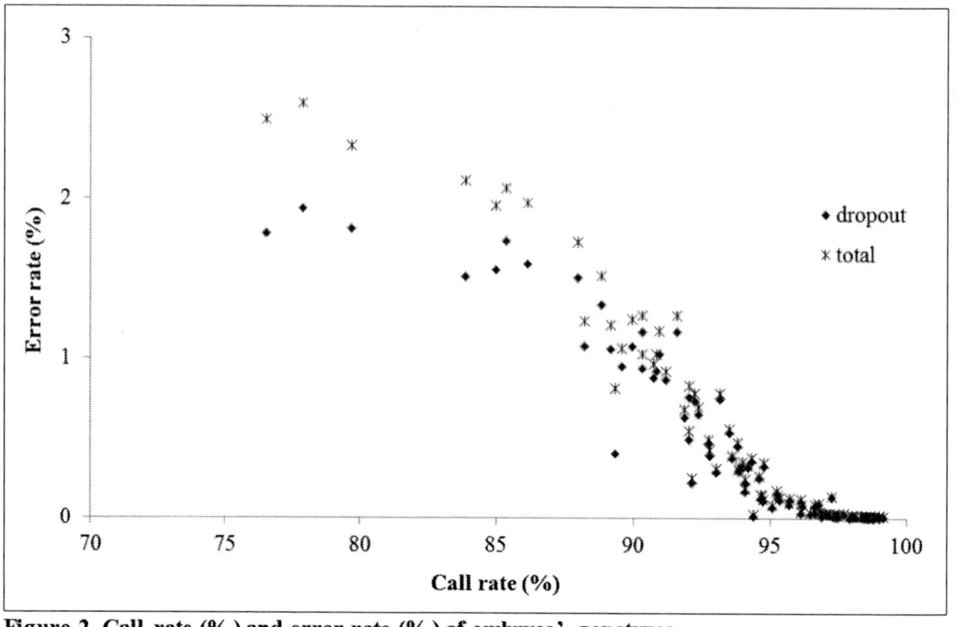

Figure 2. Call rate (%) and error rate (%) of embryos' genotypes.

It is important to emphasize that having good DNA extraction and amplification protocols remains very important. Even if imputation may improve genotype quality, it can also cause some errors and bias the genomic predictions. Pimentel *et al.* (2015) provided empirical evidence that top animals may have their genomic predictions underestimated when imputed genotypes are used, mainly due to miscalling low frequent haplotypes that could not be determined unambiguously by the imputation algorithm. As mentioned previously, the quality of the genotypes of embryos will be better assessed after the resultant calves are born and genotyped, so the comparison will be made among the embryo-calf pairs.

The genomic prediction of the embryos obtained after fixing and imputing their genotypes presented an average accuracy of 0.56 (ranging from 0.46 to 0.60), for the selection index used by the breeding program. This accuracy is equivalent to those for young bulls selected (without genomic information) for progeny testing, highlighting the potential benefit of applying genomic selection in Nellore embryos. The cost-effectiveness of this strategy is highly determined by the realized success rate of the transferred biopsied embryos.

REFERENCES

Falconer D.S. and Mackay T.F.C. (1996) Introduction to Quantitative Genetics. 4th ed. Pearson Longman, Harlow, UK.

Lauri A., Lazzari G., Galli C., Lagutina I., Genzini E., et al. (2013) *Genomics* **101**: 24.

Meuwissen T.H.E., Hayes B.J. and Goddard M.E. (2001) *Genetics* **157**: 1819.

Pimentel E.C.G., Edel C., Emmerling R. and Götz K.-U. (2015) *J. Dairy Sci.* **98**: 4131.

Saadi H.A.S., Vigneault C., Sargolzaei M., Gagné D., Fournier E., et al. (2014) *BMC Genomics* **15**: 889.

Sargolzaei M., Chesnais J.P. and Schenkel F.S. (2013) GEBV user's guide.

Sargolzaei M., Chesnais J.P. and Schenkel F.S. (2014) *BMC Genomics* **15**: 478.

Proc. Assoc. Advmt. Anim. Breed. Genet. 22:493-496

GENETICS AND GENOMICS OF BULL FERTILITY PHENOTYPES MEASURED AS PART OF THE BULL BREEDING SOUNDNESS EVALUATION

M.R.S. Fortes[1], F.S.S. Raidan[1], N. Satake[2], L.R. Porto-Neto[3], G.B. Boe-Hansen[2]

[1]The University of Queensland, School of Chemistry and Molecular Biosciences, St Lucia, Brisbane, Qld 4072.
[2]The University of Queensland, School of Veterinary Science, Gatton, Qld 4343.
[3]CSIRO Agriculture & Food, Queensland Bioscience Precinct, Brisbane, Qld 4067.

SUMMARY

Chromosomal regions that were associated ($P < 0.01$) with scrotal circumference (SC) and percentage of morphologically normal sperm (PNS) are reported according to genome-wide association studies in Tropical Composite cattle. Bulls were genotyped with Illumina SNP chips and association analyses were performed using animal models. Chromosome X had several SNP associated with SC, PNS or both traits (7,859 SNP). Polymorphisms associated with SC and PNS can contribute to new methods of estimating breeding values, which may enhance the selection of bulls with improved reproductive performance.

INTRODUCTION

Sperm concentration and morphology are important semen parameters for bull fertility, both in field situations and for artificial insemination (AI). In multiple sire mating systems, sperm morphology was considered the best indicator for calf output (Holroyd et al., 2002). Sire bulls with enhanced fertility guarantee the efficiency of transmission of favourable alleles (for any trait of economical relevance. The consequences of fertile sires are improvement in fertility rates for the herd and increase economic return. In AI centres, bulls producing high volumes of semen with appropriate sperm concentration and without significant fluctuations in semen quality are preferred. These bulls can help avoid unexpected decreases in the number of straws produced for AI, reduce economic losses and disturbances in the distribution and marketing of semen (Hering et al., 2014). Sperm concentration and morphology are typically examined as part of the Bull Breeding Soundness Evaluations (BBSE).

Bull fertility is often an overlooked component of reproductive rate in genetic studies. Bull fertility influences not only fertilization but also the viability of the preimplantation embryos and the establishment of successful pregnancy (Saacke et al., 2000). Scrotal circumference (SC) and percent of morphologically normal sperm (PNS) are indicators of bull fertility that show high heritability in beef cattle (Corbet et al., 2013). Changes in favourable allele frequencies through selection can improve SC and PNS. A limitation in the use of SC and PNS as selection tools is that these reproductive traits cannot be measured before bulls reach 12–24 months of age (Lyons et al., 2014). Identifying SNP associated with these traits could be very useful for early recognition of young sires' fertility. In this study we investigate the heritability and genetic correlations between SC and PNS measured in BBSE are investigated. Genome-wide association analyses (GWAS) were performed for SC and PNS.

MATERIAL AND METHODS

Animals, Traits and Genotypes. Blood for DNA extraction was obtained from 1,719 Tropical Composite (TC) bulls; bred by the Cooperative Research Centre for Beef Genetic Technologies. Details concerning the project design and traits measurements have been reported elsewhere (Burns et al., 2013; Corbet et al., 2013). In short, SC and PNS were measured at 24 months. BovineSNP50 chips (Matukumalli et al., 2009) were used to genotype all bulls and Bead Studio software (Illumina Inc., San Diego, CA 2006) was used to call genotypes. SNP with call rates < 80% or minor allele frequency < 0.01 were discarded. High-density (HD) genotyping of selected TC cattle was performed and genotypes were imputed using BEAGLE (Browning and Browning, 2009). Quality control and imputation resulted in 729,069 SNP genotypes for 1,719 TC.

Statistical Analyses. Allele substitution effects were estimated for each SNP separately, using an animal model. Solutions were estimated with Qxpak5 (Perez-Enciso and Misztal, 2011), using a likelihood ratio test to compare the model containing each SNP versus the model without each SNP.

RESULTS AND DISCUSSION

Descriptive statistics, genetic parameters and number of SNP associated with SC and PNS are reported (Table 1). Both traits had high heritability and so they can be used as selection tools to increase fertility. These results corroborate those reported by Corbet et al. (2013). Moreover, PNS and specially, SC are easily measured and the operating cost-benefit is agreeable since they are part of BBSE. These traits have been associated with the genetic improvement of fertility in both male and female cattle in scientific and technical studies (Silva et al., 2013).

Table 1. Descriptive statistics and genome-wide association results for percentage of normal sperm (PNS) and scrotal circumference (SC).

Traits*	N	Mean ± SD	h^2	r	P<10^6	SNP in both traits (P<10^6)
PNS	1,648	0.72 ± 0.19	0.40		8,927	
				0.11		7,859
SC	1,719	31.42 ± 2.80	0.56		14,635	

*Traits: Percentage of morphologically Normal Sperm (PNS) and Scrotal Circumference (SC). SD: standard deviation; h^2 Heritability; r: genetic correlation; P: P-value from Genome-Wide Association Study.

The genetic correlation between SC and PNS was low (Table 1) and so direct selection could be more efficient than indirect selection. However, the genetic correlation was positive indicating that the long-term increase of SC promoted by direct selection will result in an increment in PNS and consequent improvement of semen quality in young bulls. In addition, the number of SNP that were associated with the genetic variation for both traits was high: 7,859; all mapped to chromosome X. Genetic correlations are determined by either pleotropic effects or gene linkage, which depends on selection intensity, polymorphism effects, allele frequency and strength of linkage (Sheridan and Barker, 1974); factors that might vary across generations or herds. It is relevant to highlight that fixation of alleles with pleotropic effects and in linkage may reduce the genetic correlation between traits across generations (Sheridan and Barker, 1974). This might explain in part the low genetic correlation estimated between SC and PNS; selection for fertility might have been applied in this herd over generations.

The X chromosome harboured SNP that were significant for both traits, spread across millions of base pairs. Our results are evidence for polygenic regulation of these reproductive traits (Figure1).

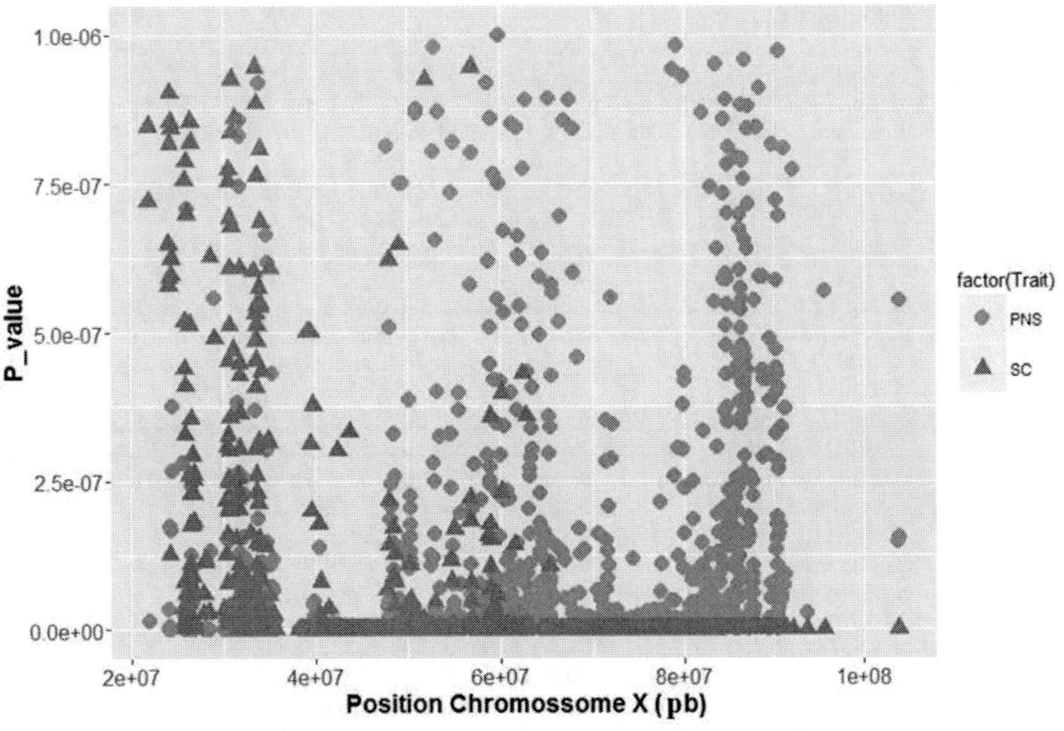

Figure 1. Polymorphisms in the X chromosome associated with percentage of morphologically normal sperm (PNS) and scrotal circumference (SC) in Tropical Composite cattle.

Associations that point to a QTL close to 32 Mb and another close to 110 Mb of chromosome X provide further evidence for results that were first reported in Holstein bulls (Blaschek et al., 2011). Candidate genes underpinning these QTL on chromosome X were proposed in Brahman bulls (Fortes et al., 2012). For example, the androgen receptor gene (*AR*) at 88 Mb of the X chromosome is a positional candidate gene, which is relevant because of its physiological role (Quigley 1998). The androgen receptor is paramount for testosterone signalling and, therefore, transcriptional regulation of genes that are critical for development and maintenance of male reproductive function.

CONCLUSION

Reported SNP associated with SC and PNS may contribute to the selection of bulls with improved reproductive performance. Direct selection for each trait would be a more efficient route than indirect selection to improve bull fertility and semen quality.

ACKNOLEDGEMENTS

Research presented in this paper was support by Meat and Livestock Australia. Project number B.NBP.0786, title "Ideal markers for tropically adapted cattle -proof of concept: causative mutations for bull fertility".

REFERENCES

Blaschek M., Kaya A., Zwald N., Memili E. and Kirkpatrick B.W. (2011) *J. Dairy Sci.* **94**:4695-4699.

Browning B.L. and Browning S.R. (2009) *A. J. Hum. Genet.* **84**:210-223.

Burns B.M., Corbet N.J., Corbet D.H., Crisp J.M., Venus B.K., Johnston D.J., Li Y., McGowan M.R. and Holroyd R.G. (2013) *Anim. Prod. Sci.* **53**:87-100.

Corbet N.J., Burns B.M., Johnston D.J., Wolcott M.L., Corbet D.H., Venus B.K., Li Y., McGowan M. R. and Holroyd R.G. (2013) *Anim. Prod. Sci.* **53**:101-113.

Fortes, M.R.S., Reverter A., Hawken R.J., Bolormaa S. and Lehnert S.A. (2012) *Bio. Reprod.* **87**:1–8.

Hering D.M., Olenski K., Rusc A. and Kaminski S. (2014) *Anim. Reprod. Sci.* **151**:126-130.

Holroyd R.G., Doogan W., De Faveri J., Fordyce G., McGowan M.R., Bertram J.D., Vankan D.M., Fitzpatrick L.A., Jayawardhana G.A. and Miller R.G. (2002) *Anim. Reprod. Sci.* **71**:67-79.

Lyons R.E., Neguyen L.T., Dierens L., Fortes M.R.S., Kelly M., McWilliam S.S., Li Y., Bunch R.J., Harrison B.E., Barendse W., Lehnert S.A. and Moore S.S. (2014) *BMC Genet.* **15**:6.

Matukumalli L.K., Lawley C.T., Schnabel R.D., Taylor J.F., Allan M.F., Heaton M.P., O'Connell J., Moore S.S., Smith T.P., Sonstegard T.S. and Van Tassell C.P. (2009) *PLoS One* **4**:e5350.

Perez-Enciso M. and Misztal I. (2011) *BMC Bioinf.* **12**:202.

Quigley C.A. (1998) In 'Testosterone', pp. 33-106, editors E. Nieschlag and H.M Behre, Springer Berlin Heidelberg, Berlin.

Saacke R.G., Dalton J.C., Nadir S., Nebel R.L. and Bame J.H. (2000) *Anim. Reprod. Sci.* **60-61**:663-677.

Sheridan A.K. and Barker J.S. (1974) *Aust. J. Biol. Sci.* **27**:89-101.

Silva M.R., Pedrosa V.B., Borges-Silva J.C., Eler J.P., Guimaraes J.D. and Albuquerque L.G. (2013) *J. Anim. Sci.* **91**:4611-4616.

Proc. Assoc. Advmt. Anim. Breed. Genet. 22:497-500

PRELIMINARY ESTIMATES OF PRODUCTIVE LIFETIME AND LIFETIME EFFICIENCY IN HOLSTEIN COWS AS AFFECTED BY AGE AT FIRST CALVING

C.J.C.Muller[1], H.L. de Waal[2] and M.M. Scholtz[2,3]

[1]Department of Animal Sciences, University of Stellenbosch, Stellenbosch 7602, South Africa;
[2]National Milk Recording, Agricultural Research Council, Stellenbosch 7602, South Africa;
[3]Natural and Agricultural Sciences, University of the Free State, Bloemfontein 9300, South Africa

SUMMARY

Longevity in dairy cows affects their lifetime performance and lifetime efficiency. Limited research has been conducted in South Africa on the effect of age at first calving (AFC) on lifetime performance. In this paper, preliminary estimates of the productive life (PL, total number of days in milk), lifetime (LT, birth date to cull date) and productive lifetime efficiency (PLE) and lifetime efficiency (LTE) for Holstein cows in South Africa as affected by AFC are presented. Lifetime production records of 509 715 cows born between 1989 and 2008 which had calved down at least once, were included in the study. Cows calving down earlier than 18 months and later than 48 months of age were removed from the data set. For each cow, the total milk, fat and protein yield, days in milk (PL) and LT of cows was determined. From this, the PLE and LTE of the milk yield for each cow was derived. Extending AFC increased LT, although decreasing PL, especially after 29 months of age. Productive lifetime efficiency peaked at an AFC of 25 months and decreased thereafter. Further analyses to estimate genetic parameters for production and ratio traits are envisaged towards identifying individual cows and sires for possible genomic analyses.

INTRODUCTION

Longevity in dairy cows is an important trait affecting the genetic progress, LT performance, LFE, and financial sustainability of a dairy herd (Fricke 2004). An early AFC increases PL (Nilforooshan & Edriss 2004) while also reducing rearing costs (Ettema & Santos 2004), being 4.3% per month less when first calving is one month earlier (Tozer & Heinrichs 2003). However, calving problems may increase when AFC is too early (Ettema & Santos 2004). For this reason most farmers rear heifers to reach first calving at an older age although not necessarily at a higher live weight. This increases the overall rearing costs because of a longer feeding period. High growth rates and longer feeding period may result in over-conditioned heifers resulting in dystocia problems (Ettema & Santos 2004). Age at first calving is therefore a benchmark that should be properly managed to increase economic returns. Torshizi *et al.* (2016) found an average AFC of 24.7±1.4 months for Iranian Holsteins which was similar for Irish seasonal calving Holsteins at 24.3 months (Berry & Cromie 2009). The heritability of AFC is generally low, ranging from 0.09 to 0.11 (Weigel & Rekaya 2000 and Changhee *et al.* 2013) indicating a large environmental effect. Limited research has been done in South Africa on factors affecting the lifetime performance of dairy cows. Mostert *et al.* (2001) found, as expected, that milk yield increased with age at calving reaching a peak at about 4th lactation. Makgahlela *et al.* (2008) showed that actual AFC decreased by 0.2 months per year. Mean breeding value for AFC also decreased by 0.06 months per year. Local breed societies present lifetime awards to cows reaching specific production milestones. Cows reach such milestones at different ages, i.e. 6 and 10 lactations, indicating differences in production and possibly production efficiency. Cows ranked for total fat and protein yield are re-ranked when lactation number is used to estimate an efficiency index (fat + protein yield/lactation number). Production efficiency should be estimated using PL or LT in days as lactations could vary being short or longer than standard lactation periods. The aim of this study is to determine the effect of AFC on the LT, PL, LF performance, PLE and LTE of Holstein cows.

MATERIALS AND METHODS

Data. Milk production records of 509 715 Holstein cows that had calved down for the first time between 1989 and 2007 were extracted from the South African National Milk Recording Scheme data base of the Agricultural Research Council (ARC). Milk production and milk composition records were compiled using standard procedures, i.e. on 10 milk recording events during the year, starting from 5 days after calving, for at least 8 milk recording events (De Waal & Heydenrych 2001). Each cow that had completed a first lactation of at least 240 days was included in the study. The milk, fat and protein yield for all subsequent lactations were added up until the end of each cow's last lactation period. Productive life was estimated for each cow totalling all the number of days-in-milk for all lactations. The LT of cows was derived from birth date to the end of the last lactation period which was regarded as the cow's cull date as exact cull dates were not available. For each cow, PLE and LTE were estimated by dividing the total milk, fat and protein yield by PL and LT. Records from cows calving down for the first time before 18 months of age and after 48 months of age were deleted from the data set. Herds with fewer than 30 records were also removed from the data set.

Statistical analyses. Analysis of variance was performed, using cows as random replicates, using GLM Procedure of SAS software (Version 9.4; SAS Institute Inc, Cary, USA) to test the effect of age at first calving (in months) on the production parameters milk, fat and protein, productive life, lifetime and productive life efficiency (total milk yield/productive life in days, PLE) and lifetime efficiency (total milk yield/lifetime in days, LTE). Fisher's least significant difference was calculated at the 5% level to compare month means (Ott 1998). A probability level of 5% was considered significant for all significance tests. The Shapiro-Wilk test was performed on the standardized residuals from the model to test for normality (Shapiro & Wilk 1965).

RESULTS AND DISCUSSION

Table 1 shows mean±standard deviation, range of records and analysis of variance results for some production parameters. Except for fat and protein percentages, the coefficient of variation was high for all traits, exceeding 30%. Average AFC was 29.6±5.1 months which is higher than general recommendations (Tozer & Heinrichs 2003). However, AFC decreased from 30.0 in 1987 to 25.3 months of age in 2006. In the present study AFC is skewed to the right with most (88%) of heifers calving down later than 24 months of age (Figure 1a). Almost 36% of heifers calved down after 30 months of age.

Table 1. The mean±standard deviation and data ranges for milk production, fat and protein percentages and lifetime parameters for Holstein cows. Least significant differences (LSD) for age at first calving intervals indicate significant differences between age at first calving intervals. (Significance **: P<0.01)

Variables	Mean	Range	LSD	Significance
Age at first calving (m)	29.6±5.1	17.8-48.0	-	-
First lactation milk yield (kg)	6217±1985	144-17329	118	**
First lactation fat (%)	3.57±0.41	0.55-7.67	0.03	**
First lactation protein (%)	3.17±0.21	0.50-5.33	0.01	**
Total lifetime milk yield (kg)	20300±14887	364-135538	813	**
Lifetime (days)	2004±720	785-4295	44	**
Productive life (days)	849±523	241-3239	31	**
Lifetime efficiency (kg/d)	9.24±4.44	0.22-43.9	0.23	**
Productive life efficiency (kg/d)	22.9±6.9	1.2-150.8	0.40	**

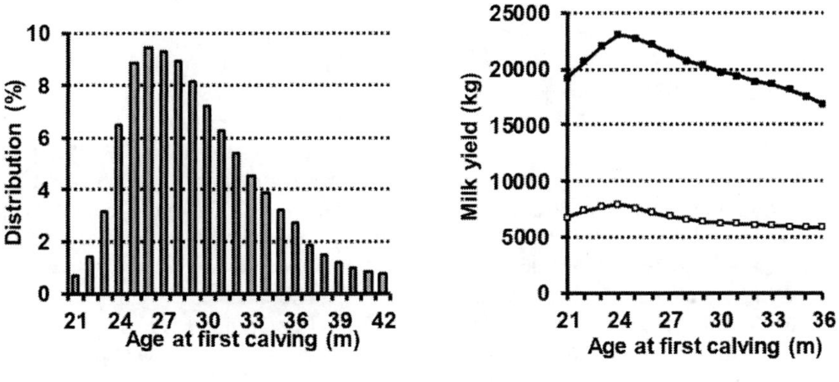

(a) (b)

Figure 1. The distribution (a) of age at first calving (AFC) records and (b) the effect of AFC on first lactation (□) and lifetime milk yield (■) for Holstein heifers.

Torshizi *et al.* (2016) found that the proportion of Holsteins cows calving down before 20 and after 30 months were low. In an earlier survey, Muller *et al.* (2014) similarly found that while only 12% of Holstein heifers calved down before 25 months of age, more than 35% of Holstein heifers calved down after 30 months of age. First lactation and LT milk production increased from 21 months of AFC (Figure 1b), peaking at 24 months of age after which production decreased indicating no advantage for a later AFC. Nilforooshan & Edriss (2004) also showed an initial increase in first lactation milk yield from 21 to 24 months of age after which production decreased. A slight negative phenotypic correlation (-0.089) was found between AFC and first lactation milk yield. In the present study, as expected, the LT of cows increased in a quadratic fashion ($P<0.05$) with an increasing AFC (Figure 2a). However, while the PL of cows increased ($P<0.05$) up to 29 months of age, it decreased ($P<0.05$) after that.

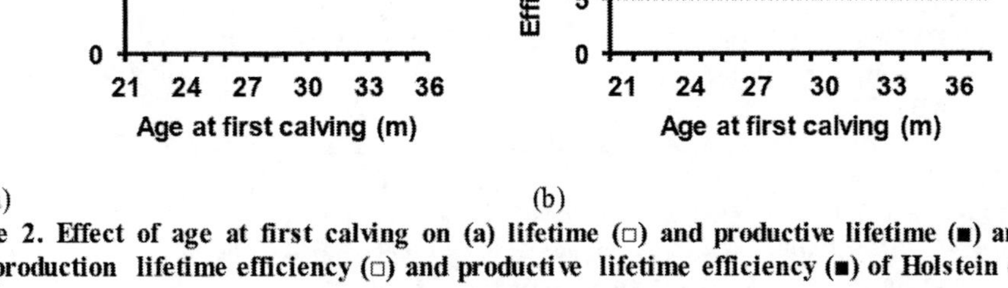

(a) (b)

Figure 2. Effect of age at first calving on (a) lifetime (□) and productive lifetime (■) and (b) milk production lifetime efficiency (□) and productive lifetime efficiency (■) of Holstein cows.

The phenotypic correlation between first lactation and lifetime milk yield levels followed a quadratic trend ($P<0.05$). Similarly, PLE and LTE peaked at about 24 to 25 months of age after which efficiencies decreased. Heinrichs & Vazquez-Anon (1993) found that AFC of Holstein heifers was 26.4 months. In that study, the distribution of AFC was also skewed towards older ages at first calving. Holstein cows calving down at more than 26 months of age produced similar amounts of milk as cows calving down at 24 months of age. Cooke *et al.* (2013) found that the total days-in-milk for heifers calving down before 23 months of age was higher ($P<0.01$) than for heifers calving down later than 30 months of age. Mostert *et al.* (2001) found that the average 305-day milk yield of Holstein cows increased when AFC increased from <24 to >32 months of age.

Changing AFC genetically would be slow as heritability (h^2) is low (<0.09) indicating that this trait is highly influenced by environmental factors (Nilforooshan & Edriss 2004). However, Makgahlela *et al.* (2008) found a moderate h^2 for AFC for South African Holsteins being 0.26±0.02.The present study showed that, over time, AFC was reduce presumably because of management improvement.

CONCLUSION

The effect of AFC on the LT (longevity) and PL of South African Holstein cows was presented. While LT increased with a later age at first calving, PL decreased especially after 29 months of age. An earlier AFC resulted in a higher LT production. Milk production PLE and LTE increased to 25 months of age after which both traits decreased. Genetic parameters should be estimated for AFC, LT, PL, LT production, PLE and LTE measures to determine relative emphasis of selection. Further analyses using alternative statistical methods, which include pedigree or genotype information, may offer additional parameters for selection in this breeding programme.

ACKNOWLEDGEMENTS

Mr. Graham Buchanan estimated total milk yields from individual lactation periods for each cow. This support is gratefully acknowledged as well as the statistical analyses provided by Mrs. Marieta van der Rijst from the Biometry Section of the ARC.

REFERENCES

Berry D.P and Cromie A.R. (2009) *Livest. Sci.* **123**: 44.

Changhee D., Nidarshani W., Kwanghyun C., Yunho C., Taejeong C., Byungo P. and Donghee L. (2013) *Asian-Australas J. Anim. Sci.* **26**: 1511.

Cooke J.S., Cheng Z., Bourne N.E. and Wathes D.C. (2013) *Open J Anim. Sci.* **3**: 1.

De Waal H. and Heydenrych H.J. (2001) *S.A. J. Anim. Sci.* **31**: 9.

Ettema J.F. and Santos J.E.P. (2004) *J. Dairy Sci.* **87**: 2730.

Fricke P.M. (2004) *Adv. Dairy Technol.* **16**: 1.

Heinrichs A.J. and Vazquez-Anon M. (1993) *J. Dairy Sci.* **76**: 671.

Makgahlela M.L., Banga C.B., Norris D., Dzama K. and Ngambi J.W. (2008) *Asian J. Anim. Vet. Adv.* **3**: 197.

Mostert B.E., Theron, H.E. and Kanfer F.H.J. (2001) *S.A. J. Anim. Sci.* **31**: 205.

Muller C.J.C., Potgieter J.P. and Cloete S.W.P. (2014) *Proc.10th Wo. Con. Gen. Appl. Livest. Prod.* Poster AB640.

Nilforooshan M.A. and Edriss M.A. (2004) *J. Dairy Sci.* **87**: 2130.

Ott R.L. (1998) An introduction to statistical methods and data analysis. Duxbury Press. 807.

Shapiro S.S. and Wilk M.B. (1965) *Biometrika.* **52**: 591.

Torshizi M. (2016) *J. Anim. Sci. Technol.* **58**: 8. doi: 10.1186/s40781-016-0089-1.

Tozer P. and Heinrichs J. (2003) *Hoards Dairyman.* **148** (5): 203.

Weigel K.A. and Rekaya R. (2000) *J. Dairy Sci.* **83**:1072.

Proc. Assoc. Advmt. Anim. Breed. Genet. 22:501-504

GENOME-WIDE MAPPING OF LOCI AFFECTING SEMEN VOLUME, SPERM CONCENTRATION AND TOTAL AND PROGRESSIVE MOTILITIES IN BOARS

T.S. Gaggini[1], R.C. Antunes[1], H. Barbosa[2], P. Charagu[2] and F.M. Rezende[1]

[1] College of Veterinary Medicine, Federal University of Uberlandia, Uberlandia, Brazil
[2] Hypor Inc., Regina, Canada

SUMMARY

This study was conducted to map genomic regions associated with semen quality traits of boars. Volume, sperm concentration and total and progressive motilities of 2,392 *in natura* ejaculates of 113 Duroc males were evaluated. Genotyping process was performed by Illumina PorcineSNP60 BeadChip (68,516 SNPs, 119 animals). After quality control (MAF <3% and call rate <90%), 118 boars, corresponding to 250 animals in pedigree, and 42,240 SNPs remained to be analysed. Genome wide-association analyses were realized by BLUPf90 using weighted single-step GBLUP method considering windows of 10 adjacent SNPs to estimate their effects. The animal model considered as fixed effect boar's litter size (except for ejaculate volume), mean age at measurement (except for total motility) and sperm concentration (only for motility evaluations) as covariates and the animal and residual random effects. It was mapped 22, 14, 10 and 11 genomic regions, distributed in 11, 9, 3 and 6 different chromosomes, explaining more than 1% of additive genetic variance of ejaculate volume, sperm concentration and total and progressive motilities, respectively. Genomic regions with a great influence on sperm quality traits' expression were identified and must be explored to understand their importance for the genetic control of these traits related to fertility.

INTRODUCTION

Artificial insemination has been extensively used in swine production and, over the years, it has promoted a considerable improvement in breeding results. However, to obtain and maintain a desirable reproduction performance it is necessary to use high quality semen originated from selected and approved boars.

Macroscopic evaluations, as ejaculate volume measurement, and microscopic exams, as sperm concentration and motility, are the most common analysis realized in boar studs (BS) routine. The objective of these evaluations is to process high quality doses of semen (Robaire and Chan 2010). Volume and sperm concentration are evaluated to determine the total number of cells in *in natura* semen and the number of possible doses to be produced per ejaculate once they are related to the number of cells per dose and the dilution rate of the doses (Flowers 1996; Waberski *et al.* 2011). Sperm motility indicates the percentage of mobile cells and, when using computer assisted sperm analysis (CASA), it can be differentiated in total motility and progressive motilities, according with the trajectory of the cells. In general, motility are positive correlated with *in vivo* fertility (Broekhuijse *et al.* 2012; Flowers *et al.* 2016) and are considered an important indicator of boar fertility (Kummer *et al.* 2013).

Despite the importance of boar semen quality traits, selection of boars to be used in BS almost does not take it into account and focuses mainly on growth and carcass characteristics (Flowers 2008). The possibility of associate sperm and genetic merit in boar selection could be interesting to reduce the number of boars required to service sows and maintain the improvement of growth and carcass quality (Oh *et al.* 2006). Although, sperm quality traits can only be evaluated on boars after puberty, the identification of genomic regions influencing their expression and their inclusion in breeding programs are an alternative to select for them in pre-pubertal stage. In that way, the aim of the study was to map chromosomal regions that potentially have association with volume, sperm concentration and total and progressive motilities in Duroc boars.

MATERIALS AND METHODS

Repeated observations of volume, concentration and total and progressive motilities of *in natura* ejaculates of 113 Duroc boars housed at the same boar stud were collected from February 2015 until May 2016. Automated semen collection system (Collectis®, IMV) was used and each ejaculate was collected in a pre-warmed (36°C) plastic container. Gel fraction of each ejaculate was filtered and discarded. After that, the ejaculate was weighted and, to a better estimation of its real volume, it was assumed that one gram corresponds to one millilitre of semen. For microscopic evaluations, samples of ejaculates were prepared (90 µL of raw semen plus 810 µL of pre-warmed extender) and submitted to CASA system (Sperm Vision® Minitüb), which determined sperm concentration and total and progressive motilities. Sperm concentration was determined through counting the cells in eight fields and establishing an average of them. Total motility corresponded to the percentage of mobile cells, independent of their trajectory, and progressive motility corresponded to progressive forward motility of the cells (>4.5 µm of distance sperm travels in straight line). It was evaluated 2,392 ejaculates and the number of ejaculates per boar was 21.17 ± 12.63. Mean values of boar's age at measurement, volume, concentration and total and progressive motilities were considered in the analysis.

Boars were genotyped with Illumina PorcineSNP60 BeadChip (Ramos *et al.* 2009), according to the manufacturer protocols (119 animals for 68,516 SNPs). The quality control of markers was made excluding those with unknown genomic position, placed in sexual chromosomes, with MAF (minor allele frequency) lower than 3% and markers and animals that presented call rate lower than 90%. After quality control, 118 animals and 42,240 SNPs remained to be analysed. Genome wide-association analysis were realized by BLUPf90 (Misztal *et al.* 2002) using weighted single-step GBLUP method (WssGBLUP, Zhang *et al.* 2014), considering windows of 10 adjacent SNPs to estimate their effects by postGSf90 (Aguilar *et al.* 2010; Wang *et al.* 2012). A total of three iterations of BLUPf90 and postGSf90 were used for the WssGBLUP. Each run of postGSf90 updated weights for SNP, whereas each run of BLUPf90 used the updated weights to constructed G matrices (Zhang *et al.* 2016). The iterations increase the weights of SNPs with large effects and decrease those with small effects.

The animal model considered as fixed effect boar's litter size (except for ejaculate volume), besides, as covariates, mean age at measurement (except for total motility) and sperm concentration (only for motility evaluations) and, as random effects, animal and residual effects. Analyses were performed using a pedigree composed by 250 animals.

The results of GWAS were reported as the percentage of the additive genetic variance explained by the windows of 10 adjacent SNPs presented in Manhattan plots drawn by R software.

RESULTS AND DISCUSSION

In Table 1 descriptive statistics for ejaculate volume, sperm concentration, total and progressive motilities were presented.

Table 1. Descriptive statistics for mean values of ejaculate volume, sperm concentration, total and progressive motilities of Duroc boars used for GWAS analysis

Semen traits	N	Mean	SD	Range
Volume (mL)	110	160.19	45.42	75.11 – 286.60
Concentration (x10^6/mL)	110	0.57	0.18	0.20 – 1.11
Total motility (%)	113	86.53	7.47	50.45 – 95.24
Progressive motility (%)	110	76.52	9.99	38.21 – 91.32

N – Number of animals evaluated; SD – Standard deviation

In Figures 1, 2, 3 and 4 were represented the genomic regions and the percentage of genetic variance explained by windows of 10 adjacent SNPs in each chromosome for ejaculate volume, sperm concentration, total and progressive motilities, respectively.

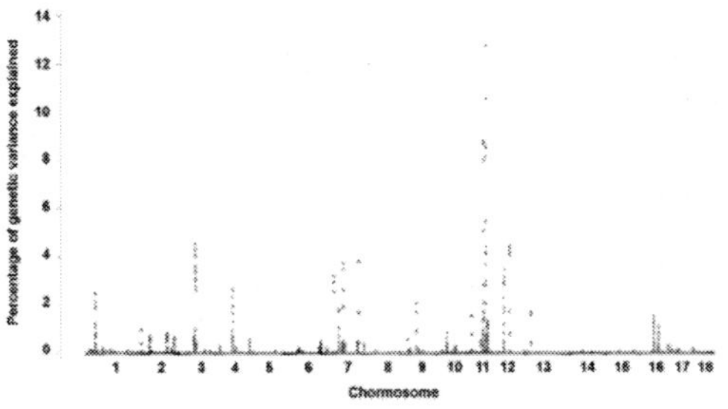

Figure 1. Manhattan plot of genomic regions associated with ejaculate volume.

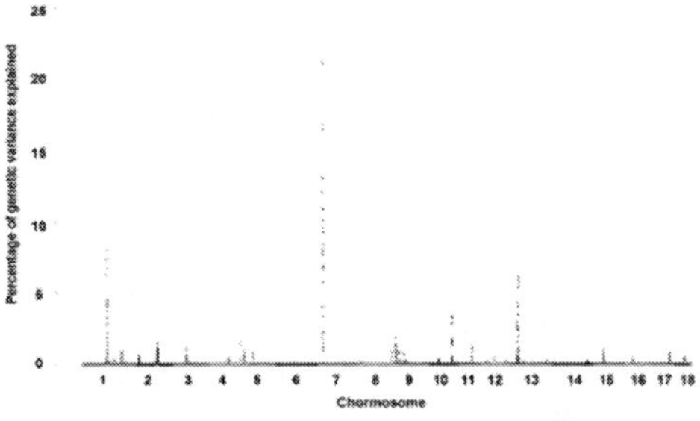

Figure 2. Manhattan plot of genomic regions associated with sperm concentration.

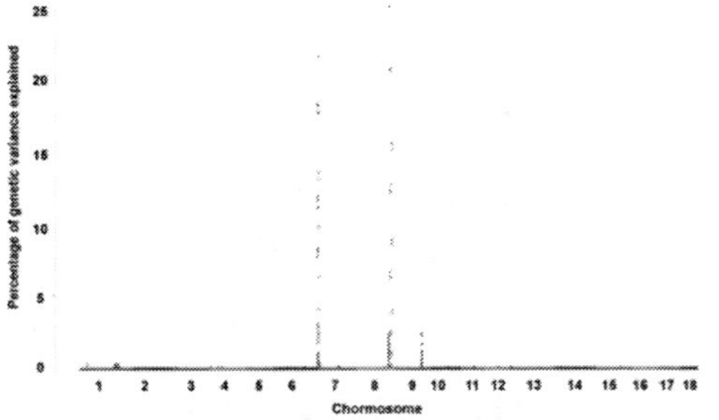

Figure 3. Manhattan plot of genomic regions associated with total motility.

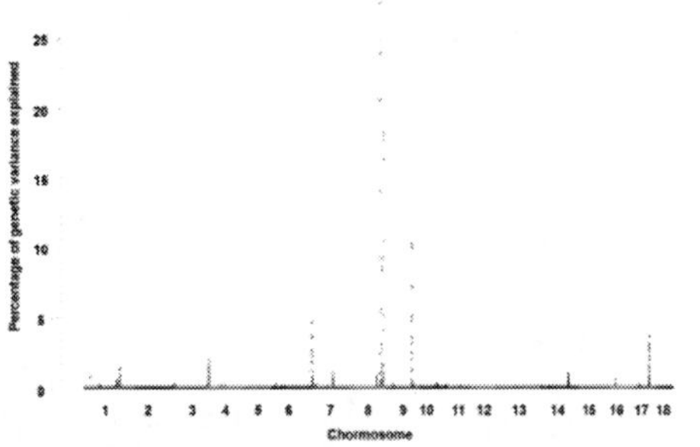

Figure 4. Manhattan plot of genomic regions associated with progressive motility.

Genomic regions that explained more than 1% of additive genetic variance of ejaculate volume were distributed in chromosomes 1, 2, 3, 4, 7, 9, 10, 11, 12, 13 and 16. Of these, chromosome 11 has two important regions explaining 8.81% (stating position 26634721 – final position 27221215) and 14.38% (starting position 49822501 – final position 50200669) of its additive genetic variance. Similarly to our results, Xing *et al.* (2009) also reported that chromosome 3 has significative quantitative trait loci (QTL) for semen volume.

Chromosomes 1, 2, 3, 4, 7, 9, 10, 11 and 13 presented genomic regions that explained more than 1% of genetic variance of sperm concentration. Windows of 10 adjacent SNPs located between 147860835-149305559 pb of chromosome 1, 8326917-8526138 pb and 8833246-8980813 pb of chromosome 7 and 22293935-22666436 pb of chromosome 13 account for 7.94%, 26.03%, 9.28% and 6.05% of its additive genetic variance, respectively. Other studies also reported significant genomic regions in chromosome 7 affecting sperm concentration, total sperm per ejaculate (Zhao *et al.* 2016) and testicular weight (Ren *et al.* 2009).

For total motility, chromosomes 7, 8 and 9 presented windows that explained more than 1% of additive genetic variation. Windows placed between 58555-740616 pb and 1637113-1806683 pb of chromosome 7 and 138036642-138190631 pb and 140609271-141043305 pb of chromosome 8 explained more than 15% of additive genetic variance (18.32%, 21.62%, 25.10% and 15.63%, in this order).

Regions that explain more than 1% of genetic variation for progressive motility were in chromosomes 1, 4, 7, 8, 9 and 17. Higher percentages of additive genetic variance (27.63%, 18.24% and 10.33%) were observed in two windows located on chromosome 8 (133531944-133969475 pb and 140609271-141043305 pb) and one on chromosome 9 (125924965-126346678 pb). There is no reference about progressive motility in literature, but Xing *et al.* (2009) and Diniz *et al.* (2014) have reported significant genomic regions in chromosome 1 affecting total motility, which can also influence sperm motility and their trajectory.

This study identified important genomic regions associated with sperm quality traits of Duroc boars. A total of 57 SNPs windows that explained more than 1% of genetic variance were identified for ejaculate volume, sperm concentration, total and progressive motilities.

Those regions must be explored to understand their importance for the genetic control of these these traits related to fertility. In the future, the markers identified in this research may be useful to improve the selection of boars to be used in boar studs.

REFERENCES

Aguilar, I., Misztal, I., Johnson, D. L., Legarra, A., Tsuruta, S., and Lawlor, T. J. (2010) *J. Dairy Sci.* **93**:743.

Broekhouijse, M.L.W.J., Sostaric, E., Feitsma, H. and Gadella, B.M. (2012) *Theriogenology* **76**:1473.

Diniz, D.B., Lopes, M.S., Broekhuijse, M.L.W.J., Lopes, P.S., Harlizius, B., Guimarães, S.E.F., Duijuvensteijn, N., Knol, E.F. and Silva, F.F. (2014) *Anim. Reprod. Sci.* **151**:201.

Flowers, W.L., Deller, F. and Stewart, K.R. (2016) *Anim. Reprod. Sci.* **174**:123.

Flowers, W.L. (1996) *Proceedings of American Association od Swine Practitioners* **1**:469.

Flowers, W.L. (2008) *Theriogenology* **70**:1297.

Kummer, A.B.H.P.; Gaggini, T.S.; Bernardi, M.L.; Mcmanus, C.; Gonçales, E.M.; Wentz, I. and Bortolozzo, F.P. (2013) *Reprod. Domest. Anim.* **48**:747.

Misztal I., Tsuruta S., Strabebl T., Auvray B., Drueta, T. and Lee, D.H. (2002) *Proceedings of 7th World Congress on Genetics Applied to Livestock Production* **28**:21.

Oh, S.H., See, M.T., Long, T.E. and Galvin, J.M. (2006) *Asian-Aust. J. Anim. Sci.* **19**:160.

Ramos, A.M., Crooijmans, R.P., Affara, N.A., Amaral, A.J., Achibald, A.L., Beever, J.E., Bendixen, C., Churcher, C., Clark, R., Dehais, P., Hansen, M.S., Hedegaard, J., Hu, Z.L., Kerstens, H.H., Law, A.S., Megens, H.J., Milan, D., Nonneman, D.J., Rohrer, G.A. Rothschild, M.F., Smith, T.P., Schnabel, R.D., Van Tassel, C.P., Taylor, J.F., Wiedmann, R.T., Schook, L.B. and Groenen, M.A. (2009) *PLoS One* **4**:6524.

Ren, D.R., Ren, J., Xing, Y.Y., Guo, Y.M., Wu, Y.B., Yang, G.C., Mao, H.R. and Huang, L.S. (2009) *J. Anim. Sci.* **87**:17.

Robaire, B. and Chan. P. (2010) 'The handbook of andrology' Schaumburg: The American Society of Andrology, Illinois. **2**.

Xing, Y., Ren, J., Ren, D., Guo, Y., Wu, Y., Yang, G., Mao, H., Brenig, B. and Huang, L. (2009) *Anim. Reprod. Sci.* **114**:210.

Zhang, X., Lourenco, D.A.L., Aguilar, I., Legarra, A. and Misztal, I. (2016) *Front. Genet.* **7**:151.

Zhang, X., Lourenco, D.A.L., Misztal, I., Aguilar, I. and Legarra, A. (2014) *Proceedings World Congress on Genetics Applied to Livestock Production* **10**.

Zhao, X., Zhao, K., Ren, J., Zhang, F., Jiang, C., Hong, Y., Jiang, K., Yang, Q., Wang, C., Ding, N., Huang, L., Zhang, Z., Xing, Y. (2016) *Anim. Sci. J.* **87**:646.

Waberski, D., Henning, H. and Pretrunkina, A.M. (2011) *Reprod. Domest. Anim.* **46**:45.

Wang, H., Misztal, I., Aguilar, I., Legarra, A. and Muir, W. M. (2012) *Genet. Res.* **94**:73.

Proc. Assoc. Advmt. Anim. Breed. Genet. 22:505-508

PRODUCTIVE LIFETIME AND LIFETIME EFFICIENCY IN HOLSTEIN COWS AS AFFECTED BY FIRST LACTATION MILK YIELD

H.L. de Waal[1], C.J.C. Muller[2] & M.M. Scholtz[1,3]

[1]National Milk Recording Scheme, Agricultural Research Council, Stellenbosch 7602, South Africa;[2]Department of Animal Sciences, University of Stellenbosch, Stellenbosch 7602, South Africa;[3]Natural and Agricultural Sciences, University of the Free State, Bloemfontein 9300, South Africa

SUMMARY

The longevity of dairy cows affects their lifetime performance and lifetime efficiency. In this paper the effect of first lactation milk yield on the lifetime (LT, in days), productive lifetime (PL, in days), total lifetime milk yield, productive lifetime milk yield efficiency (PLE) and lifetime milk yield efficiency (LTE) of South African Holstein cows are presented. Lifetime production records of 523 020 cows, born between 1989 and 2008, calving down at least once, were included in the study. For each cow, the total milk, fat and protein yield, all days in milk and lifetime of cows was determined. The PLE (total lifetime milk yield divided by PL) and LTE (total milk yield divided by LT) for milk yield were estimated for each cow. The LT and PL of Holstein cows increased up to first lactation milk yield of 6000 and 7000 kg, respectively. The milk yield LTE and PLE increased with higher milk yield levels in first lactation.

INTRODUCTION

Milk recording in South Africa has always focused on the lactation milk yield and milk composition of cows (Du Toit 2016). Local breed societies often reward lifetime performance of dairy cows by awarding special status to cows reaching milestone production levels. However, the efficiency of production has not been considered nor rewarded. A considerable re-ranking of cows occurs when lifetime production is divided by lifetime (using lactation number). Already in 1953, Leitch & Godden estimated the whole-life energy efficiency of cows at different milk production levels and different ages as indicated by lactation number. While energy efficiency increased with age, higher producing cows reached higher efficiency levels at an earlier age. Low yielding cows, regardless of a long productive life, were less efficient than higher producing cows even at a shorter productive life. As the repeatability of milk production in dairy cows is high (>0.55), it would be expected that first lactation milk yield give some indication of the future milk yield and, therefore, lifetime performance of cows. While it has been shown that lifetime milk yield and milk yield per day of productive life increases with increasing first lactation milk yield levels (Sawa & Krezel-Czopek 2009), local dairy farmers are reluctant to select for a higher production in first lactation probably because of unfavourable correlations between high milk yield levels and traits such as live weight, fertility and longevity. Little research has been done in South Africa on the lifetime performance and the efficiency of production of dairy cows. Muller & Botha (2003) showed that genetic progress in a dairy herd can be increased by selecting for higher milk yield levels in first lactation. The number of days in milk, number of completed lactations and milk yield level affects lifetime milk yield as well as economic efficiency (Heins *et al.* 2012 and Martens & Bange, 2013). The duration of each lactation is affected by calving interval which is influenced by traits like the number of days from calving to first service, first service conception rate, number of services per conception and number of days from calving to conception or days open (Muller *et al.* 2014). The aim of the study is to determine the effect of first lactation milk yield on the LT, PL, total lifetime milk yield, productive lifetime milk yield efficiency (PLE) and lifetime milk yield efficiency (LTE) of Holstein cows.

MATERIALS AND METHODS

Data. Milk production records of about 523 020 Holstein cows that had calved down for the first time between 1989 and 2008 were extracted from the South African National Milk Recording Scheme data base of the Agricultural Research Council (ARC). Milk production records were compiled using standard procedures, i.e. on 10 milk recording events during the year, starting from 5 days after calving, for at least 8 milk recording events per cow (De Waal & Heydenrych, 2001). All cows that had completed a first lactation of at least 240 days, were included in the study. The milk, fat and protein yield for all lactation periods were added up until the end of each cow's last lactation period. Productive life (PL) was estimated for each cow adding all the days in milk per lactation. The lifetime (LT) of cows was derived from birth date to the end of the last lactation period which was regarded as the cows' cull date as their actual cull dates were not recorded.

Statistical analyses. Analysis of variance, considering cows as random replicates, was performed using GLM Procedure of SAS software (Version 9.4; SAS Institute Inc, Cary, USA) to test the effect of milk yield in first lactation categories on milk, fat and protein production, PL, LT and PLE (total milk yield/productive life in days) and LTE (total milk yield/lifetime in days). Fisher's least significant difference was calculated at the 5% level to compare milk yield in first lactation category means (Ott, 1998). A probability level of 5% was considered significant for all significance tests. Shapiro-Wilk test was performed on the standardized residuals from the model to test for deviation from normality (Shapiro & Wilk, 1965).

RESULTS AND DISCUSSION

About 67% of first lactation milk production records were between 4 001 to 8 000 kg per lactation (Figure 1a). The trends for PL and LT of cows over production years remained constant until 2003 after which both traits followed a downward trend (Figure 1b). The reason for this is unclear although average age at first calving decreased from about 30.3 in 1989 to 26.9 months in 2008. It could also be related to fewer cows in milk recording or a genetic change in longevity.

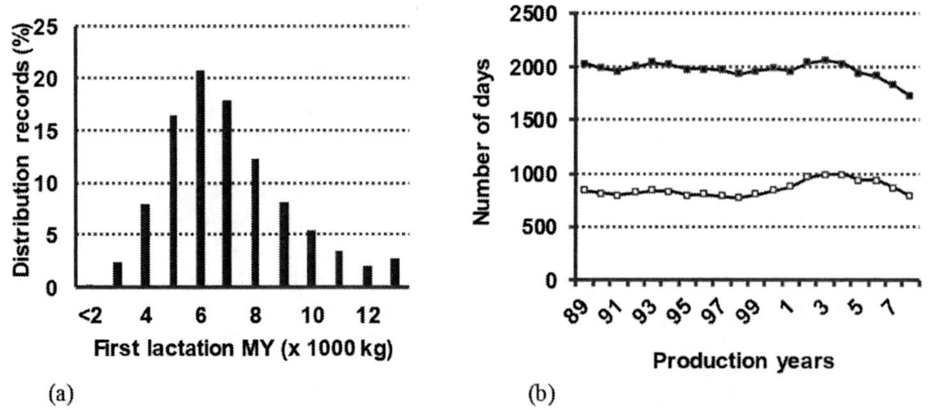

Figure 1. The (a) distribution of first lactation milk yield (MY) records in production categories and (b) the effect of production year on the productive lifetime (□) and lifetime (■) in days for South African Holstein cows.

Increasing first lactation milk yield had small linear negative (P<0.05) effects on fat and protein percentages (Table 1). Lifetime milk yield showed a linear increase (P<0.05) with increasing first lactation milk yields. Although LT and PL initially increased reaching peaks at first lactation milky yield intervals of 5001-6000 and 6001-7000 kg per lactation, respectively, traits decreased (P<0.05) after peaking. However, the PL of cows changed little at milk yield levels beyond 7000 kg per lactation being 862 and 806 days for cows producing within production intervals of 8001-9000 and 10001-11000 kg milk, respectively. In contrast to decreasing LT and PL number of days, PLE and LTE increased for cows producing at a higher level in first lactation.

Table 1. The mean (standard deviation) and data ranges for milk production, fat and protein percentages and lifetime parameters for Holstein cows in different first lactation milk yield classes. Least significant differences (LSD) for first lactation milk yield intervals indicate significant differences between first lactation milk yield classes. (PLE = Productive life efficiency; LTE = Lifetime efficiency).

	First lactation milk yield classes (kg)						
Variables	**4001-5000**	**5001-6000**	**6001-7000**	**7001-8000**	**8001-9000**	**Range (min-max)**	**LSD**
First lactation milk yield (kg)	4537 (286)	5503 (287)	6474 (286)	7465 (287)	8466 (285)	1005-17329	62.6
First lactation fat (%)	3.63 (0.38)	3.59 (0.38)	3.57 (0.39)	3.57 (0.40)	3.59 (0.40)	2.10-6.01	0.014
First lactation protein (%)	3.20 (0.21)	3.19 (0.20)	3.18 (0.19)	3.17 (0.19)	3.17 (0.19)	2.10-5.33	0.007
Total lifetime milk yield (kg)	15874 (11052)	19310 (12225)	21795 (12867)	23478 (13257)	25316 (13683)	1006-135538	438.9
Lifetime (days)	2004 (721)	2054 (713)	2041 (689)	1987 (651)	1940 (621)	785-4295	24.1
Productive life (d)	835 (500)	888 (497)	892 (474)	862 (440)	840 (412)	241-3239	16.5
PLE (kg)	18.01 (2.89)	20.96 (2.90)	23.88 (2.97)	26.82 (3.15)	29.83 (3.25)	1.2-150.8	0.22
LTE (kg)	7.02 (2.87)	8.47 (3.12)	9.74 (3.35)	10.90 (3.54)	12.14 (3.74)	0.22-43.9	0.13

Sawa & Krezel-Czopek (2009) also showed that LT milk yield in Polish Holsteins increased with increasing first lactation milk yields, the correlation coefficient being 0.44 (P<0.01). However, lifespan and PL decreased when first lactation milk yield exceeded 7000 kg milk, correlation coefficients being positive, albeit low at 0.23. In the current study, the phenotypic correlation between first lactation milk yield and LT milk showed a curvilinear trend (P<0.05). Similarly PLE and LTE were affected positively (P<0.05) with increasing first lactation milk yield levels possibly indicating that the decreases in PL and LT had a limited effect on efficiency measures. To improve efficiency measures, two possible methods can be used, i.e. improving LF milk yield while maintaining PL and LT or maintain milk yield and increasing PL and LT. Heritability estimates for cow herd life and productive life are generally, below 10% while the heritability of milk yield is moderate at about 0.25.

To improve the efficiency of production two possible methods can be used, i.e. direct selection for production traits or using ratio traits. The main advantage of most ratio traits is their ease of calculation and interpretation, as well as the ability to easily compare efficiency statistics across populations. However, according to Gunsett (1984) direct selection on ratio traits is complicated by the disproportionate fashion by which selection pressure is exerted on the component traits. This makes expected responses to selection on ratio traits difficult to determine mainly because of the poor statistical properties of ratio traits due to the antagonism between the desirable response in the numerator (i.e., increased milk yield) and the denominator (i.e., increased productive life) and the unknown relative selection pressure on each (Gunsett, 1984).

Heritability estimates for cow herd life and productive life are low. Buenger *et al.* (2001) reported heritability estimates for functional length of productive life to be 0.09 to 0.14. Increasing productive life should therefore focus on an improved environment. Although increased first lactation milk yields improves PLE and LTE, the PL and LT is decreased as shown in the present study. Juszczak *et al.* (1994) showed that the optimum first lactation milk yield regarding efficiency of production varies according to management conditions and herd milk yield levels.

CONCLUSION

This study reported on the effect of first lactation milk yield on the milk yield, PL, FT, PLE and LTE of Holstein cows. Higher first lactation milk yields resulted in an increased lifetime milk yield, productive lifetime yield, as well as PLE and LTE. Cows producing high levels of milk yield in first lactation (>8000 kg) are expected to have shorter PL and LT although the reduction is small. This resulted in increasing PL and LT efficiencies, possibly indicating that first lactation milk yield could be used as a selection tool for increased production efficiencies. To improve production efficiencies PL (days in milk) should be increased by more calving down events rather than extending lactation periods. Further work includes the estimation of genetic parameters for PL, LT and efficiency measures towards estimating breeding values for lifetime production traits. Genomic analyses may be required to identify high and low genetic merit sires for PLE and LTE.

ACKNOWLEDGEMENTS

The support provided by Mr. Graham Buchanan for estimating total milk yields from individual lactation periods for each cows is gratefully acknowledged as well as the statistical support provided by Mrs. Marieta van der Rijst from the Biometry Section of the ARC.

REFERENCES

Buenger, A., Ducrocq, V. and Swalve, H.H. (2001) *J. Dairy Sci.* **84**: 1531.
De Waal H. and Heydenrych H.J. (2001) *S.A. J. Anim. Sci.* **31**: 9.
Du Toit J. (2016) Chapt. 18. *Manual for Dairy Farming*. Section **4**: 7.
Gunsett F.C. (1984) *J. Anim. Sci.* **59**: 1185.
Heins B.J., Hansen L.B. and De Vries A. (2012) *J. Dairy Sci.* **95**: 1011.
Juszczak J., Hibner A., Zachwieja A., Tomaszewski A. and Krzyskow S. (1994) *Anim. Prod. Rev.* **4**: 3.
Leitch I. and Godden W. (1953) *Tech. Commun. Bur. Anim. Nutr.* No. 14.
Martens H. and Bange C. (2013) *J. Dairy Sci.* **89**: 3876.
Muller C.J.C. and Botha J.A. (2003) *S.A. J. Anim. Sci.* **33**: 111.
Muller C.J.C., Potgieter J.P., Cloete S.W.P. and Dzama K. (2014) *S.A. J. Anim. Sci.* **44**: 54.
Ott R.L. (1998) An introduction to statistical methods and data analysis. Duxbury Press. 807.
Sawa A. and Krezel-Czopek S. (2009) *Archiv Tierzucht.* **52**: 7.
Shapiro S.S. and Wilk M.B. (1965) *Biometrika.* **52**: 591.

Proc. Assoc. Advmt. Anim. Breed. Genet. 22:509-512

GENETIC PARAMETERS FOR ALTERNATIVE MEASURES OF FERTILITY IN A COMMERCIAL HERD OF TROPICAL COWS

F.S.S. Raidan[1,2], L.R. Porto-Neto[2], M.R.S. Fortes[1], S. Harburg[3], A. Reverter[2]

[1] The University of Queensland, School of Chemistry and Molecular Biosciences, St. Lucia Campus, Brisbane, QLD 4072
[2] CSIRO Agriculture & Food, Queensland Bioscience Precinct, Brisbane, QLD 4067
[3] North Australian Pastoral Company - NAPCO.

SUMMARY

Fertility efficiency is associated with increased profit being an important objective of breeding programs. Cow fertility is complex and difficult to improve. In this study we use the genomic relationship matrix (G) and REML approaches to investigate the genetic parameters for number of calves (NC) and fertility indices in tropical beef cattle. The fertility indices proposed were: I_1 = NC/NO, I_2 = (NC/NO)*NC and I_3 = (NC/ALC)*NC, where: NO = Number of opportunity, ALC = Age at last calve. Heritability estimates of 22% to 24% were observed for the four phenotypes. The genetic and residual correlations were close to unity, except for those pairs that included I_1 in this case, correlations were around 0.50. We conclude that NC is an efficient selection criteria for the improvement of fertility in Tropical Composite cattle.

INTRODUCTION

An important goal of tropical beef cattle breeding programs is to improve reproductive performance. However, cow fertility is complex and difficult to improve because of low heritability, delayed expression in females' life and difficulties in the routine recording of phenotypes such as pregnancy status, days open and days for first service, especially in extensive large-scale tropical beef operations. Cow longevity, an indicator of fertility, has been evaluated in some beef cattle breeding programs. However, selection for improved longevity is challenging because this trait is only available after the cow is culled or dead. Additionally, the observation results in censored data or binary distribution which requires complex statistical analyses. Alternatively, fertility indices that shows the abilities of the female to calve at a young age, to maintain the regularity of calving, and to wean heavy calves (Eler *et al.* 2008) might be advantageous because it permits the evaluation of genetic merit of females with only one or few calving events as well as the evaluation of young bulls (Santana *et al.* 2013). However, number of calves (NC, with cow age as a fixed effect in the statistical model) might be a simple and efficient predictor of cow's fertility. Thus, we propose this measure of fertility, which was less demanding and also easy to understand and can be useful for improving the fertility of the breed. In this study we make use of the genomic relationship matrix (G) to estimate genetic parameters for number of calves and fertility indices in a commercial herd of Tropical Composite cows in Australia.

MATERIAL AND METHODS

Phenotypic and genotypic data. The data set used in the present study consisted of 1,166 commercial Tropical Composite cows from North Australian Pastoral Company (NAPCO) with phenotype records including the number of calves (NC) and three alternatives fertility indices (Table 1 and Figure 1). The fertility indices explored were as follows: I_1 = NC/NO, I_2 = (NC/NO)*NC and I_3 = (NC/ALC)*NC, where: NO = Number of opportunity, ALC = Age at last calve. The I_1 index is related to reproductive efficiency of females, the value 1 or 100% was attributed for females that calving in all breeding opportunities and the indices I_2 and I_3 benefit the females that remain in the herd for longer periods of time (longevity). Genotypes were generated based on a panel with approximately 54,000 polymorphisms from the Illumina BovineSNP50. The following criteria were used for the exclusion of SNPs: minor allele frequency less than 2%, call rate less than 90% and/or duplicate samples. After quality control, 42,455 SNPs remained for analysis.

Table 1. Descriptive statistics results for reproductive traits in Tropical composite cows

Traits	Number of observation	Mean ± sd	Min	Max
Age of dam	1,166	6.17 ± 3.16	3.00	19.00
I_1 = NC/NO	1,166	0.667 ± 0.249	0.00	1.00
I_2 = (NC/NO)*NC	1,166	2.783 ± 2.94	0.00	15.00
I_3 = (NC/ALC)*NC	1,166	2.26 ± 2.62	0.00	13.24
NC	1,166	3.48 ± 3.12	0.00	15.00

NP = Number of calves, NO = Number of opportunity, ALC = Age at last calve.

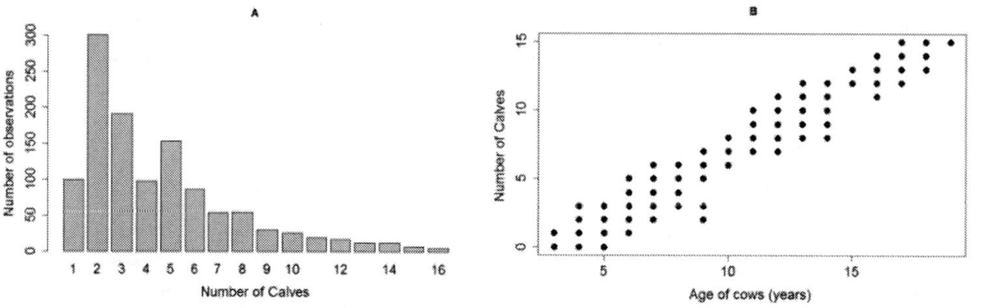

Figure 1. Distribution for number of calves (A) and distributions for number of calves by age of cows (B)

Statistical analysis. A tetra-variate analysis was performed using the general mixed model in $y_{ij} = X\beta + Zu + e_{ij}$, where: y_{ij} represents the phenotypic observations from the i-th cows (i = 1 to 1,116) at the j-th phenotype (j = 1 to 4), X is the incidence matrix relating fixed effects in $\mathbf{\beta}$ with observations in y_{ii}, Z is the incidence matrix relating random additive polygenic effects in **u** with observations in y_{ij}, and e_{ij} is the random residual effects. Fixed effects included in the model were contemporary group (i.e., cohort of cows born in the same year and raised together) and group of age of the dam. Solutions to the effects in the model as well as variance components were estimated using **G** according to Wang *et al.* (2014) in BLUPF90 programs (Misztal *et al.* 2009).

RESULTS AND DISCUSSION

Variance components, heritability and genetic and residual correlations are reported (Table 2). Moderate heritability estimates of ~ 22% were observed for the four phenotypes. The heritability estimate for number of calves was higher than those reported by Martinez *et al.* (2004) and Zhang *et al.* (2013) for number of calves born in Hereford (h^2 = 0.15), Brahman (0.15) and Tropical Composite cows (0.14). Martinez *et al.* (2004) and Zhang *et al.* (2013) evaluated the lifetime number of calves in predetermined age of cows and determined heritability estimates using **G**. These results demonstrate that **G** often explains more genetic variance than the pedigree-based estimates. In fact, heritability for number of calves at 6 years were 0.22 and 0.16 with **G** and 0.15 and 0.14 with pedigree-based matrix for Brahman and Tropical Composite cows, respectively (Zhang *et al.* 2013).

Table 2. Genetic parameters of reproductive traits in Tropical Composite cows, last 4 lines: heritability (diagonal), genetic correlation (above diagonal) and residual correlation (below diagonal) by single-step-genomic-BLUP methodology

	I_1 = NC/NO (%)	I_2 = (NC/NO)*NC	I_3 = (NC/ALC)*NC	NC
Genetic variance	0.008	0.241	0.490	0.265
Residual variance	0.029	1.922	1.639	0.921
I_1 = NC/NO (%)	0.22	0.52	0.48	0.56
I_2 = (NC/NO)*NC	0.54	0.24	0.99	0.99
I_3 = (NC/ALC)*NC	0.49	0.99	0.23	0.99
NC	0.57	0.99	0.99	0.22

NC = Number of progeny, NO = Number of opportunity, ALC = Age at last calve.

According to Chud *et al.* (2014) the low heritability estimate for fertility indices might be related to low heritability estimates for traits such as NC and ALC that compound the index. Actually, the heritability for age at calving was low, ranging from 0.05 to 0.15 over 1 to 6 calving seasons, respectively (Martinez *et al.* 2004). Furthermore, the heritability for other reproductive traits were lower than the values obtained in this study, mean and standard errors (in brackets) of 0.12 (0.07), 0.06 (0.06) and 0.11 (0.07) were obtained for conception, pregnancy and calving rates, respectively, in Tropical Composite cows (Johnston *et al.* 2013). Thus, it is possible to achieve higher genetic progress across generations through selection for NC or fertility indices evaluated here than others reproductive traits, since the heritability of NC or fertility indices were greatest.

It is important to highlight that a positive correlation between ratio values, as a fertility indices proposed here, reduces the selection response of both traits, but mainly for the trait with the weaker ratio position (Essl, 1989). That is the numerator if selection is for higher ratios and the denominator in the opposite case. Moreover, the difference between the relative selection responses for the single ratio traits becomes more diferent the closer their genetic correlation is to +1 (Essl, 1989). Thus, the ratio values can be used as a selection criteria in breeding programs however, the genetic correlation between traits included in the ratio should be strictly and routinely evaluated. Because genetic correlations change across generations, pleotropic genes may be fixed and linkage may be lost (Sheridan and Backer, 1974). However, traits can be combined in an index which included economic values (Hazel, 1947).

The genetic and residual correlations were close to unity, except for those pairs that included I_1, in this case correlation were around 0.50. Therefore, genetic progress for longevity (I_2 or I_3) can be achieved through selection for NC or fertility efficiency (I_1), which might be measured in early female's life. The length of productive life measured through 1 year after first calving in Hereford cows predicts productive life through 6 years with reasonable accuracy (Martinez *et al.* 2004). Selection for younger age at puberty leads to increase in lifetime reproductive performance of Brahmam (rg = -0.40 ± 0.20) and Tropical Composite (rg = -0.33 ± 0.28) cows (Johnston *et al.* 2013).

CONCLUSION

Based on estimates of heritability and genetic correlations, the number of calves could be a simple and useful selection criterion for improving the fertility of Tropical Composite cows in commercial operations.

ACKNOWLEDGEMENTS

Authors acknowledge Sam Harburg for contributions, North Australian Pastoral Company (NAPCO) for allowing access to their data.

REFERENCES

Chud T.C.S., Caetano S.L., Buzanskas M.E., Grossi D.A., Guidolin D.G.F., Nascimento G.B., Rosa J.O., Lôbo R.B. & Munari D.P. (2014) *Livestock Science* **170**:16-21.
Essl A. (1989) J. Anim. Breed. Genet. **106**:81-88.
Hazel L. N. (1943) Genetics. **28**:476-490.
Johnston D.J., Corbet N.J., Barwick S.A., Wolcott M.L. & Holroyd R.G. (2013) *Animal Production Science* **54**:74-84.
Martinez G.E., Koch R.M., Cundiff L.V., Gregory K.E. & Van Vleck L.D. (2004) *J Anim Sci* **82**: 1912-8.
Misztal I., Legarra A. & Aguilar I. (2009) *J Dairy Sci* **92**:4648-55.
Santana M.L., Jr., Eler J.P., Bignardi A.B. & Ferraz J.B. (2013) *J Anim Sci* **91**:2566-74.
Sheridan A.K. and Barker J.S. (1974) *Aust. J. Biol. Sci.* **27**:89-101.
Wang H., Misztal I., Aguilar I., Legarra A., Fernando R.L., Vitezica Z., Okimoto R., Wing T., Hawken R. & Muir W.M. (2014) *Front Genet* **5**:134.
Zhang Y.D., Johnston D.J., Bolormaa S., Hawken R.J. & Tier B. (2013) *Animal Production Science* **54**:16-24.

Proc. Assoc. Advmt. Anim. Breed. Genet. 22:513-516

GEOREFERENCED EVALUATION OF GENETIC PATTERNS OF MONTANA TROPICAL® CATTLE

T.G. Sousa[1], C.M. McManus[2], J.B. Ferraz[3], J.P. Eler[3], N.S. Costa[2], F.M. Rezende[1]

[1] College of Veterinary Medicine, Federal University of Uberlandia, Uberlandia, Brazil
[2] University of Brasilia, Brasilia, Brazil
[3] College of Animal Science and Food Engineering, University of Sao Paulo, Pirassununga, Brazil

SUMMARY

The influence of environmental (normalized difference vegetation index (NDVI), rainfall, minimum, maximum and average temperatures, altitude, relative humidity and temperature and humidity index) and socioeconomic (gross domestic product, number of inhabitants, municipal area and human development index) factors on the genetic distribution of Montana Tropical® beef composite cattle located in 41 counties. Expected progeny differences (EPD) of 455,175 animals for weights at weaning and at 14 months of age, post-weaning weight gain, scrotal circumference and muscle score were used to group the counties into five clusters by SAS® PROC FASTCLUS. Discriminant analysis identified significant differences between the clusters due to altitude, NDVI and maximum and minimum temperatures, by t-Student test. However, it was not possible to identify a clear relation between the cluster means for these factors and the genetic distribution of Montana® herd. Using environmental and socioeconomic information to classify the counties into the previously formed groups, the percentage of correct classifications was much lower than the classification based on EPD, which reinforces the weak influence of those factors on the genetic clustering proposed.

INTRODUCTION

The Montana Tropical® beef composite cattle, raised in Brazil since 1994, aims to exploit the benefits of heterosis and complementarity between breeds by crossing four biological groups. In this system, animal's racial composition is not fixed, which allows the customization of herds according to the conditions of each region. Hermuche *et al.* (2012, 2013) and Costa *et al.* (2014) demonstrated that environmental and socioeconomic factors can influence the dynamics of animal production and the genetic structure of populations in sheep and Holstein cattle, respectively. Manel *et al.* (2003) proposed combine molecular data and environmental conditions information for a better understanding of how geographical and environmental factors can influence the genetic structure of populations, an approach called landscape genetics. Costa *et al.* (2014) highlighted that many of these studies were done using molecular information, while few were conducted based on the genetic value of animals.

In view of racial diversity of Montana Tropical® herd and its diffusion throughout the Brazilian territory, the present study aimed to analyze the influence of environmental and socioeconomic conditions of counties in its genetic distribution.

MATERIALS AND METHODS

The analysed data consisted of 455,175 animals from 57 farms located in 41 cities placed in nine Brazilian federal states (Espírito Santo, Goiás, Minas Gerais, Mato Grosso do Sul, Mato Grosso, Pará, Rondônia, Rio Grande do Sul and São Paulo) located in Central-West, North, South and Southeast regions. Estimated progeny differences (EPD) for weights at weaning and 14 months of age, post-weaning weight gain, scrotal circumference and muscle score, which are the traits considered in selection index of Montana® breeding program, were evaluated.

These animals belong to the genetic breeding program of Montana Tropical® cattle managed by CFM-Leachman Pecuaria Ltda., whose genetic evaluations are conducted by the Animal Breeding and Biotechnology Group of the College of Animal Science and Food Engineering of University of Sao Paulo.

Environmental and socioeconomic data were obtained from different sources: NASA (National Aeronautics and Space Administration), IBGE (Brazilian Institute of Geography and Statistics) and INMET (National Meteorological Institute).

The environmental variables evaluated were: normalized difference vegetation index (NDVI), rainfall, minimum, maximum and average temperatures, altitude, relative humidity and temperature and humidity index. Additionally, the socioeconomic variables considered were: gross domestic product, number of inhabitants, municipal area and human development index.

The 41 cities were grouped into five clusters by PROC FASTCLUS procedure of SAS® software, being the cluster 1 composed by the counties with highest mean EPD values for studied traits and cluster 5 by the lowest ones. The determination of environmental and socioeconomic factors that differ between the clusters was performed by PROC GLM procedure of SAS® software. Additionally, discriminant analyses were performed by PROC STEPDISC and PROC DISCRIM procedures of SAS® software.

RESULTS AND DISCUSSION

The 41 counties, where the evaluated Montana Tropical® herd is raised, were grouped into five clusters and the mean EPD for analyzed traits per cluster were presented in Table 1. The clusters differed on the mean EPD values for weaning (WW) and 14 months of age (W14) weights, post-weaning weight gain (PWG), scrotal circumference (SC) and muscle score (MS). Cluster 1 was composed by the counties with highest mean EPD values and cluster 5 by the lowest ones. The number of counties (animals) per cluster was 3 (4,893), 4 (81,553), 16 (238,141), 12 (123,818) e 6 (6,770), in this order.Analysing the differences between clusters due the environmental and socioeconomic factors, only NDVI (p=0.08) and altitude (p=0.05) were significant by test F. However, the comparison of the mean values between clusters for the same factors, under t-Student test, revealed additional significant differences by maximum and minimum temperatures (Table 2). Although, the t-test is less rigorous than the F test, it was chosen to present these results in view of the pioneering nature of the present study in the search for environmental and socioeconomic variables that explain the genetic distribution of Montana Tropical® animals in the country. The analysis of the mean values for these four environmental factors by cluster (Table 2) didn't allow a clear identification of their influence on the genetic distribution of Montana Tropical® beef cattle.

Using all the environmental and socioeconomic information to classify the counties into the previously formed groups (Table 3), the percentage of correct classifications was much lower than the classification based on EPDs (Table 4), which reinforces the weak influence of those factors on the genetic clustering proposed.

Table 1. Descriptive statistics of EPD by cluster for weaning (WW) and 14 months of age (W14) weights, post-weaning weight gain (PWG), scrotal circumference (SC) and muscle score (MS).

Cluster	Trait	Mean	STD	Minimum	Maximum
1	WW (kg)	3.78	1.29	2.99	5.27
	W14 (kg)	6.02	1.36	4.58	7.29
	PWG (kg)	1.74	1.12	0.65	2.88
	SC (cm)	0.32	0.05	0.29	0.38
	MS (unit)	0.14	0.02	0.13	0.16
2	WW (kg)	2.63	0.65	1.93	3.45
	W14 (kg)	2.40	0.61	1.66	3.11
	PWG (kg)	0.31	0.16	0.13	0.51
	SC (cm)	0.15	0.05	0.08	0.20
	MS (unit)	0.06	0.01	0.04	0.07
3	WW (kg)	1.25	0.39	0.54	1.90
	W14 (kg)	1.06	0.33	0.53	1.62
	PWG (kg)	0.24	0.21	-0.22	0.56
	SC (cm)	0.06	0.03	0.02	0.12
	MS (unit)	0.03	0.01	0.01	0.05
4	WW (kg)	-0.08	0.42	-0.63	0.54
	W14 (kg)	-0.11	0.46	-0.89	0.47
	PWG (kg)	-0.06	0.27	-0.48	0.42
	SC (cm)	-0.01	0.04	-0.11	0.03
	MS (unit)	0.00	0.01	-0.03	0.01
5	WW (kg)	-1.33	0.14	-1.53	-1.13
	W14 (kg)	-1.01	0.40	-1.76	-0.69
	PWG (kg)	-0.11	0.07	-0.18	0.02
	SC (cm)	-0.06	0.02	-0.09	-0.04
	MS (unit)	-0.02	0.00	-0.03	-0.02

STD: standard deviation.

Table 2. Mean values of environmental factors that presented significant effects between clusters.

Cluster	Altitude (m)	NDVI	Tmin (°C)	Tmax (°C)
1	393.47 [a,b]	0.57 [a,b]	25.34 [a,b]	34.81 [a,b]
2	458.91 [b]	0.53 [a]	23.43 [a,b]	32.09 [a,b]
3	289.50 [a]	0.61 [b]	26.04 [b]	35.30 [b]
4	399.09 [b]	0.58 [a,b]	25.11 [a,b]	34.01 [a,b]
5	484.97 [b]	0.54 [a]	22.95 [a]	31.34 [a]

NDVI: normalized difference vegetation index; Tmax: maximum temperature; Tmin: minimum temperature.
Means followed by different letters in columns differ (P <0.05) by the t-Student test.

Table 3. Percentage of counties classified into each cluster using environmental and socioeconomic factors.

Cluster	1	2	3	4	5
1	33,33	0,00	33,33	0,00	33,33
2	25,00	50,00	0,00	0,00	25,00
3	6,25	6,25	75,00	0,00	12,50
4	16,67	8,33	16,67	33,33	25,00
5	0,00	16,67	0,00	0,00	83,33

Table 4. Percentage of counties classified into each cluster using the expected progeny differences of evaluated traits.

Cluster	1	2	3	4	5
1	100,00	0,00	0,00	0,00	0,00
2		75,00	25,00	0,00	0,00
3			100,00	0,00	0,00
4				100,00	0,00
5					100,00

These results can be explained by the fact that this breed is been raised in Brazil for only 22 years, which could be too short for the environmental and socioeconomic factors exert any influence on the genetic distribution of the Montana Tropical® herd. Moreover, the use of reproductive biotechnologies, such as artificial insemination (around 80% of the cows), could be masking the environmental and socioeconomic effects, since it allows the transference of genetic material between regions.

In conclusion, there were differences in the genetic merit of animals among the counties where Montana Tropical® beef composite cattle are raised. However, the only environmental factors were significantly different between clusters without a clear influence on the genetic distribution of these animals in Brazil.

REFERENCES

Costa, N.S., Hermuche, P., Cobuci, J.A.; Guimarães, R.F., Carvalho J.O.A., Costa, C.N., Gomes, R.A.T., Mcmanus, C. (2014) *Genetics and Molecular Research (GMR)* **13**:9806.

Hermuche, P., Silva, N.C, Guimarães, R.F., Carvalho J.O.A., Paiva, S.R., Gomes, R.A.T., Mcmanus, C. (2012) *Revista brasileira de cartografia* **64**:821.

Hermuche, P., Guimarães, R.F., Carvalho J.O.A.; Paiva, S.R., Gomes, R.A.T.; Mcmanus, C. (2013) *Applied Geography* **44**:172

Manel, S., Schwartz, M.K., Luikart, G.; Taberlet, P. (2003) *Trends in Ecology and Evolution* **18**:189.

Proc. Assoc. Advmt. Anim. Breed. Genet. 22:517-520

DEVELOPMENT OF A CUSTOM ION AGRISEQ GENOTYPING-BY-SEQUENCING PANEL BASED ON THE ISAG BOVINE CORE PARENTAGE MARKERS

Prasad Siddavatam[1], Angela Burrell[1], Ryan Ferretti[2], Adam Allred[1], Kristian Ridley[3]

[1]Thermo Fisher Scientific, 2130 Woodward Street, Austin, TX, USA, 78744,
[2]Neogen GeneSeek Operations, 4131 N. 48th St., Lincoln, NE 68504,
[3]ThermoFisher Scientific, 5 Caribbean Drive, Scoresby, Victoria, Australia

SUMMARY

Historically, microsatellites have been the most popular genetic feature for distinguishing cattle breeds for the purpose of determining parentage. More recently, SNP genotyping has emerged as a desirable alternative to microsatellite typing. SNPs offer several advantages over microsatellites. Perhaps the most important advantage is that there is less ambiguity in distinguishing SNP alleles in order to confidently provide a genotype call.

AgriSeq™ is a sequencing technology that can be used for targeted amplification and re-sequencing of thousands of SNP targets in a single reaction. The Ion 540™ chip allows hundreds of samples to be genotyped at thousands of loci simultaneously. Ligating a unique barcode to each sample allows samples to be sequenced together in a single run on the Ion S5™ sequencing system.

We developed a targeted sequencing panel based on 200 bovine SNP markers selected by the International Society of Animal Genetics (ISAG) for the purpose of determining parentage. We tested this panel on 96 bovine samples obtained from the USDA representing 19 different breeds. Each sample was tested in duplicate such that 192 libraries were pooled onto a single Ion 540 chip for sequencing. Variant calling was performed using the Torrent Variant Caller (TVC) plugin as part of the Torrent Suite™ software package. Mean call rate for this dataset was 98.5%, indicating that the vast majority of SNPs yielded data of sufficient quality to make a genotype call.

INTRODUCTION

SNPs are well-suited for use as genetic markers for several reasons. Some of the advantages of using SNPs relative to other types of genetic markers are that SNPs occur abundantly in the genome, are generally stable through evolution and have a low mutation rate (1).

SNP genotyping has various applications in agriculture including genetic diagnostics, germplasm identification and genomic selection for breeding purposes (1). Next-generation sequencing allows for rapid and accurate SNP genotyping. This technology, coupled with the specificity of targeted amplification using AgriSeq, enables many samples to be genotyped simultaneously without compromising sensitivity.

Here we apply SNP genotyping for assessment of bovine parentage. Using our targeted sequencing primer design pipeline, we designed primers for the amplification and subsequent sequencing of 200 SNPs related to bovine parentage. The resulting panel was tested on 96 bovine samples representing 19 different breeds of cattle in order to assess call rate and concordance with array-based genotyping methods.

MATERIALS AND METHODS

The bulk of the primers were designed using an automated process that optimizes a number of oligonucleotide properties (GC content, melting temperature, etc.) and amplicon properties (size, centering a SNP within its amplicon, etc.). Furthermore, primers were designed to avoid overlapping nearby SNPs and are subjected to an in-silico quality assessment to ensure specificity within the genome and prevent the formation of undesired PCR products.

Library prep was performed in duplicate for each of 96 samples obtained from the USDA MARC Beef Cattle Diversity Panel v2.9. All 192 libraries were pooled onto a single Ion 540 chip for template prep and sequencing on the Ion Chef™ and Ion S5 XL.

RESULTS AND DISCUSSION

Sequencing yielded over 71 million reads. Over 66 million reads were reliably assigned to one of 192 barcodes (an average of over 347,000 reads per barcode) with >99 percent of bases aligning to the bovine genome (Table 1.)

Table 1. Sequencing summary

Number of samples	96
Number of markers	200 (ISAG)
Sequencing time	2.5 hours
Analysis time	<17 hours
Output file size	10 MB (VCF)
Mean read length	140 bp
Total reads (high quality)	>66 million
Percentage of reads mapped	99.50%

Marker coverage was highly consistent. Mean coverage was 368.6 with 97% of markers falling within one standard deviation of the mean (Figure 1.).

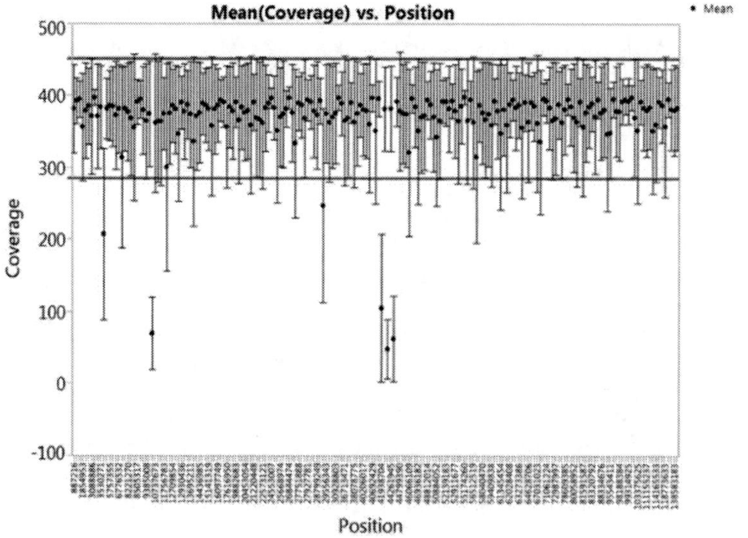

Figure 1. Marker coverage

Average call rate for these samples was 98.5%. Call rates are color-coded by breed and differences between breeds were not found to be statistically significant. Call rate was calculated for each sample as the number of markers for which data quality was high enough to make a genotype call (homozygous reference/homozygous variant/heterozygous), divided by the total number of markers in the panel (200). Seven replicates that had less than 100x coverage were excluded from analysis (Figure 2.)

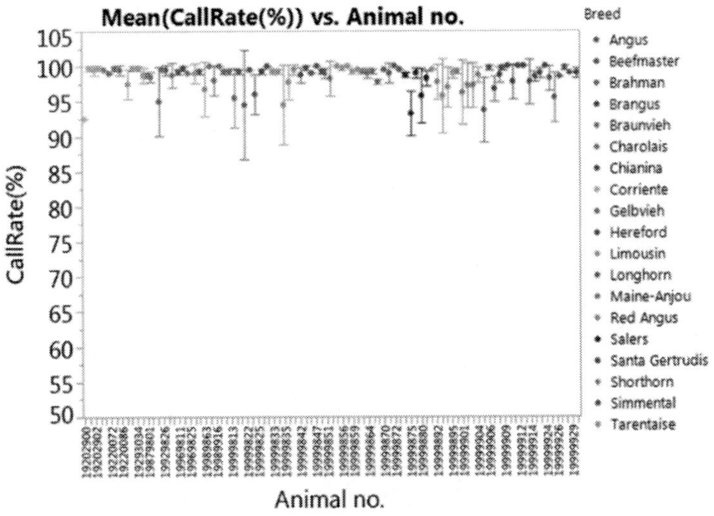

Figure 2. Sample call rates

Average marker call rate was 98.5%. 192 of the 200 markers had call rates >95% and 49 markers had 100% call rates. Only five markers had call rates <90% (Figure 3.).

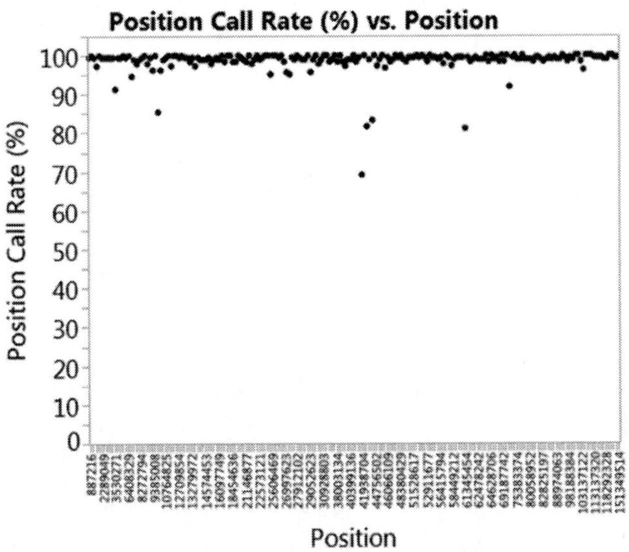

Figure 3. Marker call rates

Further investigation of the five markers that had lower call rates (<90%) revealed that performance for those markers differed greatly depending on the breed (Figure 4.).

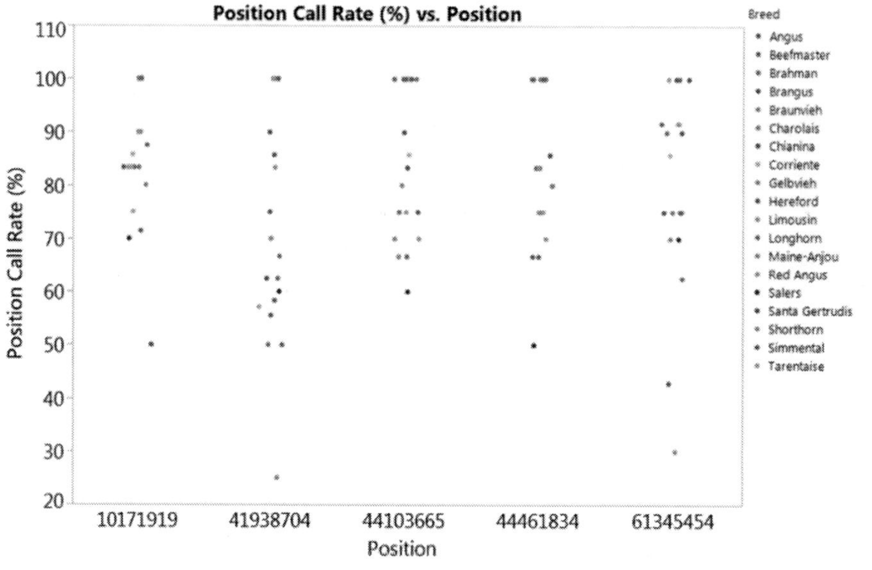

Figure 4. Markers with lower call rates show breed-specific differences in performance

Samples were hybridized to six Illumina arrays in order to obtain consensus genotype calls for the array data. Concordance was calculated as the number of times the genotype call matched between samples run on the two different technologies divided by the total number of calls. Seven replicates that had less than 100x coverage were excluded from analysis (Table 2.).

Table 2. Concordance with array data

Samples included in analysis (>100x coverage)	89
Total number of calls	36399
Number of concordant calls	35433
Concordance (%)	97.3

CONCLUSIONS

We developed a high-performing, high-throughput method for genotyping hundreds of bovine samples in a single sequencing run at hundreds of SNPs. Our method yields calls for the vast majority of markers (98.5% on average). These calls were highly concordant with array data (97.3%). While we demonstrated the utility of Ion Torrent sequencing technology for genotyping parentage markers in cattle, our approach can also be applied to other SNP genotyping problems.

REFERENCES

Patel DA, Zander M, Dalton-Morgan J, Batley J. (2015) Methods Mol Biol. 1245:1-11.

Proc. Assoc. Advmt. Anim. Breed. Genet. 22:521-524

THE USE OF MID-INFRARED SPECTROMETRY TO PREDICT MILK FATTY ACID, ENERGY BALANCE AND METHAN EMISSIONS FOR AUSTRALIAN DAIRY COWS

T. Wang[1], H.N. Phuong[1], E. Wall[2], S. Smith[2] and J.E. Pryce[1,3]

[1] Department of Economic Development, Jobs, Transport and Resources, VIC 3083, Australia
[2] Scotland's Rural College, Edinburgh, UK
[3] School of Applied Systems Biology, La Trobe University, Bundoora, VIC 3083, Australia

SUMMARY

The aim of the research was to develop prediction tools to help farmers select and manage their cows using mid-infra-red (MIR) spectroscopy. The prediction performance is evaluated on a range of complex phenotypes of energy balance, methane emissions and milk fatty acids using data from 120 Australian cows. Additionally, MIR prediction equations derived from the UK cows is also explored on the same Australia data. The accuracy (r2) of predicting energy balance using MIR from Australian cows was around 0.4, which is similar to the performance of applying the UK MIR prediction equations to Australian data of 0.39. The prediction accuracies of methane emissions and milk fatty acids were greater than 0.5. The next phase will focus on improving the accuracy and validating them against data from commercial populations.

INTRODUCTION

Mid-infrared spectral data are measures of the absorption of infrared rays at frequencies correlated to the vibration of specific chemical bonds within a molecule (Soyeurt et al., 2011). MIR prediction uses the absorbance of mid-infrared light through milk samples over a range of wavelengths to predict a given phenotype. To calculate a prediction equation that can be applied nationally, requires a reference population that has measurements of the phenotype of interest and MIR spectral data collected at the same time. The advantage of using MIR to predict these phenotypes is that the turnaround time back to farm can be fast and at little extra cost over standard milk tests, enabling reactive management decisions. To date, MIR prediction analysis has been applied to detailed milk fat and protein composition with promising prediction accuracies (Soyeurt et al., 2011). Furthermore, it has also been used to predict complex phenotypes, such as energy balance (McParland et al., 2011) and methane emissions (Dehareng et al., 2012).

As cows mobilise body fat in early lactation to sustain lactation (when feed requirements exceed intake), it is likely that special signatures of fatty acid composition are also observed in milk. There are several milk fatty acids in milk and having a greater understanding of how these are associated with energy balance may help to improve MIR predictions and could be worth breeding for in their own right.

The objective of this study was to predict a range of traits including several milk fatty acids, energy balance and methane emissions using MIR spectral data from a research herd in Victoria. In addition, a MIR prediction equation for energy balance developed using UK data and available commercially by National Milk Records (a UK milk recording organisation) was validated using this Australian data.

MATERIALS

Phenotypes and associated MIR spectral data were available from an experiment that ran from October to December 2015 consisting of 120 Australian Holstein lactating cows that calved in the spring at the research farm of the Department of Economic Development, Jobs, Transport, and Resources (DEDJTR) in Ellinbank, Victoria, Australia. The cows were divided into three batches of 40 cows to match the availability of automated feed intake equipment. Batches were balanced for days in milk (DIM), number of lactations and body weight. Cows had access to feed and water ad libitum with the diet consisting of cubes that were approximately 74% alfalfa hay, 25% crushed barley grain, and 1% minerals. The experimental duration was 32 days and the following measurements were performed: feed intake (for every meal), body weight after morning milking, body condition score 2x per week, milk yield (2x daily), milk fat percent, milk protein percent, and milk lactose percent 3 days per week at both morning and afternoon milking. MIR spectral data for each milking were obtained using a Bentley FTS instrument.

Traits. At 87-124 days in lactation, the records of 32 different types of fatty acids including saturated FA (C4:0 – C20:0), Saturated FA (10:1-18:1), Mono-unsaturated FA (18:2, 18:3) , and Trans FA (CLA) were collected for each sample twice across lactation. Across all the FA, Mahalanobis distance was calculated. Outlier samples and absorbance (with Mahalanobis distances >3) were removed.

Two equations obtained from Phuong et al (2016) were used to calculate energy balance (EB) from feed trial data. The first equation was energy intake minus energy output (EBinout), estimated using smoothed energy intake minus the energy required for maintenance, milk yield, and activity. The second equation was applied to calculate EB using milk composition only (Friggens et al., 2007).

Methane emissions were measured for each cow over a 5 day period that occurred within the 32d experiment using the SF6 tracer method of Deighton et al. (2014). Three phenotypes were calculated: 1) AvCH4, which was the mean of total methane emissions over 5 day period; 2) AvCH4yield, which was total methane yield divided by the actual feed intake over 5 day period; 3) AvCH4Intense, which was total methane yield divided by milk yield over the 5 day period.

Mid-infra-red spectral data. The MIR spectrum for each milk sample had 899 data points (absorbance) for wavelengths ranging from 649 to 3998 cm^{-1}. Using approaches developed by Grelet *et al*., (2015), several pre-processing steps including removal of outliers, standardizing, smoothing, and noise removing were applied to the raw MIR data. As reported by Hewavitharana and van Brakel (1997), and De Marchi *et al.* (2012), two spectral regions (from 1603 to 1682 cm^{-1}; from 3006 to 3998 cm^{-1}) are either water absorbance or useless chemical information, these spectra were removed leaving 620 wavelengths for analysis.

METHODS

Partial least squares regression method. Partial least squares regression method is commonly used in the analysis of MIR spectral data (De Marchi *et al*., 2012). An example is the SimPLS algorithm in R (R Development Core team, 2010) which was implemented for MIR prediction. To avoid over-fitting problems, 10 fold cross validation (10fold-CV) was used to assess the accuracy of the MIR prediction. The data sets were randomly separated into 10 subsets, and then each subset wass treated as a validation set while the other nine sets were combined as reference set. The accuracy was assessed as the coefficient of determination ($\mathbf{r^2}$) calculated as the mean of 10 fold cross validation applied to prediction equations.

RESULTS AND DISCUSSION

We found a considerable number of significant correlations between energy balance and milk fatty acids, especially C:18 and C:20 (Figure 1). The high proportion of C:18 and C:20 during periods of negative energy balance is related to a high uptake of long chain fatty acids released from the mobilisation of body fat reserves (Bastin et al., 2011). Most of the correlations between energy balance and milk fatty acids were quite small. An explanation for these findings is the late lactation stage when the cows participated in the experiment. In the current experiment, days in milk of the cows varied from 87 to 124 which were >12 weeks where milk fatty acids would change very little or even remain constant.

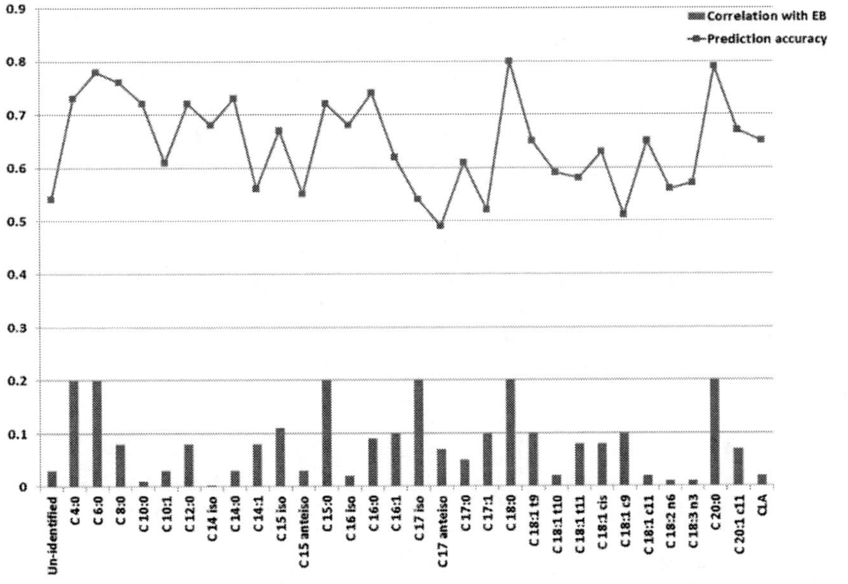

Figure 1. Pearson correlations between milk fatty acids and energy balance derived using individual cow data (blue bars) and the prediction accuracy using MIR data on the Fatty acids (red curves).

As shown in Figure 1, the accuracy of MIR prediction of milk fatty acid traits as determined by the coefficient of determination is higher than 0.50. For some fatty acids, for example, C4:0-C19:0, the prediction accuracy reached around 0.80.

Table 1. MIR prediction of two energy balance traits (predicted as the difference between energy intake and output; EBinout; energy balance using an equation applied to milk production data; EBalMilk) and three methane emission traits.

Phenotypes		Country	r^{2*}	RMSE
Energy Balance	EBinout	AU	0.42	31.27
	EBinout	UK	0.39	-
	EBalMilk	AU	0.45	31.05
Methane Emission	AvCH4	AU	0.51	21.48
	AvCH4yield	AU	0.49	31.09
	AvCH4Intense	AU	0.52	20.45

*The accuracy was assessed as the coefficient of determination (r^2) and root mean square error of calibration (RMSE) calculated as the mean of 10 fold cross validation applied to prediction equations developed using Australian (AU) or UK reference datasets.

The prediction accuracy of energy balance and methane emissions using MIR is shown in Table 1. The prediction accuracy of EBinout is 0.42 with the root mean square error of calibration (RMSE) of 31.27. For the same trait, using the equation derived from the UK (estimated with a much larger reference dataset up to thousands of animals and a similar energy balance trait), we observed a similar accuracy (around 0.39). Similarly, the prediction accuracy of energy balance based on milk composition is around 0.45. The reasons that the accuracy in Australia was lower than observed in studies such as McParland et al (2011) include the fact that the reference population was not optimised for Australia; the phenotypes were subtly different to the energy balance used for the UK and it is possible that genotype by environment interactions exist.

Similarly, the accuracy of MIR prediction of the three methane emission traits ranges from 0.49 to 0.52 with the values of RMSE ranging between 20.45 and 31.09.

All the above results show that the accuracy of MIR prediction is currently lower than other comparable studies. The most likely explanation is that most previous studies have much larger reference size. For example, McParland et al (2011) had a large reference population of 6,665 test days from 465 lactations of 277 cows matched to MIR spectra. Our reference population in comparison is small. Therefore, the strategy of improving the prediction accuracy is to increase the reference population by expanding the number of phenotypes from research herds, or devising ways in which to measure energy balance on commercial farms.

CONCLUSION

Our analyses show MIR prediction is promising but needs further improvement. In the next phase, we are investigating ways in which more energy balance phenotypes can be collected from research and commercial dairy herds.

ACKNOWLEDGEMENTS

This research belongs to the project "MIRprofit: integrating very large genomic and milk mid-infrared data to improve profitability of dairy cows" funded by the Commonwealth of Australia.

REFERENCES

Soyeurt H., Dehareng F., Gengler N., McParland S., Wall E., Berry D.P., Coffey M. and Dardenne P. (2011). J. Dairy Sci. 94:1657-67.

McParland S., Banos G., Wall E., Coffey M.P., Soyeurt H., Veerkamp R. F. and Berry D.P. (2011). Journal of Dairy Science. 94:3651-3661.

Dehareng F., Delfosse C., Froidmont E., Soyeurt H., Martin C., Gengler N., Vanlierde A. and Dardenne P. (2012). Animal. 6(10):1694-1701.

Phuong H.N., Haile-Mariam M., Garner J.B., Marett L.C., Wales W.J. and Pryce J.E. (Submitted.)

Friggens N.C., Ridder C.and Løvendahl P.. (2007). *J. Dairy Sci.* **90**:5453-5467.

Deighton M.H., Williams S.R.O., Hannah M.C., Eckard R.J., Boland T.M., Wales W.J., and Moate P.J. (2014). Animal Feed Science and Technology. 197:47-63.

Grelet C., Fernández Pierna J.A., Dardenne P., Baeten V. and Dehareng F. (2015). *J. Dairy Sci.* 98:2150–2160.

Hewavitharana, A.K. and van Brakel B. (1997). *Analyst (Lond.)*. **122**:701–704.

De Marchi, M., Toffanin V., Cassandro M., and Penasa M.. (2013). *J. Dairy Sci.* **96**:4707–4715.

Bastin, C., Gengler, N. and Soyeurt, H. (2011). *J. Dairy Sci.* **94**:4152–4163.

R Development Core Team. (2010). Vienna, Austria:R Foundation for statistical computing. Retrieved from http://www.R-project.org.

Proc. Assoc. Advmt. Anim. Breed. Genet. 22:525-528

SOMATIC CELL COUNT AND MILK UREA NITROGEN LEVELS IN HOLSTEIN AND FLECKVIEH X HOLSTEIN COWS IN A TOTAL MIXED RATION FEEDING SYSTEM

C.J.C.Muller[1,2], L. Metaxas[2,3] and S.W.P. Cloete[1,2]

[1]Directorate Animal Sciences, Western Cape Department of Agriculture, Elsenburg 7607, South Africa. [2]Department of Animal Sciences, University of Stellenbosch, Stellenbosch 7602, South Africa. [3]Free State Department of Agriculture, Glen Agricultural College, Glen 9305, South Africa

SUMMARY

Mastitis in dairy cows affects milk yield, welfare and production efficiency. In this paper the somatic cell count (SCC), milk urea nitrogen (MUN) and lactose percentage of 41 Holstein (H) and 30 Fleckvieh x Holstein (FxH) cows were compared using 1688 repeated test-day records. Production parameters did not differ between breeds except for fat and protein percentages being higher in FxH cows. Log transformed SCC did not differ between breeds. However, log transformed SCC was repeatable at 0.21 and should respond to current herd selection. Reducing SCC concentrations in milk would improve welfare of cows as cull rates for mastitis are reduced.

INTRODUCTION

Mastitis is defined as an inflammatory reaction of the udder tissue to bacterial infection. It is one of the most common diseases in dairy cows. The somatic cell count (SCC) of healthy milk is below 200,000 cells/ml of milk with an increase indicative of an infection (Robertson 2016). Mastitis is next to fertility one of the main reasons for cows being culled. Crossbreeding is gaining popularity worldwide as crossbred cows seem to be more robust (Weigel & Barlas, 2003). One aspect of robustness is the ability of cows to withstand developing mastitis under farming conditions. Heins *et al.* (2011) found that breeds differed for SCC showing that Montbéliarde x Holstein and Scandinavian Red x Holstein cows had lower ($P<0.01$) SCCs than Holstein and Normande x Holstein cows. Milk yield between these breeds also differed ($P<0.01$) with Holsteins producing the most milk. Somatic cell score (SCS) also increased from first to fifth lactation being 2.73 vs. 4.02 for Holsteins. Montbéliarde x Holstein cows were superior to the other breed groups across lactations for SCS. Prendeville *et al.* (2010) found that even though milk yield differed, udder health (SCS and the incidence of mastitis at least once per lactation) did not differ between Holstein-Friesian, Jersey and Holstein-Friesian x Jersey cows under grazing conditions. The total incidence of mastitis (accounting for repeated incidences) were higher for Jersey cows in comparison to Holstein-Friesian cows, being 1.54 vs. 1.24. Washburn *et al.* (2002) and Berry *et al.* (2007) found that the prevalence of mastitis was higher for Holstein and Holstein-Friesian (HF) cows in comparison to Jersey cows. In South Africa, dairy farmers in pasture-based systems, have attempted crossbreeding using the Fleckvieh breed, a Simmental derived dual-purpose breed from Germany. Muller *et al.* (2009) and Metaxas *et al.* (2014) have shown better fertility and higher fat and protein percentages in FxH cows in comparison to H cows. Farmers perceive a lower incidence of mastitis in Fleckvieh crossbred cows. These claims have not been tested in previous research. The aim of this study is thus to compare the SCC, mastitis incidence, MUN levels in milk of H and FxH cows in a total mixed ration feeding system.

MATERIALS AND METHODS

Data. The study was conducted at the Elsenburg Research Farm of the Western Cape Department of Agriculture. The area has a typical Mediterranean climate with short, cool, wet winters and long, warm, dry summers with an average annual rainfall of 650 mm. Milk production data of H and FxH cows in a zero-grazing system were collected over six years between 2008 and 2013. Cows from both breeds were kept together in a dry lot (open camp) with a fence-line feeding trough. Cows were fed a total mixed ration (TMR) providing 17% CP and 11 MJ ME/kg DM. Feeding was twice a day at *ad libitum* levels, i.e. orts not exceeding 5% of feed provided. Fresh drinking water was freely available. Cows were machine-milked twice a day in a milking parlour about 500m from the dry lot. The milk yield of cows at the evening and following morning's milking was recorded approximately every 35 days during the lactation period. Each cow had at least three and a maximum of nine milk recording events per lactation. At each milk production recording event, milk samples were collected at both evening and morning milking sessions. Samples were combined and analysed at the milk testing laboratory of the National Milk Recording Scheme for their fat, protein and lactose concentrations as well as SCC and milk urea nitrogen (MUN) of each sample.

Statistical analyses. Repeated test-day records (n=1688) of cross-bred (50% Fleckvieh) cows (n=30) were grouped together and compared to H cows (n = 41). Fixed effects fitted in ASReml included parity (1 to 5), genetic group (FxH or H), year (2008-2013) and the genetic group x year interaction. Days in milk were fitted as a fixed linear component as well as random cubic spline components to model deviations from linearity following a smooth trend (Gilmour *et al.* 2006). Random animal models were included to account for the repeated sampling of individual cows.

RESULTS AND DISCUSSION

Until recently, breeding programmes have put more emphasis on milk production performance without considering functional traits (Walsh *et al.*, 2009). The effect of breed (genotype) on udder health has been mostly comparing Holstein, Jersey and Jersey x Holstein cows. Although the Fleckvieh breed is the second largest dairy breed in the world, dairy farmers are not familiar with the breed, probably because of the breed's more pronounced dual-purpose characteristics. For this reason, crossbreeding studies in the USA and Ireland have used the Montbéliarde breed, a Simmental derived breed from France which shows more explicit dairy characteristics. Descriptive statistics of milk production parameters for both H and FxH cows are presented in Table 1. The coefficients of variation (CV) for production traits were in accordance with similar data. As expected, SCCs varied greatly, the appropriate CV being 213%. This is because of cows with mastitis showing extremely high SCCs. The repeatability of traits ranged from 0.02 for MUN to 0.25 for lactose percentage. All traits, except MUN, seem likely to respond to selective breeding in the current herd. Considering the relatively small sample size, it is pleasing to see that most repeatability estimates were significant ($P<0.05$), i.e. above twice the appropriate standard error.

Table 1. Descriptive statistics for the traits analysed on test day records (n=1688) for milk production traits, somatic cell count (SCC) and milk urea nitrogen (MUN), as well as the repeatability of the respective traits

Trait	Mean ± s.d.	Range	Repeatability ± s.e.
Milk yield (kg)	21.3±6.4	2.7 – 59.7	0.19±0.03
Fat (%)	4.17±0.56	2.61 – 6.53	0.16±0.03
Protein (%)	3.33±0.38	2.37 – 4.87	0.23±0.04
Lactose (%)	4.78±0.23	3.25 – 5.45	0.25±0.04
Untransformed SCC	372±795	3 – 9,233	0.21±0.04
MUN	15.5±4.8	5.9 – 34.4	0.02±0.01

Breed differences were observed, with fat and protein percentages being higher ($P<0.01$) for FxH vs. H cows (Table 2). Other traits did not differ between breed combinations. These results are in accordance with those of Metaxas *et al.* (2014). The distribution of SCCs did not differ ($P>0.05$)

for H and FxH cows over all lactation periods and lactation stages (Figure 1). Most, 63 and 67% of all H and FxH SCC records, respectively, had less than 200,000 cell/ml of milk. High SCC (>600,000 cells/ml milk) was recorded in 20 and 18% of all H and FxH records, respectively.

Table 2. Least-squares means (±s.e.) depicting differences between Holsteins (H) and Fleckvieh x Holsteins (FxH) cows for test-day milk yield (MY), fat percentage (BF), protein percentage (PP), lactose percentage (LP), the log of somatic cell count (SCC) and milk urea nitrogen (MUN) recorded either in the autumn or spring

Effect	Trait					
and level	MY (kg)	BF (%)	PP (%)	LP (%)	SCC	MUN (mg/dL)
Breed	0.54	**	**	0.97	0.54	*
FxH	21.6±0.6	4.26±0.06	3.35±0.04	4.71±0.03	5.08±0.15 (161)	16.1±0.03
H	22.1±0.6	4.08±0.06	3.22±0.04	4.71±0.03	5.17±0.15 (175)	15.5±0.03

* P<0.01; ** P<0.01; Actual significance for P>0.05. Geometric means for SCC are in brackets

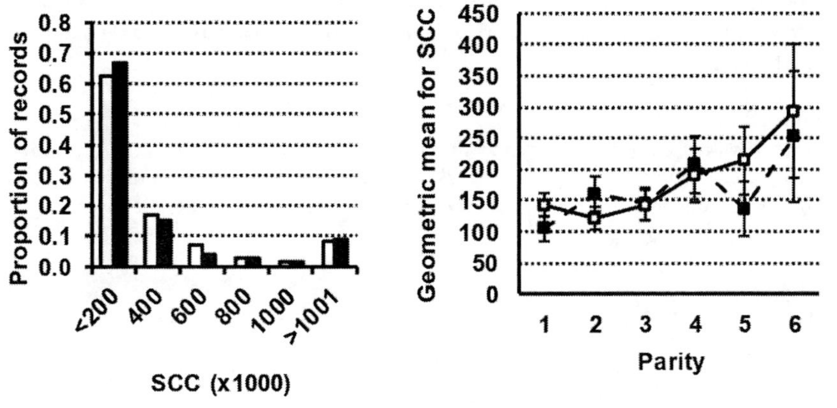

(a) (b)

Figure 1. The distribution of somatic cell count (SCC) records within categories (a) and geometric means (±s.e.) for SCC or Holstein (□) and Fleckvieh x Holstein cows (■) across parities (b)

Walsh *et al.* (2007) noted that the production of cows within a feeding system is a function of their genetic merit and environmental effects. According to Mrode & Swanson (1996) milk yield is positively correlated with SCC. Significant differences between breeds for SCC are thus expected for breeds differing in milk yield. In the present study, the milk yield of H and FxH cows did not differ (P>0.05) reflecting small differences in the average SCC and, theoretically, the number of mastitis cases. The correlation between animal effects for milk yield and for SCC was accordingly small and not significant at 0.01±0.16 in the present study. Washburn *et al.* (2002) found that high-producing HF cows had higher SCCs than Jerseys.

The MUN levels in milk can be used to assess the protein and energy status of cows. High levels (greater than 18 mg/dl of milk) indicate a diet containing high levels of easily degradable protein sources (pasture containing high levels of CP), low fermentable energy levels in the diet, high milk yield levels as well as breed. Jersey cows seem to have lower MUN levels in comparison to Holstein cows (Johnson & Young, 2003). Contrary to these results, Wattiaux *et al.* (2005) found higher test-day MUN concentrations for Jersey and Brown Swiss cows in comparison to Holstein cows on the same diet. Miglior *et al.* (2006) found that the MUN concentration in Ayrshire milk was higher than in Holstein milk.

Kgole (2013) found that non-genetic factors affecting MUN in Holstein cows were herd-test-day, lactation stage and year of calving. Herd-test-day contributed most to the observed variation in the latter study, namely 58.6 and 63.2% in parity 1 and 3, respectively. The heritability estimate for MUN was 0.09±0.01 in first parity and 0.11±0.01 in 2nd and 3rd parities. Between-animal variation in the present study was accordingly low, indicating that factors other than the animal contribute substantially to variation in MUN. Genetic correlations between MUN and milk production traits were positive, albeit low, ranging from 0.01±0.00 to 0.10±0.004 across parities (Kgole 2013). This positive association is undesirable, indicating that high-producing cows are less efficient in utilizing dietary protein.

CONCLUSION

FxH cows outperformed H cows for fat and protein percentages with no observed difference in milk yield. Significant between-animal variation suggests that current herd gains are feasible for SCC in the cows studied. With repeatability being the theoretical upper limit of heritability, these results may suggest underlying genetic variation among cows which may be exploited by selection, thus benefitting the welfare of lactating cows.

ACKNOWLEDGEMENTS

This research was partly funded by the World Simmental Fleckvieh Federation, South African Simmentaler Breed Society, Western Cape Department of Agriculture and the Western Cape Agricultural Research Trust. The inputs of the dairy management team at the Elsenburg Research Dairy are greatly appreciated.

REFERENCES

Berry D.P., Lee J.M., Macdonald K.A., Stafford K., Matthews L. and Roche J.R. (2007) *J. Dairy Sci.* **90**: 637.

Gilmour A.R., Gogel B.J., Cullis B.R. and Thompson R. (2006) ASREML - User Guide, Release 2.0. VSN International Ltd, Hemel Hempstead, HP11ES, UK.

Heins B.J. and Hansen L.B. (2011) *J. Dairy Sci.* **95**: 918.

Kgole M.L. (2013) MSc Dissertation. University of Pretoria. South Africa.

Metaxas L., Muller C.J.C., Dzama K. and Botha J.A. (2014) *47th SASAS Congress*. p. 78.

Miglior F., Sewalem A., Jamrozik J., Lefebvre D.M. and Mooré R.K. (2006) *J. Dairy Sci.* **89**: 4886.

Mrode R.A. and Swanson G.J.T. (1996) *Anim Breeding*. Abstr. **66**: 847.

Muller C.J.C., Botha J.A., Cloete S.W.P. and Potgieter J.P. (2009) *Proc. Assoc. Advmt. Anim. Breed. Genet.* **17**: 616.

Prendiville R., Pierce K.M. and Buckley F. (2010) *J. Dairy Sci.* **93**: 2741.

Robertson N.R. (2016). Section 5. Milk production and quality. *Dairy Farming Manual*.Chapt. 52.

Walsh S., Buckley F., Berry D.P., Rath M., Pierce K., Byrne N. And Dillon P. (2009) *J. Dairy Sci.* **90**: 5769.

Washburn S.P., White S.L., Green Jr. J.T. and Benson G.A. (2002) *J. Dairy Sci.* **85**: 105.

Wattiaux M.A., Nordheim E.V. and Crump P. (2005). *J. Dairy Sci.* **88**: 3020.

Weigel K.A. and Barlass K.A. (2003) *J. Dairy Sci.* **86**: 4148.

Proc. Assoc. Advmt. Anim. Breed. Genet. 22:529-532

GENETIC PARAMETERS OF PUBERTY ESTIMATED USING TWO GENETICALLY DIVERGENT GROUPS OF HOLSTEIN-FRIESIAN DAIRY HEIFERS

M. D. Price, M. D. Camara, J. R. Bryant, S. Meier and C. R. Burke

DairyNZ Limited, Private Bag 3221, Hamilton, New Zealand

SUMMARY

Genetic parameters for age at puberty were estimated for Holstein-Friesian dairy heifers in an experimental herd comprised of genetically divergent lines for fertility. Despite the non-random population structure, the estimated heritability of age at puberty was approximately 13%, based on behavioural observations, and between 41% and 76% based on blood progesterone levels. Although the phenotypic correlation between these two different measures was moderate (~0.5), the genetic correlation was much stronger (~0.9). In addition, the genetic correlations between age at puberty traits and fertility BV was ~-0.3, which suggests that age at puberty estimates may aid in the genetic evaluation of lowly heritable fertility traits.

INTRODUCTION

Previous studies of the onset of puberty, defined as age at first behavioural oestrus, estimated its heritability as approximately 0.27 in NZ beef cattle (Morris *et al.* 2000; Amyes and Morris 2009), and Martin *et al.* (1992) reported an average of 0.4 across nine beef cattle studies (ranging from 0.07 to 0.67). Fewer studies have reported the heritability of age at puberty in dairy cattle, but one available estimate was 0.09 (Morris and Hickey 2004). While the onset of puberty itself is an important trait for herd management purposes, its influence on reproductive traits is also of interest since genetic correlations may aid in the genetic improvement of lowly heritable fertility. In the NZ dairy industry, the fertility breeding value (BV) is comprised of several distinct traits such as PM21 (presented for mating within 21 days of planned start of mating; h^2=0.05) and CR42 (calving rate in 1st 42 days after planned start of calving; h^2=0.03). In NZ beef cattle, favourable genetic correlations have been reported between heifer age at puberty and several reproductive traits, such as scrotal circumference in NZ beef bulls (-0.25), pregnancy rate (-0.23) and calving date (0.57) (Morris and Amyes 2010).

Oestrus onset may instead be directly measured and defined as the age when blood progesterone (P4) concentration has reached a certain threshold. McNaughton *et al.* (2005) used a criterion of P4 >1ng/mL for 2 of 3 consecutive weekly samples in NZ dairy heifers. Age at puberty determined in this way may have an advantage over behavioural measurements due to its quantitative accuracy, and may thus provide a better estimate of heritability. No heritability estimate for P4-based age at puberty in dairy heifers is currently known.

The objective of this study was to estimate genetic parameters of several age at puberty (AP) traits in dairy cattle, including their heritabilities and genetic correlations with fertility BVs. These AP traits would be determined via either observational oestrus or several P4-based criteria.

The data used was from a physiological study of NZ dairy heifers in which the experimental herd was genetically divergent on fertility (Meier *et al.* 2017). This divergence could introduce bias into any genetic analyses, so an auxiliary objective of this study was to evaluate a basic method to account for this.

MATERIALS AND METHODS

Study animals. The study population consisted of 527 Holstein-Friesian heifers born across 379 herds between June and September 2015 and produced by mating low or high fertility BV dams and sires to generate divergent genotypes (Low BV heifers: n=252, μ=-5.12, σ=1.37; High BV heifers:

n=275, μ=5.00, σ=0.74). These heifers were reared in four mobs, each consisting of a mixture of high and low BV heifers, with the ratio per mob being no more than 40:60 either way.

Data. From April to November 2016, P4 was determined in weekly collected plasma samples. Tail-paint or heat mount detectors (Kamar Inc., Zionsville, IN USA) were used, and checked approximately weekly to quantify mounting activity associated with expression of oestrus. From these data, we defined six potential indicators of age at puberty: (AP1) when P4 first reached 1.0 ng/mL; (AP2) when P4 first reached 1.0 ng/mL for 2 consecutive weekly samplings; (AP3) when P4 first reached 1.0 ng/mL for 2 of 3 consecutive weekly samplings; (AP4) when P4 first reached 0.7 ng/mL; (AP5) when P4 first reached 0.7 ng/mL for 2 consecutive weekly samplings; and, (APK) when either the heat mount detector was activated, tail-paint was mostly worn off (scored 1 on a 1-5 rubbing scale; 5 being as new), or the animal was visibly in oestrus.

The full pedigree of these animals was extracted from the New Zealand Dairy Industry Good Animal Database (DIGAD), which consisted of 10,992 records, up to 18 generations deep. Also extracted from DIGAD were the fertility BVs for the 527 heifers estimated in the most recent national animal evaluation (January 2017). Although these BVs are already genetic estimates based on pedigree linkages rather than own data, they were used as a response variable in the models, and are referred to as the "fertility BV trait" in this study.

Variance component estimation. The data came from a population that is genetically divergent for fertility and, therefore, not normally distributed for this trait. Because fertility is likely to be genetically correlated with puberty, the herd is likely to be genetically divergent (and, therefore, not normally distributed) for puberty also, and so the genetic variance components and heritability of puberty will be overestimated by a standard univariate mixed model. To account for this, a two-model approach was used: the first (standard) model included mob as the only fixed effect; the second model included fixed effects for both mob and fertility group (low or high). Both models included a pedigree-based random animal effect. The second model, by absorbing some of the puberty variation into the fertility group effect, will underestimate the variance components for puberty. Consequently, the first and second models provide upper and lower bounds for the genetic variance of age at puberty, and thus a range for heritability. However, due to the potential for some confounding of mob effect with fertility group (as mobs were not exactly balanced), the upper heritability bound may still contain some downwards bias. If there is no genetic correlation between fertility and puberty, these upper and lower bounds ought to be approximately the same.

Bivariate mixed models (having fixed mob effect and random animal effect) were fitted with pairs of puberty definitions as response variables to estimate both the phenotypic and genetic covariances and correlations between them. Assuming linearity of genetic covariance between the fertility and age at puberty traits, the divergent population structure will not affect genetic covariance estimates, and so standard bivariate mixed models (without fertility group effect) were used.

In order to estimate genetic correlations between fertility and puberty traits, a Pearson correlation was used, in which puberty genetic variance was estimated from a (standard) univariate mixed model and fertility genetic (co)variances were estimated from a bivariate fixed effect model (mob effect; no random effect), with fertility BV and a puberty trait as response variables. As the fertility BVs are already genetic estimates, the residual variance in fertility and residual covariance from the bivariate fixed effect model are estimates of the genetic (co)variances of fertility. Standard errors are not readily available for this genetic correlation, but they ought to be of a similar magnitude to those of the other genetic correlations.

ASReml (Gilmour *et al.* 2015) was used to perform all model analyses.

RESULTS AND DISCUSSION

The lower and upper bounds for heritability estimates of the six defined puberty traits and their standard errors are presented in Table 1. Generally, the difference between lower and upper estimates was about 10% for P4-based puberty, although the high standard errors may indicate that this range is larger than calculated. Nonetheless, the two-model approach does seem to provide a useful range of potential heritabilities. Observation-based puberty (APK) unexpectedly had an upper bound which was slightly lower than its lower bound, although given the large standard errors, this is not a significant anomaly. The fact that these two heritability bounds are almost identical indicates, initially at least, a weak genetic correlation between APK and fertility.

Table 1. Upper and lower bounds of heritability estimates of puberty traits (± s.e.)

	AP1	AP2	AP3	AP4	AP5	APK
h^2_{Upper}	0.63 ± 0.17	0.76 ± 0.19	0.71 ± 0.18	0.53 ± 0.16	0.76 ± 0.19	0.13 ± 0.10
h^2_{Lower}	0.49 ± 0.16	0.66 ± 0.19	0.62 ± 0.18	0.41 ± 0.15	0.67 ± 0.19	0.14 ± 0.10

The heritability of P4-based puberty (AP1-AP5) was moderate/high (41%-76%), whereas the heritability of APK was low (~13%). The low APK heritability is consistent with previous findings (Morris and Hickey 2004), and the large difference in heritability estimates between these two types of puberty measures is likely due mostly to higher measurement error in APK. The heritability of P4-based puberty relying on when P4 first reached 1.0 or 0.7 ng/mL (i.e. AP1 or AP4) was lower than those utilising consecutive P4 data (i.e. AP2, AP3 or AP5), possibly due to a higher prevalence of initial false positives. Defining puberty onset using a 2 of 3 criterion (AP3) reduced heritability by 4-5% compared with the first of 2 consecutive weeks with elevated P4 (AP2), which may mean that this AP3 estimate has a similar problem to AP1 and AP4. Using the more sensitive P4 threshold of 0.7 ng/mL did not change heritability when considering 2 consecutive detections (AP5), although when considering the first detection only (AP4), this more sensitive measure had an even lower heritability than the AP1 estimate.

Table 2. Genetic correlations (below diagonal) and phenotypic correlations (above diagonal) between age-at-puberty (AP) and fertility traits, derived from bivariate models (± s.e.)

	AP1	AP2	AP3	AP4	AP5	APK
AP1		0.972 ± 0.003	0.972 ± 0.003	0.969 ± 0.003	0.972 ± 0.003	0.485 ± 0.037
AP2	1.000 ± 0.003		*	0.943 ± 0.006	*	0.505 ± 0.037
AP3	1.000 ± 0.003	*		0.946 ± 0.005	*	0.514 ± 0.035
AP4	1.000 ± 0.004	1.000 ± 0.007	1.000 ± 0.006		0.947 ± 0.005	0.477 ± 0.037
AP5	0.999 ± 0.003	*	*	0.999 ± 0.007		0.500 ± 0.037
APK	0.830 ± 0.217	0.925 ± 0.217	0.888 ± 0.205	0.770 ± 0.244	0.896 ± 0.200	
Fertility†	-0.333	-0.285	-0.295	-0.335	-0.279	-0.252

* Log-likelihood failed to converge
† Standard errors not available for fertility genetic correlations

Genetic and phenotypic correlations between puberty traits, and their genetic correlations with fertility BV are presented in Table 2. Correlations between P4-based puberty traits (AP1-AP5) are close to 1, as expected, with phenotypic correlations slightly lower than genetic correlations. Phenotypic correlations between APK and AP1-AP5 are moderate at ~0.5, but the genetic correlations are high at 0.77-0.92. Although P4-based puberty and observation-based puberty ought to be genetically the same or strongly associated traits, P4 fluctuations may occur prior to *corpus luteum* (CL) formation, thus resulting in false positives. As previously discussed, the AP2 trait ought to be the best measure for avoiding false positives, and the fact that it has the highest genetic correlation with APK indicates that it is succeeding in doing so, and is thus a good measure of the genetics of puberty onset (i.e. oestrus).

Genetic correlations between fertility and puberty traits are consistent at approximately -0.3, including that of APK, despite this latter measure having similar lower and upper heritability bounds. This low/moderate correlation agrees with previous findings (Morris and Amyes 2010). Of note is that this correlation is stronger for the P4-based puberty traits which are more sensitive to early P4 levels (i.e. AP1 or AP4), indicating that the fertility BV trait may have a stronger genetic association with P4 increase in general than with the stable P4 increase at oestrus.

CONCLUSIONS

Although the data is limited in size and divergent in nature, the analysis undertaken here has yielded useful preliminary results which are consistent with current literature. Age at puberty as determined via behavioural oestrus observations has a low heritability and a moderate genetic correlation with fertility. Age at puberty as determined via blood P4 measurement has a high heritability; a moderate genetic correlation with fertility; and, given appropriate threshold criteria, appears to be able to capture the genetic signal of CL formation associated with puberty onset, as behavioural oestrus does. P4-based puberty may thus be of good use for the genetic improvement of puberty and/or fertility. Although P4 measurement may have limited practicality on a national scale under traditional selection, it may be quite feasible within the reference population of a genomic selection scheme. Consequently, these findings indicate that further investigation to establish more robust genetic parameter estimates and assessment of feasibility are warranted.

ACKNOWLEDGEMENTS

This work was funded by a partnership (DRCX1302) between the New Zealand Ministry of Business, Innovation and Employment and New Zealand dairy farmers through DairyNZ Inc. A substantial contribution to establishing the research resource included in-kind support from LIC and CRV Ambreed. DairyNZ technical and farm staff are gratefully acknowledged for successfully executing the logistically challenging tasks of sampling and data capture. Plasma samples were analysed for progesterone by Angela Sheahan.

REFERENCES

Amyes N. C. and Morris C. A. (2009) *Proc. NZ Soc. Anim. Prod.* **69**: 203.

Gilmour A. R., Gogel B. J., Cullis B. R., Welham S. J. and Thompson R. (2015) 'ASReml User Guide Release 4.1' VSN International Ltd, UK.

Martin L. C., Brinks J. S., Bourdon R. M. and Cundiff L. V. (1992) *J. Anim. Sci.* **70**: 4012.

McNaughton L. R., Bennet R., Stanley G., Harcourt S. and Spelman R. J. (2005) *Proc. Assoc. Advmt. Anim. Breed. Genet.* **16**: 242.

Meier S., Fisher B., Eketone K., McNaughton L. R., Amer P. R. *et al.* (2017) *Proc. NZ Soc. Anim. Prod.* **77**:(in press).

Morris C. A. and Amyes N. C. (2010) *Proc. NZ Soc. Anim. Prod.* **70**: 204.

Morris C. A. and Hickey S. M. (2004) *Proc. NZ Soc. Anim. Prod.* **64**: 116.

Morris C. A., Wilson J. A., Bennett G. L., Cullen N. G., Hickey S. M. *et al.* (2000) *NZ J. Agric. Res.* **43**: 88.

Proc. Assoc. Advmt. Anim. Breed. Genet. 22:533-536

INTEGRATION OF IVF TECHNOLOGIES WITH GENOMIC SELECTION TO GENERATE HIGH MERIT AI BULLS: A SIMULATION STUDY

F.S.Hely[1], P.R.Amer[1], B.Oback[2] and D.N.Wells[2]

[1]AbacusBio Limited, P O Box 5585, Dunedin 9058, New Zealand,
[2]AgResearch Ruakura, Private Bag 3123, Hamilton 3240, New Zealand

SUMMARY

A stochastic simulation model was used to determine the impact of incorporating genomic selection into the IVF process to generate high merit young bull candidates, from a breeding company perspective. IVF candidates were simulated with varying selection proportions applied at the blastocyst stage, then combined with standard genomic selection scheme candidates with the top 50 bulls selected on EBV using truncation selection. Selecting a low proportion of genotyped blastocysts for cloning was key to producing high EBV young bull candidates where the cost of the IVF technologies would be offset by a large reduction in rearing costs through having fewer candidates to rear through until semen producing age.

INTRODUCTION

Dairy cattle breeding is a highly competitive business driven by commercial semen companies that have generated substantial genetic gains. Genomic selection has been widely used in dairy cattle improvement systems since 2008, with recent estimates of the reduction in generation interval ranging from 7 months for dams of cows up to 4.5 years for the sires' of bulls' pathway (Garcia-Ruiz et. al., 2016), accompanied by substantial improvements in selection pressure on key traits. These improvements are based on genomic testing of young bull candidates to predict performances earlier along with improvements in the accuracy of genomic breeding values, particularly in lower heritability traits.

Fisher et al (2012) proposed a method of genotyping bovine embryo biopsies, with rates of *in vivo* development not significantly different to fresh control embryos. Carrying out genomic selection at the blastocyst level would allow for intense selection for favourable genotypes within a set of candidate embryos. This could be exploited within a commercial breeding program, either through increased selection intensity at an earlier stage, or through reduced rearing costs to identify top young AI bull candidates because only blastocysts of sufficient predicted genetic merit would be reared through to semen production and beyond.

This study used simulation to determine the impact of incorporating genomic selection into the IVF process. A breeding company perspective was taken, whereby elite young bull candidates were generated within a genomic selection breeding scheme.

MATERIALS AND METHODS

A stochastic simulation framework was developed using the python programming language, and with a generic parameterisation based on industry statistics of ages of young bulls. The simulation framework starts by using a burn-in phase to model a base population selection pool for a genomic selection breeding scheme, followed by various scenarios incorporating varying levels of selection candidates generated via IVF and genomic selection. Bulls generated via IVF coupled with genomic selection then had to compete with a wider population of bulls generated from conventional matings. into the final pool of bulls available for selection.

An initial pool of selection candidates was generated using 25 unrelated animals of each sex expressing a single polygenic normally distributed trait with a heritability (h^2) of 0.25. The burn in phase was simulated over 9 discrete generations of random mating between equal numbers of males and females with known pedigree to generate an effective population size of 50 on average, before expanding out via factorial mating to 60,000 individuals in the 10th generation. True breeding values (TBV) in the first generation were drawn from a random normal distribution with a mean of zero and standard deviation equal to the square root of the heritability. For subsequent generations where parent information was available, the progeny TBV was calculated as the average of the parent TBV plus a Mendelian sampling term. Estimated breeding values (EBV) were simulated for the expanded population individuals as a trait correlated to their true breeding values using a Cholesky decomposition (Van Vleck and Gregory, 1992), with an accuracy of 0.8 simulated for genomic BVs.

Following the expansion stage, a genomic selection breeding scheme was simulated by selecting 50 sires and 1500 dams on EBV from the final pool of 60,000 individuals. The mating structure was weighted such that the top quintile of sires by EBV were randomly assigned the top 60% of the dams, the second quintile were randomly assigned to the next highest 20% of dams, and then the third, fourth and final quintile were assigned the remaining best ranked 10, 6 and 4% of dams respectively, to replicate a commercial industry structure. Each mating produced a single male offspring, resulting in a contribution of 1500 bull calf candidates to the final pool for selection.

Additional selection candidates resulting from IVF with prior prediction of genetic merit and selection at the pre-implantation stage were generated. The reproductive technologies processes were simulated using a series of random variates to assess the likelihood of a given cross between a selected male and female progressing through each stage from oocyte production through to survival of a semen producing bull. Top sires and dams were selected from the pool of candidates on EBV and mated using a factorial cross mating design. It was assumed that females could be flushed for oocytes multiple times, with each cross producing 10 oocytes.

Table 1. The base input parameters used to simulate the stages of the IVF process, including the unit for each input factor.

IVF input parameters	Value	Unit
Oocytes	10.7	Per cow flushed
Viability rate of oocytes	0.9	Oocytes viable per oocytes recovered
IVP development rate	0.31	Blastocysts for testing per oocyte flushed
DNA tests	1	Per blastocyst for testing
Semen sexing rate	0.5	Male blastocysts per blastocyst for testing
Selection rate	0.05 to 1	Selected blastocysts/blastocyst tested
Mortality (post biopsy, cryopreservation)	0.1	Deaths per biopsied blastocyst
Cloning factor	4	Demi-embryos per embryo cloned
Cloning success rate	0.85	Surviving embryos per demi-embryo
Embryos implanted per recipient	1	
Embryo survival rate	0.2 to 0.4	Per embryo implanted
Calf survival rate	0.95	Per calf born
Acceptable bulls	1	Per viable calf

For each oocyte simulated, a series of random standard uniform variates were generated to simulate the likelihood that each oocyte is viable and passes the in-vitro production (IVP) stage, using the base parameters shown in Table 1. Genomic EBVs were generated for the remaining blastocysts assigned to be male using the equation below

$$GEBV = C_{[2,1]} \times TBV + C_{[2,2]} \times a, \text{ where } a \sim Norm(0,1),\ C = Chol\begin{bmatrix} 1 & \rho^2 \\ \rho^2 & \rho^2 \end{bmatrix}$$

where ρ^2 is the accuracy of genomic selection. Blastocysts were then selected prior to cloning at a specified rate using truncation selection.

The cloning of the embryos occurred after the biopsy and genomic selection stage, with up to 4 demi-embryos created per embryo cloned and a cloning success rate of 85%. Finally, random standard uniform variates were used for the likelihood of both embryo and calf survival, with the assumption that all surviving calves were acceptable for potential usage as a bull, and these bulls were added to the pool of conventional selection candidates. Each of these probabilities act as linear multipliers. The values shown in Table 1 for example, a single cow flushed would produce between 0.087 calves with a blastocyst pre-selection rate of 0.05 and 1.17 calves with a pre-selection rate of 1, where all blastocysts were retained.

Costs were assigned to each stage of the IVF process, then multiplied by the input parameters shown in table 1. If there was no pre-selection of blastocysts, the estimated cost to produce a single viable bull calf was $2,647 (NZ), of which 57% was attributed to rearing costs.

Scenarios were compared based on the top 50 bulls selected from the pool of 1500 selection candidates via truncation selection, and according to their additional costs attributable to the IVF process. The proportion of IVF bulls selected in the top 50 out of those generated was compared, along with the TBV superiority of the top 50 bulls. The pre-selection rate applied to the IVF blastocysts following DNA testing was varied between 0.05 and 1 (i.e. all blastocysts selected) and the number of cows flushed increased to maintain the same number of effective calves for each selection rate. The number of effective calves generated was varied between 2 and 36, with an embryo mortality rate of 0.2.

RESULTS AND DISCUSSION

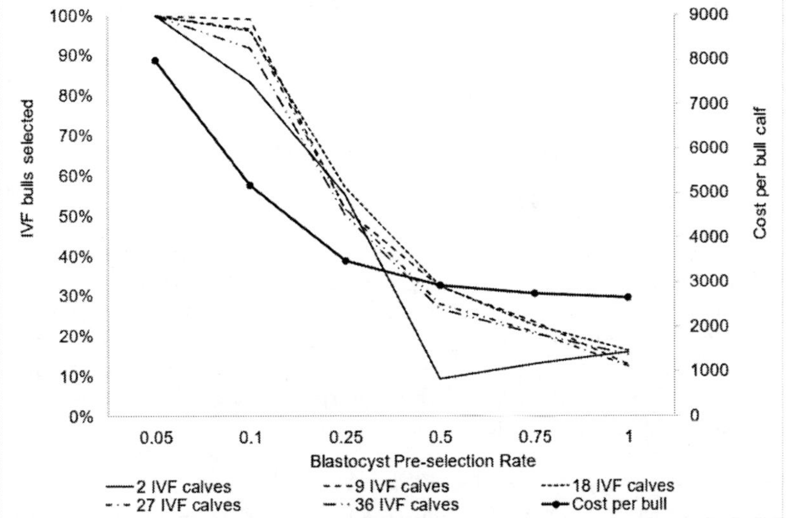

Figure 1. The average percentage of IVF bulls created that were selected in the top 50 as the blastocyst pre-selection rate increased from 0.05 to 1, with between 2 and 36 IVF calves created from the required number of cows flushed. The cost per bull calf for the given pre-selection rate is shown on the secondary axis.

Figure 1 shows the average percentage of the IVF bulls generated that were selected in the top 50 as the pre-selection rate applied to blastocysts during the IVF process was increased from 0.05 to 1, with the cost per bull calf produced. When the top 5% of blastocysts were pre-selected for implantation using genomic selection, all bull calves created via IVF were selected in the top 50 candidates in all replicates, regardless of the number of calves created, at a cost of $7,990 per calf. The percentage selected dropped to between 50 and 60% with a pre-selection rate of 0.25 and a cost of $3490 per calf, with a further reduction to between 10 and 20% without any pre-selection of blastocysts. These results show that the pre-selection rate applied to blastocysts was the key variable, where a low selection rate of blastocysts would require a larger number of cows to be flushed and blastocysts biopsied for DNA testing, (e.g. 100 cows flushed to produce on average 9 IVF calves) but in turn this would reduce the rearing costs of those young bull candidates which would not have been of sufficiently high merit to be selected in the top 50 after pooling with the calves selected at birth.

Table 2 shows the mean TBV and EBV of the pool of 1500 young bull candidates, along with the non-IVF bull and IVF bulls and IVF bull calves selected in the top 50 for a blastocyst pre-selection rate of 0.05. The TBV of the IVF bull calves was significantly ($p<0.05$) higher than the non-IVF bulls with 9, 18 or 27 IVF calves produced, although this superiority was not observed with pre-selection rates higher than 0.1.

Table 2. The mean (standard deviation) TBV and EBV of all candidates for selection, the non-IVF bull calves and IVF bull calves in the top 50 selected, with a blastocyst pre-selection rate of 0.05.

IVF calves produced	9 IVF Calves	18 IVF Calves	27 IVF Calves	36 IVF Calves
Cows Flushed	100	200	300	400
All Candidates TBV	0.86 (0.25)	0.84 (0.24)	0.84 (0.24)	0.84 (0.23)
All Candidates EBV	0.55 (0.16)	0.55 (0.15)	0.56 (0.15)	0.56 (0.15)
Non IVF bull TBV	1.23 (0.26)	1.21 (0.25)	1.22 (0.25)	1.22 (0.24)
Non IVF bull EBV	1.49 (0.17)	1.50 (0.15)	1.51 (0.15)	1.54 (0.16)
IVF bulls selected	9.12 (2.39)	17.56 (4.75)	26.72 (4.61)	36.92 (6.09)
IVF bull TBV	1.45 (0.30)	1.43 (0.34)	1.39 (0.28)	1.37 (0.30)
IVF Bull EBV	1.78 (0.19)	1.74 (0.21)	1.73 (0.20)	1.70 (0.19)

The utilisation of IVF technologies in combination with genomic selection at the blastocyst stage could be advantageous from a breeding company perspective, to carry out intensive selection on blastocysts prior to implantation, as it would reduce the rearing costs required to identify the top young bulls. While the cost of a large scale IVF program incorporating genomic selection may be prohibitive, the scenarios tested in this simulation project suggest that it could be used in combination with a more traditional genomic selection breeding scheme to increase the merit of the semen marketed from young bulls.

REFERENCES

Fisher P.J., Hyndman D.L., Bixley M.J., Oback F.C., Popovic L., McGowan L.T., Berg M.C. and Wells D.N. (2012) *Proc. NZ Soc. Anim. Prod.* **72**: 156.

Garcia-Ruiz A., Cole J.B., VanRaden P.M., Wiggans G.R., Ruiz-Lopez F.J. and Van Tassell C.P. (2016) *Proc. Natl. Acad. Sci. USA*. **113**: 3995.

Van Vleck L.D. and Gregory K.E. (1992) *J. Anim. Sci.* **70**: 57.

Proc. Assoc. Advmt. Anim. Breed. Genet. 22:537-540

DEVELOPING A GENOMIC SELECTION BREEDING PROGRAM FOR COMPLEX PEARL COLOUR TRAITS WITHIN THE SILVER-LIPPED PEARL OYSTER

D.B. Jones[1], P.B. Toole[1, 2], M.S. Khatkar[3,4], H.W. Raadsma[3,4], D.R. Jerry[1,3] and K.R. Zenger[1,3]

[1] Centre for Sustainable Tropical Fisheries and Aquaculture, College of Science and Engineering, James Cook University, Queensland, Australia
[2] Atlas South Sea Pearls Ltd., Western Australia, Australia
[3] ARC Hub for Advanced Prawn Breeding, James Cook University, Townsville QLD, Australia
[4] Sydney School of Veterinary Science, Faculty of Science, The University of Sydney, Camden, NSW, Australia

SUMMARY

Selective breeding within the pearling industry is heavily confounded by the complexity of production traits and the ability to accurately measure these phenotypes. Pearls are produced by implanting a nucleus along with a small piece of donor tissue from a sacrificed oyster into a host oyster. Unravelling trait architecture for these complex pearl quality traits is integral if genomic selection is to be implemented into established selective breeding programs. By combining simulated and exploratory datasets, this study uses genome-wide loci to better understand the genetic architecture of pearl colour which provides insights into the optimal design and implementation for a genomic selection breeding program within the pearling industry.

INTRODUCTION

The silver-lipped pearl oyster is a globally important aquaculture species with pearl production from this species currently the second most valuable Australian aquaculture export (GLOBEFISH 2016). However, like most aquaculture industries, pearling has yet to establish and implement the advanced selective breeding programs required for industry progression. Although traditional animal improvement methods have had some success with simple traits (i.e. animal growth), they are inefficient for the complex pearl traits (size, colour, lustre and shape), which are polygenic, hard to measure and have a low heritability (Jerry *et al.* 2012, Jones *et al.* 2014). For the Australian pearl industry (and aquaculture in general) to maintain international competitiveness, industry must engage in a paradigm shift in breeding practices and implement pioneering technologies that circumnavigate current limitations associated with sole reliance on phenotypic selection.

Successful incorporation of genomic data into traditional selective breeding programs depends on many factors including the diversity within a farm, the species genome size and structure, and the architecture of traits of interest. Herein, we describe the development of an optimal genomic selection approach required for rapid-genetic improvement in pearl colour in pearl oysters. In doing so, we propose a breeding system which promises to not only improve efficiency of selection within the pearling industry, but will serve as a case study for many aquaculture species.

MATERIALS AND METHODS

Experimental animals, pearl seeding and phenotypic records. To investigate the ideal design of a genomic selection breeding program for pearl colour, we utilised a dataset previously published in Jones *et al.* (2014) as a pilot dataset to explore the parameters required for robust application. Briefly, this dataset contains 2,306 individually traced commercial pearl grading phenotypes for 358 donor oysters from 6 families, as well as genotypic data for 1,146 SNPs across these individuals. Herein, we focus on the analysis of pearl colour, categorised into five sub-categories; SW.O.G: silver and white vs. gold vs. all remaining colours; G.O: gold vs. remaining colours; S.O: silver vs.

remaining colours; W.O: white vs. remaining colours; and finally SW.O: silver and white vs. remaining colours (Jones *et al.* 2014).

Variance components, heritability and genetic parameters. Previous estimates of variance components and heritability (h^2) for pearl colour have been based on pedigree information. To test the premise that genomic relationships are equally useful in calculating unbiased heritabilities (or variance explained by SNPs), we generated genomic relationship matrixes (GRMs) across the 358 oysters using GCTA v1.26.0 (Yang *et al.* 2011). Variance components and heritabilities were calculated using mixed linear models (MLM) whereby all SNPs were fit as random effects. Briefly,

$$y = X\beta + g + \varepsilon \text{ with } V = A\sigma_g^2 + I\sigma_\varepsilon^2$$

where g in an n x 1 vector of the total genetic effects of the individuals with $g \sim N(0, A\sigma_g^2)$, and **A** is interpreted as the GRM between individuals. An estimate of σ_g^2 can then be produced by the restricted maximum likelihood (REML) approach, relying on GRM estimates from all SNPs. Co-variables identified as significant in the previously published heritability estimates (i.e. seeding nucleus size and seeding technician) were also included within current the analysis described here. In addition to calculating GRMs, individual animal breeding values (best linear unbiased predictors; BLUPs) and SNP effects were calculated after incorporating in genomic relationship information. Genetic parameters generated using GRM were then compared to previously published pedigree derived results (Jones *et al.* 2014).

Optimal number of markers. To test the minimum number of SNPs necessary to produce robust estimates for GRM within the test farm data, we simulated a theoretical dataset to compare the potential benefit of including larger numbers of markers within GRM calculations. Firstly, a founder population was identified as the last generation of 1,000 historic simulated generations with 430 animals in each generation (equal to the effective population size of wild oyster populations as per Lind *et al.* 2007). Simulations were conducted for 4,200 SNPs using QMsim (Sargolzaei *et al.* 2009). From the founder population, 20 males and 20 females were used for breeding with each mating producing 50 offspring. The effective population size reflects the number of founders within the pilot dataset described above (N=50). Simulations were run for 10 discrete generations with random selection of parents from the earlier generation. For positional information, the 4,200 SNPs were placed proportionally to the length of the 14 linkage groups of a *Pinctada maxima* linkage map published in Jones *et al.* (2013). In the last three generations of the simulations, only 2,000 SNPs remained polymorphic. The resulting dataset was utilised to run comparisons of GRMs between 1,000 and 2,000 SNPs.

Optimal number of samples. Power calculations for related vs unrelated individuals based on genetic parameters related to pearl quality were conducted to estimate the minimum number of samples required to accurately identify additive genetic variance. Using methods described in Visscher *et al.* (2014), we simulated the power, defined as 'the probability of detecting $h^2 > 0$ for a quantitative trait for the given type I error rate and the SNP-heritability assumed in the population'. Heritabilities of 0.05 - 0.30 were run to reflect previous estimates of h^2 for pearl colour (0.14 - 0.36; Jones *et al.* 2014). The type 1 error rate was set at 0.01 and the variance of the SNP-derived genetic relationships was 0.00002 (for unrelated individuals) and 0.025 (for related individuals within this study population, obtained from the genetic relatedness between individuals).

RESULTS AND DISCUSSION

Pearl colour trait heritability and genetic parameters. The average difference between heritability estimates of pearl colour using GRM instead of pedigree was minor (average h^2

difference of 0.02 ± SD 0.02) indicating that the substitution of the relationship matrix provides similar power and accuracy to pedigree data when separating variance components and calculating heritability (Figure 1). The calculation of animal BLUP values based on the GRM were highly correlated to those previously published with r^2 values ranging from 0.72 to 0.88 for the different colour categories except for S.O where the r^2 was 0.46. This lower r^2 may be due to the relatively low number of silver pearls observed within the data (N = 216).

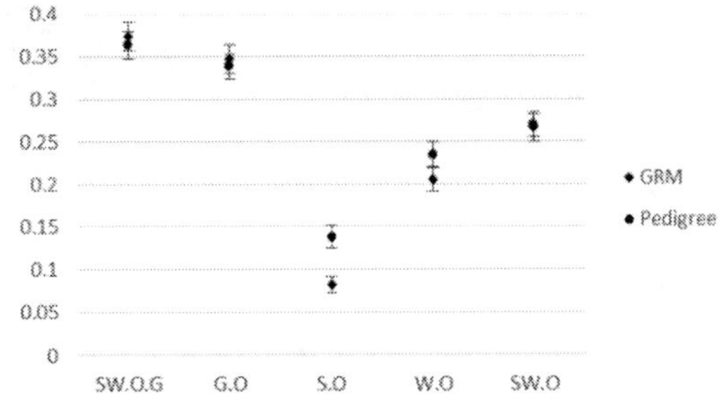

Figure 1: Heritability estimates derived from GRM (♦) as compared to pedigree (●).

The architecture of pearl colour has previously been reported to be polygenic and influenced by many genes of small effects (Jerry *et al.* 2012, Jones *et al.* 2014). One major region on linkage group 12 returned 13 significant genetic associations across the different categorisations of pearl colour which have SNP effect sizes ranging from 0.11 to 0.26 (Jones *et al.* 2014). For these SNPs, the SNP effects returned in GCTA using GRM were highly correlated to the previously published GWAS SNP effects (SW.O.G r^2 = 0.97; G.O r^2 = 0.96; S.O r^2 = 0.94; W.O r^2 = 0.98; SW.O r^2 = 0.98).

Optimal number of markers. To determine the potential effect of adding more SNPs into GRM calculation, we simulated a larger genotypic dataset containing 2,000 SNPs. The r^2 correlation between 1,000 and 2,000 SNPs was 0.98 indicating that increasing the number of SNPs with similar spacing throughout the genome yielded very little improvement to GRM accuracy in this test farm data. This indicates that 1,000 genome-wide markers is sufficient to give accurate GRM calculations for this closed farm population with limited founders (Ne previously estimated at 60). If however, if this is to be applied outside of this closed farm population, increasing the marker density would yield substantial benefit. The relative advantage of GBLUP models is at higher marker density and low heritability. In Atlantic salmon, GBLUP performed better with upwards of 4,000 SNPs (Ødegård *et al.* 2014).

Optimal number of samples. Simulations of the power to detect the unbiased heritable component of a trait from related individuals (variance of SNP-derived genetic relationships of 0.025) at sample sizes ranging from 100 – 400 and h^2 of 0.05 – 0.30 reveal that 99% power is obtained at 300 samples for a h^2 of 0.2 (Figure 2). For the pilot dataset with 2,000 phenotypic records (from 358 unique related individuals), power to detect the heritable component of a trait is estimated to range from 0.78 – 1.00 (for h^2 of 0.10 – 0.30) indicating that the current number of individuals is sufficient for estimating trait heritability. However, if this was to be expanded to unrelated individuals (i.e. variance of SNP-derived genetic relationships of 0.00002), 99% power is only reached at 8,000 samples for h^2 > 0.2.

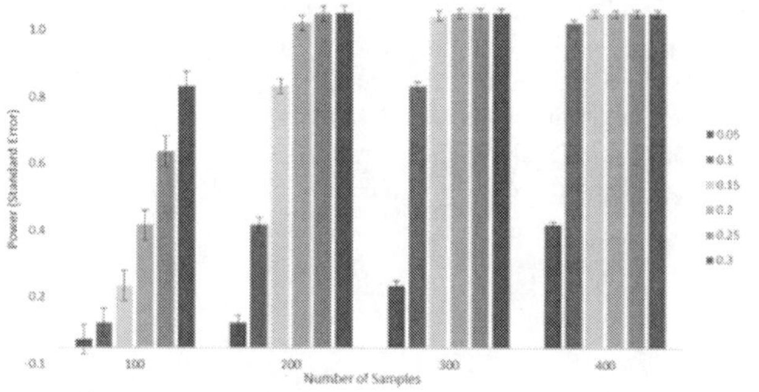

Figure 2: Power simulations for varying heritabilities across increasing sample size.

CONCLUSIONS

The re-analysis of the farm test dataset within GCTA using GRM instead of pedigree returned comparable heritability, animal BLUP and SNP effects to previously published GWAS results. This suggests that the use of GRM alone is as effective as pedigree data within a closed breeding population and adds support to the notion that pearl colour is a highly complex polygenic trait. The ability to use GRM instead of pedigree to calculate BLUP enables the inclusion of individuals for which pedigree information is not known, but more importantly the relationship between relatives can be calculated more accurately (Veerkamp *et al.* 2011). Considering this, it is hypothesised that the variance components can be estimated more precisely with GRM. Furthermore, for a genomic selection breeding program to be implemented for complex traits such as pearl colour within a closed population of pearl oysters, a minimum number of 300-400 farm data records are required to estimate the variance explained by the genome-wide SNPs for the range of heritabilities evaluated. To extend this to a breeding population with a larger number of founders (N = 300), simulations indicate that the minimum number of markers required to achieve an equivalent GRM outcome is ~3,000 SNPs and that a minimum of 8,000 samples would be required to reliably detect heritable components of pearl production traits. Based on these recommendations, the current pearl oyster breeding program has collected phenotypic data (i.e. pearl quality and growth traits) and genotypic data (a minimum of ~3,500 genome-wide SNPs) from 10,000 farm production animals to achieve these outcomes. These data simulations described herein are integral to refining the direction of ongoing research into implementing advanced genomic selection into traditional breeding programs.

REFERENCES

GLOBEFISH, F. (2016). FAO GLOBEFISH Highlights, 1/2016: 48

Jerry, D. R., R. Kvingedal, C. E. Lind, B. S. Evans, J. J. U. Taylor and A. E. Safari (2012) Aquaculture 338-341: 66-71.

Jones, D. B., D. Jerry, M. Khatkar, H. Raadsma and K. Zenger (2013) BMC Genomics 141: 810.

Jones, D. B., D. R. Jerry, M. S. Khatkar, G. Moser, H. W. Raadsma, J. J. Taylor and K. R. Zenger (2014) Aquaculture 4340: 476-485.

Lind, C. E., B. S. Evans, J. J. U. Taylor and D. R. Jerry (2007) Mol. Ecol. 1624: 5193-5203.

Ødegård, J., T. Moen, N. Santi, S. A. Korsvoll, S. Kjøglum and T. H. E. Meuwissen (2014) 5402.

Sargolzaei, M. and F. S. Schenkel (2009) Bioinformatics 255: 680-681.

Veerkamp, R. F., H. A. Mulder, R. Thompson and M. P. L. Calus (2011) 948: 4189-4197.

Visscher, P. M., G. Hemani, A. A. E. Vinkhuyzen, G.-B. Chen, S. H. Lee, N. R. Wray, M. E. Goddard and J. Yang (2014) PLoS Genet. 104: e1004269.

Yang, J., S. H. Lee, M. E. Goddard and P. M. Visscher (2011) Am. J. Hum. Genet. 881: 76-82.

Proc. Assoc. Advmt. Anim. Breed. Genet. 22:541-544

MULTI-TISSUE GENOME WIDE EXPRESSION OF MRNA AND MICRORNA IN CATTLE SELECTED FOR HIGH AND LOW RESIDUAL FEED INTAKE

Yizhou Chen[1], Cedric Gondro[2], Wijdan Al-Husseini[2], Sara De las Heras Saldana[2], Robert Herd[3], Paul Arthur[1]

[1] Elizabeth Macarthur Agricultural Institute, NSW Department of Primary Industries, Menangle, NSW, Australia

[2] Environmental and Rural Science, University of New England, Armidale, NSW, Australia

[3] Beef Industry Centre, NSW Department of Primary Industries, Armidale, NSW, Australia

SUMMARY
Residual feed intake is a measure of feed efficiency in beef cattle which is an economically important trait in beef production. Gene expression is the key determinant of cellular phenotype and genome-wide expression analysis will provide insight into the molecular events underlying biology of feed efficiency. We carried out genome-wide gene expression in liver, muscle and blood with total 199 samples and micorRNA expression with 96 samples. These datasets provide a resource of expression QTL mapping for understanding the functional consequences of genetic variation, and how it affects feed efficiency in beef cattle.

INTRODUCTION
Feed use and feed efficiency are important for the survival and selective advantage of an animal. In animal production enterprises, feed intake and feed efficiency are important factors that affect overall profitability. In a typical beef production system, about 65–85% of the feed is used to maintaining the cow breeding herd (Montaño-Bermudez and Nielsen 1990). Feed efficiency in beef cattle can be measured as Net feed intake (NFI) or residual feed intake (RFI). This is the difference between an animal's actual feed intake recorded over a test period and the expected feed intake based on the animal's size and growth rate (Koch et al. 1963).

There is strong evidence that genetic variation in RFI exists. The estimated heritability of RFI in cattle populations is moderate ranged from 0.07 to 0.62 (Berry and Crowley 2013). In Australia, heritability estimates for NFI at feedlot is 0.41 and 0.34 for post-weaning (Jeyaruban et al. 2009).

However, the accurate measurement of RFI for individual animals is an expensive process, and this has been a major limitation to the adoption of feed efficiency as an economically important trait in animal breeding. Much of the research projects were focus on to to develop genetic markers that can be used for genomic selection using high density single nucleotide polymorphism (SNP) chips and more recently, next generation sequencing technologies that enable breeders to select animals based on genomic sequences (Meuwissen et al. 2001; Barendse et al. 2007; Meuwissen and Goddard 2010; Bolormaa et al. 2011; Khansefid et al. 2014).

In the past decades, we have gained considerable knowledge of understanding of animal's development with the advance of genome sequence of human and many other species (The-ENCODE-Project-Consortium 2012). The genome sequence contains all the information necessary to develop from the initial zygote to an adult with full set of organs to respond to the environmental influence. Although all cells from an individual have the same genome sequence, there are more than 400 distinct cell types which and their cellular developments, morphology, and function are governed by precise patterns of gene expression which are regulated by the function al elements in the genome.

A number of studies of gene expression in beef cattle have been published (Chen et al. 2011; Tizioto et al. 2015; Weber et al. 2016) with a limited number of samples and a number of differentially expressed genes between high and low NFI cattle were revealed by comparing two extreme phenotype groups. However, a large gene expression dataset with sufficient number of animals and crossing multiple tissues is essential to study the patterns of transcriptome variation across individuals and tissues. We carried out genome-wide gene expression in liver, muscle, and blood tissues with total 199 samples, In addition, we also carried out genome-wide expression of micorRNA for 96 samples. These datasets provide a resource of expression QTL mapping for understanding the functional consequences of genetic variation, and how it affects feed efficiency in beef cattle.

MATERIALS AND METHODS

All animals used for the genome-wide gene expression by RNA-sequencing were from Angus divergent selection line for NFI at the Agricultural Research Centre, Trangie, NSW (Arthur *et al.* 2001). The first set of animals consisted of 48 young bulls born in 2005 which were approximately three generations of divergent selection for NFI. Liver biopsies were taken from 24 animals with the lowest RFI and 24 animals with the highest NFI at the end of the test at feedlot Tullimba. The second set of animals contained 48 young bulls born at 2008. Muscle biopsies were taken at post weaning at the Agricultural Research Centre, Trangie. The third dataset was 30 steers and 30 heifers born in 2012, the blood tissue and liver biopsies were taken at the end of NFI test.

All animals used in these experiments were recorded for average daily gain during the 70day; net feed intake, average daily feed intake; P8 fat thickness (ultrasound) at the end of NFI test; RIB fat thickness (ultrasound); eye muscle area (ultrasound).

All experiments were approved by the University of New England Animal Ethics Committee (AEC 06/123, AEC14-002 and AEC14-036) and New South Wales Department of Primary Industries (NSW DPI) Animal Research Authority ((ORA09/015, ORA 13/16/004). Male calves were castrated at 4 months of age. After weaning animals were grown on native pastures until they reached feedlot entry weight of approximately 400 kg BW. NFI was tested in the Beef Research Feedlot Tullimba, NSW with an automated recording system. During the 70-day test, the animals had ad libitum access to a barley-based feedlot ration containing 12 MJ metabolizable energy per kilogram dry matter and 15–17% crude protein.

Ninety-seven RNA-sequencing samples were obtained using HiSeq 2000 (Illumina Inc) and the RNA-sequencing libraries were created from the polyadenylated fraction of RNA from each animal by using modified protocol of Illumina sample preparation. The remaining 104 RNA-sequence samples were obtained by using HiSeq 2500 (Illumina Inc) at Beijing Genome Institute and the sequencing library were created by using non-strand specific protocol with poly-A selection of mRNA (the Illumina Tru Seq™) protocol.

Small RNA libraries were constructed for each animal using 1μg total RNA with NEXTflex™ Small RNA-Seq Kit v2 (Bioo Scientific, TX, USA) following the protocols supplied by the manufacturer. The libraries were sequenced at Ramaciotti Center, University of NSW with Illumina HiSeq 2000 Sequencing System.

The quality of the sequence was assessed with FastQC v0.11.3 (Andrews 2010) and the low quality bases and adaptor sequences were removed by Trimmomatic v0.33 (Bolger et al. 2014). We used topHat v2.1.1 (Trapnell et al. 2009) to align all paired reads against the Bos taurus reference genome (Ensembl UMD3.1). Read counts for each sample was obtained with HTSeq v 0.6.0 (Anders et al. 2014). In order to visualize the clustering of the tissues samples, batch effects, and possible outliers we performed plots from the output of Principal Component Analysis (PCA) on the raw counts.

We used R package ComBat (Johnson et al. 2007) to adjust for batch effect in liver samples of the bull and steer datasets. The differential expressed (DE) genes between tissue were obtained by Edge R (false discovery rate <0.05 and the logarithm fold change (logFC) ≥ 1.5).

RESULTS AND DISCUSSION

In total, we obtained more than 1772 million high quality paired sequence reads for RNA-sequencing and 240 million microRNA sequence reads. Among all the RNA-sequencing samples, more than 90% of the paired sequence reads from liver and muscle tissue were aligned to the bovine genome (bostau6, UMD_3.1) albeit, the mapping rate for sequence reads in blood is lower (~ 81%) (Table 1). For the microRNA sequencing, more than 80% of the sequence reads mapped to the bovine genome and 69% sequence reads mapped to known bovine mature miRNA (bostau6, UMD_3.1).

Table 1 Summary of the RNA-sequencing results

Dataset	Liver /bull	Liver /steer	Liver /heifer	Muscle /bull	Blood /steer	Blood /heifer	Liver [2] miRNA	Muscle[2] miRNA
No sample	47	25	27	48	25	27	48	48
Paired reads (10^6)	6.0	11.3	11.3	6.8	11.2	11.2	2.81	2.2
Range (10^6)	1.6-12.5	11.0-11.5	11.0-11.5	0.6-9.4	9-11.5	11-11.5	0.32-9.8	0.4-7.5
Mapped pair reads %[1]	81	90.8	90.8	88.7	80.6	80.6	88 (0.69)[3]	84 (0.69)[3]

[1] the concordant pair alignment rate; [2]:miRNA-sequencing using 1x75 single read; [3]:the sequence reads mapped to known miRNA in bovine genome sequenced (UMD3.1).

We explored gene expression similarity between tissues and across samples by principal component analysis (PCA) on the raw counts. Liver, muscle and blood tissues show a characteristic transcriptional signature (Figure 1 A). Furthermore, the expression profiles in liver of young bulls are quite distinct to steers and heifers, while there is little difference of expression pattern in liver between steer and heifer (figure 1B). Indeed, we found the largest number of differential expressed genes between tissues (>12000). In liver expression, there are more than 9000 differentially expressed genes between young bull and steer/ heifer, only 2 differentially expressed genes between steers and heifers (data not shown due page limits). This suggests that male hormones played important roles in liver expression. The liver is one of the most essential organs involved in the regulation of energy homeostasis and associated with lipid formation and breakdown, glucose production and catabolism, and cholesterol synthesis and secretion. It was well documented in mice and human that males, testosterone works via androgen receptors to increase insulin receptor expression and glycogen synthesis, decrease glucose uptake and lipogenesis, and promote cholesterol storage in the liver (Shen & Shi 2015).

Understanding the gene regulation and the identification of the functional elements in genome are important to increase the accuracy of genome selection by increase the level of linkage disequilibrium in the marker panel by including functional SNPs. These gene expression datasets provide a potential resource for mapping functional elements by eQTL mapping.

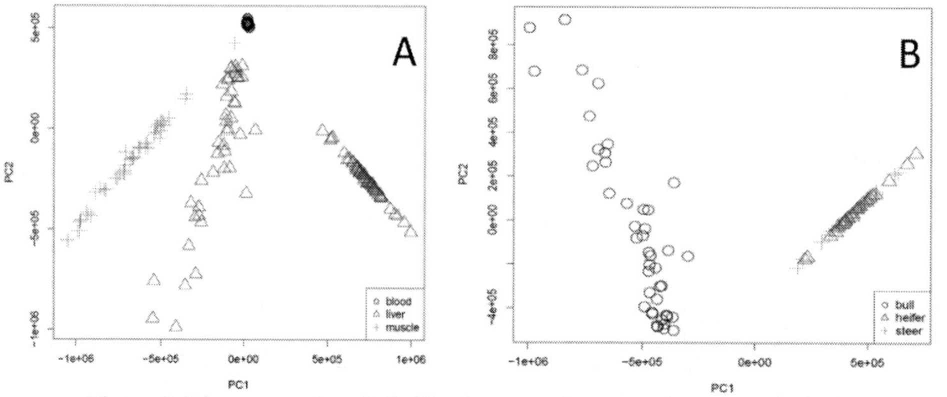

Figure 1 Gene expression similarities between tissues and sex by principal components analysis. A:gene expression between tissues across all samples. B:Liver gene expression between young bulls, steers and heifers.

REFERENCE

Anders S., Pyl P.T. & Huber W. (2014) Bioinformatics, btu638.

Andrews S. (2010). http://www.bioinformatics.babraham.ac.uk/projects/fastqc.

Archer J.A., Richardson E.C., Herd R.M. & Arthur P.F. (1999) Austra. J. Agri. Res. **50**.

Arthur P.F., Archer J.A., Johnston D.J., Herd R.M., Richardson E.C. & Parnell P.F. (2001) J. Anim. Sci. **79**, 2805.

Barendse W., Reverter A., Bunch R.J., Harrison B.E., Barris W. & Thomas M.B. (2007) Genetics 176, 1893

Bolormaa S., Hayes B.J., Savin K., Hawken R., Barendse W., Arthur P.F., Herd R.M. & Goddard M.E. (2011) J. Anim. Sci. 89, 1684

Berry D.P. & Crowley J.J. (2013) *J Anim Sci* 91, 1594-613.

Chen Y., Gondro C., Quinn K., Herd R.M., Parnell P.F. & Vanselow B. (2011) Animal Genetics 42, 475.

Johnson W.E., Li C. & Rabinovic A. (2007) Biostatistics 8, 118-27.

Jeyaruban M.G., Johnston D.J. & Graser H.U. (2009) Proc. Assoc. Advmt. Anim. Breed. Genet, **18**,584

Khansefid M., Pryce J.E., Bolormaa S., Miller S.P., Wang Z., Li C. & Goddard M.E. (2014) J. Anim. Sci. 92, 3270

Koch R.M., Swiger L.A., Chambers D. & Gregory K.E. (1963) J. Anim. Sci. 22, 486.

Montaño-Bermudez M. & Nielsen M.K. (1990) J. Anim. Sci. 68, 2297-309

Meuwissen T., Hayes B. & Goddard M. (2001) Genetics **157**, 1819 - 29.

Meuwissen T. & Goddard M. (2010) Genetics 185, 623-31.

Shen M. & Shi H. (2015) Int J Endocrinol 2015, 294278.

The-ENCODE-Project-Consortium (2012) Nature 489, 57-74.

Tizioto P.C., Coutinho L.L., Decker J.E., Schnabel R.D., Rosa K.O., Oliveira P.S., Souza M.M., Mourão G.B., Tullio R.R., Chaves A.S., Lanna D.P., Zerlotini-Neto A., Mudadu M.A., Taylor J.F. & Regitano L.C. (2015) Bmc Genomics 16, 1

Trapnell C., Pachter L. & Salzberg S.L. (2009) Bioinformatics 25, 1105-11

Weber K.L., Welly B.T., Van Eenennaam A.L., Young A.E., Porto-Neto L.R., Reverter A. & Rincon G. (2016) PLoS ONE 11, e0152274.

Proc. Assoc. Advmt. Anim. Breed. Genet. 22:545-548

MULTI-ASSEMBLER PIPELINE FOR THE *DE NOVO* TRANSCRIPTOME ASSEMBLY ON NON-MODEL ORGANISMS: THE CASE OF THE BLACK TIGER PRAWN (*Penaeus monodon*)

R. Huerlimann[1,2], L. Gordon[1,3], J. Goodall[1,4], K. Siemering[1,3], N. Wade[1,4], M. Sellars[1,4], G.E. Maes[1,5], D.R. Jerry[1,3]

[1] ARC Research Hub for Advanced Prawn Breeding
[2] Centre for Sustainable Tropical Fisheries and Aquaculture, James Cook University, Townsville, Queensland, Australia
[3] Australian Genome Research Facility, Melbourne, Victoria, Australia
[4] CSIRO Agriculture and Food, Brisbane, Queensland, Australia
[5] Laboratory of Biodiversity and Evolutionary Genomics, KU Leuven, Leuven, Belgium

SUMMARY

A high-quality transcriptome is important for genome annotation and differential gene expression studies, but a comprehensive transcriptome assembly for non-model species like prawns is still challenging. Most assemblies are carried out in a single assembler; however, recent publications have shown that while different assemblers produce a shared core of contigs, they each also produce unique contigs. Using the transcriptome assembly of the black tiger prawn (*Penaeus monodon*) as an example, we merged the assemblies generated by four transcriptome assemblers, and incorporated newly published best practices into a novel pipeline. This multi assembler approach produces an improved, less redundant assembly which is also transferable to other non-model species. Therefore, in contrast to older approaches, using multiple assemblers improves assemblies by using the strengths of different assemblers, while decreasing their weaknesses.

INTRODUCTION

Complete transcriptomes are an important resource that can be used for differential gene expression studies (Wang et al. 2009), genome annotation (Saha et al. 2002), and more recently genome scaffolding (Song et al. 2016), among other applications. The two main methods for transcriptome assembly are genome guided and *de novo*. A genome guided transcriptome assembly is computationally simpler, but depends on the completeness of the reference genome and is impeded by sequencing errors and isoforms (Grabherr et al. 2011). In contrast, the *de novo* approach is used when no reference genome is available, but is computationally more complicated, especially for large data sets. While model species often have a variety of genomic resources available, these are by definition lacking for non-model species.

Recently the number of transcriptome assemblers has exploded from the limited number that was available ten years ago. These various assemblers have different strengths and weaknesses, resulting in contigs that are unique to a specific tool (Smith-Unna et al. 2016). Trinity, one of the most popular assemblers (cited in 2865scientific articles based on Web of Science as of January 2017), can assemble most transcripts including different isoforms, or recent gene duplications (Grabherr et al. 2011), although with the drawback of the final transcriptome often including a large number of misassembled contigs. Another bias in transcriptome assembly is introduced by sequencing errors or increased heterozygosity due to sequencing multiple individuals, both leading to more fragmented assemblies.

Recently, MacManes (2016) published recommendations for the transcriptome assembly of non-model species, suggesting to only sequence tissues from one individual, which is not always possible (for example for small organisms), and to use Rcorrector to reduce sequencing errors.

BUSCO (Simão et al. 2015) and TransRate (Smith-Unna et al. 2016) are used to assess the final transcriptome and remove low coverage reads. Transfuse has been made available (github.com/cboursnell/transfuse), which can merge multiple transcriptome assemblies from different individuals or different assemblers using reads to improve the final assembly. Therefore, the aim of this study is to compare the assembly of individual samples using four assemblers (Trinity, Bridger, BinPacker and IDBA-tran) merged using Transfuse with a single assembly in Trinity.

MATERIALS AND METHODS

Samples were collected from five different individuals of *Penaeus monodon* (3 female, 2 males). Two replicates each of the following tissues were sent for sequencing: eyestalk, female gonad, male gonad, gills, haemolymph, hepatopancreas, muscle and stomach. One sample each from gills, haemolymph and stomach failed the library preparation, resulting in 13 successfully sequenced samples. Sequencing was carried out at the Australian Genome Research Facility in Melbourne, Australia, on a HiSeq 2500 using a 125 bp paired-end, strand-specific, ribo-minus protocol. On average, 20 million reads were obtained per sample with an average of 91% bases ≥Q30.

Two assembly approaches were used: one assembling all samples collectively in a single assembler (single assembly, Fig. 1a) and the other where each sample was assembled individually in four assemblers (multi assembly, Fig. 1b). The transcriptome generally followed the recommendations of MacManes (2016). For both approaches, the individual samples were collectively error corrected using RCorrector version 1.0.2 (Song et al. 2015).

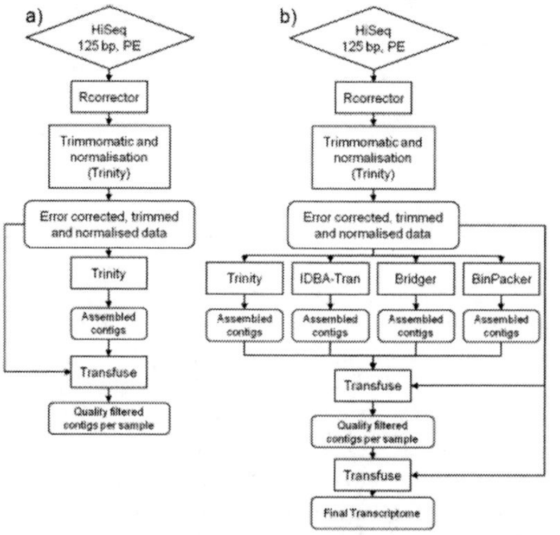

Fig. 1 Assembly pipeline for a) single assembler approach and b) multi assembler approach

Using Trinity 2.2.0 (Grabherr et al. 2011), adapter and bases with a Phred score <2 were trimmed with trimmomatic (Bolger et al. 2014) and reads were in-silico normalised. For the single assembly, the 13 samples were concatenated and assembled in Trinity. The multi assembly was carried out for each sample individually in Trinity 2.2.0, BinPacker 1.0 (Liu et al. 2016), Bridger r2014-12-01 (Chang et al. 2015) and IDBA-Tran 1.1.1 (Peng et al. 2013). For IDBA-Tran the k60 transcriptome was used for downstream processing. For both approaches transfuse version 0.5.0 (https://github.com/cboursnell/transfuse) was used to remove redundant contigs, and also merge the individual assemblies for the multi assembly approach. For the multi assembly, the transcriptomes of the four assemblers were merged by sample using transfuse in a first round. In a second round, samples were then merged with transfuse into a final transcriptome.

The two final assemblies were annotated using Blast2Go (Conesa et al. 2005) against the SwissProt database (Boeckmann et al. 2003) downloaded on 12. January 2017. The quality of both assemblies was assessed with BUSCO version 1.2 (Simão et al. (2015) using the arthropod set and TransRate version 1.0.3 (Smith-Unna et al. 2016).

RESULTS AND DISCUSSION

The aim of this study was to compare the use of multiple assemblers (multi: Trinity, BinPacker, Bridger, IDBA-Tran) on individual samples with a combined approach using only one assembler (single: Trinity). When comparing the two approaches, the multi assembly resulted in a more manageable number of contigs and lower duplication levels; however, at the price of completeness (Fig. 2a). The number of fragmented contigs was comparable in both approaches.

The raw assembly in Trinity resulted in 280,846 contigs, with 85% of the arthropod Benchmarking Universal Single-Copy Orthologs (BUSOs) complete, of which 24% were duplicated (Fig. 2a). After merging with Transfuse, this was reduced to 212,526 contigs with C:83%[D:24%] and 36,086 contigs annotated with SwissProt. In contrast, the sum of the contigs of all samples in the four assemblers added up to 2,412,355 contigs. Merging the individual assemblies by sample reduced the total number of contigs in the 13 samples combined to 392,349 with a C:70%[D:11] (Fig. 2a). The second round of merging of the individual samples into a final transcriptome resulted in 73,406 contigs with C:70%[D:10%] of which 17,885 contigs were annotated with SwissProt. The single assembly resulted in 10,470 unigenes (29% of annotated contigs), while the multi assembly resulted in 8,450 unigenes (47% of annotated contigs), with 7071 shared unigenes (Fig. 2b). The BUSCO analysis and unigene comparison shows that while the single assembly approach produces more annotated contigs, most of these contigs are duplicated.

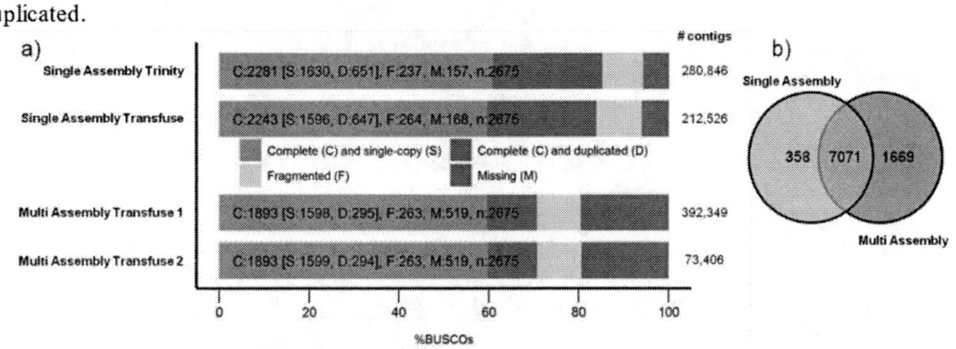

Fig. 2 a) Benchmarking Universal Single-Copy Orthologs (BUSOs) values and #contigs for the two approaches using single and multiple assemblers. b) Venn diagram showing number of shared and unique genes identified in Blast2Go.

Table 1 Quality assessment using TransRate. Scores and percentages derived from mapping reads to the assembly.

	Assembly Score	# of contigs	Assembly Size	N50	Percent mapping	Percent bases uncovered	Percent contigs low covered
Single	0.48	212,526	171.1 Mb	1571	81.8	35.6	80.2
Multi	0.36	73,406	65.5 Mb	1687	82.7	18.2	36.9

Comparing the TransRate mapping scores of the two assemblies strategies, the multi assembly exhibited higher support for the contigs. While the single assembly has a slightly higher assembly score of 0.48 compared to 0.36 in the multi assembly, the percentages of reads mapping to the transcriptome (81-83%) and N50 values (1500 bp to 1687 bp) were comparable (Table 1). However, the multi assembly had a lower proportion of bases that were not covered by reads (18.2% compared to 35.6%) and fewer contigs with low read coverage reads (36.9% compared to 80.2%).

Compared to two other multi-tissue decapods assemblies, the present assembly lies between the assemblies of the two freshwater crayfish *Astacus astacus* (Theissinger et al. 2016) and *Cherax quadricarinatus* (Tan et al. 2016). The *A. astacus* assembly combined four tissues (abdominal muscle, hepatopancreas, ovaries and green glands) and used Trinity only for the assembly. This resulted in 158,649 non-redundant contig and 45,415 contigs after filtering for lowly expressed transcripts, with a BUSCO score of C:64%[D:27%] and a TransRate assembly score of 0.20. In contrast, the *C. quadricarinatus* assembly combined five tissues (heart, kidney, hepatopancreas, central nerve cord, and testis) from a single individual and used both Trinity and IDBA-Tran for the assembly and merged the contigs using Corset (Davidson et al. 2014). This resulted in 180,635 contigs between Trinity and IDBA-Tran, and a final assembly of 44,525 contigs, with a BUSCO score of C:74%[D:7%]

Based on these results, using multiple assemblers in conjunction with a merging software like Transfuse highly reduces the number of contigs to a more realistic number by removing redundant contigs. However, while the multi assembler approach in this study also reduced the over-inflation of contigs commonly found in Trinity, it came at the cost of completeness of the assembly. While older approaches to transcriptome assembly relied on a single assembler, the field is now moving towards using multiple assemblers which improves assemblies by using the strengths of different assemblers, while decreasing their weaknesses.

REFERENCES

Boeckmann B., Bairoch A., Apweiler R., Blatter M.-C., Estreicher A., Gasteiger E., Martin M. J., Michoud K., O'Donovan C. and Phan I. (2003). *Nucleic acids research* **31**: 365-370.

Bolger A. M., Lohse M. and Usadel B. (2014). *Bioinformatics*: btu170.

Chang Z., Li G., Liu J., Zhang Y., Ashby C., Liu D., et all. (2015). *Genome biology* **16**: 1.

Conesa A., Götz S., García-Gómez J. M., Terol J., et all. (2005). *Bioinformatics* **21**: 3674-3676.

Davidson N. M. and Oshlack A. (2014). *Genome biology* **15**: 1.

Grabherr M. G., Haas B. J., Yassour M., Levin J. Z., Thompson D. A., Amit I., Adiconis X., Fan L., Raychowdhury R. and Zeng Q. (2011). *Nature biotechnology* **29**: 644-652.

Liu J., Li G., Chang Z., Yu T., Liu B., et all. (2016). *PLoS Comput Biol* **12**: e1004772.

MacManes M. D. (2016). *bioRxiv*: 035642.

Peng Y., Leung H. C., Yiu S.-M., Lv M.-J., et all. (2013). *Bioinformatics* **29**: i326-i334.

Saha S., Sparks A. B., Rago C., Akmaev V., Wang C. J., Vogelstein B., Kinzler K. W. and Velculescu V. E. (2002). *Nature biotechnology* **20**: 508-512.

Simão F. A., Waterhouse R. M., Ioannidis P., et all. (2015). *Bioinformatics*: btv351.

Smith-Unna R., Boursnell C., Patro R., et all. (2016). *Genome research*: gr. 196469.196115.

Song L. and Florea L. (2015). *GigaScience* **4**: 1.

Song L., Shankar D. S. and Florea L. (2016). *The Plant Genome*.

Tan M. H., Gan H. M., Gan H. Y., Lee Y. P., Croft L. J., Schultz M. B., Miller A. D. and Austin C. M. (2016). *Organisms Diversity & Evolution* **16**: 185-200.

Theissinger K., Falckenhayn C., Blande D., Toljamo A., Gutekunst J., Makkonen J., Jussila J., Lyko F., Schrimpf A. and Schulz R. (2016). *Marine genomics*.

Wang Z., Gerstein M. and Snyder M. (2009). *Nature reviews genetics* **10**: 57-63.

Proc. Assoc. Advmt. Anim. Breed. Genet. 22:549-552

JUVENILE IGF-I RESPONSE IN INRA RFI SELECTION LINES PARTLY REFLECTS CHANGES IN POST-WEANING ATTRIBUTES

H. Gilbert[1], I. Louveau[2], Y. Billon[3] and K.L. Bunter[4]

[1] GenPhySE, INRA, Université de Toulouse, INP, ENSAT, Castanet-Tolosan, France
[2] PEGASE, INRA, Agrocampus Ouest, Saint-Gilles, France
[3] GenESI, INRA, Surgères, France
[4] Animal Genetics and Breeding Unit*, University of New England, Armidale, NSW, Australia

SUMMARY

Performance data were recorded at INRA in two lines divergently selected for residual feed intake (RFI) for eight generations (G0 to G7). A subsample of piglets were bled shortly after weaning to establish concentrations of juvenile IGF-I. The line effect was clearly detected for juvenile IGF-I, confirming previous studies which found that juvenile IGF-I is an indirect early predictor of efficiency. The line difference in juvenile IGF-I was partly explained by differences between the RFI lines in their growth immediately after weaning, but remained significant after post-weaning growth was accounted for. Selection for efficiency has implications for post-weaning management to limit feed deprivation and growth delays during the post-weaning period.

INTRODUCTION

Selection for feed efficiency is important, but phenotyping is costly. Previous studies have demonstrated that juvenile IGF-I recorded shortly after weaning is genetically correlated with feed intake, efficiency and fatness traits in pigs (Bunter *et al.* 2005), as well as piglet birth weight (Hermesch *et al.* 2001), and that selection for lower RFI is accompanied by reduced juvenile IGF-I (Iowa State lines) (Bunter *et al.* 2010). Therefore, juvenile IGF-I is an early predictor of efficiency during growing-finishing growth stages. Postnatal IGF-I is related to growth and development (Le Roith *et al.* 2001) but literature on its role in the early post-weaning stage is scarce. In this study, an independent validation of the results reported from the Iowa State lines was investigated, and the impact of post-weaning growth for the measured levels of juvenile IGF-I were explored, using the INRA lines - divergently selected for RFI over 8 generations.

MATERIALS AND METHODS

Animals and records. Performance data were recorded at INRA in two lines divergently selected for residual feed intake (RFI) for eight generations (G0 to G7). A total of 419 pigs from lines divergently selected for low (LRFI) or high (HRFI) residual feed intake were tested in 8 batches. Male and female piglets born in generation G7 (after 7 generations of selection: 117 LRFI and 123 HRFI) and entire males from generation G8 (106 LRFI and 73 HRFI) were recorded. The difference between lines for RFI in G8 was 137 g/day ($P < 0.001$). Details of the selection of the RFI lines have been given in Gilbert *et al.* (2017a). In a given batch, pigs were born the same week, weaned on the same day at 28.3 ± 1.7 days and followed exactly the same protocol for performance testing. At weaning, pigs were penned per line in groups of 24. During the growing-finishing period (10 weeks of age (START) until slaughter weight of 115 kg), 12 pigs of the same line and sex were allotted per pen equipped with a single-place electronic feeder to record feed intake (ACEMA 64).

*AGBU is a joint venture of NSW DPI and the University of New England

Pigs were weighed at birth (BIRTH_WT), at weaning (WEAN_WT), and the week after weaning at the time of blood sampling for IGF-I measurements (35.3 ± 1.7 days) (IGF-I_WT). Average daily gain (ADG) traits between time points were calculated. The summary statistics of the traits are given in Table 1, together with the trait abbreviations.

Table 1. Summary statistics for weight (WT), growth (ADG) and log IGF-I traits after outlier elimination

Traits	unit	N	Mean	SD	Min	Max
BIRTH_WT	kg	416	1.50	0.31	0.50	2.45
WEAN_WT	kg	417	9.0	1.5	3.6	12.9
IGF-I_WT	kg	415	10.0	1.6	5.1	14.8
ADG_BIRTH_WEAN	g/day	416	265	48	99	397
ADG_BIRTH_IGF-I	g/day	415	242	40	121	365
ADG_BIRTH_START	g/day	412	353	57	194	507
ADG_WEAN_IGF-I	g/day	415	149	102	-186	570
ADG_WEAN_START	g/day	412	416	83	181	634
lIGF-I (RIA)	log(ng/mL)	178	3.99	0.39	3.00	4.98
lIGF-Ij (ELISA)	log(ng/mL)	417	4.50	0.49	3.18	5.65

IGF-I data. Animals were bled post-weaning to measure concentrations of juvenile IGF-I using radioimmunoassay (RIA) (IGF_I) or ELISA (IGF-Ij) methodology. Blood samples were obtained from the jugular using a vacutainer and deposited on the Primegro IGF-I bloodspot cards (for IGF-Ij). For 178 G7 pigs, 5 mL of blood was also collected into heparin tubes, centrifuged and plasma was aliquoted and stored at -20°C for RIA measurements. Samples extracted from Primegro bloodspot cards were assayed using an IGF-I Quantikine ELISA Kit (R&D systems). As per manufacturer's instruction, the raw values were standardized to NIBSC/WHO 02/254 values by applying a multiplication factor of 1.54. For the RIA methodology, concentrations of plasma IGF-I were determined using a double antibody RIA after an acid-ethanol extraction (Louveau and Bonneau, 1996) with recombinant human radiolabelled IGF-I (PerkinElmer).

Statistical analyses. The IGF-I and IGF-Ij measurements were log-transformed (lIGF-I and lIGF-Ij, respectively) for analysis. Outlier values for raw and log data exceeded 1.5 times the inter-quartile range based on the log transformed distribution and excluded from the analyses (2 pigs excluded). The relationships between IGF-Ij measurements, divergent selection for RFI and early growth were evaluated using a series of model comparisons. The simplest models accounted for the batch effect (production traits, Model M0) or a combination of batch and assay (BA, accounting for sampling date and generation) (lIGF-Ij, M1), along with sex within generation and line effects. For all samples but 11, effects of batch, sampling date and assay were confounded. Age at sampling was not significant for IGF-Ij traits and not included. A second level of models applied to lIGF-Ij fitted additional linear covariates across lines: early body weights until blood sampling (M2 to M4) or early ADG traits (M5 to M7, Table 2).

RESULTS AND DISCUSSION

Correlations between assay procedures for IGF-I. The correlation between the ELISA values, corrected for batch of birth and assay effects, and RIA measurements of IGF-I, corrected for batch, was 0.72. Thus, only results on IGF-Ij will be reported on the following to maximise the number of available measurements.

Line effect on lHG-Ij and early growth. The line effect was significant for weaning body weight (9.2 kg in LRFI vs 8.7 kg in HRFI, P<0.001), ADG from birth to weaning (270 vs 260 g/d, P=0.04) and ADG from weaning to blood sampling (120 vs 177 g/d, P<0.001) (Table 2). Piglets from both lines were born with similar birth weights, but LRFI piglets were 0.5 kg heavier at weaning. This is consistent with differences between these lines reported by Gilbert et al. (2017a). Due to reduced growth in the LRFI line immediately after weaning the line difference in body weight was no longer significant at one week post-weaning. Subsequently, overall growth until the end of the post-weaning period was similar in the two lines. In a study on the G9 pigs of the same selection lines, a larger line difference (-66 g/d) in ADG the week after weaning was observed in piglets having no creep feeding before weaning. This was accompanied by a 25% reduction in feed intake in the LRFI piglets (Gilbert et al. 2017b).

A very significant difference in IGF-Ij levels was found between the lines, regardless of adjustments for growth or weight measured at different time periods, with lower (untransformed) values in the LRFI line compared to the HRFI line (-35.4 ng/mL, $P < 0.001$). A larger line difference in IGF-Ij (47 ng/mL) measured in G5 of the Iowa State lines was reported by Bunter *et al.* (2010). This study confirms the association between selection for low RFI and reduced IGF-Ij.

Table 2. Model R^2 and the significance of the line effect (P(line)) on weight (WT), growth (ADG) and juvenile IGF-I (lIGF-Ij) traits, including models fitting body weight or average daily gain as linear covariates for lIGF-Ij

Model*	Trait	R^2	P(line)	P(WT)	P(ADG)
M0 Y = batch + sex + line	BIRTH_WT	0.08	0.26	-	-
M0 Y = batch + sex + line	WEAN_WT	0.07	<0.001	-	-
M0 Y = batch + sex + line	IGF-I_WT	0.04	0.54	-	-
M0 Y = batch + sex + line	START_WT	0.07	0.73	-	-
M0 Y = batch + sex + line	ADG_BIRTH_WEAN	0.06	0.04	-	-
M0 Y = batch + sex + line	ADG_BIRTH_IGF-I	0.05	0.32	-	-
M0 Y = batch + sex + line	ADG_BIRTH_START	0.07	0.60	-	-
M0 Y = batch + sex + line	ADG_WEAN_IGF-I	0.22	<0.001	-	-
M0 Y = batch + sex + line	ADG_WEAN_START	0.11	0.16	-	-
M1 Y = BA + sex + line	lIGF-Ij	0.41	<0.001	-	-
M2 Y = BA + sex + line + BIRTH_WT	lIGF-Ij	0.42	<0.001	0.04	-
M3 Y = BA + sex + line + WEAN_WT	lIGF-Ij	0.42	<0.001	0.11	-
M4 Y = BA + sex + line + IGF-I_WT	lIGF-Ij	0.49	<0.001	<0.001	-
M5 Y = BA + sex + line + ADG_BIRTH_WEAN	lIGF-Ij	0.42	<0.001	-	0.22
M6 Y = BA + sex + line + ADG_BIRTH_IGF-I	lIGF-Ij	0.49	<0.001	-	<0.001
M7 Y = BA + sex + line + ADG_WEAN_IGF-I	lIGF-Ij	0.65	<0.001	-	<0.001

*BA= combination of batch of birth and assay accounting for sampling date and generation

Line effect on lIGF-Ij when early growth measurements are accounted for. Accounting for pre-weaning WT or ADG covariates in the analysis did not change the significance of line differences for lIGF-Ij or increase coefficient of determination (R^2) of the model: 0.41 (M1) vs 0.42 (M2, M3 and M5) (Table 2). There was no evidence in these data that weaning weight significantly affected post-weaning gain within or across lines, supporting results from M1 vs M3. In contrast, including body weight at blood sampling (M4) or ADG from birth to sampling (M6) increased the model R^2 to 0.49, but with limited impact on the estimated line difference for lIGF-Ij measurements. Finally, accounting for ADG_WEAN_IGF-I decreased the line difference by 49%, as showed in Figure 1. This suggests that the line difference in lIGF-Ij is partly due to line differences in weight gain after weaning. These results are consistent with the literature on the role of IGF-I as a growth factor involved in growth and protein metabolism (Le Roith *et al.* 2001) that depends on the feed intake and nutritional status of the animal (Caroll *et al.* 1998).

The LRFI piglets had a higher growth rate before weaning compared to HRFI piglets, and all were suckled only by LRFI sows, ie no cross-fostering was allowed across lines. Therefore, line differences in piglet performance are potentially confounded with line differences in maternal effects. Weaning is a stressful event for piglets, with separation from the dam, a change of feed and mixing of litters. It generates a transient reduction of feed intake that can lead to digestive disorders. The greater difficulty of the LRFI piglets to adjust to weaning needs further examination to decipher the role of pre-weaning conditions from individual sensitivity to stress on these results. However, the absence of body weight difference between lines when growing-finishing starts also suggests a good resilience of these piglets, which return to a higher growth rate after the stress of weaning (Gilbert *et al.* 2017b).

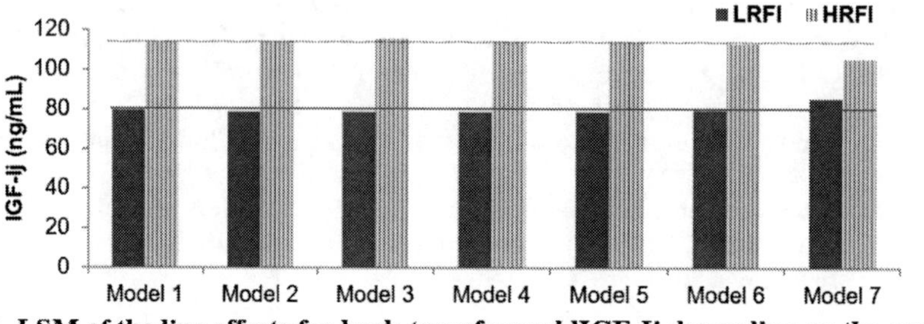

Figure 1. LSM of the line effects for back-transformed lIGF-Ij depending on the covariate included in the model – see Table 2 for details on the models.

CONCLUSION

Our study confirms that juvenile IGF-I is an indirect indicator of growing-finishing feed efficiency of the pigs. The immediate growth after weaning affected IGF-Ij, which could be considered for a better prediction of genetic merit for feed efficiency. The biological mechanisms underlying these phenomena remain to be studied for a better understanding of the relationships between post-weaning growth, juvenile IGF-I and subsequent feed efficiency. Our study confirmed that weaning creates a greater growth check in pigs selected for low RFI, not explained by variation in weaning weight, but also that LRFI pigs show good resilience to the challenge, as indicated by their overall post-weaning growth rate. Altogether, our results show that selection for reduced RFI should be combined with optimized management of the young weaned pig.

REFERENCES

Bunter K.L., Hermesch S., Luxford B., Graser H. and Crump R. (2005) *Anim. Prod. Sci.* **45**: 783.

Bunter K.L., Cai W., Johnston D.J. and Dekkers J.C.M. (2010) *J. Anim. Sci.* **88**: 1973.

Carroll J.A.,Veum T.L. and Matteri R.L. (1998) *Domest. Anim. Endocrinol.* **15**: 183.

Gilbert H., Billon Y., Brossard L., Faure J., Gatellier P., et al. (2017a) *Animal* Jan 25 : 1.

Gilbert H, Ruesche J., Muller N., Billon Y., Robert F., et al. (2017b) In '49th Journées de la Recherche Porcine'

Hermesch S., Bunter K.L. and Luxford B.G. (2001) *Proc. Assoc. Advmt. Anim. Breed. Genet.* **14**: 227.

Le Roith D., Bondy C., Yakar S., Liu J.L. and Butler A. (2001) *Endocr. Rev.* **22**: 53.

Louveau I. and Bonneau M. (1996) *Reprod. Nutr. Dev.* **36**: 301.

Proc. Assoc. Advmt. Anim. Breed. Genet. 22:553-556

A GENETIC EXPLORATION OF AUSTRALIAN LARGE WHITE PIGS

K.P. Gore, R. Banks, V. Boerner, S. Hermesch

Animal Genetics and Breeding Unit (AGBU), University of New England, Armidale, NSW 2351, Australia. AGBU is a joint venture of NSW Department of Primary Industries and the University of New England.

SUMMARY

Limited information is available about the genetic structure of Australian pigs. The genetic relationships of 20 Large White and 2 Duroc pigs sampled from one herd were explored and their genetic distance to European pigs of the same and different breeds was estimated using SNP data. On average, 36% of SNPs were heterozygous for Large White pigs. Mean correlations between Australian Large White and European breeds were highest for European Large White (0.35) and Middle White (0.34), and lowest for European Duroc (0.20) and Meishan (0.09). The analysis of breed percentages based on constrained genomic regression showed highest similarity of Australian Large White with European Large White (36.8%) followed by Middle White (15.9%) and Welsh (12.6%). Based on this small sample, the Australian pigs sampled retained significant heterozygosity and can be regarded as a distinct population to the sampled European breeds.

INTRODUCTION

Importation of porcine genetic material into Australia has been prohibited for over 2 decades. The domestic industry has had to breed for productivity while controlling inbreeding without the ability to introduce external genetics. With the number of producers shrinking by approximately two thirds during that time (Australian Pork Limited 2013), sourcing diverse off-farm genetics has become more difficult.

Genomic information being accessible on-line offers opportunities to examine relationships and genetic diversity between populations. The availability of this information vastly reduces the cost to any given individual researcher, and enables initial exploration of the genetic (genomic) structure of the Australian population to be undertaken.

MATERIAL AND METHODS

Hair samples were obtained from 22 pigs (21 boars and 1 sow) in the herd at the University of Queensland, Gatton Australia which has used boars from other Australian herds. The samples were from 20 Large White and 2 Duroc. The 2 Duroc pigs were not discussed in detail in the results presented here due to the small numbers.

Samples were genotyped using the GeneSeek Genomic Profiler HD chip. The chip originated from the original Illumina Porcine60k chip with approximately 12,000 SNPs of low informativeness in major commercial breeds removed and an extra 20,000 added to fill gaps in the chromosomes (J. Walker, personal communication, April 1 2017). Quality control consisting of minimum 85% call rate was applied resulting in 1 sample being rejected. One duplicate was also detected. SNPs were removed where there was at least 1 missing call resulting in 44,749 SNPs. The 2 possible heterozygote calls were not considered as different in any calculation.

The publically available European SNP data was already subject to quality controls from its original publication (Wilkinson *et al.* 2013a,b). The Australian genotypes were merged with the European SNP data by SNP name and SNPs were removed if there was at least 1 missing call resulting in 24,564 SNPs that were available in both SNP data sets. Pearson correlation in R (R Core Team 2015) was used to generate the correlation matrix between the genotypes.

Breed percentage was calculated for the Australian Large White pigs using the European data as the reference set with constrained genomic regression (Boerner 2017). This was repeated for the European Large White pigs with the Australian Large White pigs in the reference set for comparison. Principal component analysis on the Middle White, European Large White and Australian Large White breeds was done by generating a genetic relationship matrix (Yang *et al.* 2010) followed by singular value decomposition using the SVD function from the NumPy package (van der Walt *et al.* 2011) for python. The R functions kmean and dist (R Core Team 2015) were used to assign individuals to clusters and calculate the distance between the cluster centres.

RESULTS AND DISCUSSION

Within herd comparison. The mean correlation between the genotypes was 0.41 with a range from 0.33 to 0.62 and a standard deviation of 0.04. The maximum value of 0.62 was confirmed by pedigree records to be a parent-progeny pair.

The percentage of heterozygous SNPs for each pig ranged from 34 to 38 with a standard deviation of 1. The low standard deviation is likely to be the result of considering a single breed. Including the 2 Duroc pigs increased the standard deviation to 2 because of their lower percentage of heterozygous SNPs (31 and 32). Duroc is the smaller breed in comparison to Large White in Australia which was reflected in higher inbreeding levels and smaller effective population size for Duroc in comparison to Large White based on pedigree information (D'Augustin *et al.* 2017). The results of this study based on genomic information corresponded to the findings based on pedigree data despite the small sample size.

The percentage of heterozygous SNPs in the European and Australian breeds is shown in Table 1. The Australian Large White pigs (AULW) were the highest of all with a mean of 35.5. These means were much lower than those of Zhang and Plastow (2011), which may be the result of only considering SNPs that were called for all pigs. Imputation of these sporadic uncalled SNPs may allow more of the data to be used. Li *et al.* (2006) showed that reasonable accuracy of imputed SNPs can be achieved with as few as 90 individuals which could be achieved for this sample of pigs with additional genotyping.

European comparison. The heat map (Figure 1) of the correlation matrix indicated that the Australian animals could be considered a separate breed to the European Large Whites. The squares along the diagonal show the groups of animals of the same breed. The order of breeds from the top left to the bottom right corner was Meishan (MS), Gloucestershire Old Spots (GL), Berkshire (BK), Wild boar (WB), Large Black (LB), British Saddleback (BS), Tamworth (TA), Hampshire (HA), Mangalica (MA), Australian Duroc, Duroc (DU), Landrace(LR), Welsh (WE), Pietrain (PI), Middle White (MW), European Large White (LW), Australian Large White (AULW). The highest mean correlations to Australian Large White were European Large White (0.35) and Middle White (0.34) as shown in Table 2. The lowest correlation was Meishan(0.09).

Table 1 Mean percentage of heterozygous SNPs for breeds

AULW	MS	GL	BK	WB	LB	BS	TA	HA	MA	DU	LR	WE	PI	MW	LW
35.5	16.3	23.3	23.8	20.7	26.7	30.2	21.7	24.3	15.2	26.3	32.3	33.9	34.2	29.6	33.9

Principal component analysis of the White breeds (Middle White, European Large White and Australian Large White) showed that each breed formed separate clusters (Figure 2). The Middle White and European Large White breeds were focused on due to the higher relationship shown in the correlation heat map. The distance between the cluster centers was 14.0 from the Middle White to the European Large White, 12.6 from the Middle White to the Australian Large White and 9.0

between the European and Australian Large Whites.

There were significant proportions for European Large White, Middle White, Pietrain and Welsh breeds. Minimal proportions of Landrace, Duroc, British Saddleback, Gloucestershire Old Spots, Meishan, Tamworth, Hampshire and Large Black (Table 3). Breeds where the mean was below 1% are not shown (Berkshire 0.6%, Mangalica 0.2%, Wild Boar 0.8%).

Table 2 Mean correlations between Australian Large White and European breeds (*100)

	MS	GL	BK	WB	LB	BS	TA	HA	MA	DU	LR	WE	PI	MW	LW
Mean	9	28	28	28	29	31	30	28	29	20	29	30	29	34	35
SD	1	2	2	2	2	2	2	2	2	2	2	2	2	2	2

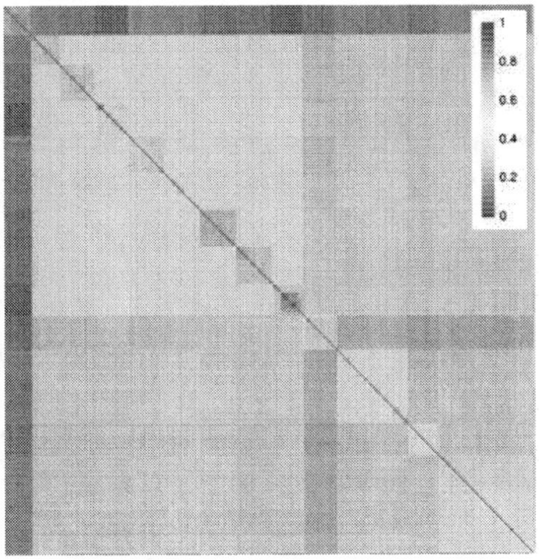

Figure 1 Heatmap of individual correlations between Australian and European Breeds.

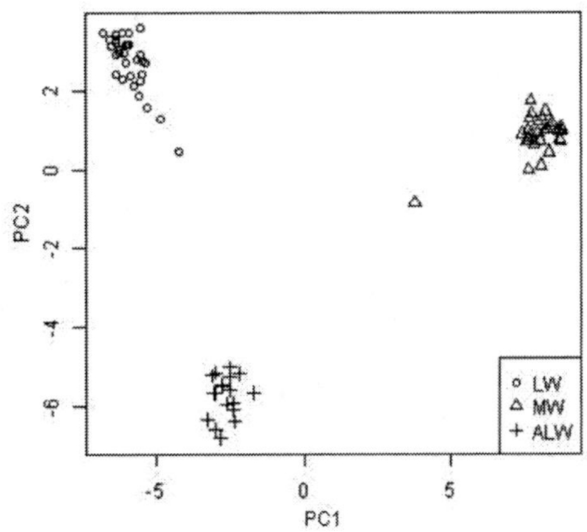

Figure 2 PC1 vs PC2 for European Large White, Middle White and Australian Large White.

Table 3 European breed percentages of Australian Large White pigs

	MS	GL	LB	BS	TA	HA	DU	LR	WE	PI	MW	LW
Mean	2.3	2.4	1.1	2.6	2.0	1.9	3.1	5.8	12.6	12.5	15.9	36.8
Min	0.9	0.0	0.0	0.0	0.0	0.0	0.0	0.0	4.0	2.6	10.3	21.1
Max	4.1	5.8	4.6	7.5	4.8	4.5	5.9	18.2	26.0	18.3	22.3	44.5
SD	1.1	2.0	1.5	2.6	1.4	1.2	1.6	4.7	5.3	3.8	3.4	5.2

The calculated breed percentages of European Large White pigs is shown in Table 4. Breeds where the mean was below 1% are not shown (Duroc 0.2%, Gloucestershire Old Spots 0.8%, Hampshire 0.3%, Mangalica 0.6%). The highest percentage was the Australian Large White at 45%. The Australian Large White pigs showed nearly 10 times the percentage of the Welsh breed (12.6%) than the European Large White pigs (1.3%).

Table 4 Breed percentages of European Large White pigs

	MS	BK	WB	LB	BS	TA	LR	WE	PI	MW	AULW
Mean	3.9	2.3	3.5	2.3	5.5	1.3	6.8	1.3	10.2	15.9	45
Min	1.4	0	0	0	0	0	1.5	0	3.1	8.2	31.3
Max	6.8	6.8	8.2	7.1	10.2	4.1	11.6	10.3	38.6	21.8	52.8
SD	1.3	1.6	2.5	1.8	2.8	1.2	2.9	2.3	6	3.1	5.2

CONCLUSIONS

This Australian Large White population has different from the European Large White population. The between-breed correlation matrix showed a higher relationship between the White breeds when compared to the other breeds but the principal component analysis showed that this sample of Australian Large White pigs was distinctly different from the European White breeds. Although the Australian genotypes originated from just 1 herd, there was a similar level of genetic diversity within this one herd as within the European Large White population, suggesting that this herd at least is maintaining diversity. Both the Australian and European Large White populations retained genetic contributions from other breeds, presumably reflecting introductions over time. This study is based on a small sample and caution should be exercised in concluding that the diversity estimated within this herd is an accurate estimate of that in the whole Australian population. Further investigation of the genomic structure of a larger sample of Australian pigs is required in order to obtain more detailed knowledge of the genetic diversity of Australian pigs.

ACKNOWLEDGEMENTS

We would like to acknowledge Tanya McKenna at UQ for the provision of hair samples, Rose Andrew for making us aware of the European data and Pam Wiener for her assistance with the European data.

REFERENCES

Australian Pork Limited (2013) Australian Pig Annual 2012-2013

Boerner V. (2017) Proc. Assoc. Adv. Anim. Breed. Genet. **22**: In prep.

D'Augustin O., Boerner V. Hermesch S. (2017) Proc. Assoc. Adv. Anim. Breed. Genet. **22**:Subm.

Li Y., Willer C, Sanna S, Abecasis G (2006) *Annu Rev Genomics Hum Genet* **10**:387

R Core Team (2015). R: A language and environment for statistical computing. R Foundation for Statistical Computing, Vienna, Austria.

Van der Walt S, Colbert S Chris, Gael Varoquaux (2011) *Computing in Science & Engineering* **13**:22-30

Wilkinson S, Lu ZH, Megens H, Archibald AL, Haley C, Jackson IJ, Groenen MAM, Crooijmans RPMA, Ogden R and Wiener P (2013) *PLoS Genetics* **9**: e1003453. doi:10.1371/journal.pgen.1003453

Wilkinson S, Lu ZH, Megens H, Archibald AL, Haley C, Jackson IJ, Groenen MAM, Crooijmans RPMA, Ogden R, Wiener P (2013) Data from: Signatures of diversifying selection in European pig breeds. Dryad Digital Repository. http://dx.doi.org/10.5061/dryad.c2124

Yang J, Benyamin B, McEvoy B, Gordon S, Henders A, Nyholt D, Madden P, Heath A, Martin N, Montgomery G, Goddard M, Visscher P (2010) *Nat Genet* **42**:565

Zhang C, Plastow G (2011) *Curr Genomics* **12**:138

Proc. Assoc. Advmt. Anim. Breed. Genet. 22:557-560

WEIGHTING OF GENOMIC AND PEDIGREE RELATIONSHIPS IN SINGLE STEP EVALUATION OF CARCASS TRAITS IN AUSTRALIAN SHEEP

A.J. McMillan[1] and A.A. Swan[1]

[1]Animal Genetics Breeding Unit*, University of New England, New South Wales, Australia

SUMMARY

An issue for implementation of single step genomic evaluations is how to weight genomic and pedigree relationships in modelling genetic co-variance. A weighting parameter lambda ranging between 0 and 1 can be used in the statistical model, with higher values corresponding to greater weighting of genomic information. We investigated appropriate values of lambda for a range of carcass traits in terminal sire sheep breeds, using the accuracy and bias of genomic prediction of breeding values as criteria. The accuracy generally increased with lambda, although the "optimal" value of lambda at the maximum accuracy varied widely, covering almost the entire range of possible values across traits. Accuracy typically approached an asymptote towards the optimal lambda, so a wide range of values could be used with minimal loss of prediction accuracy. The bias in Estimated Breeding Values (EBVs) increased with lambda, such that EBVs over-predicted phenotypic performance at high values of lambda.

INTRODUCTION

Evaluations utilising genomic information in the form of blended EBVs have been available to Australian sheep breeders since 2011 (Swan *et al.* 2012). However utilising all available information on animals including phenotypes, genotypes and pedigree information in routine Australian sheep analysis were desired but had not previously been accomplished. In 2016 large scale multi-trait single step analyses (Legarra *et al.* 2014) were implemented for carcass and live weight traits in the three major breed evaluations, Terminal sires, Maternal breeds, and Merinos. These analyses include 17 traits, with pedigrees in excess of 2 million animals, and SNP genotypes for up to 15 thousand animals.

One of the issues for the implementation of single step in routine evaluations is how to optimally combine genomic and pedigree information for genotyped animals, since it is often argued that SNP genotypes do not explain all of the genetic variation (Goddard *et al.* 2011). To accommodate this, the variance of breeding values for genotyped animals can be modelled as $(\lambda G + (1-\lambda)A_{22})\sigma_u^2$, where G is the genomic relationship matrix calculated from SNP genotypes, A_{22} is the pedigree relationship matrix between genotyped animals, σ_u^2 is the genetic variance, and λ (lambda) is a weighting factor between 0 and 1. This variance matrix can be used in single step analyses, and often a high value of lambda, between 0.95 and 0.99, is used for the pragmatic reason that the resulting modified genomic relationship matrix can be reliably inverted. However, the broader questions remain, what is an appropriate value for lambda, and does lambda vary between traits? In this paper we use cross-validation to investigate the accuracy of genomic predictions across a range of lambda values for a range of carcass traits important for the terminal sire single step evaluation.

* AGBU is a joint venture of NSW Department of Primary Industries and University of New England

MATERIALS AND METHODS

A subset of animals from the Australian Terminal sire sheep evaluation were chosen, derived from the genomic reference population first established by the Sheep CRC (Van der Werf 2010). These animals have both genotypes and phenotypes for the traits studied. Key traits from the single step carcass analysis for terminal sires were investigated, including post-weaning weight (pwt), post-weaning eye muscle depth from live animal scanning (pemd), post-weaning fat depth from live animal scanning (pfat), hot carcass weight (hcwt), carcass eye Muscle Depth (cemd) carcass C-site fat depth (ccfat), lean meat yield (lmy), intra-muscular fat (imf), and shear force at day 5 (sf5). A summary of the animals recorded per trait and total animals in the pedigree is shown in Table 1.

Table 1: Data summary for terminal sire cross-validation analyses, with size of pedigree, number of animals recorded (and genotyped), number of Poll Dorset/White Suffolk animals with records (PD/WS rec), number of sires (PD/WS sires), and number of cross-validation sets (PD/WS ncv).

Trait	Pedigree	Records	PD rec	PD sires	PD ncv	WS rec	WS sires	WS ncv
pwt	28826	7714	3764	247	12	2567	169	8
pemd	28825	7713	3764	247	12	2566	169	8
pfat	28820	7712	3763	247	12	2567	169	8
hcwt	31774	8976	4298	248	14	2981	170	9
cemd	31345	8720	4172	248	13	2896	170	9
ccfat	31191	8630	4132	248	13	2868	170	9
lmy	22752	5254	2416	85	8	1658	56	5
imf	29952	8088	3905	215	13	2770	154	9
sf5	30764	8374	4017	248	13	2814	170	9

For each of these traits, the procedure involved estimating SS-GBLUP (Single Step Genomic BLUP) REML variance components using the Wombat software package (Meyer 2007) for values of lambda ranging between 0 and 1 in increments of 0.1. Animals with phenotypes were then allocated to cross-validation groups of approximately 300, stratified within two breeds, Poll Dorset (PD) and White Suffolk (WS). Animals were allocated to breeds based on the breed content of their sires. In addition, progeny from the same sire family were always allocated to the same cross-validation group, such that no animal in a cross-validation set would have half-sibs in the training data. Within these strata animals were allocated to groups at random, and the same groupings were used for all values of lambda. Summaries of the cross-validation schemes are shown in Table 1.

SS-GBLUP analyses were carried out for each cross-validation set across the range of lambda values specified above, using the 's1step' option in Wombat. Phenotypes for animals in the cross-validation set were omitted from the training data, but their pedigree and genotype data were included in the analysis in order to obtain their EBVs. Prediction accuracy was then calculated as the correlation between these EBVs and their phenotypes (adjusted for fixed effects). To approximate the correlation between True Breeding Value and EBV, these correlations were then scaled by the square root of the heritability of the trait, which was assumed to be the heritability estimated in the absence of genomic information. EBV bias was also calculated for each cross-validation set as the slope of the regression of phenotype on EBV (the expected value of the slope is 1, and if the estimate is less than 1 then EBVs over-predict phenotypic performance). Prediction accuracies and bias was then averaged across the cross-validation sets.

RESULTS AND DISCUSSION

Table 2 shows for each trait the estimated heritability (lambda = 0), the maximum cross-validation accuracy for breeds (r_{max}), the value of lambda where the maximum cross-validation accuracy was observed, and the range in lambda values where the accuracy varied by ±0.01. Figure 1 displays accuracy across the range of lambda values by sire breed.

Table 2: Terminal sire lambda cross-validation summary, with estimated heritability (h^2), maximum cross-validation accuracy (r_{max}), $\lambda_{max} = \lambda$ at r_{max}, and range in lambda where accuracy varied by ±0.01 around r_{max} (λ_{low} to λ_{high}).

Trait	h^2	$r_{max(PD)}$	$r_{max(WS)}$	$\lambda_{max(PD)}$	$\lambda_{max(WS)}$	$\lambda_{low(PD)}$	$\lambda_{low(WS)}$	$\lambda_{high(PD)}$	$\lambda_{high(WS)}$
pwt	0.29	0.32	0.27	0.50	0.20	0.20	0.10	0.95	0.60
pemd	0.35	0.37	0.28	0.80	0.95	0.40	0.70	0.95	0.95
pfat	0.24	0.28	0.29	0.95	0.95	0.70	0.70	0.95	0.95
hcwt	0.14	0.41	0.27	0.20	0.95	0.20	0.80	0.50	0.95
cemd	0.21	0.34	0.21	0.60	0.90	0.40	0.50	0.95	0.95
ccfat	0.28	0.21	0.22	0.60	0.60	0.40	0.40	0.95	0.95
lmy	0.49	0.26	0.39	0.80	0.60	0.60	0.40	0.95	0.80
imf	0.60	0.35	0.28	0.80	0.60	0.50	0.40	0.95	0.95
sf5	0.37	0.28	0.24	0.60	0.50	0.30	0.30	0.95	0.95

Lambda values at maximum accuracy were 0.5 or greater, except pwt (WS) and hcwt (PD). As the maximum accuracy was approached, the response surface was generally asymptotic (see Figure 1), such that the range encompassing r_{max} ± 0.01 was large. Therefore, accuracy was relatively insensitive over a large range of lambda values especially beyond 0.5.

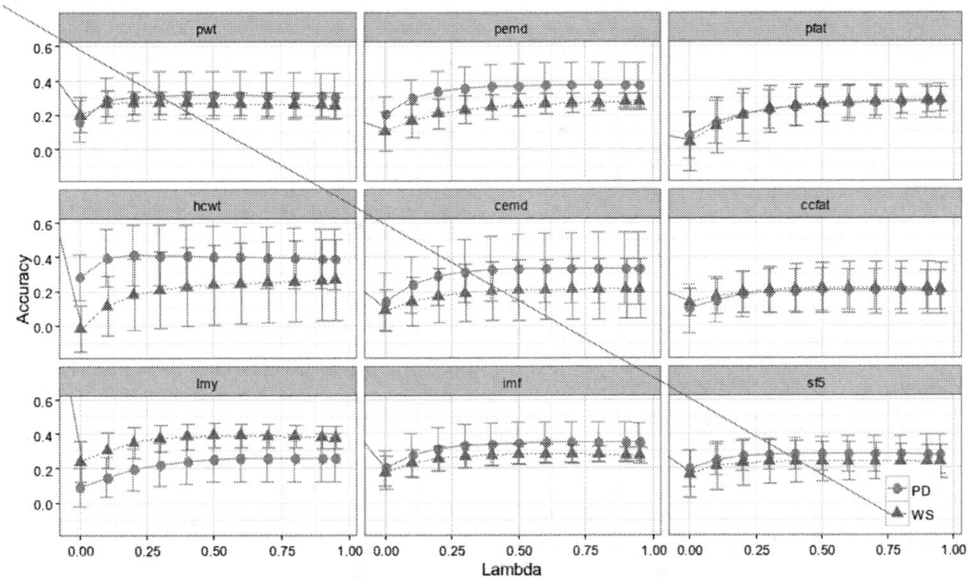

Figure 1: Accuracy versus lambda by sire breed in terminal sires (PD=Poll Dorset, WS = White Suffolk). Error bars show ± 1 standard deviation.

The slope of the regression of phenotype on EBV was used to assess the bias of EBVs across the range of lambda values and is shown in Figure 2. Results show some variation between traits and sire breeds within traits, but there is a clear general trend that the bias increases with lambda. That is, higher values of lambda lead to EBVs which over-predict phenotypic performance. In selection cohorts with a mix of genotyped and un-genotyped contemporaries, this may lead to genotyped animals being incorrectly favoured. It is uncertain to us why the bias increases with lambda, but it may be due to an increasing influence of small genomic relationships in the *G* matrix which are due to identity by state rather than identity by descent genome sharing.

Correlations between EBVs for different lambda values were also calculated for different classes of animals, including progeny tested sires, and animals with and without phenotypes. For EBVs calculated with lambda of 0.5 and 0.95, these correlations ranged between 0.96 and 0.99, demonstrating that a wide range of lambda values between 0.5 and approaching 1 can be used with minimal impact on the ranking of animals.

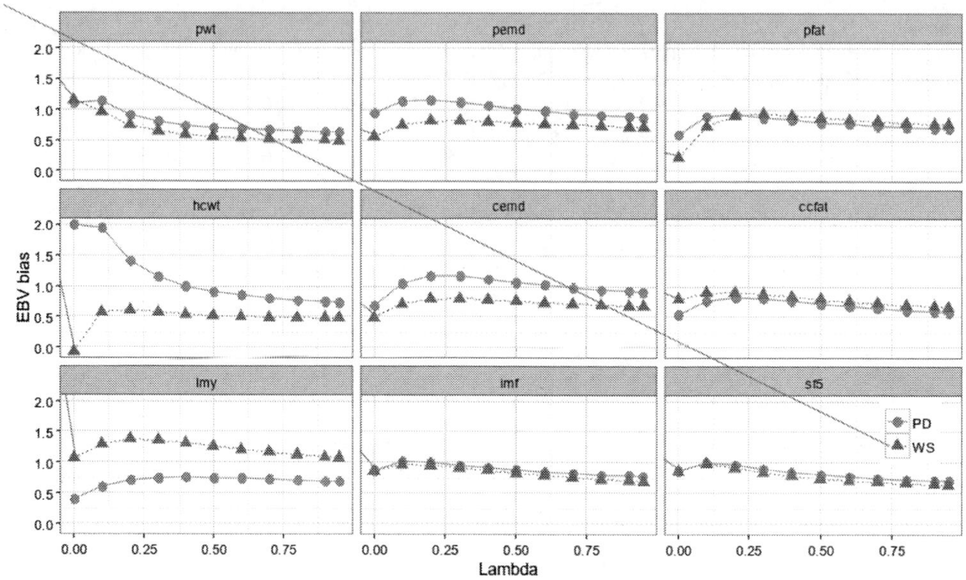

Figure 2: EBV bias versus lambda by sire breed in terminal sires (PD = Poll Dorset, WS = White Suffolk).

Given the relatively large window for insensitivity of prediction accuracy, high correlation of EBVs between lambda 0.5 and 0.95 and the levels of bias in EBVs when lambda is high we have initially used a value of 0.5 for lambda in routine industry evaluations. More research on this issue is warranted, including the impact of lambda in multi-trait models.

ACKNOWLEDGEMENTS

This research was funded by Meat and Livestock Australia. The authors acknowledge the contributions of the Sheep CRC Information Nucleus and industry-funded research flocks.

REFERENCES

Meyer K. (2007) *Journal of Zhejiang University. Science B* **8**: 815.
Swan, A.A., Johnston, D.J., Brown, D.J., Tier, B. and Graser, H-U. (2012) *Ani. Prod. Sci.* **52**: 126.
Legarra A., Christensen O.F., Aguilar I., and Misztal I. (2014) *Lives. Sci.* **166**: 54.
Goddard M.E., Hayes B.J. and Meuwissen T.H.E. (2011) *Jour. Ani. Breed. Genet.* **128**: 409.
Van der Werf J.H.J., Kinghorn B.P. and Banks R.G. (2010). *Ani. Prod. Sci.* **50**: 998.

Proc. Assoc. Advmt. Anim. Breed. Genet. 22:561-564

GENETIC VARIATION OF WEANER SURVIVAL IN MERINO SHEEP AND ITS RELATIONSHIPS WITH GROWTH AND WOOL

S.F. Walkom[1], A.N. Thompson[2], E. Bowen[3], D.J. Brown[1]

[1] Animal Genetics and Breeding Unit*, University of New England, Armidale, NSW, Australia
[2] School of Veterinary and Life Sciences, Murdoch University, Murdoch, WA, Australia
[3] Sheep Data Management, 14 Goonawarra Street, Wagga Wagga, NSW, Australia

SUMMARY

There is little evidence that mortality rates in Australian sheep during the period after weaning are improving over time. This paper explores the potential for producers to select to improve survival rates and the potential impact this may have on key production traits. A total of 114,272 weaner survival records were obtained from a wide variety of Australian Merino sheep types and production systems. Weaner survival, a binary dependent variable, was analysed as a continuous trait using a sire model. The heritability of weaner survival was estimated at 0.05 ± 0.01, significantly greater than zero. The survival of weaned lambs to yearling age was influenced by weaning weight with higher survival rates observed in heavier lambs (r_g = 0.14). Weaner survival adjusted for weaning weight was found to be antagonistically genetically correlated with fleece weight (r_g = -0.12 to -0.24). Due to antagonistic genetic correlations with greasy fleece weight and other production traits, producers should record weaner survival which will assist Sheep Genetics to produce breeding values and incorporate weaner survival in future indexes.

INTRODUCTION

The Australian sheep flock includes a significant proportion of young Merino sheep that are often characterised by poor growth, slower development and high mortality in the period follow weaning (Hatcher *et al.* 2008). There is little evidence to show that survival of Merino sheep for the post-weaning period have improved since the 1950's (Hatcher *et al.* 2008) with weaner mortality rates in Australian Merino flocks at a constant 5.2% (Campbell *et al.* 2014). Current management protocols to improve weaner survival are based on providing adequate nutrition and controlling worm burdens and fly strike to enable weaners to achieve live weight targets by weaning and maintain positive growth rates in the period following weaning (Hatcher *et al.* 2008, Campbell *et al.* 2014). It has been reported that lighter weaners were less able to cope with nutritional and or other stresses owing to lower energy reserves than heavier weaners and to improve post-weaning survival, Merinos should be managed to achieve approximately 45% of mature liveweight at weaning (Thompson *et al.* 2011). However, genetic parameters for weaner survival in sheep have not been estimated and the capacity to select for improved survival rates is unknown. The aims of this paper were to quantify the genetic variation in the Australian Merino population for survival from weaning to the yearling stage (7 to 9 months after weaning) and to estimate the genetic relationships between survival and key growth, carcass and wool traits.

MATERIALS AND METHODS

Data were obtained from 18 Merino flocks with lambs born from 1990 to 2014. The flocks included ram breeding, sire evaluation and research flocks from across Australia contributing to the MERINOSELECT database (Brown *et al.* 2007). Weaner survival was analysed as a binary trait of the lamb with animals assigned a value of 1 if alive or 0 if dead at the yearling stage. Weaner survival was verified by the presence of weight or production records provided to Sheep Genetics at or after

[1]AGBU is a joint venture of the NSW Department of Primary Industries and University of New England

the yearling stage (7 - 9 month period after weaning). Only animals with a known sire (syndicate sires removed) and a weaning weight record were included in the analyses. Contemporary groups (flock x location x year x sex) with a large number of animals which could not be assigned as dead or alive were excluded from the study. After data cleaning and the removal of uninformative contemporary groups, 104,557 weaner survival records were available for analysis with an average survival rate of 93%.

The growth and wool traits analysed included weaning, post-weaning and yearling liveweight (kg), ultrasound fat and muscle depth (mm), greasy fleece weight (kg), fibre diameter (μm), coefficient of variation (cv) in fibre diameter (%), curvature, staple length (mm), and staple strength (N/ktex). All scan and fleece traits were recorded at the yearling stage.

Statistical analysis. Genetic parameter estimates were calculated using a sire model in ASReml (Gilmour *et al.* 2009). The models fitted to the data were developed and described by Brown and Swan (2016) and was based on the linear mixed model:

$$y = X\hat{\beta} + Z_1 s + Z_2 m + Z_2 mp + sxf + e$$

where, y is a vector of observations for the trait; $\hat{\beta}$ is a vector of the fixed effects including birth type (1,2,3,4+), rearing type (1,2,3+), age of dam (as a linear and quadratic term) (mean 4.5 years of age), age of the animal (linear) and contemporary group. Contemporary group for the production traits described flock, management group, sex, and date of measurement (Brown and Swan 2016). All contemporary groups were transformed to a common mean as done routinely for Sheep Genetic analyses (Brown *et al.* 2007). The vectors s, m, and mp are the sire genetic effects, maternal genetic effects, and permanent environment due to dam effects, respectively. The incidence matrices X, Z_1, and Z_2 relate the respective effects to y; and e is a vector of random error effects. A sire by flock (sxf) term was also fitted as random for the production traits.

Weaner survival was recorded as a binary trait (0/1) but analysed as a continuous trait. A logit function was also tested, but for computational ease was not used in the bi-variate analysis. The fixed effect models fitted for weaner survival were based on the terms normally fitted for weaning weight in Sheep Genetics analyses. Data limitations (unavailable for a large proportion of individuals) meant that the contemporary group structure could not include weaning date, age at weaning or weaning management groups, all of which are fitted in the routine genetic evaluation of weaning weight (Brown and Swan 2016). Thus, contemporary groups for weaner survival described flock, flock location, year of birth and sex (male or female). The analysis was repeated with weaning weight fitted as a covariate for survival, in part to help account for the influence of weaning date, age at weaning, weaning management groups and maternal effects. The influence of weaning weight was tested by fitting weaning weight, first, as a covariate (linear effect) across the population and also as a nested covariate within contemporary group.

RESULTS AND DISCUSSION

The association of weaner weight with survival. Weaning weight exerted a strong, positive and highly significant effect on weaner survival, which is consistent with other reports in the literature (Hatcher *et al.* 2008, Thompson *et al.* 2011). In the current study, the regression of weaner survival on weaning weight predicted that on average, an 0.006 ± 0.001 (0.6%) improvement in weaner survival for every 1 kilogram increase in weaning weight, assuming a linear relationship (weaning weight; mean of 25kg, range of 6 - 49kg). However, the influence of weaning weight on weaner survival was not uniform across contemporary groups with the nested effect of weaning weight was highly significant and ranging from -0.078 to +0.050 weaner survival / kg of weaning weight. Some of the variation in survival responses to weaning weight between contemporary groups observed in the current study was likely to be due to interactions between weaning weight and post-weaning growth rate. Although not tested in this study, Thompson *et al.* (2011) concluded that post-weaning growth rates should exceed 30 g/day, and that growth rates below this level resulted in a decline in survival rates. Overall, the results suggest that the optimum weaning weight in pertaining to weaner survival is likely to differ across production systems, breeds and environments.

Variance components. Heritability of weaner survival in Merinos was low but significantly greater than zero and estimated at 0.05 ± 0.01 (Table 1) when analysed from a sire model. An animal model was also tested and estimated a heritability of 0.13 ± 0.01. Maternal genetic and permanent environmental effects were minimal and not significant whether an animal or sire model was fitted but were in part limited, since the structure of the survival trait means that all dams will have to have survived to the yearling stage. However, the significant effect of weaning weight on weaner survival is likely to be capturing some of the maternal environmental influence. Fitting weaning weight as a covariate had a small but not significant effect on the heritability and additive variance of weaner survival. Analysing weaner survival as a binary trait using a sire model with the logit-link function produced a heritability on the underlying scale of 0.19 which when transformed using the average frequency (incidence) equated to an approximate estimate of 0.01 on the observed scale.

Table 1: Heritability of weaner survival and genetic correlations for weaner survival with production traits when weaner survival is unadjusted for weaning weight, adjusted for average weaning weight (adjusted) or within each contemporary group (nested)

Trait	Records	unadjusted	adjusted	nested
Heritability of weaner survival	104,557	0.055 ± 0.005	0.053 ± 0.005	0.052 ± 0.005
Genetic correlations with:				
Weaning weight	193,784	0.14 ± 0.06	-	-
Post-weaning weight	106,968	0.30 ± 0.06	-	-
Yearling weight	110,023	0.24 ± 0.06	-	-
Yearling fat depth	39,318	0.34 ± 0.11	0.35 ± 0.10	0.36 ± 0.10
Yearling eye muscle depth	39,968	0.35 ± 0.09	0.13 ± 0.14	0.31 ± 0.09
Yearling greasy fleece weight	78,079	-0.12 ± 0.08	-0.22 ± 0.08	-0.24 ± 0.08
Yearling fibre diameter	82,293	0.16 ± 0.07	0.07 ± 0.08	0.07 ± 0.08
Yearling fibre diameter cv	81,687	-0.09 ± 0.08	-0.07 ± 0.08	-0.07 ± 0.08
Yearling curvature	74,575	0.05 ± 0.07	0.08 ± 0.07	0.07 ± 0.07
Yearling staple strength	31,131	0.10 ± 0.10	0.11 ± 0.10	0.10 ± 0.10
Yearling staple length	54,069	0.15 ± 0.08	0.10 ± 0.08	0.10 ± 0.08

Genetic relationship of weaner survival with production traits. Weaner survival (unadjusted for weaning weight) was moderately positively genetically correlated with liveweight (Table 1). This was consistent with the significant phenotypic influence of weaning weight on weaner survival observed in this study and in the literature (Hatcher *et al.* 2008, Thompson *et al.* 2011).

The genetic correlations of weaner survival with ultrasound fat depth was moderate and positive at 0.34 (Table 1). The genetic correlation for lamb survival with fat depth of the carcase at the GR site and the 5th rib has been reported as 0.34 and 0.00, respectively (Brien *et al.* 2013). These results suggest that high "genetic fat" was favourably associated with survival in lambs prior to and following weaning, and this is independent of any effect of weaning weight *per se*. The genetic correlations for weaner survival with ultrasound muscle depth were positive and ranged from 0.13 to 0.35 (Table 1). Lamb survival to weaning was lowly positively correlated with carcase eye muscle depth and area with Brien *et al.* (2013) reporting estimates of 0.17 and 0.04, respectively. Low to moderate positive correlations observed in this study suggest that weaner survival rates to the post-weaning stage will improve with selection for increased fat and muscle depth.

The genetic correlation of greasy fleece weight with weaner survival was -0.12, suggesting a weak genetic relationship (Table 1). After adjusting weaner survival for the effect of weaning weight this genetic correlation was slightly stronger at -0.24 (Table 1). These low negative genetic correlations suggest that high genetic fleece weight is associated with poorer survival rates from weaning to post-weaning at a standardised weaning weight (weight corrected). Previous research by Ferguson *et al.* (2007) and Hatcher and Atkins (2007) have both indicated unfavourable phenotypic associations of fleece weight with lamb survival. Adams *et al.* (2006) proposed that Merinos genetically superior for fleece weights have relatively smaller energy reserves which could contribute to the unfavourable genetic correlations observed in this study.

The genetic correlations for weaner survival and fleece quality traits, including mean and coefficient of variation in fibre diameter, curvature, staple length and staple strength were all low and generally not significantly different from zero (Table 1). Adjusting weaner survival for weaning weight had no significant impact on the genetic correlations between survival and wool quality traits.

CONCLUSION

Survival in Merinos from weaning to the yearling stage is lowly heritable but not zero, indicating that genetic variation exists which could be exploited. The survival of lambs from weaning to yearling was significantly influenced by weaning weight, with higher survival rates observed in genetically heavier lambs. The relationship with weight indicated that selection for heavier weaning and post-weaning weights, and in turn higher growth rates, will improve weaner survival. However, there remains genetic variation in weaner survival unrelated to weaning weight which can be selected for, and which is antagonistically associated with fleece weight. Due to antagonistic genetic correlations with key production traits, recording weaner survival would enable Sheep Genetics to calculate breeding values, and allow more balanced selection for improved survival and production traits.

ACKNOWLEDGEMENTS

This research is funded by Australian Wool Innovation. The authors acknowledge the contributions of the Sheep CRC Information Nucleus, the Australia Merino Sire Evaluation Association and industry-funded research flocks.

REFERENCES

Adams N.R., Briegel J.R., Greeff J.C., and Bermingham E.N. (2006) *Aus. J. Ag. Res.* **57**: 27.

Brien F.D., Rutley D.L., Mortimer S.I., and Edwards J.E. (2013) *Proc. Assoc. Adv. Anim. Breed. Genet.* **20**: 237.

Brown D.J., Huisman A.E., Swan A.A., Graser H.U., Woolaston R.R., *et al.* (2007) *Proc. Assoc. Adv. Anim. Breed. Genet.* **17**:187.

Brown D.J. and Swan A.A. (2016) *Anim. Prod. Sci.* **56**: 690.

Campbell A.J.D., Broekhuizen A., Curtis K., Croker K.P., Behrendt R., *et al.* (2014) *Anim. Prod. Sci.* **54**: 783.

Ferguson M.B., Adams N.R., and Robertson I.R.D. (2007) *Proc. Assoc. Adv. Anim. Breed. Genet.* **17**: 195.

Hatcher S. and Atkins K.D. (2007). *Proc. Assoc. Adv. Anim. Breed. Genet.* **17**: 260.

Hatcher S., Eppleston J., Graham R.P., McDonald J., Schlunke S., *et al.* (2008). *Aus. J. Exp. Ag.* **48**: 966.

Gilmour A.R., Gogel B.J., Cullis B.R. and Thompson R. (2009) 'ASReml User Guide Release 3.0.' (VSN International Ltd: Hemel Hempstead).

Thompson A.N., Ferguson M.B., Campbell A., Gordon D.J., *et al.* (2011) *Anim. Prod. Sci.* **51**: 784.

Proc. Assoc. Advmt. Anim. Breed. Genet. 22:565-568

PROGESTERONE IS AN INDIRECT INDICATOR OF REPRODUCTIVE OUTCOMES FOR YEARLING EWES

K.L. Bunter[1], J.E. Newton[2] and D.J. Brown[1]

[1] Animal Genetics and Breeding Unit, University of New England, Armidale, Australia
[2] Agriculture Research Division, DEDJTR, Victorian Government, Bundoora, Australia

SUMMARY

Ewes joined to first lamb as yearlings in industry ram breeding flocks were sampled for progesterone concentration (PROG) exactly 14 days after first exposure to males. The heritability of PROG was 0.22±0.06. PROG had positive genetic (P>0.05) and phenotypic (P<0.05) correlations with yearling, but not two-year old (2yo) ewe, reproductive traits affected by fertility. Pubertal status assigned using PROG was a significant (P<0.0001) factor for yearling fertility and related traits. Systematic effects, such as birth-rear type and dam age groups influenced yearling outcomes but were generally not significant for reproductive performance of 2yo ewes.

INTRODUCTION

The reproductive performance of ewes joined to lamb as yearlings is highly variable across flocks and years (Fogarty *et al.* 2007), even when weight and condition at joining are sufficient. A similar situation exists with beef heifers first joined to calve as two-year olds in Northern herds, where failure to attain puberty during the joining period has been identified as a contributing factor (Johnston *et al.* 2009). In that study, serial ovarian scanning was used to identify attainment of puberty based on the age when the first *corpus-luteum* (CL) was observed. However, this strategy is costly and time consuming and an alternative could be to evaluate physiological status based on reproductive hormone levels, such as progesterone. Circulating progesterone is potentially suitable as a marker for puberty, because it is produced post-puberty by the CL and is maintained at relatively high levels throughout most of the reproductive cycle (Foster and Jackson 2006).

In this study we investigate the use of progesterone sampling in the field during the first joining event for ewe lambs (<1 year old), under a controlled protocol of ram exposure and timing of sampling. The implications of systematic effects for progesterone and the subsequent reproductive performance of yearling and 2yo ewes are evaluated, along with the association between sire breeding values for yearling reproductive performance traits and progesterone level.

MATERIALS AND METHODS

Ewes used in this study were sourced from industry ram-breeding flocks representing a range of production environments and breeds recorded across nine sites in Southern Australia. Pedigree and birth details were available, along with accompanying growth and reproductive data. Ewes represented Merino (MER), maternal (MAT) and terminal (TERM) breed types. Flocks commenced joining predominantly in February and March when ewes averaged 7.5 months of age, but ewes ranged from 152 to 321 days of age at the commencement of joining. Blood samples from all sites were collected exactly 14 days after the introduction of ewes to males (teasers or rams). Plasma from these samples was assayed for progesterone concentration using a commercial ELISA for human samples, following the manufacturer instructions (Demeditec 2009). Ewes were classified as not pubertal (<0.95 ng/ml), of uncertain pubertal status (0.95-1.05 ng/ml) or pubertal (>1.05 ng/ml) at D14 of joining based on previous studies defining the threshold at which puberty is indicated (Sangha *et al.* 2002), while allowing for variation due to ELISA procedures. Ewes without progesterone recorded were classified as unknown status. Accompanying reproductive data from these flocks were extracted from the Sheep Genetics database for four years (2013-2016).

Reproductive traits included fertility (FERT), number of lambs born (NLB) and weaned (NLW) for ewes joined, litter size at lambing (LSIZE) and weaning (LWEAN) for ewes which lambed. Fertility and litter size traits were inferred from lambing data, or scanning data otherwise.

Reproductive traits were analysed fitting contemporary group (CG: 44 levels), which was a combination of site, year of joining and joining sub-group, and flock-dam breed group as the base model (M0). The M0 contemporary group for progesterone reflected site, date of bleeding and assay plate (PCG: 39 levels). Additional systematic effects were then investigated through a series of analyses. Dam age group (AGD: 4 levels; 1, 2, 3-5, 6+ years), month of birth (MON: 9 levels; March - November) and birth-rear type (BRT: 7 levels; 11, 2-, 21, 22, 31, 32, 33) were added simultaneously to M0 (M1). Pubertal status (PUB: 4 levels) was added to M1 for yearling traits, or after accounting for whether the ewe was previously joined as a yearling (YJOIN) for 2yo ewes traits (M2). Heritability estimates and genetic correlations between progesterone values and reproductive traits were estimated from a series of bivariate analyses under M1 fitting an animal model, using ASREML (Gilmour *et al.* 2009). Pearson correlations between sire breeding values for yearling reproductive traits (obtained from Sheep Genetics) and progesterone concentration (for sires with daughters recorded for progesterone) were calculated.

RESULTS AND DISCUSSION

Reproductive data were dominated by MAT breed types (~85% of all data) contributing to the relatively high mean litter size (Table 1). Fertility, NLB and NLW were substantially higher for ewes bred to lamb as 2yo ewes compared to yearling ewes, as expected. The heritability for progesterone concentration was moderate (0.22±0.06, Table 1). Heritability estimates for FERT, NLB and NLW were higher for yearling than 2yo ewes. Negligible heritability for yearling LWEAN (Table 1) indicates that culling on yearling LWEAN will be ineffective.

Table 1. Raw data characteristics along with heritability estimates (h^2) and the phenotypic variance (σ^2_P) for progesterone concentration and the reproductive traits (model M1)

	Yearling ewes				2yo ewes			
	N	Mean (SD)	h^2	σ^2_P	N	Mean (SD)	h^2	σ^2_P
FERT	10998	0.59 (0.49)	0.18±0.02	0.21	6494	0.92 (0.27)	0.03±0.02	0.02
NLB	10998	0.89 (0.86)	0.13±0.02	0.63	6494	1.46 (0.76)	0.08±0.02	0.54
NLW	9422	0.64 (0.78)	0.13±0.02	0.52	5913	1.07 (0.86)	0.07±0.02	0.69
LSIZE	6201	1.49 (0.54)	0.09±0.02	0.26	5972	1.58 (0.65)	0.08±0.02	0.39
LWEAN	4850	1.13 (0.67)	0.03±0.02	0.42	5421	1.17 (0.83)	0.07±0.02	0.63
PROG	1894	0.98 (0.13)	0.22±0.06	0.009	na	na	na	na

Contemporary group was very highly significant ($P<0.0001$) but explained less than 10% of variation for all reproductive traits (R^2(M0), Table 2). The addition of MON, AGD and BRT in combination increased model R^2 by up to 70% for YFERT, YNLB and YNLW (M1, Table 2), although overall R^2 remained relatively low, as expected for reproductive traits. Month of birth was the most significant factor ($P<0.0001$) affecting YFERT and therefore YNLB and YNLW, and remained significant for reproductive traits of 2yo ewes. Month of birth was more significant than month of joining when fitted concurrently (not presented). Birth-rear type was significant ($P<0.05$) for reproductive outcomes of yearling but not 2yo ewes, while dam age group was only significant for fertility (not litter size or lamb survival) outcomes and progesterone levels (M1, Table 2). Month of birth remained significant for yearling (but not 2yo ewes) reproductive traits even when age at the commencement of joining was fitted as a linear covariate (not presented), demonstrating that the effect of MON for yearling outcomes was not solely due to variation in age at joining. Pubertal status assigned using progesterone results was significantly associated with YFERT, YNLB and YNLW, but did not greatly increase model R^2 values due to both limited data for PROG and because systematic effects were common to both reproductive traits and PROG (Table 2). PUB was also significantly associated with litter size traits (but not fertility) of ewes lambing as two-year olds. This suggests that females which attain puberty early may also have higher litter size when more mature, supporting results observed by Edwards *et al.* (2015).

Table 2. The significance of systematic effects for yearling (Y) and 2yo ewes (H) reproductive traits and progesterone (PROG) under various models

Model 1 (M1)	Trait	R^2(M0)	R^2(M1)	P(AGD)	P(MON)	P(BRT)	P(PUB)
CG + AGD + MON + BRT	YFERT	8.8	14.3	0.04	<0.0001	0.006	-
CG + AGD + MON + BRT	YNLB	9.0	15.3	0.28	<0.0001	0.0002	-
CG + AGD + MON + BRT	YNLW	9.6	15.0	0.24	<0.0001	0.31	-
CG + AGD + MON + BRT	YLSIZE	7.6	9.3	0.26	<0.0001	0.007	-
CG + AGD + MON + BRT	YLWEAN	5.8	6.7	0.32	0.003	0.75	-
HCG + AGD + MON + BRT	HFERT	7.7	8.4	0.82	0.003	0.02	-
HCG + AGD + MON + BRT	HNLB	5.5	6.4	0.87	<0.0001	0.25	-
HCG + AGD + MON + BRT	HNLW	5.8	6.5	0.58	<0.0001	0.81	-
HCG + AGD + MON + BRT	HLSIZE	7.3	7.9	0.74	0.001	0.43	-
HCG + AGD + MON + BRT	HLWEAN	9.1	9.5	0.63	0.003	0.99	-
PCG + AGD + MON + BRT	PROG	46.4	51.3	0.04	<0.0001	0.07	na
Model 2 (M2)		R^2(M1)	R^2(M2)				
M1 + PUB	YFERT	14.3	14.7	0.05	<0.0001	0.01	<0.0001
M1 + PUB	YNLB	15.3	15.5	0.34	<0.0001	0.0003	<0.0001
M1 + PUB	YNLW	15.0	15.1	0.32	<0.0001	0.31	0.001
M1 + PUB	YLSIZE	9.3	9.3	0.28	<0.0001	0.007	0.70
M1 + PUB	YLWEAN	6.7	6.8	0.33	0.004	0.75	0.51
M1 + YJOIN + PUB(YJOIN)	HFERT	8.4	8.4	0.80	0.01	0.02	0.81
M1 + YJOIN + PUB(YJOIN)	HNLB	6.4	6.9	0.68	0.005	0.37	0.03
M1 + YJOIN + PUB(YJOIN)	HNLW	6.5	6.7	0.39	0.001	0.79	0.07
M1 + YJOIN + PUB(YJOIN)	HLSIZE	7.9	8.5	0.61	0.02	0.67	0.0006
M1 + YJOIN + PUB(YJOIN)	HLWEAN	9.5	9.8	0.41	0.01	0.99	0.01

Least square means show declining fertility outcomes with increasing MON, of large magnitude for yearling ewes (Y) and lesser magnitude for 2yo ewes (H). Relative to lambs reared as singles, lambs reared as multiples had reduced YFERT, but not reduced HFERT. YFERT was lower when progesterone sampling indicated that the ewe was not showing signs of puberty 14 days into the joining period (Table 3). In addition, ewe lambs born to yearling dams had both lower progesterone (0.92 vs 0.95, P=0.02) and poorer fertility outcomes (0.54 vs 0.59, P=0.01) than ewe lambs born to older dams.

Genetic and phenotypic correlations between PROG with YFERT, YNLB or YNLW suggest a positive genetic association between progesterone levels and yearling reproductive traits influenced by fertility (Table 4). Correlations between sire breeding values for PROG with ASBVs for YNLB or YNLW, derived using more extensive data, were positive in two of the three breed groups. For sires with N>10 daughters sampled for progesterone and with an accuracy >30% for the ASBV for YNLB, Pearson correlation coefficients were 0.35 and 0.51 (P=0.02) in Merino's (20 sires), 0.21 and 0.10 in MAT breeds (48 sires) and -0.16 and -0.33 in TERM breeds (12 sires) for YNLB and YNLW. However, yearling reproductive data for TERM breed ewes were affected by a delay in joining following pharmaceutical intervention, whereas MER and MAT ewes were naturally joined. The timing of sampling for progesterone was chosen to minimise false negatives (ie ewes which tested negative because of the phase of their cycle). However, it was also possible for ewes to attain puberty within the joining interval after progesterone sampling and therefore a single sample of progesterone is not a perfect predictor for the early attainment of puberty.

Table 3. Least square means for systematic factors affecting progesterone concentrations or yearling (Y) and 2yo ewes (H) reproductive traits

Factor		Month of birth					Birth-rearing group			Pubertal status			
Trait		**6**	**7**	**8**	**9**	**10**	**SS**	**MS**	**MM**	**0**	**1**	**2**	**U**
PROG	**Y**	0.98^{a}	0.96^{a}	0.96^{a}	0.85^{b}	0.83^{b}	ns	ns	ns	na	na	na	na
FERT	**Y**	72^{a}	70^{a}	55^{b}	33^{c}	37^{c}	59^{a}	59^{a}	56^{b}	50^{a}	57^{a}	67^{b}	58^{ac}
	H	96^{a}	91^{ab}	90^{ab}	87^{b}	84^{bc}	ns	ns	ns	ns	ns	ns	ns
NLB	**Y**	1.12^{a}	1.06^{a}	0.74^{b}	0.38^{c}	0.45^{c}	0.85^{a}	0.89^{a}	0.79^{b}	0.75^{a}	0.80^{a}	1.0^{b}	0.83^{ac}
	H	1.56^{abc}	1.48^{ac}	1.47^{ac}	1.38^{b}	1.50^{c}	ns	ns	ns	1.46^{a}	1.52^{ab}	1.68^{b}	1.61^{bc}
NLW	**Y**	0.84^{a}	0.76^{a}	0.49^{b}	0.21^{c}	0.11^{c}	ns	ns	ns	0.52^{a}	0.57^{ab}	0.68^{b}	0.53^{ac}
	H	1.43^{a}	1.08^{b}	1.11^{b}	0.88^{c}	1.08^{ab}	ns	ns	ns	ns	ns	ns	ns
LSIZE	**Y**	1.52^{a}	1.44^{a}	1.30^{bc}	1.26^{c}	1.25^{c}	1.35^{ab}	1.39^{a}	1.31^{b}	ns	ns	ns	ns
	H	1.71^{ac}	1.58^{a}	1.63^{a}	1.54^{b}	1.70^{ac}	ns	ns	ns	1.64^{a}	1.69^{a}	1.79^{b}	1.68^{a}
LWEAN	**Y**	1.15^{a}	1.02^{a}	0.92^{bc}	0.86^{c}	0.75^{abc}	ns	ns	ns	ns	ns	ns	ns
	H	1.60^{ac}	1.15^{a}	1.28^{a}	1.00^{b}	1.20^{ac}	ns	ns	ns	1.25^{a}	1.32^{a}	1.22^{ab}	1.12^{b}

SS: born-reared single; MS: multiple-reared single; MM: multiple-reared multiple; 0: not pubertal; 2: pubertal; 1: intermediate; U: untested; ns: P>0.05; na: not applicable; common superscripts within factor indicate P>0.05 (Month of birth and Birth-rearing type levels simplified for presentation)

Table 4. Genetic (r_g) and phenotypic (r_p) correlations between progesterone concentration and reproductive traits for ewes joined to lamb as yearlings or 2yo ewes

Trait		FERT	NLB	NLW	LSIZE	LWEAN
Yearling	r_g	0.21±0.18	0.25±0.19	0.09±0.19	0.39±0.28	0.05±0.39
	r_p	0.16±0.02	0.12±0.02	0.05±0.02	-0.02±0.04	-0.11±0.04
2yo ewes	r_g	na	0.08±0.21	0.19±0.22	-0.05±0.22	0.11±0.23
	r_p	na	0.05±0.03	0.06±0.03	0.06±0.03	0.08±0.03

CONCLUSIONS

Results from this study suggest that failure to attain puberty is a likely contributor to failed reproductive performance ewes joined to lamb as yearlings. Progesterone measured at D14 after the commencement of joining was a heritable indicator of puberty and fertility. Several systematic effects which contribute to yearling reproductive performance were not significant for outcomes of 2yo ewes, and therefore models used for the genetic evaluation of yearling reproductive outcomes requires refinement for more accurate genetic evaluation of performance in this age class.

REFERENCES

Fogarty N.M., Ingham V.M., Gilmour A.R., Afolayan R.A., Cummins L.J., Hocking-Edwards J.E. and Gaunt G.M. (2007) *Aust. J. Agric. Res.* **58**: 928.

Foster D.L. and Jackson L.M. (2006) in physiology of reproduction pp2127-2176.

Edwards S.J., Juengel J.L., O'Connell A.R., Johnstone P.D., Farquhar P.A. and Davis G.H. (2015) *Sm. Rum. Res.* **123**: 118.

Gilmour A.R., Gogel B.J., Cullis B.R. and Thompson R. (2009) '*ASREML user Guide Release 3.0*' VSN International Ltd, Hemel Hempstead, UK.

Johnston D.J., Barwick S.A., Corbet N.J., Fordyce G., Holroyde R.G., Williams, P.J. and Burrow H.M. (2009) *Anim. Prod. Sci.* **49**: 399.

Sangha G.K., Sharma R.K. and Guraya, S.S. (2002) *Sm. Rum. Res.* **43**: 53.

Proc. Assoc. Advmt. Anim. Breed. Genet. 22:569-572

GENETIC RESPONSES IN COMPONENT AND COMPOSITE REPRODUCTION TRAITS IN MERINO EWES DIVERGENTLY SELECTED FOR NUMBER OF LAMBS WEANED

S.W.P. Cloete[1,2], J.B. van Wyk[3] & J.J. Olivier[2]

[1]Department of Animal Sciences, University of Stellenbosch, Private Bag X1, Matieland 7602, South Africa; [2]Directorate Animal Sciences: Elsenburg, Private Bag X1, Elsenburg 7607, South Africa; [3]Department of Animal, Wildlife and Grassland Sciences, University of the Free State, PO Box 339, Bloemfontein 9300, South Africa.

SUMMARY

This study reports phenotypic and genetic parameters and trends for reproduction traits in a Merino flock divergently selected for number of lambs weaned per ewe joined. Three component traits (ewes conceived per ewe joined, lambs born and average lamb weaning weight per ewe reared) and three composite traits (number of lambs born and weaned as well as total weight of lamb weaned, all per ewe joined) were assessed. Most traits were variable and heritable and should respond to selection. Genetic trends suggested divergence between the lines in all traits assessed. Composite trait selection for number of lambs weaned resulted in genetic trends in the desired direction in those component traits considered here.

INTRODUCTION

It is generally accepted that reproduction traits are economically important in sheep. It has been suggested that a trait closely resembling lamb output per ewe joined, termed as net reproduction rate, most closely resembles the breeding objective strived for (Olivier 1999). However, selection decisions are complicated by the composite nature of this trait. Snowder and Fogarty (2009) argued that selection for such a composite trait will ensure that the contributing component traits remain in balance. However, approaches followed to select for reproduction vary in different sheep producing countries (Brien *et al.* 2014). Selection focuses on ewe records and number of lambs weaned in Australia and South Africa. In contrast, in New Zealand selection was based on lamb records, including ewe litter size and lamb survival as separate traits. Bunter and Brown (2015) contended that selection based on a balanced index, including reproduction rate as well as direct and maternal components of lamb weaning weight could potentially yield better and more predictable genetic gains. Against this background, data from a South African resource flock divergently selected for number of lambs weaned per ewe joined were used to study genetic and phenotypic parameters and trends for composite as well as component reproduction traits.

MATERIALS AND METHODS

Animals and selection procedures. Two lines of Merino sheep were divergently selected from the same base population from 1986 to the present, solely using maternal ranking values for number of lambs reared per joining. The selection regime resulted in two lines differing appreciably in reproduction (Cloete *et al.* 2004), termed the High (H) line for the line selected in the upward direction and the Low (L) line for the line selected in the downward direction. Details of the origin of the lines and the procedures for the selection of replacements have been reported elsewhere (Cloete *et al.* 2004; 2009). Only data recorded from 1987 to 2007 were used in this study. Outside sires were since introduced to the flock, to link this genetic resource with the broader South African Merino industry (Cloete *et al.* 2014). The lines are managed as a single flock, except at mating.

Location and data recording. The resource flock is being kept at the Elsenburg Research Farm near Stellenbosch. The climate, pastures grown as well as the management of the animals at joining in single-sire groups and at lambing were described by Cloete *et al.* (2004; 2009). The composite reproduction traits number of lambs born per ewe joined (NLB_{EJ}), number of lambs weaned per ewe joined (NLW_{EJ}) and total weigh of lamb weaned (pre-corrected for lamb age and sex) per ewe joined (TWW_{EJ}) were recorded (Cloete *et al.* 2004). The component traits number of ewes lambed per ewe joined (EL_{EJ}), number of lambs born per ewe lambed (NLB_{EL}) and average lamb weaning weight per ewe reared (AWW_{ER}) were derived additionally.

Statistical analyses. The data were analysed for fixed effects to obtain an operational model, fitting the effects of line (H or L), lambing year (1987-2007) and ewe age group (2-7+ years), as well as interactions. Random effects for each trait included additive animal (ewe) effects, ewe permanent environmental (PE) effects to accommodate repeated records and service sire PE (SS). All analyses were conducted in ASREML (Gilmour *et al.* 2015). Fixed effects significant in analyses to determine an operational model were used in downstream analyses. After the appropriate random effects for each trait were determined with Log Likelihood tests, bivariate analyses were conducted to derive correlations between traits for additive genetic, ewe PE and SS effects based on significance in univariate analyses. Genetic trends were constructed from within-line regressions of animal solutions from single-trait analyses excluding selection line (and its interactions with year) on birth year. The pedigree file included 6167 animals, the progeny of 300 sires and 1444 dams.

RESULTS AND DISCUSSION

Reproduction traits were all highly variable, coefficients of variation ranging from 37-73 % (Table 1). Lamb AWW_{ER}, in contrast, was less variable. These results are consistent with results in the literature (Cloete *et al.* 2004; Safari *et al.* 2005; Bunter and Brown 2015) although the observed coefficients of variation were on the higher end of the ranges reported.

Table 1. Descriptive statistics for the traits analysed on the ewes forming part of the study, namely ewes lambed per ewe joined (EL_{EJ}), number of lambs born per ewe lambed (NLB_{EL}), number of lambs born per ewe joined (NLB_{EJ}), number of lambs weaned per ewe joined (NLW_{EJ}), weight of lamb weaned per ewe joined (TWW_{EJ}) and average weaning weight per ewe reared (AWW_{ER})

Trait	N	Mean	s.d.	CV	Minimum	Maximum
EL_{EJ}	3790	0.86	0.35	40.7	0	1
NLB_{EL}	3256	1.38	0.51	37.0	1	3
NLB_{EJ}	3790	1.19	0.67	56.3	0	3
NLW_{EJ}	3790	0.90	0.66	73.3	0	3
TWW_{EJ}	3790	19.8	14.1	71.2	0	64.8
AWW_{ER}	2756	22.5	4.4	19.6	9.3	46.6

All traits were affected ($P<0.01$) by selection line and year (Table 2). Only NLB_{EL} was not affected by interactions between line with year or age, although line x year approached significance. Ewe age affected all traits, except AWW_{ER}. EL_{EJ} and AWW_{ER} were affected by the line x age interaction. Cloete *et al.* (2003) also reported that the composite reproduction traits were independent of this interaction. Least squares means for the component traits EL_{EJ} and NLB_{EL} and the composite trait NLW_{EJ} (Figure 1) all indicated observed divergence between the selection lines. No distinct line differences were observed in the early years of the experiment, but consistent significant differences in favour of the H Line for NLB_{EL} were observed from 1995 ($P<0.05$). Significant divergence for EL_{EJ} was first observed in 1999 ($P<0.05$). However, the H Line clearly and consistently outperformed ($P<0.05$) the L line for NLW_{EJ} from 1993. It is notable that this line difference in NLW_{EJ} were found prior to obvious divergence in the component traits reported here.

Table 2. Significance, using type III p-values, for the fixed effects fitted in the full model to the traits analysed in the study, namely EL_{EJ}, NLB_{EL}, NLB_{EJ}, NLW_{EJ}, TWW_{EJ} and AWW_{ER} (see Table 1 for trait abbreviations)

Effect	Trait					
	EL_{EJ}	NLB_{EL}	NLB_{EJ}	NLW_{EJ}	TWW_{EJ}	AWW_{ER}
Selection line (SL)	**	**	**	**	**	**
Year (Y)	**	**	**	**	**	**
Ewe age (A)	**	**	**	**	**	0.678
SL x Y	**	0.052	**	**	**	**
SL x A	**	0.378	0.223	0.370	0.739	**

* - P<0.05; ** P<0.01; Actual significance for P>0.05

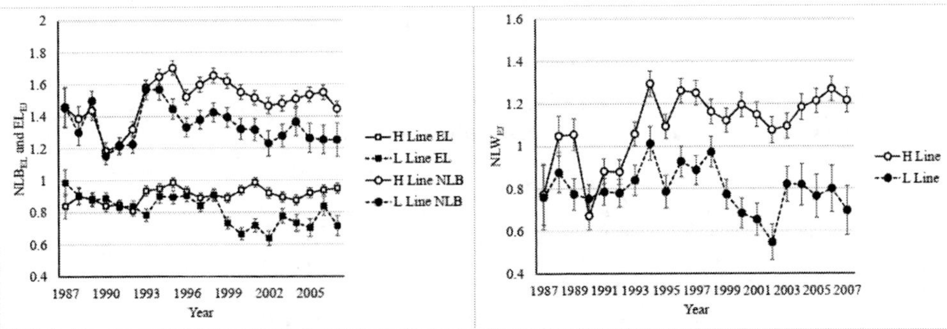

Figure 1. Least squares means (±s.e.) depicting the selection line x year interaction for the component traits EL_{EJ} and NLB_{EL} (left) and the composite trait NLW_{EJ} (right).

All reproduction traits were lowly heritable (Table 3), with estimates below 0.10 except for NLB_{EL} and NLB_{EJ}. The heritability of EL_{EJ} was not significantly different to zero. All traits were affected by animal PE. These results are not presented, but the derived estimates ranged from 0.04±0.02 for NLB_{EL} to 0.14±02 for EL_{EJ}. All traits except for NLB_{EL} were affected by SS. These estimates were small at 0.04±0.01 for EL_{EJ} and 0.02±0.01 for NLB_{EJ}, NLW_{EJ}, TWW_{EJ} and AWW_{ER}. These results were consistent with results reported in the literature (Cloete *et al.* 2004; Safari *et al.* 2005; 2007; Bunter and Brown 2015). Genetic correlations were favourable and significant, except for correlations of all traits with AWW_{ER}, where the estimates were still favourable but commonly smaller than the corresponding s.e. Animal PE and SS correlations among reproduction traits were mostly similar to genetic correlations. However, these correlations became negative with AWW_{ER} for the reproduction traits EL_{EJ}, NLB_{EJ} and NLW_{EJ}. The exception in this respect was PE correlations of TWW_{EJ} with AWW_{ER}. Previous studies also reported favourable genetic correlations among reproduction traits (Cloete *et al.* 2004; Safari *et al.* 2005; 2007; Bunter and Brown 2015) and potentially small or unfavourable correlations with AWW_{ER} (Bunter and Brown 2015).

Linear estimates of genetic trends for the respective traits are reported in Table 4. In terms of composite traits, both lines responded in the expected direction from the selection pressure applied. Expressed relative to the overall least squares mean for the first year with data (1987), the responses in the composite traits were larger in magnitude compared to component traits. Not surprisingly, the responses in the composite traits were consistent in direction and magnitude with previous results in the same resource flock (Cloete *et al.* 2004). The latter authors related the asymmetry in the responses of the H and L lines to an attempt to select against natural selection in the L Line, as well as to a reduced selection differential stemming from the reduced lamb output in the latter line.

Table 3. Phenotypic variance components (σ^2_P) and (co)variance ratios for EL_{EJ}, NLB_{EL}, NLB_{EJ}, NLW_{EJ}, TWW_{EJ} and AWW_{ER} (see Table 1 for trait abbreviations)

Component and trait	Trait					
	EL_{EJ}	NLB_{EL}	NLB_{EJ}	NLW_{EJ}	TWW_{EJ}	AWW_{ER}
σ^2_P	0.117	0.229	0.406	0.399	174.97	17.99
(Co)variance ratios*						
EL_{EJ}	**0.02±0.02**	0.78±0.43	0.78±0.23	0.64±0.32	0.57±0.10	0.16±0.58
NLB_{EL}	0.03±0.03	**0.13±0.03**	1.00±0.05	0.76±0.20	0.66±0.18	0.30±0.22
NLB_{EJ}	0.71±0.01	0.88±0.02	**0.10±0.02**	0.83±0.11	0.72±0.12	0.12±0.27
NLW_{EJ}	0.54±0.01	0.47±0.02	0.64±0.01	**0.04±0.02**	0.96±0.03	0.34±0.37
TWW_{EJ}	0.56±0.01	0.32±0.05	0.57±0.01	0.94±0.01	**0.06±0.02**	0.37±0.44
AWW_{ER}	-0.01±0.03	-0.40±0.02	-0.52±0.02	-0.59±0.01	0.26±0.02	**0.06±0.03**

* Heritability in bold on the diagonal, genetic correlations above the diagonal and phenotypic correlations below the diagonal

Table 4. Genetic trends for EL_{EJ}, NLB_{EL}, NLB_{EJ}, NLW_{EJ}, TWW_{EJ} and AWW_{ER} (see Table 1 for trait abbreviations) expressed relative to trait means in 1987

Trait	High Line*		Low Line*	
	Regression ± s.e.	As % of mean	Regression ± s.e.	As % of mean
EL_{EJ}	0.0047±0.0004	0.50	-0.0028±0.0001	-0.30
NLB_{EL}	0.0114±0.0002	0.76	-0.0035±0.0002	-0.23
NLB_{EJ}	0.0186±0.0002	1.33	-0.0053±0.0003	-0.38
NLW_{EJ}	0.0159±0.0001	1.96	-0.0064±0.0002	-0.79
TWW_{EJ}	0.430±0.004	2.35	-0.147±0.005	-0.81
AWW_{ER}	0.065±0.001	0.29	-0.042±0.001	-0.19

* All regressions were significant (P<0.01)

CONCLUSIONS

This study suggested that composite trait selection for NLW_{EJ} resulted in genetic responses in the desired direction in the component traits studied, as suggested by Snowder and Fogarty (2009). The present study did not include a measure of lamb survival or ewe rearing ability. It is thus important that further studies should also consider these traits (Bunter and Brown, 2015).

REFERENCES

Brien F.D., Cloete S.W.P., Fogarty N.M., Greeff J.C., Hebart M.L., Hiendleder S., Hocking Edwards J.E., Kelly J.M., Kind K.L., Kleeman D.O., Plush K.L. and Miller, D.R. (2014) *Anim. Prod. Sci.* **54**: 667.

Bunter K.L. and Brown D.J. (2015) *Proc. Assoc. Advmt. Anim. Breed. Genet.* **21**: 201.

Cloete S.W.P, Gilmour A.R., Olivier J.J. and Van Wyk J.B. (2003) *S. Afr. J. Anim. Sci.* **33**: 43.

Cloete S.W.P., Gilmour A.R., Olivier J.J and van Wyk J.B. (2004) *Anim. Prod. Sci.* **44**: 745.

Cloete S.W.P., Misztal I. and Olivier J.J. (2009) *J. Anim. Sci.* **87**: 2196.

Cloete S.W.P., Olivier J.J., Sandenbergh L. and Snyman M.A. (2014) *S. Afr. J. Anim. Sci.* **44**: 308.

Gilmour A.R., Gogel B.J., Cullis B.R., Welham S.J. and Thompson R. (2015) ASREML - User Guide Release 4.1. VSN International Ltd, Hemel Hempstead, HP11ES, UK.

Olivier J.J. (1999) *Proc. Assoc. Advmt. Anim. Breed. Genet.* **13**: 119.

Safari E., Fogarty N.M. and Gilmour A.R. (2005) *Livest. Prod. Sci.* **92**:271.

Safari E., Fogarty N.M., Gilmour A.R., Atkins K.D., Mortimer S.I., Swan A.A., Brien F.D., Greeff J.C. and van der Werf J.H.J. (2007) *Aust. J. Agric. Res.* **58**: 177.

Snowder G.D. and Fogarty N.M. (2009) *Anim. Prod. Sci.* **49**: 9.

Proc. Assoc. Advmt. Anim. Breed. Genet. 22:573-576

GENETIC PARAMETERS FOR DAG- AND COVER SCORES OF MERINO EWES DIVERGENTLY SELECTED FOR NUMBER OF LAMBS WEANED

A.J. Scholtz[1], J.J.E. Cloete[2,3] and S.W.P. Cloete[1,3]

[1]Directorate Animal Sciences: Elsenburg, Private Bag X1, Elsenburg 7607, South Africa
[2]Elsenburg Agricultural Training Institute, Private Bag X1, Elsenburg 7607, South Africa
[3]Department of Animal Sciences, University of Stellenbosch, Private Bag X1, Matieland 7602, South Africa

SUMMARY

Divergent selection for number of lambs weaned per ewe joined from the same base population since 1986 resulted in Merino lines that differ markedly in reproduction rate. Subjective scores for dags (DS), breech cover (BCS), crutch cover (CCS) and belly wool quality (BQS) (1 = lowest; 5 = highest) were recorded for mature ewes of these lines. Expressed relative to High (H) line performance, BCS was 19% higher in the Low (L) line ewes. Corresponding selection line differences in favour of the H line amounted to 9% for DS and 32% for CCS. In contrast, BQS were reduced by 24% in the H Line. Ewes that reared more lambs had a higher mean for DS and lower scores for the other traits than ewes not rearing any lambs (P<0.01). All traits were heritable, with estimates ranging from 0.36 for DS to 0.68 for BCS. Genetic correlations suggested that DS was not highly related to BSC, CCS or BQS. Genetic correlations among the latter three traits all exceeded 0.80, suggesting that these traits were genetically very similar. Selection for improved reproduction in Merinos resulted in favourable breech and crutch characteristics, but not in BQS. The favourable breech and crutch conformation supported the reduced susceptibility to breech strike in this line.

INTRODUCTION

Two Merino lines were divergently selected over about 30 years to differ substantially in their reproductive ability (Cloete *et al.* 2004). A difference in the susceptibility of breeding ewes to breech strike was also reported between these lines, with the line selected for increased reproduction (High or H line) being less susceptible than the line selected against reproduction (Low or L line) (Scholtz *et al.* 2010). Cloete *et al.* (2005) also reported line differences, with unmulesed hoggets from the L line being scored as more wrinkly than their H line contemporaries. Breech characteristics have been associated with the susceptibility of Merino sheep to breech blowfly strike and it has been argued that strikes could be reduced if sheep were selected to eliminate susceptible crutches (De Vries and De Klerk 1944). It has been recommended that the bare patch in the locality of the anus and vulva should be as "*large as possible*" (De Vries and De Klerk 1944). Despite these arguments or recommendations, genetic solutions to breech strike were not pursued in earlier years, most probably due to the exceptional effectiveness of surgical mulesing. Mulesing, however, has been discarded as an appropriate management strategy because of its obvious impact on animal welfare.

Selective breeding for resistance to breech strike is considered an alternative long term solution. Apart from wrinkle scores, breech cover score has been identified as an indicator trait associated with breech strike (Brown *et al.* 2010). Dag score was also found to be genetically related to breech strike (Greeff *et al.* 2014). These traits were thus studied in mature ewes recorded in the recent years of the H and L lines in an attempt to relate the proven line difference in breech strike prevalence (Scholtz *et al.* 2010) of breeding ewes to these subjective scores.

MATERIALS AND METHODS

Animals and selection procedures. Two lines of Merino sheep were divergently selected from the same base population from 1986 to the present, using maternal ranking values for number of lambs weaned per ewe joined. Details of the origin of the lines and the procedures for the selection of replacements have been reported elsewhere (Cloete and Scholtz 1998; Cloete *et al.* 2004). Briefly male and female progeny of ewes that reared more than 1 lamb per joining (i.e. reared twins at least once) were preferred as replacements in the H line. Replacements in the Low (L) line were preferably descended from ewes that reared fewer than one lamb per joining (i.e. were barren or lost all lambs at least once).

Location and recordings. The resource flock was maintained at the Elsenburg Research Farm near Stellenbosch. The climate, pastures grown and management of the animals were described by Cloete *et al.* (2004), while lambing and reproduction practices in the breeding flock were described by Cloete and Scholtz (1998). Winter lambing (June-July) was practiced routinely. Although these references are quite old, the same basic conditions still prevail at the site. All lambs had their tails docked at the third palpable joint and none of the ewes participating in the study were mulesed as lambs. Mature ewes were shorn in April/May just prior to lambing and crutched in springtime (5-6 month's wool growth) to reduce the probability of strikes over the early summer period (Scholtz *et al.* 2010). A number of commercial rams have been introduced to the flock since 2008 to link the lines to the commercial industry. These rams were selected on the same principles used for within-line selection and it was thus not attempted to account for their impact. Since the traits under consideration were not directly selected for it was assumed that the impact of these introductions would be minimal for the traits considered. From 2009 to 2016, mature reproducing ewes (2 – 7+ years) were subjectively scored for the accumulation of dags (DS), breech cover (BCS) and crutch cover (CCS) by the same experienced scorer using the Visual Breech Scoring System (Australian Wool Innovation Limited 2007). All scores were recorded in November, a month after the ewes weaned their lambs. Scores of 1 to 5 were allocated to each trait with 1 = least expression of the trait and 5 = most expression of the specific trait. Quality of belly wool (BQS) was evaluated on a linear scale from 1 to 5; where 1 = poor and 5 = excellent quality. Belly wool quality was defined as the regularity, evenness and definition of crimp, softness of handle and the absence of coarse fibres.

Statistical analyses. Systematic effects present for DS, BCS, CCS and BQS were determined in a general linear model analysis to obtain an operational model before random effects were added. The fixed effects model used included the effects of selection line (H vs. L), reproduction status (3 levels: 0, 1, 2+ lambs weaned), ewe age group (6 levels: 2 – 7+ years) and interactions among effects. Ewes dry after lambing were managed in a separate group with yearling replacements. Initial fixed effect models and the subsequent single- and four-trait genetic analyses were conducted in ASReml (Gilmour *et al.* 2015). Ewe additive genetic and ewe permanent environmental (PE) terms were included as random effects and assessed for significance by log likelihood test. Heritability estimates and genetic correlations among traits were derived from the 4-trait analyses.

RESULTS AND DISCUSSION

The scorer used the full range of scores for all subjective traits considered (Table 1). All traits were variable, with coefficients of variation ranging from 41% for DS to 52% for CCS.

Table 1. Descriptive statistics for dag score (DS), breech cover score (BCS), crutch cover score (CCS) and belly wool quality score (BQS) recorded on ewes post-weaning

Trait	Number of records	Mean	Standard deviation	Minimum	Maximum
DS	671	1.39	0.57	1	5
BCS	1107	2.27	1.03	1	5
CCS	1107	1.69	0.88	1	5
BQS	1101	2.30	0.96	1	5

When expressed relative to the relevant least squares mean for the L Line, the cover score traits were between 8% (DS) and 32% for CCS lower ($P<0.05$) (i.e. influenced in the desired direction) in the H Line (Table 2). These results are consistent with previous results reporting favourable correlated responses in dag score and perineal bare area dimensions in the H Line (Scholtz *et al.* 2011). It also supports the report that H Line ewes were less susceptible to breech strike than L Line contemporaries (Scholtz *et al.* 2011). In contrast, BQS was poorer in the H Line than in the L line. Ewes that reared lambs had a higher mean for DS and lower scores for the other traits than those ewes not rearing any lambs ($P<0.01$). The differences between ewes rearing singles and those rearing multiples were, however, small and not significant. Ewes at 4 and 6 years of age had lower means for DS than 2-, 3- and 7+-year-old ewes ($P<0.05$). BCS was unaffected by ewe age, while CCS declined with ewe age. BQS generally also declined with ewe age, but tended to stabilise from 5-year-old ewes. All traits barring BCS were affected by lambing year (results not shown). Year effects depend on climatic and managerial factors inherent to that specific year. Such effects were thus not presented, except when year interacted with selection line.

Table 2. Least-squares means (±s.e.) depicting the effects of selection line, number of lambs weaned, ewe age and lambing year on DS, BCS, CCS and BQS recorded on Merino ewes in the initial fixed model analyses (see Table 1 for abbreviations)

Effect and level	Number of observations[#]	Trait DS	BCS	CCS	BQS
Selection line		*	**	**	**
H Line	872	1.25 ± 0.03	2.17 ± 0.05	1.51 ± 0.04	2.12 ± 0.04
L Line	235	1.36 ± 0.05	2.69 ± 0.10	2.22 ± 0.07	2.77 ± 0.08
Lambs weaned		**	**	**	**
0	285	1.20 ± 0.05	2.99 ± 0.07	2.49 ± 0.05	3.04 ± 0.06
1	604	1.37 ± 0.04	2.16 ± 0.06	1.55 ± 0.04	2.23 ± 0.05
2+	218	1.36 ± 0.08	2.14 ± 0.11	1.56 ± 0.09	2.08 ± 0.09
Ewe age		*	0.17	**	**
2 Years	290	1.35 ± 0.03	2.61 ± 0.08	2.11 ± 0.06	2.76 ± 0.07
3 Years	301	1.40 ± 0.03	2.49 ± 0.08	2.03 ± 0.06	2.64 ± 0.07
4 Years	234	1.24 ± 0.03	2.34 ± 0.08	1.86 ± 0.06	2.41 ± 0.07
5 Years	180	1.33 ± 0.04	2.47 ± 0.09	1.86 ± 0.07	2.27 ± 0.08
6 Years	77	1.17 ± 0.04	2.38 ± 0.13	1.81 ± 0.10	2.26 ± 0.11
7+ Years	25	1.35 ± 0.07	2.30 ± 0.20	1.53 ± 0.15	2.37 ± 0.17

[#] - For BCS and CCS; * - $P<0.05$; ** $P<0.01$; Absolute significance shown for $P>0.05$

CCS and BQS were affected by a significant interaction between lambing year and selection line in the fixed model analyses. Expressed relative to H Line means, the mean for CCS of L Line ewes exceeded that of H Line ewes by 18% in 2016, as compared to between 40 and 64% in other years. The corresponding difference for BQS amounted to 13% in 2016, compared to 31-40% in other years. Selection line also interacted with number of lambs weaned for CCS. Although CCS was clearly lower in H Line ewes compared to their L Line contemporaries across reproduction categories, the magnitude of the difference amounted to 48% for ewes not rearing a lamb, 34% for ewes that reared singles and 60% for ewes that reared multiples (all $P<0.05$). Significant fixed effects and interactions were included in the subsequent random model analyses.

Heritability estimates in the four-trait analysis were mostly within 0.02 relative to single-trait estimates. The only slightly larger difference were for BQS where the single-trait estimate amounted to 0.39 ± 0.08 and the four-trait estimate to 0.46 ± 0.05. Some variation in this trait was repartitioned from ewe PE to the additive component in the four-trait analysis. Only four-trait results are thus presented in Table 3. All traits were heritable, estimates ranging from 0.36 for DS to 0.68 for BCS. In addition, the PE of ewes also affected DS (0.30 ± 0.09) and BQS (0.10 ± 0.03), resulting in repeatability estimates for both traits exceeding 0.50. We did not find previous results in the literature on the magnitude of heritability estimates for the indicator traits for breech strike studied in mature ewes. However, but it is well known that these traits are heritable in young animals (Brown *et al.* 2010; Scholtz *et al.* 2011; Greeff *et al.* 2014), supporting the present results on the genetic basis of indicator traits. Genetic correlations suggested that DS was not highly related to BCS, CCS or BQS (Table 3). In contrast, the genetic correlations among the latter three traits all exceeded 0.80, suggesting that they were genetically very similar. Phenotypic correlations were mostly in the same direction as genetic correlations, but smaller in magnitude.

Table 3. Four-trait phenotypic (σ^2_P), additive (σ^2_A) and ewe permanent environmental (σ^2_C) variance components and (co)variance ratios for DS, BCS, CCS and BQS of the Merino ewes studied (see Table 1 for abbreviations)

Component and trait	Trait			
	DS	BCS	CCS	BQS
σ^2_P	0.360	0.965	0.568	0.658
σ^2_A	0.128	0.661	0.358	0.300
σ^2_C	0.108	-	-	0.063
(Co)variance ratios*				
DS	**0.36 ± 0.10**	0.00 ± 0.10	0.08 ± 0.11	0.06 ± 0.15
BCS	-0.02 ± 0.05	**0.68 ± 0.03**	0.83 ± 0.03	0.82 ± 0.05
CCS	0.03 ± 0.05	0.63 ± 0.02	**0.63 ± 0.03**	0.85 ± 0.05
BQS	0.01 ± 0.05	0.60 ± 0.03	0.57 ± 0.03	**0.46 ± 0.05**

* Heritability in bold on the diagonal, genetic correlations above the diagonal and phenotypic correlations below the diagonal

CONCLUSIONS

All traits were heritable and should respond to selection if needed. Selection for improved reproduction in Merinos resulted in favourable dag, breech and crutch characteristics previously related to reductions in breech strike. In contrast, scores for BQS were compromised by selection for an improved reproduction.

REFERENCES

Australian Wool Innovation (AWI) (2007) Online: www.wool.com.au/publications.

Brown D.J., Swan A.A. and Gill J.S. (2010) *Anim. Prod. Sci.* **50**: 1060.

Cloete S.W.P. and Scholtz A.J. (1998) *Anim. Prod. Sci.* **38**: 801.

Cloete S.W.P., Gilmour A.R., Olivier J.J and van Wyk J.B. (2004) *Anim. Prod. Sci.* **44**: 745.

Cloete S.W.P., Olivier J.J., van Wyk J.B., Schoeman S.J. and Erasmus G.J. (2005) *Proc. Assoc. Advmt. Anim. Breed. Genet.* **16**: 24.

De Vries A.H. and De Klerk J.C. (1944) *Farming in South Africa*, **78**: 485.

Gilmour A.R., Gogel B.J., Cullis B.R., Welham S.J. and Thompson R. (2015) ASREML - User Guide Release 4.1. VSN International Ltd, Hemel Hempstead, HP11ES, UK.

Greeff J.C., Karlsson L.J.E. and Schlink A.C. (2014) *Anim. Prod. Sci.* **54**: 152.

Scholtz A.J., Cloete S.W.P., van Wyk J.B., Kruger A.C.M. and van der Linde T.C. de K. (2010) *Anim. Prod. Sci.* **50**: 210.

Scholtz A.J., Cloete S.W.P., Cloete J.J.E., Kruger A.C.M., van Wyk J.B. and van der Linde T.C.de K. (2011) *Proc. Assoc. Advmt. Anim. Breed. Genet.* **19**: 175.

Gap in pagination due to unavailable papers.

Pages 577-584

A GENOME-WIDE SCAN OF POSITIVE SELECTION SIGNATURE USING OVINE INFINIUM® HD SNP BEADCHIP IN TWO ROMNEY LINES, SELECTED FOR RESISTANCE OR RESILIENCE TO NEMATODES

J. Yan[1], V.S.R. Dukkipati[1], H.T. Blair[1], P.J. Biggs[1], J. C. Hamie[2] and A. Greer[2]

[1] IVABS, Massey University, Palmerston North 4442, New Zealand
[2] Agricultural and Life Sciences, Lincoln University, Lincoln 7647, New Zealand

SUMMARY

A study was undertaken to identify selection signatures associated with resistance to gastrointestinal nematodiasis in sheep. Ninety-three Romney sheep from two selection lines (resistant or resilient) were genotyped using the Ovine Infinium® HD SNP BeadChip, and extended haplotype homozygosity (EHH) and site-specific extended haplotype homozygosity (EHHS) analyses were undertaken. In total, 224 SNPs (147 in EHH and 77 in EHHS) were found to be significant (p <0.0001). Preliminary exploration of 10 SNPs found them to be located within two previously identified QTLs, namely LATRICH_2 and FECGEN, which were associated with nematode larval count and faecal egg count, respectively.

INTRODUCTION

Gastrointestinal nematodes are one of the most serious parasitic threats for sheep (Familton and McAnulty 1997; Perry and Randolph 1999), costing approximately $300 million annually to the New Zealand sheep industry (Rattray 2003). The current high dose usage of anthelmintics, owing to anthelmintic resistance, is not welcomed by the present global market, considering the increasing consumers' preference for organic products. Therefore, alternative anti-parasite strategies are necessary. Genetic selection is one of the most important ways in animal husbandry to improve the quality of domestic animals. Several studies have shown that resistance to nematodiasis in sheep is highly variable and heritable between individuals so that selective breeding can be an alternative choice for nematode control (Morris *et al.* 1995, 2000, 2005).

The advent of high-density single nucleotide polymorphism (SNP) microarray chips has facilitated detection of selection signatures based on patterns of linkage disequilibrium in selection lines. This is based on the assumption that the frequency of a novel mutation, that confers an advantage, will increase more rapidly than that of a neutral mutation (Sabeti *et al.* 2002). Consequently, long linkage disequilibrium (LD) blocks that incorporate genomic regions containing the causative genetic mutations could exist in populations undergoing artificial selection, given that they have been bred for insufficient generations to break the LD through recombination (Slatkin 2008). Hence, a high frequency and unusually long haplotype within a selected population could indicate the presence of a positive selection signature.

To detect these signatures, an algorithm called extended haplotype homozygosity (EHH) was initially introduced (Sabeti *et al.* 2002) which quantifies the decay of haplotype homozygosity within a population. Subsequently, another method known as the site-specific extended haplotype homozygosity (EHHS), was introduced to do the same purpose between populations (Sabeti *et al.* 2007). These methods have been successfully used to detect selection signatures in animals (McRae *et al* 2014; Somavilla *et al.* 2014; Zhang *et al.* 2012). Using the Ovine Infinium® HD SNP BeadChip the current study attempts to detect positive selection signatures in two Romney sheep lines selected for divergent approaches to coping with nematode infections, that being either resistance or resilience.

MATERIALS AND METHODS

This study was carried out following the guidelines of the 1999 New Zealand Animal Welfare Act and was approved by the Lincoln University Animal Ethics Committee (Permit Numbers: LUAEC#588). During early 2015, ear punch samples were collected into Allflex tissue sampling units (TSU), using an Allflex NZ tissue sampling applicator (TSU Applicator – 22134), from 93 Romney sheep belonging to two selection lines (nematode resistant, n = 42, and nematode resilient, n = 51), currently being maintained at Lincoln University, Lincoln, New Zealand. Details regarding the selection lines were described elsewhere (Morris *et al.* 2000). Sheep in the two lines were selectively bred based on faecal egg count (FEC), for at least 24 years (1985-2009) and since then have been randomly bred within each line. The tissue samples were submitted to AgResearch, Invermay Agricultural Centre, Mosgiel, New Zealand, for DNA extraction and SNP genotyping using the Ovine Infinium® HD SNP BeadChip.

The original SNP data (idat files) were converted into to PLINK format (PED/MAP) in GenomeStudio® (Illumina, San Diego CA, USA). Quality control was performed using PLINK_v1.9 (Chang *et al.* 2015; Purcell *et al.* 2007). A within individual call rate threshold of 99% was applied and SNPs with a call rate <95%, or a minor allele frequency <1%, or a p value of $<10^{-6}$ for Hardy-Weinberg equilibrium were excluded. After quality control a total of 463,392 SNPs, located on the 26 autosomes in all 93 sampled individuals were retained for further analysis.

The SNP data was reformatted in PLINK and inputted into fastPHASE_v1.4 (Scheet & Stephens 2006) in order to reconstruct the haplotypes for each autosome, using the default parameters. The resultant haplotype data (phased data) was used to detect positive selection signatures by calculating the allele-specific extended haplotype homozygosity (EHH) within populations as well as the site-specific extended haplotype homozygosity (EHHS) between populations, using an R package, REHH 2.0 (Gautier *et al.* 2017). For EHH, the test statistic was iHS (Gautier & Naves 2011), standardized ratio of the integrated allele-specific EHH (iHH), while for EHHS, two separate test statistics were employed – xp-EHH (Sabeti *et al.* 2007) and Rsb (Tang *et al.* 2007). Significance of detected signatures of selection was determined based on the p values for iHS, xp-EHH and Rsb.

RESULTS AND DISCUSSION

The purpose of this study was to identify SNPs or genes exhibiting positive selection from the pressure of nematodes. SNP genotypes pertaining to 463,392 markers covering the 26 autosomes were used. Since the two selection lines investigated in the study were selectively bred for at least 24 years (1985-2009) based on faecal egg count (FEC) using best linear unbiased prediction (BLUP) techniques and since then randomly bred within each line, these populations are suited to investigations of selection signatures of long time breeding associated with resistance and resilience.

EHH testing revealed 62 and 85 SNPs to exhibit positive selection signatures (p<0.0001) in the nematode resistant and resilient groups, respectively. None of the identified SNPs were shared between the two lines, indicating genetic differences (possibly due to selection pressure) between these two populations. An iHS plot for OAR2 for the two lines, revealing the differences between the two populations is shown in Figure 1. EHHS testing was also performed to detect positive selection signatures between the two populations. Two different algorithms, XP-EHH and Rsb were used. A total of 39 and 48 SNPs were detected exhibiting positive selection signatures, in the two algorithms, respectively. None of these SNPs were common to those detected in the within-line EHH analysis. Figure 2 depicts the results for the Rsb and XP-EHH algorithms with respect to markers on chromosome 13. So far, there is no demonstrated advantage of one over the other, and the results from these two algorithms can be slightly different (Gautier *et al.* 2017). Therefore, the SNPs shared by two algorithms should be more reliable. Ten SNPs (Table 1) were shared between the two algorithms and those were found to be located in regions of known significance and hence, can be considered as highly confident SNPs associated with nematode resistance and resilience in sheep. A QTL search (http://www.animalgenome.org/QTLdb/sheep) showed the SNP oar3_OAR11_48327544 to be located within the reported QTL:12901, which has been demonstrated to be associated with larva count (Crawford *et al.* 2006). The remaining nine SNPs were all located within a small region overlapping the gene Protein Tyrosine Phosphatase, Receptor Type T (*PTPRT*) on chromosome 13 (Oar_v4.0). *PTPRT* is a protein coding gene and GO annotations show that this gene is related to *phosphatase activity* and *beta-catenin binding*, possibly indicating that this gene could have a function in resistance to nematodes. Further, these SNPs are harboured within the previously identified QTL:16027, which has been associated with faecal egg counts (Sonstegard *et al.* 2012).

A previous study (McRae *et al.* 2014) performed on Romney and Perendale sheep, that used F_{ST} and Peddrift to detect differentiation between nematode resistant and susceptible lines, identified sixteen significant regions, which included candidate genes involved in chitinase activity and the cytokine response. The samples from nematode resistant sheep obtained for the current study came from the same line as that from McRae *et al.* (2014), but were from a different generation. In the current study, apart from the 10 significant SNPs detected by both the Rsb and XP-EHH algorithms, there were 77 other significant SNPs detected by one or either of the algorithms. These 77, together with the 147 SNPs (62 and 85 in the resistant and resilient lines, respectively) detected in the EHH analysis will be explored in detail to find out if they are located in or near to gene regions of immunological significance.

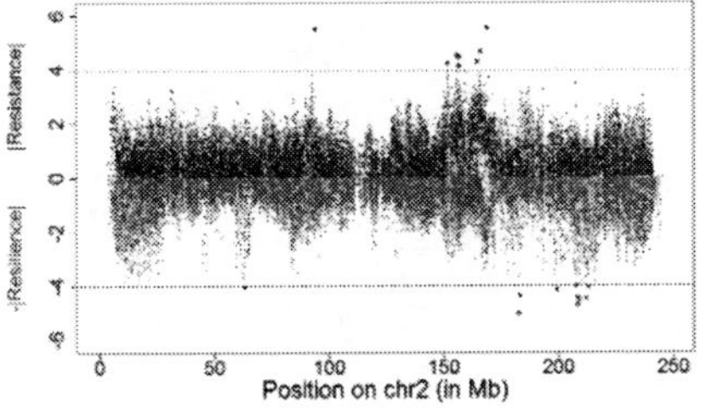

Figure 1 iHS difference for markers on chromosome 2 between lines of Romney sheep selected for either resistance or resilience to gastro-intestinal nematodes

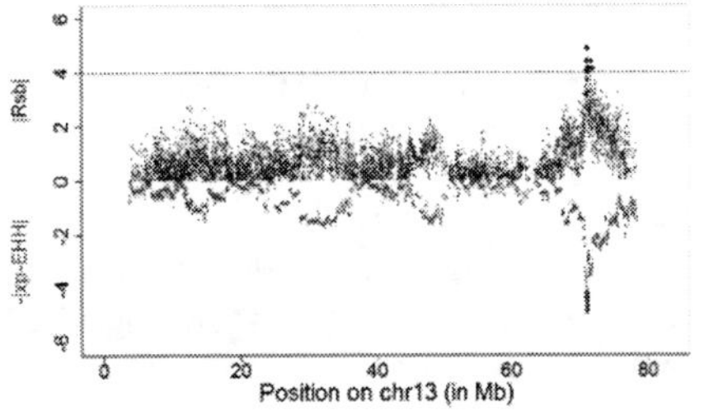

Figure 2 REHH difference between XP-EHH or Rsb methods for markers on chromosome 13

CONCLUSION

This study provided a genome-wide map of positive selection signatures in two Romney sheep lines selected for FEC. Several significant SNPs were identified and preliminary analysis of ten of the identified SNPs revealed that they were located within two previously detected QTLs associated with gastrointestinal nematodiasis in sheep. The significance of the remaining SNPs is currently being explored.

ACKNOWLEDGEMENTS

This study was funded by Massey-Lincoln and the Agricultural Industry Trust (project # 2015/5) and Massey University. Also, financial support to the primary author, in the form of a doctoral scholarship to JY from Massey University, New Zealand, is gratefully acknowledged.

Table 1. SNP markers detected by both the EHHS algorithms, XP-EHH and Rsb, to suggest evidence of positive selection in lines of Romney sheep selected for either resistance or resilience to gastro-intestinal nematodes

SNP	Chr	Position	Gene		QTL ID	
oar3_OAR11_48327544	11	48327544	none	LATRICH_2	QTL:12901	larva count
oar3_OAR13_70810243	13	70810243	PTPRT	FECGEN	QTL:16027	FEC
oar3_OAR13_70820259	13	70820259	PTPRT	FECGEN	QTL:16028	FEC
oar3_OAR13_70853062	13	70853062	PTPRT	FECGEN	QTL:16029	FEC
oar3_OAR13_70853714	13	70853714	PTPRT	FECGEN	QTL:16030	FEC
oar3_OAR13_70870621	13	70870621	PTPRT	FECGEN	QTL:16031	FEC
oar3_OAR13_70876794	13	70876794	PTPRT	FECGEN	QTL:16032	FEC
oar3_OAR13_70887333	13	70887333	PTPRT	FECGEN	QTL:16033	FEC
oar3_OAR13_70891326	13	70891326	PTPRT	FECGEN	QTL:16034	FEC
oar3_OAR13_70896117	13	70896117	PTPRT	FECGEN	QTL:16035	FEC

REFERENCES

Chang, C.C., *et al.* (2015) *GigaScience* **4**(1): 7.

Crawford A.M., *et al.* (2006) *BMC Genom.* **7**: 178.

Familton, A. and McAnulty, R. (1997) 'Sustainable control of internal parasites in ruminants', pp. 67-80.

Gautier, M., Klassmann, A. and Vitalis, R. (2017) *Mol. Ecol. Resour.* **17**(1): 78.

Gautier, M. and Naves, M. (2011) *Mol. Ecol.* **20**(15): 3128.

McRae, K.M., McEwan, J.C., Dodds, K.G. and Gemmell, N.J. (2014) *BMC Genom.* **15**(1): 1.

Morris, C., *et al.* (2000) *Anim. Sci.* **70**(1): 17.

Morris, C., Watson, T., Bisset, S., Vlassoff, A. and Douch, P. (1995) 'Breeding for resistance to infectious diseases in small ruminants', pp.77-98.

Morris, C., Wheeler, M., Watson, T., Hosking, B. and Leathwick, D. (2005) *N. Z. J. Agric. Res.* **48**(1): 1.

Perry, B. & Randolph, T. (1999) *Vet. Parasitol.* **84**(3): 145

Purcell, S., *et al.* (2007) *Am. J. Hum. Genet.* **81**(3): 559.

Rattray, P. (2003) 'Helminth parasites in the New Zealand Meat & Wool Pastoral Industries: A review of current issues.' *Final Report*. 114-117.

Sabeti, P.C., *et al.* (2002) *Nature* **419**(6909): 832.

Sabeti, P.C., *et al.* (2007) *Nature* **449**(7164): 913.

Scheet, P. and Stephens, M. (2006) *Am. J. Hum. Genet.* **78**(4): 629.

Slatkin, M. (2008) *Nat. Rev. Genet.* **9**(6): 477.

Silva, M.V.B., *et al.* (2012) *Anim Genet.* **43**(1): 63.

Somavilla, A.L., *et al.*(2014) *Anim. Genet.* **45**(6): 771.

Tang, K., Thornton, K.R. and Stoneking, M. (2007) *PLoS Biol.* **5**(7): e171.

Zhang, H., *et al.* (2012) *BMC Genom.* **13**(1): 1.

GENETIC ASSOCIATION OF SKIN THICKNESS WITH LAMB SURVIVAL FROM BIRTH TO WEANING, AND GROWTH AND WOOL TRAITS IN NEW ZEALAND ROMNEY SHEEP

M. Soltani-Ghombavani, V.S.R. Dukkipati and H.T. Blair

Institute of Veterinary, Animal and Biomedical science, Massey University, Palmerston North, New Zealand

SUMMARY

Lamb survival, as a trait of high economic importance with low heritability, might show more response to selection by considering traits of higher heritability, genetically correlated with survival, as a supplement to direct selection for the trait itself. This study aimed to estimate heritability and genetic association of skin thickness (ST), as a potential trait in indirect selection for lamb survival, with lamb survival from birth to weaning (SAW), and a few growth and wool traits including fat depth (FD), eye-muscle depth (EMD), weaning weight (WWT) and 12-month fleece weight (FWT) in New Zealand Romneys. Data for ST, FD, and EMD were collected using ultrasound scans on hoggets at 8-10 months. Appropriate animal and sire models were applied to estimate the genetic parameters using ASReml software. ST had an estimated heritability of 0.26, and showed genetic correlations of 0.27 (±0.22), 0.22 (±0.10), -0.18 (±0.12), -0.21 (± 0.12) and 0.27 (±0.12) with SAW, FD, EMD, WWT, and FWT, respectively. The preliminary estimates of heritability and genetic correlation of skin thickness with lamb survival, obtained in this study, might suggest the idea of considering this trait in selection for lamb survival, though its unfavourable correlation with other traits should also be considered.

INTRODUCTION

Lamb mortality is a major issue to sheep producers both in New Zealand and worldwide, not only due to economic losses but also as an animal welfare and management problem. Lamb survival rates of 75 to 97% has been reported in New Zealand (Hight and Jury 1970; Dalton *et al.* 1980; Gumbrell and Saville 1986), though mortality rates of up to 40% have been found on some farms (Fisher 2004). In countries like New Zealand, the UK and Australia, where lambing mostly takes place outdoor, thermoregulatory capacity of newborn lambs plays a major role in lamb survival due to its contribution to starvation-exposure mortality rates, as the second most common cause of lamb deaths in the neonatal period after dystocia (Kerslake *et al.* 2005; Everett-Hincks *et al.* 2007).

Due to a low heritability of lamb survival (Lopez-Villalobos and Garrick 1999; Brien *et al.* 2010), indirect selection, based on selection for other easy-to-measure traits of higher heritability that are genetically correlated with survival can be considered as a supplement to direct selection for the trait itself. Skin thickness as a trait of moderate to high heritability (Slee *et al.* 1991; Gregory 1982a) has been shown to be associated with cold tolerance (Samson and Slee 1981), as a component of lamb survival, which is moderately to highly heritable itself (Wolff *et al.* 1987; Slee *et al.* 1991).

Hence, selection for skin thickness might be a potential alternative to selection for cold resistance and consequently lamb survival. Unlike cold resistance, whose assessment needs laboratory-based techniques that are not feasible for breeders, skin thickness could be easily measured in the field using objective techniques like ultrasonography (Brown *et al.* 2000). Prior to implementing this trait in selection for lamb survival, it is inevitable to first estimate its heritability and genetic association with other economic traits. Although a limited number of studies were undertaken for estimating these parameters (Slee *et al.* 1991; Gregory 1982a; Gregory 1982b; Coy 1983; Hynd *et al.* 1996), the size of populations in those experiments were too small. Therefore, the objective of this study was to estimate heritability for ultrasonographically measured traits (skin thickness, subcutaneous fat depth, and eye muscle depth), lamb survival and some growth and wool traits (weaning weight, and fleece weight at 12 months). Also, genetic correlation of skin thickness (as the proposed trait influencing lamb survival) with other traits of interest was estimated.

MATERIALS AND METHODS

Data collection. Data for skin thickness, fat depth and eye muscle depth were collected by ultrasonography on four Terminal Romneys for Increased Genetic Gain (TRIGG) farms in the Manawatu region of New Zealand as part of routine farm operations using ultrasound at approximately 8 months of age, during 2011 to 2015. A commercial operator took measurements using an ultrasound scanning machine (Sonosite M Turbo) with a 38mm probe at 7.5 MHz set at a depth of 40 mm on the left dorsal loin region of the lambs around the 12th rib. Live weight was also recorded at scanning. For three out of four farms the ultrasound data were recorded only during 2011 to 2014. Additional data on date of birth, sex, flock, birth rank, rearing rank, dam age, dam and sire identities, status of lamb at weaning (alive or dead), weaning weight, weaning date, fleece weight at 12 months, and age at shearing were obtained from the Sheep Improvement Limited (SIL) database. Data cleaning was done so that records with dam age of 9 years or more (n=9), birth ranks of 4 and 5 (n=114), and rearing rank of 4 (n=20) were removed from the data because of their small numbers. Also, lambs of unknown parents in the pedigree (n=408) were excluded from the analysis. After data cleaning and editing for incorrect pedigree and outlier values, the data set had 24,097 lambs born to a total of 199 sires and 6,413 dams.

Statistical analysis. Univariate procedure in SAS software (SAS, 2015) was used to check for normality and edit the data (removing outlier observations). Data were analysed by the PROC MIXED procedure in SAS software (SAS, 2015) to identify significant fixed effects to be included in the final models. Sex, birth year, and birth flock were included as fixed effects for all the traits. Furthermore, for all the traits except skin thickness, dam age was included in the final models. Also, weight at ultrasonography was considered as a covariate for the analysis of skin thickness, fat and eye muscle depth. Birth rank was included as fixed effect in the analysis of survival at weaning, and rearing rank for the other traits excluding skin thickness. In addition, age at weaning and age at fleece weight measurement were considered as covariates in the models analysing the traits weaning weight and fleece weight at 12 months, respectively. Also, all the significant two-way interactions between these fixed effects were included in the final models. (Co)variance components were estimated by Restricted Maximum Likelihood (REML) procedure using the ASREML software (Gilmour *et al.* 2015). Appropriate animal models were used for estimation of heritability for all the traits. The random effects included direct additive genetic effect for all the traits, and also maternal genetic and maternal environmental effects for the traits survival at weaning and weaning weight.

Because survival was coded as a binary trait, a generalized linear model analysis was performed, assuming a binomial distribution for this trait and using both logit and probit link functions. For all other traits, a linear animal model was used assuming normal distribution. Genetic correlations were estimated using bivariate analyses applying the best models determined in the univariate analyses. In the bivariate model where survival was included as a trait, a sire model was used and only those sires with at least 50 and 30 records of their progeny for lamb survival and skin thickness, respectively, were included in the analysis. In those models, lamb survival was considered as a threshold trait and skin thickness as normal trait. It should be noted that the statistical analysis was performed using the skin thickness data only from those animals that were alive until ultrasound scanning (at around 8 months age) and this might have led to bias in the resulting genetic correlation of lamb survival with skin thickness.

RESULTS AND DISCUSSION

Number of observations, mean, standard deviation (SD), minimum (Min), maximum (Max), and coefficient of variation (CV) for the analysed traits are presented in Table 1. As shown, ultrasound skin thickness in this study, recorded at around 8 months of age, had a mean of 2.92 mm (Table 1), which is consistent with a report by Jopson *et al.* (2000) in new-born Coopworth lambs in New Zealand, though skin thickness was measured using skinfold callipers in their experiment. In the current study ewe lambs had significantly (P<0.01) thicker skin compared to males (3.24 vs. 2.87), while Jopson *et al.* (2000) did not find any significant difference between ewe and ram lambs. On the other hand, neither birth rank nor age of dam had any significant effect on skin thickness when adjustment was made for live weight at measurement, both of which are in agreement with the study by Jopson *et al.* (2000). Also, skin thickness was significantly affected by both birth flock and year in the present study.

Table 1. Descriptive statistics and number of records for the traits analysed

Trait	No. of records	Mean	SD	Min.	Max.	CV
Survival at weaning (%)	23976	0.81	0.39	0	1.00	47.99
Skin thickness (mm)	6082	2.92	0.50	1.50	5.00	17.20
Fat depth (mm)	6171	2.86	1.43	1.00	12.00	50.02
Eye muscle depth (mm)	4389	25.60	3.14	4.00	38	12.25
Weaning weight (kg)	18657	28.68	6.10	10.00	57.00	21.25
Fleece weight at 12 months (kg)	5426	3.32	0.67	1.60	5.80	20.29

Table 2 presents heritability estimates for the traits of interest and genetic correlation of skin thickness with other traits. As expected, lamb survival at weaning had low direct and maternal heritability estimates, which is in line with several other studies (Lopez-Villalobos and Garrick 1999; Brien *et al.* 2010). As mentioned at the outset, this finding shows that direct genetic selection for this trait is not promising. On the other hand, skin thickness as the main trait of interest considered for indirect selection for lamb survival showed a moderate heritability of 0.26±0.04, which confirms the results from previous studies showing this trait to be heritable (Slee *et al.* 1991; Gregory 1982a). Furthermore, skin thickness showed a positive genetic correlation (0.27±0.22) with lamb survival at weaning, which is favourable and consistent with the results from a study by Jopson *et al.* (2000) that showed 2.7% increase in lamb survival from tagging until weaning for each millimetre of increase in skin thickness in Coopworth sheep. This finding could be attributed to the effect of skin thickness on improved thermoregulation.

Table 2. Estimates (±SE) of the direct (h_a^2) and maternal (h_m^2) heritabilities and maternal environmental (me^2) effects for each trait, and genetic correlations (r_g) with skin thickness

Trait	h_a^2	h_m^2	me^2	r_g
Survival at weaning (using probit link)	0.033 ± 0.01	0.061 ± 0.02	0.008 ± 0.02	0.27 ± 0.22
Survival at weaning (using logit link)	0.035 ± 0.01	0.053 ± 0.02	0.016 ± 0.017	-
Skin thickness	0.26 ± 0.04	-	-	-
Fat depth	0.36 ± 0.04	-	-	0.22 ± 0.10
Eye muscle depth	0.39 ± 0.05	-	-	-0.18 ± 0.12
Weaning weight	0.33 ± 0.04	0.17 ± 0.03	0.14 ± 0.02	-0.21 ± 0.12
Fleece weight at 12 months	0.50 ± 0.04	-	-	0.27 ± 0.12

There was also a favourable positive genetic correlation of 0.27±0.12 between skin thickness and fleece weight at 12 months. Similarly, Gregory (1982b) found a significant genetic correlation of 0.39 between skin thickness and clean fleece weight in South Australian Merino sheep. Contrary to this, Hynd *et al.* (1996) indicated a slight negative correlation between skin weight (as an indicator of thickness) and clean fleece weight. Unfavourably, the genetic correlation of skin thickness and fat depth was positive with a value of 0.22±0.10. In agreement with this, Jopson *et al.* (2000) showed that lambs from lines selected for high backfat depth had thicker skins than those selected for low backfat depth. Also, unfavourable genetic correlations of -0.18±0.12 and -0.21±0.12 were found between skin thickness and the traits eye muscle depth and weaning weight, respectively.

CONCLUSION

The preliminary estimates of heritability of skin thickness, together with its favourable genetic correlation (although with a high standard error) with lamb survival at weaning obtained in this study, suggests the idea of considering this trait as a likely attribute in indirect selection of lamb survival in selection programs. However, its inclusion should be with caution due to its unfavourable genetic correlation with fat depth, eye muscle depth, and weaning weight, as well as high standard errors associated with them. Otherwise, selection of animals with thicker skin might result in lambs with improved survival as well as increased fleece weight, but with a greater fat depth, and less muscle depth and lower weaning weight.

ACKNOWLEDGMENTS

The authors would like to thank the SIL and TRIGG for providing the data used in this study. We also are grateful to New Zealand Education for supporting the first author financially through New Zealand International Doctoral Research Scholarship.

REFERENCES

Brien F.D., Hebart M.L., Smith D.H., *et al.* (2010) *Anim. Prod. Sci.* **50**: 1017.

Brown D.J., Wolcott M.L. and Crook B.J. (2000) *Wool Tech. Sheep Breed.* **48**: 269.

Coy J. (1983) Proc. Workshop. 'Breeding Productive Sheep'. (Vie. Stud Merino Sheep Breeders Association).

Dalton D.C., Knight T.W., and Johnson D.L. (1980) *N. Z. J. Agric. Res.* **23**: 167.

Everett-Hincks J.M., Dodds K.G. and Wilson T. (2007) *Proc. Soc. Sheep & beef Cattle Veterinarians NZVA*: 21.

Fisher M.W. (2004) *Livest. Prod. Sci.* 85: 165.

Hight G.K. and Jury K.E. (1970) *N. Z. J. Agr. Res.* **13**: 735.

Gilmour A.R., Gogel B.J., Cullis B.R., Welham S.J. and Thompson R. (2015) ASReml User Guide Release 4.1, VSN International Ltd, Hemel Hempstead, HP1 1ES, UK.

Gregory I.P. (1982a) *Aust. J. Agric. Res.* **33**: 355.

Gregory IP. (1982b) *Aust. J. Agric. Res.* **33**: 363.

Gumbrell R.C. and Saville D.J. (1986) *Proc. N. Z. Soc. Anim. Prod.* **46**: 263.

Hynd P.I., Ponzoni R.W., Grimson R., Jaensch K.S., *et al.* (1996) *Wool Tech. Sheep Breed.* **44**: 167.

Jopson N.B., Greer G.J., Bain W.E., Findlay J.A. et all. (2000) *Proc. N. Z. Soc. Anim. Prod.* **60**: 61.

Kerslake J.I., Everett-Hincks J.M. and Campbell A.W. (2005) *Proc. N. Z. Soc. Anim. Prod.* **65**: 13.

Lopez-Villalobos N. and Garrick D.J. (1999) *Proc. N. Z. Soc. Anim. Prod.* **59**: 121.

SAS (2015) Statistical Analysis System. SAS Institute, version 9.4, Cary, NC, USA.

Samson D.E. and Slee J. (1981) *Anim. Prod.* **33**: 59.

Slee J. and Simpson S.P. (1991) *Res. Vet. Sci.* **51**: 34.

Wolff J.E., Baker R.L., Dobbie P.M., Ford A.J., Jordan R.B. (1987) *Proc. N. Z. Soc. Anim. Prod.* **47**: 93.

ACCOUNTING FOR POPULATION STRUCTURE IN GENOMIC PREDICTION OF AUSTRALIAN MERINO SHEEP

N. Moghaddar[1,3], D.J. Brown[2,3], A.A. Swan[2,3] and J.H.J. van der Werf[1,3]

[1] School of Environmental and Rural Science, University of New England, Armidale, Australia.[2] Animal Genetics and Breeding Unit (AGBU), University of New England, Armidale, Australia. [3] Cooperative Research Centre for Sheep Industry Innovation, Armidale, Australia.

SUMMARY

The aim of this study was to compare different ways of accounting for population structure for genomic prediction of three economic traits in an Australian Merino sheep population. Population structure was accounted for either by fitting genetic groups (GG) derived from pedigree, or fitting principal components (PCs) calculated from the genomic relationship matrix based on 50k density SNP marker genotypes. Genomic breeding values (GBV) were calculated using genomic best linear unbiased prediction (GBLUP) and the GBV accuracy was evaluated based on 5 fold cross-validation across half-sib families. Best linear unbiased estimation (BLUE) of GG or PC effects were added to the GBV. Results showed that accounting for population structure either by fitting GG or PCs improved the accuracy of genomic prediction. Furthermore, fitting the first two PCs gave a similar accuracy to fitting GG derived from pedigree. The improvement in GBV accuracy after accounting for population structure in studied traits was not high (3.8% when averaged across traits) which may be because the genomic relationship matrix will implicitly account for some of the population structure effect when the GG or PCs are not fitted in analysis. In the case of missing or incomplete pedigrees, PCs can be used to account for population structure and to improve the prediction accuracies.

INTRODUCTION

Differences in average genetic effects of breeds or strains within breeds (population structure) may affect the accuracy of genetic merit evaluation of selection candidates. Population structure could bias the genomic estimated breeding values (GBV) and hence affect the realized selection response. Australian Merino sheep is a highly diverse population due to different breeding objectives within the various types of Merino, and due to different production environments. The Merino breed consists of many sub-populations according to wool quality, e.g. strong wool, fine wool and ultra-fine wool Merinos. Accounting for population structure is a very importance feature of MERINOSELECT which is the national genetic evaluation of Australian Merino sheep (Brown *et al.* 2015; Swan *et al.* 2014)

The effect of population structure can be accounted for in the estimation of breeding values (based on phenotype and pedigree), according to genetic groups derived from pedigrees. However, in the case of incomplete pedigree information, population structure can be derived from genotypes by using Principal Components (PCs) from the genomic relationships matrix (GRM) (Price *et al.* 2006). Fitting PCs explicitly in the model is likely more accurate than accounting for the structure implicitly through the GRM (Van der Werf *et al.* 2013). The aim of this study was to compare fitting genetic groups based on pedigree with fitting PCs based on the genomic relationship matrix when accounting for population structure in genomic prediction of Australian Merino sheep.

MATERIALS AND METHODS

Reference population, phenotypes and validation population. The traits studied were post weaning weight (PWW, 6,388 records), ultrasound scanned eye muscle depth (PEMD, 4,012 records) measured between 150 and 290 days from birth and yearling greasy fleece weight (YGFW, 5,200 records) on Merino sheep. Animals originated from the "Sheep Cooperative Research Centre Information Nucleus Flock" (INF) and the Resource Flock (RF) which consisted of eight sites located across different regions of Australia and these were linked to each other by using common sires through artificial insemination between 2007 and 2015. More information on the scope and design of the INF is provided by Van der Werf *et al.* (2010). The accuracy of genomic prediction was evaluated based on the average of 5-fold cross-validation, where whole half sib families were sampled such that half sibs could not appear in training as well as validation set. The accuracy was calculated as the correlation between the GBV and the corrected phenotype, divided by the square root of the trait heritability.

Genotypes. Genotypes were available based on real 50K Ovine marker panel (Illumina Inc., San Diego, CA, USA) or 12K which was imputed to 50K. The 50K and 12K marker panel provided respectively 48,559 and 12,646 SNP genotypes after applying quality control. The sporadic missing genotypes were imputed first using Beagle 3.0 (Browning 2009). Animals genotyped with 12K marker density then were imputed to 50K density using Beagle 3.0 and using all Merino animals genotyped with 50K marker density as reference set. Accuracy of imputation was shown to be high (on average 0.96).

Statistical methods. Genomic best linear unbiased prediction (GBLUP) was used to calculate the Genomic Breeding Values (GBV) using the ASReml (Gilmour *et al.* 2009) program. The model fitted for each trait was: $y = Xb + Z_1g + Z_2m + e$ where y is a vector of phenotypes, b is a vector with fixed effects, g is the random additive genetic effect of the animal, m is a vector with maternal effects and e is vector of random residual effects, X, Z_1 and Z_2 are incidence matrices relating effects to animals. The parameters g, m and e are considered normally distributed as: $g \sim N(0, G\sigma_g^2)$, $m \sim N(0, I\sigma_m^2)$ and $e \sim N(0, I\sigma_e^2)$, respectively and G was the genomic relationship matrix calculated based on 50k markers genotypes using the VanRaden (2008) method. The common fixed effects in all models were birth type, rearing type, gender, age at measurement and contemporary group which was flock × birth year × management group. In the GG models 5 genetic groups were fitted as a regression (fixed continuous variable) on proportion of Merino sub-population (strains) where the proportions for individual animals were derived from a deep pedigree. In the PC models principal components were fitted by regression on up to ten eigenvectors associated with the largest 10 principal components.

RESULTS AND DISCUSSION

Tables 1, 2 and 3 compare the accuracy of genomic prediction between different models of fitting GG or PCs to account for population structure for PWW, PEMD and YGFW, respectively. Results show higher prediction accuracy for three different traits studied when population structure was accounted for in the model and then solutions for GG or PCs' effects were added to the GBV. This result was in line with a previous study by Daetwyler *et al.* (2013) who showed higher genomic prediction accuracy within Australian sheep breeds by accounting for population structure using PCs. However, the improvement in accuracy compared to only fitting the GRM in this study was not very high and on average 3.4% in absolute value.

Results showed fitting the first two largest PCs resulted in similar prediction accuracy to fitting GG from pedigree. Brown *et al.* (2015) and Swan *et al.* (2014) also showed strong correlation between using GG derived from pedigree and PCs calculated from genomic relationship matrix to correct the impact of population structure on estimation of genetic merits of animals. In this study the accuracy of GBV (GG/PC effect inclusive) was not increased by fitting more PCs. Results also showed a continuous decrease in GBV accuracy if the GG or PC effect solution was not added to GBV (Tables 1-3).

Table 1. Variance components, (SE) and average accuracy of genomic predictions from 5 fold cross-validation for PWW based on fitting genetic groups (GG) or Principal Components (PCs).

Model	Ve[1]	Va[2]	Vdam[3]	r(GBV1,Res)[4]	r(GBV2,Res+GG)[5]
No GG	13.83 (0.73)	12.05 (0.91)	2.08 (0.61)	NA	0.348
GG	14.61 (0.74)	10.22 (0.89)	2.28 (0.61)	0.243	0.368
1PC	14.24 (0.73)	10.98 (0.90)	2.23 (0.61)	0.218	0.342
2PC	14.41(0.73)	10.55 (0.89)	2.30 (0.61)	0.215	0.355
3PC	14.96 (0.74)	9.33 (0.88)	2.44 (0.61)	0.194	0.322
4PC	14.94 (0.74)	9.36 (0.88)	2.43 (0.61)	0.194	0.322
5PC	14.93 (0.74)	9.40 (0.88)	2.43 (0.61)	0.191	0.322
10PC	14.99 (0.74)	9.24 (0.88)	2.45 (0.61)	0.178	0.316

[1]Residual variance, [2]Additive genetic variance, [3]Dam permanent environmental effect, [4]Average of correlation between GBV (corrected for GG or PC effects) and corrected phenotypes (adjusted for GG effects). [5]Average of correlation between GBV (plus solution for GG or PCs) and corrected phenotypes (not adjusted for GG effect).

Table 2. Variance components, (SE) and accuracy of genomic prediction for PEMD based on fitting genetic groups (GG) or Principal Components (PCs).

Model	Ve[1]	Va[2]	r(GBV1,Res)[3]	r(GBV2,Res+GG)[4]
GG not fitted	5.066 (0.22)	2.251 (0.25)	NA	0.384
GG fitted	5.398 (0.23)	1.728 (0.25)	0.348	0.420
1PC	5.146 (0.22)	2.121 (0.25)	0.341	0.412
2PCs	5.237 (0.22)	1.976 (0.25)	0.320	0.422
3PCs	5.504 (0.22)	1.565 (0.25)	0.317	0.394
4PCs	5.496 (0.23)	1.552 (0.25)	0.316	0.393
5PCs	5.510 (0.23)	1.550 (0.25)	0.316	0.393
10PCs	5.524 (0.23)	1.550 (0.25)	0.311	0.387

[1]Residual variance, [2]Additive genetic variance, [3]Average of correlation between GBV (corrected for GG or PC effects) and corrected phenotypes (adjusted for GG effects). [4]Average of correlation between GBV (plus solution for GG or PCs) and corrected phenotypes (not adjusted for GG effect).

Tables 1, 2 and 3 also show the additive genetic, residual and dam variance (for PWW and YGFW only) for different models. Results show a continuous decrease in additive genetic variance and an increase in residual variance by fitting GG or fitting 1 to 10 PCs. The change in dam effect was very small in PWW and YGFW.

Results of this study showed that accounting for population structure according to pedigree or genomic information improves the total genetic merit prediction accuracy. However, the increase in prediction accuracy in traits studied was not very high compared to fitting only the GRM. This indicate that it is likely that the GRM could account for only part of the effect of population structure implicitly as was indicated before (Van der Werf *et al.* 2013).

The reason for lower accuracy of GBVs (corrected for PCs) by fitting more PCs would be because fitting more PCs can capture part of the total additive genetic variance between different flocks and between half-sib families within flocks.

In term of estimating the total genetic merits for animals with pedigree information the results show the GG model seems to work slightly better than PCs model. However, fitting the first two largest PCs derived from the GRM can also sufficiently account for population structure. This shows that in the case of missing, incomplete or not reliable pedigree information and if the animals were genotyped, PCs could be used to account for population structure to obtain higher prediction accuracies within a breed. This could be more important in prediction of unbiased breeding values on the national scale such as Australian Sheep Breeding values (ASBV) with probable larger impact of genetic groups.

Table 3. Variance component, (SE) and accuracy of genomic prediction for YGFW based on fitting genetic groups (GG) or Principal Components (PCs).

Model	Ve[1]	Va[2]	V(dam)[3]	r(GBV1,Res)[4]	r(GBV2,Res+GG)[5]
GG not fitted	0.160 (0.01)	0.128 (0.01)	0.016 (0.01)	NA	0.564
GG fitted	0.163 (0.01)	0.121 (0.01)	0.017 (0.01)	0.532	0.611
1PC	0.153 (0.01)	0.131 (0.01)	0.020 (0.01)	0.524	0.562
2PCs	0.156 (0.01)	0.127 (0.01)	0.021 (0.01)	0.519	0.604
3PCs	0.157 (0.01)	0.122 (0.01)	0.021 (0.01)	0.509	0.569
4PCs	0.161 (0.01)	0.122 (0.01)	0.021 (0.01)	0.509	0.566
5PCs	0.163 (0.01)	0.121 (0.01)	0.022 (0.01)	0.508	0.566
10PCs	0.167 (0.01)	0.116 (0.01)	0.021 (0.01)	0.487	0.560

[1]Residual variance, [2]Additive genetic variance, [3]Dam permanent environmental effect, [4]Average of correlation between GBV (corrected for GG or PC effects) and corrected phenotypes (adjusted for GG effects), [5]Average of correlation between GBV (plus solution for GG or PCs) and corrected phenotypes (not adjusted for GG effect).

ACKNOWLEDGEMENTS

The authors would like to extend their gratitude to Klint Gore (University of New England, Armidale, NSW, 2351, Australia) for preparing and cleaning genotype data and managing the CRC information nucleus database, and all staff involved at the Sheep CRC Information Nucleus sites across Australia.

REFERENCES

Brown D.J., Swan A.A., Gill J.S., *et al.* (2015). *Proc. Assoc. Advmt. Anim. Breed. Genet.* **20**: 66.

Browning B.L. and Browning S.R. (2009) *Am. J. Hum. Genet.* **84**: 210.

Gilmour A.R., Gogel B.J., Cullis B.R., *et al.* (2009) ASReml User Guide Release 3.0.

Price A.L., Patterson N.J., Plenge R.M., *et al* (2006) *Nature Genet.* **38**: 904.

Swan A.A., Brown D.J., Daetwyler H.D *et al.* (2014) *Proc. 10th World Congress Gen. Appl. Livest. Prod.*, Vancouver, Canada.

Van der Werf J.H.J., Kinghorn B.P. and Banks R.G. (2010) *Anim. Prod. Sci.* **50**: 998.

Van der Werf J.H.J., Brown D.J. and Swan A.A. (2013) *Proc. Assoc. Advmt. Anim. Breed. Genet.* **20**: 340.

VanRaden P.M. (2008) *J. Dairy Sci.* **91**: 4414.

PRELIMINARY RESULTS ON THE INHERITANCE OF EARLY LAMB GROWTH AS AN INDICATOR FOR MILK PRODUCTION IN MERINO SHEEP?

J.C. Greeff[1], L.J.E. Karlsson[2] and A.C. Schlink[3]

[1]Department of Agriculture and Food WA, South Perth WA 6151
[2]RMB 314, Bridgetown WA 6255
[3]University of Western Australia, Crawley, WA

SUMMARY

The inheritance of early growth rate of the litter from birth to 30 days of age was investigated as an indirect trait for milk production of the dam in a medium type Merino flock in a production system where ewes were mated over a 35 day period. Body weight at marking was used to calculate a standardised cumulative weight gain for each litter to an average age of 30 days. The standardised 30 day weight gain had a low heritability (0.17 ± 0.04). It was also genetically positively correlated with hogget body weight (r_g = 0.38 ± 0.09). The genetic correlations of 30 day weight gain with clean fleece weight, fibre diameter and staple strength at hogget age, did not differ significantly from zero. It was suggested that the heritability may be higher in a production system where artificial insemination is generally used, as lamb age would be less variable in such a system. These results indicate that it should be possible to develop an indirect breeding value for milk production in Merino sheep.

INTRODUCTION

Milk production is crucial for lamb survival and early lamb growth. Four different measurement techniques have been developed to measure milk production, i.e. (i) use of exogenous oxytocin and directly milking the ewe, (ii) weighing the lamb before and after suckling, (iii) measuring water turnover rate using tritiated water, and (iv) using lamb weight gain as an indirect estimate of milk yield (Geenty 2010). However, it is difficult and expensive to measure milk production using methods (i) to (iii) on individual sheep. Afolayan *et al.* (2009) measured milk production in crossbred meat sire x Merino ewes and estimated the genetic correlations between milk production and average lamb weaning weight in the litter and reported a genetic correlation of 0.44. Geenty (1979) found that the phenotypic correlation between milk production of the dam and the early growth rate of their lambs varies between 0.22 to 0.76 in different meat breeds and their crosses. This implies that early growth rate of the lamb(s) may qualify as a potential indicator trait of milk production for the dam. However, very little information is available on early weight gain prior to the lamb starting to consume solids and this could be affected by age of the dam, litter size, sex of the lambs, and age of the lamb(s) when first weighing occurs at marking. Snyman *et al.* (2016) published genetic parameters of milk production using the oxytocin method where milk production was measured at 3 and 12 weeks of lactation. They found high genetic correlations between milk production and the maternal effect of early body weight. Thus early growth rate of the lamb can be used as an indicator of milk production in the dam. This study investigates the inheritance of early growth rate up to 30 days of age of the lamb as a trait of the dam, and its genetic relationship with body weight, wool production and fibre traits of the dam at hogget age.

MATERIAL AND METHODS

The Australian Wool Innovation Breech strike flock of the Department of Agriculture and Food Western Australia was used in this study (Greeff *et al.* 2014). Ewes were naturally mated in single sire groups over a 35 day period. Maternal pedigree, birth weight, litter size, birth date and sex of the lambs were recorded at birth during July/August. Marking weight was recorded at an average

age of 30 days of age on 1954 lambs that were born in 2014, 2015 and 2016 from 954 dams mated to 143 sires. Weight gain from birth to marking for the litter was standardised by calculating the daily rate of gain from birth to first weighing and predicting the total litter weight at 30 days of age.

Wool production and fibre traits produced over a 12 month growth period and the body weights were recorded on the dams and their contemporaries at approximately 18 months of age. This dataset consisted of 7956 sheep that were the progeny of 194 sires mated to 3120 dams that were between 2 and 8 years of age. All flystrike information was available on each sheep.

STATISTICAL ANALYSIS

ASREML (Gilmour *et al.* 2009) was used to analyse the data. A univariate analysis was carried out to obtain estimates of the heritability of weight gain of the litter up to 30 days of age as a trait of the dam. A repeatability model was fitted to the data with dam as random genetic effect, and also fitting dam as a permanent maternal environmental effect. An animal model was fitted for wool yield and the fibre traits and body weight at hogget age (see Table 1). Year of birth (2014-2016), age of the dam (2-8 years), and year of observation were fitted as fixed effects for the 30 day weight gain trait. For the wool, fibre and body weight traits at hogget age, year of birth (2006 to 2014), birth status (single or multiple), sex (male or female), and whether the animal was struck by blowflies were fitted as fixed effects. Sex was confounded with management group as males and females were managed in separate groups. All two way interactions were initially fitted for both models. Statistically non-significant factors ($P>0.05$) were dropped from the final model. This was followed by a bivariate analysis to obtain genetic covariances to estimate the genetic correlations between 30-day-weight gain as a trait of the dam and her production traits at hogget age

RESULTS AND DISCUSSION

Table 1 shows that the average and standard deviation for the different traits in this study. The averages of the wool and fibre traits in this Merino flock shows that this flock is representative of a typical medium wool type in Western Australia (Greeff and Cox 2006). The weight gained from birth to 30 days of age was on average 10.5kg. Year of birth of the ewe, year of measurement, age of the ewe were significant ($P<0.01$) environmental effects of 30 day weight gain.

Table 1. Number of records, raw means and standard deviation (SD) of the different traits

Trait	Abbreviation	n	Mean	SD
30 day weight gain (kg)	30d_WT	1954	10.5	3.85
Greasy fleece weight (kg)	GFW	7965	3.87	0.83
Yield (%)	Yld	7956	70.9	4.23
Clean fleece weight (kg)	CFW	7965	2.74	0.61
Fibre diameter (micron)	FD	7965	19.2	1.59
Coefficient of variation of Fibre diameter (%)	FDCV	7956	20.5	2.42
Fibre curvature (deg)	CUR	7965	93.7	11.1
Standard deviation of CUR	SDCUR	7965	55.7	6.08
Staple strength (N/Ktex)	SS	7953	26.0	10.83
Hogget body weight (kg)	HWT	7952	47.9	17.0

Significant effects ($P<0.05$) of fixed effects were observed on various traits, as follows:

Year of birth of the ewe: GFW, CUR, FD, FDCV, SS, YLD and HWT.

Age of the dam: GFW, CFW and FD.
Sex of the lamb: GFW, FDCV, SS, Yld, FD and HWT.
Birth status: GFW and CFW.
Flystrike: GFW, CFW and HWT.
A year of birth by sex interaction was also significant for GFW, CUR, FDCV, SS, CFW, GFW, YLD, FD and HWT.

Table 2 shows that the heritability estimates found in this study for body weight and the wool traits at hogget age are very similar to that which has been widely reported in the literature (Safari *et al.* 2005).

A low heritability of 0.17 (± 0.04) was found for weight gain to 30 days of age as a trait of the dam. The permanent maternal environmental component effect explained an additional 8.4% (± 4.9) of the phenotypic variation in 30-day weight gain.

Standardised 30-day weight gain was also genetically positively correlated (0.38 ± 0.09) with the dam's body weight at hogget age. For the wool traits, fibre curvature and clean yield had a small but significant positive (0.18 ± 0.09) and negative genetic correlation (-0.18 ± 0.09) with 30-day weight gain, respectively. None of the other wool production traits (clean fleece weight, fibre diameter and staple strength) showed a significant relationship with 30-day body weight.

Table 2. The phenotypic variation (Vp), heritability (h^2) of 30 day weight gain of the litter as a trait of the dam and her production traits at hogget age, and the genetic correlations (r_g) between 30 day weight gain of the litter and the production traits at hogget age.

Trait	Vp	h^2	SE	r_g (30 day weight gain and trait)	SE
30 day weight gain	13.7	0.17	0.04		
GFW	0.40	0.46	0.02	0.23	0.09
CFW (kg)	0.22	0.47	0.02	0.15	0.09
YLD (%)	15.1	0.64	0.02	-0.18	0.08
FD (micron)	2.02	0.62	0.02	0.06	0.08
FDCV (%)	5.47	0.38	0.02	0.03	0.09
CUR (deg)	112	0.65	0.02	0.18	0.08
CURVSD (deg)	33	0.63	0.02	0.10	0.08
SS (N/Ktex)	67	0.52	0.02	-0.17	0.10
HWT (kg)	118	0.42	0.02	0.38	0.09

CONCLUSIONS

Standardised 30-day weight gain was lowly heritable (0.17 ± 0.04) in a naturally mating Merino flock. However, it is not surprising that the heritability estimate is relatively low, considering the large variation that exists in lamb age at marking due to the long lambing period. Furthermore, ewes with singletons and ewes with lambs younger than 30 days of age would not be adequately challenged to express their full milk production potential. Thus, as expected this estimate is lower than that of directly measured milk production of 0.32 (Barillet and Boichard 1987) in Lacaune sheep. Snyman *et al.* (2016) reported that the heritability of milk production using the oxytocin method was 0.02 (Grootfontein Merino), 0.21 (Afrino), 0.10 (Cradock Merino) and 0.29 (Elsenburg Merino) in four different flocks. They also found relatively high genetic correlations (Afrino, 0.76; Elsenburg Merino – not reported; Cradock Merino 0.83; Grootfontein Merino 0.62) between directly measured milk production and early growth rate up to 42 days of age in the four flocks. The heritability of this study agree well with the results of Snyman *et al.* (2016) which indicates that there may be opportunities to improve milk production indirectly by selecting on early growth rate.

No unfavourable genetic relationships were found between early weight gain and any of the wool production traits. Further research is necessary to estimate the heritability of the trait especially in a flock that predominantly uses artificial insemination, where the age of the lambs at marking is less variable than in a naturally mated flock. Various options should be evaluated to identify an optimum age when lambs should be weighed to obtain a more accurate indirect measurement of milk production as older and bigger lambs will challenge the dam more to obtain a better indication of her milk production potential. It may offer opportunities to select indirectly for milk production in sheep.

ACKNOWLEDGEMENT

The researchers acknowledge the generous financial support provided by Australian woolgrowers and the Commonwealth through Australian Wool Innovation Limited. The authors also thank the Research station staff and Technical Officers, especially Nicola Stanwyck for managing the flock and for the collection of the data.

REFERENCES

Afolayan R.A., Fogarty N.M., Morgan J.E., Gaunt G.M., Cummins L.J. and Gilmour A.R. (2009) *Small Rumin. Res.* **82**: 27.

Barillet F. and Boichard D. (1987) *Genet. Select. Evol.* **19**: 459.

Geenty K.G. (1979) *Proc. NZ Soc. Anim Prod* **39**: 202.

Geenty K.G. (2010) Lactation and Lamb growth. In International Sheep and Wool Handbook. Edited by D.J. Cottle, Nottingham University Press.

Greeff J.C. and Cox G. (2006) *Aust. J. Exp. Agric.* **46**: 803.

Greeff J.C., Karlsson L.J.E. and Schlink A.C. (2014) *Anim. Prod. Sci.* **54**: 125.

Safari E., Fogarty N.M. and Gilmour A.R. (2005). *Livest. Prod. Sci.* **92**: 271.

Snyman M.A., Cloete S.W.P. and Olivier W.J. (2016) Accessed from: gadi.agric.za/Agric/Vol16No1_2016/melk.php.

THE EFFECT OF INCLUDING IMMUNE COMPETENCE IN MERINO SHEEP BREEDING PROGRAMS

A. Alexander [1,2], B. Hine[1], J. Smith[1], and S. Dominik[1]

CSIRO Agriculture and Food, Armidale, Australia
[2] University of Melbourne, Melbourne, Australia

SUMMARY

The effect of including general immune competence (IC) as a novel trait in a fibre production (FP) and a dual purpose selection (DP) index was investigated. Two levels of economic values were assumed for IC and the sensitivity of index responses to these were tested. The results showed that the addition of IC to a selection index requires careful consideration in order to achieve the envisaged improvements in health and welfare outcomes expected while addressing primary production objectives.

INTRODUCTION

Selection for production traits with little or no emphasis on health and welfare traits can lead to increased susceptibility to disease (Rauw *et al.* 1998). For example in sheep, Shaw *et al.* (2012) reported that production focused selection has led to a decrease in parasite resistance. In an effort to improve general disease resistance, methods to assess immune competence were first developed in both pigs and dairy cattle (Wilkie and Mallard 1999; Mallard and Wagter 2001). In Australia, methods for assessing immune competence in beef cattle (Hine *et al.* 2016) and sheep have recently been evaluated, providing estimates for this study (Hine and Smith, CSIRO Agriculture and Food, 2016, preliminary estimates). This study assumed economic values for immune competence and explored the effect of these on fibre and dual purpose selection indexes for Merino sheep.

MATERIAL AND METHODS

Immune competence. Responses of the immune system can be broadly classified as being innate or adaptive with innate responses providing the first line of defence, which arise quickly and are broad in action, while adaptive responses provide a second line of defence and are slower to develop but more specific in their action. Further, adaptive immune responses are specifically tailored to the type of pathogen being encountered with antibody-mediated immune responses (AMIR) predominating upon exposure to extracellular pathogens and cell-mediated immune responses (CMIR) predominating upon exposure to intracellular pathogens. Overall immune competence, defined as a combination of AMIR and CMIR, has been demonstrated to be correlated with infectious and metabolic diseases in dairy cattle (Thompson-Crispi *et al.* 2012). Overall immune competence (IC) has been used as breeding objective trait in this study.

Selection indexes. Breeding objectives were derived from the Sheep Genetics (2014) Dual Purpose Plus and Fibre Production Plus indexes. The breeding objective traits in the dual purpose (DP) index include the adult (a) expression of clean fleece weight (aCFW), fibre diameter (aFD), bodyweight (aWT), yearling eye muscle depth (yEMD) and number of lambs weaned (NLW). The breeding objective traits in the fibre index (FP) include aCFW, aFD, aWT, NLW and adult staple Strength (aSS). Selection criteria for both indexes include NLW and yWT, yCFW and yFD. Yearling staple strength (ySS) was a selection criterion for FP only, and yearling eye muscle depth (yEMD) for DP only. Traits were recorded on the selection candidates, sire and dam and half-sibs.

To test the effect of including immune competence as a novel trait in the FP and DP indexes, IC was added as a breeding objective trait (DP+IC and FP+IC). Additional selection criteria were IC and its component traits CMIR and AMIR.

Table 1. Heritabilities (on the diagonal in bold), genetic (below the diagonal) and phenotypic (above the diagonal) correlations, phenotypic trait variance (V_P), the economic value (EV in $) and standardized economic values (stdEV = EV* σp in $) for breeding objective traits of the dual purpose and fibre indexes (DP/FP and DP/FP+IC)

Traits	aWT	aCFW	aFD	aSS	NLW	yEMD	yWT	yCFW	yFD	ySS	IC	AMIR	CMIR
V_P	28.77	0.26	1.35	83.93	0.27	3.27	23.3	0.16	3.00	83.8	2.36	0.24	0.04
DP EV	0.08	0.74	-3.44	--	125.29	3.83	--	--	--	--	1.40/2.02	--	--
FP EV	-0.03	1.55	-13.74	1.57	126.96	--	--	--	--	--	1.40/2.02	--	--
DP stdEV*	0.43	0.38	-3.99	--	65.50	6.93	--	--	--	--	2.15/8.98	--	--
FP stdEV*	-0.16	0.79	-15.39	14.38	66.12	--	--	--	--	--	2.15/8.98	--	--
aWT	**0.44**	0.29	0.08	-0.13	0.01	-0.08	0.56	0.50	0.17	-0.13	0.00	-0.02	0.01
aCFW	-0.15	**0.50**	0.22	0.28	0.00	-0.20	0.26	0.50	0.22	0.26	-0.01	-0.01	0.01
aFD	0.02	0.28	**0.67**	0.01	0.02	0.02	0.13	0.22	0.7	0.35	0.05	0.06	0.04
aSS	-0.31	0.37	-0.03	**0.35**	0.00	--	0.09	0.03	0.03	0.40	0.00	0.02	-0.02
NLW	0.33	-0.47	0.01	0.15	**0.07**	0.10	0.02	-0.07	0.03	0.00	0.001	0.001	0.001
yEMD	-0.20	-0.11	-0.08	--	0.28	**0.33**	0.83	-0.03	0.06	--	-0.01	0.03	0.00
yWT	0.77	-0.09	0.22	0.17	-0.1	0.85	**0.43**	0.42	0.13	0.09	0.00	-0.02	0.01
yCFW	-0.15	0.80	0.15	0.14	-0.65	0.10	0.23	**0.36**	0.22	0.03	-0.01	-0.01	0.01
yFD	0.03	0.15	0.80	0.26	-0.07	0.17	0.22	0.15	**0.77**	0.48	0.05	0.06	0.04
yss	-0.31	-0.14	0.15	0.80	-0.45	--	0.17	0.16	0.27	**0.40**	0.00	0.02	-0.02
IC	0.06	-0.15	0.31	0.21	0.001	0.00	0.06	-0.15	0.31	0.21	**0.53**	0.76	0.76
AMIR	0.02	0.03	0.28	0.20	0.001	0.11	0.02	0.03	0.28	0.20	0.82	**0.47**	0.16
CMIR	0.02	-0.24	0.20	0.12	0.001	0.00	0.02	-0.24	0.20	0.12	0.79	0.29	**0.42**

stdEV=EVσp

Index calculations were performed using the MTIndex software (http://www.personal.une.edu.au/~jvanderw). The total dollar response and individual trait responses to selection were calculated per animal and per round of selection.

Genetic and phenotypic variances, heritabilities and correlations for breeding objective traits and selection criteria are shown in Table 1 (Brown and Swan 2015; Purvis and Swan 1999; Huisman *et al.* 2008; Huisman and Brown 2008; Huisman and Brown 2009a; Huisman and Brown 2009b; Swan *et al.* 2008; Dominik and Swan 2016).

Economic values. The economic value for IC was developed on the basis of the strong favourable correlation between IC with dag score post-weaning (r_g=-0.55; Hine and Smith, CSIRO Agriculture and Food, 2016, preliminary estimates). Dags (faecal soiling of the breech) cause hygiene and contamination issues at shearing and slaughter. Correlations suggested that lower IC is associated with higher dag score, potentially leading to the need for an extra crutch throughout the year and prior to shearing and potential penalties when selling lambs, which can result in extra costs for the producer. Based on costs for crutching obtained from High Voltage Shearing Pty Ltd. in Armidale, NSW (pers. comm., 15 December 2016), two economic values were used. A simple 'market crutch' that requires only the area around the breech to be shorn was valued at $1.40/head, marked in the index abbreviation as "a". A full crutch on a non-mulesed sheep was valued at the highest price $2.02/head due to the tendency of the extra wool on those animals to be more soiled and difficult to remove ("b"). Using these two values, the sensitivity of index responses to different emphasis on IC was tested (DP/FP+ICa and DP/FP+ICb). Economic values for the other breeding objective traits were obtained from Brown and Swan (2015). All economic values are summarised in Table 1.

RESULTS AND DISCUSSION

The inclusion of IC in the DP index (Table 2), increased total dollar response per animal and round of selection from $11.75 (DP) by 27% (DP+ICa) and 26% (DP+ICb) respectively.

With a low economic value placed on IC (DP+ICa), an 8 micron decrease in aFD and an over 30% increase in NLW was observed, which led to the substantial increase in total dollar response. The response in IC was slightly unfavorable. When the economic value for IC was increased, the moderate unfavourable genetic correlations between IC with FD and CFW significantly influenced index responses. As a result, with increasing economic weight on IC the response in IC increased only slightly and the response in aFD, which has a high economic value, was maximised. The response in aWT was unfavorable, but NLW was still greatly improved compared to DP.

The total dollar response showed a small increase of $0.02 for the FP index with the inclusion of IC at a low economic weight (FP+ICa) and $0.21 at a high economic weight (FP+ICb) (Table 2). Compared to DP these increases were lower due to low individual trait responses. The inclusion of IC increased the emphasis on a FD, which has a high economic value in the FP index. However, this increased emphasis on aFD was balanced by aSS, which is unfavourably correlated with FD. Staple strength is as economically important as FD, but is not as heritable. The IC trait and its components (AMIR and CMIR) are correlated to both FD and SS. These competing interests are reflected in only small changes in all traits and a small increase in the total dollar response.

The results showed that with the assumed economic values, no major changes were achieved in IC. The assumptions on the economic value for IC were simplistic but could be considered conservative as it did not take into account any decrease in animal health treatment costs associated with a variety of common diseases which may be realised as a consequence of improved IC. Also the influence of improved consumer confidence that could be expected from improving IC, and as a consequence animal welfare by reducing disease incidence and deaths, was not considered. The influence of these factors on the economic value of IC could be substantial.

Table 2. Standard deviation of the breeding objective (SD_{BO}), total dollar response (SD_{Index}), index accuracy (Acc) and trait responses per animal per round of selection in the dual purpose (DP) and fibre production (FP) index without and with immune comptence at EV $1.40 (DP/FP+ICa) and EV $2.02 (DP/FP+ICb)

	SD_{BO}	SD_{Index} ($)	Acc	aWT (kg)	aCFW (kg)	aFD (micron)	NLW (no of lambs)	yEMD (cm)	aSS (Nktex)	IC (stddev)
DP	21.28	11.75	0.55	1.31	-0.19	-0.17	0.09	-0.12	--	--
DP+ICa	21.27	14.87	0.70	1.03	-0.13	-0.25	0.12	-0.12	--	-0.04
DP+ICb	21.30	14.86	0.70	1.04	-0.13	-0.24	0.12	-0.12	--	0.00
FP	27.42	14.90	0.54	0.55	-0.11	-0.35	0.09	--	-0.56	--
FP+ICa	27.73	14.92	0.54	0.55	-0.11	-0.32	0.09	--	-0.50	0.00
FP+ICb	27.73	15.11	0.55	0.55	-0.11	-0.32	0.09	--	-0.52	0.02

CONCLUSION

Improvement in overall immune competence in sheep is desirable for future production to improve welfare and reduce health costs. Here it was shown, that the inclusion of this novel trait in a sheep breeding framework that is highly production focused requires a full economic evaluation of immune competence to integrate it effectively in genetic improvement programs.

ACKNOWLEDGEMENTS

This research was supported by CSIRO, the Grains Research and Development Cooperation and Bayer. Access to data has kindly been provided by Meat and Livestock Australia

REFERENCES

Brown D.J. and Swan A.A. (2015) *Anim. Prod. Sci.* **56** (4): 690.

Dominik S. and Swan A.A. (2016) *Anim. Prod. Sci.* doi: 10.1071/AN15738.

Hine B., Ingham A., Dominik S. and Colditz I. (2016) Final Report. Meat and Livestock Australia Limited, North Sydney, Australia. (online 7 December 2016).

Huisman A.E., Brown D.J., Ball A.J. and Graser H.U. (2008) *Aust. J. Exp. Agr.* **48** (9) 1177.

Huisman AE and Brown DJ (2008) *Aust. J. Exp. Agr.* **48**(9) 1186.

Huisman A.E. and Brown D.J. (2009a) *Aust. J. Exp. Agr.* **48** (9) 1186.

Huisman A.E. and Brown D.J. (2009b) *Aust. J. Exp. Agr.* **49** (4) 289.

Mallard BA and Wagter L (2001) Method of identifying high immune response animals. University of Guelph, Canada, assignee. US Pat. No. 6,287,564. (online 5 January 2017).

Purvis IW and Swan AA (1999) *Proc. Assoc. Advmt. Anim. Breed. Genet.* **13**: 177.

Rauw W.M., Kanis E., Noordhuizen-Stassen E.N., Grommers F.J. (1998). *Livest. Prod. Sci.* **56**: 15–33.

Shaw R.J., Morris C.A., Wheeler M., Tate M. and Sutherland I.A. (2012) *Vet. Parasitol.* **186**: 109.

Swan A.A., Purvis I.W. and Piper L.R. (2008) *Aust. J. Exp. Agr.* 48(9): 1168.

Thompson-Crispi, K.A., Hine, B., Quinton M., Miglior F. and Mallard B.A. (2012) *J. Dairy Sci.* **95** :3888–3893.

Wilkie B. and Mallard B. (1999) *Vet. Immunol. Immunopathol.* **72**(1-2):23.